# ADVANCES IN BIOMECHANICS AND TISSUE REGENERATION

# ADVANCES IN BIOMECHANICS AND TISSUE REGENERATION

*Edited by*

MOHAMED H. DOWEIDAR

ELSEVIER

ACADEMIC PRESS
An imprint of Elsevier

Academic Press is an imprint of Elsevier
125 London Wall, London EC2Y 5AS, United Kingdom
525 B Street, Suite 1650, San Diego, CA 92101, United States
50 Hampshire Street, 5th Floor, Cambridge, MA 02139, United States
The Boulevard, Langford Lane, Kidlington, Oxford OX5 1GB, United Kingdom



**Notices**

Knowledge and best practice in this field are constantly changing. As new research and experience broaden our understanding, changes in research methods, professional practices, or medical treatment may become necessary.

Practitioners and researchers must always rely on their own experience and knowledge in evaluating and using any information, methods, compounds, or experiments described herein. In using such information or methods they should be mindful of their own safety and the safety of others, including parties for whom they have a professional responsibility.

To the fullest extent of the law, neither the Publisher nor the authors, contributors, or editors, assume any liability for any injury and/or damage to persons or property as a matter of products liability, negligence or otherwise, or from any use or operation of any methods, products, instructions, or ideas contained in the material herein.

**Library of Congress Cataloging-in-Publication Data**
A catalog record for this book is available from the Library of Congress

**British Library Cataloguing-in-Publication Data**
A catalogue record for this book is available from the British Library

ISBN 978-0-12-816390-0

For information on all Academic Press publications visit our website at https://www.elsevier.com/books-and-journals

*Publisher:* Mara Conner
*Acquisition Editor:* Fiona Geraghty
*Editorial Project Manager:* Thomas Van Der Ploeg
*Production Project Manager:* R.Vijay Bharath
*Cover Designer:* Miles Hitchen

Typeset by SPi Global, India

# Contents

## 8. Towards the Real-Time Modeling of the Heart

R.R. RAMA AND S. SKATULLA



## 9. Computational Musculoskeletal Biomechanics of the Knee Joint

HAFEDH MAROUANE, ABOULFAZL SHIRAZI-ADL AND MASOUD SHARIFI



## 10. Determination of the Anisotropic Mechanical Properties of Bone Tissue Using a Homogenization Technique Combined With Meshless Methods

MARCO MARQUES, JORGE BELINHA, ANTÓNIO F. OLIVEIRA AND RENATO M. NATAL JORGE



## 11. Analysis of the Biomechanical Behavior of Osteosynthesis Based on Intramedullary Nails in Femur Fractures

SERGIO GABARRE, JORGE ALBAREDA, LUIS GRACIA, SERGIO PUÉRTOLAS, ELENA IBARZ AND ANTONIO HERRERA



## 12. Biomechanical Study in the Calcaneus Bone After an Autologous Bone Harvest

JAVIER BAYOD LÓPEZ, RICARDO BECERRO DE BENGOA VALLEJO AND MARTA E. LOSA IGLESIAS



# Part II

# MECHANOBIOLOGY AND TISSUE REGENERATION

## 13. Multidimensional Biomechanics Approaches Though Electrically and Magnetically Active Microenvironments

S. RIBEIRO, C. GARCIA-ASTRAIN, M.M. FERNANDES, S. LANCEROS-MENDEZ AND C. RIBEIRO



## 14. Using 3-D Printing and Bioprinting Technologies for Personalized Implants

JULIEN BARTHES, EDWIN-JOFFREY COURTIAL, ESTEBAN BRENET, CELINE BLANDINE MULLER, HELENA KNOPF-MARQUES, CHRISTOPHE MARQUETTE AND NIHAL ENGIN VRANA



## 15. Computational Simulation of Cell Behavior for Tissue Regeneration

S. JAMALEDDIN MOUSAVI AND MOHAMED H. DOWEIDAR

# Contributors

**Jorge Albareda** Department of Orthopaedic Surgery and Traumatology, Lozano Blesa University Hospital; Aragón Health Research Institute; Department of Surgery, University of Zaragoza, Zaragoza, Spain

**Miguel Ángel Ariza-Gracia** Instituto de Investigación en Ingeniería de Aragón, Universidad de Zaragoza, Zaragoza, Spain; Institute for Surgical Technology and Biomechanics, Universität Bern, Bern, Switzerland

**Stéphane Avril** Mines Saint-Etienne, University of Lyon, INSERM, U1059 Sainbiose, Saint-Etienne, France

**Jacobo Ayensa-Jiménez** Aragon Institute of Engineering Research (I3A), University of Zaragoza; Institute of Health Research of Aragon (IIS), Zaragoza, Spain

**Eduardo Bajador** Department of Gastroenterology, Lozano Blesa University Hospital, Zaragoza, Spain

**Julien Barthes** INSERM UMR 1121, 11 rue Humann; Protip Medical, 8 Place de l'Hôpital, Strasbourg, France

**Jose A. Bea** Aragon Institute of Engineering Research, University of Zaragoza, Zaragoza, Spain

**Jorge Belinha** Institute of Science and Innovation in Mechanical and Industrial Engineering (INEGI); School of Engineering, Polytechnic of Porto (ISEP), Porto, Portugal

**Esteban Brenet** INSERM UMR 1121, 11 rue Humann, Strasbourg, France

**Begoña Calvo Calzada** Instituto de Investigación en Ingeniería de Aragón, Universidad de Zaragoza, Zaragoza, Spain; Centro de Investigación Biomédica en Red en el área temática de Bioingeniería, Biomateriales y Nanomedicina (CIBER-BBN), Madrid, Spain

**Carmen Carda** Department of Pathology, University of Valencia and INCLIVA Health Research Institute; Biomedical Research Networking Center—Bioengineering, Biomaterials and Nanomedicine (CIBER-BBN), Valencia, Spain

**A.T.A. Castro** Faculty of Engineering of University of Porto (FEUP), Porto, Portugal

**Myriam Cilla** Aragón Institute of Engineering Research (I3A), University of Zaragoza; Centro de Investigación en Red en Bioingeniería, Biomaterialesy Nanomedicina, CIBER-BBN; Centro Universitario de la Defensa, Academia General Militar, Zaragoza, Spain

**Edwin-Joffrey Courtial** 3dFAB Université Lyon 1—CNRS 5246 ICBMS, Lyon, France

**Lucília P. da Silva** I3Bs—Research Institute on Biomaterials, Biodegradables and Biomimetics of University of Minho, Headquarters of the European Institute of Excellence on Tissue Engineering and Regenerative Medicine, University of Minho, Guimarães, Portugal; ICVS/3B's—PT Government Associate Laboratory, Guimarães, Portugal

**Ricardo Becerro de Bengoa Vallejo** Department of Nursing, School of Nursing, Physiotherapy and Podiatry, Complutense University, Madrid, Spain

**Manuel Doblaré** Aragon Institute of Engineering Research (I3A), University of Zaragoza; Institute of Health Research of Aragon (IIS), Zaragoza, Spain; Biomedical Research Networking Center in Bioengineering, Biomaterials and Nanomedicine (CIBER-BBN), Madrid, Spain; Mechanical Engineering Department, School of Engineering and Architecture (EINA), University of Zaragoza, Zaragoza, Spain

**Mohamed H. Doweidar** Mechanical Engineering Department, School of Engineering and Architecture (EINA), University of Zaragoza; Biomedical Research Networking Center in Bioengineering, Biomaterials and Nanomedicine (CIBER-BBN); Aragón Institute of Engineering Research (I3A), University of Zaragoza; Institute of Health Research of Aragon (IIS), Zaragoza, Spain

**M.M. Fernandes** Center/Department of Physics; CEB—Centre of Biological Engineering, University of Minho, Braga, Portugal

**Maria G. Fernandes** I3Bs—Research Institute on Biomaterials, Biodegradables and Biomimetics of University of Minho, Headquarters of the European Institute of Excellence on Tissue Engineering and Regenerative Medicine, University of Minho, Guimarães, Portugal; ICVS/3B's—PT Government Associate Laboratory, Guimarães, Portugal

**Luis. J. Fernández** Aragon Institute of Engineering Research (I3A), University of Zaragoza; Institute of Health Research of Aragon (IIS), Zaragoza, Spain; Biomedical Research Networking Center in Bioengineering, Biomaterials and Nanomedicine (CIBER-BBN), Madrid, Spain; Mechanical Engineering Department, School of Engineering and Architecture (EINA), University of Zaragoza, Zaragoza, Spain

**Julio Flecha-Lescún** Instituto de Investigación en Ingeniería de Aragón, Universidad de Zaragoza, Zaragoza, Spain

**Sergio Gabarre** Vlaams Instituut voor Biotechnologie, Leuven, Belgium

**Alberto García** Laboratori de Calcul Numeric, Universitat Politecnica de Catalunya, Barcelona, Spain

**C. Garcia-Astrain** BCMaterials, Basque Center for Materials, Applications and Nanostructures, UPV/EHU Science Park, Leioa, Spain

**Fernanda Gentil** School of Health - P. Porto, Porto, Portugal

**H.I.G. Gomes** Faculty of Engineering of University of Porto (FEUP), Porto, Portugal

**Luis Gracia** Department of Mechanical Engineering, University of Zaragoza; Aragón Institute for Engineering Research, Zaragoza, Spain

**Antonio Herrera** Aragón Institute for Engineering Research; Aragón Health Research Institute; Department of Surgery, University of Zaragoza, Zaragoza, Spain

**Elena Ibarz** Department of Mechanical Engineering, University of Zaragoza; Aragón Institute for Engineering Research, Zaragoza, Spain

**Helena Knopf-Marques** INSERM UMR 1121, 11 rue Humann; Protip Medical, 8 Place de l'Hôpital; Université de Strasbourg, Faculté de Chirurgie Dentaire, Fédération de Médecine Translationnelle de Strasbourg, Fédération de Recherche Matériaux et Nanosciences Grand Est, Strasbourg, France

**S. Lanceros-Mendez** BCMaterials, Basque Center for Materials, Applications and Nanostructures, UPV/EHU Science Park, Leioa; IKERBASQUE, Basque Foundation for Science, Bilbao, Spain

**Hin Lee Eng** NUS Tissue Engineering Program, Life Sciences Institute; Department of Orthopaedic Surgery, National University of Singapore, Singapore

**Javier Bayod López** Group Applied Mechanics and Bioengineering, School of Engineering and Architecture, University of Saragossa, Zaragoza, Spain

**Marta E. Losa Iglesias** Faculty of Health Sciences, Rey Juan Carlos University, Madrid, Spain

**Mauro Malvè** Department of Engineering, Public University of Navarra, Pamplona; Centro de Investigación en Red en Bioingeniería, Biomaterialesy Nanomedicina, CIBER-BBN; Aragón Institute of Engineering Research (I3A), University of Zaragoza, Zaragoza, Spain

**Hafedh Marouane** Division of Applied Mechanics, Department of Mechanical Engineering, Polytechnique, Montréal, QC, Canada

**Alexandra P. Marques** I3Bs—Research Institute on Biomaterials, Biodegradables and Biomimetics of University of Minho, Headquarters of the European Institute of Excellence on Tissue Engineering and Regenerative Medicine, University of Minho, Guimarães, Portugal; ICVS/3B's—PT Government Associate Laboratory, Guimarães, Portugal; The Discoveries Centre for Regenerative and Precision Medicine, Headquarters at University of Minho, Guimarães, Portugal

**Marco Marques** Institute of Science and Innovation in Mechanical and Industrial Engineering (INEGI); Faculty of Engineering of University of Porto (FEUP), Porto, Portugal

**Christophe Marquette** 3dFAB Université Lyon 1—CNRS 5246 ICBMS, Lyon, France

**José Javier Martín de Llano** Department of Pathology, University of Valencia and INCLIVA Health Research Institute, Valencia, Spain

**Miguel Angel Martínez** Aragón Institute of Engineering Research (I3A), University of Zaragoza, Zaragoza, Spain; Department of Mechanical Engineering, University of Zaragoza, Zaragoza, Spain; Centro de Investigación en Red en Bioingeniería, Biomaterialesy Nanomedicina, CIBER-BBN, Zaragoza, Spain

**Manuel Mata Roig** Department of Pathology, University of Valencia and INCLIVA Health Research Institute, Valencia, Spain

**Andres Mena** CIBER, Zaragoza, Spain

**Lara Milián** Department of Pathology, University of Valencia and INCLIVA Health Research Institute, Valencia, Spain

**Judith Millastre** Department of Gastroenterology, Lozano Blesa University Hospital, Zaragoza, Spain

**S. Jamaleddin Mousavi** Mines Saint-Etienne, University of Lyon, INSERM, U1059 Sainbiose, Saint-Etienne, France; Mechanical Engineering Department, School of Engineering and Architecture (EINA); Aragón Institute of Engineering Research (I3A), University of Zaragoza; Biomedical Research Networking Center in Bioengineering, Biomaterials and Nanomedicine (CIBER-BBN), Zaragoza, Spain

**Celine Blandine Muller** INSERM UMR 1121, 11 rue Humann; Protip Medical, 8 Place de l'Hôpital, Strasbourg, France

**Renato M. Natal Jorge** Institute of Science and Innovation in Mechanical and Industrial Engineering (INEGI); Faculty of Engineering of University of Porto (FEUP), Porto, Portugal

**Ignacio Ochoa** Aragon Institute of Engineering Research (I3A), University of Zaragoza; Institute of Health Research of Aragon (IIS), Zaragoza, Spain; Biomedical Research Networking Center in Bioengineering, Biomaterials and Nanomedicine (CIBER-BBN), Madrid, Spain; Human Anatomy and Histology Department, Faculty of Medicine, University of Zaragoza, Zaragoza, Spain

**Sara Oliván** Aragon Institute of Engineering Research (I3A), University of Zaragoza; Institute of Health Research of Aragon (IIS), Zaragoza, Spain; Biomedical Research Networking Center in Bioengineering, Biomaterials and Nanomedicine (CIBER-BBN), Madrid, Spain; Human Anatomy and Histology Department, Faculty of Medicine, University of Zaragoza, Zaragoza, Spain

**António F. Oliveira** ICBAS—Abel Salazar Institute of Biomedical Sciences, Porto, Portugal

**Marco Parente** INEGI, Institute of Mechanical Engineering and Industrial Management, Faculty of Engineering of University of Porto, FEUP, Porto, Portugal

**Estefania Peña** Aragón Institute of Engineering Research (I3A); Department of Mechanical Engineering, University of Zaragoza; Centro de Investigación en Red en Bioingeniería, Biomaterialesy Nanomedicina, CIBER-BBN, Zaragoza, Spain

**Juan A. Peña** Aragón Institute of Engineering Research (I3A); Department of Management and Manufacturing Engineering, University of Zaragoza, Zaragoza, Spain

**Claudie Petit** Mines Saint-Etienne, University of Lyon, INSERM, U1059 Sainbiose, Saint-Etienne, France

**M.M.A. Peyroteo** Institute of Science and Innovation in Mechanical and Industrial Engineering (INEGI); Faculty of Engineering of University of Porto (FEUP), Porto, Portugal

**Sergio Puértolas** Department of Mechanical Engineering, University of Zaragoza; Aragón Institute for Engineering Research, Zaragoza, Spain

**José A. Puértolas** Department of Material Science, University of Zaragoza, Zaragoza, Spain

**R.R. Rama** Center for Research in Computational and Applied Mechanics; Department of Civil Engineering, Computational Continuum Mechanics Research Group, UCT, Cape Town, South Africa

**Teodora Randelovic** Aragon Institute of Engineering Research (I3A), University of Zaragoza; Institute of Health Research of Aragon (IIS), Zaragoza, Spain

**S. Ribeiro** Center/Department of Physics, University of Minho; Centre of Molecular and Environmental Biology (CBMA), Universidade do Minho, Braga, Portugal

**C. Ribeiro** Center/Department of Physics; CEB—Centre of Biological Engineering, University of Minho, Braga, Portugal

**José Félix Rodríguez Matas** Chemistry, Materials and Chemical Engineering Department "Giulio Natta", Politecnico di Milano, Milan, Italy

**Pablo Sáez** Laboratori de Calcul Numeric, Universitat Politecnica de Catalunya, Barcelona, Spain

**María Sancho-Tello** Department of Pathology, University of Valencia and INCLIVA Health Research Institute, Valencia, Spain

**Carla F. Santos** Faculty of Engineering of University of Porto, (FEUP), Porto, Portugal

**Masoud Sharifi** Division of Applied Mechanics, Department of Mechanical Engineering, Polytechnique, Montréal, QC, Canada

**Aboulfazl Shirazi-Adl** Division of Applied Mechanics, Department of Mechanical Engineering, Polytechnique, Montréal, QC, Canada

**S. Skatulla** Center for Research in Computational and Applied Mechanics; Department of Civil Engineering, Computational Continuum Mechanics Research Group, UCT, Cape Town, South Africa

**Nihal Engin Vrana** INSERM UMR 1121, 11 rue Humann; Protip Medical, 8 Place de l'Hôpital, Strasbourg, France

**Yingnan Wu** NUS Tissue Engineering Program, Life Sciences Institute; Department of Orthopaedic Surgery, National University of Singapore, Singapore

**Zheng Yang** NUS Tissue Engineering Program, Life Sciences Institute; Department of Orthopaedic Surgery, National University of Singapore, Singapore

**Lu Yin** NUS Tissue Engineering Program, Life Sciences Institute; Department of Orthopaedic Surgery, National University of Singapore, Singapore

# PART I

# BIOMECHANICS

CHAPTER

# 1

# Personalized Corneal Biomechanics


*Miguel Ángel Ariza-Gracia**,†, *Julio Flecha-Lescún**, *José Félix Rodríguez Matas*‡, *Begoña Calvo Calzada**,§

*Instituto de Investigación en Ingeniería de Aragón, Universidad de Zaragoza, Zaragoza, Spain †Institute for Surgical Technology and Biomechanics, Universität Bern, Bern, Switzerland ‡Chemistry, Materials and Chemical Engineering Department "Giulio Natta", Politecnico di Milano, Milan, Italy §Centro de Investigación Biomédica en Red en el área temática de Bioingeniería, Biomateriales y Nanomedicina (CIBER-BBN), Madrid, Spain


## 1.1 INTRODUCTION

About 90% of incoming information reaches the brain through the eyes. According to the World Health Organization (WHO), about 285 million people are visually impaired worldwide. Globally, the first cause of visual impairment is uncorrected refractive error: myopia, hyperopia, astigmatism, and age-related presbyopia represent 43% of the total (not including presbyopia). Cataracts, with 33%, and glaucoma, with 2%, are the second- and third-leading causes of visual impairment.[1] Refractive errors in Western Europe and the United States affect one-third of people over 40 years old.

Nowadays, refractive surgeries are applied to change the curvature of the corneal surface and to modify its optical power. Despite the surgical breakthroughs over the last decades (radial keratotomy [RK], photorefractive keratotomy [PRK], and laser in situ keratomileusis [LASIK]), the unpredictability of the surgical outcomes remains. This unpredictability is manifested inside effects that can lead to unexpected results in visual acuity after an intervention. Sometimes, undercorrection (≈11.9%) or overcorrection (≈4.2%) may occur and a second "enhancement" procedure is required. In many cases, additional surgery may be used to refine the result. According to the Food and Drug Administration, close to one million LASIK procedures are performed annually in the United States, positioning it as one of the most common surgeries.

Regarding ectatic disorders, Keratoconus (KC) shows major incidence in the general population (1–430/2000),[2] but official statistics do not include those who have been misdiagnosed or lately diagnosed. KC has a negative impact on the patient's life because it decreases visual acuity and has a lasting negative impact on all aspects of a patient's life. Keratoconus affects three million patients worldwide with a higher prevalence among females [1]. Also, South Asian ethnicity has an incidence probability 4.4 times higher than Caucasians, and they are also more prone to be affected earlier [2]. Not only that, but advanced Keratoconus can cause corneal blindness, which is responsible for 40,000 people needing a corneal transplant in Europe every year.[3]

The corneal shape is the result of the equilibrium between its mechanical stiffness (related to the corneal geometry and the intrinsic stiffness of the corneal tissue), intraocular pressure (IOP), and the external forces acting upon it such as external pressure. An imbalance between these parameters, for example, an increment of IOP, a decrement of the corneal thickness induced by refractive surgery, or a corneal material weakening due to a disruption of collagen fibers, can produce ocular pathologies (ectasias) that seriously affect a patient's sight.

[1] World Health Organization, Visual Impairment and Blindness; Fact Sheet No. 282 (http://www.who.int/mediacentre/factsheets/fs282/en/index.html, WHO link to digital version).

[2] The Global Keratoconus Foundation (http://www.kcglobal.org/, the Global Keratoconus Foundation link to digital version).

[3] See http://cordis.europa.eu/news/rcn/32213_en.html, CORDIS Europe link to digital version.


Advances in Biomechanics and Tissue Regeneration
https://doi.org/10.1016/B978-0-12-816390-0.00001-7

Consequently, it is important to understand how these ocular factors are related to pathologies in order to improve treatments. In order to do that different corneal features must be properly characterized:

- physiological conditions of the eye: IOP and interaction of the eyeball with the surrounding media;
- patient-specific corneal geometry; and
- patient-specific mechanical properties of the eye.

To date, IOP can be measured using contact tonometers (e.g., Goldmann Applanation Tonometry) [3, 4], whereas corneal topography is obtained with corneal topographers (e.g., Pentacam or Sirius [5]). The availability of high-resolution topographical data and a patient's IOP have made it possible to reconstruct a patient's specific geometric model. In this regard, some patient-specific models have already been reported in the literature [6, 7]. However, the workflow described in these studies cannot be automated in a straightforward manner so as to permit personalized analysis on large populations in order to, for example, characterize the mechanical properties of the corneal tissue.

Noncontact tonometry (e.g., CorVis ST, Oculus Optikgeräte GmbH [8]) has recently gained interest as a diagnostic tool in ophthalmology as an alternative method for characterizing the mechanical behavior of the cornea. In a noncontact tonometry test, a high-velocity air jet is applied to the cornea for a very short time (less than 30 ms), causing the cornea to deform while the corneal motion is recorded by a high-speed camera. A number of biomarkers associated with the motion of the cornea, that is, maximum corneal displacement and the time between the first and second applications, among others, have been proposed to characterize preoperative and postoperative biomechanical changes [8–16].

As the dynamic response is the result of the interplay between different corneal features (IOP, geometry, material), it is reasonable to argue that a misunderstanding of the diagnostic tools is likely to be the cause of the unexpected clinical results already occurring (e.g., a softer cornea with a higher IOP could show the same behavior as a stiffer cornea with a lower IOP). Although geometry and IOP can already be measured accurately, the mechanical behavior of the cornea cannot be directly characterized in vivo.

Precise knowledge about the underlying factors that affect the corneal mechanical response will allow establishing better clinical diagnoses, monitoring the progression of different diseases (e.g., Keratoconus), or designing a priori patient-specific surgical plans that may reduce the occurrence of unexpected outcomes.

The construction of predictors for real-time clinical applications must rely on mathematical tools that, given a set of clinical biomarkers, can return the material parameters of a given constitutive model [17]. In the present study, a K-nn (nearest neighbor) approach is used to determine the corneal material parameters using three clinical biomarkers: the maximum corneal displacement measured during a noncontact tonometry test ($U$), the patient's IOP, and the geometrical features of the cornea.

This chapter explores methodologies to determine the patient-specific geometry and mechanical properties of the cornea. Shedding light on patient-specific corneal biomechanics will allow performing a personalized assessment in ocular surgeries and treatments. This is further demonstrated by two applications: the prediction of a patient-specific refractive surgery (astigmatic keratotomy [AK]) in an animal model, and the qualitative assessment in the level of stresses induced by an intracorneal ring segment implantation in humans that, clinically, is impossible to measure.

## 1.2 EYE ANATOMY

The eye is composed of different structures and layers (see Fig. 1.1). Among the most important macroscopic structures, those providing the eyeball's shape are the cornea (i.e., the outermost transparent layer), the sclera (i.e., the white layer protecting and shaping the eye), and the limbus (i.e., the transition between the cornea and sclera). Besides, the cornea, which represents ≈45 of the 60 diopters of the optical power of a relaxed eye, and the crystalline, ciliary muscles, retina, and optical nerve are the optical elements in charge of vision quality. Generally, ocular structures present three main layers: the fibrous layer that protects and gives the shape (tunica externa bulbi), the vascular layer that perfuses the organ (tunica vasculosa bulbi), and the nervous layer that provides the sensorial faculties (tunica interna bulbi). The mechanical compliance of the human eye is mainly associated with the collagen fibrils embedded in the fibrous layer (cornea, sclera, limbus, and lamina cribosa). Although human eye dimensions vary significantly between patients, average measures can be set. Generally, the main dimensions of an emmetropic eye (nonrefractive errors) are

- an axial (sagittal) diameter of 24–25 mm (i.e., the distance between the corneal apex and the sclera);
- a transversal (i.e., nasal-temporal plane) diameter of 23.5 mm;
- a vertical (i.e., superior-inferior plane) diameter of 23 mm;
- a mean corneal diameter of 11–12 mm;

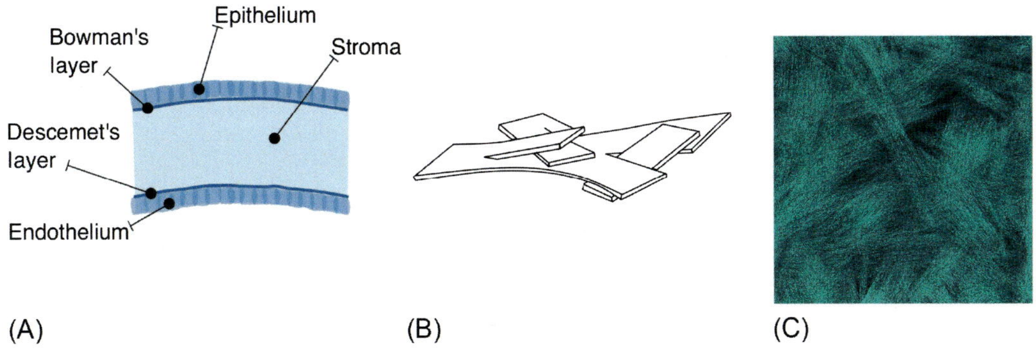


FIG. 1.1 Structure of the cornea. (A) Conceptual diagram of the different corneal layers, from the most external layer (epithelium) to the most internal layer (endothelium). The stroma represents almost 90% of the corneal thickness; (B) 3D diagram of the out-of-plane collagen interwoven; (C) second-harmonic generation imaging of a porcine cornea. In-plane collagen fibers distribution ($z$-depth of 75 μm, $x-y$ resolution of 100 μm). *(A, B) Adapted from K. Anderson, A. El-Sheikh, T. Newson, Application of structural analysis to the mechanical behaviour of the cornea, J. R. Soc. Interface 1 (April) (2004) 3–15; (C) taken by D. Haenni, M.A. Ariza-Gracia, P. Büchler at ZMB, University of Zurich.*

- an increasing thickness from the center to the periphery (550–750 μm);
- a volume of ≈6 $cm^3$; and
- a weight of ≈7.5 g.

Finally, the eye is inserted in the ocular socket, surrounded by fat tissue, held by the extraocular muscles and the optical nerve, and protected from external agents by the lids and eyelids. To preserve the shape, the eyeball is filled with the aqueous humor (anterior chamber) and the vitreous humor (posterior chamber), and is subjected to a typical IOP ranging between 12 and 22 mmHg in healthy patients.

Tissue-speaking, the cornea is a highly porous tissue formed by a laminar structure. Apart from the high water content (around 80%), there are three main layers: the epithelium, the endothelium, and the central stroma (see Fig. 1.1A). Apart from these main layers, there are specialized extracellular structures called Bowman and Descemet membranes [18, 19]. The constitution of each layer is vastly different. However, the most important is the stroma, which represents 90% of the corneal thickness. Its structure presents several overlapping collagen lamellae composed of bundles of collagen fibrils (see Fig. 1.1B and C) surrounded by a gelatinous matrix mostly composed of glycoproteins. The microstructure of the stroma is highly heterogeneous, depending on the specific region and corneal layer being evaluated [20–24]. The anterior stromal lamellae are more closely packed and less hydrated than the posterior stroma, with stronger junctions between collagen lamellas. Thus, the anterior stroma is suggested to hold the main role in maintaining the corneal strength and curvature. This anisotropy in the stromal architecture is also suggested to result in an anisotropic mechanical behavior of the corneal tissue, being supported by experimental and clinical studies [25–28]. Furthermore, collagen fibers are differently distributed over the surface and thickness. This leads to a complex behavior, exhibiting different zone-wise mechanical (no time-dependent) and dynamical (time-dependent) properties.

## 1.3 PATIENT-SPECIFIC GEOMETRY

Different imaging techniques have been developed in recent to evaluate the geometry of the cornea [29, 30], but the most common and important is corneal topography [31–34], a noninvasive imaging technique for mapping the anterior and posterior surfaces of the cornea. Nowadays, there are two technologies used to measure corneal topographies: Placido-based systems (reflection-based) and Scheimpflug photography-based (projection-based) [35–37] systems.

Sirius and Pentacam are among the most-used devices in clinics. Sirius enables retrieving 25 radial sections of the cornea and anterior chamber in a few seconds, measuring 35,632 points on the anterior surface and 30,000 on the posterior surface (in high-resolution mode). Furthermore, it provides consistent measurements of curvatures (anterior and posterior), pachymetry, and anterior chamber depth [38–40]. The Oculus Pentacam calculates a three-dimensional (3D) topographical model of the anterior eye segment using as many as 25,000 true elevation points. Pentacam corneal topographies are represented as point cloud surfaces in the form of two 141 × 141 matrices. The first matrix contains the coordinates $(x, y, z)$ of the anterior corneal surface, whereas the second matrix represents the available pachymetry (corneal thickness) data at each $(x, y)$ point. Because pachymetry data are sometimes not available at all points in the anterior surface point cloud, the number of nonzero elements in the pachymetry matrix determines the total number of available data points for surface reconstruction. The posterior surface is the result of a point-to-point subtraction between the anterior surface and the pachymetry data.

The availability of high-resolution topographical data and the patient's IOP have made it possible to reconstruct a patient's specific geometric model of the cornea, which makes it possible to study specific treatments and pathologies, develop a robust methodology to incorporate a patient's specific corneal topology into a finite element (FE) model of the eyeball, and account for the stress-free configuration of the eyeball.

### 1.3.1 Corneal Surface Reconstruction

A reliable patient-specific FE model of the cornea must incorporate the patient's topographical data as much as possible. In this regard, the proposed framework makes use of actual patient data where available, minimizing the amount of extrapolated data required to build a full 3D FE model amenable for numerical simulations. Current topographers provide topographical data limited to a corneal area between 8 and 9 mm in diameter due to patient misalignment, blinking, or eyelid aperture (see Fig. 1.2A). However, a corneal diameter of 12 mm (average human size) is needed to build a 3D FE model [6, 7].

In order to overcome this limitation, a surface continuation algorithm is proposed. Data extrapolation is performed by means of a quadric surface given in matrix notation as

$$\mathbf{x}^T\mathbf{A}\mathbf{x} + 2\mathbf{B}^T\mathbf{x} + c = 0, \tag{1.1}$$

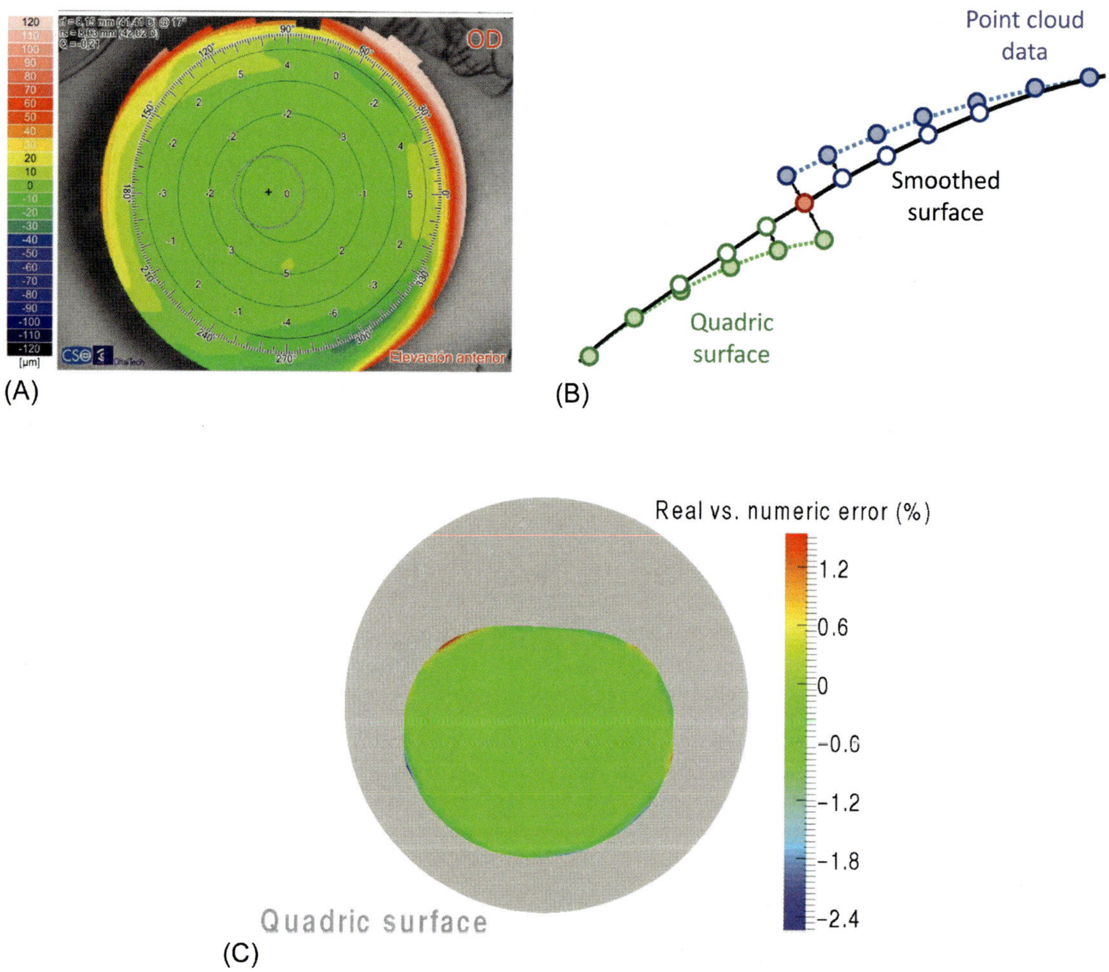


FIG. 1.2 Corneal surface reconstruction. (A) Anterior elevation of healthy cornea measured with Sirius; (B) surface smoothing at the joint between the extended surface and the patient's corneal surface; (C) projection of the 12 mm diameter corneal surface in the optical axis plane. *Gray area* corresponds to the extended surface required in order to achieve a 12-mm diameter (approximating surface). Contour map of the error between the point cloud data prior and after smoothing (less than 5% at the corneal periphery). *Adapted from M.Á. Ariza-Gracia, J. Zurita, D.P. Piñero, B. Calvo, J.F. Rodríguez Matas, Automatized patient-specific methodology for numerical determination of biomechanical corneal response, Ann. Biomed. Eng. 44 (5) (2016) 1753–1772.*

where A is a $3 \times 3$ constant matrix, B is a $3 \times 1$ constant vector, and $c$ is a scalar, which defines the parameters of the surface. Eq. (1.1) is fitted to the topographical data by means of a nonlinear regression analysis.

To extend the corneal surface, the quadric surface, Eq. (1.1) should properly approximate the periphery of the patient's topographical data. For this reason and before fitting Eq. (1.1), the central corneal part is removed using a level-set algorithm based on the relative elevation of each corneal point with respect to the apex (for further details see Ref. [12]).

When using an analytical surface such as Eq. (1.1) to extend the corneal surface, there will always be a jump at the joint between the approximating surface and the point cloud surface (see Fig. 1.2B). This discontinuity in the normal of the surface may lead to convergence problems or to nonrealistic stress distributions on the cornea during FE analysis. Hence, a smoothing algorithm based on the continuity of the normal between the quadric surface and the point cloud data is applied, as shown in Fig. 1.2B, producing local alterations in the patient's topographic data near the border. However, these alterations are very small (less than 3%) as outlined in the contour map of the error between the topographic point cloud data prior and after smoothing (Fig. 1.2C), where the depicted data corresponds to an extreme post-LASIK patient.

## 1.3.2 Corneal Surface Finite Element Model

Once the corneal surface fitting is completed, it is introduced in the 3D model of the anterior half ocular globe, which accounts for three different parts: the cornea, the limbus, and the sclera. Because only the cornea can be partially measured by a topographer and neither the sclera nor the limbus can be measured with this procedure, average parts are used instead. The sclera was assumed as a 25 mm in diameter sphere with a constant thickness of 1 mm, whereas the limbus is a ring linking both the sclera and the cornea. The geometry has been meshed using hexahedral elements by means of an in-house C program, as shown in Fig. 1.3A, thus allowing precise control of the mesh size as well as generating meshes with trilinear (8 nodes) or quadratic (20 nodes) hexahedral elements. Pachymetry data measured with the topographer are accurately mapped onto the 3D FE model during mesh generation. Finally, the FE model of the eyeball is completed by defining the corneal fibers over the two preferential orientations (a nasal-temporal and superior-inferior directions) and one single circumferential orientation embedded in the limbus (Fig. 1.3B).

Symmetry boundary conditions have been defined at the scleral equator ($\Pi$ plane in Fig. 1.3A) [41, 42], that is, the base of the semisclera. The optical nerve insertion was neglected as it is not necessary for the present simulation. Hence, the boundary nodes are allowed to move on the symmetry plane $\Pi$ but not normal to the plane, resulting in a much less restrictive boundary condition than fixing all nodal degrees of freedom [6, 7]. In addition, the inner surface of the eyeball is subject to the actual patient's IOP, which was previously measured by means of Goldmann Applanation Tonometry. The FE model is generated with quadratic hexahedra and 5 elements through the thickness (11 nodes), resulting in an eyeball with 62,276 nodes (186,828 degrees of freedom) and 13,425 quadratic elements. All FE simulations and methodologies presented in this chapter are carried out using the commercial FE software Abaqus (Dassault Systemes Simulia Corp.) and MATLAB (MathWorks).

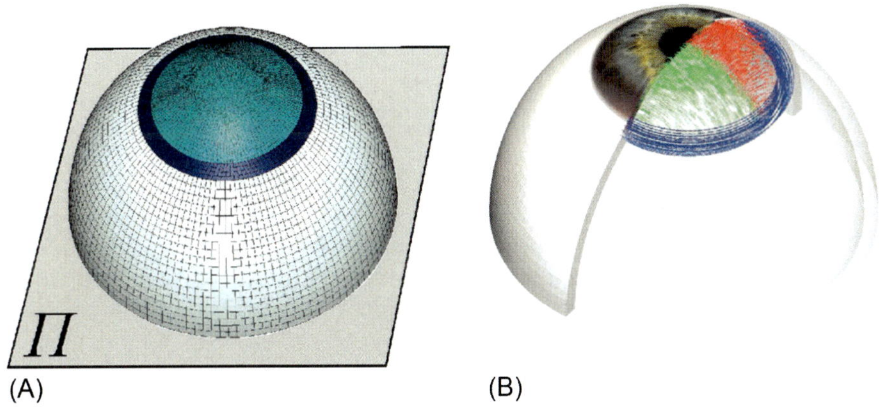


FIG. 1.3 Numerical model of the eyeball. (A) Finite element mesh of the eyeball: Sclera (*white region*), limbus (*dark blue region*), cornea (*light blue region*); (B) direction of collagen fibers. Two orthogonal directions for the cornea (*red and green fibers*), and one circumferential direction of the limbus (*blue fibers*). *Adapted from M.Á. Ariza-Gracia, J. Zurita, D.P. Piñero, B. Calvo, J.F. Rodríguez Matas, Automatized patient-specific methodology for numerical determination of biomechanical corneal response, Ann. Biomed. Eng. 44 (5) (2016) 1753–1772.*

**GIVEN**

$\boldsymbol{X}_{REF}$, fibers direction $\boldsymbol{M}$ and $\boldsymbol{N}$, itemax, TOL

**INITIALIZE**

$k=0,\ \boldsymbol{X}_k = \boldsymbol{X}_{REF},\ \boldsymbol{M}_k = \boldsymbol{M},\ \boldsymbol{N}_k = \boldsymbol{N}$

**REPEAT**

(i) FE analysis: internal pressure $(IOP*)$
$\boldsymbol{X}_k^d \rightarrow \text{ABAQUS}(\boldsymbol{X}_k, \boldsymbol{M}_k, \boldsymbol{N}_k)$
$\mathbf{E}_k = \boldsymbol{X}_k^d - \boldsymbol{X}_{REF}$
$e_k = \|\boldsymbol{X}_k^d - \boldsymbol{X}_{REF}\|_\infty$

(ii) Update zero-pressure geometry
$\boldsymbol{X}_{k+1} = \boldsymbol{X}_k - \mathbf{E}_k$

(iii) Fibre pull—back (cornea & limbus)
deformation gradient $\boldsymbol{F}^{k+1} = \frac{\partial \boldsymbol{X}_{REF}}{\partial \boldsymbol{X}_{k+1}}$
$\boldsymbol{M}_{k+1} = (\boldsymbol{F}^{k+1})^{-1}\boldsymbol{M}$
$\boldsymbol{N}_{k+1} = (\boldsymbol{F}^{k+1})^{-1}\boldsymbol{N}$

(iv) $k = k+1$

**UNTIL** OR($e_k \leqslant$ TOL, $k >$ itemax)

(A)

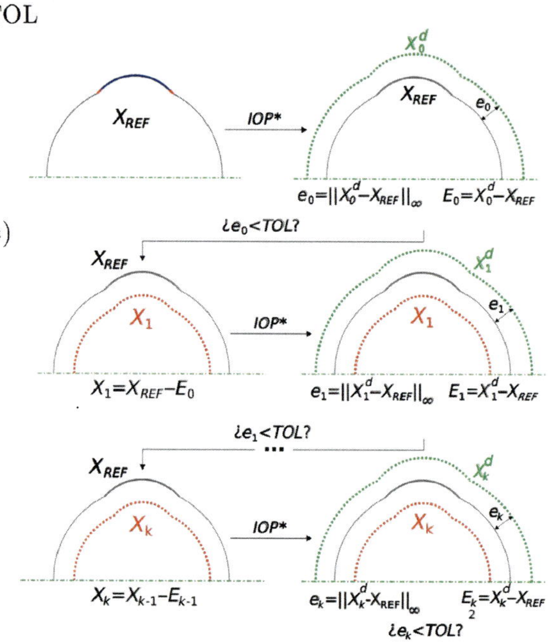


(B)

FIG. 1.4 Zero-pressure algorithm. (A) Zero-pressure algorithm accounting for the pull-back algorithm with a consistent mapping of the fibers onto the current unloaded state; (B) iterative scheme of the algorithm. *Adapted from M.Á. Ariza-Gracia, J. Zurita, D.P. Piñero, B. Calvo, J.F. Rodríguez Matas, Automatized patient-specific methodology for numerical determination of biomechanical corneal response, Ann. Biomed. Eng. 44 (5) (2016) 1753–1772.*

### 1.3.3 Stress-Free Configuration of the Eyeball: Reference Geometry

When an eye is measured by a topographer, the identified geometry belongs to a deformed configuration due to the effect of the IOP (hereafter referred to as the image-based geometry) but the corneal prestress is neglected. Hence, an accurate stress analysis of the cornea starts by identifying the initial state of stresses due to the physiological IOP present on the image-based geometry, or equivalently, the actual geometry associated with the absence of IOP (hereafter referred to as the zero-pressure geometry), as shown in Fig. 1.4A. Consequently, an iterative algorithm is used to find the zero-pressure configuration of the eye [43] (see in the algorithm in Fig. 1.4B) that keeps the mesh connectivity unchanged and iteratively updates the nodal coordinates. Moreover, the local directions of anisotropy (orientation of collagen fibers) are also consistently pulled back to the current zero-pressure configuration.

In Fig. 1.4, $X_{REF}$ stands for the patient's geometry reconstructed from the topographer's data, where $\mathbf{X}$ represents an $N_n \times 3$ matrix that stores the nodal coordinates of the FE eyeball, with $N_n$ the number of nodes in the FE mesh; $X_k$ is the zero-pressure configuration identified at iteration $k$; and $\mathbf{X}_k^d$ is the deformed configuration obtained when inflating the zero-pressure configuration $X_k$ at the IOP pressure. The iterative algorithm updates the zero-pressure geometry, $X_k$, until the infinite norm of the nodal error between $X_{REF}$ and $X_k^d$ is less than a tolerance, *TOL*.

## 1.4 PATIENT-SPECIFIC MATERIAL BEHAVIOR

A number of material models have been proposed to reproduce the behavior of the cornea, ranging from simple hyperelastic isotropic materials [44] to more complex models coupling the hyperelastic isotropic response for the matrix (i.e., neo-Hookean models) with the anisotropic response of the collagen fibers of the eye [7, 25, 41, 42, 45–48]. These material models have been incorporated into computer models of the eye to simulate surgical interventions and tonometry tests in an effort to demonstrate the potential of these in silico models [6, 7, 11, 12, 49–54]. Once a constitutive model is chosen, it must be particularized for each patient. Unfortunately, many of the methodologies to retrieve material parameters require a high computational effort and cannot be used in clinics. The proposed predictive tool relies on a dataset generated by the results of FE simulations of the noncontact tonometry test. The simulations are based on combinations of patients of a real clinical database (the patient-specific corneal geometry and the Goldmann IOP [12]) and of corneal material properties of the numerical model to predict the corneal apical displacement.

In brief, the FE model is used to perform a Monte Carlo (MC) simulation in which the material parameters and the IOP uniformly vary within an established range. The range of the material parameters was determined by considering the experimental results from an inflation test reported in the literature [48, 55] and the physiological response of the cornea to an air-puff device (i.e., displacement of the cornea using a CorVis device).

First, the inflation tests were used to initially screen the model parameters, to constrain the search space of the optimization, and to avoid an ill-posed solution [56]. Second, the range of each material parameter was then determined such that the in silico inflation curve was within the experimental window. In this way, both physiological behaviors of the cornea are simultaneously fulfilled: the response to an inflation test (biaxial stress) and the response to an air-puff test (bending stress). Subsequently, the generated dataset was used to implement different predictors for the mechanical properties of the patient's corneal model in terms of variables that are identified in a standard noncontact tonometry test.

### 1.4.1 Material Model

One feasible form of the strain energy function for modeling the cornea corresponds to a modified version of that proposed by Gasser et al. [57] for arterial tissue, where the neo-Hookean term has been substituted by an exponential term

$$\begin{aligned}\psi(\mathbf{C},n_\alpha) &= D_1\cdot\{\exp[D_2\cdot(\tilde{I}_1-3)]-1\}+\frac{k_1}{2\cdot k_2}\cdot\sum_{\alpha=1}^{N}\{\exp[k_2\langle\tilde{E}_\alpha\rangle^2]-1\}\\ &+K_0\cdot\left(\frac{J_{el}^2-1}{2}-\ln(J_{el})\right),\\ \text{with } \tilde{E}_\alpha &\overset{\text{def}}{=}\kappa\cdot(\tilde{I}_1-3)+(1-3\kappa)\cdot(\tilde{I}_{4(\alpha\alpha)}-1),\end{aligned} \tag{1.2}$$

where C is the right Cauchy-Green tensor; $J_{el}=\sqrt{\det\mathbf{C}}$ is the elastic volume ratio; $D_1$, $D_2$, $k_1$, and $k_2$ are material parameters; $K_0$ is the bulk modulus; $N$ is the number of families of fibers; $\tilde{I}_1$ is the first invariant of the modified right Cauchy-Green Tensor $\tilde{\mathbf{C}}=J_{el}^{-2/3}\mathrm{C}$; and $\tilde{I}_{4(\alpha\alpha)}=n_\alpha\cdot\tilde{\mathbf{C}}\cdot n_\alpha$ is the square of the stretch along the fiber's direction $n_\alpha$. The parameter $\kappa$ describes the level of dispersion in the fiber's direction and has been assumed to be zero because it has been reported that a dispersion in the fibers of $\pm 10$ degrees about the main direction results in a maximum variation of 0.03% on the maximum corneal displacement [12].

The strain-like term $\tilde{E}_\alpha$ in Eq. (1.2) characterizes the deformation of the family of fibers with preferred direction $n_\alpha$. The model assumes that collagen fibers bear load only in tension while they buckle under compressive loading. Hence, only when the strain of the fibers is positive, that is, $\tilde{E}_\alpha>0$, do the fibers contribute to the strain energy function. This condition is enforced by the term $\langle\tilde{E}_\alpha\rangle$, where the operator $\langle\cdot\rangle$ stands for the Macauley bracket defined as $\langle x\rangle=\frac{1}{2}(|x|+x)$. The model has been implemented in a *UANISOHYPER* user subroutine (Abaqus, Dassault Systèmes).

Due to the random distribution of the fibers, far from the optic nerve insertion, the sclera has been assumed to be an isotropic hyperelastic material [58] (Eq. 1.3).

$$\psi_Y=\sum_{i=1}^{3}K_i(J_{el}-1)^{2\cdot i}+\sum_{i=1}^{3}C_{i0}\cdot(\tilde{I}_1-3)^i, \tag{1.3}$$

where $C_{10}=810$ [kPa], $C_{20}=56,050$ [kPa], $C_{30}=2,332,260$ [kPa], and $K_i$ [kPa] is automatically set by the FE solver during execution.

### 1.4.2 Monte Carlo Simulation

In order to obtain the personalized corneal material parameters for a given patient, it is necessary to build a reliable dataset on which to fit or train a predictive model. In the present case, we chose to construct our dataset using an MC analysis. First, the upper and lower boundaries of the material parameters were searched to restrict the number of combinations. This prescreening experiment used ex vivo inflation experiments [48, 55] to establish a reliable range of material parameters that made our simulations behave physiologically under membrane tension. A total of

81 combinations were simulated to mimic the inflation experiments. The in silico inflation curves were then compared with experiments [48, 55] and the range of material parameters leading to curves within the experimental window was determined. The identified range of parameters was set to $D_1$[kPa] $\in$ (0.0492, 0.492), $D_2$[−] $\in$ (70, 144), $k_1$[kPa] $\in$ (15, 130), and $k_2$[−] $\in$ (10, 1000).

Afterward, the MC analysis was used to generate the dataset. A uniform distribution of the material parameters was assumed because there are no a priori data on the dispersion of the mechanical parameters in the human cornea and, therefore, a total ignorance about the population is assumed. Otherwise, a bias could be introduced on the outcome of the system. Additionally, to account for the physiological diurnal variations in the IOP [59], variations in the IOP ranging from 8 to 30 mmHg along with the patient's IOP at the moment of the examination were also considered in the MC simulation. Hence, for each available geometry in the clinical database, 72 different samples of the material parameters and the IOP, uniformly distributed in their respective ranges, were used to conduct 72 simulations of the noncontact tonometry test. Consequently, a total of 9360 computations (i.e., 72 combinations × 130 geometries) was scheduled. The generated dataset consisted of the following variables: classification (healthy, KC, and LASIK), computation exit status (failed or successful), material parameters ($D_1$, $D_2$, $k_1$, and $k_2$), IOP, CCT, nasal-temporal curvature ($R_h$), superior-inferior curvature ($R_v$), and the computed maximum displacement of the cornea ($U_{num}$).

After the dataset was generated, an ANOVA analysis was done to identify the most influential model parameters (geometry, pressure, and material) on the numerical displacement, $U_{num}$, obtained with the noncontact tonometry simulation. The results from this analysis were used to identify the geometric parameters to be included in the construction of the predictor functions for the material parameters. ANOVA was conducted on the global dataset without differentiation between the populations and for each of the populations (healthy, Keratoconus or KC, and LASIK). Because the dataset is randomly generated, ANOVA cannot be conducted directly on the data. Instead, a quadratic response surface was first fitted to $U_{num}$ (e.g., $U_{num} = f$(geometry, pressure, material)). Then, a Pareto analysis (i.e., it states the most influential parameters on an objective variable, arranging them in decreasing order by taking into account the cumulative sum of the influence until reaching a 95% variation on the objective variable) was used to determine the most influential parameters on the dependent variable, $U_{num}$.

The simulations show that the proposed material model is adequate to reproduce both the inflation and the bending response of the cornea when subjected to an air puff for different levels of the IOP (see Fig. 1.5A). In particular, the range of parameters used for the MC simulation is able to accommodate the experimental response to corneal inflation tests reported in the literature (see Fig. 1.5B). Note that traditional model development of corneal mechanics has mainly considered inflation tests to identify the model parameters. However, when the response to an air puff is considered, we found that there are a number of combinations for which the inflation response is within the experimental range but the corneal displacement due to the air puff is not. An example of this situation is given by the red and blue lines in Fig. 1.5A. In both cases, the response to the inflation test is identical, but the response to the air puff is not physiological for the red line. Therefore, from the total number of samples in the MC simulation, only those samples that reconcile the response to an inflation and to an air-puff test to be within the experimental ranges were considered [9, 13, 60]. After including this exclusion criterion, only 29% (1127 of 3855) of the healthy cases, 30.5% (1327 of 4344) of the KC cases, and 21.5% (219 of 1017) of the LASIK cases were included in the training dataset (see Fig. 1.5C for a healthy population).

The empirical distribution of the material parameters related to the matrix ($D_1$ and $D_2$) did not follow a uniform distribution, whereas those related to the fibers ($k_1$ and $k_2$) were found to be uniformly distributed (the results are shown in Ariza-Gracia et al. [17]). A Kolmogorov-Smirnov test showed nonsignificant differences between the material parameters of the healthy LASIK and the KC LASIK populations (see Table 1.1). By contrast, significant differences were found for $D_1$ and $D_2$ between the healthy KC populations.

When the cornea is under the action of the IOP (i.e., its physiological stress state), the cornea is in a membrane stress state where the full cornea works in tension (i.e., both extracellular matrices and both families of collagen fibers), and therefore, no bending effects exist. However, during an air puff, the cornea experiences bending. Whereas the anterior surface goes from a traction state of stress to a compression state of stress, the posterior surface works in tension. Hence, in the anterior corneal stroma, the collagen fibers are not contributing to load bearing because they do not support buckling and the stiffness of the cornea mainly relies on the extracellular matrix. At the same time, the collagen fibers on the posterior stroma suffer from a higher elongation, resulting in an overall nonphysiological state of stress. In this regard, due to the action of the IOP, no significant differences in the maximum principal stress and in the maximum principal stretch were observed between the different populations for both the anterior and posterior corneal surfaces. In contrast, when the maximum principal stress and stretch are compared at the instant of the maximum corneal displacement, significant statistical differences between all populations were found on the posterior surface (see Table 1.2). However, at the anterior surface, significant differences were found only for the maximum principal stretch, whereas for the maximum principal stress, differences were found only between the healthy and KC populations (see Table 1.2).

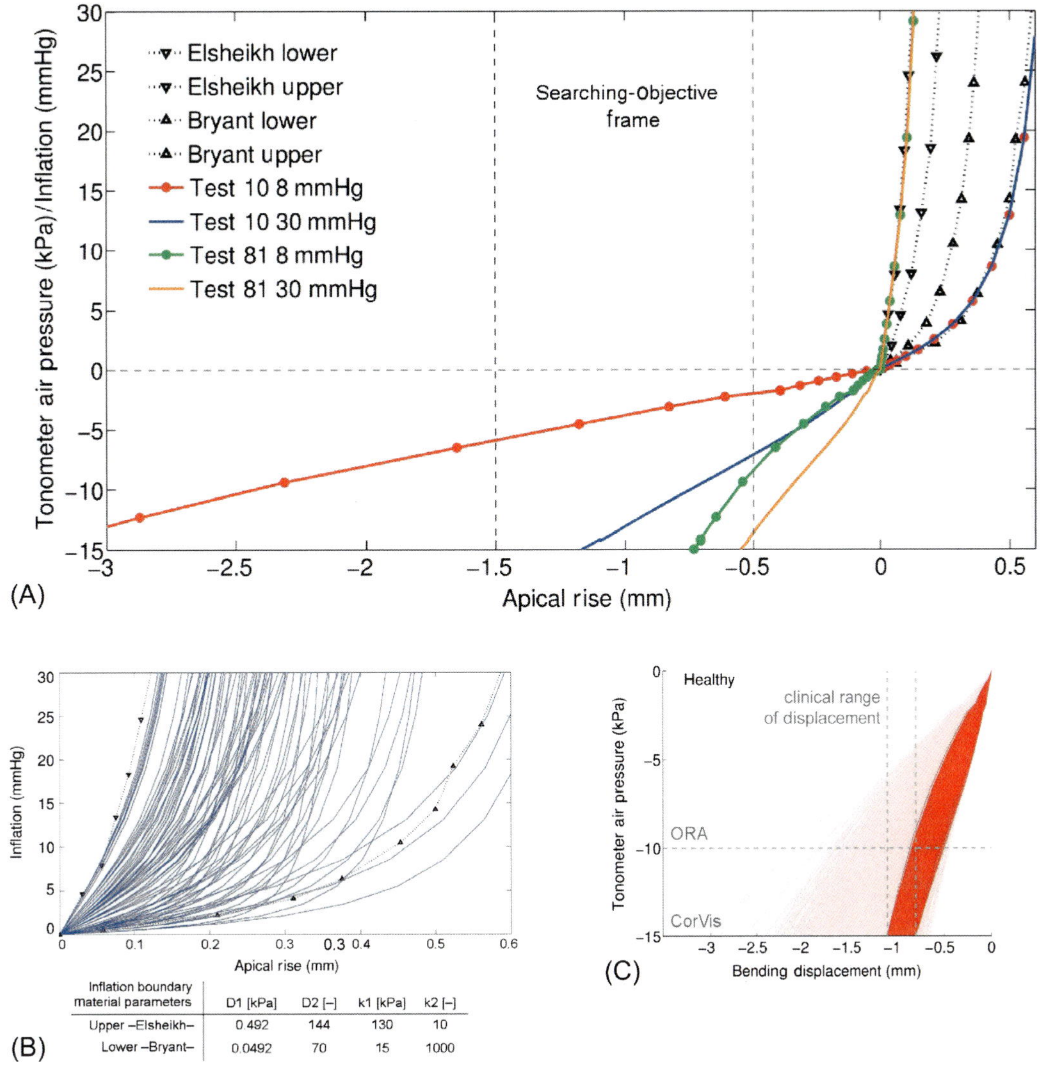


FIG. 1.5 Results of the Monte Carlo simulation. (A) Mechanical corneal response to both experiments: inflation and air puff. The physiological range for the inflation is limited by the inflation real curves reported in the literature [48, 55] (see *black dashed lines* and *triangles*), whereas the physiological range of the air-puff behavior must lie within the searching objective frame (i.e., the reported experimental displacement to CorVis [9]); (B) prescreening of the material parameters within the physiological inflation range; (C) results of the Monte Carlo simulation for a healthy population (i.e., those whose topography and IOP were diagnosed as healthy by an optometrist). *Dark red curves* belong to the simulations that cast a numerical displacement that is contained within the experimental range ($U_{Healthy}$[mm] $\in (0.8, 1.1)$). *Adapted from M.Á. Ariza-Gracia, J. Zurita, D.P. Piñero, B. Calvo, J.F. Rodríguez Matas, Automatized patient-specific methodology for numerical determination of biomechanical corneal response, Ann. Biomed. Eng. 44 (5) (2016) 1753–1772.*

TABLE 1.1 Kolmogorov-Smirnov Hypothesis Test Between Populations Regarding the Material Parameters

| | $D_1$ | | $D_2$ | | $k_1$ | | $k_2$ | |
|---|---|---|---|---|---|---|---|---|
| **Comparison** | ***h*** | ***P*-value** | ***h*** | ***P*-value** | ***h*** | ***P*-value** | ***h*** | ***P*-value** |
| Healthy KC | 1 | <0.001 | 0 | 0.058 | 0 | 0.328 | 0 | 0.983 |
| Healthy LASIK | 0 | 0.869 | 0 | 0.779 | 0 | 0.584 | 0 | 0.482 |
| KC LASIK | 0 | 0.098 | 0 | 0.161 | 0 | 0.681 | 0 | 0.725 |

*Notes:* $h$ indicates the result of the hypothesis test (i.e., $h = 1$ rejects the null hypothesis that both populations come from the same continuous probability distribution); *P*-value indicates the asymptotic *P*-value of the test (i.e., *P*-value $< 0.05$ means that the null hypothesis can be rejected at a 5% significance level).

TABLE 1.2 Kolmogorov-Smirnov Hypothesis Test Between Populations Regarding the Stress-Strain Apical Behavior

| | Anterior | | | | Posterior | | | |
|---|---|---|---|---|---|---|---|---|
| | Stretch | | Stress | | Stretch | | Stress | |
| Comparison | *h* | *P*-value | *h* | *P*-value | *h* | *P*-value | *h* | *P*-value |
| Healthy KC | 1 | <0.001 | 1 | <0.001 | 1 | <0.001 | 1 | <0.001 |
| Healthy LASIK | 1 | <0.001 | 0 | 0.073 | 1 | <0.001 | 1 | <0.001 |
| KC LASIK | 1 | <0.001 | 0 | 0.083 | 1 | <0.001 | 1 | 0.049 |

*Notes*: *h* indicates the result of the hypothesis test (i.e., $h = 1$ rejects the null hypothesis that both populations come from the same continuous probability distribution); *P*-value indicates asymptotic *P*-value of the test (i.e., *P*-value $< 0.05$ means that the null hypothesis can be rejected at a 5% significance level).

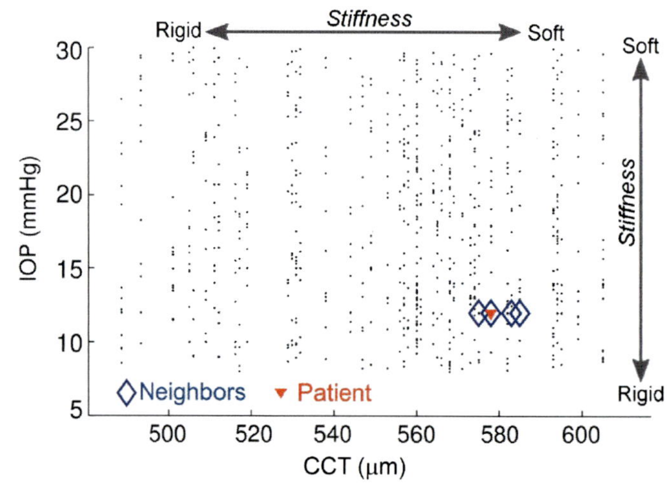


FIG. 1.6 Coupled effect of the corneal response. *Black dots* represent healthy patients with an apical displacement of 1 mm for different combinations of IOP and central corneal thickness (CCT). *Adapted from M.Á. Ariza-Gracia, S. Redondo, D.P. Llorens, B. Calvo, J.F. Rodríguez Matas, A predictive tool for determining patient-specific mechanical properties of human corneal tissue, Comput. Methods Appl. Mech. Eng. 317 (2016) 226–247.*

### 1.4.3 Neighborhood-Based Protocol (K-nn Search)

Once a reliable dataset was built using the MC analysis, a K-nn search method was used to determine the patient-specific material parameters. A K-nn search does not require fitting a particular mathematical function to predict the material parameters in terms of the corneal patient's geometric data and the mechanical response to the air puff because it simply searches for the closest point in the database to the patient's data (IOP, CCT, and *U*).

Furthermore, this method also helps to demonstrate the inherent coupling that exists between CCT, IOP, and *U* (see Fig. 1.6) [10]. For a given value of the IOP or CCT, different combinations of the material properties and corneal thickness lead to the same corneal displacement, *U*. When patient-specific information (IOP, CCT, and *U*) is used as an input to the dataset (see the red triangle), it is possible to define a neighborhood of feasible points around the patient's data (see blue diamonds) from which the material parameters can be estimated. Although this method is straightforward in terms of searching and implementation, it is also very expensive in terms of computation because the accuracy of the method is highly affected by the resolution of the grid used for the dataset (number of samples present in the dataset).

### 1.4.4 Validation With Clinical Data

To validate the predictive tool, five unknown patients were considered to predict their material parameters and to check the accuracy of the method (Table 1.3). Afterward, the patient-specific material parameters were used to simulate the patient-specific noncontact tonometry test using the clinical data available for each case, that is, the topography of the cornea and IOP. In most cases, the predicted displacements ($U_{num}$) were in close proximity to the measured displacement (*U*), with the largest error difference, $\epsilon(\%)$, being 10% for the KC eye (patient kc2). In addition, although local minima exist and we are aware of them, material predictions associated with local minima also lead to a

TABLE 1.3 Validation Using A Priori Unknown Clinical Patient Data

| | | Output | Validation | |
|---|---|---|---|---|
| $L$ | Input | $D_1$ \|$D_2$ \|$k_1$ \|$k_2$ [kPa] \| [–] \| [kPa] \| [–] | $U_{num}$ [mm] | $\epsilon(\%)$ [–] |
| $h_0$ | IOP = 12 mmHg<br>CCT = 578 μm<br>U = 1.00 mm | 0.277 \| 120.6 \| 20.8 \| 516.9 | 1.007 | 0.726 |
| $kc_0$ | IOP = 15 mmHg<br>CCT = 545 μm<br>U = 1.12 mm | 0.267 \| 103.5 \| 17.9 \| 525.3 | 1.153 | 2.968 |
| $kc_1$ | IOP = 15 mmHg<br>CCT = 544 μm<br>U = 1.03 mm | 0.330 \| 109.0 \| 17.6 \| 374.5 | 1.025 | 0.529 |
| $kc_2$ | IOP = 15 mmHg<br>CCT = 464 μm<br>U = 1.05 mm | 0.385 \| 126.7 \| 20.8 \| 267.5 | 1.161 | 10.565 |
| $kc_3$ | IOP = 16 mmHg<br>CCT = 460 μm<br>U = 1.12 mm | 0.388 \| 120.5 \| 18.5 \| 592.7 | 1.131 | 0.940 |

*Notes*: ($D_1$ [kPa] |$D_2$ [–] | $k_1$ [kPa] | $k_2$ [–]): Parameters of the Demiray + G-H-O energy strain function; $U_{num}$ [mm]: maximum deformation amplitude provided by the numerical simulation of the noncontact tonometer; $\epsilon(\%) = (|U_{num} - U|/U) \times 100$: percentage difference between numerical and clinical displacement.

predicted corneal displacement close to the actual measurements. In patient kc2, for which the material predictions led to the worst corneal displacement predictions, it was found that the closest neighbor to the patient's data was located at a distance that was an order of magnitude higher than for the other patients. This result indicates the need for a higher number of samples in the dataset, that is, a denser sampling of the parameter space.

## 1.5 SURGERY SIMULATION

The biomechanical response of the cornea plays a significant role in the final corneal curvature and hence, in the success of any refractive surgery. FE-based biomechanical models of the eye have been presented as a powerful tool for a better assessment of refractive surgeries, aiming at providing personalized mechanical and optical information to the surgeons. To stress the usefulness of these methods, two examples of common ocular surgeries are introduced.

### 1.5.1 Simulation of Refractive Surgery: Astigmatic Keratotomy

AK is an ocular refractive surgery that aims to reduce corneal astigmatism (i.e., the difference between the maximum and minimum curvature), modifying a more elliptic into a more spherical cornea. To validate the simulation, four New Zealand rabbits were subjected to AK surgery performed by the same corneal surgeon. Two straight (transverse) relaxing incisions of 6 mm were made and were located in a 7-mm diameter optical zone along the most curved meridian (see Fig. 1.7B). The tissue was cut to 80% in depth (320 μm) using a guarded diamond blade. A week after the surgery, the follow-up was carried out by measuring the corneal topography.

To retrieve the animal-specific material properties of the cornea, indentation and inflation experiments were carried out on corneas of the same specimen. The material parameters of the constitutive model were subsequently obtained by means of an inverse FE method (see further details in Ariza-Gracia et al. [61]).

As New Zealand rabbits have an almost spherical cornea with circumferential collagen fibers (see Fig. 1.7A) [62], in our experiments, the surgery had the opposite effect of increasing astigmatism. To control the corneal optics, presurgical and postsurgical topographies of the anterior corneal surface were acquired using a MODI corneal topographer (Construzione Strumenti Oftalmici, CSO, Florence, Italy). An in-house optical software was developed to obtain the refractive power and the wavefront aberration of the optical system. The optical outcomes were predicted using both material parameters and patient-specific geometries.

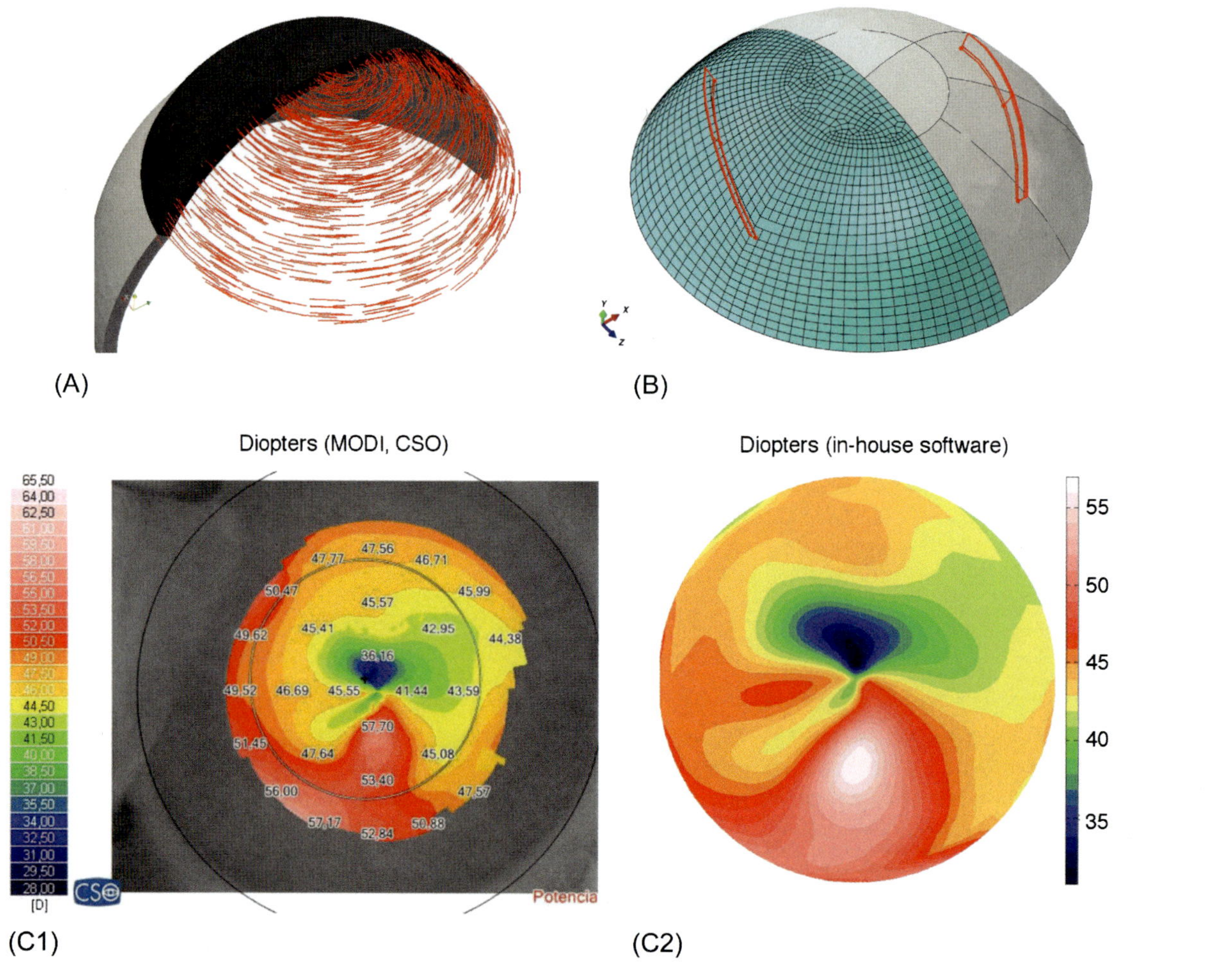


FIG. 1.7 Simulation of astigmatic keratotomy in New Zealand rabbits. (A) Collagen distribution in New Zealand rabbits; (B) FE model of the surgery (relaxing incisions are outlined in *red*); (C.1) Sagittal refractive power for specimen 3 measured with MODI; (C.2) Numerical sagittal refractive power of the FE simulation. *Adapted from M.Á. Ariza-Gracia, Á. Ortillés, J.Á. Cristobal, J.F. Rodríguez Matas, B. Calvo, A numerical-experimental protocol to characterize corneal tissue with an application to predict astigmatic keratotomy surgery, J. Mech. Behav. Biomed. Mater. 74 (2017) 304–314.*

From the corneal topography (see Fig. 1.7C), the topographic axis was used to determine the treatment position, and the topographic cylinder measurement was used to define the length of the incisions. The patient-specific model was built by transforming a geometrical template into the point cloud of the actual corneal coordinates [12]. The template was composed of 63,361 nodes (190,083 D.O.F.) and 50,466 C3D8H elements. The chosen mesh is fine enough to properly capture the contact between the indenter and the cornea and to provide an accurate record of the pressure, area, and forces measured at the tip. Regarding the boundary conditions, a restrained displacement was imposed at the scleral rim to mimic the clamping of the sample, and a uniform pressure was imposed on the inner surface to reproduce the IOP. The contact between the indenter and the cornea was modeled as a frictionless hard contact allowing separation.

The optical quality of the cornea was used as a criterion to validate the model. Thus, the cylindrical (*Cyl*) and spherical (*Sph*) power of the wavefront aberration of the ocular system was used to assess the optical quality.

Optics are purely determined by the geometry of the cornea while geometry depends on IOP and material stiffness [10]. Hence, only patient-specific models are reliable in terms of refractive outcomes. Although all specimens presented the same trend, for the sake of simplicity all results correspond to the specimen with the most representative bow-tie pattern associated with astigmatism (see Fig. 1.7C).

The optical features of the actual geometry are given to a circular area of 3 mm in radius (i.e., the optical zone corresponding to the physical pupil of the system). The models are able to reproduce the overall optical quality of the cornea after the AK surgery. The prediction of the cylindrical power (*Cyl*) is in good agreement with experiments

TABLE 1.4 Experimental (Exp.) and Numerical (Patient-Specific, PS) Corneal Optics Before (Pre.) and After (Post.) the AK Surgery

| | $z_2^{-2}$ | $z_2^0$ | $z_2^2$ | *Sph* | *Cyl* | $\phi$ |
|---|---|---|---|---|---|---|
| Exp. Pre. | −0.97 | −22.79 | −0.95 | 10.8 | −0.9 | 23 |
| Exp. Post. | 0.44 | −25.5 | −7.6 | 13.7 | −4.8 | 2 |
| PS. Pre. | −1.94 | −23.01 | −0.96 | 10.7 | −1.3 | 32 |
| PS. Post. | −6.76 | −26.45 | 1.99 | 13.7 | −4.4 | −37 |

*Notes*: Zernike coefficients ($z_m^n$) in microns. Cylindrical (*Cyl*) and spherical power (*Sph*) in diopters. Astigmatic angle ($\phi$) measured with respect to the nasal-temporal axis (horizontal axis) in degrees.

(+0.4D and −0.4D preoperatively and postoperatively). Similarly, the spherical power (*Sph*) is in good agreement with the experiments (−0.1D and 0D preoperatively and postoperatively). Surprisingly, despite using a patient-specific geometry, the optical features of the presurgical numerical model do not match the actual optical parameters of the presurgical experimental eye. This is explained by the impact of the prestress algorithm on the initial geometry. Due to its iterative nature, the patient-specific geometry is recovered with a maximum error of approximately 2 μm (i.e., the convergence limit of the algorithm). Because the optical features are very subtle, small perturbations of the geometry will lead to different optical values. In this vein, the simulated Zernike coefficients do not match the actual postsurgical Zernike coefficients of the experiment. Although not all the optical features are accurately predicted, numerical models predict the overall dioptric correction, showing the same trend as the experiments: an increase in spherical and cylindrical power, as qualitatively expected due to the corneal physiology of the rabbit. Because the initial configuration of the cornea of the rabbit is almost spherical, a clear astigmatism is induced (Table 1.4).

## 1.5.2 Simulation of ICRS Implantation

KC is an idiopathic, noninflammatory, and degenerative corneal disease that typically develops in the inferior-temporal and central zones. Corneas with Keratoconus present a loss of organization in the corneal collagen fibrils that results in a localized thinning and conical protrusion. Although its incidence is low, 0.05%–2.5%, the absence of a cure and its long-term blinding effects put KC on the spot [63, Chapter 19]. Advanced KC is managed by implanting intrastromal corneal ring segments (ICRS) [64, 65] to avoid penetrating the keratoplasy or the corneal graft [66, 67].

The patient-specific model is built from clinical data (see Section 1.3). The geometry is meshed with 284,184 eight-node linear hybrid hexahedral elements (C3D8H). As the problem is dominated by the biaxial stress state induced by the IOP and the bending is negligible [12], the mesh size at the center of the cornea is not critical. However, areas close to the insertion of the ICRS undergo great deformation and, thus, a finer mesh is required. Hence, the cornea is overmeshed in some regions to avoid a bad mesh transition between the fine mesh surrounding the ICRS and the ideal coarse mesh in the rest of the cornea. The presurgical IOP of the patient is set to a 12 mmHg (1.6 kPa) and a condition of symmetry on the equatorial plane of the sclera. In addition, the initial prestress of the corneal tissue is introduced. The constitutive behavior of the cornea is the same as aforementioned (see Section 1.4).

Based on corneal nomograms, the surgeon decided to implant two triangular ICRSs (AFR20140) of 200 μm, which covered 140 degrees, at the superior-nasal plane and the inferior-temporal plane, respectively, and at a distance of 5 mm with respect to the corneal center. ICRSs are made of polymethyl methacrylate and modeled as an elastic material with a Young's modulus of 3300 MPa and a Poisson ratio of 0.4 (data from Addition Technology, Inc., Sunnyvale, CA).

The simulation of the surgery is split into three stages: (i) the creation of the laser incision in the stroma, (ii) the widening of the incision to introduce the rings, and (iii) the insertion of the ICRS. Once the incision is empty and the new equilibrium state is achieved, the periphery of the surgery is morphed to the shape of the ICRS by applying a displacement boundary condition such that, in the end, the gap takes the exact form of the ICRS's cross-section. Subsequently, ICRSs are inserted into contact with the corneal stroma, enabling the contact from the beginning.

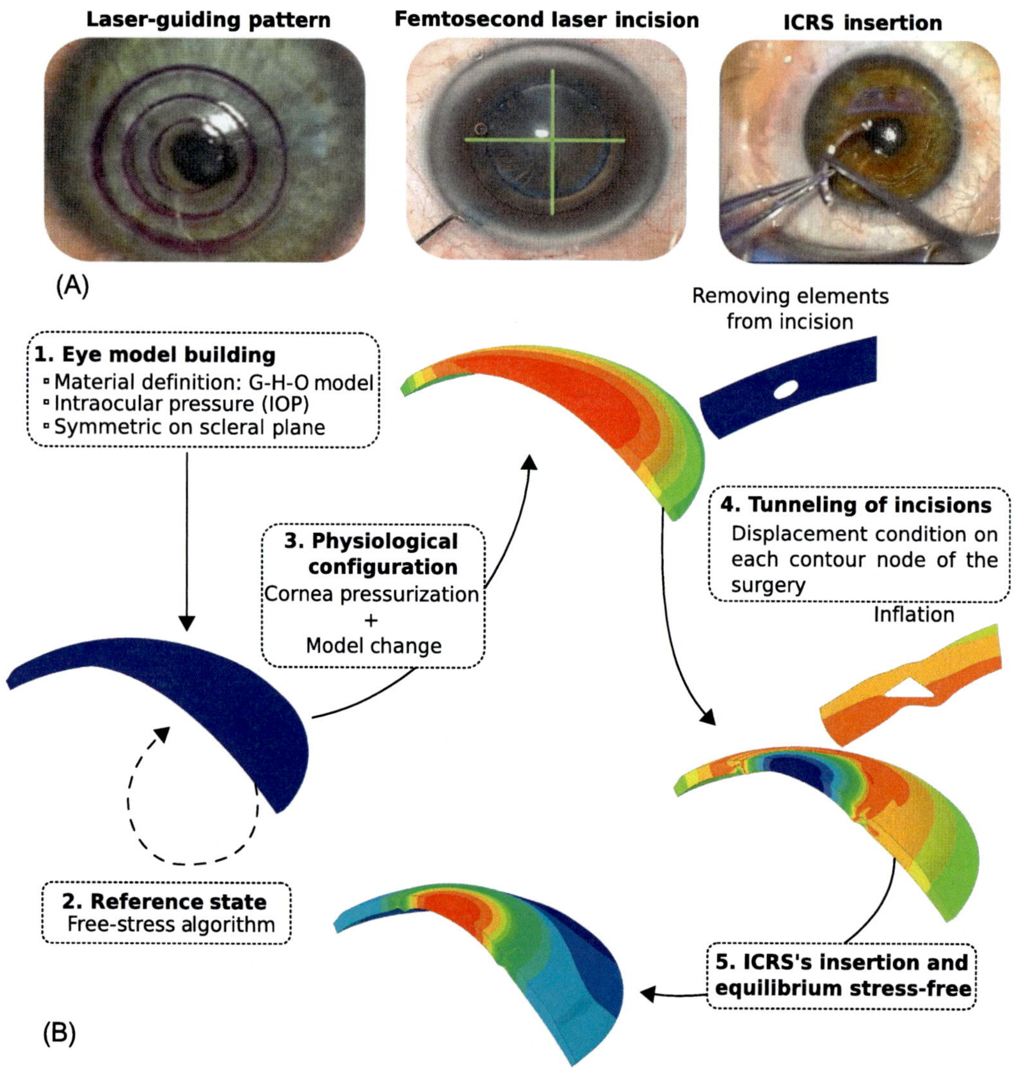


FIG. 1.8 Clinical and computational surgical procedure. (A) Clinical procedure: a pattern is marked so the laser can perform the incision and the rings can be manually inserted; (B) computational procedure. *Adapted from J. Flecha-Lescun, B. Calvo, J. Zurita, M.Á. Ariza-Gracia, Template-based methodology for the simulation of intracorneal segment ring implantation in human corneas, Biomech. Model. Mechanobiol. 17 (4) (2018) 923–938.*

The complete procedure (see Fig. 1.8) consists of the following steps: (1) building the patient-specific geometry of the cornea; (2) obtaining the stress-free configuration of the cornea; (3) physiological prestressing of the cornea due to the IOP and removing the elements inside the laser incision; (4) morphing the incision to the ICRS's section; and (5) registering the ICRS to the incision, forcing contact with the ICRS, and achieving the mechanical stability (equilibrium step).

Inserting ICRS on the cornea relaxes the stress on the stromal tissue, relieving the stretch on those areas affected by KC. As KC's growth is hypothesized to be stress-driven, relaxing the stress will prevent its progression. Not only that, but ICRSs regularize the anterior corneal surface, improving visual acuity (see Fig. 1.9A and B).

When comparing the elevation corresponding to the same circumference (diameter of 5 mm), numerical models were able to match exactly the presurgical geometry (see left panel in Fig. 1.9C). After introducing the ICRS, numerical models presented a fair approximation of the clinical cases (see right panel in Fig. 1.9C). The main differences can be associated to two main factors: first, an average material model was used because patient-specific properties were not available and, second, in real surgeries there is an unknown based on the final positioning of the ICRS in the corneal stroma.

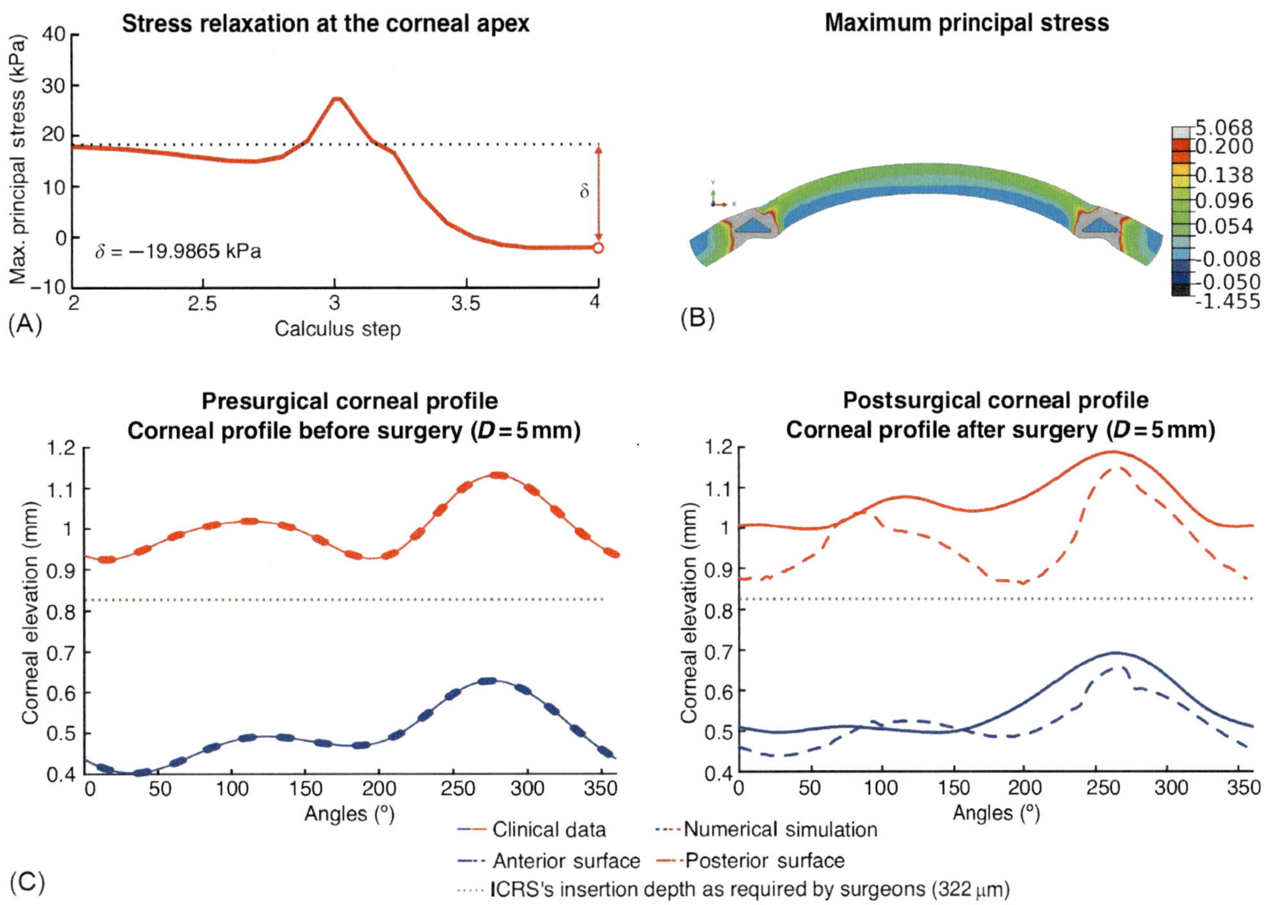


FIG. 1.9 Simulation of ICRS implantation. (A) Stress relaxation at the corneal apex after the introduction of ICRS; (B) main principal stress in MPa; (C) profile for corneal elevation at the same circumference (5 mm optical zone). (*Left*) Presurgical corneal elevation. (*Right*) Postsurgical corneal elevation.

## 1.6 CONCLUSIONS

The determination of the individualized mechanical properties of the cornea has gained interest thanks to the availability of new imaging and diagnostic techniques that allows for a more complex examination in clinics. Still, unexpected clinical outcomes continue forcing clinicians to use secondary "refining" treatments to achieve the desired optical outcomes with the consequent repercussions on a patient's welfare. Unfortunately, commercial devices cannot determine the mechanical properties of the stroma by themselves. Today, certain information about the tissue, such as stress, strain, or material behavior, cannot be determined in vivo without using numerical methods in conjunction with clinical data.

In ophthalmology, topographers have been widely used to recover the deformed corneal geometry. However, these devices give no information about the tensional state of the tissue. To obtain information about the tissue behavior in physiological conditions, it is necessary to use, for example, FE models so as to recover the reference (Lagrangian) configuration of the eyeball. Once this reference geometry is recovered by means of iterative algorithms, the deformed geometry of the eyeball can be recovered including information of its tensional state. Noteworthy, if the definition of the direction of the collagen fibers has been imposed on the deformed (Eulerian) configuration, it is necessary to carry out a consistent pullback of the fibers to the Lagrangian configuration.

Clinicians could benefit from these personalized models to plan surgeries in advance, testing different clinical scenarios without harm to the patient. However, these models may require patient-specific material properties of, at least, the corneal stroma in order to provide a reliable assessment. Also, numerical tools are often time-consuming and cannot be applied effectively in clinics, where a real-time application is usually desired.

In this vein, the use of numerical tools such as artificial neural networks, response surfaces, or K-nn methods is highly useful. The main caveat of these tools, however, is the need for a reliable and extensive dataset to train or fit the models. Among these tools, K-nn showed a satisfactory trade-off between accuracy and prediction time, being amenable for a real-time clinical application. It should be noted, though, that for using K-nn, the dataset must present a fine grid size in which neighbors are not quite far from the patient under analysis or, otherwise, it could introduce higher errors.

Only when both patient-specific geometry and materials are set should these models be encouraged for clinical use. Currently, many surgical applications are under analysis, such as laser refractive surgeries, relaxing incisions, cataract surgery, or other techniques such as the insertion of intracorneal segment rings for preventing the evolution of Keratoconus. But, at this point, it is essential to remark that unless patient-specific optical validation is provided, it is impossible to assess the predictive abilities of these models.

In this chapter, we showed two examples of how in silico models can be used for predicting clinical outcomes if calibrated properly. The first example showed that given patient-specific geometry, material, and optics, we could assess qualitatively and quantitatively the postsurgical visual acuity of the patient, being below the error of measurement of commercial devices. In the second example, we explored the possibility of predicting the effective relaxation of the corneal stroma affected by Keratoconus after the implantation of intracorneal ring segments. As Keratoconus has been suggested to be promoted by tensile stresses on corneal stroma, this assessment is extremely important and cannot be done by any commercial device. These two examples are just a peek into the vast world of possibilities that in silico models could open in modern ophthalmology.

Still, the cornea presents a clear interplay between different features such as geometry (thickness), material stiffness, and IOP. Corneal stability, which is ruled by corneal biomechanics, is the one leading the final visual acuity of the patient. Hence, it is not possible to assess the final visual acuity of the patient without first thoroughly assessing the patient's corneal biomechanics. Devices such as noncontact tonometers are a great contribution to the ophthalmological field because, currently, there is no alternative to induce a mechanical deformation in vivo. However, the diagnosis should be handled with care as they are not measuring the physiological response of the cornea, nor the total contribution of the fibers as the anterior stroma is compressed during bending. Still, the use of these commercial devices in conjunction with in silico approaches is an appealing alternative to assessing corneal biomechanics and, hopefully, to improving unexpected postsurgical outcomes.

Although further research must be done in this field such as exploring fluid-structure interaction simulations so as to carry out more precise material optimizations or instructing models with microstructural features essential for some specific surgeries, we are facing exciting days that will eventually end up with a full coupling between in silico models and clinical practice.

## Acknowledgments

This work was supported by the Spanish Ministry of Economy and Competitiveness (Projects DPI2014-54981-R and DPI2017-84047-R), Department of Industry and Innovation (Government of Aragón) and European Social Fund 2014–2020 (FSE-DGA group T24_17R). J. Flecha was supported by the Spanish Ministry of Economy and Competitiveness (BES-2015-073630).

## References

[1] B.A. Fink, H. Wagner, K. Steger-May, C. Rosenstiel, T. Roediger, T.T. McMahon, M.O. Gordon, K. Zadnik, Differences in Keratoconus as a function of gender, Am. J. Ophthalmol. 140 (3) (2005). 459.e1–459.e12.

[2] A.R. Pearson, B. Soneji, N. Sarvananthan, J.H. Sandford-Smith, Does ethnic origin influence the incidence or severity of Keratoconus? Eye (Lond.) 14 (4) (2000) 625–628.

[3] P.R. Ruiseñor Vázquez, J.D. Galletti, N. Minguez, M. Delrivo, F. Fuentes Bonthoux, T. Pförtner, J.G. Galletti, Pentacam Scheimpflug tomography findings in topographically normal patients and subclinical Keratoconus cases, Am. J. Ophthalmol. 158 (1) (2014). 32–40.e2.

[4] K.C. Ogbuehi, U.L. Osuagwu, Corneal biomechanical properties: precision and influence on tonometry, Cont. Lens Anterior Eye 37 (3) (2014) 124–131.

[5] J.-L. Bourges, N. Alfonsi, J.-F. Laliberté, M. Chagnon, G. Renard, J.-M. Legeais, I. Brunette, Average 3-dimensional models for the comparison of Orbscan II and Pentacam pachymetry maps in normal corneas, Ophthalmology 116 (11) (2009) 2064–2071.

[6] A.S. Roy, W.J. Dupps Jr, Patient-specific modeling of corneal refractive surgery outcomes and inverse estimation of elastic property changes, J. Biomech. Eng. 133 (1) (2011) 11002.

[7] H.P. Studer, H. Riedwyl, C.A. Amstutz, J.V.M. Hanson, P. Büchler, Patient-specific finite-element simulation of the human cornea: a clinical validation study on cataract surgery, J. Biomech. 46 (4) (2013) 751–758.

[8] J. Hong, J. Xu, A. Wei, S.X. Deng, X. Cui, X. Yu, X. Sun, A new tonometer-the Corvis ST tonometer: clinical comparison with noncontact and Goldmann Applanation Tonometers, Invest. Ophthalmol. Vis. Sci. 54 (1) (2013) 659–665.

[9] M. Lanza, S. Iaccarino, M. Bifani, In vivo human corneal deformation analysis with a Scheimpflug camera, a critical review, J. Biophotonics 9 (2016) 464–477.
[10] M.Á. Ariza-Gracia, J.F. Zurita, D.P. Piñero, J.F. Rodríguez Matas, B. Calvo, Coupled biomechanical response of the cornea assessed by non-contact tonometry. A simulation study, PLoS ONE 10 (3) (2015) e0121486.
[11] A. Sinha Roy, M. Kurian, H. Matalia, R. Shetty, Air-puff associated quantification of non-linear biomechanical properties of the human cornea in vivo, J. Mech. Behav. Biomed. Mater. 48 (April) (2015) 173–182.
[12] M.Á. Ariza-Gracia, J. Zurita, D.P. Piñero, B. Calvo, J.F. Rodríguez Matas, Automatized patient-specific methodology for numerical determination of biomechanical corneal response, Ann. Biomed. Eng. 44 (5) (2016) 1753–1772.
[13] Z. Hassan, L. Modis, E. Szalai, A. Berta, G. Nemeth, Examination of ocular biomechanics with a new Scheimpflug technology after corneal refractive surgery, Cont. Lens Anterior Eye 37 (5) (2014) 337–341.
[14] F. Faria-Correia, I. Ramos, B. Valbon, A. Luz, C.J. Roberts, R. Ambrósio Jr, Scheimpflug-based tomography and biomechanical assessment in pressure-induced stromal keratopathy, J. Refract. Surg. 29 (5) (2013) 356–358.
[15] Y. Hon, A.K.C. Lam, Corneal deformation measurement using Scheimpflug noncontact tonometry, Optom. Vis. Sci. 90 (1) (2013) e1–e8.
[16] S. Kling, S. Marcos, Contributing factors to corneal deformation in air puff measurements, Invest. Ophthalmol. Vis. Sci. 54 (7) (2013) 5078–5085.
[17] M.Á. Ariza-Gracia, S. Redondo, D.P. Llorens, B. Calvo, J.F. Rodríguez Matas, A predictive tool for determining patient-specific mechanical properties of human corneal tissue, Comput. Methods Appl. Mech. Eng. 317 (2016) 226–247.
[18] T. Seiler, M. Matallana, S. Sendler, T. Bende, Does Bowman's layer determine the biomechanical properties of the cornea? Refract. Corneal Surg. 8 (2) (1992) 139–142.
[19] E.L. Sancho, Modelado Biomecánico de los componentes refractivos del ojo humano y tratamientos refractivos asociados (Ph.D. thesis), Universidad de Zaragoza, 2010.
[20] M. Winkler, G. Shoa, Y. Xie, S.J. Petsche, P.M. Pinsky, T. Juhasz, D.J. Brown, J.V. Jester, Three-dimensional distribution of transverse collagen fibers in the anterior human corneal stroma, Invest. Ophthalmol. Vis. Sci. 54 (12) (2013) 7293–7301.
[21] M. Winkler, G. Shoa, S.T. Tran, Y. Xie, S. Thomasy, V.K. Raghunathan, C. Murphy, D.J. Brown, J.V. Jester, A comparative study of vertebrate corneal structure: the evolution of a refractive lens, Invest. Ophthalmol. Vis. Sci. 56 (4) (2015) 2764–2772.
[22] S.J. Petsche, D. Chernyak, J. Martiz, M.E. Levenston, P.M. Pinsky, Depth-dependent transverse shear properties of the human corneal stroma, Invest. Opthalmol. Vis. Sci. 53 (2) (2012) 873.
[23] M. Winkler, D. Chai, S. Kriling, C.J. Nien, D.J. Brown, B. Jester, T. Juhasz, J.V. Jester, Nonlinear optical macroscopic assessment of 3-D corneal collagen organization and axial biomechanics, Invest. Ophthalmol. Vis. Sci. 52 (12) (2011) 8818–8827.
[24] A. Benoit, G. Latour, S.-K. Marie-Claire, J.-M. Allain, Simultaneous microstructural and mechanical characterization of human corneas at increasing pressure, J. Mech. Behav. Biomed. Mater. 60 (2016) 93–105.
[25] A. Pandolfi, G.A. Holzapfel, Three-dimensional modeling and computational analysis of the human cornea considering distributed collagen fibril orientations, J. Biomech. Eng. 130 (6) (2008) 61006.
[26] J.B. Randleman, D.G. Dawson, H.E. Grossniklaus, B.E. McCarey, H.F. Edelhauser, Depth-dependent cohesive tensile strength in human donor corneas: implications for refractive surgery, J. Refract. Surg. 24 (1) (2008) S85–S89.
[27] M. Lombardo, G. Lombardo, D.J. Friend, S. Serrao, M.A. Terry, Long-term anterior and posterior topographic analysis of the cornea after deep lamellar endothelial keratoplasty, Cornea 28 (4) (2009) 408–415.
[28] M. Lombardo, G. Lombardo, G. Carbone, M.P. De Santo, R. Barberi, S. Serrao, Biomechanics of the anterior human corneal tissue investigated with atomic force microscopy, Invest. Opthalmol. Vis. Sci. 53 (2) (2012) 1050.
[29] S. Dorairaj, J.M. Liebmann, R. Ritch, Quantitative evaluation of anterior segment parameters in the era of imaging, Trans. Am. Ophthalmol. Soc. 105 (2007) 99–108. discussion 108–110.
[30] M.Q. Salomão, A. Esposito, W.J. Dupps, Advances in anterior segment imaging and analysis, Curr. Opin. Ophthalmol. 20 (4) (2009) 324–332.
[31] D.P.S. O'Brart, D.C. Saunders, M.C. Corbett, E.S. Rosent, The corneal topography of Keratoconus, Eur. J. Implant Refract. Surg. 7 (1) (1995) 20–30.
[32] A. Agarwal, A. Argawal, S. Jacob, Textbook on Corneal Topography. Including Pentacam and Anterior Segment OCT, second ed., Jaypee Brothers Medical Publishers (P) Ltd Corporate, New Delhi, 2010, p. 397.
[33] M.M. Sinjab, Corneal Topography in Clinical Practice (Pentacam System). Basics & Clinical Interpretation, second ed., Jaypee Brothers Medical Publishers (P) Ltd Corporate, New Delhi, 2012, p. 233.
[34] A. Garg, J.L. Alio, B. Pajic, C.K. Mehta, Mastering the Techniques of LASIK, EPILASIK, and LASEK: Techniques and Technology, Jaypee Brothers Medical Publishers, New Delhi, 2007, p. 394.
[35] D.P. Piñero, N. Alcón, In vivo characterization of corneal biomechanics, J. Cataract Refract. Surg. 40 (6) (2014) 870–887.
[36] D.P. Piñero, N. Alcón, Corneal biomechanics: a review, Clin. Exp. Optom. 98 (2) (2015) 107–116.
[37] M. Kaschke, K.-H. Donnerhacke, M.S. Rill, Optical Devices in Ophthalmology and Optometry. Technology, Design Principles, and Clinical Applications, Wiley-VCH Verlag GmbH & Co. KGaA, Weinheim, 2014, p. 625.
[38] A. Bedei, I. Appolloni, A. Madesani, A. Pietrelli, S. Franceschi, L. Barabesi, Repeatability and agreement of 2 Scheimpflug analyzers in measuring the central corneal thickness and anterior chamber angle, volume, and depth, Eur. J. Ophthalmol. 22 (suppl. 7) (2012) 29–32.
[39] G. Savini, P. Barboni, M. Carbonelli, K.J. Hoffer, Accuracy of corneal power measurements by a new Scheimpflug camera combined with Placido-disk corneal topography for intraocular lens power calculation in unoperated eyes, J. Cataract Refract. Surg. 38 (5) (2012) 787–792.
[40] R. Montalbán, D.P. Piñero, J. Javaloy, J.L. Alió, Intrasubject repeatability of corneal morphology measurements obtained with a new Scheimpflug photography-based system, J. Cataract Refract. Surg. 38 (6) (2012) 971–977.
[41] P.M. Pinsky, D. van der Heide, D. Chernyak, Computational modeling of mechanical anisotropy in the cornea and sclera, J. Cataract Refract. Surg. 31 (1) (2005) 136–145.
[42] E. Lanchares, B. Calvo, J.A. Cristóbal, M. Doblaré, Finite element simulation of arcuates for astigmatism correction, J. Biomech. 41 (4) (2008) 797–805.
[43] F. Riveros, S. Chandra, E.A. Finol, T.C. Gasser, J.F. Rodriguez, A pull-back algorithm to determine the unloaded vascular geometry in anisotropic hyperelastic AAA passive mechanics, Ann. Biomed. Eng. 41 (4) (2013) 694–708.

[44] M.A. Lago, M.J. Rupérez, F. Martínez-Martínez, C. Monserrat, E. Larra, J.L. Güell, C. Peris-Martínez, A new methodology for the in vivo estimation of the elastic constants that characterize the patient-specific biomechanical behavior of the human cornea, J. Biomech. 48 (1) (2015) 38–43.
[45] A. Pandolfi, F. Manganiello, A model for the human cornea: constitutive formulation and numerical analysis, Biomech. Model. Mechanobiol. 5 (4) (2006) 237–246.
[46] F. Bao, M. Deng, Q. Wang, J. Huang, J. Yang, C. Whitford, B. Geraghty, A. Yu, A. Elsheikh, Evaluation of the relationship of corneal biomechanical metrics with physical intraocular pressure and central corneal thickness in ex vivo rabbit eye globes, Exp. Eye Res. 137 (2015) 11–17.
[47] A. Elsheikh, C.W. McMonnies, C. Whitford, G.C. Boneham, In vivo study of corneal responses to increased intraocular pressure loading, Eye Vis. 2 (2015) 20.
[48] C. Whitford, H. Studer, C. Boote, K.M. Meek, A. Elsheikh, Biomechanical model of the human cornea: considering shear stiffness and regional variation of collagen anisotropy and density, J. Mech. Behav. Biomed. Mater. 42 (2015) 76–87.
[49] E. Lanchares, M.A. del Buey, J.A. Cristóbal, L. Lavilla, B. Calvo, Biomechanical property analysis after corneal collagen cross-linking in relation to ultraviolet A irradiation time, Graefes Arch. Clin. Exp. Ophthalmol. 249 (8) (2011) 1223–1227.
[50] R. Navarro, F. Palos, E. Lanchares, B. Calvo, J.A. Cristóbal, Lower- and higher-order aberrations predicted by an optomechanical model of arcuate keratotomy for astigmatism, J. Cataract Refract. Surg. 35 (1) (2009) 158–165.
[51] S. Kling, N. Bekesi, C. Dorronsoro, D. Pascual, S. Marcos, Corneal viscoelastic properties from finite-element analysis of in vivo air-puff deformation, PLoS ONE 9 (8) (2014) e104904.
[52] I. Simonini, A. Pandolfi, Customized finite element modelling of the human cornea, PLoS ONE 10 (6) (2015) e0130426.
[53] I. Simonini, A. Pandolfi, The influence of intraocular pressure and air jet pressure on corneal contactless tonometry tests, J. Mech. Behav. Biomed. Mater. 58 (2016) 75–89.
[54] I. Simonini, M. Angelillo, A. Pandolfi, Theoretical and numerical analysis of the corneal air puff test, J. Mech. Phys. Solids 93 (2016) 118–134.
[55] M.R. Bryant, P.J. McDonnell, Constitutive laws for biomechanical modeling of refractive surgery, J. Biomech. Eng. 118 (4) (1996) 473–481.
[56] S. Kok, N. Botha, H.M. Inglis, Calibrating corneal material model parameters using only inflation data: an ill-posed problem, Int. J. Numer. Methods Biomed. Eng. 30 (12) (2014) 1460–1475.
[57] T.C. Gasser, R.W. Ogden, G.A. Holzapfel, Hyperelastic modelling of arterial layers with distributed collagen fibre orientations, J. R. Soc. Interface 3 (6) (2006) 15–35.
[58] A. Eilaghi, J.G. Flanagan, I. Tertinegg, C.A. Simmons, G.W. Brodland, C.R. Ethier, G. Wayne Brodland, C. Ross Ethier, Biaxial mechanical testing of human sclera, J. Biomech. 43 (9) (2010) 1696–1701.
[59] M.Á. Ariza-Gracia, D.P. Piñero, J.F. Rodriguez, R.J. Pérez-Cambrodí, B. Calvo, Interaction between diurnal variations of intraocular pressure, pachymetry, and corneal response to an air puff: preliminary evidence, JCRS Online Case Rep. 3 (1) (2014) 12–15.
[60] T. Huseynova, G.O. Waring IV, C. Roberts, R.R. Krueger, M. Tomita, Corneal biomechanics as a function of intraocular pressure and pachymetry by dynamic infrared signal and Scheimpflug imaging analysis in normal eyes, Am. J. Ophthalmol. 157 (4) (2014) 885–893.
[61] M.Á. Ariza-Gracia, Á. Ortillés, J.Á. Cristobal, J.F. Rodríguez Matas, B. Calvo, A numerical-experimental protocol to characterize corneal tissue with an application to predict astigmatic keratotomy surgery, J. Mech. Behav. Biomed. Mater. 74 (2017) 304–314.
[62] S. Hayes, C. Boote, J. Lewis, J. Sheppard, M. Abahussin, A.J. Quantock, C. Purslow, M. Votruba, K.M. Meek, Comparative study of fibrillar collagen arrangement in the corneas of primates and other mammals, Anat. Rec. Adv. Integr. Anat. Evol. Biol. 290 (12) (2007) 1542–1550.
[63] J.L. Alió, F. Pierre, B. Ramez, T. Sherwin, L. Buzzonetti an Francesco Versaci, M.W. Belin, A. Renato, F. Cavas-Martínez Jr, F. Bao, in: Alió, J.L. (Ed.), Keratoconus: Recent Advances in Diagnosis and Treatments, Springer International Publishing, 2017.
[64] L. Akaishi, P.F. Tzelikis, I.M. Raber, Ferrara intracorneal ring implantation and cataract surgery for the correction of pellucid marginal corneal degeneration, J. Cataract Refract. Surg. 30 (11) (2004) 2427–2430.
[65] M.A. Zare, H. Hashemi, M.R. Salari, Intracorneal ring segment implantation for the management of Keratoconus: safety and efficacy, J. Cataract Refract. Surg. 33 (11) (2007) 1886–1891.
[66] J.A.P. Gomes, D. Tan, C.J. Rapuano, M.W. Belin, J. Renato Ambrósio, J.L. Guell, F. Malecaze, K. Nishida, V.S. Sangwan, Global consensus on Keratoconus and ectatic disease, Cornea 34 (4) (2015) 359–369.
[67] A. Vega-Estrada, J. Alio, The use of intracorneal ring segments in Keratoconus, Eye Vis. 3 (2016) 8.

## Further Reading

[68] K. Anderson, A. El-Sheikh, T. Newson, Application of structural analysis to the mechanical behaviour of the cornea, J. R. Soc. Interface 1 (April) (2004) 3–15.
[69] J. Flecha-Lescun, B. Calvo, J. Zurita, M.Á. Ariza-Gracia, Template-based methodology for the simulation of intracorneal segment ring implantation in human corneas, Biomech. Model. Mechanobiol. 17 (4) (2018) 923–938.

CHAPTER

# 2

# Biomechanics of the Vestibular System: A Numerical Simulation

*Carla F. Santos*, Jorge Belinha†, Fernanda Gentil‡, Marco Parente§, and Renato M. Natal Jorge**

***Faculty of Engineering of University of Porto (FEUP), Porto, Portugal †Institute of Science and Innovation in Mechanical and Industrial Engineering (INEGI), School of Engineering, Polytechnic of Porto (ISEP), Porto, Portugal ‡School of Health - P. Porto, Porto, Portugal §INEGI, Institute of Mechanical Engineering and Industrial Management, Faculty of Engineering of University of Porto, FEUP, Porto, Portugal**

## 2.1 INTRODUCTION

The vestibular system is located in the posterior portion of the inner ear. It is a key component to our sense of balance and movement. Any changes in this system can cause effects or symptoms such as dizziness, blurred vision, imbalance, and nausea, which are vertiginous syndrome indicators. Vertigo is reported as one of the most common symptoms in the world [1]. It is considered the third most frequent complaint in medicine, transmitting a sense of inadequacy and insecurity [2]. The main roles of the vestibular system, from a functional point of view, include the correction of any involuntary movement of the body mass center and balance position to avoid falls; the accurate perception of the body position in the environment and the perception of the direction and acceleration movements; and the control of eye movement to maintain a clear visual field when the individual, the environment, or both are in movement [3].

Gait and balance functions need the activation of three different sensory afferent systems, vestibular, visual, and proprioceptive, which integrate the information and lead to body and view stabilization. A healthy and active lifestyle demands a permanent stimulation of the vestibular system, which takes place during reactive adaptation, determining a head position in the space relative to the fundamental posture, as in relation to the displacement [3, 4].

The static balance maintenance is a complex process, which has a vestibular intervention. The postural control system results in the interaction between the input from the sensory system, the coordination of the central nervous system (CNS), and the motor output. In this way, the sensory system provides information about the corporal position segment in relation to other segments and the environment. These are transmitted and integrated with the CNS and sent to the motor system, which is responsible for the correct and appropriate activation of muscles for movements [5, 6].

The interaction between the different sensory systems happens as a consequence of the stimulation of the vestibular receptors by any head movement, which leads to a neural information transition to the cerebellum and the vestibular nuclei located in the brain stem [5, 6].

The vestibular system nuclei send signals primarily to the neural central structures that control eye movements and to the muscles that contribute to keeping an upright position. The projections to the former provide anatomical basis of the vestibulo-ocular reflex (VOR), which is required for clear vision, and the projections to the muscles that control our posture are necessary to keep us standing [4], maintain balance, and allow the visual control of motion events [3, 6].

Due to some vestibular disorders, characterized by dizziness symptoms triggered by some head positions (positional vertigo) or during the movement to one specific position of the cephalic segment (positioning vertigo), there

*Advances in Biomechanics and Tissue Regeneration*
https://doi.org/10.1016/B978-0-12-816390-0.00002-9

are some authors such as Powers and Howley [7] stating that the receptors inside the vestibular system are sensitive to any movement change position. The head movements excite the hair cells placed in the sensitive structures, clarified in the next paragraphs, generating nervous impulses, which are sent to the CNS to recognize the new position.

Patients with vestibular dysfunction exhibit a measurable impairment in motor behavior controlled by the vestibular system (postural control and oculomotor and spatial orientation) and perceptual illusions, such as vertigo [4]. This and other clinical symptomatology will be detailed in the next section.

The reduced dimensions of such a system are an aspect that has to be reported, since the vestibular system controls all the balance function comprised in just 8 mm.

The global structure of the vestibular system consists of three semicircular canals (SCCs) placed orthogonally (Fig. 2.1) connected to a central chamber called the vestibule, comprising the saccule and utricle. The cochlea, which is responsible for the auditory part, is attached to the vestibular system.

The material constituents of the vestibular system include the membranous labyrinth enclosed in a bony labyrinth with the same shape. The tubular part of the membranous labyrinth within the SCCs is called the semicircular ducts (SCDs). In Fig. 2.1 the yellow part corresponds to the membranous labyrinth, and the green one is the bony labyrinth.

There are also two distinct fluids in the vestibular system: the perilymph, which is located between the labyrinths, and the endolymph, which is located inside the membranous labyrinth. This second fluid introduces the connection of the five sensory structures of this system: the cupulas of the three SCCs, shown in red in Fig. 2.1, and the two maculae located in the utricle and the saccule [4], shown in blue in Fig. 2.1.

Each canal is composed of a circular path of fluid continuity, discontinued by a swallow small pathway called the ampulla, where the hair cell sensory epithelium is placed, the cupula. The three cupulas in the SCCs, arranged in three orthogonal planes, are sensitive to angular (rotational) acceleration [8, 9]. The sensory cells exhibit constant discharge of neurotransmitters that are modified by the direction of cupula deflection. This output signal has an origin in the velocity of head rotation, which is why these cells can be named "rate sensors" in engineering terms. The signal response of the three canals during movement or standing balance is merged by the brain, leading to a vector representation of the instantaneous speed of head rotation relative to space. Since there are three canals on each inner ear orthogonal to each other, they can decompose a movement into three components. Each canal has a partner on the other side of the head. When one sensor of a canal is excited, the other correspondent is inhibited.

The vestibule is constituted by the otolithic structures, saccule and utricle; it comprises two macula portions, one in each sac, which support the otoconia. The macula is the correspondent to the cupula but with regard to linear accelerations. It is composed of a gel layer membrane covered with calcium carbonate crystals called otoconia and has substantially more mass than the cupula. The mass of the otolithic membrane causes the maculae to be sensitive to gravity and linear acceleration. Dedicated hair cells are one of the most functional structures contained in the gel layer and are the biological sensors that convert displacement due to head motion into neural firing [3].

When the head moves, the inertia of the otoconia bends the hair cells in the opposite direction. Bending all the hairs toward the kinocilium depolarizes the cell, inducing an increase in action potential frequency in the afferent nerve. Bending away from the kinocilium causes hyperpolarization and reduced action potential frequency [10].

The vestibular system's mechanical properties are essential to the understanding and interpretation of vestibular neurophysiological behavior. Due to the reduced dimensions of the SCCs, a computational numerical model appears

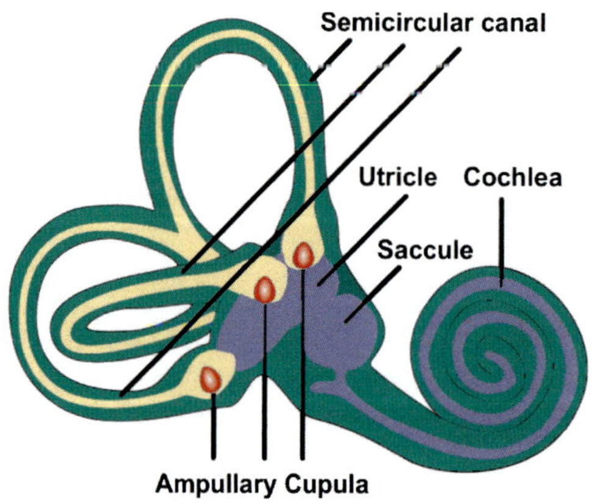


FIG. 2.1 Vestibular system with the three cupulas in each semicircular duct and both maculae in the saccule and utricle.

to be a valid research opportunity to test and validate hypotheses, such as the orientation and shape of the otolith maculae.

The patient's early evaluation also takes place in biomechanical concerns. In the last decades, research and study in the field of otoneurology of patients suffering have been developed. The outcome of such research led to quick balance recovery strategies development, preventing the onset or recurrence of vestibular disorders and consequently resulting in a fast reintegration to the daily routine.

## 2.2 DIAGNOSING VESTIBULAR DYSFUNCTIONS

The final goal of any health research project is to take a step into the quality improvement of the patient's daily life. A deep comprehension of the development of vestibular dysfunctions is a desirable start. In the present chapter, some of the main diseases affecting the vestibular system and respective therapies will be discussed.

The literature shows that 85% of the balance dysfunctions could be related to the inner ear disorders [11], mainly from the vestibular segment. People suffering from vertigo and dizziness have a higher risk of falls, so it is important to develop mechanisms to decrease or even eliminate these symptoms. There are high mortality and morbidity rates associated with falls in patients above 65 years old. The accidents constitute the sixth cause of death in this age group; falls are responsible for half of these episodes [12, 13]. Additionally, falls are one of the main causes of accidental bone fractures in older patients. Furthermore, the healthcare related to these kinds of fracture involves high costs, which are another strong motivation to decrease the number of people who suffer from dizziness [14].

Vertigo is one of the most common medical complaints, affecting approximately 20%–30% of the population worldwide [15]. Vertigo may be present in patients of all ages, and its prevalence increases with age, being the most frequent complaint in people over 70 years old. Additionally, vertigo is about two to three times more predominant in women than in men [16]. These numbers lead us to fathom the conditions behind the vestibular system failure. The vestibular system degeneration associated with the decrease of the CNS capacities and the aging of the human body contribute to balance changes. Other factors such as multiple and excessive medications expand the balance alterations, responsible for the vertigo occurrence and for the imbalance of the geriatric population.

The circumstances regarding the orthostatic balance correspond to the body and view stabilization, ensuring the maintenance of the gravitational center (GC) inside the sustentation polygon, either at rest or in motion [5, 17].

There are several reasons that could lead to a mismatch information between the systems in control of the balance function. When it occurs, all the system breaks, leading to equilibrium loss, until the trigger is identified and starts to compensate the central mechanisms to return balance. Consequently, vertigo can be defined as an illusionary sensation of movement. However, it is often referred to the broadest sense of dizziness, imbalance, instability, or special disorientation [5, 17].

The classification of dizziness was first performed by Drachman and Hart following [18, 19], a categorization driven by complaints of the patient, defined four categories of dizziness: presyncope (described as a feeling of imminent loss of consciousness), imbalance (or apparent sense of falling, not exclusively associated with the movement), numbness (described in Anglo-Saxon as light-headedness that does not present a clear definition or associated diagnosis), and vertigo (defined as an illusion of movement, often with sense of rotation and with nausea and vomiting). The last is the most prevalent cause of dizziness.

Vertigo can be caused by ear disorders, by the nerve connecting the ear to the brain, or even by the brain. Vertigo can also be related to visual disorders or sudden changes in blood pressure. Brain-related causes are less commonly associated with isolated vertigo and nystagmus but can still produce signs and symptoms. Many other conditions can affect the inner ear and cause vertigo, like bacterial or viral infections, cancer, abnormal pressure, nervous inflammation, or toxic substances.

Benign paroxysmal positional vertigo (BPPV) is the most common cause of vertigo, and it is described as a brief, intense sensation of spinning that occurs when there are changes in the head orientation with respect to gravity [9].

The cause of BPPV is the presence of otoconia in the SCC, which are normally located in the macula of the otolith organs, as explained in the previous section. If they fall from the utricle and become loose in the SCC, it will induce a false spinning sensation with no head rotational motion actually occurring. The sensation is generated by the distinct information sent to the brain by the two different systems, visual and vestibular [9, 20].

Therefore the BPPV diagnostic is obtained by the Dix-Hallpike maneuver, which consists of a specific set of movements performed by an audiologist, where the presence of nystagmus confirms the disorder [21].

The main effective therapy used in the reported condition is the vestibular rehabilitation composed of particle repositioning maneuvers and habituation therapy. Despite the success rates of such therapy, there are some key details that should be better analyzed. Therefore, due to the vestibular system conditions, a computational biomechanical approach seems to be a suitable analysis methodology.

## 2.3 NUMERICAL METHODS APPLIED TO HUMAN MORPHOLOGY

The reduced dimensions of the vestibular system hinder the experimental analysis of such a system, being the computational simulation an alternative approach to study the rehabilitation process.

In 1933, when the primary investigations of the vestibular system were performed, Steinhausen formulated a mathematical description of the SCC that considered the dynamics of the cupula endolymph system as a highly damped torsion pendulum for the sensation of the angular motion [22].

The fluid–structure interaction simulation of the endolymph and cupula during head rotation allows the measurement of the fluid interactions between the three ducts and the displacement of the cupula during the movement. This model could be useful to understand the physiological and mechanical aspects of SCC, and it could be described as a band-pass filter relating the displacement of the cupula to the angular velocity of the head [23]. Simulation methods for fluid-structure interaction include the finite element method (FEM), which have been gradually developed. FEM is a computational tool for performing mathematical analysis. It includes the use of mesh generation techniques for dividing a complex problem into small elements.

Simultaneously, meshless methods have been under strong development and are continuously extending their application field [24]. Additionally, due to the efficiency and accuracy of their discretization formulation, meshless methods are more flexible discretization techniques and a competitive and alternative numerical method in computational mechanical analysis [25].

The meshless methods discretize the domain based on a cloud of nodes [25, 26], instead of the rigid element concept used in FEM. In the early years, the solution of partial differential equations was the main focus of interest. However, today, meshless methods are applied to a wide range of applications [27].

Meshless methods can be divided into many classes or even subclasses; one of the most common classifications used is the "not-truly meshless method" or "truly meshless method" classification [27].

A meshless method is labeled "not truly" when a background mesh is required to perform the numerical integration of the integrodifferential equations ruling the studied physical phenomenon.

On the other hand, "truly" meshless methods only require an unstructured cloud of nodes to discretize the problem domain, because the influence domain, integration points, shape functions, and other necessary mathematical constructions are obtained directly from the spatial coordinates of the nodes.

Thus truly meshless methods are capable to obtain the cloud of nodes using just CAT scan or the MR images by considering the pixels (or voxels) position. Afterward, using only the nodal spatial information, these truly meshless methods are able to obtain directly the nodal connectivity, the integration points, and the shape functions. Furthermore, using a gray range of medical images, truly meshless methods are even capable of recognizing distinct biological structures and then attribute to each node the corresponding material properties [27, 28].

The natural neighbor radial point interpolation method (NNRPIM) is an example of a truly meshless method, which allows discretizing the problem domain using only a nodal cloud coming from the voxel position of the micro-CT scan (no other information is required). Then, using the natural neighbor concept, the Voronoï diagram discretizing the problem domain can be constructed. From the Voronoï diagram, it is possible to establish directly the nodal connectivity and define the position and weight of background integration points. The NNRPIM formulation and its extension to free vibration analysis were used in the present work, and some applications will be detailed in the next section.

Another popular and accurate meshless method used in this work is the radial point interpolation method (RPIM). The RPIM is an interpolator meshless method that enforces nodal connectivity using the influence-domain concept. To solve the integrodifferential equations governing the physical phenomenon, the RPIM uses a background cloud of integration points, constructed using integration cells and the Gauss-Legendre quadrature rule. This numerical integration process represents a significant percentage of the total computational cost of the analysis.

There are other meshless methods used in the present work that could be used to simulate biological structures, such as the smoothed-particle hydrodynamics (SPH), which is a computational method used also for simulating fluid flows.

It uses a Lagrangian formulation in which the particles are constant mass particles and particle mesh is attached to the material and particles deform as the material starts to deform. The SPH method works by dividing a continuous field into a set of discrete particles. These particles have a spatial distance over which their properties are "smoothed" by a *kernel function*. The particles are identified with some characteristics, such as mass, position, and velocity, but particles can also carry estimated physical properties depending on the problem, like mass density, temperature, and pressure. This method is best suited for problems in which the mesh deformations are large [29].

The SPH was first developed to simulate astrophysical fluid dynamics. Since then, it has been successfully applied to a vast range of problems, as fluid flows. It was developed by Gingold and Monaghan (1977) and Lucy (1977). A computational domain of this method is represented by computational points—particles.

This method has some advantages over grid-based techniques because its concept is simple and it is relatively easy to incorporate complicated physical effects into the SPH formalism [30, 31].

During 1995 Liu and his team [32], after some tests with SPH conditions, proposed a reproducing kernel particle method that improves the accuracy of SPH approximation.

SPH method was initially developed as a probabilistic meshless particle method and was later modified to a deterministic meshless method. Over the years the method has been optimized to increase the accuracy solution and to be applied in most of the cases.

In the next subchapter, there will be detailed numerical models and simulations performed with the meshless methods on the vestibular system of the inner ear.

## 2.4 BIOMECHANICAL MODEL OF THE SEMICIRCULAR DUCTS

In this section the computational model developed within the scope of this work is presented.

Regarding the computational analysis software, it was found that ABAQUS possesses enough flexibility to build and analyze such complex systems. ABAQUS is a commercial software, widely used in several computational mechanic fields, to build and simulate numerical models. Furthermore, it allows combining the FEM with the SPH meshless method. The first circular model built to study the vestibular system using the SPH method to simulate the endolymph is represented in Fig. 2.2. This model comprises the membrane of the duct discretized with shell elements (S3R) and the inside endolymph discretized with particle (PC3D) for the SPH method [33].

In the present model, the shell was defined as a rigid body because it does not present significant deformations. The fluid particle distribution is regular, and the sum of the volume of the particles is equal to the volume of the SCC with the cupula. The dimensions of the duct represented in Fig. 2.2 were obtained from a 3-D model of the complete vestibular system that can be found in the work of O.W. Henson et al. [34], which was constructed based on magnetic resonance imaging. The considered endolymph material properties were obtained from the literature, being $1.0 \times 10^{-3}\,\text{kg/m}^3$ and $4.8 \times 10^{-3}\,\text{Pa s}$, the density and viscosity of the endolymph, respectively [35]. Additionally, to

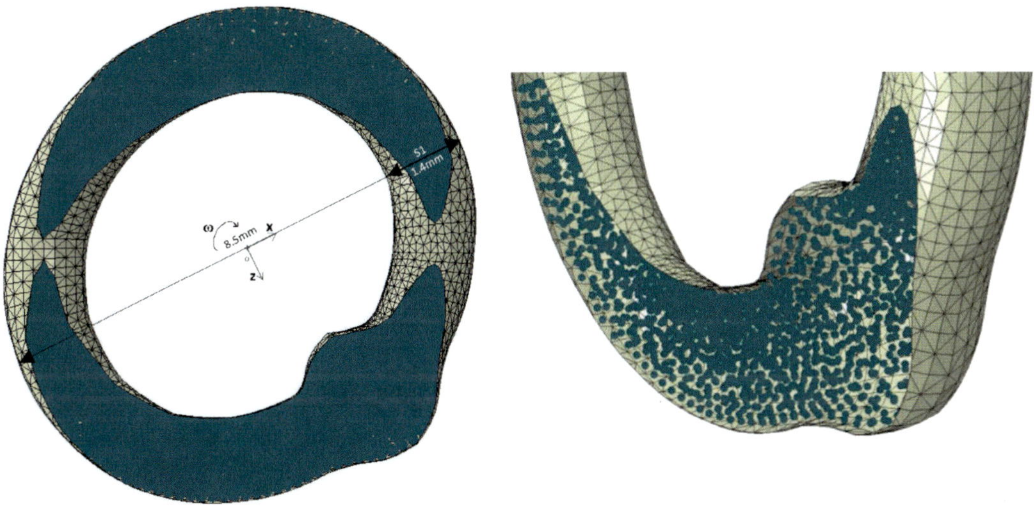


FIG. 2.2 Three-dimensional model of the semicircular duct of the vestibular system.

validate the model, three distinct particle discretization meshes were considered and analyzed, named as follows: Mesh1 (1790 particles), Mesh2 (7410 particles), and Mesh3 (13,637 particles).

Regarding the essential boundary conditions, to each node of the elements belonging to the membrane, a prescribed time-dependent angular velocity ($\omega = \pi/2$ rad/s) is enforced, with respect to point O in Fig. 2.2. To analyze the biomechanics of the SCC model, two distinct angular velocity functions were considered: profile ω1 and profile ω2 (Fig. 2.3A); both were previously used by Selva et al. [36] and WU. Cai-qin et al. [37].

Thus this model considers the gravity acceleration along the zz-axis. The general contact between the membrane elements and the fluid particles is performed explicitly by ABAQUS. To compare and validate the results of the several meshes considered (Mesh1, Mesh2, and Mesh3) with similar works of other authors [36, 37], instantaneous and local discharge variable was analyzed in Section S1 of Fig. 2.2 (Fig. 2.3B and C). The discharge is the volume rate of fluid flow transported through a given cross-sectional area.

The results obtained in this work are closed to the solution obtained by P. Selva et al. [36] and WU. Cai-qin et al. [37] when a similar angular velocity is applied to the SCC model, regardless of the considered particle mesh. Additionally, the convergence of the analysis was confirmed.

To perceive the model behavior during the imposed angular velocity, the velocity of the fluid along time was analyzed. The fluid velocity, at some instants of the simulation, is presented in Fig. 2.4A. As it is perceptible, the inner fluid velocity increases with time, as expected. After the general analysis of the global domain, a section of the duct was analyzed in details. The velocity field results correspond to Section S1 (see Fig. 2.2) obtained when profile ω1 was applied. The Fig 2.4C shows the 2-D sectional view on the left side and the velocity 3-D distribution on the right side.

The velocity maps (Fig. 2.4A and B), representing the canal fluid flow, permitted to observe the centrifugal force and the consequent higher velocity in the cupular part, due to the duct shape. The use of a three-dimensional model allowed the visualization of evidence of a 3-D velocity field in a section of the canal in some instants of the simulation. The analysis of the fluid in the section analyzed showed, as expected, that the velocity is higher in the center of the canal (Fig. 2.4C).

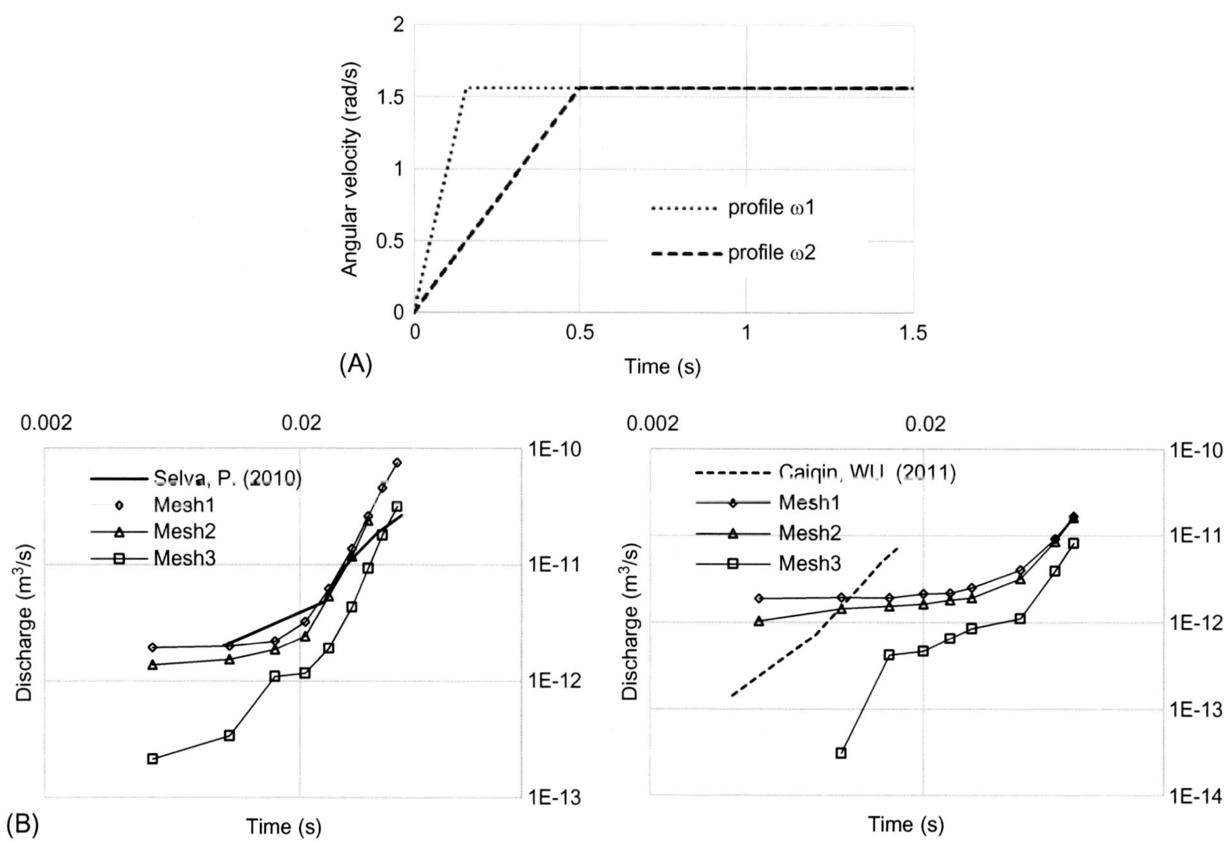


FIG. 2.3 (A) Two angular velocity functions imposed in the model, (B) discharge in the three meshes from both time steps. Profile ω1 *(left side)* and profile ω2 *(right side)*.

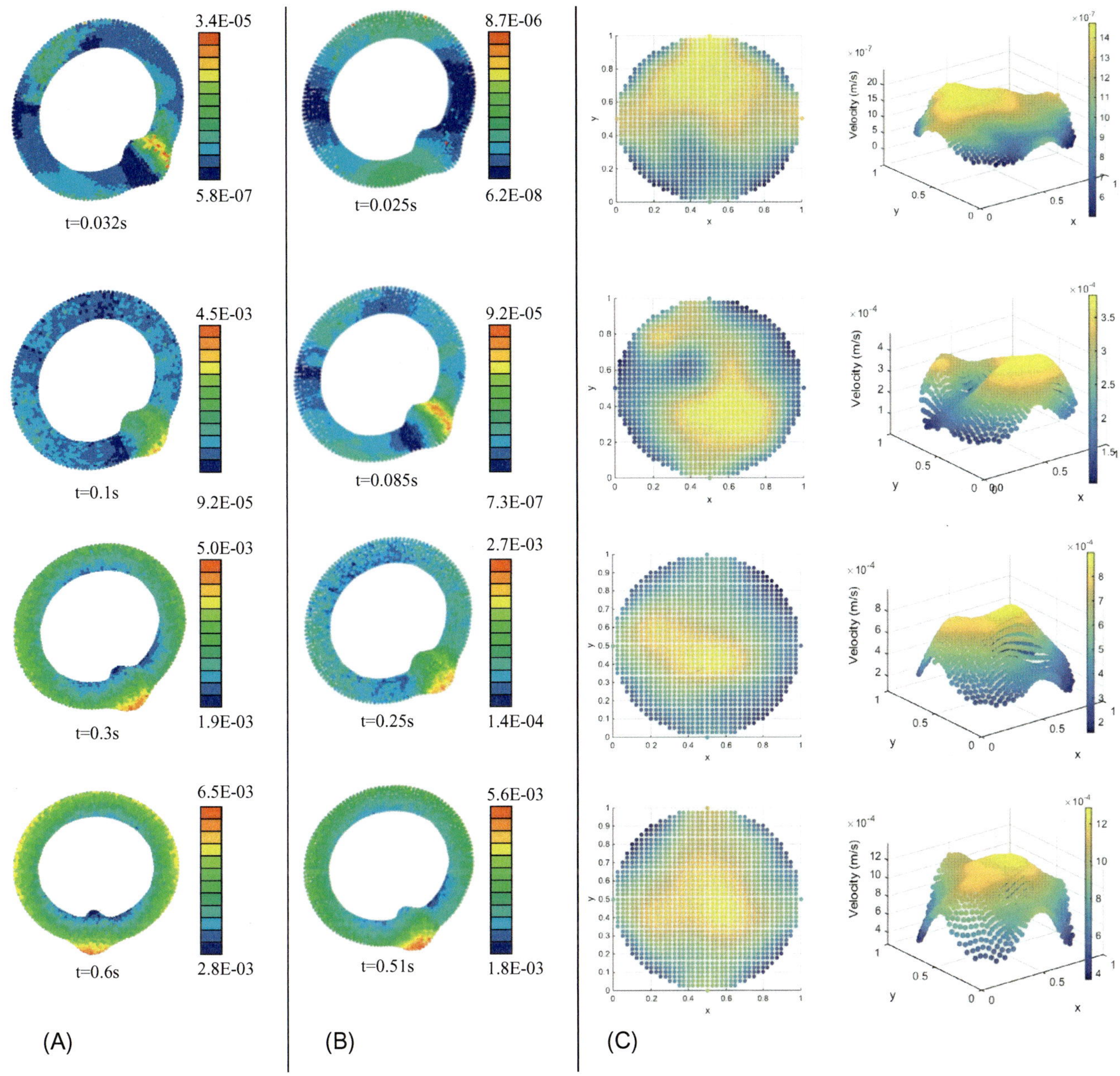


FIG. 2.4 (A) Fluid velocity (m/s) along time with profile ω1, (B) fluid velocity (m/s) along time with profile ω2, and (C) 3-D velocity field in section S1.

It was found that the use of SPH allows a more realistic representation of the fluid behavior. The applied methodology permitted to obtain promising results, according to the biomechanical properties of the vestibular system components available in the literature.

After the promising results, two more complex models of the vestibular system were built (Fig. 2.5) with the same methodology. The second model (Fig. 2.5A) was a semicircular duct with a closed shape to the real duct and with the cupula. The third model (Fig. 2.5B) was the organic vestibular system with three SCCs, a vestibule, and cupulas.

The results obtained with both models were similar to the first model, allowing us to perform new inner ear disorder simulations mainly due to the new structures.

One particular case of BPPV is cupulolithiasis, which occurs when the otoconia get lost in the SCC and attach to the cupula, inducing a false sensation of movement and leading to vertigo [38]. In the present chapter, generalized displacements of the cupula (free vibrations) will be applied to the obtained modal. The main goal was to understand the mechanical implications of the attached otoconia to the cupula.

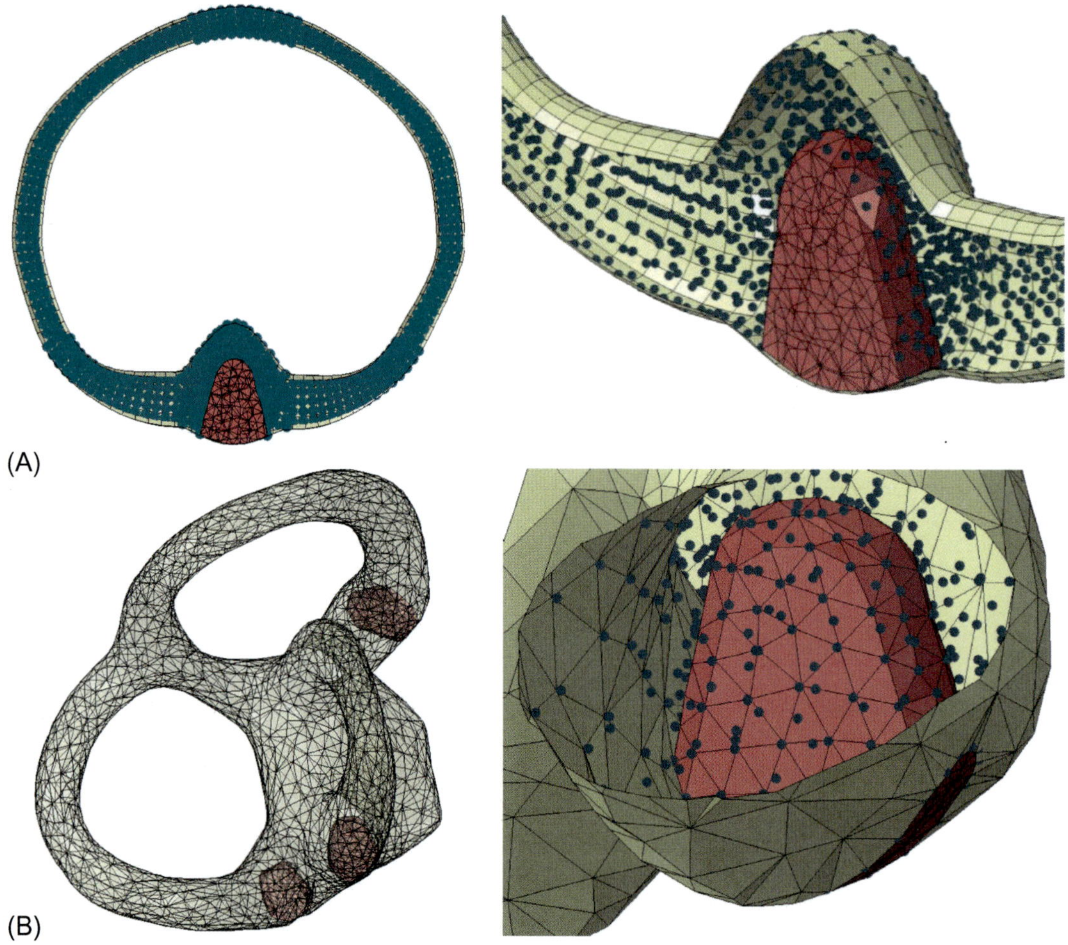


FIG. 2.5 (A) Three-dimensional model of the semicircular duct with a cupula and (B) three-dimensional model of the vestibular system.

To obtain numerically the free vibration modes of the cupula, 2-D and 3-D discrete models were built using the standard dimensions and the approximated mechanical properties found in the literature [39]; the applied boundary conditions were correlated with a real situation. Three different scenarios were simulated: the cupula itself, the cupula with the surrounded fluid, and the cupula with the attached otoconia and the surrounded fluid.

The free vibrations of the cupula were obtained with the software FEMAS, which is a finite element and meshless analysis software fully developed in an academic environment [27], and it possesses a graphical user interface (GUI) running in a MATLAB environment.

The formulations used to study the discrete models of the cupula were the FEM and two distinct meshless methods detailed in the previous section: RPIM and NNRPIM.

The results presented in this chapter include the vibration modes (Fig. 2.6) and the corresponding frequencies (Table 2.1) of the cupula 2-D with 417 nodes and 382 elements.

To visualize automatically the vibration mode, the results of the figures correspond to a fictitious displacement field. With this visualization technique, it is possible to visualize in each figure the magnitude of the fictitious displacement corresponding to the indicated vibration mode, since the red color corresponds to the higher displacement values and the blue color to the lower ones. In Table 2.1 the corresponding 10 first vibration frequencies of the cupula are presented.

The results indicate that the first vibration frequency is between 51.48 and 54.70 Hz. Note that the following vibration frequencies are much higher, which allows identifying with enough precision the magnitude of frequency of the first mode.

All the vibration frequencies obtained are contained in the human hearing range, since it goes from 20 to 20,000 Hz. In general, the vibration frequencies obtained with both meshless methods used (RPIM and NNRPIM) are lower than the ones obtained with FEM.

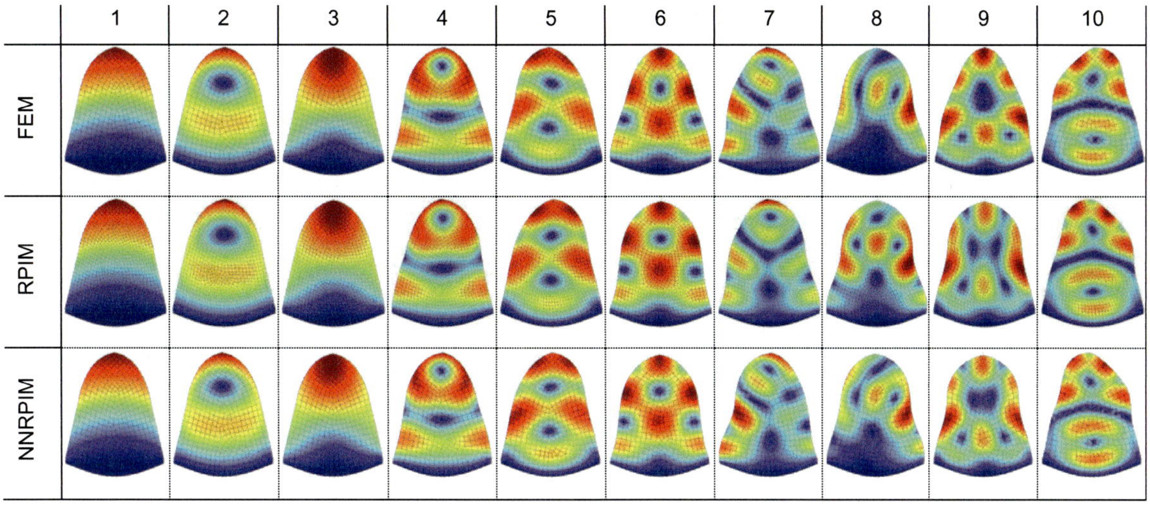


FIG. 2.6 Ten first modes of vibration of the two-dimensional cupula with FEM, RPIM, and NNRPIM.

TABLE 2.1 The 10 First Vibration Frequencies of the Two-Dimensional Cupula with FEM, RPIM, and NNRPIM

| | Natural frequency (Hz) | | | | | | | | | |
|---|---|---|---|---|---|---|---|---|---|---|
| | **1** | **2** | **3** | **4** | **5** | **6** | **7** | **8** | **9** | **10** |
| FEM | 54.70 | 121.32 | 131.62 | 226.17 | 259.80 | 265.13 | 345.26 | 350.33 | 362.51 | 386.60 |
| RPIM | 51.78 | 117.27 | 126.98 | 216.15 | 249.26 | 251.89 | 324.66 | 329.32 | 340.28 | 359.79 |
| NNRPIM | 51.48 | 117.36 | 126.35 | 213.64 | 249.12 | 251.08 | 321.02 | 330.41 | 338.08 | 361.29 |

Thus, to observe the effect of the endolymph around the cupula, the first five vibration modes of the cupula are presented in Fig. 2.7, for the FEM and both meshless methods. Note that the first two vibration modes obtained with the FEM resemble the first vibration mode obtained in Fig. 2.6. Furthermore, the third, fourth, and fifth vibration modes obtained now are very similar to the second vibration mode obtained in the analysis of the cupula without surrounding fluid.

The same effect is verified for the RPIM and NNPRIM approaches, the first vibration mode of the present RPIM analysis resembles the first vibration mode obtained in the cupula without fluid, and the second and third vibration modes of the present analysis are very similar to the second vibration mode obtained in Fig. 2.6.

Regarding the natural frequencies of the cupula surrounded by endolymph, in this study, the results obtained were 59 and 75 Hz in FEM, 61 Hz in the RPIM analysis, and 60 and 63 Hz in NNRPIM analysis, for the first resembling mode. Note that these values are closed to the ones observed in Table 2.1, indicating that the fluid has a reduced influence on the magnitude of the cupula's natural frequency.

In the next step of the present study, another possible cause of vertigo was simulated, the cupulolithiasis, which happens when the otoconia get attached to the cupula. Then, two different situations were simulated, one with one otoconium and another with three otoconia, as could be observed in the first line of Fig. 2.8.

The goal is to understand if the inclusion of the otoconia changes significantly the natural frequency of the cupula. Since the results obtained in the previous analyses with the three formulations were very similar, in Fig. 2.8, only the results regarding the NNRPIM formulation are presented.

When one otoconium (or one otoconium group) is considered, the frequencies of the cupula resembling its natural frequency (observed in Fig. 2.8) using the NNPRIM formulation are 61.42 Hz, which is also a very close result with the three-otoconium situation, which obtained 61.80 Hz for the same conditions. This may indicate that the size of the otoconium particles is not enough to change significantly the vibration frequency of the cupula [40].

The global results showed that the three different methodologies applied are capable to achieve similar results. However, most importantly, this work opens a new research branch in the computational analysis of the vestibular system.

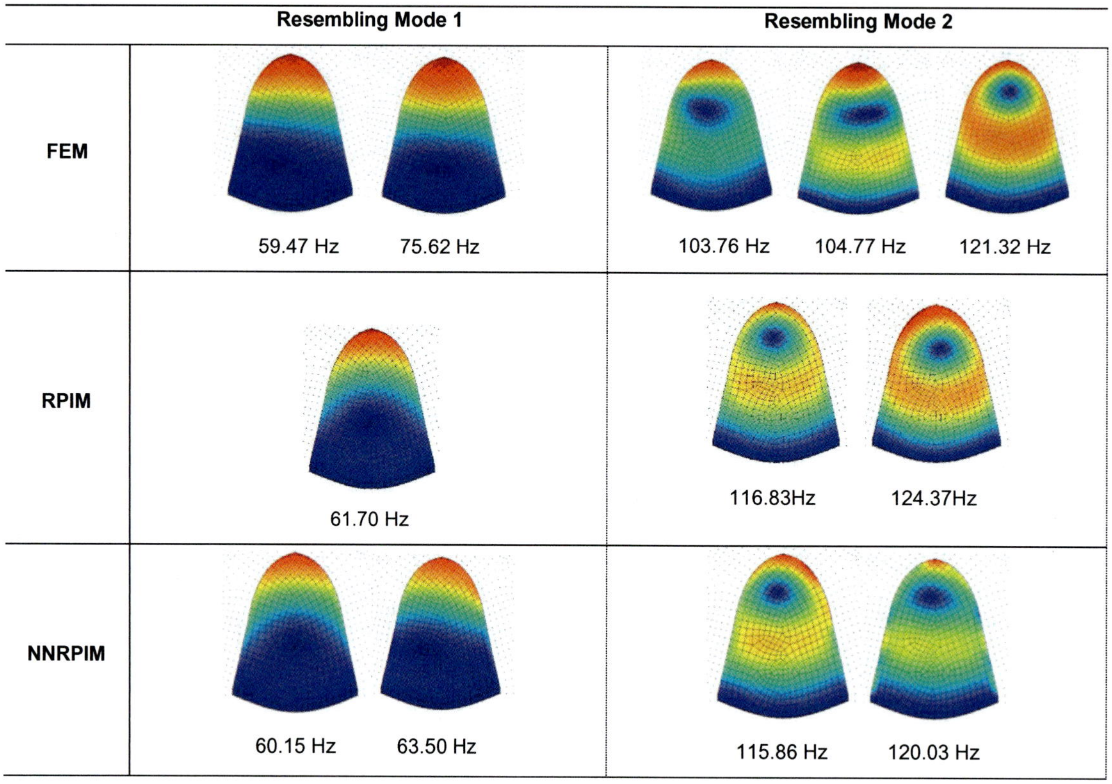


FIG. 2.7 Two first modes and natural frequencies of the two-dimensional cupula with endolymph using FEM and RPIM and NNRPIM.

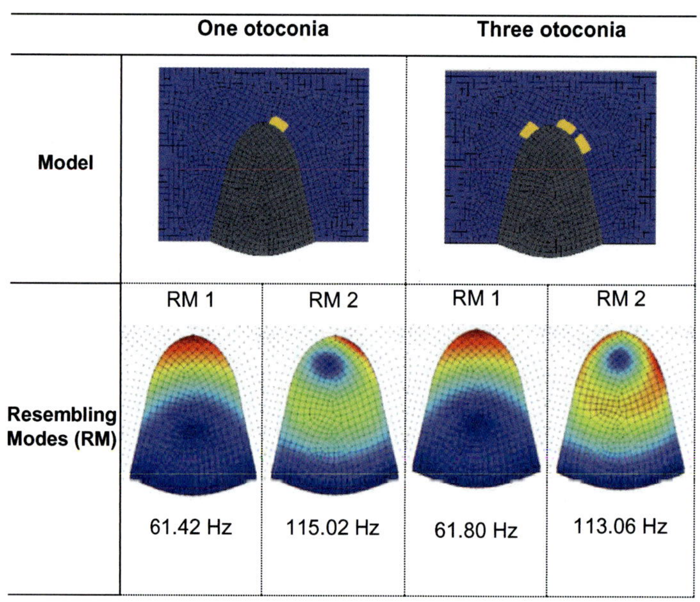


FIG. 2.8 Two first modes and natural frequencies of the cupula with endolymph and otoconia.

In conclusion, it was possible to understand that the natural frequency of the cupula ranges between 51 and 61 Hz for all the analyzed situations. Moreover, all the other vibration frequencies possess very distinct values. Since theoretically the resonance phenomena will force the cupula to vibrate, the induced vibration will allow detaching the otoconium particles from the cupula, reducing the vertigo symptoms. This could be achieved with noninvasive ways, such as listening to music with a predominant set of bass sounds.

## 2.5 CONCLUSIONS

The vestibular system plays an important role in the maintenance of the body balance. It is imperative to study this system considering that vertigo affects a huge part of the older population. The research work presented in this chapter has demonstrated once again the value of computational numerical simulation to study biological structures and phenomena.

In conclusion, a better comprehension of the biomechanics of the vestibular system is the first step to plan different kinds of simulations with numerical models. The vestibular system model built using several numerical methods, as FEM and meshless methods, aims to be the first step to reinvent the rehabilitation process. The development of a tool, which can help the audiologist in their daily practice, is an important step in such scientific field that can contribute to a fast recovery from vestibular disorders.

## Acknowledgments

The authors acknowledge the funding by Ministério da Ciência, Tecnologia e Ensino Superior, Fundação para a Ciência e a Tecnologia, Portugal, and Programa Operacional Capital Humano (POCH), comparticipado pelo Fundo Social Europeu e por fundos nacionais do MCTES under research grants SFRH/BD/108292/2015, IF/00159/2014, and by project funding MITEXPL/ISF/0084/2017 and UID/EMS/50022/2013 (funding provided by the interinstitutional projects from LAETA). Additionally, the authors gratefully acknowledge the funding of Project NORTE-01-0145-FEDER-000022, SciTech, Science and Technology for Competitive and Sustainable Industries, co-financed by Programa Operacional Regional do Norte (NORTE2020), through Fundo Europeu de Desenvolvimento Regional (FEDER).

## References

[1] H. Neuhauser, T. Lempert, Vertigo: epidemiologic aspects, Semin. Neurol. 29 (5) (Nov. 2009) 473–481.
[2] R. Humphriss, Clinical outcomes of vestibular rehabilitation, Physiotherapy 87 (7) (2001) 368–373.
[3] J.E. Hall, Guyton and Hall Textbook of Medical Physiology, 13th ed., Elsevier Inc, 2015.
[4] S.J. Herdman, Vestibular rehabilitation, Curr. Opin. Neurol. 26 (1) (Mar. 2013) 96–101.
[5] D.A. Winter, Human balance and posture control during standing and walking, Gait Posture 3 (4) (Dec. 1995) 193–214.
[6] H.C. Diener, J. Dichgans, On the role of vestibular, visual and somatosensory information for dynamic postural control in humans, Prog. Brain Res. 76 (1988) 253–262.
[7] S.K. Powers, E.T. Howley, Exercise Physiology: Theory and Application to Fitness and Performance, McGraw-Hill Humanities/Social Sciences/Languages, 2012.
[8] S. Herdman, Vestibular Rehabilitation, F.A. Davis Co., 2007.
[9] H. Reisine, J.I. Simpson, V. Henn, A geometric analysis of semicircular canals and induced activity in their peripheral afferents in the rhesus monkey, Ann. N. Y. Acad. Sci. 545 (1988) 10–20.
[10] R. Jaeger, A. Takagi, T. Haslwanter, Modeling the relation between head orientations and otolith responses in humans, Hear. Res. 173 (1–2) (Nov. 2002) 29–42.
[11] M.J. Gámiz, J.A. Lopez-Escamez, Health-related quality of life in patients over sixty years old with benign paroxysmal positional Vertigo, Gerontology 50 (2) (Feb. 2004) 82–86.
[12] P.B. Salzman, Gait and balance disorders in older adults, Am. Fam. Physician 82 (1) (2010) 61–68.
[13] U. Olsson Möller, P. Midlöv, J. Kristensson, C. Ekdahl, J. Berglund, U. Jakobsson, Prevalence and predictors of falls and dizziness in people younger and older than 80 years of age—a longitudinal cohort study, Arch. Gerontol. Geriatr. 56 (1) (Jan. 2013) 160–168.
[14] R.E. Gans, Overview of BPPV: Treatment Methodologies – Hearing Review, [Online]. Available: http://www.hearingreview.com/2000/09/overview-of-bppv-treatment-methodologies/, 2000. Accessed 16 April 2018.
[15] M. von Brevern, H. Neuhauser, Epidemiological evidence for a link between vertigo and migraine, J. Vestib. Res. 21 (6) (2011) 299–304.
[16] H.K. Neuhauser, T. Lempert, Vertigo: epidemiologic aspects, Semin. Neurol. 29 (5) (2009) 473–481.
[17] F.B. Horak, Clinical assessment of balance disorders, Gait Posture 6 (1) (Aug. 1997) 76–84.
[18] K. Hanley, T. O'Dowd, N. Considine, A systematic review of vertigo in primary care, Br. J. Gen. Pract. 51 (469) (Aug. 2001) 666–671.
[19] R.W. Baloh, Vertigo, Lancet 352 (9143) (Dec. 1998) 1841–1846.
[20] T.D. Fife, et al., Practice parameter: therapies for benign paroxysmal positional vertigo (an evidence-based review): [RETIRED], Neurology 70 (22) (May 2008) 2067–2074.
[21] A. Katsarkas, Benign paroxysmal positional vertigo (BPPV): idiopathic versus post-traumatic, Acta Otolaryngol. 119 (7) (1999) 745–749.
[22] M. Kassemi, D. Deserranno, J.G. Oas, Fluid–structural interactions in the inner ear, Comput. Struct. 83 (2–3) (Jan. 2005) 181–189.
[23] C. Wu, C. Hua, L. Yang, P. Dai, T. Zhang, K. Wang, Dynamic analysis of fluid-structure interaction of endolymph and cupula in the lateral semicircular canal of inner ear, J. Hydrodyn. Ser. B 23 (6) (Dec. 2011) 777–783.
[24] P.K.T. Belytschko, Y. Krongauz, D. Organ, M. Fleming, Meshless methods: an overview and recent developments, Comput. Methods Appl. Mech. Eng. 139 (1–4) (Dec. 1996) 3–47.
[25] G. Chen, et al., A new approach for assigning bone material properties from CT images into finite element models, J. Biomech. 43 (5) (Mar. 2010) 1011–1015.
[26] T.Q.B.T.-V. Vu, N.-H. Nguyen, A. Khosravifard, M.R. Hematiyan, S. Tanaka, A simple FSDT-based meshfree method for analysis of functionally graded plates, Eng. Anal. Bound. Elem. 79 (Jun. 2017) 1–12.

[27] J. Belinha, Meshless Methods in Biomechanics – Bone Tissue Remodelling Analysis, Lecture Notes in Computational Vision and Biomechanics, vol. 16, Springer Netherlands, 2014.
[28] T.Q. Bui, M.N. Nguyen, C. Zhang, An efficient meshfree method for vibration analysis of laminated composite plates, Comput. Mech. 48 (2) (Aug. 2011) 175–193.
[29] G.R. Liu, M.B. Liu, Smoothed Particle Hydrodynamics, World Scientific, 2003.
[30] L. Lobovský, J. Křen, Smoothed particle hydrodynamics modelling of fluids and solids, Appl. Comput. Mech. 1 (2007) 521–530.
[31] J.J. Monaghan, Smoothed particle hydrodynamics, Reports Prog. Phys. 68 (8) (Aug. 2005) 1703–1759.
[32] W.K. Liu, S. Jun, S. Li, J. Adee, T. Belytschko, Reproducing kernel particle methods for structural dynamics, Int. J. Numer. Methods Eng. 38 (10) (May 1995) 1655–1679.
[33] C.F. Santos, J. Belinha, F. Gentil, M. Parente, R.N. Jorge, An alternative 3D numerical method to study the biomechanical behaviour of the human inner ear semicircular canal, Acta Bioeng. Biomech. 19 (1) (2017) 3–15.
[34] O.W. Henson, et al., Department of Cell and Developmental Biology, University of North Carolina, Chapel Hill and The Center for In Vivo Microscopy, Duke University, Durham, NC, Copyright, 2000.
[35] J.L. Davis, J. Xue, E.H. Peterson, J.W. Grant, Layer thickness and curvature effects on otoconial membrane deformation in the utricle of the red-ear slider turtle: static and modal analysis, J. Vestib. Res. 17 (4) (Jan. 2007) 145–162.
[36] P. Selva, J. Morlier, Y. Gourinat, Development of a dynamic virtual reality model of the inner ear sensory system as a learning and demonstrating tool, Model. Simul. Eng. 2009 (2009) 1–10.
[37] C. Wu, K. Wang, L. Yang, P. Dai, Three-dimensional models of the membranous vestibular labyrinth in the guinea pig inner ear, in: Proc. 2011 4th Int. Conf. Biomed. Eng. Informatics, BMEI 2011, vol. 1, 2011, pp. 541–544.
[38] T.M. Squires, M.S. Weidman, T.C. Hain, H.a. Stone, A mathematical model for top-shelf vertigo: the role of sedimenting otoconia in BPPV, J. Biomech. 37 (8) (Aug. 2004) 1137–1146.
[39] J.L. Davis, J. Xue, E.H. Peterson, J.W. Grant, Layer thickness and curvature effects on otoconial membrane deformation in the utricle of the red-ear slider turtle: static and modal analysis, J. Vestib. Res. 17 (4) (2007) 145–162.
[40] C.F. Santos, J. Belinha, F. Gentil, M. Parente, R.N. Jorge, The free vibrations analysis of the cupula in the inner ear using a natural neighbor meshless method, Eng. Anal. Bound. Elem. 92 (Jul. 2018) 50–63.

CHAPTER

# 3

# Design, Simulation, and Experimentation of Colonic Stents

*Sergio Puértolas**,†, Antonio Herrera*†,‡,§, *Eduardo Bajador*¶, *Judith Millastre*¶, *Elena Ibarz**,†, *Sergio Gabarre*‖, *José A. Puértolas*#, *Luis Gracia**,†

***Department of Mechanical Engineering, University of Zaragoza, Zaragoza, Spain †Aragón Institute for Engineering Research, Zaragoza, Spain ‡Aragón Health Research Institute, Zaragoza, Spain §Department of Surgery, University of Zaragoza, Zaragoza, Spain ¶Department of Gastroenterology, Lozano Blesa University Hospital, Zaragoza, Spain ‖Vlaams Instituut voor Biotechnologie, Leuven, Belgium #Department of Material Science, University of Zaragoza, Zaragoza, Spain**

## 3.1 INTRODUCTION

Stents are medical devices that scaffold or brace the inside circumference of tubular passages in the human body. They are endoprostheses with a flexible tubular mesh structure that are manufactured mainly with biocompatible metallic alloys. Stenting technology has become an alternative to recover the lumen and restore flow or transit in many vital ducts.

The first applications were for the treatment of coronary atherosclerosis. In 1964 Charles Dotter [1] performed the first angioplasty in a peripheral artery of the legs. Since then stents have been widely used for the treatment of cardiovascular ischemia, and their use has been extended to the treatment of stenosis in ducts of the respiratory and digestive system: tracheobronchial, biliary duct, esophageal, urethral or urethral/prostatic, and colon.

Many different stent designs are currently being marketed or evaluated for the foregoing applications with different characteristics due mainly to the materials used and their structure. Biocompatibility is maintained using materials such as stainless steel 316L, 316LVM, tantalum, shape memory alloy NiTi [2–4], or more recently biodegradable materials based on polymers from the lactic acid, glycolic, and caprolactone families [5, 6], and metals [7], either Mg-based [8, 9] or Fe-based alloys [10]. In fact, metals have superior mechanical properties over polymers for replicating the properties of SS316L, the reference material for coronary stents.

Usually, the final geometry is a coil, helical, spiral, woven, individual, or sequential ring, and a closed or open cell, which are manufactured from wires, tubes, ribbons, or sheets by laser cutting, photochemical etching, or wire forming such as braiding, knitting, etc. Deployment, efficiency, and stability of treatments for stenosis are obtained by balloon-expandable or self-expanding methods.

In the majority of coronary and noncoronary applications, the combination of different types of biomaterials and geometries allows stents to be designed with both deployment methods for stenosis: balloon-expandable and self-expanding stents with different efficiencies and stabilities.

Colonic strictures are a common complication of enteral diseases. Benign causes of obstructions like Crohn's disease and ulcerative colitis present a group of strictures scattered throughout the colon, since they are not caused by local growth of malignant tumor and in most cases evolve favorably under pharmacological treatment. Therefore the use of stents is not a common practice in benign pathologies. However, colonic stents have been used for more than 20 years for preoperative and palliative relief of malignant colonic obstruction [11, 12].

Large bowel cancer is the most common cancer in the digestive system. Colorectal cancer is the third most common cancer in men, after lung and prostate cancer, and the second most common in women, after breast cancer, and represents the second leading cause of cancer death in the world. It is the third most common malignant tumor in


https://doi.org/10.1016/B978-0-12-816390-0.00003-0

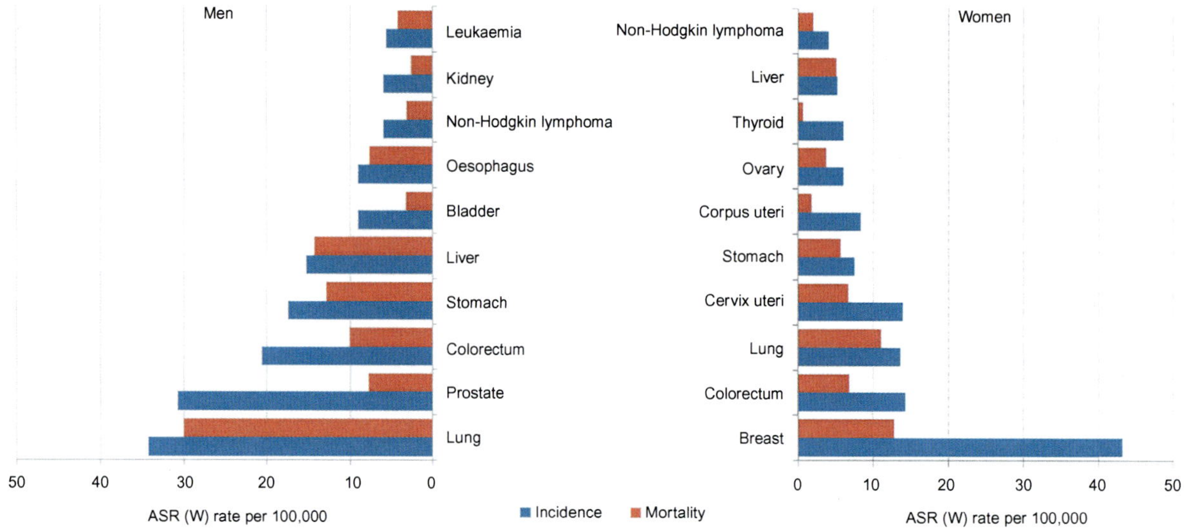


FIG. 3.1 Malignant tumors with the highest ratio of incidence and mortality in the world (age-standardized rates for world standard population, ASR(W) per 100,000 population) [13].

incidence and the fourth in mortality in the world, after lung cancer [13] (Fig. 3.1). Due to its endoluminal growth, 29% of patients with this tumor developed intestinal obstruction [14].

Colonic occlusion is the primary reason for emergency colorectal surgery [15]. Traditional treatment of colonic obstruction is surgery, which is usually a colon resection with primary anastomosis and is more often associated (two-stage procedure) with bowel resection and colostomy (Hartmann's procedure) [16] and restoration of gastro-intestinal continuity in a subsequent intervention. However, both possibilities are limited, especially when it comes to elderly patients with associated serious diseases, causing high morbidity, prolonged hospitalization, and a mortality rate between 5% and 11% [17]. Moreover, in many cases it is not possible to restore intestinal continuity, so the patient is left with a permanent colostomy and consequent deterioration in their quality of life [18].

Sixty percent of patients with colorectal cancer present with a tumor in the left colon, and up to 25% have complete or partial occlusions at the time of diagnosis [19]. As is well known, metallic self-expanding stents are an option widely used to treat acute obstruction of the colon, either for palliation or bridging or transition to surgical intervention, thus avoiding emergency surgery rates of morbidity and mortality greater than 30% [19]. The stent allows decompression, complete preparation of the colon, and resection surgery in one step [20].

Colonic stenting in acute large bowel obstruction is more likely to be successful in shorter, malignant strictures with less angulation distal to the obstruction. Longer benign strictures are less likely to be successful and may be associated with an increased risk of perforation [21]. Although the placement of colonic stents is much less invasive, morbidity can be high due to the complications associated with this procedure. The most common complications include perforation, wrong positioning, migration, and reobstruction. To prevent or minimize these complications, it is necessary to make a correct choice of stent to use [22]. Since the technical success of a stent for a particular obstructive lesion depends mainly on its mechanical behavior, mechanical modeling is required to design the stent structure through a predictive model as in Kim et al. [23] for braiding stents and Tzamtzis et al. [24] for an aortic valve. This will allow custom-designed stents for each type of obstruction and patient.

Colonic stents were introduced in the early 1990s. In 1990 Dohmoto et al. [25] presented the treatment of a malignant rectal obstruction with the placement of an expandable metal stent, and in 1994 Tejero et al. [26] reported two cases of acute malignant obstruction treated successfully with the placement of a stent prior to elective surgery. In 1996 the first prototype a self-expanding colonic stent made of Nitinol was presented [27]. Since then, the field of colorectal stent placement has evolved substantially with the introduction of new stent models and more versatile deployment systems. Self-expandable metallic stents (SEMSs) are widely used to treat malignant colonic obstruction [28]. They may be used both as a palliative measure and as a preoperative bridge to facilitate a one-stage surgical resection of primary colonic tumors [29, 30]. Increasing experience in the management of colorectal cancer with colonic stents decreases morbidity and mortality, as well as cost, in comparison with surgical intervention for acute colonic obstruction.

Despite the plethora of stent designs, diameters, and materials from which they are fabricated, the technology is imperfect. Uncovered SEMSs allow tissue ingrowth and fully covered SEMSs migrate. New designs and materials to minimize these risks as well as those associated with occlusion and device-associated perforation are needed [22].

Stent design principally has two different manufacturing strategies: braided or knitted wire-based stents, and tube-based stent designs [31] to which laser cuts are made to shape their final repose configuration. They can also be classified according to whether they are covered or uncovered. Although the tendency is to design stents covered to prevent tumor growth and thus reduce the risk of restenosis, in practice these stents have not shown clear advantages over other types [32]. Tumor ingrowth occurs more frequently in the uncovered group, while late migration is more common in covered stents. In [33], an increase in late migration (40%, $P = .005$) and greater loss of stent function (60%, $P = .0018$) with covered stents are reported.

Different designs of colorectal stents are available for clinical use. The characteristics of some of the most extended and used such as Ultraflex, Wallstent, and colonic Wallflex are included in Table 3.1.

In the literature, little quantitative data are found to characterize the behavior of the different types of colonic stents. A limited number of studies have investigated the influence of design parameters for helical coils [34] and braided stents [23, 35] on mechanical behavior. Experimental bench testing to understand the properties of self-expandable metallic stents for different biliary and airway endoprostheses have been reported [36, 37]. Finite element analysis is commonly used to contrast the main mechanical parameters (radial and axial strength, flexibility, etc.) of different stents [38–42].

The combination of different geometries and materials provides stents with a wide range of mechanical properties related to radial strength, trackability, elastic recoil, longitudinal flexibility, etc. For example, an expanded balloon together with a slotted tube provides great radial strength, while a self-expanding coil stent is highly flexible with good trackability. In the next section, some of the most relevant mechanical and functional characteristics are defined.

Since the first case of colorectal stent placement reported by Dohmoto et al. [25] in 1991, different stent designs have been widely used to relieve various forms of malignant gastrointestinal obstructions. In fact, the first colonic stent was primarily designed for vascular or esophageal use. Nowadays, different configurations (braided mesh, knitted, cell wire, etc.) and materials (stainless steel, Nitinol, polymers) have been developed and marketed with clear differences between them in their clinical performance, such as those observed in comparative studies of clinical trials [44–47].

The following configurations strive to fulfill the criteria of the ideal stent for colonic obstructions:

- Mechanical properties such as flexibility, high expansion rate, mechanical stability, and radial compression resistance.
- Properties related to the stent delivery system, such as tractability, crossability, and pushability, needed to advance the stent to the point of the obstruction.
- Clinical features such as the prevention of the migration of propulsive forces of colonic mobility, perforation, good radiopacity, and tumor- related ingrowth or overgrowth.

**TABLE 3.1** Main Available Commercial Colorectal Stents [43]

| Stent name | Material | Length (mm) | Diameter flare (mm) | Diameter body (mm) | Manufacturer |
|---|---|---|---|---|---|
| Ultraflex Precision Colonic Stent | Nitinol | 57, 87, 117 | 30 | 25 | Boston Scientific |
| Wallflex Colonic TTS | Nitinol | 60, 90, 120 | 25 | 22 or 25 | Boston Scientific |
| Wallstent Enteral TTS | Elgiloy | 60, 90 | NA | 20 or 22 | Boston Scientific |
| Colonic Z stent | Stainless steel | 40, 60, 80, 100, 120 | 35 | 25 | Wilson-Cook |
| Niti-S Colorectal Stent | Nitinol | 60, 80, 100 covered, uncovered | 28–30 | 20, 22, 24 | Taewoon-Medical Co., Ltd. |
| TTS Niti-S Colorectal Stent | Nitinol | 60, 80, 100 | 28 | 20 | Taewoon-Medical Co., Ltd. |
| Hanarostent colorectal (Choo stent) | Nitinol | Covered: 60, 90, 110, 130 | NA | 18, 22 | MI Tech Co. Ltd. |
| | | Uncovered: 80, 110, 140 | NA | 18, 22 | |

*NA*, not available.

## 3.2 IDEAL MECHANICAL PROPERTIES FOR COLONIC STENTS

Colonic stents present the following characteristics:

- Diameters are two to three times greater than other types of stent (30–35 mm vs. 10, 12 mm in vascular stents and 15, 18 mm in esophageal stents).
- Great opening capacity (a lumen between 15 and 20 mm must be obtained, at least).
- Recovery force as uniform as possible, independent of the opening diameter.

Therefore the mechanical performance of colonic stents presents a number of differences compared to the usual low expansion rate coronary stents, needing high rates of elastic recovery (superelasticity) and enough radial expansion force in a wide deployment diameter range.

At present, there are two design alternatives: the self-expanding stent, based on the superelastic properties of certain metallic alloys (e.g., Nitinol), and the balloon-expandable stent, which uses plastic deformations in materials like stainless steel 316L, with the help of a balloon, and remains the reference method in vascular applications.

However, for colonic stents, with final diameters in the range 25–30 mm and expansion rates higher than 5, self-expanding stents have become a clear alternative to traditional balloon-expandable deployment.

### 3.2.1 Mechanical Parameters

The following are relevant aspects concerning large diameter stents, including qualitative evaluation of colorectal applications, and should be considered to evaluate the performance of different models:

- **Expansion rate** (Fig. 3.2A): The ratio between the unexpanded and expanded diameter of the stent. A higher value is better.
- **Stent shortening** (Fig. 3.2B): Reduction of the overall length of the device during expansion (commonly expressed in percentage). A lower value is better.

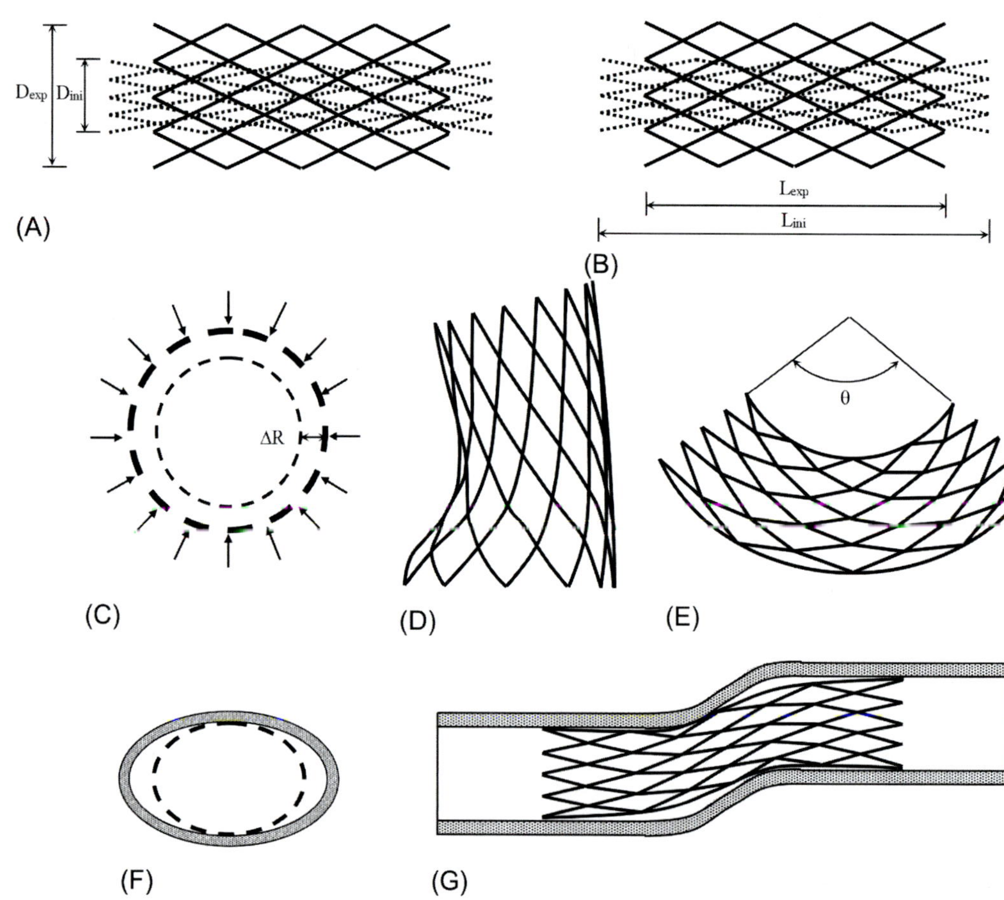


FIG. 3.2 Mechanical parameters of stents [48]: (A) expansion rate; (B) stent shortening; (C) radial compression resistance; (D) buckling resistance; (E) longitudinal flexibility; (F) perimetral adaptability; (G) longitudinal adaptability.

- **Radial compression resistance** (Fig. 3.2C): Stiffness of the stent when compressed by uniformly distributed radial pressure. A higher value is better. In the case of use materials that present hysteresis loops in their strain-stress curve (e.g., Nitinol), two parameters related to compression and expansion are defined. In such a case, the radial compression resistance is controlled by two terms. The first is the radial compression resistance (RCR): stiffness of the stent when compressed by uniformly distributed radial pressure (crimping process). A higher value is better. The second is the chronic radial expansion force (CEF): radial force exerted by the stent against the duct wall in deployment (release process). A higher and progressive (low slope) value evolution is better.
- **Buckling resistance** (Fig. 3.2D): Stent resistance to inelastic collapse when compressed by pressure above a critical value. A higher value is better.
- **Longitudinal flexibility** (Fig. 3.2E): Flexibility of the stent when globally bent as a tubular beam. A higher value (lower stiffness) is better.
- **Perimetral adaptability** (Fig. 3.2F): Ability to tightly adapt to the shape of the conduit cross- section for whose patency the device is maintained. A higher value is better.
- **Longitudinal adaptability** (Fig. 3.2G): Ability to tightly adapt to the shape of the conduit longitudinal section for whose patency the device is maintained. A higher value is better.
- **Mesh opening**: Total amount of conduit tissue area not covered by metal on each of the deformed cells of the expanded stent. A lower value is better.
- **Migration tendency**: Probability of stent displacement after correct implantation due to peristaltic movement in combination with poor mechanical performance. A lower value is better.

## 3.2.2 Commercial Stents

### 3.2.2.1 *Self-Expanding Stainless Steel Stents*

#### 3.2.2.1.1 WALLSTENT (FIG. 3.3)

This stent is based on braided wires in a tubular configuration with two families of helicoidal springs rolled up in inverse directions. There is no connection between the wires in the helicoidal shape, which results in the formation of

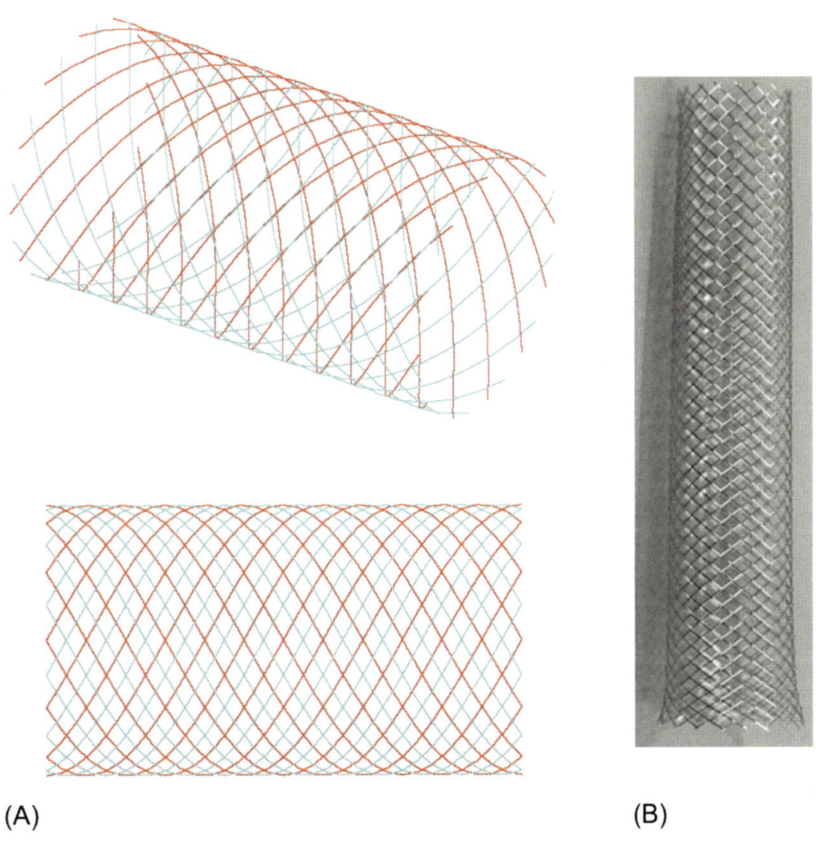

(A) (B)

FIG. 3.3 Wallstent: (A) geometrical scheme; (B) real view. *(Source: Boston Scientific.)*

sliding knots. This stent always operates in elastic mode since the deformation from its work shape is achieved by pulling each wire from the helicoidal mode to an almost straight line. The high number of wires in the Wallstent provides a small cell size.

##### 3.2.2.1.2 GIANTURCO STENT (FIG. 3.4)

Rings of zigzag wire segments formed by Z-shaped radial springs, joined by fine nylon or stainless steel wires, form the structure of this stent. In contrast to the version for the esophagus, which is covered by a double layer of polyethylene, the version for the colon, the colonic Z-stent, does not present any covering. The stent presents a bell shape at the extremes of the stents. The cells in the stent are extremely open, due to the high length of the wire segments.

##### 3.2.2.1.3 SONG AND CHOO-Z-MODIFIED GIANTURCO STENTS (FIG. 3.5)

These are variants of the Gianturco type with a curved Z-shaped radial spring. In the Song model (Fig. 3.5B) the end rings present a greater diameter than the rest of the rings. This configuration introduces a right-angle discontinuity, which cannot be supported by fine wires, as in the basic Gianturco stent. In the transition zone, the segments are embedded in polyurethane. The Choo-Z stent (Fig. 3.5C) introduces certain modifications in this design: the number of rings in the central region increases and the length of the wires decreases.

#### 3.2.2.2 *Self-Expanding Nitinol Stents*

##### 3.2.2.2.1 ESOPHACOIL STENT (FIG. 3.6)

This stent presents a spiral that forms a unique ribbon helicoidal spring configuration and originally was applied to the esophagus. The helix is manufactured from a ribbon with a square transversal section and minimum helix pitch so that it contacts the turns of the spiral. This structure reduces the space between the turns in the central region and can be considered a stent with a practically closed cell. In general, the manufacturing process is simple and the design process does not need any modification, except the section.

##### 3.2.2.2.2 ULTRAFLEX STENT (FIG. 3.7)

The model for the colon is a braided and fabric Nitinol wire structure following the typical pattern used in cardiovascular applications, known as the Strecker stent, which is a balloon-expandable operating mode stent. It is based on a

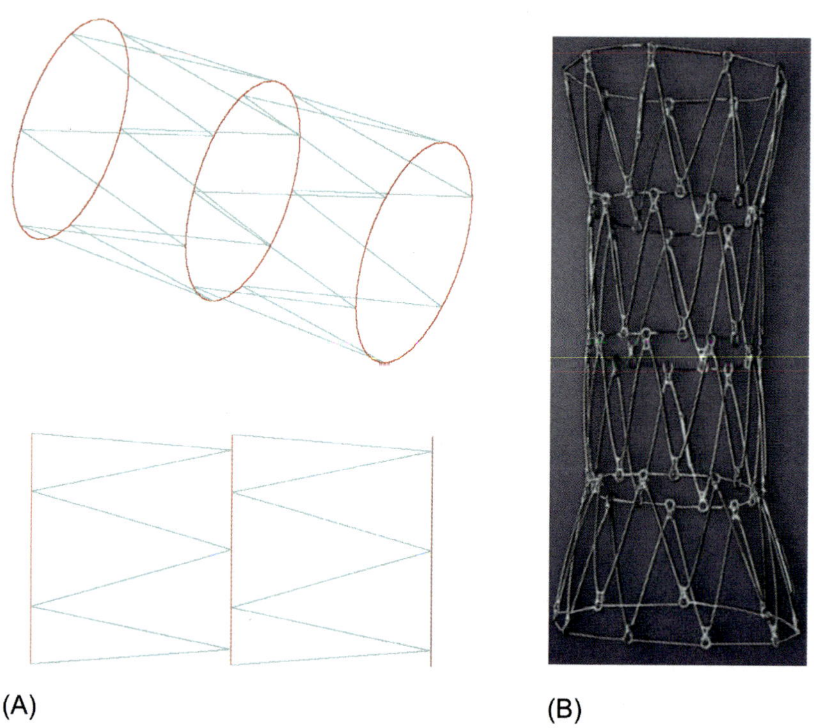


FIG. 3.4 Gianturco: (A) geometrical scheme; (B) real view. *(Source: colonic Z-stent, Cook Medical.)*

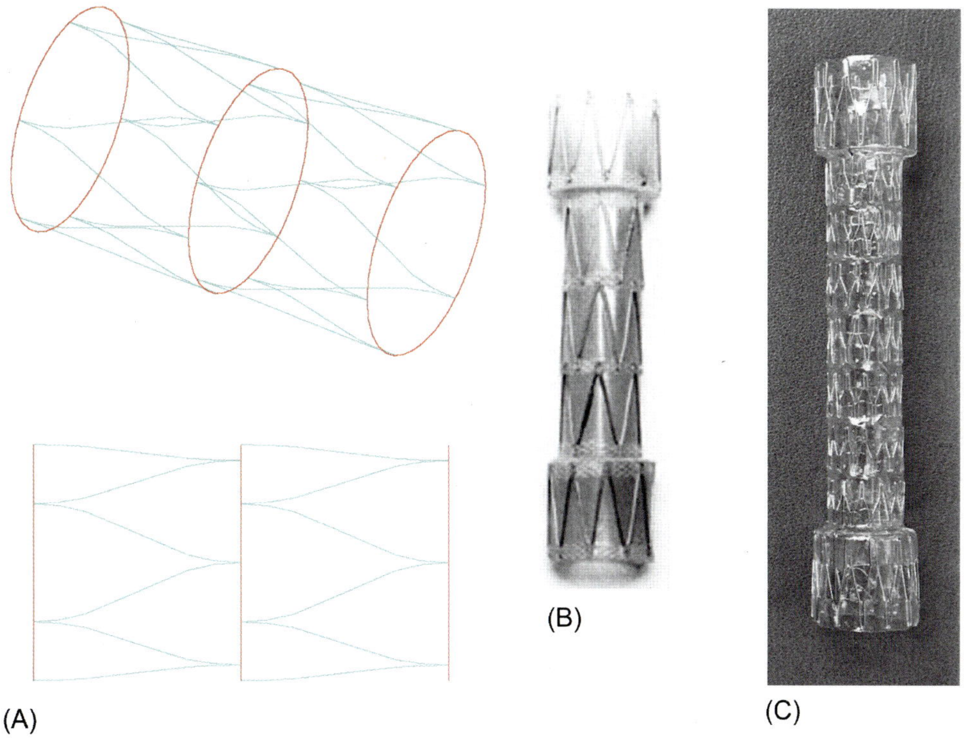


FIG. 3.5 Modified Gianturco stents: (A) geometrical scheme; (B) Song Z-stent real view; (C) Choo-Z stent real view. *(Sources: (B) MI-Tech. (C) Wilson-Cook.)*

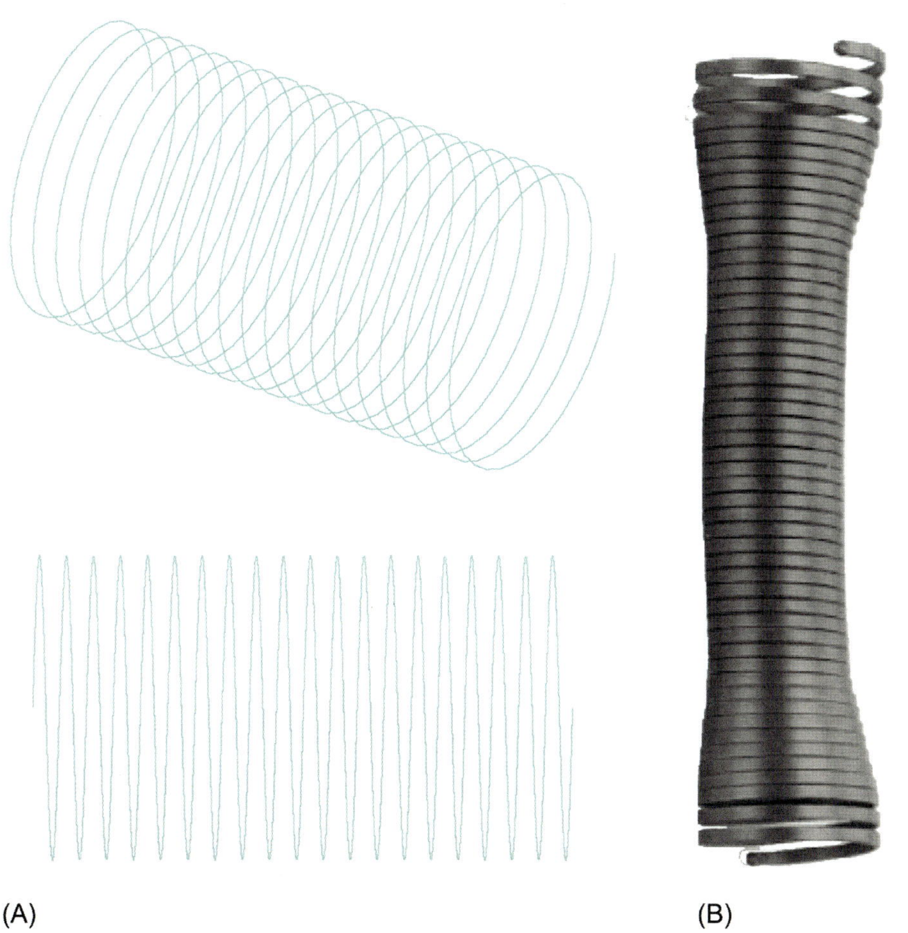


FIG. 3.6 Esophacoil stent: (A) geometrical scheme; (B) real view. *(Source: Medtronic/Instent.)*

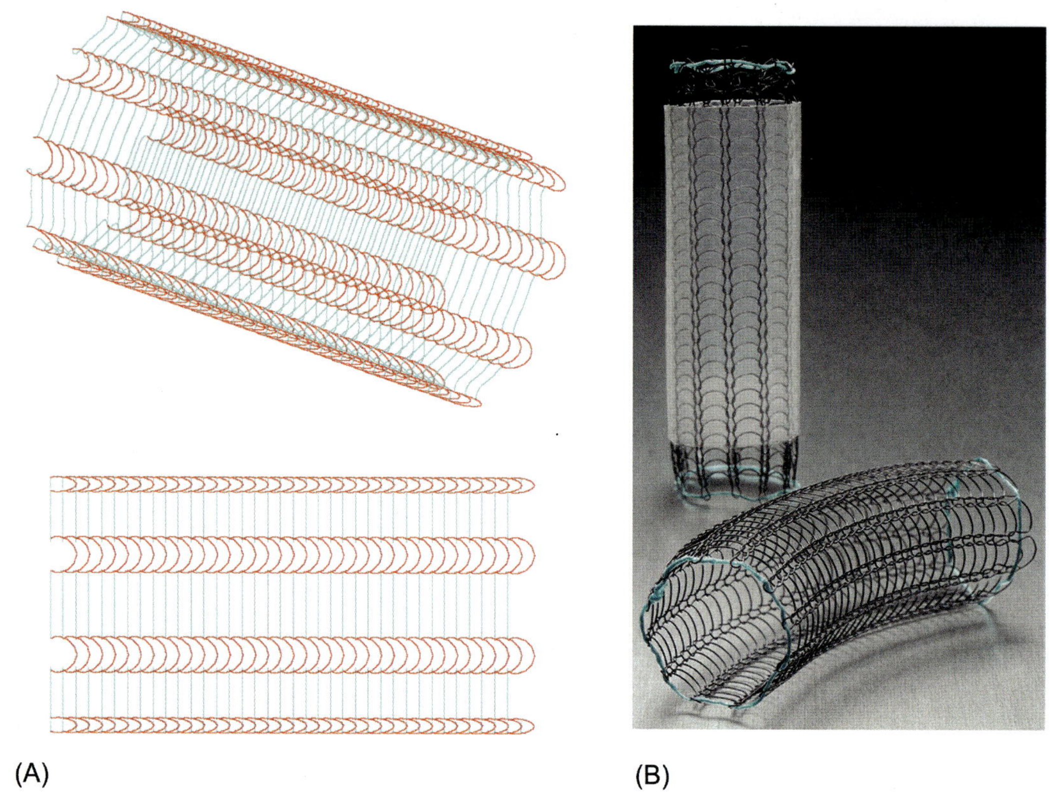

FIG. 3.7 Ultraflex stent: (A) geometrical scheme; (B) real view. *(Source: Boston Scientific.)*

concatenation of radial springs with multiple arcs and sliding joints between the rings. A bell shape is introduced in half of the stent by means of a thermal treatment for fixing this expanded form. In a rigorous classification, this stent belongs to the sliding knots design. Its particular braiding provides flexible longitudinal bands alternating with narrow rigid bands.

#### 3.2.2.2.3 CHOO STENT (FIGS. 3.8 AND 3.9)

To take advantage of all the Nitinol power, the stent must have fixed knots. Therefore the loss of flexibility is compensated by the superelasticity. This idea is carried out in cardiovascular stents like the Cook (Fig. 3.8) and Symphony (Fig. 3.9) stents by braided knots or laser, respectively. The Choo colorectal stent introduces an alternative fixation of the knots. In a peripheral ring, the knots are crossed by torsion of the wires and in the next ring the knots are free.

## 3.2.3 Mechanical Behavior

### 3.2.3.1 *Resistance Mechanisms*

Each type of stent bases its mechanical behavior on several distinct resistance mechanisms that will be analyzed in this section.

#### 3.2.3.1.1 HELICOIDAL SPRING

The design parameters corresponding to a coil spring are: helix pitch ($h$), diameter ($D$), wire diameter ($d$) or section dimensions if not circular, and number of wires ($n$). Due to the change of stiffness with expansion, a nonlinear inverse problem must be solved to determine the stiffness of the stent corresponding to every expanded configuration.

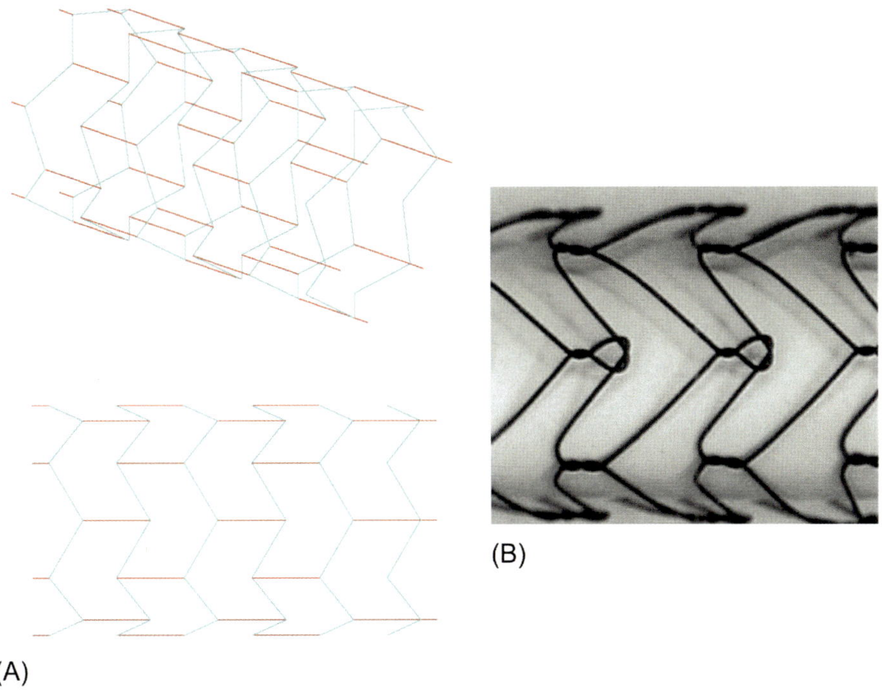


FIG. 3.8 Cook ZA stent: (A) geometrical scheme; (B) real view. *(Source: Cook Inc.)*

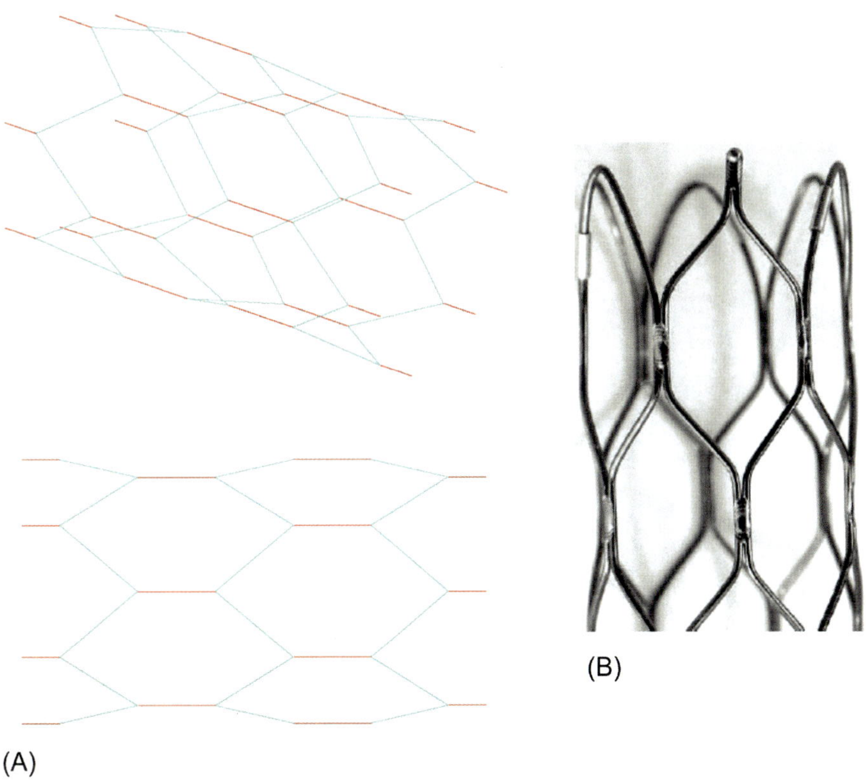


FIG. 3.9 Symphony stent: (A) geometrical scheme; (B) real view. *(Source: Boston Scientific.)*

By assuming that under small loads the total length of the wire ($L$) remains unchanged, then changes in $D$ must be accommodated by changes in $h$ according to the constraint:

$$L = \sqrt{h^2 + \pi^2 D^2}$$

*KINEMATICS RELATIONS*

Undeformed spiral length: $L_0 = \sqrt{h^2 + \pi^2 D^2}$
Reference angle: $\tan\alpha_0 = \frac{h}{D}$
Deformed spiral length: $L = \sqrt{(h - \Delta h)^2 + \pi^2 (D + \Delta D)^2}$
Deformed reference angle: $\tan\alpha = \frac{h - \Delta h}{D + \Delta D}$
Angle increment: $\Delta\alpha = \alpha_0 - \alpha$
Angle per unit length: $\theta = \frac{\Delta\alpha}{L}$
Undeformed spiral slope: $\tan\beta_0 = \frac{h}{\pi D}$
Deformed spiral slope: $\tan\beta = \frac{h - \Delta h}{\pi(D + \Delta D)}$

*STATIC EQUILIBRIUM*

Equilibrium in half spiral (Fig. 3.10B), where $p_0$ is nominal pressure.
Axial force: $T = \frac{p_0 (h - \Delta h) L}{2\pi \cos\beta}$
Torsion moment: $M_t = \frac{p_0 (h - \Delta h) L \ (D + \Delta D) \tan\beta}{4\pi \cos\beta}$

*BEHAVIOR EQUATIONS*

Deformation due to axial stress is neglected, therefore $L = L_0$.
Torsion moment: $M_t = GJ\theta$
where $J$ is the torsion constant ($I_0$, polar inertia if circular section).

Considering a free-body diagram of one-half of one spiral wire in the stent (Fig. 3.10B), taking into account that the stent is unloaded along its axis, equilibrium of forces, axial $F_a$ and shear $F_s$, moments, bending $M_b$, and torsion $M_t$ derived from the Bernoulli-Euler and Coulomb theories for slender beams (along the directions of the tangent and normal to the helix), the relationship between the effective pressure, $p$, and the diameter of the stent is:

$$p = \frac{4\cos^2\alpha}{D^2 \sin\alpha}\left[EI\Delta\chi \sin\alpha - GJ\Delta\vartheta \cos\alpha\right]$$

where $E$ and $G$ are the elasticity and shear moduli, respectively; $I$ and $J$ are the inertial constants for bending and torsion; $\Delta\chi$ and $\Delta\vartheta$ are the changes in principal curvature and twist of the spring induced by the pressure; and $\alpha$ is the angle of the spring, tan $\alpha = h/\pi D$.

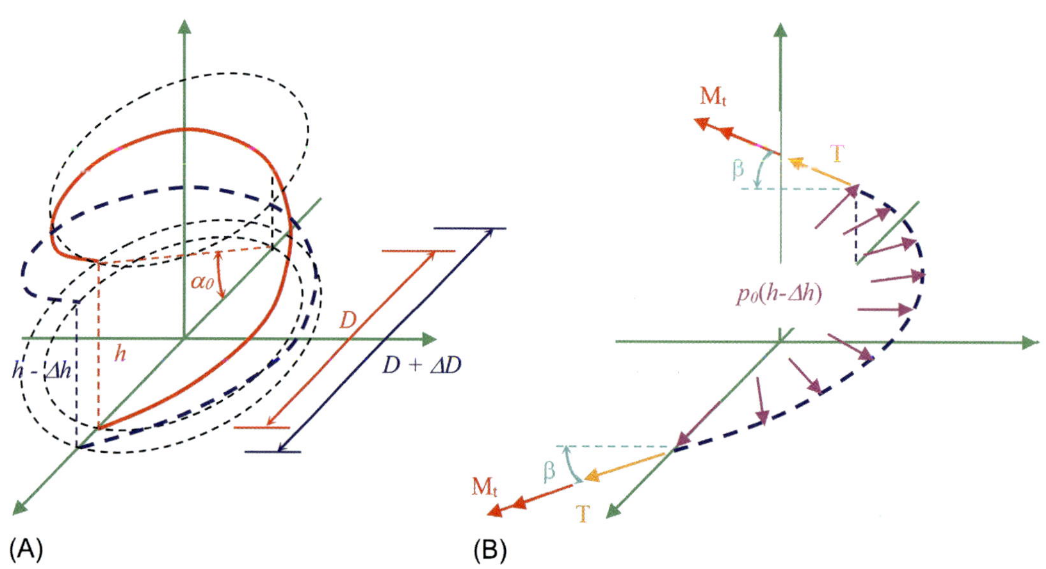


FIG. 3.10 Helicoidal spring: (A) design parameters; (B) free-body diagram.

Wallstent, braided, Song, and Esophacoil stents are based on this mechanism. The Esophacoil stent is, in fact, a simple helicoidal spring, while the others are two layers of helicoidal springs turned in reverse with sliding knots between the wires. In this case, the number of wires per layer is another design parameter.

#### 3.2.3.1.2 RADIAL SPRING

In Fig. 3.11A, the analyzed case corresponds to a Z shape. The design parameters are: pitch ($h$), diameter ($D$), wire diameter ($d$), and number of circumference divisions ($n$). Again, a nonlinear inverse problem must be solved to determine the stiffness of the stent corresponding to every expanded configuration. By assuming that under small loads the total length of the arm between knots ($L$) remains unchanged, then changes in $D$ must be accommodated by changes in $h$ according to the kinematic constraint

$$L=\sqrt{h^2+\frac{\pi^2 D^2}{4n^2}}$$

Considering the free-body diagram of one-half of one wire ring in the stent (Fig. 3.11B), and taking into account that the stent is unloaded along its axis, the next equations are:

*KINEMATICS RELATIONS*

Undeformed truss length: $L_0=\sqrt{h^2+\frac{\pi^2D^2}{4n^2}}$

Undeformed truss angle: $\tan\alpha_0=\frac{\pi D}{2nh}$

Deformation equation: $y=\delta\left[1-3\left(\frac{x}{L}\right)^2+2\left(\frac{x}{L}\right)^3\right]$

where $\delta$ is perpendicular displacement to the truss; $L$ is the distance between the ends of truss; and $x$ is the coordinate in the direction of the length.

Deformed truss length: $L_d=\int_0^L\sqrt{1+y'^2}dx$

Deformed truss angle: $\tan\alpha=\frac{\pi(D+\Delta D)}{2n(h-\Delta h)}$

Perpendicular displacement to truss: $\delta=\frac{\pi\Delta D}{2n}\cos\alpha$

Distance between ends of truss: $L=\sqrt{(h-\Delta h)^2+\frac{\pi^2(D+\Delta D)^2}{4n^2}}$

*STATIC EQUILIBRIUM*

Equilibrium in half spring (Fig. 3.11B).

Shear force: $V=p_0(h-\Delta h)\frac{D+\Delta D}{2}$

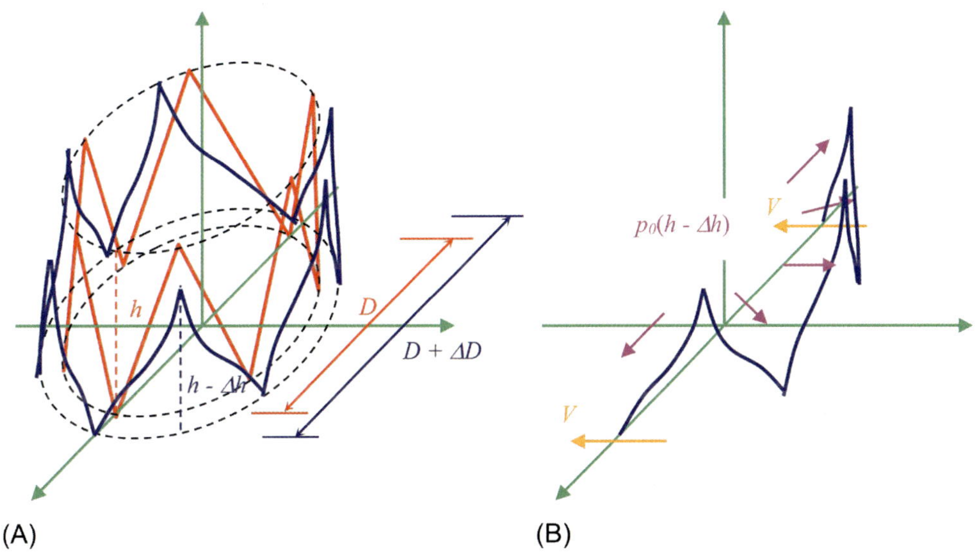


FIG. 3.11 Radial spring: (A) design parameters; (B) free-body diagram.

*BEHAVIOR EQUATIONS*

Bending moment: $M=\frac{6EI\delta}{L^2}$; shear force: $V=\frac{2M}{L}=\frac{12EI\delta}{L^3}$

The relationship between the pressure and the diameter of the stent may be written as:

$$p=\frac{12EI\pi\Delta D}{nL^3D}\cos\alpha$$

where $\tan\alpha=\pi D/2nh$.

Gianturco, Song, Choo, Cook, and Symphony stents correspond to this mechanism, with some variants. Therefore, on the one hand, some of them present sliding knots among modules (Gianturco, Song, and Choo), while others present rigid knots (Cook and Symphony); on the other hand, the knots in the different modules of the stents can be corresponding (Symphony) or not corresponding (Gianturco, Song, Choo, and Cook). Finally, Gianturco, Song, and Cook stents have straight arms, while Choo and Symphony stents have curved trusses.

### 3.2.3.1.3 RADIAL MULTIPLE ARCS SPRING

The Ultraflex Tracheobronchial Stent (Boston Scientific) is based on the radial spring with multiple arcs and sliding knots between wires (Fig. 3.12A), with design parameters: pitch ($h$), diameter ($D$), wire diameter ($d$), number of arcs along the circumference ($n$), arc radius ($r$), arc angle with respect to the stent axis ($\alpha$), and arc semiangle ($\theta_0$).

The behavior of this mechanism is much more complex, including multiple deformations at different levels. However, the theory of slender arcs can be used to obtain a simplified model, depending on the bending deformations of the smaller arcs, which are the most significant. According to Fig. 3.12B, considering the bending deformations for an arc induced by pressure, the relationship between the pressure and the deformed shape of the stent may be obtained.

Therefore we can write:

*KINEMATICS RELATIONS*

Undeformed circumference length: $L_0=\pi D$

Deformed circumference length: $L=\pi(D+\Delta D)$

Increasing circumference length: $\Delta L=\pi\Delta D=n\delta_h$

Increase in step: $\Delta p=\delta_v\cos\alpha$

*STATIC EQUILIBRIUM*

Balancing forces in half spring (Fig. 3.12A), we have:

Axial force: $T=p_0(p-\Delta p)\frac{D+\Delta D}{2}$

*BEHAVIOR EQUATIONS*

Horizontal displacement: $\delta_h=\frac{2Tr^2}{EI}\left(\theta_0+\frac{1}{2}\cos2\theta_0-\frac{3}{4}\sin2\theta_0\right)$

Vertical displacement: $\delta_v=\frac{Tr^2}{EI}\left(\frac{1}{2}\theta_0\sin2\theta_0-\frac{3}{4}\cos2\theta_0\right)$

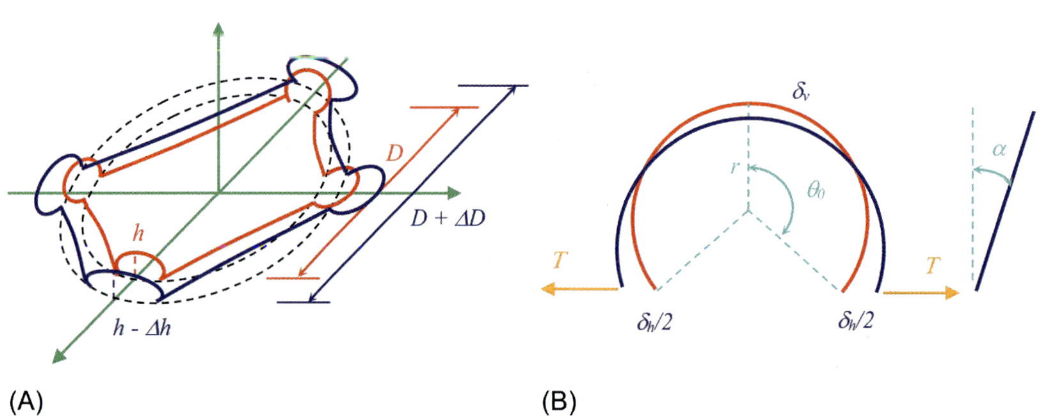


FIG. 3.12 Radial multiple arcs spring: (A) design parameters; (B) free-body diagram.

The relationship between the pressure and change of diameter $D$ and $h$ of the stent may be written as:

$$p = \frac{\pi EI}{nr^2 hf(\theta)} \frac{\Delta D}{D}$$

and

$$p = \frac{2EI}{r^2 g(\theta)} \frac{\Delta h}{h}$$

where $f(\theta)$ and $g(\theta)$ are geometric functions.

The three mechanisms analyzed, with their respective equations, must be considered as a simplified approach to the actual mechanical behavior of the stents, which presents complex effects not included in this approach such as plastic deformations and residual stresses appearing in some conformation processes, frictional resistance in sliding configurations, etc.

## 3.3 DESIGN METHODOLOGY

### 3.3.1 Stent Material

Nitinol is a superelastic alloy capable of reaching very high strain ranges without yielding, allowing complete elastic recovery. The thermomechanical behavior of this alloy is associated with austenite-martensite phase transformation. In the martensite phase, the material is easily deformable, and during martensite-austenite transformation under heating, the material will recover its original shape prior to deformation. Additionally, when the material is deformed in the austenitic state up to 8%, the stress-induced martensite phenomenon allows the alloy to undergo complete recoverable deformation. This property is known as superelasticity or rubber-like behavior.

This property makes it possible to design stents that present a soft CEF, allowing a gradual relaxation of the constriction, and a high RCR, which prevents stent collapse due to external actions (Fig. 3.13).

The alloy used for stent simulation and manufacturing corresponds to an NiTi alloy (50.8 at.% Ni, 49.2 at.% Ti) supplied by Memory Metalle GmbH (Germany).

Mechanical properties of the material are determined by means of uniaxial tensile tests. In this case, a universal machine test Instron 5565 has been used, testing standard specimens obtained from sheets 1 mm thick. Every test is done considering complete loading and unloading with displacement control at a reference temperature $T_0 = 22°C$.

To simulate material behavior, a phenomenological constitutive model is used, based on the formulation developed by Auricchio and Taylor [49, 50]. In this respect, the material can be modeled in finite element software (e.g., Abaqus 6.11) as an embedded user material subroutine (UMAT).

Fig. 3.14 shows the average curve of the tensile tested samples versus strain-stress curves resulting from a uniaxial virtual test simulated and adjusted using the material subroutine implemented in Abaqus 6.11 code for $T_0 = 22°C$ and $T_w = 37°C$, respectively. Thermomechanical parameters, necessary to obtain material behavior at body temperature $T_w = 37°C$, are retrieved from the literature [51]. Table 3.2 includes NiTi material mechanical properties fitted to average values obtained from experimental testing.

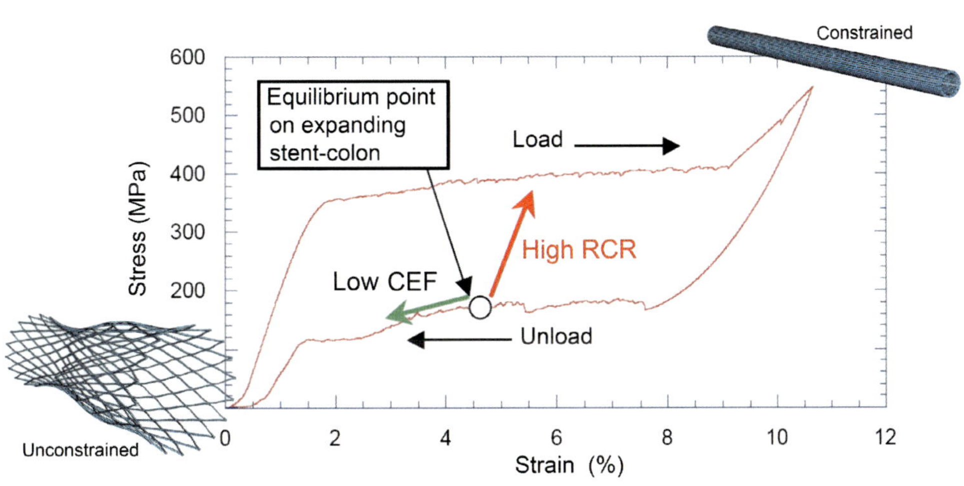


FIG. 3.13 Strain-stress curve of Nitinol alloys. Application of superelastic property to colonic stent designs. *CEF*, chronic expansion force; *RCR*, radial compression resistance.

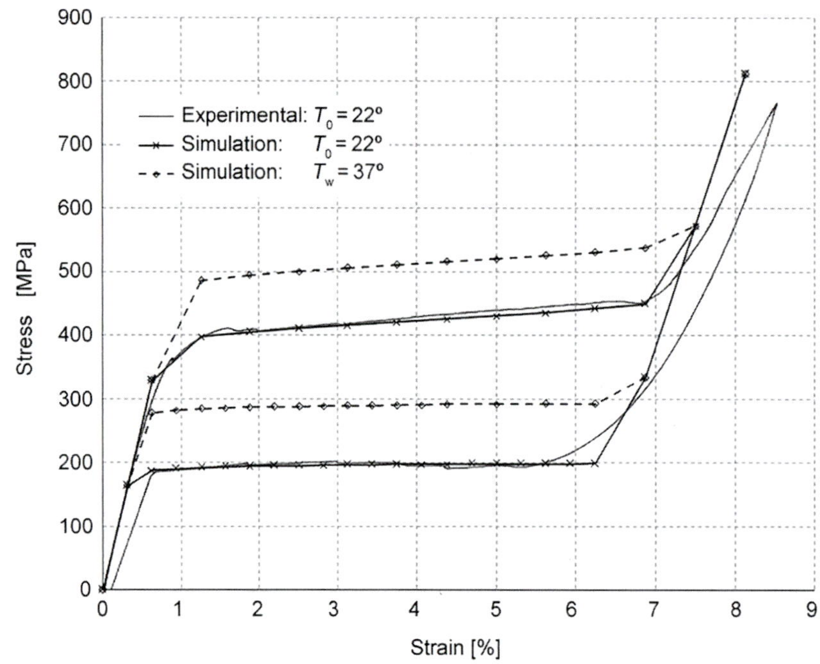


FIG. 3.14 Strain-stress curves of the used Nitinol alloy [52]. Experimental curve obtained from mechanical test at 22°C, adjusted curve in Abaqus 6.11 using the parameters in Table 3.2, and extrapolated curve at body temperature (37°C).

TABLE 3.2 NiTi Material Parameters Defining the Constitutive Stress-Strain Curve of the Alloy Used in the Present Study

| Parameter | Description | Value |
|---|---|---|
| $E_A$ | Austenite elasticity module | 52,650 MPa |
| $\nu_A$ | Austenite Poisson's ratio | 0.33 |
| $E_M$ | Martensite elasticity module | 38,250 MPa |
| $\nu_M$ | Martensite Poisson's ratio | 0.33 |
| $\varepsilon_L$ | Maximum transformation strain | 6% |
| $\sigma_s^{AM}$ | Start of transformation austenite martensite stress | 385 MPa |
| $\sigma_f^{AM}$ | End of transformation austenite martensite stress | 460 MPa |
| $\sigma_s^{MA}$ | Start of transformation martensite austenite stress | 200 MPa |
| $\sigma_f^{MA}$ | End of transformation martensite austenite stress | 180 MPa |
| $T_0$ | Reference temperature | 22°C |
| $C^{AM}$ | $\frac{\partial \sigma_s^{AM}}{\partial T}$ | 6.1 MPa/°C |
| $C^{MA}$ | $\frac{\partial \sigma_s^{MA}}{\partial T}$ | 6.1 MPa/°C |
| $\varepsilon_{L\_Vol}$ | Volumetric transformation strain | 0.06 |

The loading and unloading rate of change of plateau with the temperature $C^{AM}$ and $C^{MA}$ were obtained from the literature [51].

### 3.3.2 Stent Geometry

Considering habitual diameters for colonic stenting of the order of 30 mm, resistance to radial compression and buckling are determining factors, and stents that show better behavior in this sense are those of rigid knots. Therefore, initially, a configuration of Z-shaped radial spring with rigid knots may be suitable for the development of a new design.

The geometrical pattern adopted allows the manufacturing of the initial shape performed by the cutting laser technique, creating a longitudinal nonhomogeneous distribution of grooves on the tube surface, with a given width and length. In the next step, the slotted tube is expanded until it reaches the stent configuration. The chosen design allows a precision manufacturing process for obtaining controlled custom designs for each pathology and patient in an efficient and economically viable manner.

The geometric typology proposed is a tube-based stent with closed diamond-shaped cells (Palmaz-Schatz type), which correspond to the basic geometry module used by the authors to design and manufacture a prototype colonic stent [48, 53]. In a previous comparative analysis of stents for the treatment of colorectal obstruction [48] and by means of experimental animal testing [54], it was shown that stents based on closed cells with arms joined by rigid nodes develop higher radial force response, necessary for reopening occlusive colonic strictures where greater strength is required.

### 3.3.3 Finite Element Model

To get an accurate geometric model, the previous deformation process is simulated, starting with the slotted tube, expanding and fixing its final shape. In this respect, a 3D finite element model is developed (Fig. 3.15) from a solid 3D CAD model.

The total number of elements in the model was 184,770 (180,450 brick type, 4320 wedge type). The forming process was simulated in the way described in [52] using a rigid surface for the forming tool inserted in the tube. This is a realistic assumption, taking into account that its stiffness is several orders of magnitude greater than the stent structure. The geometric model and subsequent meshing were made using pyFormex [55] and I-Deas [56] software.

### 3.3.4 Cell Model

A unique module, which corresponds to a cell based on radial spring mechanisms, is used in preliminary simulations (Palmaz-Schatz design with diamond-shaped geometry).

The geometry consists of two mesh types:

- Hexahedral mesh (brick type). This mesh shapes the grooved tube except elements that are on the corners.
- Tetrahedral mesh (wedge type). This mesh models the corners and softens them avoiding stress gathering.

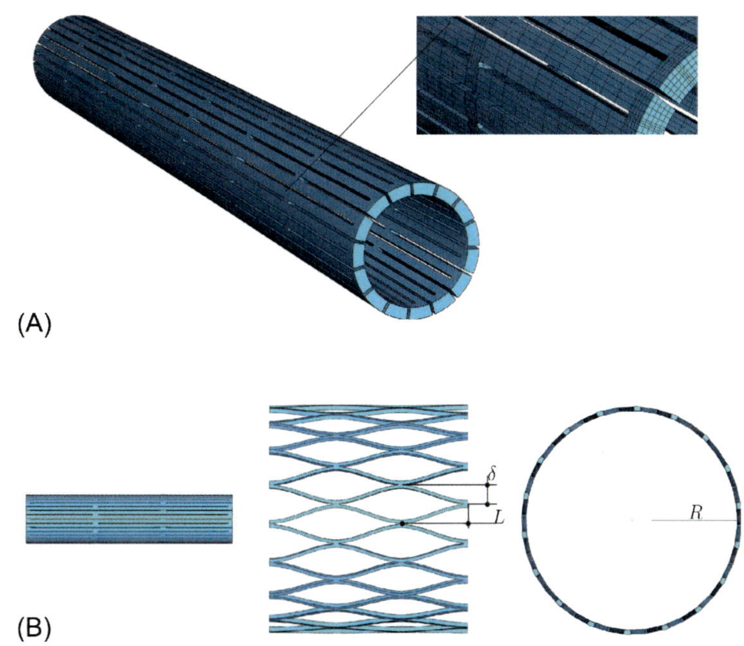


FIG. 3.15 3D model of a Palmaz-Schatz-type stent: (A) raw model and finite element mesh detail; (B) scheme of shaping process: initial configuration and repose final geometry.

The model is 3D, using BRICK- and WEDGE-type elements, with linear interpolation. Only elements that are wedge type are used to model the groove's corners and avoid stress concentration (Fig. 3.16).

Using just a module it is possible to determine the final shape geometry reached by truss cells, without spending a lot of time on computation. In Fig. 3.17 we can see the curvature adopted for each truss, which conforms to the ring spring.

Nonetheless, this reduced model is validated by comparison with previous results of a complete stent simulation, and is found to be valid without significant loss of accuracy. Comparing deformations between Fig. 3.17 with Fig. 3.18A we can see very similar expanded geometries and both reproduced geometries obtained by the manufacturing process (Fig. 3.18B).

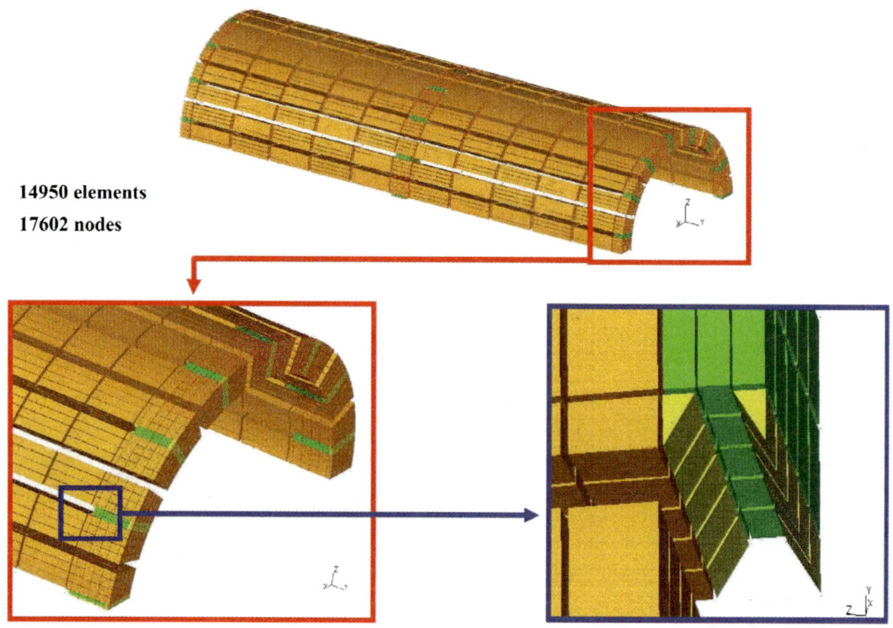


FIG. 3.16 General view of the simplified one-module model.

FIG. 3.17 Simulation of one stent ring cell expansion to determine configuration after the shaping process.

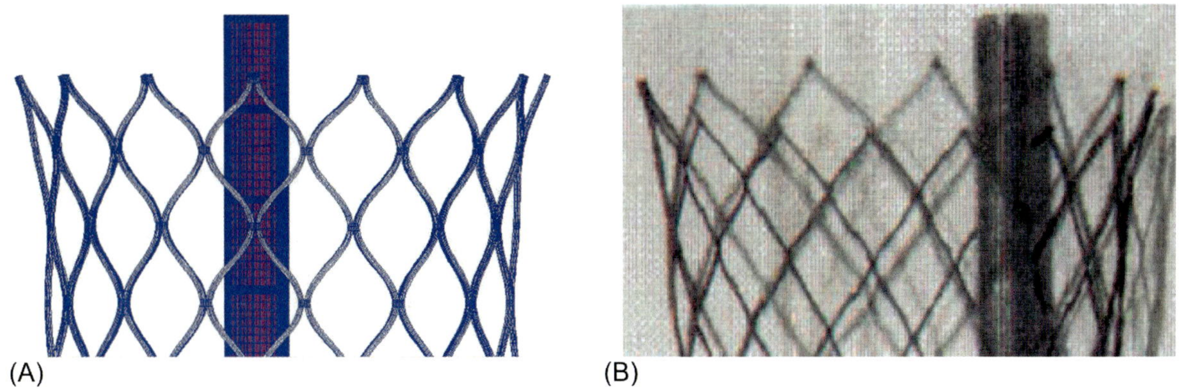


FIG. 3.18 Opening at the end of the stent (dog-boning effect): (A) geometry after shaping process simulation; (B) geometry obtained by the manufacturing process.

## 3.4 SIMULATION METHODOLOGY

The analyses are carried out using commercial FEM software Abaqus 6.11, which takes into account the nonlinear behavior of the material constitutive equations associated with Nitinol alloys treated in the previous section.

The model has been applied to simulate different processes by mean of the Abaqus program, such as:

- Steps of the forming process for the final shape.
- Surgical handling. Compression process and deployment (implantation process).
- Application of a pressure wave on the external surface of the stent structure to simulate the response to peristaltic movement, which is the most demanding mechanical requirement.

### 3.4.1 Shaping Process

The shaping process is simulated using a rigid surface for the forming tool inserted in the tube (Fig. 3.19). This is a realistic assumption, taking into account that its stiffness is several orders of magnitude greater than the stent initial structure.

Contact conditions (frictionless) are imposed between the stent-node-based surface (inner node-set) and the outer surface of the forming tool. A user-defined material subroutine (UMAT) for NiTi mechanical behavior, based on the developments included in [57], is employed using Abaqus Standard 6.11 software [58], using the material parameters in Table 3.2. Once the final shape is achieved, using a specific bell-shaped tool (Fig. 3.20), the stress and strain fields were removed.

The final obtained shape is in accordance with the initial requirements and is considered to be the definitive design of the stent.

### 3.4.2 Surgical Handling: Crimping and Releasing From Catheter

Usually, the stent is mounted in a preloaded constrained position on a delivery catheter. When the obstruction has been localized, a guide is passed through the lumen beyond the stenosis and then the stent is advanced over it and positioned across the stricture. Then, the constraining mechanism is released and the expansive forces cause radial opening of the lumen. After that, to get a more realistic simulation, both catheter introduction and liberation were simulated in two steps:

**(a)** Crimping (Fig. 3.21): the stent was compressed into a 14 Fr catheter (1 Fr = 0.333 mm) defined as a rigid surface. The stent was fixed in the axial direction and free in the radial direction allowing it to deform freely under the contact pressure of a crimping tool and catheter with no contact friction.

**(b)** Releasing (Fig. 3.22): while holding the stent fixed in the axial direction and the rigid surface that model the delivery system is sliding with frictionless contact, the stent progressively recovered its shape but was limited with the compressive force exerted by the colonic tissue.

The definitive design of the stent was able to crimp and release without any problems.

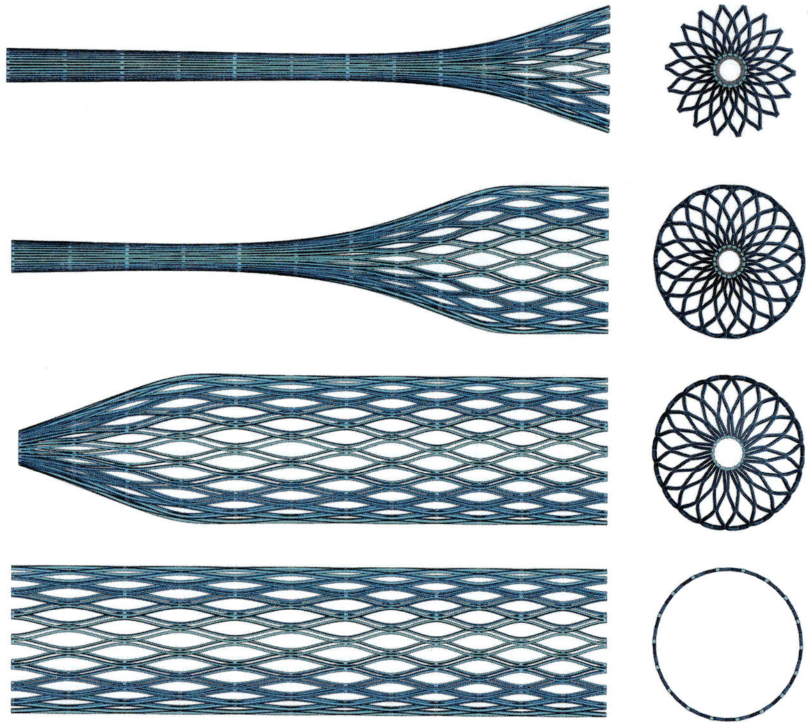

FIG. 3.19 Simulation of the forming-expanding process of the grooved tube. Expansion until a cylindrical shape with 20 mm inner diameter is obtained.

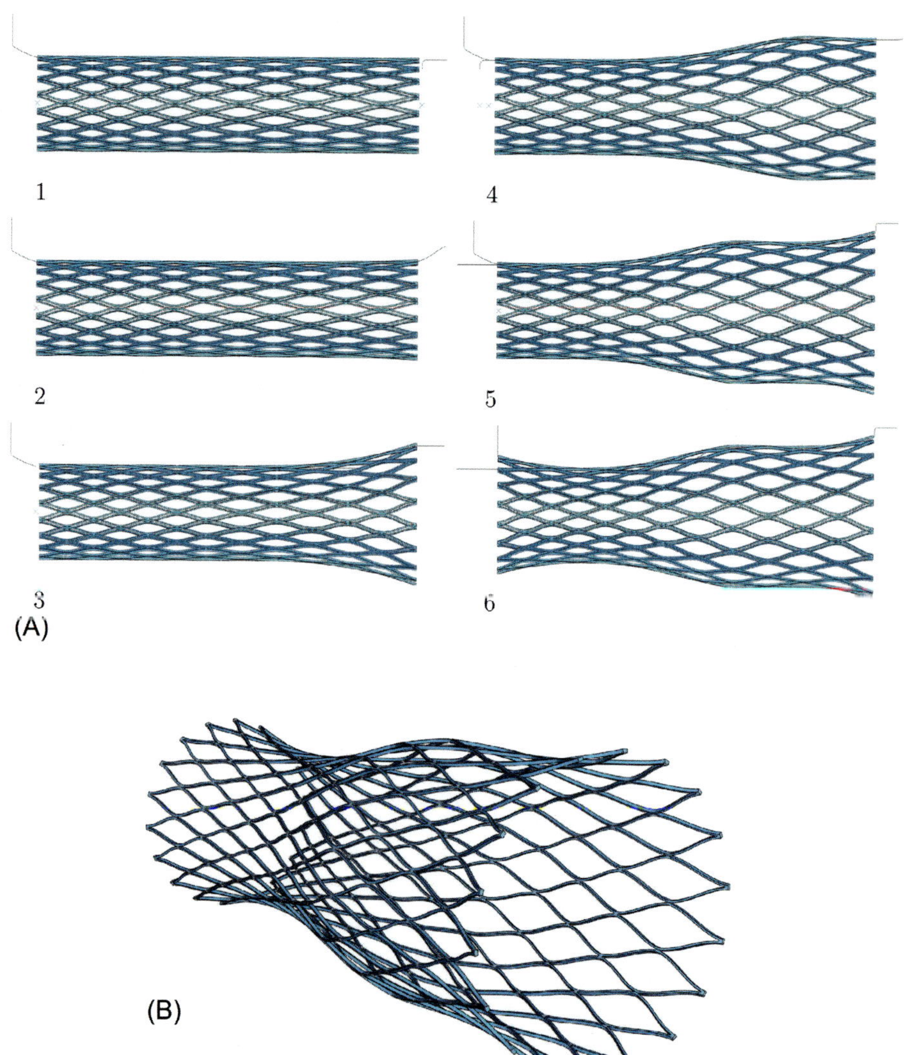


FIG. 3.20 Simulation of forming sequence of final stent bell-shaping process: (A) the final shape forming tool is modeled by a rigid surface, represented by a line corresponding to the external profile; (B) finite element model of the expanded stent [54].

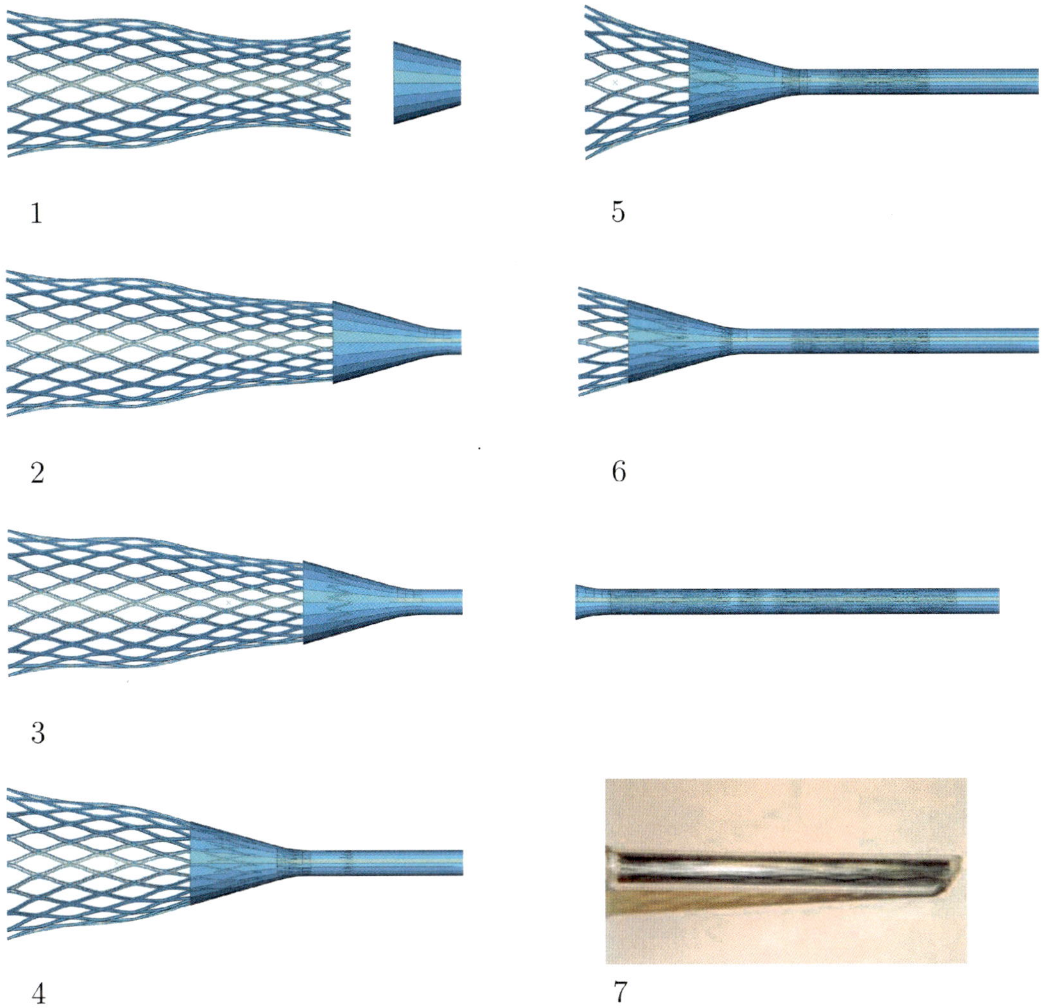


FIG. 3.21 Step simulation of stent introduction into the catheter. A funnel shape tool is modeled by a rigid surface. It is used to crimp the stent into catheter.

### 3.4.3 Peristaltic Motion

While stents work well in vascular structures, the biliary tract, or ureter, its operation is more difficult in the colon, because of the special peristalsis and fecal content characteristics, favoring its displacement or obstruction. The large intestine presents a complex peristalsis, consisting of deep circular and circular muscle contractions, causing a wave that progresses longitudinally along the intestine and leads to feces in the rectal ampulla.

Since the colon transports solid material, it is necessary to generate its movement through local contraction of great strength, which spreads in wavy stretches that resemble the motion of a worm. Therefore after positioning the stent, the application of a pressure wave on the external surface of the duct stent structure is modeled to simulate the response to peristaltic movement, which is the most demanding mechanical requirement. The loading conditions consider a wave motion with an amplitude that produces a uniform constriction in the section of duct stent that has a diameter at maximum compression of 6 mm, according to clinical experience. A ring compression is modeled as a displacement-driven process. This option is chosen after studying different alternatives (displacement wave, pressure wave) and after verifying in clinical experimentation that the radial recovering force of the stent is appropriate when it is reached by the peristaltic wave, suffering maximum radial compression, and being able to achieve complete recovery when the peristaltic pressure disappears (Fig. 3.23).

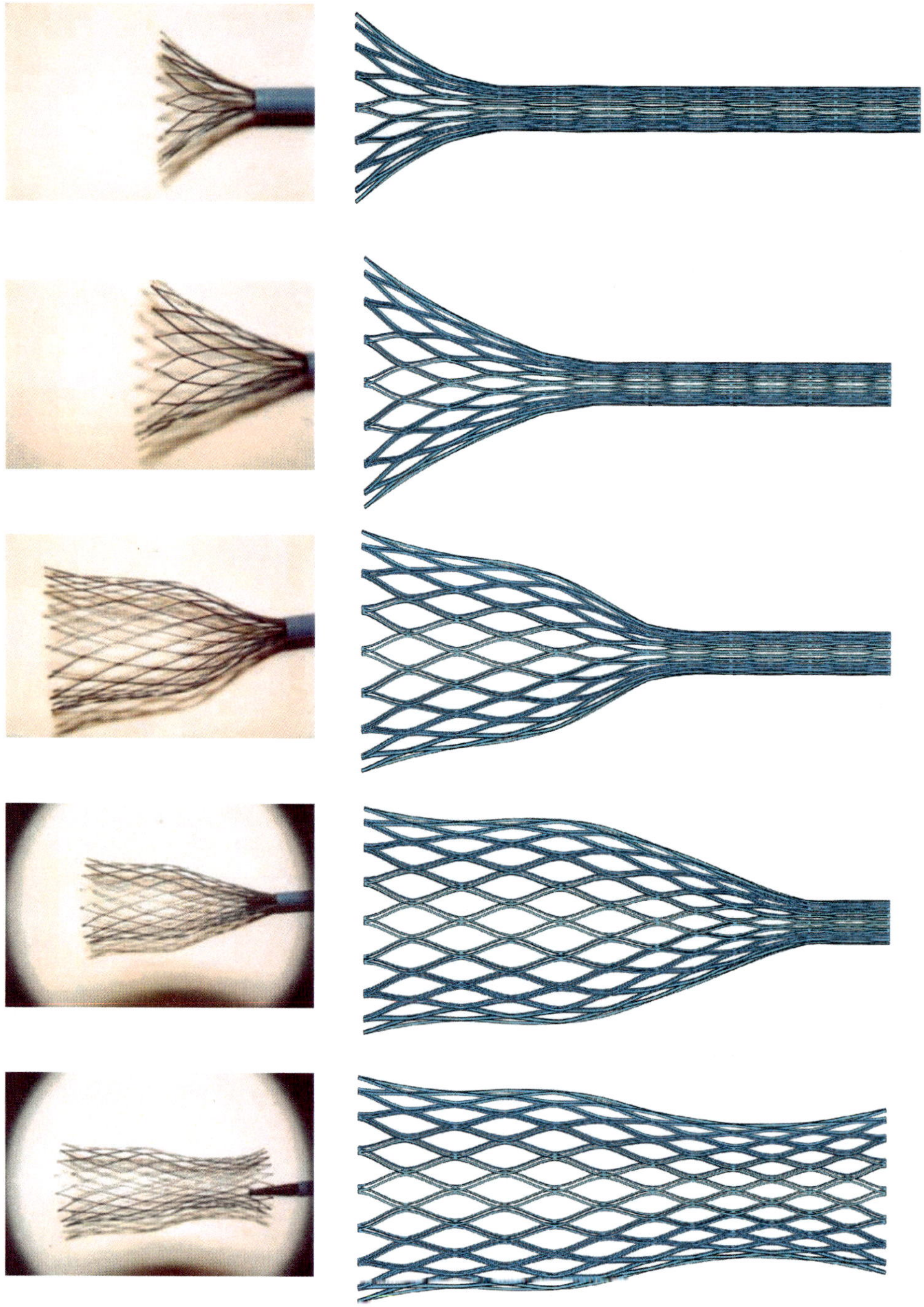

FIG. 3.22 Simulation of stent releasing from catheter. Images of the test *(left)* versus those obtained by numerical simulation *(right)*.

## 3.5 MANUFACTURING AND ANIMAL EXPERIMENTATION

### 3.5.1 Material

The stent was manufactured from NiTi tube with a 4.5 mm diameter supplied by Minitubes Ltd., Grenoble, France. The composition was the standard superelastic 50.8 at.% Ni and 49.2 at.% Ti. In particular, for the shape memory alloy used, and prior to any thermomechanical process, the transformation temperatures were $M_f = -55°C$, $M_s = -38°C$, $A_s = -17°C$, and $A_f = -2°C$, so, at body temperature, the alloy was in a superelastic state.

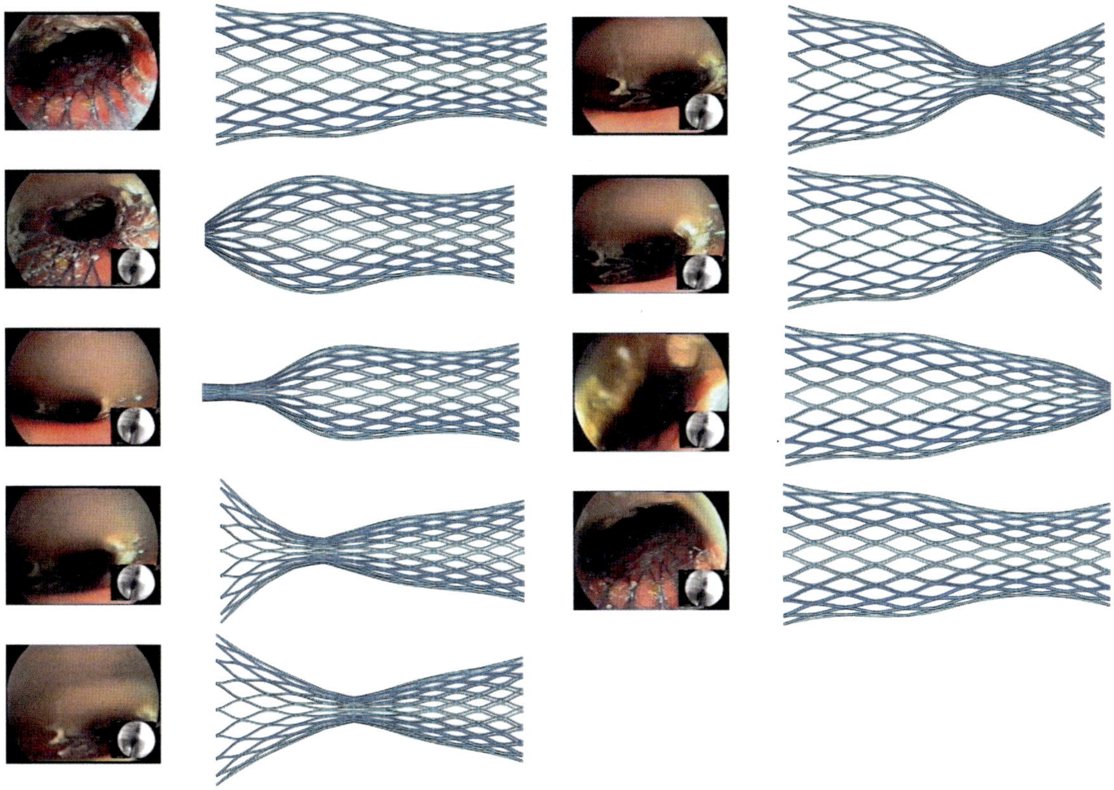

FIG. 3.23 Peristaltic movement. Comparison between finite element simulation and endoscopic images [54].

## 3.5.2 Stent Manufacturing Process

Here, we present a manufacturing process for tube-based stents. In the proposed design, the stent is manufactured from an NiTi thin-walled tube of 4.5 mm outer diameter with a wall thickness of 400 μm, with a composition of 50.8 at.% nickel and 49.2 at.% titanium (supplied by Minitubes Ltd., Grenoble, France). A laser-cutting technique is used to obtain the initial shape, creating a longitudinal nonhomogeneous distribution of grooves on the surface, in this case with a width of 100 μm and a length ranging from 6.15 to 16 mm, separated by a length of 400 μm. Therefore a scaffold preform of struts linked by solid hinges is obtained (Fig. 3.24A).

The expansion process from the slotted tube shape to the final operating form is performed by means of permanent deformation in the martensite phase. In this NiTi alloy, this phase is reached below $M_s = -38°C$. Therefore the deformation process had to be carried out at cryogenic temperatures. Different metallic cylinders with diameters in the range from 7.5 to 30 mm are used in the intermediate expansion steps. In the last step, a component with a nonsymmetric double bell-shaped profile is used to obtain the final shape of the stent (Fig. 3.24B). This shape is fixed by means of further thermal treatment at 660°C for 7 min. The process ends with quenching in water after the annealing process.

Therefore the final bell-shaped stent is obtained by the plastic deformation process at cryogenic conditions (Fig. 3.24B). The repose shape obtained after thermal annealing treatment and electropolishing [53] (Fig. 3.24C) gives a bell-shaped profile similar to the commercial device Medtronic-CoreValve, which is used for the treatment of aortic valves, presenting high expansion rates with values of 5 and 7 at the ends of the stent.

## 3.5.3 Instrumental Adaptation Test

Once the stent is manufactured, its suitability for surgical handling is checked. For this, the following experimental testing stages are carried out (Fig. 3.25):

- Introduction into the conveyor using a funnel. The stent is crimped into a conveyor using a funnel.
- Introduction into the catheter by pushing a cylinder tool.
- Gradual release of the stent from the catheter.

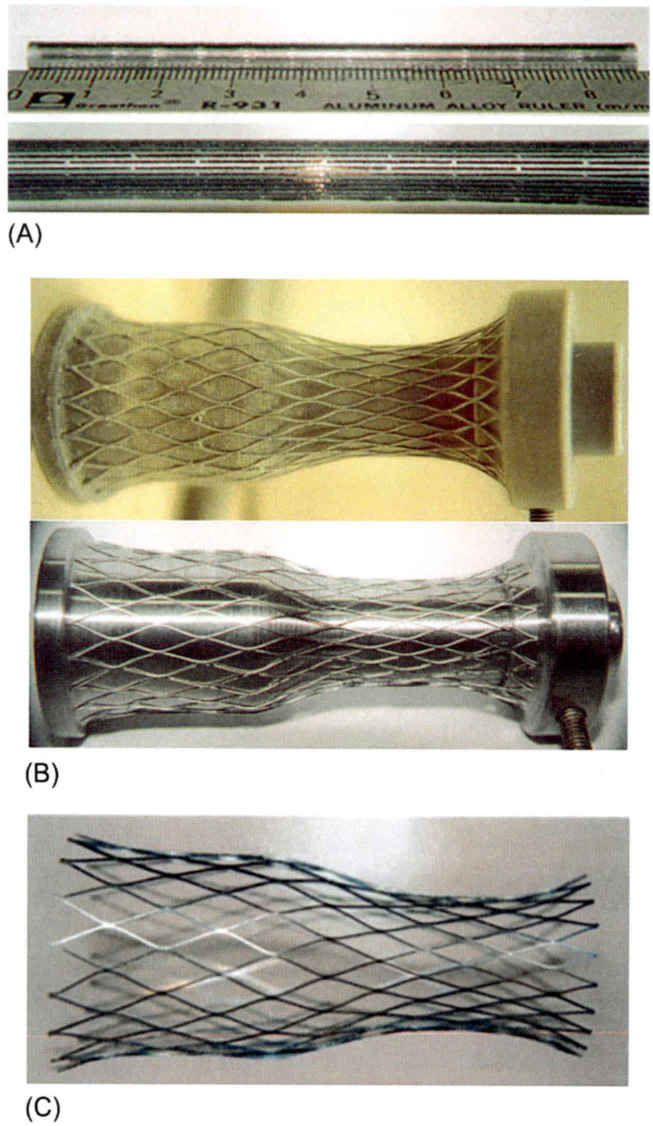

FIG. 3.24 Stent manufacturing: (A) slotted tube [54]; (B) stent after expansion process; (C) final free expanded shape of the stent, with a maximum diameter of 35 mm.

### 3.5.4 Animal Experimentation

Sooner or later the process of validating a new design will go through in vivo experimentation. Normally, it is carried out in animals of porcine species, between 40 and 60 kg, due to anatomical similarities of the descending colon with humans.

In this case, the in vivo test is described for validation of a stent geometry corresponding to the prototype proposed by the same authors in [53, 54]. It is a self-expanding stent with diamond cell type and bell-shaped profile with longitudinal variable radial strength.

The in vivo testing procedure includes:

1. Surgical creation of a model of acute colon stenosis.
2. Implantation of the stent after the appearance of the first symptoms.
3. Radiological and endoscopic follow-up for 10 days (minimum).
4. Control of physical conditions, nutritional tolerance, depositional habit, and weight gain.

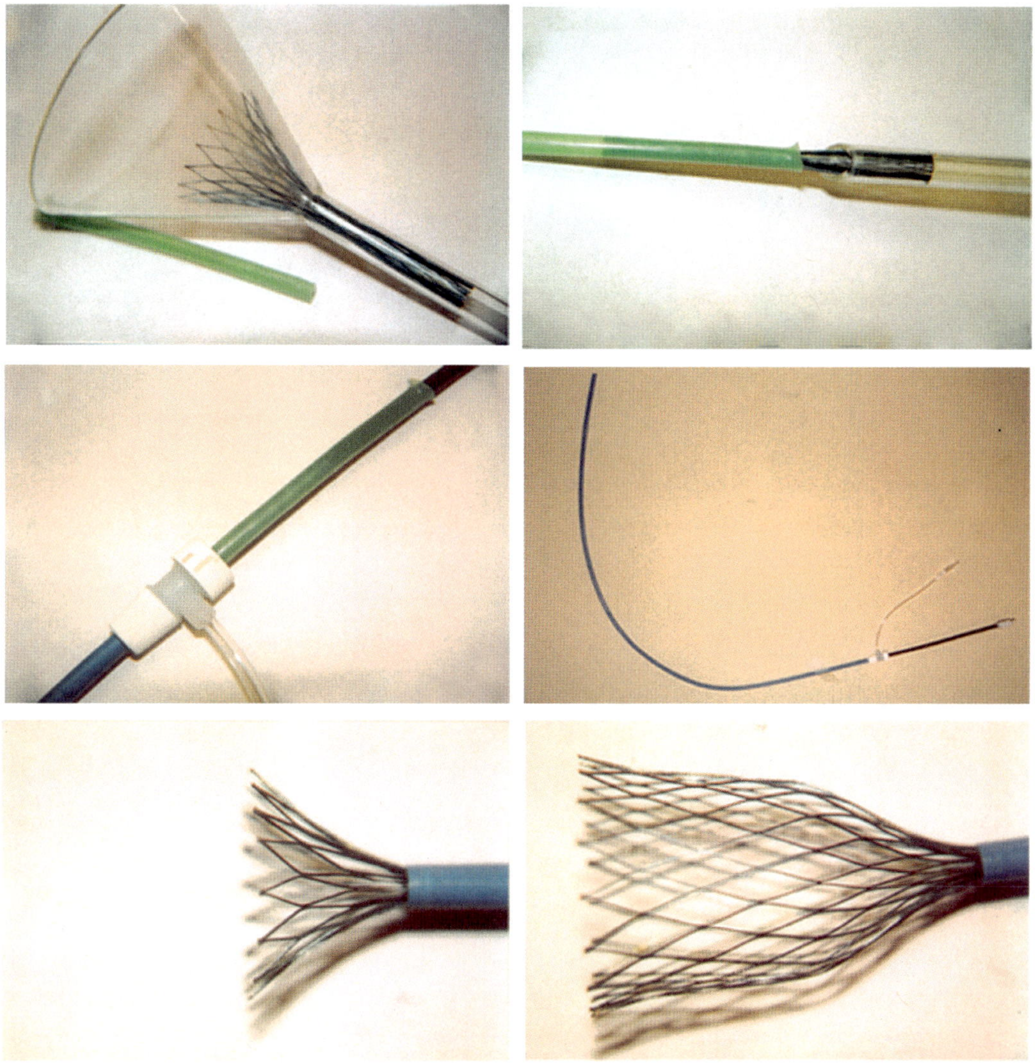

FIG. 3.25 Preliminary surgical handing tests.

### 3.5.4.1 Stenosis Generation

In vivo testing is carried out in porcine specimens. Through laparotomy, an autologous peritoneal patch measuring 4–6cm in length and 4cm in width is harvested and folded into a double-layer band (6 × 2cm). The peritoneal band is wrapped around the sigmoid colon approximately 20cm from the anus (Fig. 3.26). A 6mm diameter probe is placed inside the colon and the band is then tightened. The band is sutured onto the colon in an interrupted fashion. A contrast enema is then performed to verify and document the presence of stenosis.

### 3.5.4.2 Insertion Process

Stent deployment is carried out by means of a 6 Fr multipurpose catheter and an introducer sheath is placed inside the colon, carefully crossing the stenotic site. The stent is mounted on a 14 Fr carriage and pushed out of the sleeve into the sheath (Fig. 3.27A). Later, a 14 Fr pusher is used to place the stent at the tip of the carriage. Injecting contrast medium checks the exact deployment site. After the deployment site is chosen, the self-expandable stent is deployed by keeping it in place with the pusher and gently sliding the introducer sheath backwards (Fig. 3.27B).

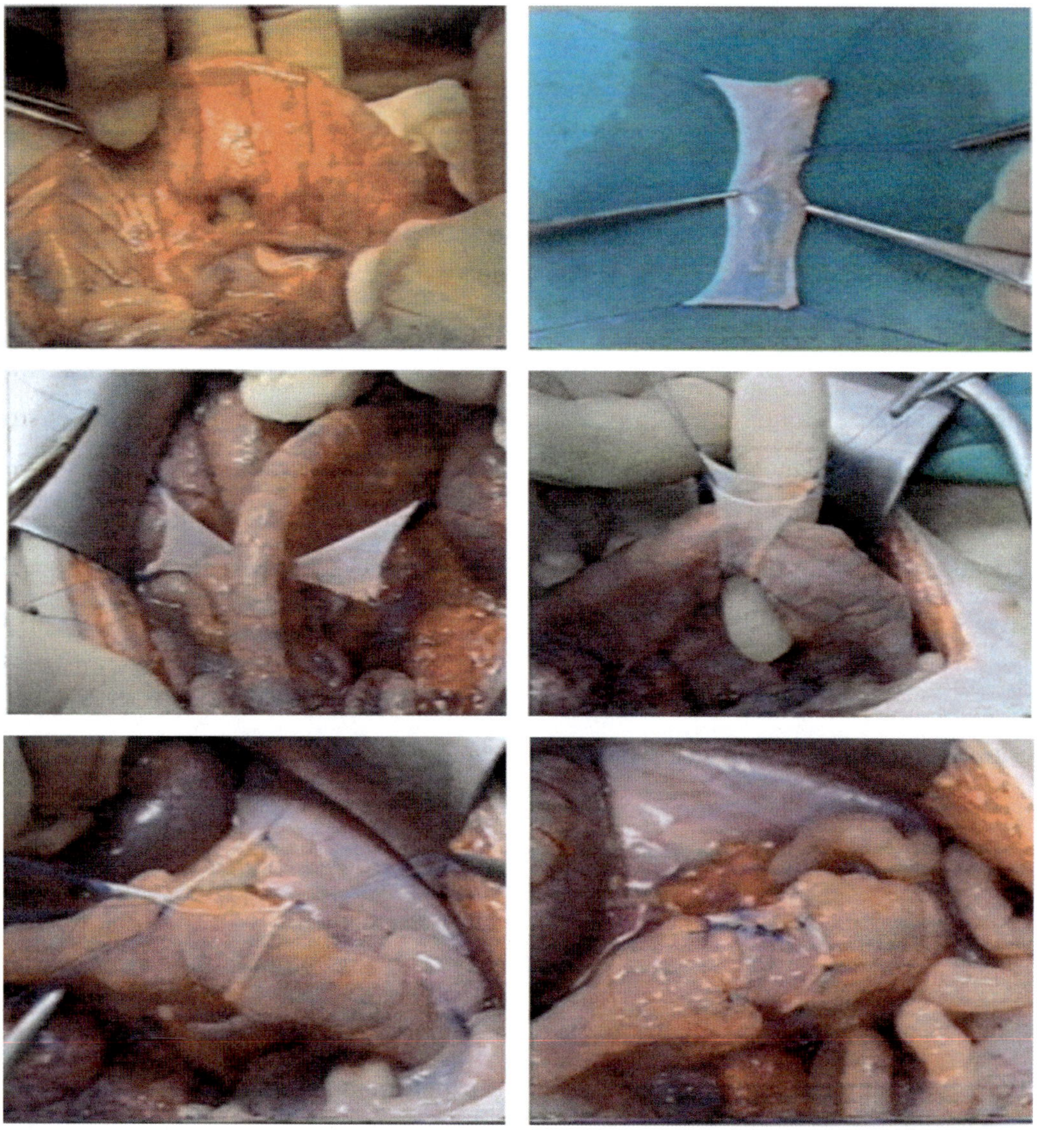

FIG. 3.26 Surgical intervention using endothelial tissue from the animal itself. Formation of a compressive collar in the area of the descending colon.

## 3.6 CUSTOMIZED PARAMETRIC DESIGN

In this section, a strategy and methodology for the design of customized self-expandable NiTi colonic stents adapted to each patient-specific case are presented, using a laser cutting tube stent design, by controlling geometry parameters. The choice of necessary stent properties depends mainly on the length of the stenosis and degree of occlusion. Therefore a stent design with variable radial stiffness along its longitudinal axis would be a good option. By acting independently on different geometric factors, variable geometries can be obtained with different radial force reactions. The design adjustment according to specific requirements is possible, getting a better fit to ill-duct and reducing complications.

Therefore finite element simulations are used to analyze the behavior of a predefined set of geometric configurations to establish an extrapolation technique that allows the designing of new stents adapted to the requirements of specific patients.

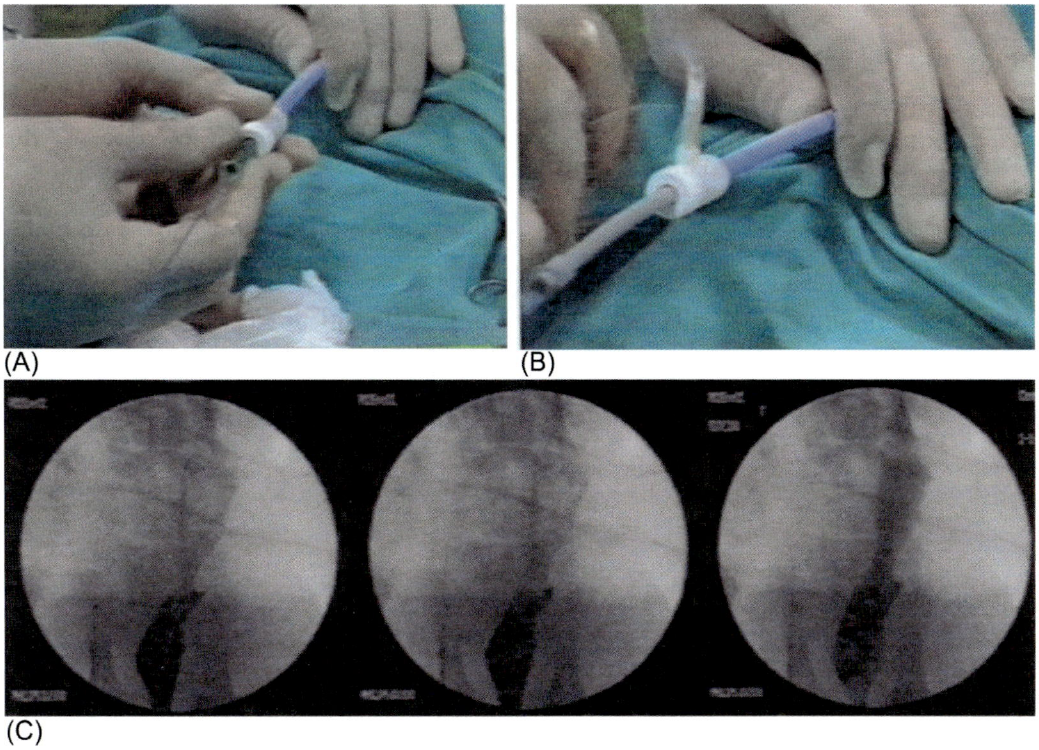

FIG. 3.27 Surgical intervention: (A) introduction rectally of the catheter, guide wire, and transporter; (B) stent release and removal of guide and transporter; (C) radiological monitoring of stent deployment.

The geometric typology used for this purpose corresponds to a tube-based stent with closed diamond-shaped cells (Palmaz-Schatz type), which correspond to a basic geometry module used by the authors to design a colonic stent [53, 54]. In a previous comparative analysis of stents for the treatment of colorectal obstruction [48], it was shown that stents based on closed cells with arms joined by rigid nodes develop higher radial force responses necessary for reopening occlusive colonic strictures where greater strength is required.

The repose shape obtained after the subsequent shaping process, thermomechanical treatment (numerically: stress field to zero), and electropolishing [53] (Fig. 3.20B) gives a bell-shaped profile similar to the commercial device Medtronic-CoreValve used for the treatment of aortic valves.

By giving each stent a constant number of circumferential cells, it is possible to obtain areas with different stiffnesses, which vary the pattern cut along the longitudinal axis, and adjust the expansion ratio factor imposed on each ring of cells until it achieves the desired shape.

Acting over these two design parameters, a shape profile with a longitudinally variable radial strength to facilitate optimal adaptation to the duct walls and prevent migration risk can be set [54]. The pressure exercised against duct walls by each annular cell may be adjusted independently of the profile adopted to avoid migration. Therefore it is possible to obtain a stent with a shape adapted to each type of stenotic duct and variable radial stiffness according to the needs of each patient and pathology.

As an example, a distribution of estimated required net radial forces is presented in Fig. 3.28 in three different areas of the same acute stenosis, which by surgical prescription are considered suitable for restoring transit on the section of affected colon. Those values might be estimated by balloon catheter via colonoscopy in subacute strictures; by performing a minimal local stricture dilatation and recording the pressure to inflate the balloon, an estimation of radial force needed in each area can be obtained.

Among the different possibilities (i.e., a single solution does not exist), by varying just length slots and keeping the rest of the parameters constant, for example, the stent configuration can be obtained. The preliminary dimensioning of geometry gives rise to a cutting pattern as illustrated in Fig. 3.29A, with zones 1–3 with slot length according to the given specifications, and a transition zone 4. After the forming process simulation, the stent configuration shown in Fig. 3.29B is obtained.

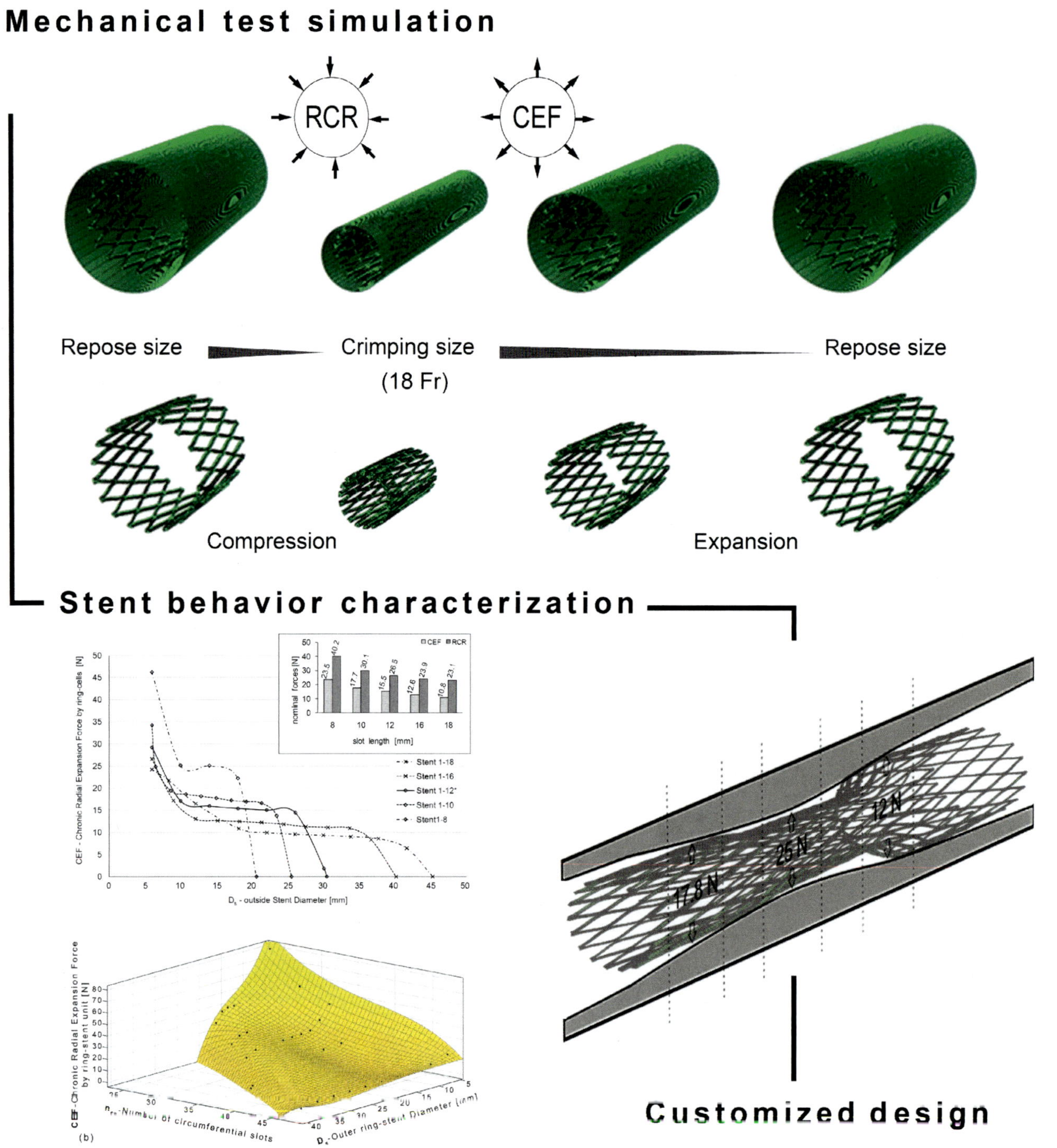


FIG. 3.28 Methodological sequence to characterize the influence of design parameters on stent behavior. *CEF*, chronic expansion force; *RCR*, radial compression resistance.

## 3.7 CONCLUSIONS

A finite element simulation methodology for colonic stents has been presented. Every step in the design and implantation process can be simulated (shaping process, catheter introduction, releasing at the position, and the effect of peristaltic waves) in an accurate way. These aspects present high mechanical requirement and hence are essential in colonic stent design.

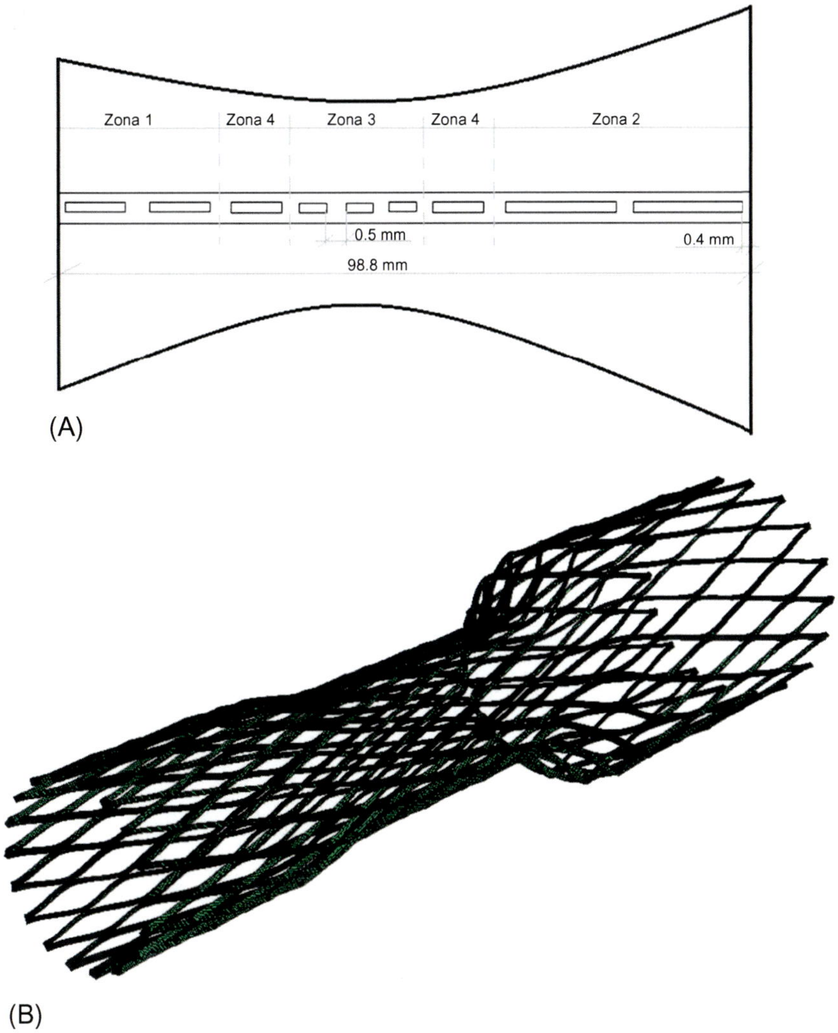


FIG. 3.29 Customized colonic stent: (A) scheme of cutting pattern; (B) profile adopted imposing a constant shape factor in every lattice cell ring, which constitutes the whole stent [52].

A custom-made stent design has been developed and simulated, exhibiting suitable mechanical performance. Moreover, the proposed design allows customized stent manufacture and makes possible the adaptation of endoprosthesis to the particularities of each case to be treated. Therefore we can decrease the contact radial force in the healthy parts of the colon, while, in the stenotic zone, the radial expansion reaction may be higher.

The superelasticity of NiTi alloys is ideal to support the mechanical requirement in colonic stents, avoiding strain values reaching the elastic limit, so that no yielding occurs. However, it is necessary to have good dimensional calibration to ensure that the lattice structure develops enough radial force to recover the previous stent/duct equilibrium stage, guaranteeing proper intestinal transit. Therefore RCR magnitude is essential to prevent the collapse of the stent once implanted, when it is subjected to pressures generated by peristalsis or any other kind of external compressive load.

The results of the study show that NiTi self-expanding bell-shaped customized stents may be an effective alternative to relieve colonic obstruction, either as a bridge to elective single-stage surgery avoiding a stoma, or as a definitive palliative solution in patients with an irresectable tumor or poor estimated survival.

## References

[1] C.T. Dotter, M.P. Judkins, Transluminal treatment of arteriosclerotic obstruction. Description of a new technic and a preliminary report of its application, Circulation 30 (1964) 654–670.

[2] T. Duerig, A. Pelton, D. Stockel, An overview of nitinol medical applications, Mater. Sci. Eng. A: Struct. Materials Propert. Microstruct. Process. 273 (1999) 149–160.

[3] A.R. Pelton, V. Schroeder, M.R. Mitchell, X.-Y. Gong, M. Barney, S.W. Robertson, Fatigue and durability of Nitinol stents, J. Mech. Behav. Biomed. Mater. 1 (2008) 153–164.

[4] T.W. Duerig, A.R. Pelton, An overview of superelastic stent design, Mater. Sci. Forum 394-395 (2001) 1–8.
[5] S. McMahon, N. Bertollo, E.D.O. Cearbhaill, J. Salber, L. Pierucci, P. Duffy, T. Dürig, V. Bi, W. Wang, Bio-resorbable polymer stents: a review of material progress and prospects, Prog. Polym. Sci. 83 (2018) 79–96.
[6] T. Hu, C. Yang, S. Lin, Q. Yu, G. Wang, Biodegradable stents for coronary artery disease treatment: recent advances and future perspectives, Mater. Sci. Eng. C 91 (2018) 163–178.
[7] H. Hermawan, D. Dubé, D. Mantovani, Developments in metallic biodegradable stents, Acta Biomater. 6 (2010) 1693–1697.
[8] B. Heublein, R. Rohde, V. Kaese, M. Niemeyer, W. Hartung, A. Haverich, Biocorrosion of magnesium alloys: a new principle in cardiovascular implant technology? Heart 89 (2003) 651–656.
[9] R. Waksman, R. Pakala, P.K. Kuchulakanti, R. Baffour, D. Hellinga, R. Seabron, F.O. Tio, E. Wittchow, S. Hartwig, C. Harder, R. Rohde, B. Heublein, A. Andreae, K.H. Waldmann, A. Haverich, Safety and efficacy of bioabsorbable magnesium alloy stents in porcine coronary arteries, Catheter. Cardiovasc. Interv. 68 (2006) 607–617. discussion 618-9.
[10] H. Hermawan, A. Purnama, D. Dube, J. Couet, D. Mantovani, Fe–Mn alloys for metallic biodegradable stents: degradation and cell viability studies, Acta Biomater. 6 (2010) 1852–1860.
[11] T.H. Baron, Colonic stenting: technique, technology, and outcomes for malignant and benign disease, Gastrointest. Endosc. Clin. N. Am. 15 (2005) 757–771.
[12] T.H. Baron, Technique of colonic stenting, Colon. Endoprosthet. 16 (2014) 108–111.
[13] J. Ferlay, I. Soerjomataram, R. Dikshit, S. Eser, C. Mathers, M. Rebelo, D.M. Parkin, D. Forman, F. Bray, Cancer incidence and mortality worldwide: sources, methods and major patterns in GLOBOCAN 2012, Int. J. Cancer 136 (2015) E359–E386.
[14] G.T. Deans, Z.H. Krukowski, S.T. Irwin, Malignant obstruction of the left colon, Br. J. Surg. 81 (1994) 1270–1276.
[15] R. Frago, E. Kreisler, S. Biondo, E. Alba, J. Domínguez, T. Golda, D. Fraccalvieri, M. Millán, L. Trenti, Complications of distal intestinal occlusion treatment with endoluminal implants, Cir. Esp. 89 (2011) 448–455.
[16] R.D. Gongaware, C.A. Slanetz, Hartmann procedure for carcinoma of the sigmoid and rectum, Ann. Surg. 178 (1973) 28–30.
[17] C.S. McArdle, D.J. Hole, Emergency presentation of colorectal cancer is associated with poor 5-year survival, Br. J. Surg. 91 (2004) 605–609.
[18] S. Biondo, J. Martí-Ragué, E. Kreisler, D. Parés, A. Martín, M. Navarro, L. Pareja, E. Jaurrieta, A prospective study of outcomes of emergency and elective surgeries for complicated colonic cancer, Am. J. Surg. 189 (2005) 377–383.
[19] S. Adamsen, S. Meisner, Expandable metal stents for malignant colorectal obstruction, in: Expandable Metal Stents in the Gastrointestinal Tract, vol. 3, 2001, pp. 103–107.
[20] R. Castaño, J.D. Puerta, O.A. Alvarez, J. Lopera, E. Sanin, F. Erebrie, E. NuñEz, L.E. GarcíA, Self-expanding metal stents in malignant and benign colonic obstructions, in: DDW Abstract Issue 2008Digestive Disease Week 2008, vol. 67, 2008, p. AB151.
[21] D.J. Boyle, C. Thorn, A. Saini, C. Elton, G.K. Atkin, I.C. Mitchell, K. Lotzof, A. Marcus, P. Mathur, Predictive factors for successful colonic stenting in acute large-bowel obstruction: a 15-year cohort analysis, Dis. Colon Rectum 58 (2015) 358–362.
[22] N. Srinivasan, R.A. Kozarek, Stents for colonic strictures: materials, designs, and more, Colon. Endoprosthet. 16 (2014) 100–107.
[23] J.H. Kim, T.J. Kang, W.-R. Yu, Mechanical modeling of self-expandable stent fabricated using braiding technology, J. Biomech. 41 (2008) 3202–3212.
[24] S. Tzamtzis, J. Viquerat, J. Yap, M.J. Mullen, G. Burriesci, Numerical analysis of the radial force produced by the Medtronic-CoreValve and Edwards-SAPIEN after transcatheter aortic valve implantation (TAVI), Med. Eng. Phys. 35 (2013) 125–130.
[25] M. Dohmoto, K.D. Rupp, G. Hohlbach, Endoscopically-implanted prosthesis in rectal carcinoma, Dtsch. Med. Wochenschr. 115 (1990) 915.
[26] E. Tejero, A. Mainar, L. Fernández, R. Tobío, M.A. De Gregorio, New procedure for the treatment of colorectal neoplastic obstructions, Dis. Colon Rectum 37 (1994) 1158–1159.
[27] R.M. Bashir, D.E. Fleischer, T.J. Stahl, S.B. Benjamin, Self-expandable nitinol coil stent for management of colonic obstruction due to a malignant anastomotic stricture, Gastrointest. Endosc. 44 (1996) 497–501.
[28] R. Moroi, K. Endo, R. Ichikawa, H. Nagai, H. Shinkai, T. Kimura, F. Ishiyama, K. Yaguchi, S. Kayaba, T. Shimosegawa, The effectiveness of self-expandable metallic stent insertion in treating right-sided colonic obstruction: a comparison between SEMS and decompression tube placement and an investigation of the safety and difficulties of SEMS insertion in right colons, Gastroenterol. Res. Pract. 2014 (2014) 372918.
[29] Z. Lackberg, M.A. Abbas, Colonic stenting: when and how, Semin. Colon Rectal Surg. 28 (2017) 34–40.
[30] D.G. Adler, Colonic strictures: dilation and stents, Gastrointest. Endosc. Clin. N. Am. 25 (2015) 359–371.
[31] D. Stoeckel, A. Pelton, T. Duerig, Self-expanding nitinol stents: material and design considerations, Eur. Radiol. 14 (2004) 292–301.
[32] Y. Zhang, J. Shi, B. Shi, C.Y. Song, W.F. Xie, Y.X. Chen, Comparison of efficacy between uncovered and covered self-expanding metallic stents in malignant large bowel obstruction: a systematic review and meta-analysis, Color. Dis. 14 (2012) e367–e374.
[33] K.M. Lee, S.J. Shin, J.C. Hwang, J.Y. Cheong, B.M. Yoo, K.J. Lee, K.B. Hahm, J.H. Kim, S.W. Cho, Comparison of uncovered stent with covered stent for treatment of malignant colorectal obstruction, Gastrointest. Endosc. 66 (2007) 931–936.
[34] J. Huo, R. Rojas, J. Bohlin, J. Hilborn, E.K. Gamstedt, Parametric elastic analysis of coupled helical coils for tubular implant applications: experimental characterization and numerical analysis, J. Mech. Behav. Biomed. Mater. 29 (2014) 462–469.
[35] R. Rebelo, N. Vila, R. Fangueiro, S. Carvalho, S. Rana, Influence of design parameters on the mechanical behavior and porosity of braided fibrous stents, Mater. Des. 86 (2015) 237–247.
[36] H. Isayama, Y. Nakai, Y. Toyokawa, O. Togawa, C. Gon, Y. Ito, Y. Yashima, H. Yagioka, H. Kogure, T. Sasaki, T. Arizumi, S. Matsubara, N. Yamamoto, N. Sasahira, K. Hirano, T. Tsujino, N. Toda, M. Tada, T. Kawabe, M. Omata, Measurement of radial and axial forces of biliary self-expandable metallic stents, Gastrointest. Endosc. 70 (2009) 37–44.
[37] A. Ratnovsky, N. Regev, S. Wald, M. Kramer, S. Naftali, Mechanical properties of different airway stents, Med. Eng. Phys. 37 (2015) 408–415.
[38] J.A. Grogan, S.B. Leen, P.E. McHugh, Comparing coronary stent material performance on a common geometric platform through simulated bench testing, J. Mech. Behav. Biomed. Mater. 12 (2012) 129–138.
[39] D.J. McGrath, B. O'Brien, M. Bruzzi, P.E. McHugh, Nitinol stent design – understanding axial buckling, J. Mech. Behav. Biomed. Mater. 40 (2014) 252–263.
[40] L. Petrini, F. Migliavacca, F. Auricchio, G. Dubini, Numerical investigation of the intravascular coronary stent flexibility, J. Biomech. 37 (2004) 495–501.

[41] F. Migliavacca, L. Petrini, M. Colombo, F. Auricchio, R. Pietrabissa, Mechanical behavior of coronary stents investigated through the finite element method, J. Biomech. 35 (2002) 803–811.

[42] F. Etave, G. Finet, M. Boivin, J.C. Boyer, G. Rioufol, G. Thollet, Mechanical properties of coronary stents determined by using finite element analysis, J. Biomech. 34 (2001) 1065–1075.

[43] L. Feo, D.M. Schaffzin, Colonic stents: the modern treatment of colonic obstruction, Adv. Ther. 28 (2011) 73–86.

[44] G. Costamagna, A. Tringali, J. Spicak, M. Mutignani, J. Shaw, A. Roy, E. Johnsson, E.G.H. De Moura, S. Cheng, T. Ponchon, M. Bittinger, H. Messmann, H. Neuhaus, B. Schumacher, R. Laugier, J. Saarnio, F.I. Ariqueta, Treatment of malignant gastroduodenal obstruction with a nitinol self-expanding metal stent: an international prospective multicentre registry, Dig. Liver Dis. 44 (2012) 37–43.

[45] Y. Hori, I. Naitoh, K. Hayashi, T. Ban, M. Natsume, F. Okumura, T. Nakazawa, H. Takada, A. Hirano, N. Jinno, S. Togawa, T. Ando, H. Kataoka, T. Joh, Predictors of outcomes in patients undergoing covered and uncovered self-expandable metal stent placement for malignant gastric outlet obstruction: a multicenter study, Gastrointest. Endosc. 85 (2017) 340.

[46] J. Rayhanabad, M.A. Abbas, Long-term outcome of endoscopic colorectal stenting for malignant and benign disease, Am. Surg. 75 (2009) 897–900.

[47] N. Suzuki, B.P. Saunders, S. Thomas-Gibson, C. Akle, M. Marshall, S. Halligan, Colorectal stenting for malignant and benign disease: outcomes in colorectal stenting, Dis. Colon Rectum 47 (2004) 1201–1207.

[48] S. Domingo, S. Puértolas, L. Gracia-Villa, J.A. Puértolas, Mechanical comparative analysis of stents for colorectal obstruction, Minim. Invasive Ther. Allied Technol. 16 (2007) 126–136.

[49] F. Auricchio, R.L. Taylor, J. Lubliner, Shape-memory alloys: macromodelling and numerical simulations of the superelastic behavior, Comput. Methods Appl. Mech. Eng. 146 (1997) 281–312.

[50] J. Arghavani, F. Auricchio, R. Naghdabadi, A finite strain kinematic hardening constitutive model based on Hencky strain: general framework, solution algorithm and application to shape memory alloys, Int. J. Plasticity 27 (2011) 940–961.

[51] F. Iannaccone, N. Debusschere, S. De Bock, M. De Beule, D. Van Loo, F. Vermassen, P. Segers, B. Verhegghe, The influence of vascular anatomy on carotid artery stenting: a parametric study for damage assessment, J. Biomech. 47 (2014) 890–898.

[52] S. Puertolas, D. Navallas, A. Herrera, E. Lopez, J. Millastre, E. Ibarz, S. Gabarre, J.A. Puertolas, L. Gracia, A methodology for the customized design of colonic stents based on a parametric model, J. Mech. Behav. Biomed. Mater. 71 (2017) 250–261.

[53] S. Domingo, S. Puertolas, L. Gracia-Villa, M. Mainar, J. Uson, J.A. Puertolas, Design, manufacture and evaluation of a NiTi stent for colon obstruction, Biomed. Mater. Eng. 15 (2005) 357–365.

[54] S. Puértolas, E. Bajador, J.A. Puértolas, E. López, E. Ibarz, L. Gracia, A. Herrera, Study of the behavior of a bell-shaped colonic self-expandable NiTi stent under peristaltic movements, Biomed. Res. Int. 2013 (2013) 370582.

[55] pyFormex, Available: www.nongnu.org/pyformex/.

[56] Siemens, I-Deas - NX11, Available: http://www.plm.automation.siemens.com/.

[57] F. Auricchio, A robust integration-algorithm for a finite-strain shape-memory-alloy superelastic model, Int. J. Plast. 17 (2001) 971–990.

[58] Dassault Systèmes, Available: http://www.3ds.com/.

C H A P T E R

# 4

# Mechanical and Microstructural Behavior of Vascular Tissue

*Estefanía Peña*[*,†,‡], *Alberto García*[§], *Pablo Sáez*[§], *Juan A. Peña*[*,¶], *Myriam Cilla*[*,‡,|], *Miguel Angel Martínez*[*,†,‡]

**[*]Aragón Institute of Engineering Research (I3A), University of Zaragoza, Zaragoza, Spain [†]Department of Mechanical Engineering, University of Zaragoza, Zaragoza, Spain [‡]Centro de Investigación en Red en Bioingeniería, Biomaterialesy Nanomedicina, CIBER-BBN, Zaragoza, Spain [§]Laboratori de Calcul Numeric, Universitat Politecnica de Catalunya, Barcelona, Spain [¶]Department of Management and Manufacturing Engineering, University of Zaragoza, Zaragoza, Spain [|]Centro Universitario de la Defensa, Academia General Militar, Zaragoza, Spain**

## 4.1 INTRODUCTION

The study of the mechanical factors that induce vascular pathologies, especially in arteries, has been one of the main research lines in biomechanics. This is not surprising because according to the World Health Organization (WHO), cardiovascular chronic diseases are the leading cause of death. In this regard, a wide number of constitutive relations describing the mechanical response have been proposed for cardiovascular tissue. Therefore, the mechanical variables that strongly influence the mechanobiology on the vascular tissue may be computed [1]. Accurate mechanical models and appropriate numerical approaches can be an asset in the study of cardiovascular dysfunctions and the simulation of surgical interventions, for example, carotid stenting.

The arterial wall is composed of three distinct elements: the vascular smooth muscles (VSMs) that form the cellular part of the vessel, and the extracellular matrix major components elastin and collagen. Collagen is the main load-bearing component within the tissue while the elastin provides elasticity to the tissue. The mechanical behavior of an arterial wall is governed mainly by smooth muscle cells (SMC), the matrix material (consisting mainly of water, elastin, and proteoglycans), and the collagen fibers. Their three-dimensional (3D) organization is vital to accomplish proper physiological functions. The combined contribution of these constituents determines the mechanical response of the tissue, described as highly nonlinear and anisotropic, due to the existence of clear preferential orientations of the fiber bundles. The microstructural composition and thus the mechanical properties of the arterial wall vary along the cardiovascular tree and also for different species [2–6]. García et al. [3] found differences in the stress-strain curves for the circumferentially oriented swine carotid samples depending on the longitudinal position, becoming stiffer when increasing the distance from the aorta. This finding was directly correlated with a significant variation on the tissue composition and the presence of the different microconstituents. Peña et al. [4] compared ascending thoracic aorta, descending thoracic aorta (DTA), and infrarenal abdominal aorta. Abdominal tissues were found to be stiffer and highly anisotropic. They found that the aorta changed from a more isotropic to a more anisotropic tissue and became progressively less compliant and stiffer with the distance to the heart. They also observed substantial differences in the anisotropy parameter between aortic samples where abdominal samples were more anisotropic and nonlinear than the thoracic samples.

Numerous constitutive models have been proposed to describe the arterial mechanical response. The preferred methodology to describe and reproduce its complex mechanical response is the definition of a strain energy function (SEF) from which the stress response is derived; see, for example, Refs. [7–12] and references therein. Although phenomenological models (PMs) may reproduce the biomechanical properties of the vascular tissue, their material



*Advances in Biomechanics and Tissue Regeneration*
https://doi.org/10.1016/B978-0-12-816390-0.00004-2

parameters lack a clear physical meaning. Moreover, these models are unreliable for predictions beyond the strain range used in parameter estimation [13]. However, structurally motivated material models may provide increased insights into the underlying mechanics and physics of arteries and could overcome this drawback [14].

Identification of an appropriate SEF is the preferred method to describe the complex nonlinear elastic properties of vascular tissues [15]. An ideal SEF should be based on histological analysis to provide a better description of wall deformation under load [11, 16]. Early SEFs were purely phenomenological functions where parameters involved in the mathematical expression have not physiological meaning [17, 18]. Later, structure-based or constituent-based SEF was developed, where the parameters mean some physical and structural properties of the different components of the vessel wall [9, 12]. In several tissues, there is a strong alignment of the collagen fibers with little dispersion in their orientation. In other cases, such as the artery wall, there is significant dispersion in the orientation, which has a significant influence on the mechanical response. Proposed structure-based models were used by taking into account the spatial dispersion or distribution or waviness of collagen fiber directions [7, 8, 16, 19, 20].

In this chapter, we study some constitutive models, some with a phenomenological approach and others microstructurally and physically oriented, that describe more accurately the features of the arterial wall. The mechanical behavior and structural organization of the DTA and the carotid artery are studied together in order to overcome some limitations of previous models. We aim to examine the advantages and limitations of phenomenological versus more microstructural-oriented approaches.

## 4.2 MICROSTRUCTURAL MODELING OF THE CAROTID ARTERY

The carotid artery has been widely studied because it is prone to develop atherosclerosis, stenosis, collagen remodeling, etc. An extensive range of experimental results for the carotid artery can be found in the literature. Experimental procedures such as inflation tests on animals [6, 21, 22] and humans [23, 24], simple tension tests [3, 25], and nano-indentation tests [26] have been used to determine the mechanical properties of carotids. However, these studies are usually fitted with phenomenological and microstructure-based SEFs without taking into account the histological data from different regions. Therefore, the constitutive laws commonly consider independent material parameters for each position.

### 4.2.1 Experimental Findings for the Porcine Carotid Artery

To analyze the microstructure of the porcine carotid artery, we used three kinds of experimental data of a swine carotid artery obtained previously in our group [3, 27]. Tissue samples from 14 specimens were processed in a histological laboratory and images were used to quantify their microconstituents [3]. We also analyzed the collagen bundle distribution means of using polarized light microscopy techniques [27]. We used the average collagen distribution to calibrate the presented microstructural model (MM) in order to obtain general constitutive parameters for the porcine carotid. Finally, we also carried out several cyclic uniaxial tension tests [3] on swine carotid tissue. The details are illustrated in the following sections.

#### *4.2.1.1 Histological Analysis*

García et al. [3] reported the composition of the porcine carotid depending on the location of the sample by image analysis techniques. In order to quantify the percentage areas of elastin fibers and SMC, a segmentation procedure was followed by means of Software ImageJ. Several histologies of different samples were analyzed. The samples were grouped depending on the type of stain (Orcein's elastin, Masson's trichrome, and antismooth muscle actin immunohistochemical study) and the elastin, collagen, and muscular tissue composition were determined (Fig. 4.1). Quantification of elastin areas for distal and proximal locations was, respectively, $19.6 \pm 4.1$ and $52.6 \pm 6.7$. The area percentages of SMC achieved by means of the antismooth muscle actin stains were $44.3 \pm 4.2$ and $31.0 \pm 3.3$ for distal and proximal samples. Finally, collagen quantification by using Masson's trichrome stain reported $23.5 \pm 3.9$ and $14.9 \pm 2.3$ for swine samples in the two different locations, respectively [3].

Regarding the collagen fiber distribution, segments of each vessel were fixed and prepared for histological analysis. Slides were stained for birefringence enhancement with a Picrosirius Red stain, which causes collagen to appear in a brighter orange-yellow when viewed through polarized light [28–30]. Samples were analyzed in a BX50 microscope (Olympus, Melville, NY) equipped with an Achromat UD 16/0.17 objective and a CMEX-1300x camera (Euromex microscopen B.V., Arnhem, The Netherlands) equipped with a Universal Rotary Stage (Carl Zeiss GmbH, Jena, Germany) [30].

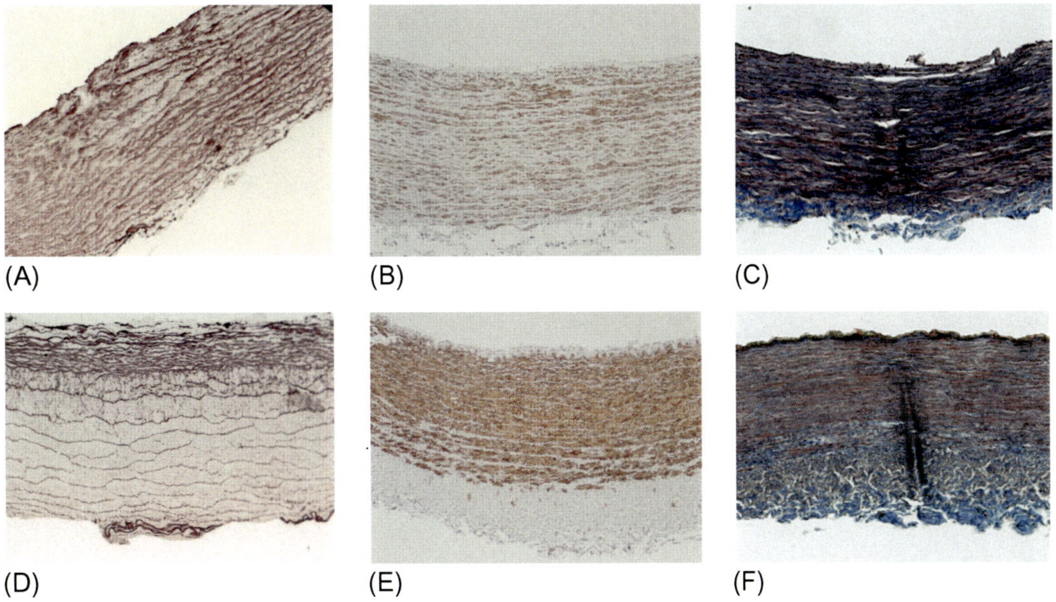


FIG. 4.1 Histological and immunohistochemical images of porcine carotid tissue (10×). (A) Van Giesson elastin stain, proximal location. (B) Antismooth muscle actin, proximal location. (C) Masson's trichrome stain, proximal location. (D) Van Giesson elastin stain, distal location. (E) Antismooth muscle actin, distal location. (F) Masson's trichrome stain, distal location.

The orientation of a collagen fiber in 3D space was uniquely defined by its elevation angle $\Phi$ and its azimuthal angle $\Theta$, which were measured by (in-plane) rotating and (out-of-plane) tilting the stage. In Sáez et al. [27], two almost-symmetric families of fibers were found (one with positive and the other with negative elevation angles). Furthermore, with respect to the azimuthal angle, the dispersion showed only a family of fibers could be observed. Due to the assumption of two families of fibers, separate Bingham ODFs were introduced to capture each of them independently (Fig. 4.2). The fit Bingham parameters, mean elevation angle $\phi_r$, mean azimuthal angle $\theta_r$, and $\kappa_{1, 2, 3}$ are given in Table 4.1.

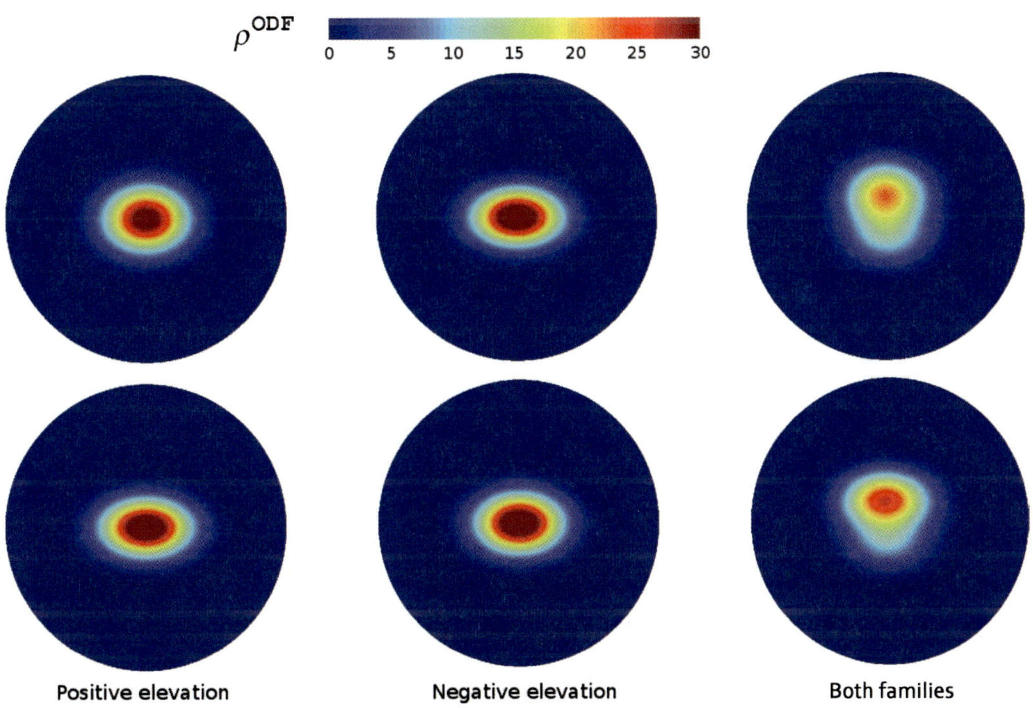


FIG. 4.2 Collagen fiber orientation distribution for all proximal (*up*) and distal (*down*) samples represented by the Bingham orientation density function (ODF) [27].

TABLE 4.1 Mean and Standard Deviation of Collagen Fiber Orientation in Swine Carotid Artery Samples [27]

| | $\bar{\phi}_r$ | SD | $\bar{\theta}_r$ | SD | $\kappa_{1,\,2,\,3}$ |
|---|---|---|---|---|---|
| **POSITIVE ELEVATION FAMILY** | | | | | |
| Proximal | 90.0 | ±8.5 | 8.7 | ±5.0 | [9.6, 0.0, 22.4] |
| Distal | 91.8 | ±7.8 | 8.8 | ±3.8 | [18.6, 0.0, 30.6] |
| **NEGATIVE ELEVATION FAMILY** | | | | | |
| Proximal | 89.0 | ±7.5 | 8.8 | ±5.0 | [18.6, 0.0, 30.3] |
| Distal | 90.8 | ±8.3 | 7.7 | ±0.5 | [15.0, 0.0, 28.4] |

#### 4.2.1.2 *Uniaxial Mechanical Test*

Concerning the mechanical behavior, simple tension tests of the carotid strips were performed in a high precision drive Instron Microtester 5548 system adapted for biological specimens [3]. Different loading and unloading cycles were applied corresponding to approximately 60, 120, and 240 [kPa] stress levels at 30%/min of strain rate. Three preliminary cycles at those load levels were applied in order to precondition the sample. Fig. 4.3 shows the mean and standard deviations obtained at several points for all individuals. The third cycle of the highest level of stress (240 [kPa]) was used in the subsequent stress-stretch analysis. In that work, we reported important differences in the mechanical response between the proximal and distal behaviors for the circumferential direction. By contrast, the longitudinal tests showed quite similar stress-stretch curves for both locations.

### 4.2.2 Material Models for the Carotid Artery

Let $\mathcal{B}_0 \subset \mathbb{E}^3$ be a reference or rather material configuration of a body $B$ of interest. The notation $\varphi: \mathcal{B}_0 \times \mathcal{T} \rightarrow \mathcal{B}_t$ represents the one-to-one mapping, continuously differentiable, transforming a material point $\mathbf{X} \in \mathcal{B}_0$ to a position $\mathbf{x} = \varphi(\mathbf{X}, t) \in \mathcal{B}_t \subset \mathbb{E}^3$, where $\mathcal{B}_t$ represents the deformed configuration at time $t \in \mathcal{T} \subset \mathbb{R}$. The mapping $\varphi$ represents a motion of the body $B$ that establishes the trajectory of a given point when moving from its reference position X to x. The two-point deformation gradient tensor is defined as $\mathrm{F}(\mathrm{X}, t) := \nabla_{\mathrm{X}} \boldsymbol{\varphi}(\mathrm{X}, t)$, with $J(\mathbf{X}) = \det(\mathbf{F}) > 0$ the local volume variation.

The free energy density function is given by a scalar-valued function $\Psi$ defined per unit reference volume in the reference configuration and, for isothermal processes, decoupled in volumetric and isochoric parts. Consistent with the constrained mixture approach [31], we assume a free energy function of the form

$$\Psi = \phi_{\text{elas}} \Psi_{\text{elas}} + \phi_{\text{vsmc}} \Psi_{\text{vsmc}} + \phi_{\text{coll}} \Psi_{\text{coll}}, \tag{4.1}$$

where the subscripts elas, vsmc, and coll refer to elastic fiber, VSM cell, and collagen fiber contributions, respectively, where $\phi_i$ and $\Psi_i$ describe the volume fractions and the passive free energy associated with each constituent.

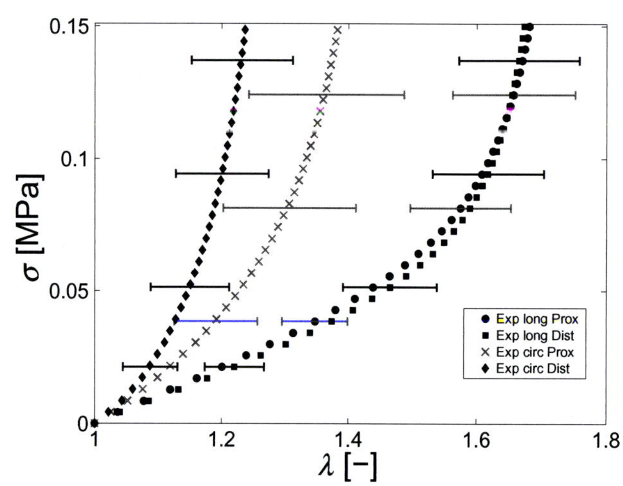


FIG. 4.3 Experimental stress-stretch (Mean ±SD) for the proximal and distal positions for all tensile tests extracted from García et al. [3].

For the elastin, we use the classical neo-Hookean SEF

$$\Psi_{\text{elas}} = \mu_{\text{elas}}[I_1 - 3] \tag{4.2}$$

with $I_1$ the first invariant of the symmetric Cauchy-Green tensor and $\mu_{\text{elas}}$ is a stress dimensional material parameter.

Following Bellin et al. [32], the behavior of SMC is modeled using an exponential function

$$\Psi_{\text{vsmc}} = \frac{c_{1\text{vsmc}}}{2c_{2\text{vsmc}}}[\exp(c_{2\text{vsmc}}[\lambda_{\text{vsmc}}^2 - 1]^2) - 1] \tag{4.3}$$

with $\lambda_{\text{vsmc}} = \|\mathbf{F} \cdot \mathbf{m}\|$ the stretch experienced by the smooth muscle where m is the unit vector associated with the SMC orientation, and $c_{1\text{vsmc}}$ and $c_{2\text{vsmc}}$ are stress dimensional and dimensionless material parameters, respectively. The contribution of the smooth muscle to the SEF was only considered in the media of the vessel and the SMCs were supposed to be circumferentially oriented [33].

$\Psi_{\text{coll}}$ defines the anisotropic contribution of the collagen fiber. Based on the previous literature and histological results, four SEFs representing collagen fibers are chosen: a PM [34], a cross-linked phenomenological model (CLPM) [35], an MM [36], and a cross-linked microstructural model (CLMM) [27].

#### 4.2.2.1 Phenomenological Model

Holzapfel et al. [34] proposed a multilayer fiber-reinforced composite model for the arterial wall that considers the histological structure of arteries with a preferential direction for the fiber orientation and with an exponential form that represents the prominent stiffening characteristics of the arterial collagen

$$\Psi_{\text{coll}} = \sum_{i=4,6} \frac{c_{1\text{coll}}}{2c_{2\text{coll}}}[\exp(c_{2\text{coll}}[\rho[I_i - 1]^2 + [1-\rho][I_1 - 3]^2]) - 1]. \tag{4.4}$$

In this equation, $I_1$ represents the first invariant of the Cauchy-Green tensor [37] characterizing the isotropic mechanical response of the elastin [38, 39] while $I_4 \geq 1$ and $I_6 \geq 1$ characterize the mechanical response in the preferential directions of the fibers/cells [34]. $c_{1\text{coll}} > 0$ is the stress-like parameter and $c_{2\text{coll}} > 0$ is the dimensionless parameter. The $\rho \in [0, 1]$ parameter is also dimensionless and accounts for the fiber dispersion. The SEF represents the strain energy stored in a composite material reinforced in two preferred directions represented by the invariants $I_4$ and $I_6$. Both invariants can also be expressed as a function of the main stretches, Eq. (4.5). Due to each family of fibers represents the main direction of collagen bundles ($\theta_1$ and $\theta_2$) that are orientated in a helicoidal manner and both families of fibers were assumed to have the same mechanical response. The anisotropy directions were assumed to be oriented at $\pm\theta$ degrees with respect to the longitudinal axis [9], so $\theta_1 = \theta$ and $\theta_2 = -\theta$ in Eq. (4.5), see Holzapfel et al. [9].

$$I_4 = \lambda_1^2 \cos^2\theta_1 + \lambda_2^2 \sin^2\theta_1, \quad I_6 = \lambda_1^2 \cos^2\theta_2 + \lambda_2^2 \sin^2\theta_2. \tag{4.5}$$

This model was used by García et al. [3] to fit the mechanical properties of the porcine carotid arteries by uniaxial tensile tests where $\theta$ was treated as an unknown phenomenological variable included in the minimization algorithm. Proximal and distal samples were fitted separately considering independent material parameters for each position. It is emphasized that the model defined by Eq. (4.17) should be considered as a phenomenological approach and cannot be regarded as a structural model. Note that the parameter $\rho$ has no histological meaning.

#### 4.2.2.2 Cross-Linked Phenomenological Model

O'Connell et al. [40] in rat aortas and Sáez et al. [27] in pig carotids have shown that collagen fibers are bundled around the SMC with some collagen fibrils linking the main fiber in a predominant perpendicular direction. In the case of the media layer of carotid arteries, previous studies reported that SMC keep a highly predominant circumferential direction [27]. With this orientation obtained experimentally, the PM presented in Holzapfel et al. [34] was unable to fit the uniaxial test. The CLPM was initially proposed by Sáez et al. [35] and used to fit the mechanical properties of the porcine carotid arteries from experimental data presented in García et al. [3], where the $\theta$ parameter was directly fixed to the circumferential direction [27, 40].

A linear interpolation of a well-known isotropic and anisotropic SEF was used by Sáez et al. [35] to recover this behavior as

$$\Psi_{\text{coll}} = \sum_{i=4,6} [1 - 2\alpha] \frac{c_{1\text{coll}}}{2c_{2\text{coll}}}[\exp(c_{2\text{coll}}[I_i - 1]^2) - 1] + \alpha \frac{c_{1\text{coll}}}{2c_{2\text{coll}}}[\exp(c_{2\text{coll}}[I_1 - 3]^2) - 1], \tag{4.6}$$

where $\alpha$ represents the amount of cross-links, being $\alpha \in [0, 0.5]$. For the extreme cases, $\alpha = 0$ means no cross-links on the tissue while for $\alpha = 0.5$, the degree of links is high enough to consider a fully isotropic distribution of the fibers. The physical meaning of this parameter $\alpha$ is the inclusion in the model of the attachments between the main fibers of collagen, which provides a certain degree of stiffening in the transverse direction of these fibers.

#### 4.2.2.3 Microstructural Model

Alastrué et al. [36] proposed a microsphere-based model to account for the dispersion of the collagen fibers around a preferential direction, overcoming the one-dimensional (1D) limitation of the previous characterization of the collagen fiber. $\Psi_{\text{coll}}$ is defined as the sum of the contributions of each collagen family of fibrils as

$$\Psi_{\text{coll}} = \sum_{j=1}^{N} [\Psi_{\text{coll}}]^j = \sum_{j=1}^{N} \langle n\rho\psi_{\text{coll}} \rangle^j = \sum_{j=1}^{N} \frac{1}{4\pi} \int_{\mathbb{U}^2} (n\rho[\psi_{\text{coll}}])^j dA, \tag{4.7}$$

where $n$ is the chain density, $N$ denotes the number of families of collagen fibers, $N = 2$ according to the experimental results of orientation of collagen fibers [9], and applying a discretization to the continuous orientation distribution on the unit sphere $\mathbb{U}^2$, $[\Psi_{\text{coll}}]^j$ corresponds to the expression

$$[\Psi_{\text{coll}}]^j = \sum_{i=1}^{m} n\rho(\mathbf{r}^i; \mathbf{Z}, \mathbf{Q}) \psi^i_{\text{coll}}(\lambda^i_{\text{coll}}), \tag{4.8}$$

where $\mathbf{r}^i$ are the unit vectors associated with the discretization on the microsphere over the unit sphere $\mathbb{U}^2$, $m$ is the number of discrete orientation vectors [7], $\lambda^i_{\text{coll}} = \|\mathrm{F} \cdot \mathrm{r}^i\|$ the stretch in the $\mathrm{r}^i$ direction, and $\psi^i_{\text{coll}}(\lambda^i_{\text{coll}})$ the strain energy density function associated with the $\mathrm{r}^i$ direction. Using Eqs. (4.7), (4.8) it results in

$$\Psi_{\text{coll}} = \sum_{j=1}^{N} \sum_{i=1}^{m} (w^i n\rho[\psi^i_{\text{coll}}])^j, \tag{4.9}$$

where $w^i$, $i = 1, \ldots, m$ denote related weighting factors and $\rho$ is the orientation density function (ODF) to take into account the fibril dispersion [7].

In this work, we used the Bingham ODF [41] initially proposed by Alastrué et al. [36] for the incorporation of anisotropy in a microsphere-based model with application to the modeling of the thoracic aorta. One of the main advantages of the Bingham ODF is the possibility of considering three different concentration parameters in three orthogonal directions of the space. These orientations can be easily correlated with the three main directions of a blood vessel: circumferential, radial, and axial. That function is expressed as

$$\rho(\mathbf{r}; \mathbf{Z}, \mathbf{Q}) \frac{dA}{4\pi} = [F_{000}(\mathbf{Z})]^{-1} \exp(\mathrm{tr}(\mathbf{Z} \cdot \mathbf{Q}^t \cdot \mathbf{r} \cdot \mathbf{r}^t \cdot \mathbf{Q})) \frac{dA}{4\pi}, \tag{4.10}$$

where Z is a diagonal matrix with eigenvalues $\kappa_{1,2,3}$, $\mathbf{Q} \in \mathbb{Q}^3$ defines the orientation of the three principal orthogonal directions with respect to the reference basis, and

$$F_{000}(\mathbf{Z}) = \frac{1}{4\pi} \int_{\mathbb{U}^2} \exp(\mathrm{tr}(\mathbf{Z} \cdot \mathbf{r} \cdot \mathbf{r}^t)) dA. \tag{4.11}$$

Thus, the probability of finding a find in a specific direction is controlled by the eigenvalues of Z, which might be interpreted as concentration parameters. Specifically, the difference between pairs of $\kappa_{1,2,3}$—that is, $[\kappa_1 - \kappa_2]$, $[\kappa_1 - \kappa_3]$, and $[\kappa_2 - \kappa_3]$—determines the shape of the distribution over the surface of the unit sphere. Therefore, the value of one of these three parameters may be fixed to a constant value without reducing the versatility and different distributions of a family of fibers achieved for a constant value of $\kappa_2$ and varying values of $\kappa_1$ and $\kappa_3$.

The exponential-like SEF proposed by Holzapfel et al. [9] was used to approach the fiber response

$$n\psi^i_{\text{coll}}(\lambda^i_{\text{coll}}) = \frac{c_{1\text{coll}}}{2c_{2\text{coll}}} \left( e^{c_{2\text{coll}}((\lambda^i_{\text{coll}})^2 - 1)^2} - 1 \right) \quad \text{if } \lambda_i \geq 1 \text{ otherwise } \psi_{f\,i}(\lambda_i) = 0, \tag{4.12}$$

because it is usually considered that collagen fibers only affect the global mechanical behavior in tensile states; see Holzapfel et al. [9]. The affine kinematics define the collagen fiber stretch $\lambda^i_{\text{coll}} = \|\mathrm{t}^i\|$ in the fiber direction $\mathrm{r}^i$. Finally, $c_{1\text{coll}}$ and $c_{2\text{coll}}$ are stress dimensional and dimensionless material parameters, respectively.

#### 4.2.2.4 *Cross-Linked Microstructural Model*

In an attempt to bring together the advantages from the MM and the CPM, Sáez et al. [27] proposed an anisotropic contribution of the collagen fibers to the SEF as

$$\Psi_{\text{coll}} = \langle [\rho + \alpha\hat{\rho}]\psi_{\text{f}}(\lambda)\rangle = \frac{1}{4\pi}\int_{\mathbb{U}^2} [\rho + \alpha\hat{\rho}]\psi_{\text{f}} dA, \tag{4.13}$$

where $\rho$ denotes the Bingham ODF; see Eq. (4.10). In addition, $\hat{\rho} = [\max(\rho) - \rho]/\max(\rho)$ is an effective orientation density that accounts for cross-links between the main collagen fibers. $\hat{\rho}$ represents the physical space where the main collagen fibers do not exist, and therefore collagen cross-links can be interpreted similarly to the one described in the CLPM. Specifically, $\alpha \in [0, 1]$ represents the relative amount of cross-links with $\alpha = 0$ and $\alpha = 1$ denoting no cross-links and fully cross-linked collagen, respectively. The fully cross-linked state results in an isotropic response of the collagen.

Finally, the numerical integration of Eq. (4.13), using $m$ integration points over the unit sphere $\mathbb{U}^2$, gives

$$\Psi_{\text{coll}} = \sum_{i=1}^{m} w_i [\rho_i + \alpha\hat{\rho}_i]\psi^i_{\text{coll}}(\lambda^i_{\text{coll}}), \tag{4.14}$$

with $w_i$ denoting integration point weights and $\psi^i_{\text{coll}}$ was defined in Eq. (4.12).

This model was formerly used by Sáez et al. [27] for fitting the mechanical properties of the porcine carotid arteries by uniaxial tensile tests where proximal and distal samples were adjusted separately considering independent material parameters for each position. Again, the physical meaning of this parameter $\alpha$ is the inclusion of the attachments between the main fibers, which provides an additional reinforcement to the collagen tissue in the transverse direction of the main fibers [35] described previously in the MM.

### 4.2.3 Results on Modeling the Porcine Carotid Artery

The fitting of the experimental mechanical data (Fig. 4.3) was developed by using a Levenberg-Marquardt minimization algorithm [42], by defining the objective function represented in Eq. (4.15). In this function, $\sigma_{\theta\theta}$ and $\sigma_{zz}$ are the Cauchy stress data obtained from the tests, $\sigma^{\Psi}_{\theta\theta}$ and $\sigma^{\Psi}_{zz}$ are the Cauchy stresses for the $i$th point computed using the SEF (Eq. 4.21) under the hypothesis of incompressibility [43], and $k$ is the number of data points.

$$\chi^2 = \Sigma^k_{i=1}[(\sigma_{\theta\theta} - \sigma^{\Psi}_{\theta\theta})^2_i + (\sigma_{zz} - \sigma^{\Psi}_{zz})^2_i], \tag{4.15}$$

where

$$\sigma^{\Psi}_{\theta\theta} = \lambda_\theta \frac{\partial \Psi_{\text{iso}}}{\partial \lambda_\theta} \quad \sigma^{\Psi}_{zz} = \lambda_z \frac{\partial \Psi_{\text{iso}}}{\partial \lambda_z}. \tag{4.16}$$

The coefficient of determination of the normalized mean square root error $\varepsilon \in [0, 1]$ was computed for each fitting $\varepsilon = \frac{\sqrt{\frac{\chi^2}{k-q}}}{\varpi}$. In this equation, $\varpi$ is the mean value of the measured stresses, $\varpi = \frac{\Sigma^k_{i=1}(\sigma)_i}{k}$, $q$ is the number of parameters of the SEF, so $k - q$ is the number of degrees of freedom, and $\varpi$ the mean stress already defined earlier. $\varepsilon \leq 0.1$ typically represents a good fit to the experimental data.

Both data (proximal and distal, see Fig. 4.3) were fitted at the same time considering only one set of parameters. That is, we considered that the mechanical properties of the elastin, VSMCs, and collagen do not change along the carotid. The values of $\phi_r$, $\theta_r$, and $\kappa_{1,2,3}$ were taken from Sáez et al. [27] and are given in Table 4.1. The values of $\phi_{\text{elas}}$, $\phi_{\text{vsmc}}$ and $\phi_{\text{coll}}$ were taken from the mean values for distal and proximal positions of the porcine carotids from García et al. [3]. For PM and CLPM approaches, a total of nine elastic parameters ($\mu_{\text{elas}}$, $c_{1\text{vsmc}}$, $c_{2\text{vsmc}}$, $c_{1\text{coll}}$, $c_{2\text{coll}}$, $\rho_{\text{prox}}$, $\rho_{\text{dist}}$, $\theta_{\text{prox}}$, $\theta_{\text{dist}}$) and seven elastic parameters ($\mu_{\text{elas}}$, $c_{1\text{vsmc}}$, $c_{2\text{vsmc}}$, $c_{1\text{coll}}$, $c_{2\text{coll}}$, $\alpha_{\text{prox}}$, $\alpha_{\text{dist}}$), respectively, should be fitted. Furthermore, the number of adjusting variables for the MM and the CLMM is five elastic parameters ($\mu_{\text{elas}}$, $c_{1\text{vsmc}}$, $c_{2\text{vsmc}}$, $c_{1\text{coll}}$, $c_{2\text{coll}}$) and seven ($\mu_{\text{elas}}$, $c_{1\text{vsmc}}$, $c_{2\text{vsmc}}$, $c_{1\text{coll}}$, $c_{2\text{coll}}$, $\alpha_{\text{prox}}$, $\alpha_{\text{dist}}$), respectively.

The results of the fitting of the different SEFs proposed to the experimental data for the distal and proximal curves are shown in Fig. 4.4. The material parameters are summarized in Table 4.2. In terms of fitting errors, both phenomenological models, PM and CLPM, present relatively low values, $\varepsilon \sim 0.05-0.11$ and $\varepsilon \sim 0.08-0.12$, respectively. However, the values of $\theta_{\text{prox}} = 43.4$ degrees for PM contradicts experimental findings [27]. Contrarily, the MM resulted in very large error values ($\varepsilon_p = 0.74$ and $\varepsilon_d = 0.37$) for the proximal and distal zones, respectively, failing to reproduce

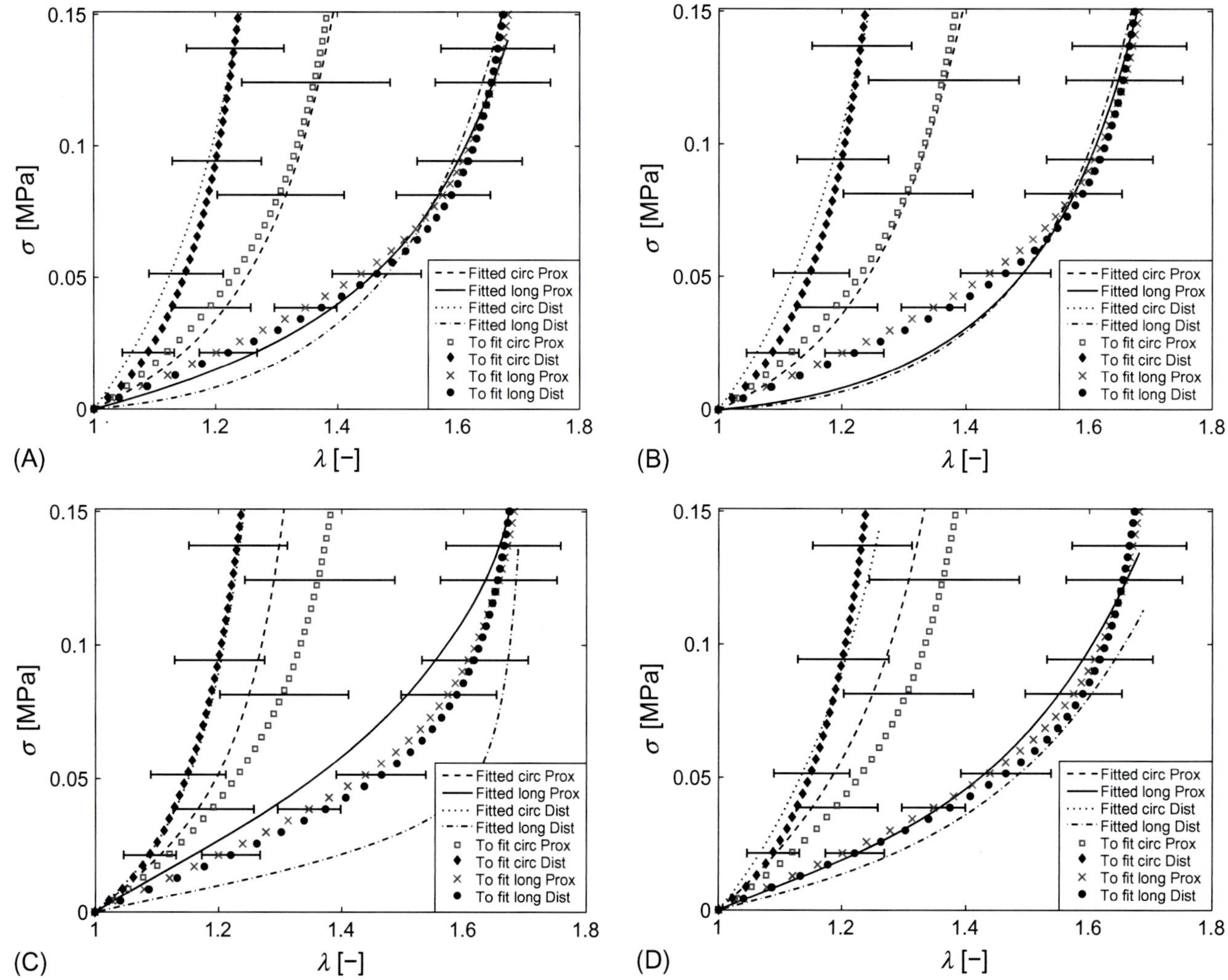


FIG. 4.4 Simulation results of mean uniaxial tension tests from proximal and distal samples obtained with the proposed constitutive laws. Experimental stress-stretch (Mean ±SD) in the proximal (×) and distal (•) positions for all tests obtained from García et al. [3]. (A) Phenomenological model; (B) cross-linked phenomenological model; (C) microstructural model; (D) cross-linked microstructural model.

TABLE 4.2 Estimated Mechanical Constitutive Model Parameters for Phenomenological Model (PM), Cross-Linked Phenomenological Model (CLPM), Microstructural Model (MM), and Cross-Linked Microstructural Model (CLMM)

| | | | | | | | | | | | |
|---|---|---|---|---|---|---|---|---|---|---|---|
| PM | $\mu_{elas}$ | $c_{1vsmc}$ | $c_{2vsmc}$ | $c_{1coll}$ | $c_{2coll}$ | $\rho_{prox}$ | $\rho_{dist}$ | $\theta_{prox}$ | $\theta_{dist}$ | $\varepsilon_p$ | $\varepsilon_d$ |
| | 0.0175 | 0.0493 | 0.9718 | 0.1401 | 1.4266 | 0.9881 | 0.6164 | 43.4281 | 89.9999 | 0.0526 | 0.1199 |
| CLPM | $\mu_{elas}$ | $c_{1vsmc}$ | $c_{2vsmc}$ | $c_{1coll}$ | $c_{2coll}$ | $\alpha_{prox}$ | $\alpha_{dist}$ | | | | |
| | 0.0072 | 0.0061 | 0.7737 | 0.3536 | 0.5668 | 0.3012 | 0.1559 | – | – | 0.0862 | 0.1250 |
| MM | $\mu_{elas}$ | $c_{1vsmc}$ | $c_{2vsmc}$ | $c_{1coll}$ | $c_{2coll}$ | | | | | | |
| | 0.0527 | 0.0290 | 1.6454 | 0.0350 | 3.3930 | – | – | – | – | 0.7433 | 0.3739 |
| CLMM | $\mu_{elas}$ | $c_{1vsmc}$ | $c_{2vsmc}$ | $c_{1coll}$ | $c_{2coll}$ | $\alpha_{prox}$ | $\alpha_{dist}$ | | | | |
| | 0.0528 | 0.0260 | 1.6770 | 0.0846 | 0.3187 | 0.3150 | 0.7161 | – | – | 0.3605 | 0.2029 |

*Notes*: The quality of the model representation is characterized by $\varepsilon_p$ and $\varepsilon_d$ for proximal and distal curves, respectively. $\mu_{elas}$, $c_{1vsmc}$, and $c_{1coll}$ are in MPa, $\theta_{prox}$ and $\theta_{dist}$ in degrees, $c_{2vsmc}$, $c_{2coll}$, $\rho_{prox}$, $\rho_{dist}$, $\alpha_{prox}$, $\alpha_{dist}$, $\varepsilon_p$, and $\varepsilon_d$ dimensionless.

accurately the experimental data over the entire stress range. As the errors show, this model overestimated the circumferential stress and underestimated the longitudinal data for proximal samples. Also, it is worth noting that this model presents some similarities to the PM, but it fixes the orientation of the fibers and includes the dispersion of the collagen bundles. The CLMM predictions improve the fitting accuracy from the MM ($\varepsilon_p \sim 0.36$ and $\varepsilon_d \sim 0.20$) while phenomenological models (PM and CLPM) capture better the proximal behavior, and the microstructural approaches (MM and CMM) are reproduced better than the distal one. The material parameter values better fit to the stress-stretch curve around the medium stress zone 60 ÷ 140 [kPa] that corresponds to the estimated physiological stress state for an artery.

The mechanical response of arteries depends significantly on the tissue structure, in particular on the percentage of their microconstituents and the fiber distribution. One of the main goals of our previous work was to determine and model the passive mechanical properties of the swine carotid artery in both the proximal and distal regions [3, 27, 35], but only from a phenomenological perspective. However, until now, we did not include the information concerning the amount of microconstituents in the constitutive equations. This missing piece of the puzzle motivates the study of the volume fractions $\phi_{elas}$, $\phi_{vsmc}$, and $\phi_{coll}$ to introduce them into all these models along with the fiber orientation (in the CLPM, MM, and CLMM) and dispersion (in the MM and CLMM) retrieved by experimental analysis [3, 27].

Therefore, the final goal of this work was: (i) to investigate the capabilities of the phenomenological [34] and microstructural approaches [27, 35, 36], including experimental information of the collagen fiber distribution [3, 27] and the volume fractions of the arterial constituents; and (ii) to compare these models between each other. In this regard, we not only propose a new, highly accurate constitutive model for the artery, but also look at the big picture comparing the performance of different constitutive models in the literature.

The more accurate fitting was achieved by the phenomenological ($\varepsilon_p = 0.0526$ and $\varepsilon_d = 0.1199$) and the CLPM ($\varepsilon_p = 0.0862$ and $\varepsilon_d = 0.125$). The MM did not capture the proximal and distal behavior at the same time, producing the worst results ($\varepsilon_p = 0.74$) for the longitudinal direction of the proximal samples. The CLMM captured better the behavior on both directions and positions ($\varepsilon_p = 0.36$ and $\varepsilon_d = 0.20$), but was still far from the errors reached by the PM.

However, despite the error results, it is worth mentioning the fundamental fact of physically motivated results. In this regard, the PM predicted a fiber orientation of $\theta_{\text{prox}} = 43.42$ degrees and $\theta_{\text{dist}} = 89.99$ degrees. This fiber orientation does not match with the experimental observations in Sáez et al. [27], where collagen fibers were observed mainly along the circumferential direction with very low dispersion for both distal and proximal locations. Additionally, the measure of the fiber dispersion, $\rho$, is not in agreement with the highly concentrated distribution obtained in our experimental results. Including the experimental observations explicitly into the PM conducted to complete useless results since the longitudinal compliance was extremely low. This can be observed in the MM where this approach gives very high fitting errors. Although the model also includes information on collagen dispersion, the imposition of the angle of the collagen bundles invalidates the predictability features of this model. The CLPM overcomes the limitations observed in the PM by including the collagen cross-links that are attached among the main fibers. However, the collagen fibers did not gather information on the collagen fiber dispersion. The CLMM improved the fitting outcome at the low and large strain regimes, considering both the real distribution and orientation of the collagen fibers. It also predicts mechanical parameters that are in agreement with the compliance reported in the literature.

The PM, with a larger number of parameters, was unable to reproduce the mechanical tests when microstructural information of the tissue was included in the model, although the best results were obtained when the parameters were considered as free variables to fit. The constitutive models described in Sáez et al. [27, 35] were able to reproduce the mechanical test by physical motivated data, even after including microstructure information from experimental results. Concerning this, we conclude that the CLMM, even though it does not produce the best results in terms of residual errors, was the only model capable of appropriately reproducing the mechanical test comprising the actual orientation and distribution of the experimental results.

## 4.3 MECHANICAL CHARACTERIZATION AND MODELING OF THE AORTA

The prevalence of aortic disease in the worldwide population has led to much research into computational modeling of cardiovascular interventions. Noninvasive techniques and especially stenting for endovascular thoracic or abdominal aneurysm repair are increasingly relevant due to the numerous advantages they offer. Before the final human clinical trials, research has been conducted on animal models, with swine the most common due to the similarities between the human and swine cardiovascular systems.

### 4.3.1 Experimental Findings for the Porcine Aorta

To analyze the microstructure of the descending thoracic and abdominal aorta, we used two kinds of experimental data of the swine aortic tree obtained previously in our group [4, 5]. Porcine aortas ($n = 7$) were harvested postmortem from approximately $3.5 \pm 0.6$-month-old female pigs, sacrificed for other studies that did not interfere with the aorta or the circulation system and processed in a histological laboratory to analyze their layered microconstituents [5]. Finally, we also carried out several cyclic biaxial tension tests [4]. Details are illustrated in the following sections.

#### 4.3.1.1 Biaxial Mechanical Test

Concerning the mechanical behavior, biaxial tests of the DTA samples were developed in Peña et al. [4]. Porcine thoracic aortas ($n = 7$) were harvested postmortem and square specimens, approximately $25 \times 25$ [mm], with their sides aligned in the circumferential and axial directions were cut using a punch cutter and a scalpel. Tests were carried out in an Instron BioPuls low-force planar-biaxial testing system. Square specimens were mounted in the planar-biaxial machine by connecting four carriages by noddle clamps. Load-controlled tests were performed at peak tension ratios in circumferential and longitudinal directions ($P_c$:$P_l$ first Piola Kirchhoff stress tensor P) of 60:30, 30:60, and 60:60 kPa at stress rates of approximately 2 kPa/s [44]. Samples were preconditioned through 10 loading cycles for each stress ratio, and the last cycle (10th) was used for subsequent analysis [4]. For the deformation measurements during biaxial testing, several randomized markers were placed on the surface of the vessel and the lengths between the two markers in each direction were measured by a Digital Image Correlation Strain Master LaVision System. Shear strains were computed and were small; consequently, they were not accounted in the constitutive model.

Finally, Fig. 4.5 shows the mean and standard deviation of the equibiaxial 2:2 experimental data for whole samples in the circumferential (A) and longitudinal (B) directions. The standard deviation is lower for the circumferential direction than the longitudinal one. The aorta revealed nonlinear and anisotropic behavior with a stiffer behavior in the circumferential than in the longitudinal direction [4].

#### 4.3.1.2 Histology and Confocal Laser Scanning Microscopy Imaging

One part of each wall specimen from the DTA (approximately $0.5 \times 1.0$ cm) was used for microscopic investigation. Segments of each vessel were fixed in formaldehyde for 24 h and then moved to 70% alcohol. The segments were dehydrated and embedded in paraffin. The histology blocks were cross-sectioned at 5 μm and stained with hematoxylin-eosin in order to study the general structure of the sample (Fig. 4.6) [5].

Confocal imaging was performed at the CIBER Confocal Microscopy Service at the University of Alcalá. Scanning was performed on a Leica TCS-SP5 Confocal Laser Scanning Microscope. Elastin and collagen naturally autofluorescence, with slight overlaps in excitation and emission wavelengths. We used the identified optimal imaging regions for elastin and collagen reported by O'Connell et al. [40]; elastin excites maximally at 488 nm (20% laser power) and emits in the region of 500–550 nm (band-pass filter). Isolated collagen excites at 514 nm (30% laser power) with emission and reflection optimal in the region of 500–530 nm (band-pass filter) [40]. DAPI (4′,6-diamidino-2-phenylindole) was used to stain cell nuclei, allowing nuclei to be excited with the same 514-nm wavelength as collagen while the emission was recorded above 615 nm (low-pass filter). Specimens were prepared by longitudinally slicing open the posterior side of the cylindrical specimen and laying it flat between the microscope slide and coverslip. Cross-sections of $10 \div 20$ μm thickness were imaged with slices taken, each 0.37 μm. Fig. 4.7 shows a series of confocal laser scanning microscopy

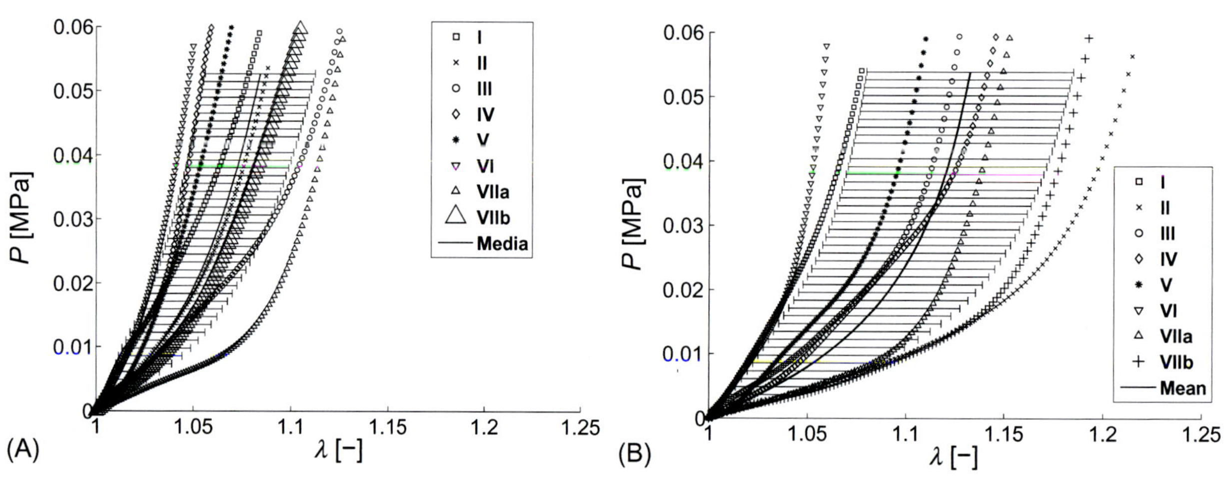


FIG. 4.5 Equibiaxial 2:2 experimental data samples in circumferential (A) and longitudinal (B) directions. *Adapted from J.A. Peña, V. Corral, M.A. Martínez, E. Peña, Over length quantification of the multiaxial mechanical properties of the ascending, descending and abdominal aorta using Digital Image Correlation, J. Mech. Behav. Biomed. 77 (2018) 434–445.*

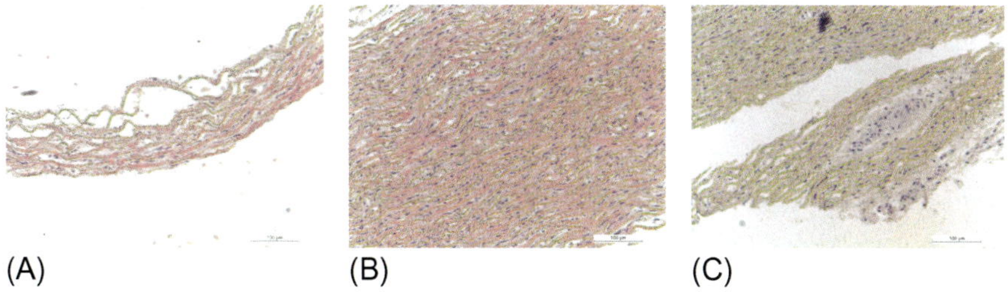

FIG. 4.6 Hematoxylin-and-eosin-stained sections 5 μm from DTA. (A) Intima; (B) media; (C) adventitia.

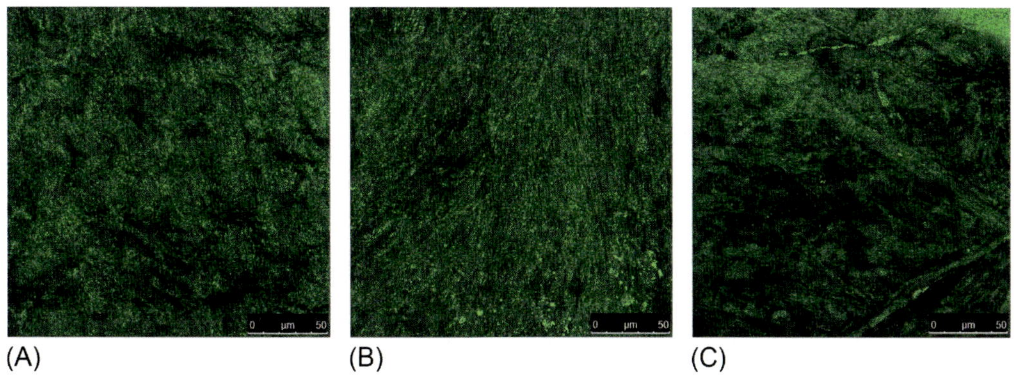

FIG. 4.7 Collagen fibers in the intima, the media, and the adventitia of the DTA. The longitudinal and circumferential directions are along horizontal and vertical directions and the artery radius is along the perpendicular one. Image extracted from an acquired image stack and corresponding to a thickness of 15 μm. (A) Intima DTA; (B) media DTA; (C) and adventitia DTA.

images that reveals that the collagen orientation distribution changed significantly across the wall thickness. In the intima, the collagen fiber shows a practically isotropic distribution. In the media, collagen fibers are dominantly aligned along the circumferential direction (vertical orientation in the images) and the dispersion of the orientation gradually increases toward the adventitia.

### 4.3.2 Material Models for the Porcine Aorta

Stress-stretch curves resulting from the tensile tests were used to fit several constitutive models. The fitted data were restricted to the elastic range, so data acquired after noticeable variations in the curve slope were neglected for the fitting procedure. Arterial tissue subject to loading exhibits strong nonlinearity, large strains, and anisotropy, so the SEF is used to reproduce the mechanical behavior of the aortas. In order to generate constitutive mechanical parameters that could be used on computational simulations, each of the biaxial specimens tested was fitted to four constitutive models proposed in the literature.

### 4.3.3 Phenomenological Model

The strong nonlinearity motivated the use of an exponential function for describing the strain energy stored in the collagen fibers that was previously proposed by Holzapfel et al. [9]. The main hypothesis was that each family of fibers represents the main direction of collagen bundles that are orientated in a helicoidal manner. Both families of fibers were assumed to have the same mechanical response and the anisotropy directions were assumed to be helically oriented at $\pm\theta$ degrees relative to circumferential direction [9]. Here, $\theta$ is treated as a phenomenological variable. The SEF proposed by the authors was

$$\Psi = \mu(I_1 - 3) + \sum_{i=4,6} \left\{ \frac{k_1}{2k_2} \exp\left(k_2[I_i - 1]^2\right) - 1 \right\}, \tag{4.17}$$

where

$$I_4 = \lambda_\theta^2 \cos^2\theta_1 + \lambda_z^2 \sin^2\theta_1, \quad I_6 = \lambda_\theta^2 \cos^2\theta_2 + \lambda_z^2 \sin^2\theta_2. \tag{4.18}$$

In this equation, $I_1$ represents the first invariant of the Cauchy-Green tensor [37] characterizing the isotropic mechanical response of the elastin [38, 39]. $\mu > 0$, $k_1 > 0$, and $k_3 > 0$ are stress-like parameters and $k_2 > 0$ and $k_4 > 0$ are dimensionless. The SEF represents the strain energy stored in a composite material reinforced in two preferred directions represented by the invariants $I_4 > 1$ and $I_6 > 1$, where it has been assumed that the strain energy corresponding to the anisotropic terms only contributes to the global mechanical response of the tissue when stretched. A total of four elastic parameters ($\mu$, $k_1$, $k_2$, $\theta$) should be fitted.

## 4.3.4 Structural Model

The Gasser, Ogden, and Holzapfel (GOH) model [8] extended the model of Holzapfel et al. [9] by the application of generalized structure tensor $\mathrm{H} = \kappa 1 + (1 - 3\kappa)\mathrm{M}_0$ (where 1 is the identity tensor and $\mathrm{M}_0 = \mathrm{m}_0 \otimes \mathrm{m}_0$ is a structure tensor defined using unit vector $\mathrm{m}_0$ specifying the mean orientation of fibers) and proposed a new constitutive model

$$\Psi = \mu(I_1 - 3) + \sum_{i=4,6} \left[ \frac{k_1}{2k_2} (\exp\{k_2 \hat{E}_i\} - 1) \right], \tag{4.19}$$

where

$$\hat{E}_i = \kappa I_1 + (1 - 3\kappa) I_i - 1 \quad i = 4,6 \tag{4.20}$$

and $\kappa \in [0, 1/3]$ is a dispersion parameter (the same for each collagen fiber family); when $\kappa = 0$, the model is the same as the one published in Holzapfel et al. [9], and when $\kappa = 1/3$, it recovers an isotropic potential similar to that used in Demiray [45]. Note that the parameter $\kappa = \frac{1}{4}\int_0^\pi \rho \sin^3\theta d\theta$ could have histological meaning due to the fully characterized distribution [8]. A total of five elastic parameters ($\mu$, $k_1$, $k_2$, $\kappa$, $\theta$) should be fitted.

## 4.3.5 Microfiber Model

As commented on in the carotid section, Alastrué et al. [36] proposed a microfiber model (microsphere-based model) to account for the dispersion of the collagen fibers around a preferential direction, overcoming the 1D limitation of previous characterizations of the collagen fiber. Consistent with the constrained mixture approach [31]

$$\Psi = \mu(I_1 - 3) + \Psi_{\mathrm{coll}}, \tag{4.21}$$

where the subscript coll refers to collagen fiber contribution. $\Psi_{\mathrm{coll}}$ is defined as the sum of the contributions of each collagen family of fibrils as

$$\Psi_{\mathrm{coll}} = \sum_{j=1}^{N} [\Psi_{\mathrm{coll}}]^j = \sum_{j=1}^{N} \langle n\rho\psi_{\mathrm{coll}} \rangle^j = \sum_{j=1}^{N} \frac{1}{4\pi} \int_{\mathbb{U}^2} (n\rho[\psi_{\mathrm{coll}}])^j dA, \tag{4.22}$$

where $N$ denotes the number of families of collagen fibers, $N = 2$ according to the experimental results of the orientation of collagen fibers [9], and applying a discretization to the continuous orientation distribution on the unit sphere $\mathbb{U}^2$, $[\Psi_{\mathrm{coll}}]^j$ corresponds to the expression

$$[\Psi_{\mathrm{coll}}]^j = \sum_{i=1}^{m} n\rho(\mathbf{r}^i)\psi^i_{\mathrm{coll}}(\lambda^i_{\mathrm{coll}}), \tag{4.23}$$

where $\mathbf{r}^i$ are the unit vectors associated with the discretization on the microsphere over the unit sphere $\mathbb{U}^2$, $m$ is the number of discrete orientation vectors [7], $\lambda^i_{\mathrm{coll}} = \|\mathrm{F} \cdot \mathrm{r}^i\|$ the stretch in $\mathrm{r}^i$ direction, and $\psi^i_{\mathrm{coll}}(\lambda^i_{\mathrm{coll}})$ the SEF associated with $\mathrm{r}^i$ direction. Using Eqs. (4.22), (4.23), this results in

$$\Psi_{\mathrm{coll}} = \sum_{j=1}^{N} \sum_{i=1}^{m} (w^i n\rho[\psi^i_{\mathrm{coll}}])^j, \tag{4.24}$$

where $w^i_{i=1,\ldots,m}$ denotes related weighting factors and $\rho$ is the ODF to take into account the fibril dispersion [7].

The exponential-like SEF proposed by Holzapfel et al. [9] was used to deal with the fiber response

$$n\psi^i_{\mathrm{coll}}(\lambda^i_{\mathrm{coll}}) = \frac{c_{1\mathrm{coll}}}{2c_{2\mathrm{coll}}} \left( e^{c_{2\mathrm{coll}}((\lambda^i_{\mathrm{coll}})^2 - 1)^2} - 1 \right) \quad \text{if } \lambda_i \geq 1 \text{ otherwise } \psi_{\mathrm{f}\,i}(\lambda_i) = 0, \tag{4.25}$$

because it is usually considered that collagen fibers only affect global mechanical behavior in tensile states [9]. The affine kinematics define the collagen fiber stretch $\lambda^i_{\mathrm{coll}} = \|\mathrm{t}^i\|$ in the fiber direction $\mathrm{r}^i$.

Two ODFs were used to model for the incorporation of anisotropy

- One of the ODFs applied most frequently is 3D bi-$\pi$-periodic von Mises ODF for the incorporation of anisotropy in a microsphere-based model with application to the modeling of the thoracic aorta [7]. This function is expressed as

$$\rho(\theta) = \rho_1(\theta) + \rho_2(\theta), \tag{4.26}$$

where $\theta = \arccos(\mathbf{m} \cdot \mathbf{r})$ is the so-called mismatch angle and m the preferred mean orientation of the collagen distribution, and

$$\rho_i(\theta) = 4\sqrt{\frac{b}{2\pi}} \frac{\exp(b[\cos(2\theta) + 1])}{\mathrm{erfi}(\sqrt{2b})}, \tag{4.27}$$

where the positive concentration parameter $b$ constitutes a measure of the degree of anisotropy. Moreover, erfi($x$) = $-i$ erf($x$) denotes the imaginary error function. Finally, $c_{1\mathrm{coll}}$ and $c_{2\mathrm{coll}}$ are stress dimensional and dimensionless material parameters, respectively. A total of five elastic parameters ($\mu$, $k_1$, $k_2$, $\kappa$, and $\theta$) should be fitted.
- We also used the Bingham ODF [41] initially proposed by Alastrué et al. [36] for the incorporation of anisotropy in a microsphere-based model with application to the modeling of the thoracic aorta and presented in Eq. (4.10).

### 4.3.6 Results on Modeling the Porcine Carotid Artery

The results of the fitting to the SEFs are shown in Table 4.3. Our results on the descriptive capacity of SEF models indicated that the worst fitting was with the HGO SEF, showing a mean RMSE of $\varepsilon = 0.2668$. On the contrary, the best

TABLE 4.3 Material Constants Obtained for the DTA Curves

| HGO model | | | | | | |
|---|---|---|---|---|---|---|
| **Specimen** | $\mu$ | $k_1$ | $k_2$ | $\theta$ | $R^2$ | $\varepsilon$ |
| I | 0.0531 | 0.0051 | 16.4048 | 63.31 | 0.9382 | 0.1458 |
| II | 0.001 | 0.0175 | 1.7351 | 80.15 | 0.6592 | 0.3459 |
| III | 0.0145 | 0.0117 | 3.0238 | 78.11 | 0.8329 | 0.2626 |
| IV | 0.0314 | 0.01363 | 8.5495 | 72.82 | 0.7947 | 0.3048 |
| V | 0.0263 | 0.0134 | 11.2069 | 79.88 | 0.8750 | 0.2486 |
| VI | 0.0435 | 0.0037 | 93.9559 | 57.43 | 0.8295 | 0.2316 |
| VIIa | 0.0129 | 0.0038 | 6.7227 | 67.58 | 0.6670 | 0.4269 |
| VIIb | 0.0010 | 0.0150 | 2.7861 | 79.14 | 0.2827 | 0.1686 |
| Mean | 0.0229 | 0.0104 | 18.0481 | 72.3025 | 0.7349 | 0.2668 |
| SD | 0.01908 | 0.0054 | 31.0638 | 8.6528 | 0.2063 | 0.0919 |

| GOH model | | | | | | | |
|---|---|---|---|---|---|---|---|
| **Specimen** | $\mu$ | $k_1$ | $k_2$ | $\kappa$ | $\theta$ | $R^2$ | $\varepsilon$ |
| I | 0.0262 | 0.0117 | 26.06 | 0.1125 | 59.83 | 0.9415 | 0.1422 |
| II | 0.0054 | 0.0654 | 7.6320 | 0.2726 | 20 | 0.6858 | 0.3431 |
| III | 0.0078 | 0.0947 | 4.6290 | 0.2885 | 1.5 | 0.8680 | 0.1701 |
| IV | 0.0146 | 0.1906 | 11.1502 | 0.2848 | 17.69 | 0.8200 | 0.2875 |
| V | 0.001 | 0.3552 | 0.0014 | 0.2742 | 74 | 0.8840 | 0.2256 |
| VI | 0.0210 | 0.0862 | 660.0371 | 0.2712 | 0.0 | 0.8494 | 0.2139 |
| VIIa | 0.0064 | 0.0258 | 18.9642 | 0.2531 | 0.0 | 0.6668 | 0.4268 |
| VIIb | 0.0025 | 0.1035 | 6.7001 | 0.2895 | 0.0 | 0.8558 | 0.2663 |
| Mean | 0.0106 | 0.1166 | 91.8967 | 0.2558 | 21.6275 | 0.8214 | 0.2594 |
| SD | 0.0090 | 0.1107 | 229.7130 | 0.0590 | 29.3472 | 0.0961 | 0.0931 |

*Continued*

TABLE 4.3 Material Constants Obtained for the DTA Curves—cont'd

| Microfiber von Mises model | | | | | | | |
|---|---|---|---|---|---|---|---|
| **Specimen** | $\mu$ | $k_1$ | $k_2$ | $b$ | $\theta$ | $R^2$ | $\varepsilon$ |
| I | 0.0021 | 0.1742 | 0.0010 | 0.2155 | 48.31 | 0.9362 | 0.0995 |
| II | 0.0015 | 0.0511 | 0.0016 | 1.2059 | 42.66 | 0.7827 | 0.3385 |
| III | 0.0016 | 0.0630 | 0.5198 | 0.5402 | 12.60 | 0.9670 | 0.1305 |
| IV | 0.0016 | 0.1224 | 0.8446 | 0.6631 | 11.34 | 0.9048 | 0.2371 |
| V | 0.0010 | 0.1033 | 3.0416 | 0.8904 | 41.82 | 0.9516 | 0.1806 |
| VI | 0.0028 | 0.1318 | 13.0384 | 0.6237 | 38.01 | 0.96281 | 0.1192 |
| VIIa | 0.0037 | 0.0426 | 0.0012 | 0.9761 | 40.56 | 0.6912 | 0.4821 |
| VIIb | 0.0010 | 0.0565 | 0.0011 | 0.8537 | 12.83 | 0.8936 | 0.2918 |
| Mean | 0.0019 | 0.0931 | 2.1811 | 0.74607 | 33.6471 | 0.8862 | 0.2349 |
| SD | 0.0009 | 0.0472 | 4.5067 | 0.3029 | 15.0599 | 0.0987 | 0.1312 |

| Microfiber Bingham model | | | | | | | |
|---|---|---|---|---|---|---|---|
| **Specimen** | $\mu$ | $k_1$ | $k_2$ | $\kappa_1$ | $\kappa_2$ | $R^2$ | $\varepsilon$ |
| I | 0.0342 | 0.0604 | 7.2772 | 0.5158 | 0.0005 | 0.9821 | 0.0812 |
| II | 0.0010 | 0.0288 | 1.2392 | 8.7994 | 6.6073 | 0.8435 | 0.1734 |
| III | 0.001 | 0.0503 | 0.5371 | 2.4282 | 1.4002 | 0.9696 | 0.1308 |
| IV | 0.0011 | 0.1153 | 0.8676 | 1.6350 | 0.3303 | 0.9013 | 0.2220 |
| V | 0.001 | 0.1129 | 2.8115 | 1.2928 | 0.0 | 0.9508 | 0.1806 |
| VI | 0.0010 | 0.1365 | 11.6631 | 5.4277 | 4.5481 | 0.9568 | 0.1230 |
| VIIa | 0.001 | 0.0361 | 1.7047 | 1.9521 | 0.0 | 0.7633 | 0.3207 |
| VIIb | 0.0011 | 0.0198 | 2.2892 | 14.9106 | 12.7635 | 0.9503 | 0.1823 |
| Mean | 0.0051 | 0.0700 | 3.5487 | 4.6202 | 3.2062 | 0.9147 | 0.1767 |
| SD | 0.0117 | 0.0449 | 3.9072 | 4.9679 | 4.5837 | 0.0756 | 0.0726 |

*Notes:* Constants $\mu$ and $k_1$ in MPa, $\theta$ in degrees, $k_2$, $\rho$, $\kappa$, $b$, $\kappa_1$, and $\kappa_2$ are dimensionless.

*Source: Adapted from J.A. Peña, V. Corral, M.A. Martínez, E. Peña, Over length quantification of the multiaxial mechanical properties of the ascending, descending and abdominal aorta using Digital Image Correlation, J. Mech. Behav. Biomed. 77 (2018) 434–445.*

fitting was with the microstructured model with the Bingham ODF showing a mean RMSE of $\varepsilon = 0.1767$. The RMSE of the GOH model and the microstructured model with the von Mises ODF function were similar. Regarding the predictive capacity of the material models, the fitted material constants using only the equibiaxial test (2:2) demonstrated a good predictive result for the biaxial tests (2:1 and 1:2), data not shown (see Peña et al. [4]), with a "predictive" error, $\varepsilon_{error} < 10\%$ for the Bingham microstructured model only.

However, despite the error results, it is worth mentioning the fundamental fact of physically motivated results. The PM [9] predicted a mean fiber orientation of $\theta_{DTA} = 72.3025$ degrees without dispersion. This fiber orientation does not match the experimental observations in Schriefl et al. [46], where collagen fibers were observed mainly along $\theta \approx 45$ degrees of circumferential direction with high dispersion for both descending aortas. Furthermore, with regard to the measure of the fiber dispersion, $\kappa \approx 0.3$ for the GOH model and $b \approx 0$, the microstructured model with the von Mises ODF function and mean fiber orientation is in keeping with the dispersed distribution obtained in the experimental results of Schriefl et al. [46]. In accordance with Schriefl et al. [46], the MM with the Bingham ODF showed $\kappa_1 - \kappa_2 \approx 0$, meaning high dispersion around the circumferential direction.

## 4.4 CONCLUSIONS

It is well known that vascular tissues are subject to finite deformations and that their mechanical behavior is highly nonlinear, anisotropic, and essentially incompressible with nonzero residual stress. The nonphysiological domain presents viscous and damage behavior and with a significant dispersion in the orientation, which has a significant influence on the mechanical response. The high complexity of biological tissues requires mechanical models that include information about the underlying constituents and that look for the physics of the whole processes within the material. This behavior of the microconstituents can be taken into macroscopic models by means of computational homogenization. It is in this context where the microsphere-based approach acquires high relevance. Regarding the parameter estimation analysis, the larger the number of parameters, the more flexible and the better fitting (i.e., concerning the residual error) is reached, as could be expected. However, too many parameters not only increase the complexity of the model [47], but also increment the disadvantages of ill-posed problems. In this regard, we agree that the main goal in constitutive models should be to include physically motivated aspects and, as much as possible, to feed these models with experimental data obtained from histological analysis, polarized light microscopy [46], or other quantitative experimental techniques [48].

## Acknowledgments

The authors gratefully acknowledge research support from the Spanish Ministry of Science and Technology through research project DPI2016-76630-C2-1-R and CIBER initiative. Part of the work was performed by the ICTS "NANBIOSIS" specifically by the Tissue and Scaffold Characterization Unit (U13) and High Performance Computing Unit (U27), of the CIBER in Bioengineering, Biomaterials and Nanomedicine (CIBER-BBN at the University of Zaragoza).

## References

[1] J.D. Humphrey, Continuum biomechanics of soft biological tissues, Proc. R. Soc. Lond. A 175 (2002) 1–44.
[2] Y.C. Fung, Biomechanics: Mechanical Properties of Living Tissues, Springer-Verlag, New York, NY, 1993.
[3] A. García, E. Peña, A. Laborda, F. Lostalé, M.A de Gregorio, M. Doblaré, M.A. Martínez, Experimental study and constitutive modelling of the passive mechanical properties of the porcine carotid artery and its relation to histological analysis. Implications in animal cardiovascular device trials, Med. Eng. Phys. 33 (2011) 665–676.
[4] J.A. Peña, V. Corral, M.A. Martínez, E. Peña, Over length quantification of the multiaxial mechanical properties of the ascending, descending and abdominal aorta using Digital Image Correlation, J. Mech. Behav. Biomed. 77 (2018) 434–445.
[5] J.A. Peña, M.A. Martínez, E. Peña, Layer-specific residual deformations and uniaxial and biaxial mechanical properties of thoracic porcine aorta, J. Mech. Behav. Biomed. 50 (2015) 55–69.
[6] D.P. Sokolis, S. Sassani, E.P. Kritharis, S. Tsangaris, Differential histomechanical response of carotid artery in relation to species and region: mathematical description accounting for elastin and collagen anisotropy, Med. Biol. Eng. Comput. 49 (2011) 867–879.
[7] V. Alastrué, M.A. Martínez, M. Doblaré, A. Menzel, Anisotropic micro-sphere-based finite elasticity applied to blood vessel modelling, J. Mech. Phys. Solids 57 (2009) 178–203.
[8] T.C. Gasser, R.W. Ogden, G.A. Holzapfel, Hyperelastic modelling of arterial layers with distributed collagen fiber orientations, J. R. Soc. Interface 3 (2006) 15–35.
[9] G.A. Holzapfel, T.C. Gasser, R.W. Ogden, A new constitutive framework for arterial wall mechanics and a comparative study of material models, J. Elasticity 61 (2000) 1–48.
[10] G.A. Holzapfel, R.W. Ogden, Constitutive modelling of arteries, Phil. Trans. R. Soc. A 466 (2010) 1551–1597.
[11] D.P. Sokolis, A passive strain-energy function for elastic and muscular arteries: correlation of material parameters with histological data, Med. Biol. Eng. Comput. 48 (2010) 507–518.
[12] M. Zullinger, A. Rachev, N. Stergiopulos, A constitutive formulation of arterial mechanics including vascular smooth muscle tone, Am. J. Physiol. Heart Circ. Physiol. 287 (2004) H1335–H1343.
[13] S. Polzer, T.C. Gasser, K. Novak, V. Mana, M. Tichy, P. Skacel, J. Bursa, Structure-based constitutive model can accurately predict planar biaxial properties of aortic wall tissue, Acta Biomater. 14 (2015) 133–145.
[14] H. Weisbecker, M.J. Unterberger, G.A. Holzapfel, Constitutive modelling of arteries considering fiber recruitment and three-dimensional fiber distribution, J. R. Soc. Interface 12 (2015) 20150111.
[15] J.D. Humphrey, Mechanics of the arterial wall: review and directions, Crit. Rev. Biomed. Eng. 23 (1995) 1–162.
[16] Y. Lanir, A structural theory for the homogeneous biaxial stress-strain relationship in flat collageneous tissues, J. Biomech. 12 (1979) 423–436.
[17] Y.C. Fung, K. Fronek, P. Patitucci, Pseudoelasticity of arteries and the choice of its mathematical expression, Am. J. Physiol. 237 (1979) H620–H631.
[18] K. Takamizawa, K. Hayashi, Strain-energy density-function and uniform strain hypothesis for arterial mechanics, J. Biomech. 20 (1987) 7–17.
[19] C.N. van den Broek, A. van der Horst, M.C.M. Rutten, F.N. van de Vosse, A generic constitutive model for the passive porcine coronary artery, Biomech. Mod. Mechanobiol. 10 (2011) 249–258.
[20] M. Zullinger, P. Fridez, K. Hayashi, N. Stergiopulos, A strain energy function for arteries accounting for wall composition and structure, J. Biomech. 37 (2004) 989–1000.
[21] P.B. Dobrin, T.R. Canfield, Identification of smooth muscle series elastic component in intact carotid artery, Am. J. Physiol. Heart Circ. Physiol. 232 (1977). HI22–HI30.
[22] P.B. Dobrin, T.R. Canfield, Elastase, collagenase, and the biaxial elastic properties of dog carotid artery, Am. J. Physiol. Heart Circ. Physiol. 247 (1984) H124–H131.
[23] G.V. Guinea, J.M. Atienza, M. Elices, P. Aragoncillo, K. Hayashi, Thermomechanical behavior of human carotid arteries in the passive state, Am. J. Physiol. Heart Circ. Physiol. 288 (2005) H2940–H2945.

[24] G. Sommer, P. Regitnig, L. Költringer, G.A. Holzapfel, Biaxial mechanical properties of intact and layer-dissected human carotid arteries at physiological and supra-physiological loadings, Am. J. Physiol. Heart Circ. Physiol. 298 (2010) H898–H912.
[25] F.H. Silver, P.B. Snowhill, D.J. Foran, Mechanical behavior of vessel wall: a comparative study of aorta, vena cava, and carotid artery, Ann. Biomed. Eng. 31 (2003) 793–803.
[26] D.M. Ebenstein, D. Coughlin, J. Chapman, C. Li, L.A. Pruitt, Nanomechanical properties of calcification, fibrous tissue, and hematoma from atherosclerotic plaques, J. Biomed. Mater. Res. A 91 (2009) 1028–1037.
[27] P. Sáez, A. García, E. Peña, T.C. Gasser, M.A. Martínez, Microstructural quantification of collagen fiber orientations and its integration in constitutive modeling of the porcine carotid artery, Acta Biomater. 33 (2016) 183–193.
[28] P.B. Canham, H.M. Finlay, D.R. Boughner, Contrasting structure of the saphenous vein and internal mammary artery used as coronary bypass vessels, Cardiovasc. Res. 34 (1997) 557–567.
[29] P.B. Canham, H.M. Finlay, J.G. Dixon, D.R. Boughner, A. Chen, Measurements from light and polarised light microscopy of human coronary arteries fixed at distending pressure, Cardiovasc. Res. 23 (1989) 973–982.
[30] J.F. Smith, P.B. Canham, J. Starkey, Orientation of collagen in the tunica adventitia of the human cerebral artery measured with polarized light and the universal stage, J. Ultrastruct. Res. 77 (1981) 133–145.
[31] J.D. Humphrey, K.R. Rajagopal, A constrained mixture model for arterial adaptations to a sustained step change in blood flow, Biomech. Model Mechanobiol. 2 (2003) 109–126.
[32] C. Bellin, J. Ferruzzi, S. Roccabianca, E.S. Di Martino, J.D. Humphrey, A microstructurally motivated model of arterial wall mechanics with mechanobiological implications, Ann. Biomed. Eng. 42 (2013) 488–502.
[33] G.A. Holzapfel, T.C. Gasser, M. Stadler, A structural model for the viscoelastic behaviour of arterial walls: continuum formulation and finite element analysis, Eur. J. Mech. A/Solids 21 (2002) 441–463.
[34] G.A. Holzapfel, C.T. Gasser, G. Sommer, P. Regitnig, Determination of the layer-specific mechanical properties of human coronary arteries with non-atherosclerotic intimal thickening, and related constitutive modelling, Am. J. Physiol. Heart. Circ. Physiol. 289 (2005) H2048–H2058.
[35] P. Sáez, E. Peña, M.A. Martínez, A structural approach including the behavior of collagen cross-links to model patient-specific human carotid arteries, Ann. Biomed. Eng. 42 (2014) 1158–1169.
[36] V. Alastrué, P. Saez, M.A. Martínez, M. Doblaré, On the use of Bingham statistical distribution in microsphere-based constitutive models for arterial tissue, Mech. Res. Commun. 37 (2010) 700–706.
[37] A.J.M. Spencer, Theory of invariants, in: Continuum Physics, Academic Press, New York, NY, 1971, pp. 239–253.
[38] N. Gundiah, M.B. Ratcliffe, L.A. Pruitt, The biomechanics of arterial elastin, J. Mech. Behav. Biomed. Mater. 2 (2009) 288–296.
[39] M.A. Lillie, R.E. Shadwick, J.M. Gosline, Mechanical anisotropy of inflated elastic tissue from the pig aorta, J. Biomech. 43 (2010) 2070–2078.
[40] M.K. O'Connell, S. Murthy, S. Phan, C. Xu, J. Buchanan, R. Spilker, R.L. Dalman, C.K. Zarins, W. Denk, C.A. Taylor, The three-dimensional micro- and nanostructure of the aortic medial lamellar unit measured using 3D confocal and electron microscopy imaging, Matrix Biol. 27 (2008) 171–181.
[41] C. Bingham, An antipodally symmetric distribution on the sphere, Ann. Stat. 2 (1974) 1201–1225.
[42] D.W. Marquardt, An algorithm for least-squares estimation of nonlinear parameters, SIAM J. Appl. Math. 11 (1963) 431–441.
[43] T.E. Carew, R.N. Vaishnav, D.J. Patel, Compressibility of the arterial wall, Circ. Res. 23 (1968) 61–86.
[44] J. Tong, T. Cohnert, P. Regitnig, G.A. Holzapfel, Effects of age on the elastic properties of the intraluminal thrombus and the thrombus-covered wall in abdominal aortic aneurysms: biaxial extension behaviour and material modelling, Eur. J. Vasc. Endovasc. Surg. 42 (2011) 207–219.
[45] H. Demiray, A note on the elasticity of soft biological tissues, J. Biomech. 5 (1972) 309–311.
[46] A.J. Schriefl, G. Zeindlinger, D.M. Pierce, P. Regitnig, G.A. Holzapfel, Determination of the layer-specific distributed collagen fiber orientations in human thoracic and abdominal aortas and common iliac arteries, J. R. Soc. Interface 9 (2012) 1275–1286.
[47] S. Zeinali-Davarani, J. Choi, S. Baek, On parameter estimation for biaxial mechanical behavior of arteries, J. Biomech. 42 (2009) 524–530.
[48] R. Rezakhaniha, A. Agianniotis, J.T.C. Schrauwen, A. Griffa, D. Sage, C.V.C. Bouten, F.N. van de Vosse, M. Unser, N. Stergiopulos, Experimental investigation of collagen waviness and orientation in the arterial adventitia using confocal laser scanning microscopy, Biomech. Model. Mechanobiol. 11 (2012) 461–473.

CHAPTER

# 5

# Impact of the Fluid-Structure Interaction Modeling on the Human Vessel Hemodynamics

*Mauro Malvè*[*,†,‡], *Myriam Cilla*[§], *Estefanía Peña*[†,‡,¶], *Miguel Angel Martínez*[†,‡,¶]

*Department of Engineering, Public University of Navarra, Pamplona, Spain †Centro de Investigación en Red en Bioingeniería, Biomaterialesy Nanomedicina, CIBER-BBN, Zaragoza, Spain ‡Aragón Institute of Engineering Research (I3A), University of Zaragoza, Zaragoza, Spain §Centro Universitario de la Defensa, Academia General Militar, Zaragoza, Spain ¶Department of Mechanical Engineering, University of Zaragoza, Zaragoza, Spain

## 5.1 CLINICAL BACKGROUND

It is estimated that approximately 60% of all human deaths are caused by cardiovascular disorders [1]. The most relevant disease in this sense is atherosclerosis, due to the increasing number of affected persons [1]. Atherosclerosis is a pathological obstruction of blood vessels consisting of the formation, growth, and development of a plaque that is deposited from different origins. The narrowing caused by this progressive deposit on the artery limits the oxygen-rich blood flow to the heart and other vessels, promoting heart attack and stroke among other pathologies and leading even to death. The lesion begins with some accumulation in the intima layer and progresses, forming a fatty cap that can eventually become calcified. There is much evidence that atherosclerosis is nonuniformly distributed in the human body. On the contrary, it occurs at certain specific locations of the cardiovascular system, especially the arterial bifurcations [2, 3].

It is known that many factors affecting the whole body such as smoking, high cholesterol, hypertension, and lifestyle, among others, promote cardiovascular diseases. In recent years, it has been suggested that perturbed blood flow and/or abnormal stresses and strains of the artery may play a considerable role in atherogenesis. Extensive numerical studies centered on the modeling of human and animal hemodynamics have shown that bifurcations are peculiar geometrical regions submitted to highly perturbed flows. The latter promotes complex flow patterns characterized by unsteadiness, high local Reynolds numbers, flow recirculation with a variety of flow structures, and local peaks of high and low endothelial shear stress. In some arteries such as the aorta, blood flow may even become locally turbulent. In this context, it is clear that near the bifurcation, the endothelium is loaded with nonuniform forces generated by the complexity of the aforementioned flow patterns. The smooth muscle cells of the artery are certainly influenced and stimulated by the blood flow acting on the vessel walls by means of the shear stress. For this reason, many investigations into atherosclerosis have been focused on searching for a surrogate marker for atherogenesis in the computational hemodynamics. The most accepted theory in respect to this is that the oscillating or low average shear stress is responsible for the appearance of the plaque. However, there are a few studies that propose high shear stress as an important factor for atherogenesis, and there is no agreement in this sense. Due to the intervariability of the arterial morphology among patients, perturbed flow and nonuniform wall shear stress (WSS) have often been correlated with geometrical factors that may considerably change among the different parts of the blood vessels [4–6].

Atherogenesis has been associated with arterial bifurcation, also considering the vessel walls. Geometrically speaking, arteries are compliant cylindrical tubes that at the bifurcation, due to the changes promoted by stretch and dilatation, may show important changes and complex morphologies. An increase or decrease of curvature as well as changes of diameter and of shape and variations of wall thickness are shown to promote the concentration of stresses

*Advances in Biomechanics and Tissue Regeneration*
https://doi.org/10.1016/B978-0-12-816390-0.00005-4

and stretches. Due to the pulsatile character of blood pressure, these regions of concentration show oscillating peaks where the highest values are usually located at the bulb of the bifurcation. As for hemodynamics, the structural problem has been usually analyzed by means of computational methods that allow studying a considerable number of geometries, situations, and conditions, yet resulting in the possibility of performing statistical analyses. Fluid dynamics and structural mechanics perspectives are not in conflict [7], as this is about changing the main focus of analysis from fluid to solid variables. Traditionally, these two fields have been often treated separately. This is mainly due to the high complexity problem that involves unsteady, pulsatile, turbulent, and non-Newtonian flow with anisotropic, nonlinear, hyperelastic, and fiber-reinforced vascular tissue. Both fluid and solid parts may even change their properties due to large-scale modifications such as cardiovascular diseases. Blood flow may show a local increase in peak velocity, an accentuation of recirculation, and a change of the resultant endothelial shear stresses. Vessel wall properties may vary in case of lipid accumulation and partial occlusion of the artery such as in the case of atherosclerosis or a loosening of elasticity and wall thickening in the case of an aneurysm. In any case, there is a mutual interaction between blood flow and compliant vessels because the cardiovascular flow exerts blood pressure on the walls. The latter is accumulated as potential energy and transferred to the blood flow as kinetic energy. Under this perspective, it is a fluid-structure interaction (FSI) problem [8]. For these reasons, a considerable number of studies have been focused on the analysis of coupled fluid-solid problems with application to the cardiovascular field. Among others, the aorta and the carotid artery have been frequently considered in healthy and diseased conditions [9–19]. These works have been focused on the aorta and on the carotid artery due to their intrinsic tendency to develop cardiovascular diseases such as atherosclerosis. Commercial and in-house software have been used on idealized and patient-specific data with the aim of quantifying physical variables not evaluable in vivo. Instantaneous, average, and oscillatory endothelial stresses are the variables mostly computed, correlated with structural variables and geometrical factors and used as a marker for the considered pathology.

In this chapter, we present two FSI models based on medical images. The aorta and the carotid artery have been analyzed, including the most important flow features, with the aim of showing the impact of the distensible walls on the fluid dynamics variables. A large number of computational fluid dynamics (CFD) studies have been proposed for hemodynamics evaluations. However, these works have the intrinsic limitation that the vessel is considered rigid. Structural models used with computational solid mechanics (CSM) have been demonstrated to be useful for quantifying and localizing peaks of stresses and strains. The latter completely neglects the effect of the blood flow that is considered just a boundary condition through the blood pressure. The main differences that can be found using the FSI approximation in the cardiovascular field are related to the amplitude and the locations of WSS as well as its intensity. While the provided results are more accurately obtained with respect to the CFD or CSM computations, the main limitation of the FSI approach is the drastic increase in computational costs that limit its applicability to clinical daily practice, contrary to the CFD approach.

## 5.2 FINITE ELEMENT MODELING OF THE HUMAN BLOOD VESSELS

### 5.2.1 Image-Based Geometrical Reconstruction

Medical images are usually required for building the vessel geometry that corresponds to the computational domain of the numerical model. Due to the high intervariability of the different parts of the blood vessels among subjects, it is preferable to use patient-specific geometries for numerical studies. Another approach is to use average data coming from various patients and provide a parametric model in which geometrical variations can be imposed. In all cases, the vessel lumen (fluid domain) and the corresponding wall thickness (solid domain) are generated. Different techniques are currently available. In the present study, for all the performed reconstructions, we have used computerized tomography (CT) images that allow acquiring three-dimensional (3D) images with a spatial resolution of less than 1 mm. The standard format for storing, transmitting, and handling medical imaging is the DICOM format (Digital Imaging and Communication in Medicine). The images containing the thoracic slices of two different patients were imported into the commercial software MIMICS (Materialise Software, Leuven, Belgium). Here, a manual segmentation was performed with the aim of extracting the vessel lumen. For this scope, the black cavity that represents the lumen was manually filled in each image. As a result, a stereolithography (STL) file with the 3D model and an Initial Graphics Exchange Specification (IGES) file of the cross-sectional slices belonging to each geometry of the two handled patients was exported. Prior to this, the obtained data were smoothed for reducing the unavoidable noise included in the acquisition of the images. Both formats can be easily imported and treated in commercial computer-aided design (CAD) software. Here, the STL file was only used as a reference for the 3D models that have been created

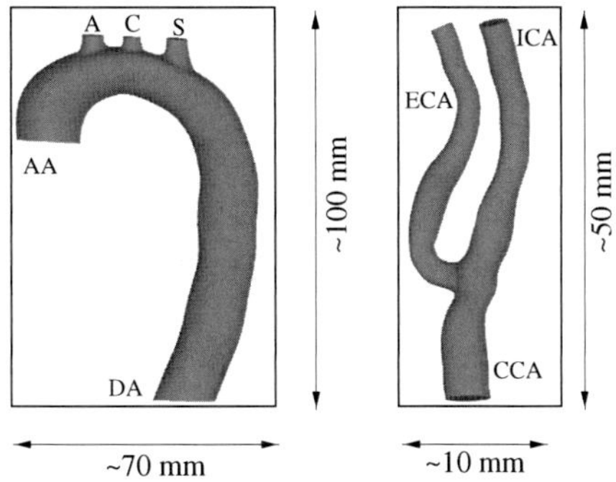


FIG. 5.1 Geometry of the considered models along with their main dimensions: aorta (*left panel*) and carotid artery (*right panel*).

generating a loft from the existing splines. The latter has been carried out using the commercial CAD software Rhinoceros (Robert McNeel & Associates, Seattle, WA). During this step, the geometries have been cropped to allow smoothed and planar surfaces where, in a second step, the boundary conditions have been imposed. Later on, the obtained 3D models of the aorta and the carotid were exported again as IGES files. The geometrical reconstruction was only performed with the aim of generating the fluid domain. For the vessel wall thickness, in fact, no data were available. For this reason, a constant value of 1.5 mm was assigned to the aorta [20, 21]. The carotid wall thickness was also found in the literature and a constant value of 0.6 mm was given [22]. In Fig. 5.1, the models are represented along with their main dimensions. The aorta includes the ascending (AA) tract, the anonyma or brachiocephalic trunk (A), the carotid (C) and the subclavian (S) outlets, and the descending trunk (DA). The carotid artery includes the main bifurcation: the common carotid artery (CCA) divides into the external carotid artery (ECA) and internal carotid artery (ICA). At the inlet and at the outlets of the models, 5-inlet and outlets diameter-long straight inlet and outlets extensions were added for facilitating the hemodynamics and dumping the effect of the imposition of the flow and pressure boundary conditions. These extensions were also considered in the solid domains that, as explained later, were also subjected to boundary conditions.

### 5.2.2 Generation of the Computational Grids

The IGES files coming from the geometry generation were imported into the commercial software Ansys IcemCFD (Ansys Inc., Canonsburg, PA). The fluid tetrahedral mesh was generated for the presented cases starting from the internal shell that represents the numerical fluid-solid interface domain. Due to the intrinsic complexity of the human vessels, an unstructured tetrahedral-based morphology was selected for discretizing the geometries. In order to establish the adequate element size for the computations and to guarantee that the provided results were grid-independent, a sensitivity study was carried out. Different grids were evaluated, increasing progressively the number of elements. For reducing the necessary computational time, this analysis was performed using the CFD, that is, imposing rigid walls for the arteries. Flow velocity profiles at different arterial sections were plotted as a function of the number of elements. Because the WSS plays a central role in atherogenesis and its evaluation is one of the goals of this study, the mesh independence analysis was also based on this variable. As detailed in Prakash and Ethier [23], mesh-independent velocity fields are not very difficult to obtain. However, WSS fields, and, in particular, WSS gradient fields are much more difficult to be accurately resolved. Achieving mesh-independence in computed WSS fields requires a considerably large number of nodes, and shows appreciable errors even on meshes that appear to produce mesh-independent velocity fields. For these reasons, and with the WSS being a very sensible variable, the adequate or "good enough" grid should be selected as a compromise between computational costs, solver efficiency and requirements, the quality of input data, and especially the needs of the scientist. In this work, we have made a compromise between the percentage of error on the WSS computation (less than 8%) and the computational requirements (one CFD analysis took about 96 h using a 16-node, Dual Nehalem (64 bits), 16-processor cluster with a clock speed of 2.33 GHz and 32 GB of memory for each node).

For the present study, the number of elements of the fluid computational meshes is of about $1 \times 10^6$ elements for the aorta and $0.5 \times 10^6$ for the carotid artery. These grids are depicted in Fig. 5.2.

The solid models of the considered arteries were meshed using hexahedral elements. The latter is, in fact, usually the best choice for structural analysis. The grids were created using the aforementioned commercial software. In particular, the blocking tool of Ansys IcemCFD was used for building "O-Grid" structures inside the arterial volume. Both considered arteries were first divided into subvolumes in which structured "O-Grid" blocks have been inserted. The computational grids used for the solid domains, shown in Fig. 5.2, are composed of $0.2 \times 10^6$ and $0.1 \times 10^6$ hexahedral

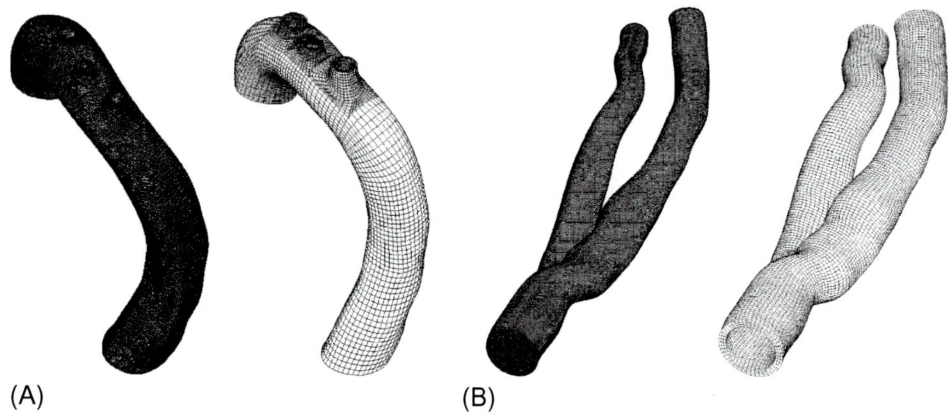

FIG. 5.2 Grids of the considered models used for the computations: aorta (A) and carotid artery (B).

elements for the aorta and the carotid artery, respectively. These sizes have been selected after a mesh-independence study realized by means of CSM analyses. These are based on the computational displacements and strains because the maximum principal stress is difficult to be accurately resolved in analogy with the WSS for the fluid computation. Also, in this case, the number of elements composing the final solid grids has been chosen as a compromise between the necessary requirements (error percentage on the strain computations) and the computational costs. With the aforementioned cluster, the final FSI computations required about 336 h to be completed with the computational requirements previously detailed.

## 5.2.3 Boundary Conditions Dilemma

For FSI problems, it is known that mixed conditions are required for correctly computing the intravascular flow and pressure. In particular, pressure information has to be given to the model for adequately computing the arterial stresses and strains. Theoretically, patient-specific measured blood flow and pressure are the perfect candidates in this sense. Unfortunately, while measured flow can be obtained in a noninvasive way by means of the laser Doppler technique, pressure measures are normally very invasive and required the use of a probe that may negatively influence the measure. In addition, flow measurements are clinically standardized while pressure measures are nonstandard and performed only in specific cases.

The impedance-based method may help to overcome this problem, providing a powerful computational tool that, starting from flow measurements, allows the computation of pressure waveforms. In the following sections, this attractive method originally proposed by Olufsen [24] will be explained. This method will be used for both the computations of the aorta and the carotid artery.

### 5.2.3.1 *Aortic and Carotid Inflow*

Patient-specific data of blood flow and blood pressure were not available for this study. Therefore, intravascular Doppler ultrasonic measured flows for the two considered arteries were taken from the literature [24, 25]. Murray's law was then used to impose the flow rate in each outlet branch. Using this law, the flow rate in each outlet is inversely proportional to the third power of the diameter of each branch:

$$\frac{Q_{root}}{D^3{}_{root}} = \frac{Q_{p1}}{D^3{}_{p1}} + \frac{Q_{p2}}{D^3{}_{p2}}, \tag{5.1}$$

where $Q$ is the flow rate and $D$ the diameter of each arterial branch [26]. In other words, for the child branches, the flow coming from the mother branch (root) splits following the cubic power of the parent diameters $D_{p1}$ and $D_{p2}$. Each branch outflow was then calculated starting from the root flow waveform using Murray's law and later used for the computation of the impedance-based pressure waveforms. For the carotid artery, we used the averaged flow given in Lee et al. [5] for the CCA. Then, this flow was divided following Murray's law at the ICA and ECA.

### 5.2.3.2 *The Impedance-Based Method*

Following the recursive approach developed by Olufsen et al. [24, 25] and extended by Steele et al. [27] for calculating the impedance of the human vascular tree, we first modeled the circulatory system as a structured fractal network in which we predicted physiological blood flow and pressure waveforms. This model consists of two parts: the large arteries (for this study, separately the aorta and the carotid artery) and the small arteries. In both, the blood flow and pressure are calculated using the incompressible axisymmetric Navier-Stokes equations for a Newtonian fluid. A one-dimensional (1D) model is obtained by integrating these equations over the cross-sectional area of each vessel [24, 25].

#### 5.2.3.3 The Vascular Fractal Network

As commented, the small arteries were modeled as a binary asymmetric-structured tree in which each vessel was assumed as a straight compliant segment. For computing the fractal tree, the relevant parameters are radii, bifurcation relationships, asymmetry, area ratios, lengths, and compliance. The vascular network resulted in a series of bifurcating segments composed of a parent and daughter vessels, as shown in Fig. 5.2. Asymmetry factors $\alpha$ and $\beta$ guide the scaling of each parent vessel in two daughter vessels according to:

$$r_{i,j} = \alpha^i \beta^{j-i} r_{root}. \tag{5.2}$$

Assuming as a root of the tree the radii of the arterial branches to which the fractal network has to be attached, we created an asymmetric vascular network (see Fig. 5.3). The structured tree continues branching until the radius of any vessel is less than a given minimum value $r_{min}$, that is the goal of the recursive method. This is normally a capillary radius because at this location the pressure can be approximately set to 0. Asymmetry ratios of the vessels were first defined as:

$$\eta = \frac{(r_0)_{d_1}{}^2 + (r_0)_{d_2}{}^2}{(r_0)_{pa}{}^2} \tag{5.3}$$

and

$$\gamma = \left(\frac{r_{0d_2}}{r_{0d_1}}\right)^2, \tag{5.4}$$

where $\eta$, the area ratio, and $\gamma$, the asymmetry ratio, are related to each other through the expression [25]:

$$\eta = \frac{1+\gamma}{(1+\gamma^{\xi/2})^{2/\xi}}. \tag{5.5}$$

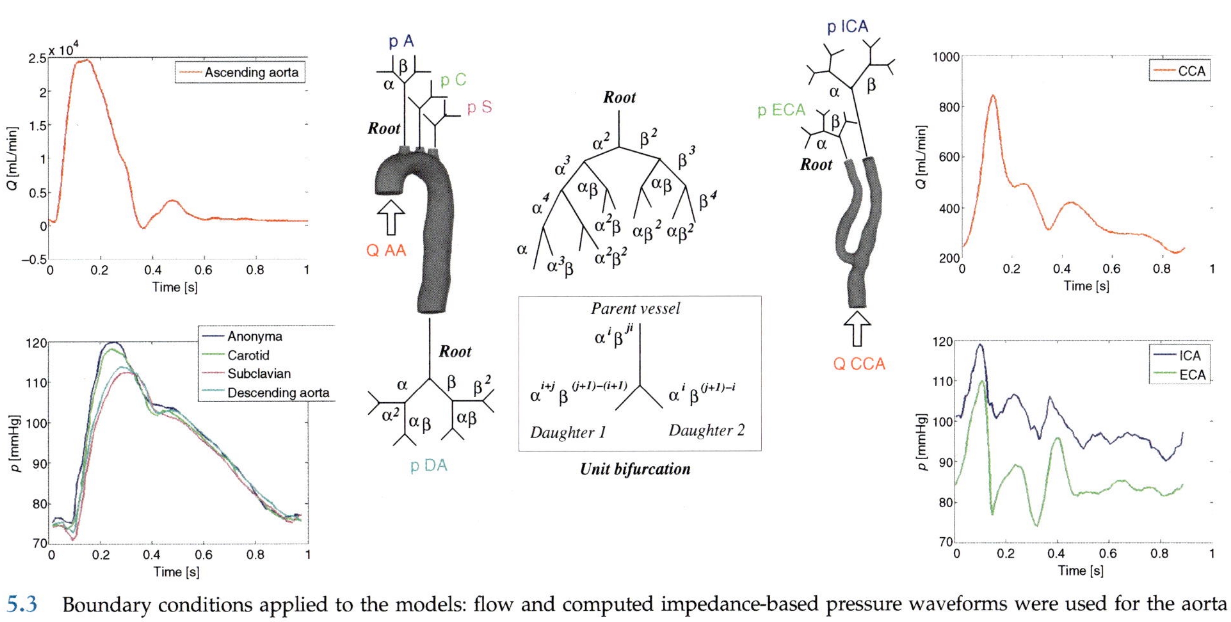


FIG. 5.3 Boundary conditions applied to the models: flow and computed impedance-based pressure waveforms were used for the aorta and carotid artery.

TABLE 5.1 Parameters Used to Describe the Structured Tree

| Level | Radius [μm] | $\alpha$ | $\beta$ | $\xi$ | $\gamma$ | $\eta$ |
|---|---|---|---|---|---|---|
| Small arteries | $250 < r$ | 0.895 | 0.566 | 2.5 | 0.4 | 1.12 |
| Resistance vessels | $50 < r < 250$ | 0.864 | 0.67 | 2.76 | 0.6 | 1.20 |
| Capillaries | $r < 50$ | 0.807 | 0.766 | 2.90 | 0.9 | 1.24 |

*Notes:* The binary network is divided into three levels as a function of the vessel radius. For each level, the main parameters, such as, for instance, the power exponent $\xi$ and the asymmetry ratio $\gamma$, were varied.

In Eq. (5.5), the exponent $\xi$ is known from the literature [26] so that the scaling parameters are obtained from the following expressions:

$$\alpha = (1+\gamma^{\xi/2})^{-1/\xi}, \quad \beta = \alpha\sqrt{\gamma}. \tag{5.6}$$

The length of each vessel is related to the radius using a special constant called length-to-radius ratio $l_{rr}$. This constant is well known in the literature [24, 27], and it was adjusted to control the outlet pressure waveform.

Following the extension of the Olufsen model performed by Steele et al. [27], we have divided the entire vascular bed in three different levels as a function of the vessel radius $r_{root}$ and the length-to-radius ratio $l_{rr}$, to mimic in more detail the structure of the human circulatory system. For each level, the parameters describing the asymmetry ratio $\gamma$ and the exponent $\xi$ were varied. The minimum radius was set to 3 μm where, as aforementioned, the blood pressure was set to 0. Table 5.1 shows the parameters used to describe each fractal tree (one for each outlet of the two models).

Different values of $l_{rr}$ can be found in the literature. Based on the studies of Iberall et al. [28] on small arteries, Olufsen used the value 50 while Steele used a multilevel approach also followed in this study. Zamir [29] suggested that the mean $l_{rr}$ is 20 with a maximum of 70. Other studies showed that this parameter widely varies in the vascular tree, being also organ-specific. In this work, the $l_{rr}$ pair has been set to 75/25 [27].

#### 5.2.3.4 *Computation of the Vascular Impedance*

The vascular impedance represents the resistance to the blood flow through the vascular network. Impedance was computed from the structured tree and used as the outlet boundary condition for large arteries. The impedance at the root of the vascular tree is recursively computed from the linear, axisymmetric, 1D Navier-Stokes equations starting from the terminal branch [24, 27]. The input impedance at the beginning of each vessel $z = 0$ was evaluated as a function of the impedance at the end of a vessel $z = L$ according to Eq. (5.7):

$$Z(0,\omega) = \frac{ig^{-1}\sin(\omega L/c) + Z(L,\omega)\cos(\omega L/c)}{\cos(\omega L/c) + igZ(L,\omega)\sin(\omega L/c)}, \tag{5.7}$$

where $L$ is the vessel length, $c = \sqrt{s_0(1-F_J)/(\rho C)}$ is the wave-propagation velocity, and

$$Z(0,0) = \lim_{\omega \to 0} Z(0,\omega) = \frac{8\mu l_{rr}}{\pi {r_0}^3} + Z(L,0), \quad F_J = \frac{2J_1(w_0)}{w_m J_0(w_0)}, \tag{5.8}$$

where $J_0(x)$ and $J_1(x)$ are the zeroth- and the first-order Bessel functions with $w_0 = i^3 w$ and $w^2 = {r_0}^2\omega/\nu$. The compliance $C$ can be estimated through the following equation:

$$C = \frac{3A_0 r_0}{2Eh}, \quad \frac{Eh}{r_0} = k_1 \exp(k_2 r_0) + k_3, \tag{5.9}$$

where $k_1$, $k_2$, and $k_3$ are known constants originally obtained by Olufsen [25], $s_0$ is the cross-sectional area, and $r_0$ is the root vessel.

### 5.2.4 Inflow and Outflow Conditions for the Aortic and Carotid Hemodynamics

The computation of the pressure waveforms was performed by means of the commercial software MatLab (The MathWorks, Natick, MA) prior to the FSI and CFD analyses. The computed waveforms are shown in Fig. 5.3 for the aorta and the carotid artery. These pressure waveforms were used as outflow conditions for the aortic and carotid hemodynamics. Flow waveforms found in the literature and used as the input of the impedance-based recursive computations were applied as inflow conditions in both cases (see Fig. 5.3).

### 5.2.5 Boundary Conditions for the Solid Domain

Some parts of the structural domains of the considered arterial models were clamped to avoid the movement as a rigid body. In particular, the extremities of the inlet and outlet extensions were constrained by fixing their surface rotations and axial translations, allowing only in-plane movement of each section. In the aorta, the mesh nodes of the extremity of the ascending aorta extension and that of the antonyma, subclavian, carotid, and descending trunk extensions were fixed. In the carotid artery, no axial or transaxial motion was permitted at the extremities of the CCA, ICA, and ECA extensions. These conditions, even nonphysiological as the inlet and outlets should theoretically be allowed to deform radially for simulating the arterial tethering, are usually assumed in the literature of the field [10, 11, 18].

### 5.2.6 Blood Flow Modeling

Arterial blood flow was modeled as laminar [30–32] incompressible and non-Newtonian. In particular, we utilized the Carreau-Yasuda model for modifying the blood viscosity as a function of the shear rate. The constitutive equation that represents this model is given by the following equation:

$$\mu = \mu_\infty + (\mu_0 - \mu_\infty)[1 + (\lambda\dot{\gamma})^a]^{\frac{n-1}{a}}, \tag{5.10}$$

where $\mu_0 = 0.056$ is the viscosity at zero shear rate expressed in [Pa s], $\mu_\infty = 0.00345$ is the viscosity for an infinity shear rate expressed in [Pa s], $\lambda = 3.313$ is the relaxation time expressed in [s], $n = 2$ is the power exponent, and $a = 0.64$ is the Yasuda exponent [33]. The blood density was set to 1060 $kg/m^3$.

### 5.2.7 Quantification of Hemodynamic Indices

Because the WSSs are biomarkers for vascular diseases, from the simulated models, we have evaluated velocity, pressure, and the most common WSS-related indicator such as time average wall shear stress (TAWSS). All hemodynamic variables were registered at every time step. The TAWSS was computed starting from the instantaneous WSS vector $\vec{\tau_w}$ registered at each time instant of the cardiac cycle period $T$. The TAWSS for pulsatile flow represents the spatial distribution of the tangential, frictional stress caused by the action of blood flow on the vessel wall temporally averaged on the entire cardiac cycle. It is often used in the computational cardiovascular hemodynamics [16, 18, 30, 31], and it can be calculated by integrating the WSS vector over the cardiac cycle:

$$TAWSS = \frac{1}{T}\int_0^T |\vec{\tau_w}| dt \tag{5.11}$$

### 5.2.8 Structural Modeling

#### 5.2.8.1 Aortic Structural Modeling

The material properties of the aortic structural model were based on the experimental data of Holzapfel and coworkers [34, 35]. The material constants were fitted by the strain energy density function (SEDF) proposed by Holzapfel et al. (see Table 5.2):

$$\psi = \mu[I_1 - 3] + \frac{k_1}{2k_2}\sum_{i=4,6}\left[\exp\left(k_2\left[[1-\rho][I_1-3]^2 + \rho[I_i - 1]^2\right]\right) - 1\right], \tag{5.12}$$

TABLE 5.2 Material Constants of the SEDF Developed in Holzapfel [34] Used for Modeling the Adventitia, the Media, and the Intima Layer of the Aorta

| | $\mu$ [kPa] | $k_1$ [kPa] | $k_2$ [−] | $\rho$ [−] |
|---|---|---|---|---|
| Adventitia | 1.8 | 7.03 | 6.04 | 0.15 |
| Media | 12.8 | 17.37 | 34.92 | 0.17 |
| Intima | 12.8 | 17.37 | 34.92 | 0.17 |

TABLE 5.3 Material Constants of the SEDF Developed in Kiousis et al. [36] Used for Modeling the Adventitia, the Media, and the Intima Layer of the Carotid Artery

| | $\mu$ [kPa] | $k_1$ [kPa] | $k_2$ [−] | $\rho$ [−] |
|---|---|---|---|---|
| Adventitia | 0.44 | 0.146 | 105 | 0.8 |
| Media | 0.7 | 0.023 | 16.9 | 0.8 |
| Intima | 0.7 | 0.023 | 16.9 | 0.8 |

where $\mu > 0$ and $k > 0$ are stress-like parameters. $k_2 > 0$ and $0 \leq \rho \leq 1$ are dimensionless parameters (when $\rho = 1$ the fibers are perfectly aligned and when $\rho = 0$ the fibers are randomly distributed so that the material is considered as isotropic), $I_1$ is the first invariant, and $I_4$ and $I_6$ are invariants that depend on the direction of the family of fibers at a material point. Two different material parameters were used for modeling the presented cases. The aortic wall was modeled as an anisotropic hyperelastic material with two families of fibers, oriented at ±30.28 degrees, with respect to the circumferential direction, for the adventitia and media layer, respectively.

#### 5.2.8.2 *Carotid Structural Modeling*

The material properties of the carotid structural model were based on the experimental data of Kiousis et al. [36]. The material constants were fitted by the SEDF defined in [36] (see Table 5.3). The carotid artery wall was modeled as an anisotropic hyperelastic material with two families of fibers, oriented at ±17.22 degrees [36].

### 5.2.9 FSI Coupling and Numerical Modeling

The simulations were run using the commercial software Adina (Adina R&D Inc., Watertown, MA). In this software, the FSI coupling can be performed after the creation of two models that separately include the fluid and the solid domain. The fluid domain was solved using a standard ALE formulation [37] while the solid domain used a typical Lagrangian formulation [38, 39]. Taking into account the moving reference velocity, the Navier-Stokes equations for the fluid domain become

$$\rho_F \frac{\partial \mathbf{v}_F}{\partial t} + \rho((\mathbf{v}_F - \mathbf{w}) \cdot \nabla)\mathbf{v}_F - \nabla \cdot \boldsymbol{\sigma}_F = \mathbf{f}_F^B, \tag{5.13}$$

where the term $\mathbf{w}$ denotes the moving mesh velocity vector [38], $\mathbf{v}_F$ is the velocity vector of the fluid, $\mathbf{f}_F^B$ is the body force per unit volume, and $\rho_F$ is the fluid density.

The governing equation of the solid domain is the momentum conservation equation:

$$\nabla \cdot \boldsymbol{\sigma}_S + \mathbf{f}_S^B = \rho_S \ddot{\mathbf{u}}_S, \tag{5.14}$$

where $\rho_S$ is the solid density, $\boldsymbol{\sigma}_S$ is the solid stress tensor, $\mathbf{f}_S^B$ is the body force per unit volume, and $\ddot{\mathbf{u}}_s$ is the local acceleration of the solid.

The domains described by Eqs. (5.13), (5.14) are coupled in the aforementioned software using a displacement compatibility and a traction equilibrium described by the following equations:

$$\mathbf{u}_S = \mathbf{u}_F \quad (x,y,z) \in \Gamma_{wall}^F \cap \Gamma_{wall}^S, \tag{5.15}$$

$$\boldsymbol{\sigma}_S \cdot \mathbf{n}_S + \boldsymbol{\sigma}_F \cdot \mathbf{n}_F = \mathbf{0} \quad (x,y,z) \in \Gamma_{wall}^F \cap \Gamma_{wall}^S, \tag{5.16}$$

where $\Gamma_{wall}^F$ and $\Gamma_{wall}^S$ are the boundaries of the fluid and solid domains, respectively, and $\mathbf{n}_S$, $\mathbf{n}_F$ the corresponding outer-pointing normals. Eq. (5.16) is an equilibrium condition between both domains $\Gamma_{wall}^F$ and $\Gamma_{wall}^S$ on the boundary surfaces. Because this condition is applied in weak form, the grids between the two domains can but are not required to match. For establishing the equilibrium, a mapping equation is provided:

$$\mathbf{F}_S(\mathbf{v},p) = \int (\mathbf{H}^S)^T \mathbf{M}_{sf} \boldsymbol{\tau}_f \cdot d\mathbf{S}, \tag{5.17}$$

where $\mathbf{M}_{sf}$ is a mapping operator used to interpolate variables at the solid and fluid nodes or vice versa [38, 39], $\mathbf{F}_S$ are the solid nodal forces, and $\mathbf{H}^S$ the interpolation functions of the solid elements.

Cardiac cycles of about 1 s were discretized in time steps of 0.0001 s. To dump the effect of initial transients, three complete cardiac cycles were computed and data from the last one was stored and postprocessed. Because the stress-free configuration was not known and the medical images in all the presented cases were given at a specific time instant, a ramp was used for reaching the diastolic pressure that was finally used as a starting configuration of the cardiac cycle. As discussed in Section 5.2.2, the simulations were carried out using a 16-node, Dual Nehalem (64 bits), 16-processor cluster with a clock speed of 2.33 GHz and 32 GB of memory for each node. Convergence was considered reached when the residuals of error for momentum and continuity fell below $10^{-4}$.

## 5.3 RESULTS

### 5.3.1 Arterial Hemodynamics

The hemodynamic characteristics of the considered human arteries are determined by the arterial morphologies that imply changes in vessel curvature, diameter, and tortuosity. Especially, bifurcation regions are predisposed to disturbed flow as visible by means of the blood flow recirculation. In Fig. 5.4, secondary flows are shown for both the aortic and carotid arteries. These main vortexes correspond to recirculation regions in the descending aorta and in the ECA, respectively. Of course, the evidenced structures are physiological and agree with the main features well established in the literature of the two arteries. The flow separation at the descending aorta and at the ICA and ECA is expected due to the abrupt change in curvature and the associated adverse pressure gradient along the outer walls of the arteries, as found also in Younis et al. [12, 13]. The blood flow is incapable of instantaneously changing the direction to accommodate the curvature. A comparison between the flow features computed by means of the CFD and the FSI techniques for the aorta and the carotid artery, respectively, shows similarities in the qualitative structures but also differences related especially with the intensity of the secondary flows. The latter has a special importance when computing the endothelial shear stresses because this may promote differences in the low and high regions, as discussed in the following sections. Recirculation regions are evidenced as well in the other location of the considered arteries such as near the anonyma, carotid, and subclavian bifurcations for the aorta and the ICA (results not shown). However, these zones present similar characteristics as those visualized in Fig. 5.4.

### 5.3.2 Instantaneous Wall Shear Stress Comparison

The WSS distributions computed by means of the CFD and FSI techniques are similar in the two considered arteries. These distributions that, as discussed earlier, are due to the flow structures inside the arteries, match well with the other results found in the literature studies [5, 13, 31]. Reverse flow and low or oscillating WSS correlate with atherosclerosis development in the carotid artery bifurcation, as documented in Ku et al. [40]. In Fig. 5.5, a comparison between the temporal minimal shear stress for the aorta (upper panel) and for the carotid artery (lower panel), respectively, for rigid wall and FSI simulations is shown. In the figure, the value of the WSS at the DA and at the ECA are plotted as a function of the cardiac cycle. These locations have been selected and postprocessed because they registered the minimum values for the variable. The computed curves for both arteries show the same trend as the inlet condition

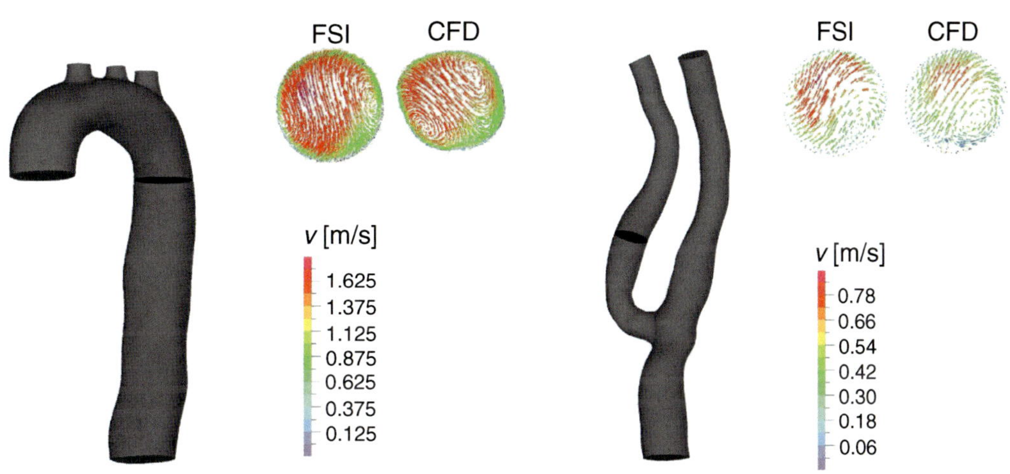


FIG. 5.4 Secondary flows represented by means of velocity *arrows* at peak flow during systole. The case of the descending aorta is represented in the *left panel*. The case of the external carotid artery is represented in the *right panel*.

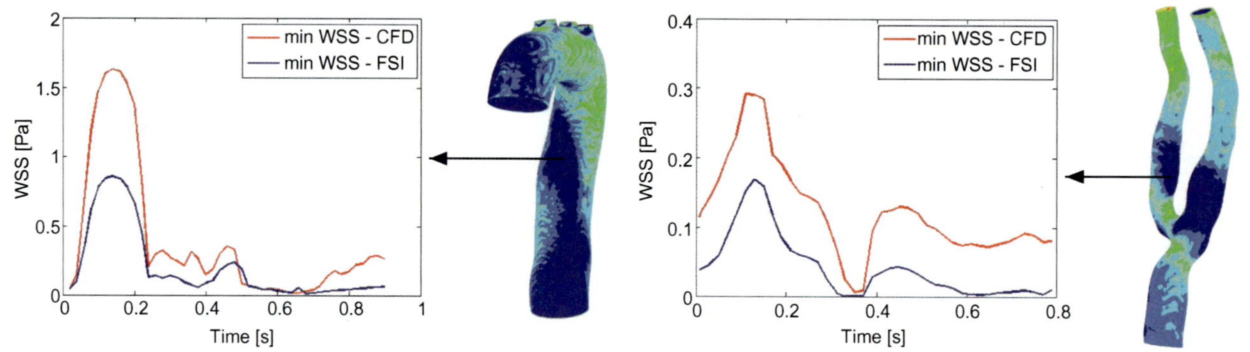


FIG. 5.5 Temporal history of the minimum WSS in specific locations related to flow recirculation for the aorta (*left panel*) and carotid artery (*right panel*).

(see Fig. 5.3), but no negligible differences were found in the WSS values. The temporal variations of the WSS show higher values if computed with CFD simulation rather than the FSI analysis. This tendency is reflected by the two cases.

For the aorta, at the DA (see Fig. 5.5, left panel) a gap between the temporal variations of the WSS computed by means of the FSI and the CFD techniques is clearly visible. The maximal difference is reached at time $t = 0.14$ s. The WSS values obtained with the CFD technique reflect higher values along the cardiac cycle. In particular, the comparison between the CFD and FSI techniques shows that the maximal value of the WSS at peak flow computed with the CFD technique is almost two times that computed with the FSI approach. At peak systole ($t = 0.14$ s), the CFD-computed WSS is 1.631 Pa while the FSI-computed value is limited to 0.8541. The latter highlights that the CFD not only tends to overestimate the WSS computation, but provides a value that is twice that computed with a compliant artery. In the diastolic region of the cycle (see Fig. 5.5), a reduction of the gap between the computed curves is visible. However, only between $t = 0.55$ s and $t = 0.65$ s do the two temporal histories overlap. For the carotid artery, at the ECA (see Fig. 5.5, bottom panel), we can observe a gap between the minimal WSS values computed with FSI and with CFD. The maximal gap between both curves is reached at the systolic peak ($t = 0.16$ s) where the value computed through rigid wall analysis is approximately twice that calculated with FSI. The gap attenuates at time $t = 0.38$ s but recovers a considerable difference (0.1 Pa) until the end of the cardiac cycle. In Fig. 5.6A and B, the spatial distributions of the WSS at peak flow during systole are depicted for the aorta and the carotid artery, respectively. The two figures highlight nonuniform distributions and a part of the aforementioned differences registered in the numerical values.

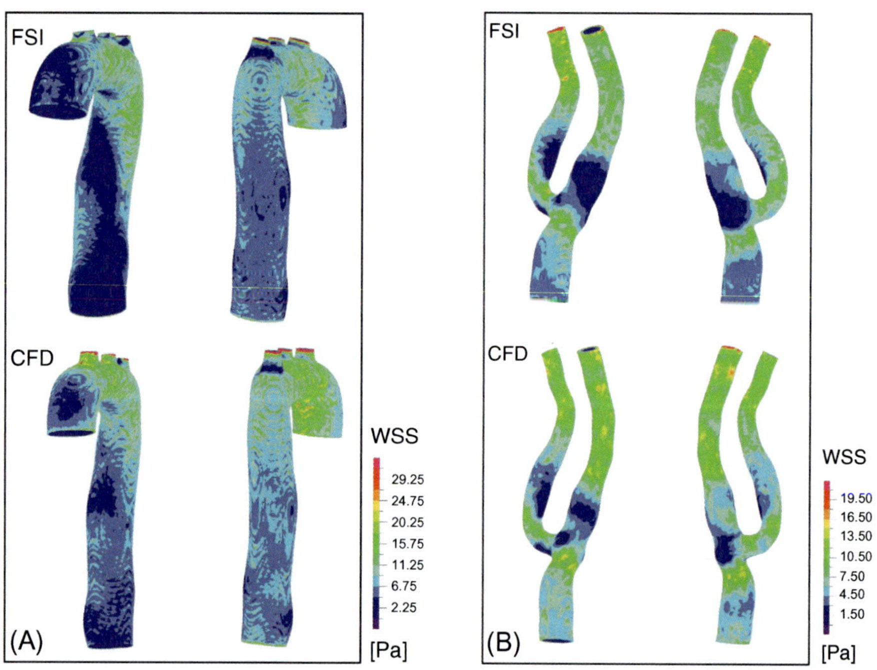


FIG. 5.6 Comparison between FSI and CFD-computed WSS distribution at peak systolic flow for the aorta (A) and the carotid artery (B).

In both cases, it is clearly visible that the amplitude of the low WSS regions is reduced when computed using the CFD technique. The reduction is visible in the descending trunk of the aorta (see Fig. 5.6A), both in the frontal and dorsal view as well as in the respective views of the carotid artery (see Fig. 5.6B). In the latter, the low WSS regions registered at the ECA, ICA, and CCA are strongly spatially less extended than those computed, including the compliance.

### 5.3.3 Time Average Wall Shear Stress Comparison

Endothelial cells subjected to low or oscillatory WSS are circular in shape without any preferred flow alignment pattern [41]. These cells, coupled with the blood stagnation usually observed in regions of low WSS, lead to increased uptake of blood-borne particles to the arterial wall, which is prevalent in atherosclerosis. This is a result of increased residence time and increased permeability of the endothelial layer [42].

In Figs. 5.7 and 5.8, the TAWSS computed using the instantaneous WSS vector $\vec{\tau_w}$ is depicted on the unloaded aorta and carotid artery, respectively. Because the TAWSS is an average variable, the differences registered by the instantaneous WSS are reduced and, qualitatively speaking, the spatial distributions of the computed TAWSS for both arteries are more similar with respect to that depicted in Fig. 5.6A and B. However, the regions of the low WSS found for both the aorta and the carotid artery, which are still lower when visualized by the FSI technique with respect to the

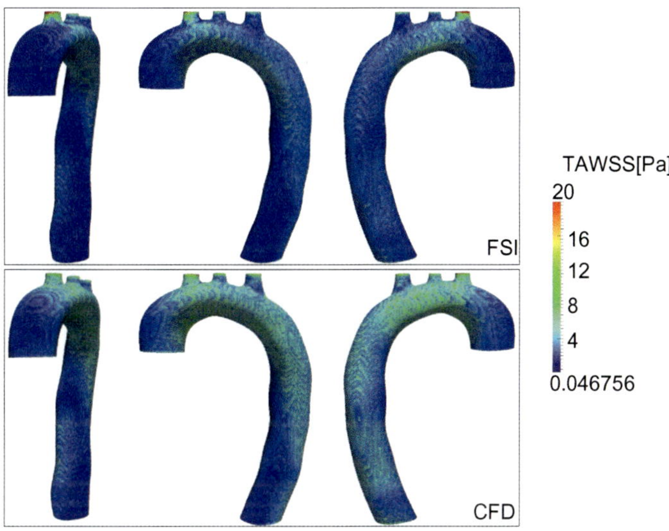


FIG. 5.7 Comparison of the CFD- and FSI-computed time average wall shear stress for the aorta.

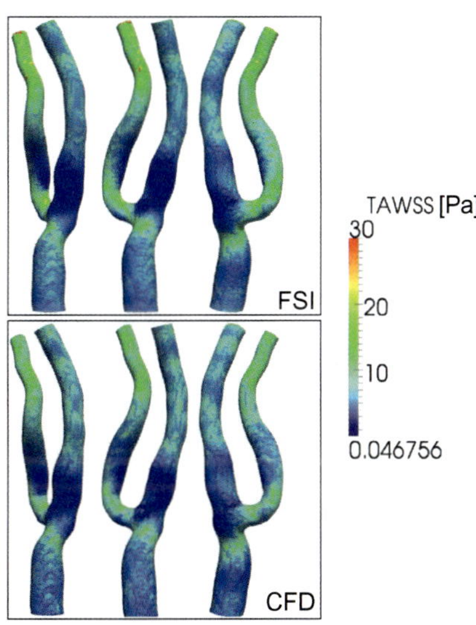


FIG. 5.8 Comparison of the CFD- and FSI-computed time average wall shear stress for the carotid artery.

CFD technique, and the differences in the spatial distribution are quite evident. Between the two considered cases, the differences are more pronounced in the aorta while the carotid artery is similar for both techniques. The regions with high TAWSS along the artery are, in fact, quite similar for the carotid artery in both cases, although the peak of low TAWSS is slightly different. On the contrary, the aorta shows considerable differences in the spatial distribution of the high WSS in the two considered cases. These distributions once again highlight that the CFD technique tends to overestimate the WSS and its related variables with respect to the FSI technique, even in the averaged values.

From the presented results, we can conclude that the arterial compliance strongly affects the WSS evaluation. The compliance of the arterial walls dilates the vessel, and the WSS is consequently altered, as observed in other works for both aorta and carotid hemodynamics [13, 16]. Generally speaking, the CFD is used more frequently with respect to the FSI technique due to its reduced computational costs. In this work, we suggest that the arterial hemodynamics should include compliant vessels. For assessing the risk of atherogenesis, in the computation of the WSS and its related indices, the compliance can be neglected only as a first approximation because important differences in the spatial amplitude and intensity can be found.

### 5.3.4 Arterial Compliance

In Fig. 5.9, the temporal variation of the compliances of the AA and DA for the aorta and of the ECA and CCA for the carotid artery, respectively, are shown. For the aorta (Fig. 5.9A), the maximal diameter variations during a cardiac cycle are 0.55 and 0.38 mm on the AA and DA, respectively. By normalizing these changes with their corresponding

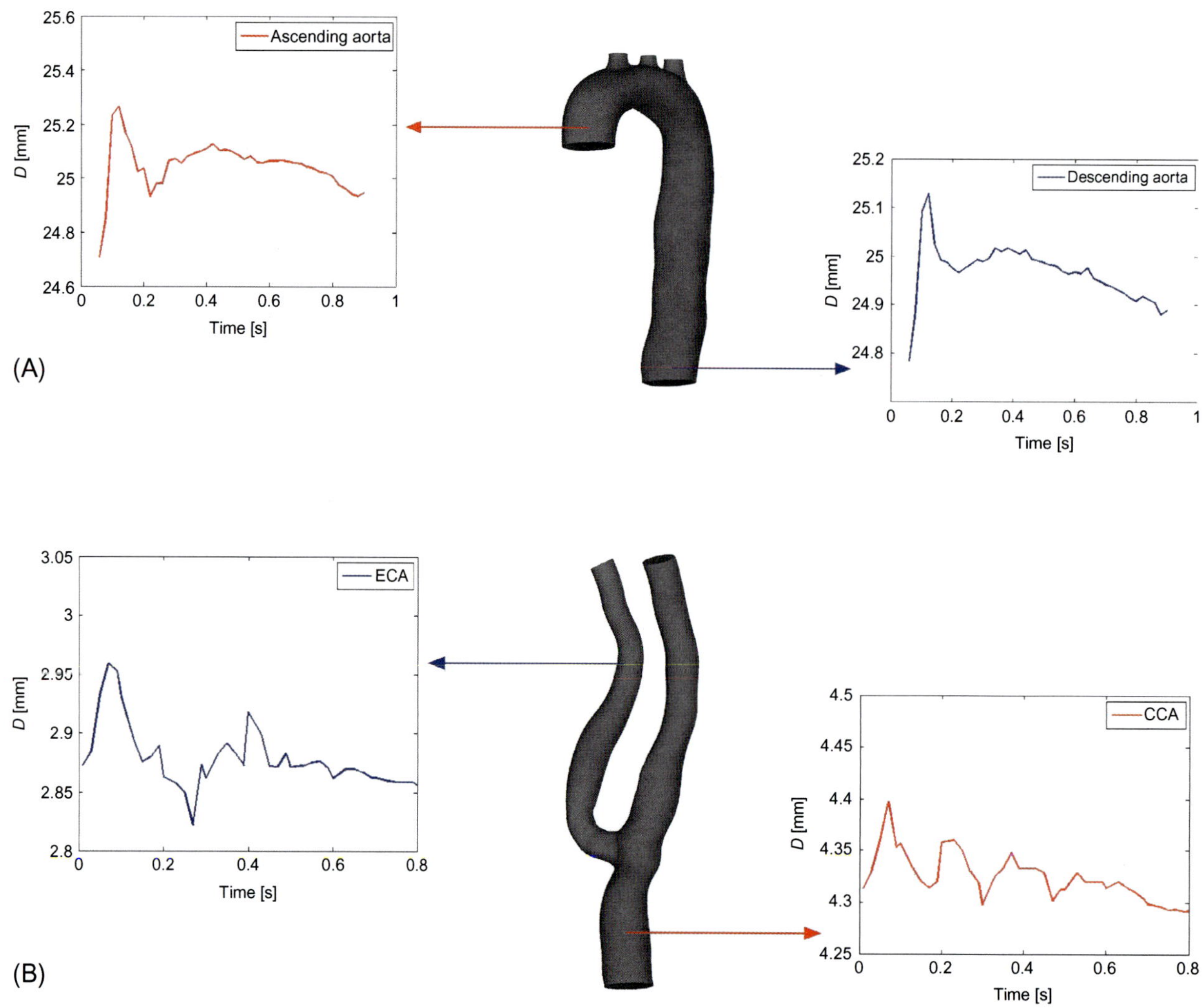


FIG. 5.9 Temporal history of the aorta (A) and carotid compliance (B) at different locations along the arteries. (A) Aortic compliance at ascending and descending trunk; (B) carotid compliance at ECA and CCA.

diameters, we obtained 2.7% and 1.9%, respectively. These values are computed with respect to the loaded configuration, that is, when the diastolic pressure has been reached. During the load of the artery until diastolic pressure, the diameter variations of the locations, shown in Fig. 5.9, are of about 21.8% for the AA and 15.1% for the DA. For the carotid artery (Fig. 5.9B), the maximal diameter variations during a cardiac cycle are 0.1 and 0.13 mm on the CCA and ECA, respectively. By normalizing these changes with their corresponding diameters, we obtained 2.4% and 3.1%, respectively. The obtained values are smaller than those measured by Studinger et al. [43]. However, the latter have been obtained using healthy and young patients during intense exercise conditions while the results presented here are obtained using rest conditions. It has to be noted that again, these values refer to the diameter variation during the cardiac cycle. The maximal diameter variation of each branch from the initial configuration to the beginning of the cycle is, in fact, much higher (about 20% and 17%), respectively. As for the case of the aorta, this is due to the pressurization of the artery until the diastolic pressure and it has not been considered for comparison purposes.

The presented results agree with those obtained in other works, which show that the arterial wall compliance for the carotid remains limited during the cardiac cycle [9]. Evidently, the WSS magnitude is reduced in the distensible model so that the compliance may play a crucial role when computing WSS and related indices for assessing atherosclerotic risk. The results of the flow study have been compared as far as possible with the numerical results found in the literature. However, this comparison can be performed only in a qualitative manner due to the different conditions in which the compared results have been obtained. These differences include pressure pulse wave, wall parameter, Reynolds number, and rheological blood property, among others. The qualitative comparison shows agreements in the essential features. For instance, during diastole, a reduction of the internal axial flow velocity is registered so that the pressure and, thus, the vessel lumen increase, as previously reported by Perktold and Rappitsch [9]. On the contrary, during systole, local accelerations and pressure decreases cause vessel lumen contractions so that the internal axial flow velocity tends to increase.

### 5.3.5 Limitations

The main limitation of the provided work that is based on FSI simulations is related to the increased computational costs with respect to the CFD and CSM techniques. This aspect may limit the introduction of the FSI to the clinical practice, contrary to the CFD, for instance. Furthermore, patient-specific boundary conditions would be of advantage for more precisely studying the presented hemodynamics. The computed impedance-based boundary conditions are specific to measured data that unfortunately have not been obtained from the same patients as those of the CT data. Additionally, the conditions are imposed as flat profiles on the inlet and outlet surfaces. However, the entrance and exit profiles are more complicated. The boundary conditions of the solids model also play a very important role. In this work, we have constrained the extremities of the structural domains. In this way, even the movement as a rigid body is impeded; the models tend to provide a certain displacement that could be avoided with the use of elastic springs. We are currently working on this issue for dumping this nonphysiological movement. Finally, the structural material models, even hyperelastic, anisotropic, and fiber-reinforced, are considered as passive, neglecting the active behavior of the muscular tissue.

However, with all these assumptions, the presented computational models shed light on the role and the arterial compliance for vessel hemodynamics, quantifying the differences with the usual rigid wall models.

## 5.4 CONCLUSION

Human cardiovascular hemodynamics is a complex problem that is usually treated separately regarding the analysis of vessel structural behavior and the arterial hemodynamics. In this chapter, a comprehensive FSI model of two human arteries is presented. The model, which considers the aorta and the carotid artery, includes the most important flow and structural features of human hemodynamics and presents a fully coupled approached between the fluid and solid domain. Arterial flow has been computed as quasisteady and non-Newtonian while the vessel wall has been modeled as anisotropic, hyperelastic, and fiber-reinforced. Arterial compliance has been evaluated in the analyzed cases as well as the WSS. As previously shown by experimental and computational studies [44, 45], the temporal variation of the vessel lumen is in phase with the pressure waveform while the temporal variation of the WSS is in phase with the flow waveform. The WSS, being a well-known biomarker for atherosclerosis, was used for computing derived variables that are frequently used for assessed atherosclerotic risk such as the time average WSS. A comparison between CFD and FSI analyses has been carried out in order to quantify the impact of the compliant vessel wall when simulating human hemodynamics. The results have shown that while the blood flow structure is relatively unaffected, the instantaneous WSS presents discrepancies in all considered cases, as previously found by other authors [9, 16]. These differences are especially related to the minimum and maximum value of the WSS history along the cardiac cycle. The comparison between the maximum values of the low WSS computed through the FSI technique exceeds twice that computed using rigid walls in both arteries while the arterial compliance between systole and diastole is limited to 3%, comparing diastolic and systolic conditions. Generally speaking, the CFD analysis tends to overestimate the instantaneous WSS computation while the TAWSS, which is an average variable, tends to dump the discrepancies yet provides more uniform spatial distribution.

The presented approach, even computationally more expensive than CFD, provides a significant insight into the role that compliance plays and also allows the computation of structural stresses and strains that affect the vessel walls. The latter, while computed within the FSI simulation and hence including the effect of hemodynamics, may provide important information for the analysis of atherosclerosis. Structural variables, which were not shown in this work, may also be considered in a large number of patients. Regions with high stresses that are related to tissue inflammation in the literature can be correlated, for instance, to low or oscillating WSS. The presence of high stress and simultaneous low or oscillatory WSS is submitted to an elevated risk of atherosclerosis, as reported by Thubrikar [46].

## Acknowledgments

This study is supported by the Spanish Ministry of Economy, Industry, and Competitiveness through the research projects DPI2017-83259-R and DPI2016-76630-C2-1-R. The support of the Instituto de Salud Carlos III (ISCIII) through the CIBER-BBN initiative and through the project "Patient-Specific Modelling of the Aortic valve replacement: Advance towards a Decision Support System (DeSSaValve)" is highly appreciated.

## References

[1] World Health Organization, World Health Organization regional office for Europe. European Health Report for 2016, (2016). European Series.
[2] T. Asakura, T. Karino, Flow patterns and spatial distribution of atherosclerotic lesions in human coronary arteries, Circ. Res. 66 (4) (1990) 1045–1066.
[3] C.G. Caro, J.M. Fitz-Gerald, R.C. Schroter, Atheroma and arterial wall shear: observation, correlation and proposal of a shear dependent mass transfer mechanism for atherogenesis, Proc. R. Soc. Lond. 177 (1046) (1971) 109–159.
[4] J.B. Thomas, L. Antiga, S.L. Che, J.S. Milner, D.A. Steinman, J.D. Spence, B.K. Rutt, D.A. Steinman, Variation in the carotid bifurcation geometry of young versus older adults: implications for geometric risk of atherosclerosis, Stroke 36 (11) (2005) 2450–2456.
[5] S.W. Lee, L. Antiga, J.D. Spence, D.A. Steinman, Geometry of the carotid bifurcation predicts its exposure to disturbed flow, Ann. Biomed. Eng. 39 (8) (2008) 2341–2347.
[6] M. Malvè, A.M. Gharib, S.K. Yazdani, G. Finet, M.A. Martínez, R. Pettigrew, J. Ohayon, Tortuosity of coronary bifurcation as a potential local risk factor for atherosclerosis: CFD steady state study based on in vivo dynamic CT measurements, Ann. Biomed. Eng. 43 (1) (2015) 82–93.
[7] Y.C.W. Fung, Biomechanics, Circulation, second ed., Springer, New York, NY, 1996 (Chapter 3).
[8] A. Quarteroni, A. Manzoni, C. Vergara, The cardiovascular system: mathematical modeling, numerical algorithms and clinical applications, Acta Numer. 26 (2017) 365–590.
[9] K. Perktold, G. Rappitsch, Computer simulation of the local blood flow and vessel mechanics in a compliant carotid artery bifurcation model, J. Biomech. 28 (1995) 845–856.
[10] D. Tang, C. Yang, S. Mondal, F. Liu, G. Canton, T.S. Hatsukami, C. Yuan, A negative correlation between human carotid atherosclerosis plaque progression and plaque wall stress: in vivo MRI-based 2D/3D FSI models, J. Biomech. 41 (2008) 727–736.
[11] H. Gao, Q. Long, M. Graves, J.H. Gillard, Z.Y. Li, Carotid arterial plaque stress analysis using fluid-structure interactive simulation based on in-vivo magnetic resonance images of four patients, J. Biomech. 42 (2009) 1416–1423.
[12] H.F. Younis, M.R. Kaazempur-Mofrad, C. Chung, R.C. Chan, A.G. Isasi, D.P. Hinton, A.H. Chau, L.A. Kim, R.D. Kamm, Hemodynamics and wall mechanics in human carotid bifurcation and its consequences for atherogenesis: investigation of inter-individual variation, Biomech. Model. Mechanobiol. 3 (2003) 17–32.
[13] H.F. Younis, M.R. Kaazempur-Mofrad, C. Chung, R.C. Chan, R.D. Kamm, Computational analysis of the effects of exercise on hemodynamics in the carotid bifurcation, Ann. Biomed. Eng. 31 (2003) 995–1006.
[14] P. Crosetto, P. Reymond, S. Deparis, D. Kontaxakis, N. Stergiopulos, A. Quarteroni, Fluid-structure interaction simulation of aortic blood flow, Comput. Fluids 43 (2011) 46–57.
[15] S.H. Lee, S. Kang, N. Hur, S.K. Jeong, A fluid-structure interaction analysis on hemodynamics in carotid artery based on patient-specific clinical data, J. Mech. Sci. Technol. 26 (12) (2012) 3821–3831.
[16] P. Raymond, P. Crosetto, S. Deparis, A. Quarteroni, N. Stergiopulos, Physiological simulation of blood flow in the aorta: comparison of hemodynamic indices as predicted by 3-D FSI, 3-D rigid wall and 1-D models, Med. Eng. Phys. 35 (2013) 784–791.
[17] N. Xiao, J.D. Humphrey, C.A. Figueroa, Multi-scale computational model of three-dimensional hemodynamics within a deformable full body arterial network, J. Comput. Phys. 244 (2013) 22–40.
[18] M. Malvè, S. Chandra, A. García, A. Mena, M.A. Martínez, E.A. Finol, M. Doblaré, Impedance-based outflow boundary conditions for human carotid haemodynamics, Comput. Methods Biomech. Biomed. Eng. 17 (11) (2014) 1248–1260.
[19] S. Roccabianca, C.A. Figueroa, G. Tellides, J.D. Humphrey, Quantification of regional differences in aortic stiffness in the aging human, J. Mech. Behav. Biomed. Mater. 29 (2014) 618–634.

[20] E.B. Rosero, R.M. Peshock, A. Khera, P. Clagett, H. Lo, C.H. Timaran, Sex, race, and age distributions of mean aortic wall thickness in a multiethnic population-based sample, J. Vasc. Surg. 53 (4) (2011) 950–957.

[21] J.D. Humphrey, G.A. Holzapfel, Mechanics, mechanobiology, and modeling of human abdominal aorta and aneurysm, J. Biomech. 45 (2012) 805–814.

[22] V. Alastrue, A. García, E. Peña, J.F. Rodríguez, M.A. Martínez, M. Doblaré, Numerical framework for patient-specific computational modelling of vascular tissue, Int. J. Numer. Methods Biomed. Eng. 26 (2010) 35–51.

[23] S. Prakash, C.R. Ethier, Requirements for mesh resolution in 3D computational hemodynamics, J. Biomech. Eng. 123 (2001) 134–144.

[24] M.S. Olufsen, C.S. Peskins, W.Y. Kim, E.M. Pedersen, A. Nadim, J. Larsen, Numerical simulation and experimental validation of blood flow in arteries with structured tree outflow conditions, Ann. Biomed. Eng. 28 (2000) 1281–1299.

[25] M.S. Olufsen, A structured tree outflow condition for blood flow in larger systemic arteries, A. J. Physiol. 276 (1999) H257–H268. Heart and Circulatory Physiology.

[26] C.D. Murray, The physiological principle of minimum work, the vascular system and the cost of blood volume, Proc. Natl Acad. Sci. USA 12 (1926) 207–214.

[27] B.N. Steele, M.S. Olufsen, C.A. Taylor, Fractal network model for simulating abdominal and lower extremity blood flow during resting and exercise conditions, Comput. Methods Biomech. Biomed. Eng. 10 (1) (2007) 37–51.

[28] A.S. Iberall, Anatomy and steady flow characteristics of the arterial system with an introduction to its pulsatile characteristics, Math. Biosci. 1 (1967) 375–385.

[29] M. Zamir, The Physics of Pulsatile Flow, in: Biological Physics Series, second ed., Springer, New York, NY, 2000. Chapter 3.

[30] U. Morbiducci, R. Ponzini, D. Gallo, C. Bignardi, G. Rizzo, Inflow boundary conditions for image-based computational hemodynamics: impact of idealized versus measured velocity profiles in the human aorta, J. Biomech. 46 (1) (2013) 102–109.

[31] D. Gallo, U. Gülan, A. Di Stefano, R. Ponzini, B. Lüthi, M. Holzner, U. Morbiducci, Analysis of thoracic aorta hemodynamics using 3D particle tracking velocimetry and computational fluid dynamics, J. Biomech. 47 (12) (2014) 3149–3155.

[32] S. Pirola, Z. Cheng, O.A. Jarral, D.P. O'Regan, J.R. Pepper, T. Athanasiou, X.Y. Xu, On the choice of outlet boundary conditions for patient-specific analysis of aortic flow using computational fluid dynamics, J. Biomech. 60 (2017) 15–21.

[33] A. Valencia, M. Villanueva, Unsteady flow and mass transfer in models of stenotic arteries considering fluid-structure interaction, Int. Commun. Heat Mass Transfer 33 (1) (2006) 966–975.

[34] G.A. Holzapfel, Determination of material models for arterial walls from uniaxial extension tests and histological structure, J. Theor. Biol. 238 (2006) 290–302.

[35] A.J. Schriefl, G. Zeindlinger, D.M. Pierce, P. Regitnig, G.A. Holzapfel, Determination of the layer-specific distributed collagen fibre orientations in human thoracic and abdominal aortas and common ilia arteries, J. R. Soc. Interface 9 (2012) 1275–1286.

[36] D.E. Kiousis, S.F. Rubinigg, M. Auer, G.A. Holzapfel, A methodology to analyze changes in lipid core and calcification onto fibrous cap vulnerability: the human atherosclerotic carotid bifurcation as an illustratory example, J. Biomech. Eng. 131 (2009). 121002-1.

[37] J. Donea, S. Giuliani, J.P. Halleux, An arbitrary Lagrangian-Eulerian finite element method for transient dynamic fluid-structure interaction, Comput. Methods Appl. Mech. Eng. 33 (1982) 689–723.

[38] K.J. Bathe, H. Zhang, Finite element developments for general fluid flows with structural interactions, Int. J. Numer. Methods Eng. 60 (2004) 213–232.

[39] K.J. Bathe, H. Zhang, S. Ji, Finite element analysis of fluid flows fully coupled with structural interactions, Comput. Struct. 72 (1999) 1–16.

[40] D.N. Ku, D.P. Giddens, C.K. Zarins, S. Glagov, Pulsatile flow and atherosclerosis in the human carotid bifurcation: positive correlation between plaque location and low and oscillating shear stress, Atherosclerosis 15 (1985) 293–302.

[41] A.M. Malek, S.L. Alper, S. Izumo, Hemodynamic shear stress and its role in atherosclerosis, J. Am. Med. Assoc. 282 (1999) 2035–2042.

[42] E.A. Murphy, F.J. Boyle, Reducing in-stent restenosis through novel stent flow field augmentation, Cardiovasc. Eng. Technol. 3 (2012) 353–373.

[43] P. Studinger, Z. Lenard, Z. Kovats, L. Kocsis, M. Kollai, Static and dynamic changers in carotid artery diameter in humans during strenuous exercises, J. Physiol. 550 (2) (2003) 565–583.

[44] B.A. Haluska, L. Jeffriess, P.M. Mottram, S.G. Earlier, T.H. Marwick, A new technique for assessing arterial pressure wave forms and central pressure with tissue Doppler, Cardiovasc. Ultrasound 5 (2007) 6.

[45] S. Tada, J.M. Tarbell, A computational study of flow in a compliant carotid bifurcation-stress phase angle correlation with shear stress, Ann. Biomed. Eng. 33 (2005) 1202–1212.

[46] M.J. Thubrikar, Vascular Mechanics and Pathologies, second ed., Springer, New York, NY, 2007.

CHAPTER

# 6

# Review of the Essential Roles of SMCs in ATAA Biomechanics

*Claudie Petit*, S. Jamaleddin Mousavi*, Stéphane Avril**

***Mines Saint-Etienne, University of Lyon, INSERM, U1059 Sainbiose, Saint-Etienne, France**

## 6.1 INTRODUCTION

Aortic aneurysms (AAs) are among the most critical cardiovascular diseases [1, 2]. Although their detection is difficult, prevention and monitoring of AA are essential as large AAs present high risks of dissection or rupture, which are often fatal complications [2, 3]. Monitoring consists of measuring the aneurysm diameter using medical imaging methods such as echography or CT scan [2, 3].

The present study is focused on ascending thoracic AAs (ATAAs). The risk of rupture of ATAAs is estimated clinically with the maximum aneurysm diameter, which consists of considering surgical repair for ATAA diameters larger than 5.5 cm. Other factors such as growth rate, gender, or smoking can be taken into account [2–4]. It is known that the criterion of maximum diameter relies on statistics of the global ATAA population. On an individual basis, many ruptures or dissections have been reported for aneurysms with diameters below the critical value [5]. Other criteria based on biomechanics were suggested [5], but they still need to be validated clinically [6–8].

The main causes of ATAAs are summarized in Table 6.1. ATAAs are a very specific class of AA due to the particularity of the ascending thoracic aorta. First, it contains the highest density of elastic fibers of all the vasculature, and these have to resist the mechanical fatigue induced by the wearing combination of pulsed pressure and axial stretching repeated every cardiac beat. As elastic fibers cannot be repaired in mature tissue [9], the ascending aortic tissue is highly prone to mechanical damage [3, 6]. Second, a major role of the contractile function in smooth muscle cells (SMCs) is evident in the ascending aorta more than anywhere else as heterozygous mutations in the major structural proteins or kinases controlling contraction lead to the formation of aneurysms of the ascending thoracic aorta [10]. Moreover, the outer curvature of the ascending thoracic aorta is constituted of a mix of cardiac neural crest- and second heart field-derived SMCs, distributed over the different medial lamellar units (MLUs) (Fig. 6.1) [11]. This may be correlated with the observation that dilatations are more often located on the outer curvature of the ascending thoracic aorta [12]. Third, the ascending thoracic aorta experiences very complex flow profiles, with significant alterations (vortex, jet flow, eccentricity, peaks of wall shear stress) in case of bicuspid aortic valves [13–15] or aortic stenosis [15–17]. It was shown that these complex hemodynamic patterns have major interactions with the aortic wall and correlate with local inflammatory effects or variations of oxidative stress in the aortic tissue [15, 18, 19].

Research studies dedicated to ATAA have always invoked one of the three previous particularities of the ascending thoracic aorta to account for the intrinsic mechanism leading to the development of an ATAA, even if recent studies tend more and more to invoke multifactorial effects. In this review, we show that all these effects converge toward a single paradigm relying upon the crucial biomechanical role of SMCs in controlling the distribution of mechanical stresses across the different components of the aortic wall. The chapter is organized as follows. In Section 6.2, we introduce the basics of arterial wall biomechanics and how the stresses are distributed across its different layers. In Section 6.3, we introduce the biomechanical active role of SMCs and its main regulators and show how this can control the distribution of stresses across the aortic wall. In Section 6.4, we review the different pathways of SMC mechanotransduction and their mechanisms at the cellular and tissue level in the aortic wall. Finally, we review studies showing that SMCs tend to have a preferred homeostatic tension. We show that mechanosensing can be understood as

*Advances in Biomechanics and Tissue Regeneration*
https://doi.org/10.1016/B978-0-12-816390-0.00006-6

TABLE 6.1 Main Causes of ATAAs Affecting Both the ECM and the SMCs

| Causes of ATAAs | Effects | Ref. |
|---|---|---|
| | *GENETIC MUTATIONS AFFECTING THE ECM* | |
| fnb1 (Marfan syndrome) | Microfibrils anomalies => wrong force transmission and alteration of the mechanotransduction | [3, 20] |
| Types I and III collagen | Anomalies of collagen fibers | [21] |
| | *GENETIC MUTATIONS AFFECTING THE SMC* | |
| ACTA2 ($\alpha$-SMA) | Dysfunction of the contractile apparatus. This mutation represents about 12% of ATAAs | [20, 22] |
| MYH11 (myosin light chain) | Dysfunction of the contractile apparatus | [20, 23] |
| TGFB (TGF-$\beta$) | Anomalies of TGFB receptors TGBFR1/2. Wrong regulation of traction forces | [20, 24, 25] |
| MYLK (myosin light chain kinase) | Alteration of myosin RLC (regulatory light chains) phosphorylation, and thus force generation | [26] |
| PRKG1 | Kinase activation resulting in SMC relaxation | [26] |
| MMP genes | Alteration of myosin regulatory light chains (RLC) phosphorylation, and thus force generation | [26] |
| | ***PHENOTYPIC SWITCHING: CONTRACTILE (C) => SYNTHETIC (S)*** | |
| Stiffening and weakening of the arterial wall *Pathologies*: atherosclerosis, arteriosclerosis, arteritis, aging | The SMCs move on to synthetic phenotype (S) (hypertrophy), they lose their quiescence (hyperplasia), they order wall remodeling by the synthesis of MMPs (degradation) and ECM (renewal). Moreover, atheroma plaques contain many SMCs | [3, 27–30] |
| Chronic overstress *Pathologies*: hypertension, dissection, ATAAs | The (C) SMCs move on to (S): remodeling, hypertrophy, hyperplasia | [9, 31] |
| Contact with blood flow *Pathologies*: intimal injury, porosity of the wall | The (C) SMCs move on to (S), formation of a neointima containing SMCs and GAGs through hyperplasia | [32] |
| | Blood-borne components interacts with SMCs | [15] |
| Change in ECM chemical composition Laminin/fibronectin ratio Elastin/collagen ratio | The (C) phenotype may be favored on laminin or matrigel (collagen + laminin) in vitro A high elastin concentration may activate actin polymerization and thus the development of the contractile apparatus | [33–36] [37] |
| Cell culture in vitro High passage Substrate (physical properties) | Wrong development of the contractile apparatus => more (S) SMCs as cell passage increases Necessity to use some stimuli such as vasoactive agonists or suitable substrates | [35, 38–41] |
| | *PARTIALLY IDENTIFIED CAUSES* | |
| Biochemical imbalance Signaling pathways involved in cell contraction | Angiotensin II, growth factors: TGF-$\beta$, PDGF | [2, 21, 26, 34, 42–45] |
| | $Ca^{2+}$ ionic channel | [39, 46–51] |
| Intercellular interactions Vasoactive agonists, neurotransmitters, hormones, ions, mechanical stimuli | Interaction with endothelial cells from the intima | [27, 29, 35, 52–54] |
| | Synchronization of several SMCs | [27, 46, 55] |
| Local changes in hemodynamics Bicuspid aortic valve, dissection, ATAAs | Disturbance of the mechanotransduction through endothelial cells and SMCs | [3, 15, 27, 56] |
| Embryonic origin of the SMCs Transition area between aortic root and arch: the media combines SMCs from different origin | Outermost SMCs are from the second heart field and the innermost ones from the neural crest. This area is prone to ATAAs and dissections | [11] |

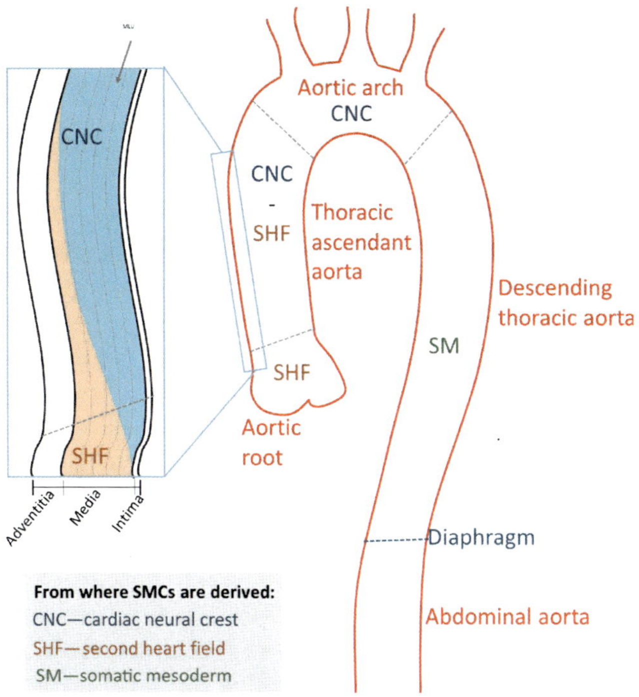


FIG. 6.1 Anatomy of the whole aorta: The thoracic part is separated from the abdominal one by the diaphragm. The thoracic aorta is divided into four parts from different embryological origins. The ascending aorta is particularly subject to ATAAs, and contains a mix of CNC- and SHF-derived SMCs. *Based on the works of E.M. Isselbacher, Thoracic and abdominal aortic aneurysms, Circulation 111 (6) (2005) 816–828 and H. Sawada, D.L. Rateri, J.J. Moorleghen, M.W. Majesky, A. Daugherty, Smooth muscle cells derived from second heart field and cardiac neural crest reside in spatially distinct domains in the media of the ascending aorta—brief report, Arterioscler. Thromb. Vasc. Biol. 37 (9) (2017) 1722–1726.*

the reaction to the homeostasis unbalance of SMC tension. The review reveals, though, that the quantification of the SMC homeostatic tension in the ascending thoracic aorta is still an open question, and the chapter closes with possible directions for research in measuring this tension at the tissue level and at the cellular level.

## 6.2 BASICS OF AORTIC WALL MECHANICS AND PASSIVE BIOMECHANICAL ROLE OF SMCS

### 6.2.1 Composition of Arteries

#### *6.2.1.1 The Extracellular Matrix*

The extracellular matrix (ECM) of the aortic tissue is made of two main fibrous proteins participating in the passive response: collagen and elastin, which are responsible for 60% in dry weight of the entire wall [21]. There are several types of collagen, but types I, III, and V are primarily found in the media layer (see section below describing the layers of the aorta), where the SMCs are located, representing about 35% of the global aortic wall in dry weight [21, 27]. Collagen fibers are not extensible and ensure the mechanical resistance of the tissue in case of overloading [10, 21, 57]. If collagen fibers can be produced over the lifespan, elastic fibers are actively synthesized in early development, and there is a loss of efficiency for the ones created during adulthood [9]. Elastin has a 40-year estimated half-life. It should last for an entire life in optimal conditions, but some pathological states or natural aging will necessarily affect it. Fibroblasts and SMCs can produce new ECM components but also matrix metalloproteases (MMPs), which degrade the current ECM. If the action of MMPs is not well regulated, the ECM may be remodeled, yielding a different mechanical behavior with possible ATAA development [9, 58]. Likewise, the loss of elasticity may be related to an anomaly of elastic fibers. Elastic fibers are mainly composed of a core of amorphous elastin surrounded by microfibrils. The microfibrils comprise collagen VI and fibrillin [59], a polymer encoded by the *fbn1* gene, whose mutation is involved in Marfan syndrome. The genetic mutations affecting the ECM in ATAAs are summarized in Table 6.1.

Another important constituent, although with lesser mass fractions, are glycosaminoglycans (GAGs), which can contribute to the compressive stiffness of the aortic tissue. As they represent about 3%–5% of the total wall by dry weight [21], they do not participate markedly to the passive response except in specific cases, such as for atherosclerosis, where GAGs are piled up during lesion development and increase the wall stiffness. GAGs refer to different types of nonsulfated (hyaluronic acid) and sulfated (keratan sulfate, dermatan sulfate, and heparan sulfate) polysaccharides [60]. The ECM contains also some glycoproteins that bind to cell membrane receptors, the integrins, and allow for cellular adhesion. Among these binding proteins, the fibronectin can also bind to collagen and heparan sulfate, and laminin is a major component of basal lamina, which influences cell responses.

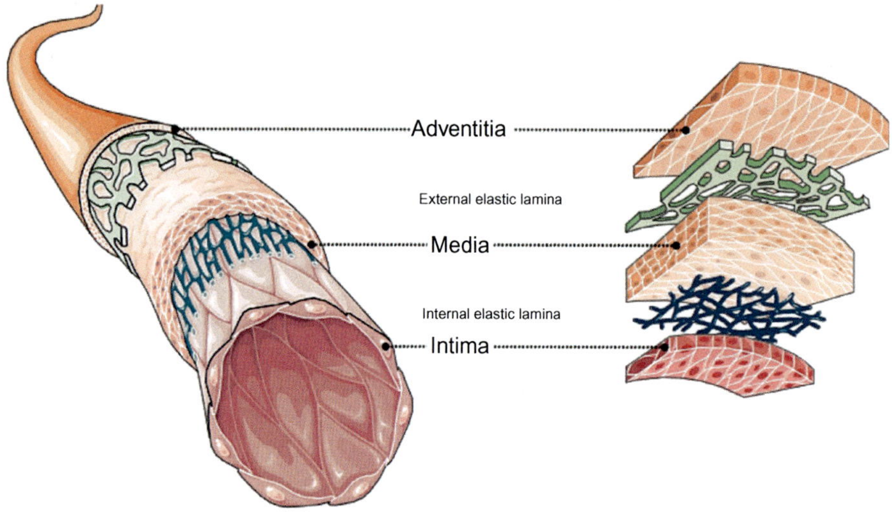


FIG. 6.2 Structure of the arterial wall. *Courtesy of T.C. Gasser, Structure and Basic Properties of the Arterial Wall, 2017 (Indisponible en Accès Libre).*

#### 6.2.1.2 A Multilayered Wall Structure

The aortic wall is divided into three main layers surrounding the lumen where the blood flow circulates (Fig. 6.2). Each layer has its function and proper mechanical properties [6, 14, 27, 57, 61, 62]. The adventitia, which is the most external layer, contains fibroblasts and is particularly collagen-rich, according to its protective role for the entire wall against high stress. The internal layer, called the intima, is directly in contact with the blood flow. It also constitutes a selective barrier of endothelial cells for preventing the wall from blood product infiltration and delivering oxygen and nutrients from the blood to the internal wall. The inner medial layer is separated from adventitia and intima by two elastic laminae, and represents about two-thirds of the whole thickness of the wall. All these layers have a passive mechanical response to the loading induced by the blood flow, but only the media can also act actively due to the presence of contractile SMCs. The media is structured into several MLUs (Fig. 6.1) [21, 62], where a layer of SMCs is tight between two thin elastin sheets through a complex network of interlamellar elastin connections [57]. The SMCs are oriented in the direction of the ECM fibers in order to better transmit the forces to each other and to successive MLUs. The number of MLUs varies according to the diameter of the artery [62] and the size of the organism: 6–8 for mice and 40–70 for the human body [21].

### 6.2.2 Basics of Aortic Biomechanics

It is commonly assumed that only the adventitia and the media are involved in the mechanical response of the entire wall, neglecting the mechanical role of the intima. This assumption is not valid in the case of pathologies resulting in a thickening of the intima such as atherosclerosis.

The aorta is submitted to four types of mechanical stresses (Fig. 6.3). The two main components are the axial one, $\sigma_z$, and the circumferential one, $\sigma_\theta$. The two other components are, namely $\sigma_r$ (radial stress) and $\tau_w$ (wall shear stress). The wall shear stress results from the friction of the blood onto the wall. The circumferential stress is related to the distension of the aorta with the variation of the blood pressure. It can reach about 150 kPa under normal conditions [21]. It can be approximated by the Laplace law according to:

$$\sigma_\theta = \frac{P \cdot r}{t} \tag{6.1}$$

where $P$ is the blood pressure, $r$ the internal aortic radius, and $t$ the thickness of the wall. If the number of MLUs varies according to the arterial diameter and across species [63], the average tension per MLU was shown to remain constant at $T = 2\ \text{N/m}$ [21], and its average circumferential stress can be determined by

$$\sigma_\theta = \frac{T}{t_{MLU}} \tag{6.2}$$

As the mean thickness of an MLU is about $t_{MLU} \simeq 15\,\mu\text{m}$, it was estimated that the average normal circumferential stress across the aorta is $\sigma_\theta = 133$ kPa [21].

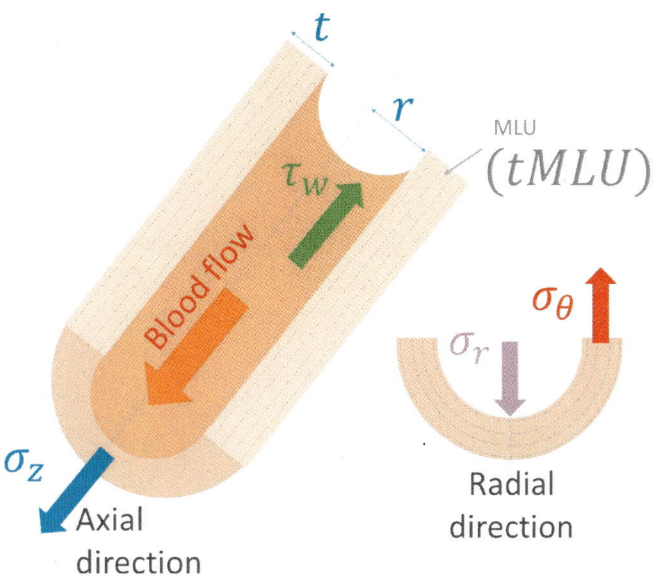


FIG. 6.3 Schematic representation of the mechanical stresses in the aortic wall, and particularly in the media. The intima is neglected in the case of aneurysms, but it cannot be the case for pathologies resulting in an intimal thickening; $t$ is the thickness of the wall, $r$ is the internal aortic radius, $P$ is the blood pressure, and $t_{MLU}$ is the mean thickness of an MLU.

### 6.2.3 Passive Mechanics of the Aortic Tissue

The passive behavior refers to the behavior of the aortic wall in the absence of vascular tone. It is mainly due to ECM components, namely elastin and collagen fibers. If the elastin is responsible for the wall elasticity, the collagen fibers are progressively tightened from their initial wavy configuration while the wall stress is increasing, and they tend to protect the other components from overstress [10, 21, 57].

Given that the tissue contains about 70%–80% of water, it is often assumed as incompressible. As a heterogeneous composite material comprising a fluid part (i.e., water) and a solid part (i.e., ECM and cells) [14, 27], divided into several layers with different mechanical properties (see Section 6.2.1.2), the aortic wall has a complex anisotropic mechanical behavior. To predict the rupture risk of ATAAs [14, 27, 57], the passive mechanical behavior of the ECM is relevant. Numerous in vitro tests using the bulge inflation device [7, 8, 64–67] confirmed that elastin in the media is the weak element of the wall toward rupture.

### 6.2.4 Multilayer Model of Stress Distribution Across the Wall

Single-layered homogenized models of arterial wall mechanics have provided important visions of arterial function. For example, Bellini et al. [68] proposed a bilayer model with different material properties for the media and adventitia layers. They split the passive contributions of elastin, SMC, and collagen fibers (modeled with four different families). Eventually, the strain-energy function (SEF) at every position may be written as [68, 69]:

$$W=\rho^{\mathrm{e}}W^{\mathrm{e}}(I_1^{\mathrm{e}})+\sum_{i=1}^{n}\rho^{\mathrm{c}_i}W^{\mathrm{c}_i}(I_4^{\mathrm{c}_i})+\rho^{\mathrm{m}}W^{\mathrm{m}}(I_4^{\mathrm{m}}) \tag{6.3}$$

where superscripts e, $\mathrm{c}_i$, and m represent, respectively, the elastin fiber constituent, the constituent made of each of the $n$ possible collagen fiber families, and the SMC constituent, with all these constituents making the mixture. In Eq. (6.3), $\rho^j$ refers to mass fraction, and $W^j$ stands for the stored elastic energy of each constituent, depending on the first ($I_1^j$) and fourth ($I_4^j$) invariants of the related constituents of the mixture ($j\in\{\mathrm{e}, \mathrm{c}_i, \mathrm{m}\}$). Let the mechanical behavior of the elastin constituent be described by a neo-Hookean SEF as in Refs. [68, 70–72]

$$W^{\mathrm{e}}(I_1^{\mathrm{e}})=\frac{\mu^{\mathrm{e}}}{2}(I_1^{\mathrm{e}}-3) \tag{6.4}$$

where $I_1^{\mathrm{e}}=tr(\mathbf{C}^{\mathrm{e}})$ and $\mu^{\mathrm{e}}$ is a material parameter with a stress-like dimension. $\mathbf{C}^{\mathrm{e}}=\mathbf{F}^{\mathrm{e}T}\mathbf{F}^{\mathrm{e}}$ denotes the right Cauchy-Green tensor where $\mathbf{F}^{\mathrm{e}}=\mathbf{F}\mathbf{G}_{\mathrm{h}}^{\mathrm{e}}$ is the deformation gradient of the elastin constituent. $\mathbf{F}$ is the corresponding deformation gradient of the arterial wall mixture and $\mathbf{G}_h^{\mathrm{e}}$ is the deposition stretch of elastin with respect to the reference

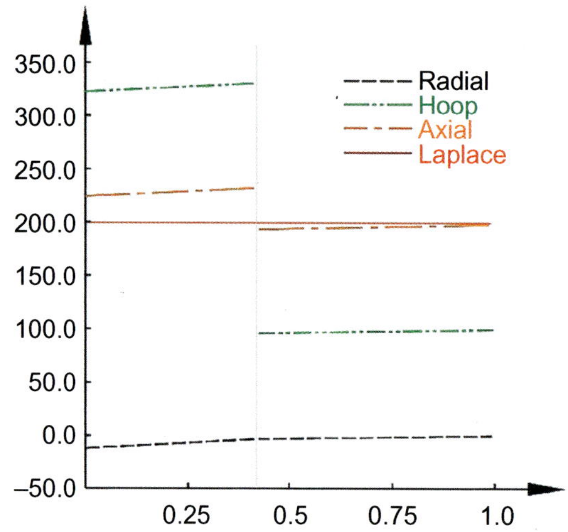


FIG. 6.4 Predicted transmural distributions of different components of Cauchy stress at mean arterial pressure (MAP ~93 mmHg) by bilayered model of Bellini et al. [68]. The mean circumferential stress, as obtained from Laplace's relation, is shown for comparison. All components of stress are plotted versus the normalized current radius, with 0 and 1 corresponding to the inner and outer radii, respectively.

configuration [68, 70]. Therefore, using the concept of constrained mixture theory, it is assumed that all constituents in the mixture deform together in the stressed configuration while each constituent has a different "total" deformation gradient based on its own deposition stretch.

The SEF of passive SMC and collagen contributions is described using an exponential expression such as [68, 70, 73, 74]:

$$W^k(I_4^k) = \frac{D_1^k}{4D_2^k}\left[\exp\left(D_2^k(I_4^k - 1)^2\right) - 1\right] \tag{6.5}$$

where $k \in \{c_i, m\}$. $D_1^k$ and $D_2^k$ are stress-like and dimensionless material parameters, respectively, and can take different values when fibers are under compression or tension [75]. $I_4^k = G_h^{k\,2}\mathbf{C}:\ \mathbf{M}^k \otimes \mathbf{M}^k$ where $G_h^k$, $k \in \{c_i, m\}$, is the specific deposition stretch of each collagen fiber family or SMCs, with respect to the reference configuration. $\mathbf{M}^k$, $k \in \{c_i, m\}$, denotes a unit vector along the dominant orientation of anisotropy in the reference configuration of the constituent made of the $i$th family of collagen fibers or of SMCs. For SMCs, $\mathbf{M}^m$ coincides with the circumferential direction of the vessel in the reference configuration while for the $i$th family of collagen fibers $\mathbf{M}^{c_i} = [0\ \sin\alpha^i\ \cos\alpha^i]$, where $\alpha^i$ is the angle of the $i$th family of collagen fibers with respect to the axial direction. $\mathbf{C} = \mathbf{F}^T\mathbf{F}$ is the right Cauchy-Green stretch tensor of the arterial wall mixture [68, 70].

This model can capture the stress "sensed" by medial SMCs and adventitial fibroblasts. The model shows interestingly that the stresses spit unevenly between the media and the adventitia (Fig. 6.4). For physiological pressures, the stress is significantly larger in the media but when the pressure increases, the stress increases faster in the adventitia. As this chapter is dedicated to SMCs, the model permitted estimating that stresses taken by SMCs remain less than a modest 40 kPa for normal physiological pressures [68].

## 6.3 ACTIVE BIOMECHANICAL BEHAVIOR

On top of its passive mechanical behavior, the aortic tissue exhibits an active component thanks to the tonic contraction of SMCs, permitting fast adaptation to sudden pressure variation during the cardiac cycle.

### 6.3.1 Smooth Muscle Cells

#### *6.3.1.1 SMC Structure*

SMCs have an elongated, fiber-like shape. Their length is about 50–100 μm and their mean diameter is 3 μm, reaching 5 μm around the nucleus [27, 76–78]. SMCs have an axial polarity. Their longest axis tends to align with the direction of the principal stress applied to the ECM (Fig. 6.2). Each MLU in the aorta contains a layer of SMCs that is connected to the elastic laminae thanks to microfibrils [21, 57]. They are circumferentially arranged throughout the media [27, 57, 76] and are particularly sensitive to $\sigma_z$ and $\sigma_\theta$ components of the wall stress [42]. This specific structure may also explain the fact that the media has been revealed as stronger circumferentially than longitudinally [14] and that the forces produced by the SMCs are maximized in this direction [76]. This ability of endothelial cells and SMCs to align along the direction of the applied stress has been confirmed by a number of in vitro studies [79–81].

The arrangement of SMCs in the media used to be controversial [82]. The most recent studies (1980s [63, 83], 1990s [84], and 2000 [85]) describe SMC orientation as circumferential whereas a helical and oblique disposition was reported earlier (1960s [86], 1970s [87]). Fujiwara and Uehara showed an oblique orientation in 1992 [76]. Likewise, data are controversial about alignment parallel to the vessel surface: Clark and Glagov [83] agree with this statement, unlike Fujiwara and Uehara [76]. Furthermore, some authors mention a change of SMC orientation in each subsequent MLU, creating a herringbone-like layout [85, 88]. Humphrey suggested that the SMCs are oriented helically, closer to a circumferential direction [27], but O'Connell suggested the SMCs may also be slightly radially tilted [82].

A recent study pointed out the importance of the helical disposition, suggesting that SMCs are oriented according to two intermingled helices [89]. This disposition was assumed in several tissue models [90, 91]. Moreover, a tissue model for coronaries taking into account the orientation of SMCs suggested they contribute both to circumferential and axial stresses and tend to reorient toward the circumferential direction when blood pressure is increased [92]. Other studies [83, 93] suggested that the almost circumferential orientation is only valid for inner MLUs of the ascending thoracic aorta because SMCs seem to orient more axially close to the adventitia. This pattern was also confirmed by Fujiwara and Uehara [76].

#### 6.3.1.2 Principle of SMC Contractility

The contractility of SMCs is their defining feature, thanks to a strongly contractile cytoskeleton. SMCs have a well-developed contractile apparatus organized in cross-linked actin bundles, regularly anchored into the membrane with dense bodies [94] (Fig. 6.5). This layout implies a bulbous morphological aspect during contraction [95]. There may be two types of actin filaments in the same bundle. The thick filament serves as a support for myosin heads and permits sliding of thin filaments during contraction, defining a so-called "contractile unit." Thin filaments are made of alpha smooth muscle actin ($\alpha$-SMA), an actin isoform specialized in the increase of cellular traction forces [24, 96, 97]. This isoform is specific to certain cell types, namely SMCs and myofibroblasts [45]. The $\alpha$-SMA filaments are created from their rod-like form, synthesized, and assembled when focal adhesions (FAs) undergo high stresses [24, 96]. Genetic mutations may affect the genes encoding the components of the contractile apparatus (Fig. 6.6) and lead to ATAAs (Table 6.1).

The main signaling pathways controlling SMC contraction are summarized in Fig. 6.6. More details about these pathways may be found in Refs. [26, 34, 46, 48, 51, 98]. However, it is important to mention that SMC contractility is controlled by the modulation of intracellular ionic calcium concentration $[Ca^{2+}]_i$. The SMC membrane has many invaginations called caveolae where extracellular $Ca^{2+}$ ions can enter the cell [99]. The increase of $[Ca^{2+}]_i$ triggers the contraction above a certain threshold, activating myosin chains [48]. Some studies revealed that $[Ca^{2+}]_i$ is a reliable indicator of SMC contractility because it increases from 100 nM in the relaxed state to 600–800 nM once fully contracted [27]. But Hill-Eubanks et al. [48] underlined later that a 400 nM concentration is sufficient to cause a complete contraction.

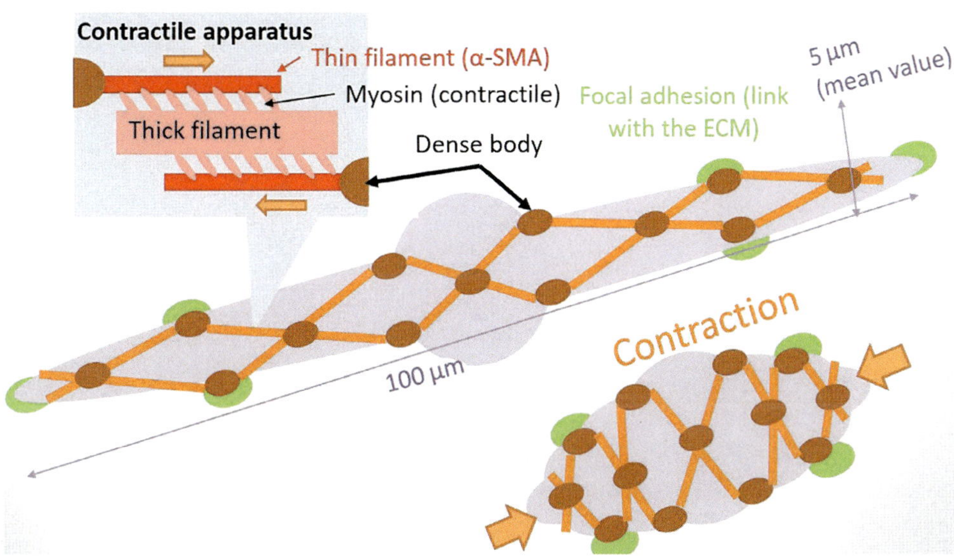


FIG. 6.5 Cellular and subcellular architecture of the SMC.

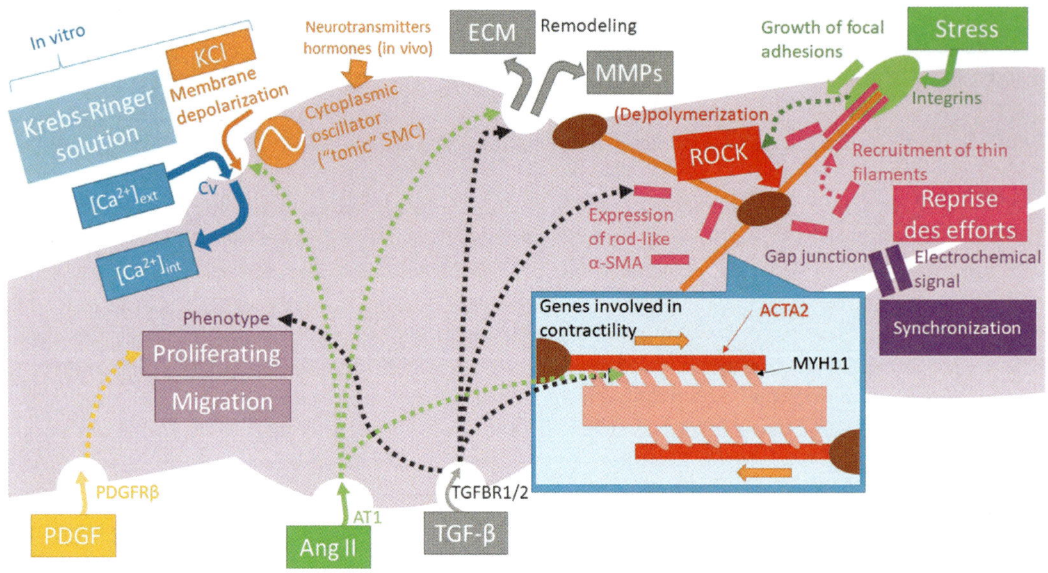


FIG. 6.6 Summary of the main signaling pathways involved in SMC contractility and phenotypic switching. *α*-SMA, alpha smooth muscle actin; Ang II, angiotensin II; AT1, angiotensin receptor; $C_v$, caveola; *PDGF*, platelet-derived growth factor; PDGFR*β*, PDGF receptor; ROCK, Rho kinase involved in cytoskeleton turnover; TGF-*β*, beta transforming growth factor; TGFBR1/2, TGF-*β* receptor. Specific genes mainly involved in the loss of contractility—ACTA2: encoding the *α*-SMA, MYH11: encoding the myosin heavy chains. Shape and orientation of an SMC according to its ECM. The SMCs can synchronize their contraction along their strongest axis thanks to intercellular interactions: between several SMCs through gap junctions, or between endothelial cells from the intima and innermost MLUs of SMCs through vasoactive agonists, neurotransmitters, or secreted GAGs.

Calcium entries in the SMC after some stimuli resulting in membrane depolarization are widely studied in vitro with electrical [95], electrochemical [39, 50, 100], chemical, or even mechanical stimulation [47, 101]. In fact, some of these studies suggested that SMCs undergo a progressive membrane depolarization as intraluminal pressure increases under normal conditions [47, 48, 101]. But when the mechanical stimulation becomes higher than normal, some studies have also highlighted that SMCs undergo more depolarization, resulting in an alteration of their reactivity [102].

The most common protocol used to control SMC contraction in vitro remains the addition of potassium ions $K^+$ from a KCl solution with a 50–80 mM concentration that depolarizes the membrane [39, 49, 50, 100, 102]. The extracellular media must also contain a calcium concentration $[Ca^{2+}]_e$ to cause the activation of myosin heads by calcium entry into the cell. This is the reason for adding the $CaCl_2$ solution to the media [49], or immersing the cells in a physiological Krebs-Ringer solution [39]. The latter has the advantage of keeping biological tissues alive. Calcium entry is also regulated thanks to a cytosolic oscillator that allows a periodic release of calcium from intracellular reservoirs (i.e., endoplasmic reticulum) [46, 103]. The frequency is highly dependent on external stimuli such as neurotransmitters, hormones, or growth factors. If its primary role is to induce a single cell contraction, the secondary role of the cytosolic oscillator is also responsible for membrane depolarization of neighboring cells in order to synchronize the contraction of several SMCs [46, 55]. The $Ca^{2+}$ signaling pathway was included in the mechanical cellular model of Murtada et al. [104] to model SMC contractility.

The angiotensin II (Ang II) signaling pathway has been widely developed in mice models and its link with aneurysms is well explained by Malekzadeh et al. [105]. It may lead to SMC contraction and may be used as a vasoconstrictor agonist in mice models [15, 105] or for isolated cells by addition in a bath [52]. But the review of Michel et al. suggests that angiotensin II may also damage the intima [15]. In this case, intimal degradation leads to the activation of other signaling pathways that have an influence on SMC tone. Moreover, some studies suggested a very active biological role of the intima through the secretion of nitric oxide (NO) that is involved in a pathway controlling cell relaxation [52, 102]. Accordingly, the intimal integrity seems to have a strong influence on cell contractile response.

The filament overlap involved in SMC contraction creates a "cross-bridge" whose function has been previously described by some subcellular models based on the sliding-filament theory [106, 107]. The cross-bridges have been assumed to be based on contractility activation/deactivation cycles through phosphorylation of the contractile unit. That is what Dillon et al. [108] have called the "latch state," which was used later in association with the

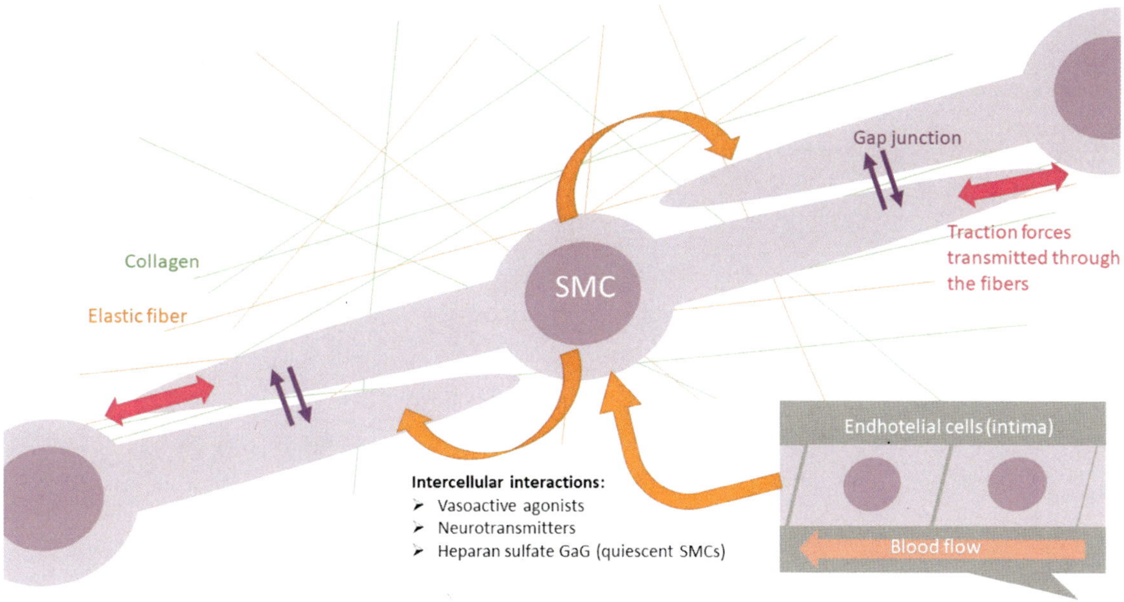


FIG. 6.7 Shape and orientation of an SMC according to its ECM. The SMCs can synchronize their contraction along their strongest axis thanks to intercellular interactions: between several SMCs through gap junctions, or between endothelial cells from the intima and innermost MLUs of SMCs through vasoactive agonists, neurotransmitters, or secreted GAGs.

sliding-filament theory to develop another subcellular model for the SMC contractile unit [39]. Several other cellular models combine the proper active behavior of SMCs with the passive behavior of its ECM [92, 109, 110]. The SMC is protected from a too high lengthening thanks to the intermediate filaments (made of desmine) linking dense bodies together [24].

#### 6.3.1.3 Intracellular Connections

Each SMC is covered by a basal lamina, a thin ECM layer (40–80 nm [27]) comprising type IV collagen, glycoproteins and binding proteins ensuring cell adhesion: the fibronectin and the laminin. The basal lamina represents about 12%–50% of the volume of SMCs. This lamina is open around the gap junctions to allow cell communication [94]. These junctions allow the cells to exchange electrochemical stimuli required to synchronize the contraction of the whole MLU and to match with the successive MLUs [111]. SMCs are linked together thanks to thin collagen microfibrils permitting the transmission of cell forces.

Interactions between the media and the other layer (intima and adventitia) also have to be considered. The synchronization of the contraction is induced in the outermost MLU by their innervation thanks to the vasa vasorum present in the adventitia, and the nervous signal is transmitted to inner MLUs thanks to gap junctions. The vasa vasorum also provides nutrients in the thickest arteries to complete the action of the intima for the innermost SMCs [27]. Moreover, endothelial cells communicate with the innermost SMCs (Fig. 6.7) secreting vasoactive agonists, neurotransmitters, and GAGs, notably heparan sulfate, which seems to influence the quiescence of the SMCs [29, 53]. Further information about this topic can be found in Lilly [54].

### 6.3.2 Multiscale Mechanics of SMC Contraction

#### 6.3.2.1 Subcellular Behavior

Many experiments were developed to characterize SMC traction forces thanks to $[Ca^{2+}]_i$ measurements [108] or traction force microscopy (TFM) techniques, from common substrate deformation methods [96, 101, 112] to uncommon specific microdevices [113–115]. SMC stiffness is closely linked to their contractile state [52, 116]. The reported values depend strongly on the measurement method. Common magnetic twisting cytometry gives a range of $[10^0-10^2]$ Pa against $[10^3-10^5]$ Pa for atomic force microscopy (AFM) [117]. Published stiffness and traction force values for SMCs are reported in Table 6.2.

If the AFM was mainly used on the ECM of aneurysm samples [33, 121], only Crosas-Molist et al. [118] characterized aortic SMCs using AFM and showed an increase of their stiffness in Marfan syndrome (from 3 kPa for a healthy tissue

TABLE 6.2 Stiffness and Traction Force Values for In Vitro SMCs

| SMC mechanical properties | Value [Pa] | Description | Ref. |
|---|---|---|---|
| **STIFFNESS** | | | |
| *$10^3-10^5$ PA* | *AFM TECHNIQUE* | | *[117]* |
| Viscoelastic properties | *In response to a vasoactive agonist (serotonin)* | | [116] (airways SMCs) |
| Storage modulus G′ | → | 150% increase | |
| Loss modulus G″ | → | 67% increase | |
| Hysterisis | → | 28% decrease after AFM stimulus: The cell elasticity prevails gradually more ("latch state") | |
| Elastic properties | *Comparison between control and Marfan-induced aneurysm tissue* | | |
| | 3k<br>7k | Increase in SMC and ECM stiffness in the pathological case | [118] (aortic SMCs) |
| | ↪ | Increase in focal adhesions size | |
| Young modulus | *Comparison between control and stimulated tissue with angiotensin II (vasoconstrictor)* | | |
| | 13.5k<br>18.5k | Increase in SMC stiffness after having their contraction induced (after 2 min) | [52, 116] |
| | 22k | After 30 min (actin polymerization dynamics) | |
| | ↪ | Increase in focal adhesions size (stronger adhesion to functionalized AFM tip with type I collagen) | |
| *$10^0-10^2$ PA* | *MAGNETIC TWISTING CYTOMETRY* | | *[117]* |
| *Increase in SMCs stiffness with substrate rigidity* | | | [119] |
| $12.6 \pm 1.6\ N/m^2$ | 12.6 | "Hard" substrate: high-density collagen | |
| $4.3 \pm 0.3\ N/m^2$ | 4.3 | "Soft" substrate: low-density collagen | |
| *Increase in SMC stiffness with contraction* | | | [119] |
| | | More effect on "Soft" substrate | |
| | | Increase linked to myosin head activation and actin polymerization | [120] (airways SMCs) |
| $9.91 \pm 0.75\ N/m^2$ | 9.9 | Unstimulated | |
| $14.27 \pm 0.85\ N/m^2$ | 14.3 | Vasoconstrictor agonist: serotonin | |
| **TRACTION FORCES** | | | |
| *TRACTION FORCES MEASURED ACCORDING TO THE CALCIUM CONCENTRATION OF THE KCL BATH* | | | |
| $2.9 \pm 0.4 \times 10^5\ N/m^2$ | 290k | $[Ca^{2+}] = 1.6$ mM | [108] (carotid SMCs) |
| $3.9 \pm 0.2 \times 10^5\ N/m^2$ | 390k | $[Ca^{2+}] = 25$ mM | |
| *TFM ON A PDMS MICRONEEDLE ARRAY WITH A FIBRONECTIN COATING, SIMULATING A SOFT MATERIAL: CELLULAR STRESS APPLIED BY THE ENTIRE CELL* | | | |
| SMCs applied stress, adhering to the pattern, unstimulated | 4.6k | Inhibition of the myosin contractility and the actin polymerization | [114] (airway SMCs) |
| Increase in the applied stress per needle with cell spreading | 10k | Weakly spread (440 $\mu m^2$) | |
| | 30k | Strongly spread (1520 $\mu m^2$) | |

TABLE 6.2 Stiffness and Traction Force Values for In Vitro SMCs—cont'd

| SMC mechanical properties | Value [Pa] | Description | Ref. |
|---|---|---|---|
| *TFM ON A GEL SUBSTRATE INCLUDING FLUORESCENT AND MAGNETIC MICROBEADS WITH FIBRONECTIN COATING: TRACTION FORCES MEASUREMENT AND MECHANICAL STIMULUS* | | | |
| 1 N/m$^2$ | 1 | Unstimulated SMCs | [101] (renal vSMCs) |
| 1.6 N/m$^2$ | 1.6 | Stimulated SMCs: 60% increase | |
| *STANDARD TFM ON GEL SUBSTRATE WITH FLUORESCENT MICROSPHERES: MEASUREMENT OF THE DEFORMATION FIELD AFTER A CHEMICAL STIMULUS* | | | |
| Increase in mean traction force (mean vector of the deformation field) | 50 | Unstimulated SMCs | [112] (airway SMCs) |
| | 100 | Vasoconstrictor agonist: histamine | |
| *THE CONTRACTILE APPARATUS IS MADE OF NONCONTRACTILE THICK FILAMENTS, LINKED TO CLASSIC FOCAL ADHESIONS (FA), AND HIGHLY CONTRACTILE α-SMA FILAMENTS, LINKED TO SUPER FOCAL ADHESIONS (SUFA)* | | | |
| Increase in mean traction force (mean vector of the deformation field) | 8.5k | Stress produced by suFAs | [96] (myofibroblasts) |
| | 3.1k | Stress produced by classic FAs | |

to 7 kPa for a pathological one). Interestingly, another team focused on rat vascular SMCs (without aneurysm) and tested them by AFM indentation with a functionalized tip to measure the adhesion forces to type I collagen. This work suggested that contracted (with Ang II) or relaxed SMCs regulate their FAs [52].

#### 6.3.2.2 (Sub)cellular Models for the SMC

A common mechanical model of the SMCs and their contractile apparatus is the sliding filament theory. The original sliding filament theory permitted modeling the $\alpha$-SMA filaments sliding on myosin heads. It was initially published by Huxley and Huxley in 1953 [107]. The filament overlap (i.e., thin filaments linked to thick filaments by myosin heads, see Section 6.3.1.2) creates a "cross-bridge" modeling the contractile unit of a single SMC.

Another important study was that of Dillon et al. [108], where the latch state was introduced to describe the activation/blockage cycles of the contractile unit through the phosphorylation process. In low phosphorylation states, the active force can be maintained by the SMCs [106]. Dillon et al. [108] also highlighted that SMCs generate a maximal force when stretched at an optimal length. Gradually, further models took into account the orientation of the SMCs in the media [92] and the interaction between the cell and its ECM [109, 110, 122]. Only Murtada's model [104] has integrated the regulation of $[Ca^{2+}]_i$ controlling SMC contraction (see Section 6.3.1.2).

### 6.3.3 Effect of SMC Contraction on the Distribution of Stresses Across the Aortic Wall

The effects of SMC contraction on the stress distribution across the wall were investigated in several studies, whose the experimental work that consists in measuring the opening angle on radially cut aortic rings. This permitted assessing the intramural stress induced by SMC contractility [122, 123]. Indeed, the opening angle experiment reveals residual stresses that can be related to passive ECM mechanics and to SMC active contraction [122, 123].

It was shown that at physiological pressure, the pure passive response of the wall does not ensure uniform stress distributions, suggesting an essential role of the basal tone of SMCs to maintain a uniform stress distribution [122, 123]. But under the effects of a vasoactive agonist, SMCs contract through myogenic response and can provoke a rise of intraluminal pressure up to 200 mmHg. As they change the intraluminal pressure, SMCs may also induce nonuniform stress distributions across the wall [123].

In summary, SMCs are very sensitive to mechanical stimuli. They tend to keep the intramural circumferential stress as uniform as possible for physiological variations of the blood pressure, but the stress becomes nonuniform for higher pressures [123]. They adapt their myogenic response, which ranges from 50 kPa for the basal tone under normal physiological conditions to 100 kPa for maximal SMC contraction [122].

## 6.4 MECHANOSENSING AND MECHANOTRANSDUCTION

Given their highly sensitive cytoskeleton and FAs (Fig. 6.8), SMCs represent real sensors of the local mechanochemical state of the ECM. Many experimental models have permitted the investigation of this mechanosensing role and how it is involved in ATAAs and dissections [15]. One of the main responses to stimuli mechanosensing is mechanotransduction [43, 124–127], which is the process of transducing wall stress stimuli into tissue remodeling [21, 27, 32, 37, 42].

### 6.4.1 Mechanosensing

Many recent studies have highlighted the effects of the environment on the SMC response, in terms of protein synthesis, proliferation, migration, differentiation, or apoptosis, thanks to its mechanosensitive architecture [43, 124–128]. Mechanosensing relies on links between the ECM, FAs, and the cytoskeleton. The microfibrils provide an adhesive support to the SMCs through collagen VI [59]. Because of this, when microfibrils are damaged, SMCs sense an increase in stiffness and are no longer able to transmit forces to each other through elastic fiber. According to several studies, the elastin acts for the maturation of the contractile apparatus of the SMCs and may encourage their quiescent phenotype [9, 41, 129]. In the ECM, two proteins are mainly involved in mechanosensing: fibronectin and laminin. Fibronectin is known for being mainly present in the ECM of blood vessels during early development and seems to encourage SMC proliferation and migration in order to build the tissue [34, 36]. On the contrary, the laminin may be required later for SMC maturation toward a contractile phenotype [130].

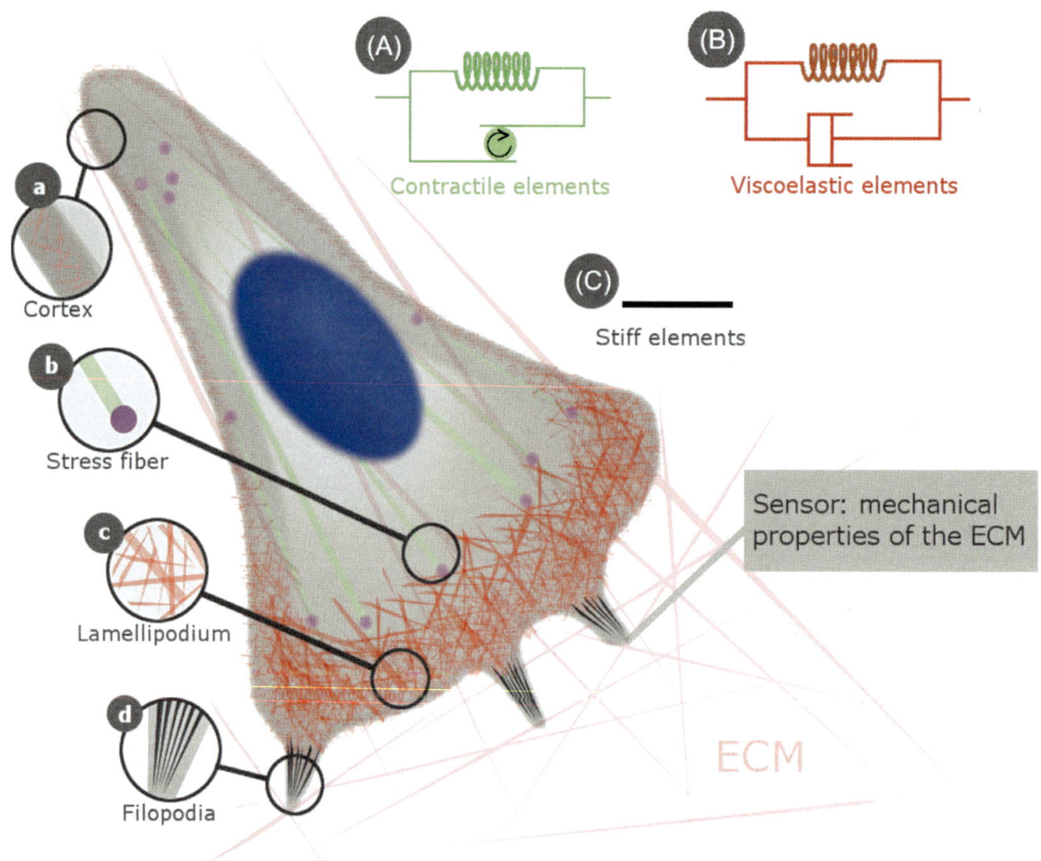


FIG. 6.8 The cytoskeleton of the migratory cell is composed of a complex network of actin bundles where three specific structures are observed: (a) The cortex: a cross-linked network surrounding the cell and ensuring the modulation of cell shape by rapid turnover during migration. (b) Stress fiber: contractile structures made of antiparallel actin bundles linked to a molecular motor, the myosin. (c) Lamellipodium: large membrane extension made of cross-linked and branched bundles, pushing the cell forward. (d) Filopodia: thin membrane extension made of parallel bundles, and projected forward to sense the mechanical properties of the substrate. Each of these structures has its proper mechanical behavior: (A) Contractile elements activated by the myosin motor. (B) Viscoelastic elements based on a Kelvin-Voigt model (viscous damper and elastic spring connected in parallel). (C) Stiff elements associated with filopodium. Here the cell is represented on a 2D substrate instead of the 3D real ECM. That is why the cell has adopted an apico-basal polarity. The apical side refers to the unattached membrane above the nucleus, and the basal one to the contact with the substrate, through focal adhesions. *Based on the lecture of E. Planus, Cours de M1, 2017 (non-disponible en accäs libre).*

Because SMCs are dynamic systems, their cytoskeleton remains in constant evolution during cellular processes. It consists in a dense fibrous actin structure that allows the cell for shape maintenance and generation of traction forces required notably during migration (Fig. 6.8). The cytoskeleton of SMCs is particularly rich in contractile $\alpha$-SMA thin filaments (see Section 6.3.1) that are used to enhance the traction forces required for cell function. SMC contraction involves a quick remodeling of its cytoskeleton in order to recruit contractile thin filaments in the direction of applied forces [24, 96] and to follow its change of shape while renewing noncontractile cortical structures [52].

In summary, the SMC may be considered a powerful sensor of the mechanical state across the aortic wall. The high sensitivity of SMCs led many research teams to point out their implication in arterial disease, including aortic aneurysms [20, 22, 27, 36, 42, 77, 129].

## 6.4.2 The Key Role of SMCs in ATAAs

The role of SMCs in the development of ATAAs is now well accepted [57, 76].

Several studies have already mentioned the change of SMC behavior in cardiovascular disease, and the consequences on the arterial wall. It was shown that hypertension is perceived by SMCs as permanent stimuli through the increase of wall stress, which induces collagen synthesis to reinforce the wall resulting in an increasing thickness [15, 27, 98, 131].

In atherosclerosis and restenosis, the growth of plaques between the media and the intima is due to SMC proliferation and migration toward the intima, forming a neointima [40, 129]. The neointimal SMCs are also able to gather lipids, increasing the stiffness and weakening the wall. Intimal integrity may also control the quiescence of SMCs thanks to heparan sulfate [29, 53] or vasoactive agonist [27, 52] synthesis. Hence, the degradation of the endothelial cell layer leads to SMC proliferation and ECM synthesis until whole intima repair [35].

All these changes suggest that SMCs can switch to another phenotype in order to repair the damaged tissue through migration toward the injured region, proliferation, and ECM synthesis [35]. Under normal conditions, mature SMCs acquire a "contractile" (C) phenotype from an immature "synthetic" (S) one, which is mainly present in early development [40, 77, 129]. But SMCs demonstrate a high plasticity as they are not fully differentiated cells, and they can return to an (S) phenotype in response to many stimuli. The phenotypic switching is due to a number of factors, summarized in Table 6.3.

The cytoplasm of (S) SMCs has more developed synthetic organites such as endoplasmic reticulum and Golgi apparatus, leading to hypertrophy [27]. The phenotypic switching does not radically change the cytoskeleton as

TABLE 6.3 SMC Phenotypic Switching Characteristics

| SMCs phenotypic switching | Effects | |
|---|---|---|
| High plasticity | The SMCs are not entirely differentiated when they reach maturity through the contractile (C) phenotype, and can move on to a synthetic (S) phenotype | [21, 27, 32, 35, 36, 56, 77] |
| | (The (S) phenotype is mainly present in the aorta during early development) | |
| ECM synthesis and degradation (through MMP synthesis) | The SMCs undergo an increase in volume (hypertrophy), with the development of their synthetic organites (Golgi apparatus and endoplasmic reticulum) | [27, 29, 40, 56] |
| Loss of quiescence: hyperplasia | (S) SMCs tend to proliferate and migrate | [77] |
| Loss of contractility | Stress produced into the wall: | [21] |
| | (C) SMCs: 100 kPa; (S) SMCs: 5–10 kPa | |
| Degradation of the contractile apparatus | The cytoskeleton is not entirely remodeled (undamaged microtubules), but there are weaker actin and myosin concentrations (contractile fibers) in (S) SMCs | [29, 35, 41, 129] |
| Modification of the basal side | Regulation of the focal adhesions | [41, 52, 56] |
| | (They grow according to the traction force direction, ensuring a strong adhesion to the ECM in response to high stress) | [24, 96] |
| Decrease in $\alpha$-SMA concentration | Degradation of the thin filaments that are responsible for amplifying and regulating the cell traction forces | [24, 97] |
| Reversible process | Once the tissue is repaired, the SMCs return to a contractile phenotype | [32, 41, 56] |
| General apoptosis | Decrease of SMC number and degradation of the ECM => loss of wall elasticity and resistance | [28, 42, 132] |

microtubules remain intact, but the contractile apparatus (i.e., $\alpha$-SMA thin filaments) is clearly affected [24, 29, 35, 41, 97, 129]. SMC contractility involves a reorganization of their contractile apparatus. In other words, high traction forces require high adhesion to the ECM; hence, it is suggested that SMCs undergo a regulation of their FAs [24, 52, 96, 133], evolving toward super focal adhesions (suFAs) in the direction of the applied stress [24, 96].

Hyperplasia concerns the loss of SMC quiescence in favor of a proliferating and migrating behavior [27, 77]. During hypertension, the increase in wall thickness has been shown to result more from hypertrophy than hyperplasia [31, 98], but the two phenomena are involved in several pathological states [27]. ATAAs also involve a reduction of the elastin/collagen ratio in the aortic wall, inducing a stiffness increase and leading to phenotypic switching of SMCs [28]. But the whole thickness is not uniformly affected: Tremblay et al. [12] have assessed SMC densities across ATAAs and deduced that it was greater in the outer curvature. The reduced contractile behavior suggests more phenotypic switching in this area.

### 6.4.3 SMC Mechanotransduction

As previously highlighted in several studies [21, 68], SMCs tend to remain in a specific mechanical state called homeostasis. It is considered a reference value for the stress they undergo in the wall under normal physiological pressure. During any cardiac cycle, SMCs do not activate suddenly their contractile apparatus according to the short variations of blood pressure. In fact, they always remain partially contracted and tend to adapt gradually to any constant increase of the mean pressure (Fig. 6.9).

Facing a constant rise of wall stress, the SMC response may be divided into two main categories according to time. In the short term, SMCs react in a progressive contraction until they reach maximal contraction, permitting the regulate of the blood flow through arterial diameter control. But beyond a given stress threshold, collagen fibers from the adventitia are recruited to protect the cells and the medial elastic fibers from higher stress values [27, 68]. In the long term, the remaining mechanical stimuli of SMCs lead to phenotypic switching or apoptosis. In this way, SMCs tend to coordinate the renewal of ECM, particularly through the synthesis of new collagen fibers to increase the wall resistance to high stress.

### 6.4.4 Mechanical Homeostasis in the Aortic Wall

Mechanical homeostatis means that SMCs try to regulate their contractile apparatus and their surrounding ECM to maintain a target wall stress corresponding to a certain mechanobiological equilibrium. The presence of a mechanobiological equilibrium was first proposed by the constant mean tension of a single MLU in a stressed aorta in spite of different species and aortic diameter [21, 63]. Humphrey [21] estimated that the circumferential stress per MLU is about $\sigma_\theta = 133$ kPa. It is assumed that SMCs and fibroblasts tend to maintain a preferred mechanical state through homeostasis. Kolodney et al. showed that cultured fibroblasts on unloaded gel substrates generate a steady tension of 3.2 kPa [134]. Moreover, Humphrey [21] suggested that homeostasis expression is similar throughout scales, from

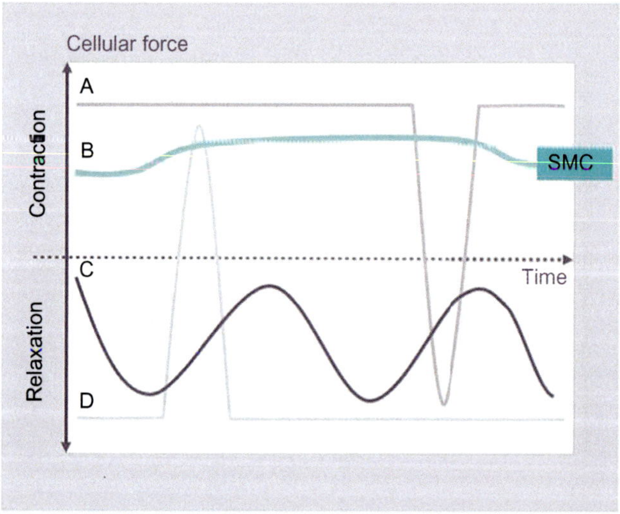


FIG. 6.9 All the smooth muscles do not have the same behavior. The SMCs are normally partially contracted and adapt their contractile response and maintain it for a long time. (A) Normally contracted: Sphincters. (B) Normally partially contracted (tone): blood vessels, airways. (C) Phasically active: stomach, intestines. (D) Normally relaxed: esophagus, urinary bladder. *Modified from B.M. Koeppen, B.A. Stanton, Berne and Levy Physiology, updated edition E-Book, Elsevier Health Sciences, Amsterdam, 2009.*

the organ level (vessel mechanoadaptation), the tissue level (ECM prestressing and synthesis/degradation), the cellular level (traction forces applied onto the ECM, see Section 6.3.2.1), the subcellular level (FAs and actin/myosin bundles), and even the molecular level ($[Ca^{2+}]_i$). These findings suggest that the cell is able to adapt its proper stress state through the regulation of $[Ca^{2+}]_i$, the cytoskeleton and FA turnover, and by controlling its surrounding ECM as well. Experimental studies of Matsumo et al. [123] showed a change of the intramural strain distribution in response to SMC contraction (and relaxation) on radially cut aortic rings and confirmed that SMCs actively adapt their contractile state to keep the intramural stress uniform. In summary, SMCs can both work actively (through contraction/relaxation) and passively by deposition and organization of the ECM [21].

That is why SMCs may undergo a phenotypic switching toward a synthetic one under several stimuli (see Section 6.4.2). Through phenotypic switching, SMCs tend to remodel their ECM to go back to a preferred state and face the variations of their environment. Humphrey has well described this equilibrium state saying: "When a homeostatic condition of the blood vessel is disturbed, the rate of tissue growth is proportional to the increased stress" [27]. But SMCs lose their contractility in return and may irrevocably affect the wall vasoactivity in which they may have a key role [21, 23, 50]. In fact, Humphrey suggests in his review that fully contractile SMCs can react mainly to circumferential wall stress (150 kPa in physiological conditions) with 100 kPa equivalent traction forces exerted on their ECMs while synthetic SMCs may only apply 5–10 kPa [21].

### 6.4.5 Consequences for Aortic Tissue

Reduction or loss of SMC contractility alters the stress distribution across the aortic wall [41, 52, 56, 97]. In reaction, the development of synthesis abilities ensures recovery processes by ECM remodeling. SMCs keep a key role in the aortic wall remodeling. In ATAAs, they tend to adapt their response through complex signaling pathways. An important one is Rho kinase (ROCK), which is mainly involved in cytoskeleton turnover for the control of cell shape and movement during migration [23]. The Rho kinase seems to influence the formation of $\alpha$-SMA thin filaments and the regulation of FAs that are involved in SMC contractility and anchoring to the ECM [24, 44, 96, 133]. Moreover, the oxidative stress induced by ATAAs enhances the inflammatory response of SMCs, increasing MMP synthesis and the further disruption of elastin fibers [18].

Remodeling was shown to be uneven in human [13] and porcine [12] aortic tissues. The authors highlighted that the outer curvature of the ATAAs is more affected. Remodeling implies phenotypic switching toward a synthetic phenotype able to synthesize both ECM compounds (i.e., collagen and glycoproteins) and MMPs to degrade the "dysfunctional" ECM, leading to ECM wear [27, 29, 40, 56]. Likewise, elastin degradation results in a permanent decrease of the elastin/collagen ratio because elastic fibers cannot be regenerated in adulthood [9]. On top of the induced stiffening, the ability of SMCs to restore a healthy state is altered as well, as it was shown that elastin is also important for activating actin polymerization [37]. Moreover, SMCs undergo a general apoptosis to reduce their number when they sense an inappropriate chemomechanical state, inducing further reduction of elasticity and mechanical resistance through a vicious circle loop [18, 28, 42, 132].

### 6.4.6 Toward an Adaptation of SMCs in ATAAs?

Any disruption of the mechanical or chemical homeostasis is interpreted by the SMCs as a distress signal, and several recovery processes can be activated in reaction, but the regulation loop is similar to a vicious circle because of the complexity to return naturally to equilibrium (Fig. 6.10). Interestingly, in hypertension, the increase of wall stress results in an increase in the arterial diameter [27, 131]. Conversely, a decrease in mean wall stress leads to atrophy [27]. Because the (*S*) SMCs can recover their (*C*) phenotype once the tissue returns to its original homeostatic stress, the phenotypic switching seems to be a reversible process [32, 41, 56]. These observations suggest a two-way mechanoadaptive process. But once affected by ATAAs, remodeled aortic ECM is known not to reach complete recovery, particularly because disrupted elastic fibers cannot be rebuilt in adulthood [9]. Aortic tissue would, therefore, evolve more or less quickly according to some factors that may slow it down. As the review study of Michel et al. [15] has already pointed out, ATAAs may result in some epigenetic modifications that have an influence on the cellular response. It could be defined as the acquisition of new constant and heritable traits without requiring any change in the DNA sequence, which results, for instance, in gene modulation. The suggested theory explains that SMC reprogramming is likely to induce a progressive dilatation of the aorta without dissection, whereas no SMC reprogramming promotes acute rupture of the wall [15]. Finally, it is well accepted that SMCs play a major role in controlling the wall evolution after aortic injury, either toward partial recovery of initial mechanical properties or fatal rupture through dissection.

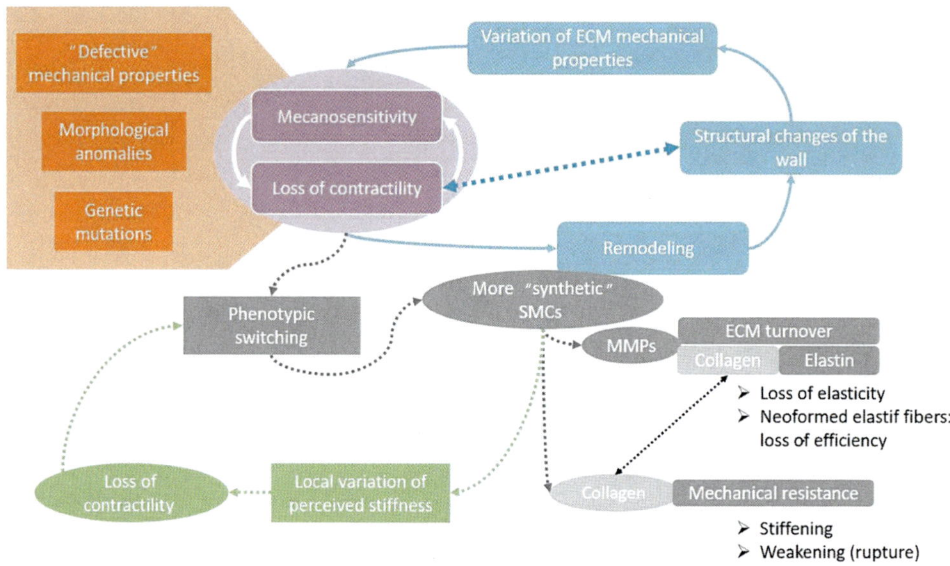


FIG. 6.10 A vicious circle: under some stimuli (*orange box*), the homeostatic state is endangered and the cell (*violet circle*) sets up complex chain reactions. Several regulation loops cross each other (*blue, gray, green*) and thus the equilibrium is difficult to reach again, especially as the elastin degradation leads to a permanent loss of elasticity and this effect is amplified with the duration of the pathological sate. Hence, although the phenotypic switching may be reversible, the permanent alteration of the ECM may prevent the cells from going back to homeostatic conditions that enhance the pathology.

However, the quantification of levels of SMC contractility that results in one type of evolution or another is still an open issue. There is still a pressing need to characterize the basal tone of SMCs in healthy aortas and ATAAs at the cellular scale.

## 6.5 SUMMARY AND FUTURE DIRECTIONS

The mechanobiology and physiopathology of the aorta have received much attention so far, but there is still a pressing need to characterize the roles of SMCs at the cellular scale. Despite the difficulties of characterizing cells having complex dynamics and fragility, techniques such as AFM or traction force microscopy could permit important progress to be made about SMC nanomechanics and provide relevant information on how SMC biomechanics are related to the irreversible alteration of stress distribution in ATAAs. This will also imply the development of new biomechanical models of the aortic wall, taking into account the contractility of SMCs.

## Acknowledgments

The authors are grateful to the European Research Council for grant ERC-2014-CoG BIOLOCHANICS. They are also thankful to E. Planus and T.C. Gasser for the courtesy of the corresponding figures.

## References

[1] M.J. Bown, A.J. Sutton, P.R.F. Bell, R.D. Sayers, A meta-analysis of 50 years of ruptured abdominal aortic aneurysm repair, Br. J. Surg. 89 (6) (2002) 714–730.
[2] J.Z. Goldfinger, J.L. Halperin, M.L. Marin, A.S. Stewart, K.A. Eagle, V. Fuster, Thoracic aortic aneurysm and dissection, J. Am. Coll. Cardiol. 64 (16) (2014) 1725–1739.
[3] E.M. Isselbacher, Thoracic and abdominal aortic aneurysms, Circulation 111 (6) (2005) 816–828.
[4] M. Fillinger, Who should we operate on and how do we decide: predicting rupture and survival in patients with aortic aneurysm, Semin. Vasc. Surg. 20 (2) (2007) 121–127.
[5] M.F. Fillinger, M.L. Raghavan, S.P. Marra, J.L. Cronenwett, F.E. Kennedy, In vivo analysis of mechanical wall stress and abdominal aortic aneurysm rupture risk, J. Vasc. Surg. 36 (3) (2002) 589–597.
[6] A.R. Marquez, Caractérisation Biomécanique des Anévrismes de L'aorte Thoracique Ascendante (PhD thesis), École Nationale Supérieure des Mines de Saint-Etienne, 2014.
[7] O. Trabelsi, F.M. Davis, J.F. Rodriguez-Matas, A. Duprey, S. Avril, Patient specific stress and rupture analysis of ascending thoracic aneurysms, J. Biomech. 48 (10) (2015) 1836–1843.
[8] A. Duprey, O. Trabelsi, M. Vola, J.-P. Favre, S. Avril, Biaxial rupture properties of ascending thoracic aortic aneurysms, Acta Biomater. 42 (suppl C) (2016) 273–285.
[9] S.M. Arribas, A. Hinek, M.C. Gonzlez, Elastic fibres and vascular structure in hypertension, Pharmacol. Ther. 111 (3) (2006) 771–791.
[10] H. Wolinsky, S. Glagov, Structural basis for the static mechanical properties of the aortic media, Circ. Res. 14 (5) (1964) 400–413.

[11] H. Sawada, D.L. Rateri, J.J. Moorleghen, M.W. Majesky, A. Daugherty, Smooth muscle cells derived from second heart field and cardiac neural crest reside in spatially distinct domains in the media of the ascending aorta—brief report, Arterioscler. Thromb. Vasc. Biol. 37 (9) (2017) 1722–1726.

[12] D. Tremblay, R. Cartier, R. Mongrain, R.L. Leask, Regional dependency of the vascular smooth muscle cell contribution to the mechanical properties of the pig ascending aortic tissue, J. Biomech. 43 (12) (2010) 2448–2451.

[13] N. Choudhury, O. Bouchot, L. Rouleau, D. Tremblay, R. Cartier, J. Butany, R. Mongrain, R.L. Leask, Local mechanical and structural properties of healthy and diseased human ascending aorta tissue, Cardiovasc. Pathol. 18 (2) (2009) 83–91.

[14] S. Pasta, A. Rinaudo, A. Luca, M. Pilato, C. Scardulla, T.G. Gleason, D.A. Vorp, Difference in hemodynamic and wall stress of ascending thoracic aortic aneurysms with bicuspid and tricuspid aortic valve, J. Biomech. 46 (10) (2013) 1729–1738.

[15] J.-B. Michel, G. Jondeau, D.M. Millewicz, From genetics to response to injury: vascular smooth muscle cells in aneurysms and dissections of the ascending aorta, Cardiovasc. Res. 114 (4) (2018) 578–589.

[16] A.J. Barker, M. Markl, J. Bürk, R. Lorenz, J. Bock, S. Bauer, J. Schulz-Menger, F. von Knobelsdorff-Brenkenhoff, Bicuspid aortic valve is associated with altered wall shear stress in the ascending aorta, Circ. Cardiovasc. Imaging 5 (4) (2012) 457–466.

[17] S. Schnell, D.A. Smith, A.J. Barker, P. Entezari, A.R. Honarmand, M.L. Carr, S.C. Malaisrie, P.M. McCarthy, J. Collins, J.C. Carr, M. Markl, Altered aortic shape in bicuspid aortic valve relatives influences blood flow patterns, Eur. Heart J. Cardiovasc. Imaging 17 (11) (2016) 1239–1247.

[18] F.J. Miller, W.J. Sharp, X. Fang, L.W. Oberley, T.D. Oberley, N.L. Weintraub, Oxidative stress in human abdominal aortic aneurysms: a potential mediator of aneurysmal remodeling, Arterioscler. Thromb. Vasc. Biol. 22 (4) (2002) 560–565.

[19] M.M. Dua, R.L. Dalman, Hemodynamic influences on abdominal aortic aneurysm disease: application of biomechanics to aneurysm pathophysiology, Vascul. Pharmacol. 53 (1–2) (2010) 11–21.

[20] A.J. Brownstein, B.A. Ziganshin, H. Kuivaniemi, S.C. Body, A.E. Bale, J.A. Elefteriades, Genes associated with thoracic aortic aneurysm and dissection: an update and clinical implications, Aorta (Stamford, CN) 5 (1) (2017) 11–20.

[21] J.D. Humphrey, M.A. Schwartz, G. Tellides, D.M. Milewicz, Role of mechanotransduction in vascular biology: focus on thoracic aortic aneurysms and dissections, Circ. Res. 116 (8) (2015) 1448–1461.

[22] D.-C. Guo, H. Pannu, V. Tran-Fadulu, C.L. Papke, R.K. Yu, N. Avidan, S. Bourgeois, A.L. Estrera, H.J. Safi, E. Sparks, D. Amor, L. Ades, V. McConnell, C.E. Willoughby, D. Abuelo, M. Willing, R.A. Lewis, D.H. Kim, S. Scherer, P.P. Tung, C. Ahn, L.M. Buja, C. S. Raman, S.S. Shete, D.M. Milewicz, Mutations in smooth muscle alpha-actin (ACTA2) lead to thoracic aortic aneurysms and dissections, Nat. Genet. 39 (12) (2007) 1488–1493.

[23] S.-Q. Kuang, C.S. Kwartler, K.L. Byanova, J. Pham, L. Gong, S.K. Prakash, J. Huang, K.E. Kamm, J.T. Stull, H.L. Sweeney, D.M. Milewicz, Rare, non-synonymous variant in the smooth muscle-specific isoform of myosin heavy chain, MYH11, R247C, alters force generation in the aorta and phenotype of smooth muscle cells, Circ. Res. 110 (11) (2012) 1411–1422.

[24] J. Chen, H. Li, N. Sundarraj, J.H.-C. Wang, Alpha-smooth muscle actin expression enhances cell traction force, Cell Motil. Cytoskeleton 64 (4) (2007) 248–257.

[25] E. Gillis, L.V. Laer, B.L. Loeys, Genetics of thoracic aortic aneurysm: at the crossroad of transforming growth factor-$\beta$ signaling and vascular smooth muscle cell contractility, Circ. Res. 113 (3) (2013) 327–340.

[26] D.M. Milewicz, K.M. Trybus, D. Guo, H.L. Sweeney, E. Regalado, K. Kamm, J.T. Stull, Altered smooth muscle cell force generation as a driver of thoracic aortic aneurysms and dissections, Arterioscler. Thromb. Vasc. Biol. 37 (1) (2016).

[27] J.D. Humphrey, Cardiovascular Solid Mechanics, Springer, New York, NY, 2002.

[28] N. Mao, T. Gu, E. Shi, G. Zhang, L. Yu, C. Wang, Phenotypic switching of vascular smooth muscle cells in animal model of rat thoracic aortic aneurysm, Interact. Cardiovasc. Thorac. Surg. 21 (1) (2015) 62–70.

[29] L. Rubbia, G. Gabbiani, Phénotype des cellules musculaires lisses artérielles et athérosclérose, Médecine/Sciences 5 (6) (1989) 389.

[30] A. Tsamis, J.T. Krawiec, D.A. Vorp, Elastin and collagen fibre microstructure of the human aorta in ageing and disease: a review, J. R. Soc. Interface 10 (83) (2013) 20121004.

[31] G.K. Owens, P.S. Rabinovitch, S.M. Schwartz, Smooth muscle cell hypertrophy versus hyperplasia in hypertension, Proc. Natl Acad. Sci. USA 78 (12) (1981) 7759–7763.

[32] J. Thyberg, K. Blomgren, U. Hedin, M. Dryjski, Phenotypic modulation of smooth muscle cells during the formation of neointimal thickenings in the rat carotid artery after balloon injury: an electron-microscopic and stereological study, Cell Tissue Res. 281 (3) (1995) 421–433.

[33] Z. Chen, Y. Xu, P. Bujalowski, A.F. Oberhauser, P.J. Boor, N-(2-aminoethyl) ethanolamine-induced morphological, biochemical, and biophysical alterations in vascular matrix associated with dissecting aortic aneurysm, Toxicol. Sci. 148 (2) (2015) 421–432.

[34] P. Reusch, H. Wagdy, R. Reusch, E. Wilson, H.E. Ives, Mechanical strain increases smooth muscle and decreases nonmuscle myosin expression in rat vascular smooth muscle cells, Circ. Res. 79 (5) (1996) 1046–1053.

[35] J. Thyberg, K. Blomgren, J. Roy, P.K. Tran, U. Hedin, Phenotypic modulation of smooth muscle cells after arterial injury is associated with changes in the distribution of laminin and fibronectin, J. Histochem. Cytochem. 45 (6) (1997) 837–846.

[36] J. Thyberg, U. Hedin, M. Sjölund, L. Palmberg, B.A. Bottger, Regulation of differentiated properties and proliferation of arterial smooth muscle cells, Arteriosclerosis (Dallas, TX) 10 (6) (1990) 966–990.

[37] J.H.-C. Wang, J.-S. Lin, Cell traction force and measurement methods, Biomech. Model. Mechanobiol. 6 (6) (2007) 361–371.

[38] T.R. Murray, B.E. Marshall, E.J. Macarak, Contraction of vascular smooth muscle in cell culture, J. Cell Physiol. 143 (1) (1990) 26–38.

[39] S.-I. Murtada, J. Ferruzzi, H. Yanagisawa, J.D. Humphrey, Reduced biaxial contractility in the descending thoracic aorta of Fibulin-5 deficient mice, J. Biomech. Eng. 138 (5) (2016) 051008.

[40] E.W. Raines, R. Ross, Smooth muscle cells and the pathogenesis of the lesions of atherosclerosis, Br. Heart J. 69 (1 suppl) (1993) S30–S37.

[41] E. Stadler, J.H. Campbell, G.R. Campbell, Do cultured vascular smooth muscle cells resemble those of the artery wall? If not, why not? J. Cardiovasc. Pharmacol. 14 (suppl 6) (1989) S1–S8.

[42] C. Li, Q. Xu, Mechanical stress-initiated signal transduction in vascular smooth muscle cells in vitro and in vivo, Cell. Signal. 19 (5) (2007) 881–891.

[43] J.S. Park, J.S. Chu, A.D. Tsou, R. Diop, Z. Tang, A. Wang, S. Li, The effect of matrix stiffness on the differentiation of mesenchymal stem cells in response to TGF-$\beta$, Biomaterials 32 (16) (2011) 3921–3930.

[44] A.P. Somlyo, A.V. Somlyo, $Ca^{2+}$ sensitivity of smooth muscle and nonmuscle myosin II: modulated by G proteins, kinases, and myosin phosphatase, Physiol. Rev. 83 (4) (2003) 1325–1358.
[45] J.J. Tomasek, G. Gabbiani, B. Hinz, C. Chaponnier, R.A. Brown, Myofibroblasts and mechano-regulation of connective tissue remodelling, Nat. Rev. Mol. Cell Biol. 3 (5) (2002) 349–363.
[46] M.J. Berridge, Smooth muscle cell calcium activation mechanisms, J. Physiol. 586 (Pt 21) (2008) 5047–5061.
[47] M.J. Davis, J.A. Donovitz, J.D. Hood, Stretch-activated single-channel and whole cell currents in vascular smooth muscle cells, Am. J. Physiol. Cell Physiol. 262 (4) (1992) C1083–C1088.
[48] D.C. Hill-Eubanks, M.E. Werner, T.J. Heppner, M.T. Nelson, Calcium signaling in smooth muscle, Cold Spring Harb. Perspect. Biol. 3 (9) (2011) a004549.
[49] U. Malmqvist, A. Arner, Kinetics of contraction in depolarized smooth muscle from guinea-pig taenia coli after photodestruction of nifedipine, J. Physiol. 519 (Pt 1) (1999) 213–221.
[50] L.A. Schildmeyer, R. Braun, G. Taffet, M. Debiasi, A.E. Burns, A. Bradley, R.J. Schwartz, Impaired vascular contractility and blood pressure homeostasis in the smooth muscle alpha-actin null mouse, FASEB J. 14 (14) (2000) 2213–2220.
[51] A.P. Somlyo, A.V. Somlyo, Signal transduction and regulation in smooth muscle, Nature 372 (6503) (1994) 231.
[52] Z. Hong, K.J. Reeves, Z. Sun, Z. Li, N.J. Brown, G.A. Meininger, Vascular smooth muscle cell stiffness and adhesion to collagen I modified by vasoactive agonists, PLoS ONE 10 (3) (2015) e0119533.
[53] K. Tran-Lundmark, P. Tannenberg, B.H. Rauch, J. Ekstrand, P.-K. Tran, U. Hedin, M.G. Kinsella, Perlecan heparan sulfate is required for the inhibition of smooth muscle cell proliferation by all-trans-retinoic acid, J. Cell. Physiol. 230 (2) (2015) 482–487.
[54] B. Lilly, We have contact: endothelial cell-smooth muscle cell interactions, Physiology 29 (4) (2014) 234–241.
[55] F.V. Brozovich, C.J. Nicholson, C.V. Degen, Y.Z. Gao, M. Aggarwal, K.G. Morgan, Mechanisms of vascular smooth muscle contraction and the basis for pharmacologic treatment of smooth muscle disorders, Pharmacol. Rev. 68 (2) (2016) 476–532.
[56] G.K. Owens, M.S. Kumar, B.R. Wamhoff, Molecular regulation of vascular smooth muscle cell differentiation in development and disease, Physiol. Rev. 84 (3) (2004) 767–801.
[57] J.E. Wagenseil, R.P. Mecham, Vascular extracellular matrix and arterial mechanics, Physiol. Rev. 89 (3) (2009) 957–989.
[58] C.L. Papke, Y. Yamashiro, H. Yanagisawa, MMP17/MT4-MMP and thoracic aortic aneurysms: OPNing new potential for effective treatment, Circ. Res. 117 (2) (2015) 109–112.
[59] C.M. Kielty, S.P. Whittaker, M.E. Grant, C.A. Shuttleworth, Attachment of human vascular smooth muscles cells to intact microfibrillar assemblies of collagen VI and fibrillin, J. Cell Sci. 103 (2) (1992) 445–451.
[60] Y. Akgul, M. Mahendroo, Cervical changes accompanying birth, in: B.A. Croy, A.T. Yamada, F.J. DeMayo, S.L. Adamson (Eds.), The Guide to Investigation of Mouse Pregnancy, Academic Press, Boston, MA, 2014, pp. 391–401.
[61] G. Finet, A. Tabib, L'arthérosclérose, in: CardiologiePresses Universitaires de Lyon, Lyon, 1999, pp. 491–499.
[62] J. Rapp, Oscillatory Flow Effects on Rat Aortic Smooth Muscle Cells, diplom.de, 1997. ISBN 978-3-8324-0048-4.
[63] H. Wolinsky, S. Glagov, A lamellar unit of aortic medial structure and function in mammals, Circ. Res. 20 (1) (1967) 99–111.
[64] A. Brunon, K. Bruyére-Garnier, M. Coret, Characterization of the nonlinear behaviour and the failure of human liver capsule through inflation tests, J. Mech. Behav. Biomed. Mater. 4 (8) (2011) 1572–1581.
[65] J.-H. Kim, S. Avril, A. Duprey, J.-P. Favre, Experimental characterization of rupture in human aortic aneurysms using a full-field measurement technique, Biomech. Model. Mechanobiol. 11 (6) (2012) 841–853.
[66] A. Romo, P. Badel, A. Duprey, J.P. Favre, S. Avril, In vitro analysis of localized aneurysm rupture, J. Biomech. 47 (3) (2014) 607–616.
[67] C. Cavinato, C. Helfenstein-Didier, T. Olivier, S.R. du Roscoat, N. Laroche, P. Badel, Biaxial loading of arterial tissues with 3D in situ observations of adventitia fibrous microstructure: a method coupling multi-photon confocal microscopy and bulge inflation test, J. Mech. Behav. Biomed. Mater. 74 (2017) 488–498.
[68] C. Bellini, J. Feruzzi, S. Roccabianca, E.S. Di Martino, J.D. Humphrey, A microstructurally motivated model of arterial wall mechanics with mechanobiological implications, Ann. Biomed. Eng. 42 (3) (2014) 488–502.
[69] J.D. Humphrey, C.A. Taylor, Intracranial and abdominal aortic aneurysms: similarities, differences, and need for a new class of computational models, Annu. Rev. Biomed. Eng. 10 (2008) 221–246.
[70] L. Cardamone, A. Valentin, J.F. Eberth, J.D. Humphrey, Origin of axial prestretch and residual stress in arteries, Biomech. Model. Mechanobiol. 8 (6) (2009) 431–446.
[71] K.L. Dorrington, N.G. McCrum, Elastin as a rubber, Biopolymers 16 (6) (1977) 1201–1222.
[72] G.A. Holzapfel, T.C. Gasser, R.W. Ogden, A new constitutive framework for arterial wall mechanics and a comparative study of material models, J. Elast. Phys. Sci. Solids 61 (1–3) (2000) 1–48.
[73] F. Riveros, S. Chandra, E.A. Finol, T.C. Gasser, J.F. Rodriguez, A pull-back algorithm to determine the unloaded vascular geometry in anisotropic hyperelastic AAA passive mechanics, Ann. Biomed. Eng. 41 (4) (2013) 694–708.
[74] J.F. Rodriguez, C. Ruiz, M. Doblaré, G.A. Holzapfel, Mechanical stresses in abdominal aortic aneurysms: influence of diameter, asymmetry, and material anisotropy, J. Biomech. Eng. 130 (2) (2008) 21–23.
[75] M.R. Bersi, C. Bellini, P. Di Achille, J.D. Humphrey, K. Genovese, S. Avril, Novel methodology for characterizing regional variations in the material properties of murine aortas, J. Biomech. Eng. 138 (7) (2016) 0710051–07100515.
[76] T. Fujiwara, Y. Uehara, The cytoarchitecture of the medial layer in rat thoracic aorta: a scanning electron-microscopic study, Cell Tissue Res. 270 (1) (1992) 165–172.
[77] R.P. Mecham, S.M. Schwartz, Vascular Smooth Muscle Cell: Molecular and Biological Responses to the Extracellular Matrix (Biology of Extracellular Matrix Series), Academic Press, Boston, MA, 1995.
[78] B.G. Miller, V.H. Gattone, J.M. Overhage, H.G. Bohlen, A.P. Evan, Morphological evaluation of vascular smooth muscle cell: length and width from a single scanning electron micrograph of microvessels, Anat. Rec. 216 (1) (1986) 95–103.
[79] G. Bao, S. Suresh, Cell and molecular mechanics of biological materials, Nat. Mater. 2 (11) (2003) 715–725.
[80] P.C. Dartsch, H. Hämmerle, Orientation response of arterial smooth muscle cells to mechanical stimulation, Eur. J. Cell Biol. 41 (2) (1986) 339–346.

[81] K. Kanda, T. Matsuda, Mechanical stress-induced orientation and ultrastructural change of smooth muscle cells cultured in three-dimensional collagen lattices, Cell Transplant. 3 (6) (1994) 481–492.
[82] M.K. O'Connell, S. Murthy, S. Phan, C. Xu, J. Buchanan, R. Spilker, R.L. Dalman, C.K. Zarins, W. Denk, C.A. Taylor, The three-dimensional micro- and nanostructure of the aortic medial lamellar unit measured using 3D confocal electron microscopy imaging, Matrix Biol. 27 (3) (2008) 171–181.
[83] J.M. Clark, S. Glagov, Transmural organization of the arterial media. The lamellar unit revisited, Arteriosclerosis (Dallas, TX) 5 (1) (1985) 19–34.
[84] M.A. Gaballa, C.T. Jacob, T.E. Raya, J. Liu, B. Simon, S. Goldman, Large artery remodeling during aging: biaxial passive and active stiffness, Hypertension (Dallas, TX: 1979) 32 (3) (1998) 437–443.
[85] K.P. Dingemans, P. Teeling, J.H. Lagendijk, A.E. Becker, Extracellular matrix of the human aortic media: an ultrastructural histochemical and immunohistochemical study of the adult aortic media, Anat. Rec. 258 (1) (2000) 1–14.
[86] M.K. Keech, Electron microscope study of the normal rat aorta, J. Biophys. Biochem. Cytol. 7 (1960) 533–538.
[87] M.J. Osborne-Pellegrin, Some ultrastructural characteristics of the renal artery and abdominal aorta in the rat, J. Anat. 125 (Pt 3) (1978) 641–652.
[88] E.C. Davis, Smooth muscle cell to elastic lamina connections in developing mouse aorta. Role in aortic medial organization, Lab. Invest. J. Tech. Methods Pathol. 68 (1) (1993) 89–99.
[89] Z. Tonar, P. Kochova, R. Cimrman, J. Perktold, K. Witter, Segmental differences in the orientation of smooth muscle cells in the tunica media of porcine aortae, Biomech. Model. Mechanobiol. 14 (2) (2015) 315–332.
[90] G.A. Holzapfel, T.C. Gasser, M. Stadler, A structural model for the viscoelastic behavior of arterial walls: continuum formulation and finite element analysis, Eur. J. Mech. A. Solids 21 (3) (2002) 441–463.
[91] G.A. Holzapfel, T.C. Gasser, A viscoelastic model for fiber-reinforced composites at finite strains: continuum basis, computational aspects and applications, Comput. Methods Appl. Mech. Eng. 190 (34) (2001) 4379–4403.
[92] H. Chen, T. Luo, X. Zhao, X. Lu, Y. Huo, G.S. Kassab, Microstructural constitutive model of active coronary media, Biomaterials 34 (31) (2013) 7575–7583.
[93] K. Hayashi, A. Kamiya, K. Ono, Biomechanics: Functional Adaption and Remodeling, Springer Science and Business Media, New York, NY, 2012.
[94] D. Balas, P. Philip, Tissus musculaires, (2002). http://www.db-gersite.com/HISTOLOGIE/HISTGENE/histgen1/histgen6/histgen6.htm. Accessed 2 November 2017.
[95] F.S. Fay, C.M. Delise, Contraction of isolated smooth muscle cells—structural changes, Proc. Natl Acad. Sci. USA 70 (3) (1973) 641–645.
[96] J.M. Goffin, P. Pittet, G. Csucs, J.W. Lussi, J.-J. Meister, B. Hinz, Focal adhesion size controls tension-dependent recruitment of $\alpha$-smooth muscle actin to stress fibers, J. Cell Biol. 172 (2) (2006) 259–268.
[97] O. Skalli, P. Ropraz, A. Trzeciak, G. Benzonana, D. Gillessen, G. Gabbiani, A monoclonal antibody against alpha-smooth muscle actin: a new probe for smooth muscle differentiation, J. Cell Biol. 103 (6) (1986) 2787–2796.
[98] P. Lacolley, V. Regnault, A. Nicoletti, Z. Li, J.-B. Michel, The vascular smooth muscle cell in arterial pathology: a cell that can take on multiple roles, Cardiovasc. Res. 95 (2) (2012) 194–204.
[99] M.J. Taggart, Smooth muscle excitation–contraction coupling: a role for caveolae and caveolins? Physiology 16 (2) (2001) 61–65.
[100] H. Karaki, N. Urakawa, P. Kutsky, Potassium-induced contraction in smooth muscle, Nihon Heikatsukin Gakkai Zasshi 20 (6) (1984) 427–444.
[101] L. Balasubramanian, C.-M. Lo, J.S.K. Sham, K.-P. Yip, Remanent cell traction force in renal vascular smooth muscle cells induced by integrin-mediated mechanotransduction, Am. J. Physiol. Cell Physiol. 304 (4) (2013) C382–C391.
[102] S. De Moudt, A. Leloup, C. Van Hove, G. De Meyer, P. Fransen, Isometric stretch alters vascular reactivity of mouse aortic segments, Front. Physiol. 8 (2017) 157.
[103] M.J. Berridge, A. Galione, Cytosolic calcium oscillators, FASEB J. 2 (15) (1988) 3074–3082.
[104] S.C. Murtada, A. Arner, G.A. Holzapfel, Experiments and mechanochemical modeling of smooth muscle contraction: significance of filament overlap, J. Theor. Biol. 297 (2012) 176–186.
[105] S. Malekzadeh, R.A. Fraga-Silva, B. Trachet, F. Montecucco, F. Mach, N. Stergiopulos, Role of the renin-angiotensin system on abdominal aortic aneurysms, Eur. J. Clin. Invest. 43 (12) (2013) 1328–1338.
[106] C.M. Hai, R.A. Murphy, Cross-bridge phosphorylation and regulation of latch state in smooth muscle, Am. J. Physiol. 254 (1988) C99–C106.
[107] H.E. Huxley, Electron microscope studies of the organisation of the filaments in striated muscle, Biochim. Biophys. Acta 12 (1) (1953) 387–394.
[108] P.F. Dillon, M.O. Aksoy, S.P. Driska, R.A. Murphy, Myosin phosphorylation and the cross-bridge cycle in arterial smooth muscle, Science (New York, NY) 211 (4481) (1981) 495–497.
[109] T. Tan, R. De Vita, A structural constitutive model for smooth muscle contraction in biological tissues, Int. J. Nonlinear Mech. 75 (suppl. C) (2015) 46–53.
[110] M.A. Zulliger, A. Rachev, N. Stergiopulos, A constitutive formulation of arterial mechanics including vascular smooth muscle tone, Am. J. Physiol. Heart Circ. Physiol. 287 (3) (2004) H1335–H1343.
[111] P. Challande, M. Briet, Y. Bézie, P. Boutouyrie, Méthodes d'exploration fonctionnelle des gros troncs artériels, in: Biologie et Pathologie du Coeur et des VaisseauxJohn Libbery Eurotext, Paris, 2007, pp. 427–433.
[112] I.M. Tolic-Norrelykke, J.P. Butler, J. Chen, N. Wang, Spatial and temporal traction response in human airway smooth muscle cells, Am. J. Physiol. Cell Physiol. 283 (4) (2002) C1254–C1266.
[113] A. Hall, P. Chan, K. Sheets, M. Apperson, C. Delaughter, T.G. Gleason, J.A. Phillippi, A. Nain, Nanonet force microscopy for measuring forces in single smooth muscle cells of the human aorta, Mol. Biol. Cell 28 (14) (2017) 1894–1900.
[114] J.L. Tan, J. Tien, D.M. Pirone, D.S. Gray, K. Bhadriraju, C.S. Chen, Cells lying on a bed of microneedles: an approach to isolate mechanical force, Proc. Natl Acad. Sci. USA 100 (4) (2003) 1484–1489.
[115] Y. Zhang, S.S. Ng, Y. Wang, H. Feng, W.N. Chen, M.B. Chan-Park, C. Li, V. Chan, Collective cell traction force analysis on aligned smooth muscle cell sheet between three-dimensional microwalls, Interface Focus 4 (2) (2014) 20130056.
[116] B.A. Smith, B. Tolloczko, J.G. Martin, P. Grütter, Probing the viscoelastic behavior of cultured airway smooth muscle cells with atomic force microscopy: stiffening induced by contractile agonist, Biophys. J. 88 (4) (2005) 2994–3007.
[117] M. Laurent, S. Hénon, E. Planus, R. Fodil, M. Balland, D. Isabey, F. Gallet, Assessment of mechanical properties of adherent living cells by bead micromanipulation: comparison of magnetic twisting cytometry vs optical tweezers, J. Biomech. Eng. 124 (4) (2002) 408–421.

[118] E. Crosas-Molist, T. Meirelles, J. López-Luque, C. Serra-Peinado, J. Selva, L. Caja, D. Gorbenko del Blanco, J.J. Uriarte, E. Bertran, Y. Mendizabal, V. Hernandez, C. Garca-Calero, O. Busnadiego, E. Condom, M. Toral, D. Castell, A. Forteza, D. Navajas, E. Sarri, F. Rodrógues-Pascual, H.C. Dietz, I. Fabregat, G. Egea, Vascular smooth muscle cell phenotypic change in patients with Marfan syndrome significance, Arterioscler. Thromb. Vasc. Biol. 35 (4) (2015) 960–972.

[119] R.D. Hubmayr, S.A. Shore, J.J. Fredberg, E. Planus, R.A. Panettieri, W. Moller, J. Heyder, N. Wang, Pharmacological activation changes stiffness of cultured human airway smooth muscle cells, Am. J. Physiol. Cell Physiol. 271 (5) (1996) C1660–C1668.

[120] S.S. An, R.E. Laudadio, J. Lai, R.A. Rogers, J.J. Fredberg, Stiffness changes in cultured airway smooth muscle cells, Am. J. Physiol. Cell Physiol. 283 (3) (2002) C792–C801.

[121] J.H.N. Lindeman, B.A. Ashcroft, J.-W.M. Beenakker, M. van Es, N.B.R. Koekkoek, F.A. Prins, J.F. Tielemans, H. Abdul-Hussien, R.A. Bank, T. H. Oosterkamp, Distinct defects in collagen microarchitecture underlie vessel-wall failure in advanced abdominal aneurysms and aneurysms in Marfan syndrome, Proc. Natl Acad. Sci. USA 107 (2) (2010) 862–865.

[122] A. Rachev, K. Hayashi, Theoretical study of the effects of vascular smooth muscle contraction on strain and stress distributions in arteries, Ann. Biomed. Eng. 27 (4) (1999) 459–468.

[123] T. Matsumoto, M. Tsuchida, M. Sato, Change in intramural strain distribution in rat aorta due to smooth muscle contraction and relaxation, Am. J. Phys. Anthropol. 271 (4 Pt 2) H1711–H1716.

[124] K.K. Parker, A.L. Brock, C. Brangwynne, R.J. Mannix, N. Wang, E. Ostuni, N.A. Geisse, J.C. Adams, G.M. Whitesides, D.E. Ingber, Directional control of lamellipodia extension by constraining cell shape and orienting cell tractional forces, FASEB J. 16 (10) (2002) 1195–1204.

[125] A. Saez, M. Ghibaudo, B. Ladoux, A. Buguin, P. Silberzan, Les cellules vivantes répondent á la rigidité de leur substrat, Images de la Physique (2007) 94–100.

[126] W.W. Ahmed, E. Fodor, T. Betz, Active cell mechanics: measurement and theory, Biochim. Biophys. Acta Mol. Cell Res. 1853 (11, Part B) (2015) 3083–3094.

[127] A. Livne, B. Geiger, The inner workings of stress fibers: from contractile machinery to focal adhesions and back, J. Cell Sci. (7) (2016) 1293–1304.

[128] C. Petit, E. Planus, F. Marchi, I.A. Ivan, Caractérisation combinée par Microscopie á Force Atomique en mode PeakForce et par Microscopie Confocale á Fluorescence de la cellule animale isolée fixée sur un micro motif adhésif á destination de futurs Travaux Pratiques de nanobiophysique, 2017 (non publié).

[129] S.K. Karnik, A critical role for elastin signaling in vascular morphogenesis and disease, Development 130 (2) (2003) 411–423.

[130] T. Tran, K.D. McNeill, W.T. Gerthoffer, H. Unruh, A.J. Halayko, Endogenous laminin is required for human airway smooth muscle cell maturation, Respir. Res. 7 (2006) 117.

[131] K. Hayashi, T. Naiki, Adaptation and remodeling of vascular wall; biomechanical response to hypertension, J. Mech. Behav. Biomed. Mater. 2 (1) (2009) 3–19.

[132] K. Riches, T.G. Angelini, G.S. Mudhar, J. Kaye, E. Clark, M.A. Bailey, S. Sohrabi, S. Korossis, P.G. Walker, D.J.A. Scott, K.E. Porter, Exploring smooth muscle phenotype and function in a bioreactor model of abdominal aortic aneurysm, J. Transl. Med. 11 (2013) 208.

[133] S. Anderson, L. DiCesare, I. Tan, T. Leung, N. SundarRaj, Rho-mediated assembly of stress fibers is differentially regulated in corneal fibroblasts and myofibroblasts, Exp. Cell Res. 298 (2) (2004) 574–583.

[134] M.S. Kolodney, R.B. Wysolmerski, Isometric contraction by fibroblasts and endothelial cells in tissue culture: a quantitative study, J. Cell Biol. 117 (1) (1992) 73–82.

C H A P T E R

# 7

# Multiscale Numerical Simulation of Heart Electrophysiology

*Andres Mena*, Jose A. Bea†*

***CIBER, Zaragoza, Spain †Aragon Institute of Engineering Research, University of Zaragoza, Zaragoza, Spain**

## 7.1 CARDIAC ELECTROPHYSIOLOGY: INTRODUCTION

In the recent decades, mathematical modeling and computer simulations have become a useful tool for tackling problems in science and engineering. In this regard, modeling the electric activity of the heart, under physiological and pathological conditions, has attracted the attention of researchers [1] because ventricular tachycardia and fibrillation are among the major causes of sudden death [2]. Because direct measurements are many times limited to only surface signals, multiscale numerical simulations where the electrical activity at the surface as well as in the myocardium can be related to the underlying electrochemical behavior of the cell, help to gain further insights into the problem.

The electric activity of the heart is usually studied using the well-known bidomain model [3, 4]. It consists of an elliptic partial differential equations and a parabolic partial differential equation coupled to a system of stiff nonlinear ordinary differential equations (ODEs) describing the ionic current through the cellular membrane. This model can be simplified to the so-called anisotropic monodomain equation [3], a parabolic reaction-diffusion equation describing the propagation of the transmembrane potential coupled to a system of ODEs describing the cellular ionic model. The monodomain model represents a much less computationally expensive model for the electric activity of the heart, and has been extensively used [5–8].

The high computational cost of the bidomain and monodomain models is due to the stiffness of the system of ODE describing the transmembrane ionic current, which introduces different space and time scales. The depolarization front is localized in a thin layer of less than a millimeter. Therefore, this requires discretizations of the order of tenths of millimeters in order to accurately resolve the depolarization front, implying models with millions of degrees of freedom to simulate the heart. The time scale is another fundamental issue in cardiac simulations. The time constants involved in the kinetics of cellular models range from 0.1 to 600 ms, requiring in some phases of the process the use of time steps of the order of a hundredth of a millisecond. Hence, solving a single heartbeat requires thousands of time steps.

A number of alternatives have been proposed to solve this problem. In this particular, the multilength scale nature of the problem has inspired the development of adaptive techniques, where the mesh is allowed to change with time coupled with adaptive time integration schemes, to improve the computational performance [9–11]. However, dynamic loading for these adaptive schemes is still cumbersome, limiting their application in massively parallel architectures. Recent efforts [12–15] suggest the use of multilevel meshes, fixed in time, along with adaptive time schemes that take advantage of the different kinetics of the ionic currents. This allows reductions of up to two orders of magnitude in CPU time with respect to traditional explicit algorithms. However, these techniques require a fine mesh (lower level mesh) for solving the partial differential equations (responsible for the propagation of the depolarization front).

Despite the efforts at designing more efficient schemes, the solution of the electrophysiology problem requires the use of algorithms with higher levels of parallelism in multicore platforms. In this regard, the next generation of


**https://doi.org/10.1016/B978-0-12-816390-0.00007-8**

high-performance computing platforms promise to deliver better performance in the PetaFLOPS range. However, achieving high performance on these platforms relies on the fact that strong scalability can be achieved, something challenging due to the performance deterioration caused by the increasing communication cost between processors as the number of cores increases. That is, with an increasing number of cores, the load assigned to each processor decreases, but the communication between different processors associated with the boundaries of a given partitioned domain increases. Therefore, when communication costs dominate, no further benefits are obtained from adding additional processors. An alternative to the multicore platforms is emerging in the newer programmable graphics processing units (GPUs), which in recent years have become highly parallel, multithreaded, many-core processors with tremendous computational horsepower [16, 17]. GPUs outperform multicore CPU architectures in terms of memory bandwidth, but underperform in terms of double precision floating point arithmetic. However, GPUs are built to schedule a large number of threads, thus reducing latencies in their multicore architecture.

Sanderson et al. [18] proposed a general purpose, graphics processing unit (GP-GPU)-based approach for the solution of advection-reaction-diffusion models. They report an increase of performance of up to 27 times for an explicit solver when used on 3D problems. Regarding cardiac electrophysiology, previous studies have reported speedups by a factor of 32 for the monodomain model [19] using an explicit finite difference scheme with a rather simple transmembrane ionic model. In their study, Sato et al. [19] established the solution of the partial differential equation (PDE) as the bottleneck of the computation with GPU. However, in their studies, older NVidia GT8800 and GT9800 GX2 cards that only supported single precision floating point operations were used, which greatly limited the computations of the parabolic system. Chai et al. [20] successfully solved a 25 million node problem on a multi-GPU architecture using the monodomain model and a four-state variable model. Bartocci et al. [21] have performed an implementation of a finite difference explicit solver for cardiac electrophysiology. They evaluated the effect of the ionic model size (number of state variables) on the performance in simulating two-dimensional (2D) tissues, and compared single precision and double precision implementation. They provided acceleration with respect to real time. For small ionic models and the single precision implementation, they reported simulations faster than real time for small problems, whereas for highly detailed models with a larger number of state variables, they reported simulation times between 35 and 70 times larger than real time. Rocha et al. [22] implemented an implicit method on the GPU. Spatial discretization of the parabolic equation was performed by means of the finite element methods (FEM), keeping full stiffness matrices. Promising acceleration ratios were achieved with 2D bidomain tissue models using an unpreconditioned conjugate gradient (CG) method. However, with unstructured 3D bidomain simulations, the number of iterations required for convergence became prohibitive. In a more recent work, Neic et al. [23] showed that 25 processors were equivalent to a single GPU when computing the bidomain equations. This new capability to solve the governing equations on a relatively small GPU cluster makes it possible to 1 day introduce simulations using patient-specific computer models into a clinical workflow. In a more recent work, Vigueras et al. [24] ported to the GPU a number of components of a parallel c-implemented cardiac solver. They report accelerations of 164 times of the ODE solver and up to 72 times for the PDE solver. They have also achieved accelerations of up to 44 times for the mechanics residual/Jacobian computation in electromechanical simulations.

When dealing with the pathological heart, ventricular tachycardia and fibrillation are known to be two types of cardiac arrhythmias that usually take place during acute ischemia and frequently lead to sudden death [25]. Even though these arrhythmias arise from different conditions, ischemia is the most important perpetrator among them. During ischemia, the delivery of nutrients to the myocardium diminishes, causing metabolic changes that result in a progressive deterioration of the electric activity in the injured region [26]. These metabolic changes are mainly hypoxia, increased concentrations of the extracellular potassium $[K^+]_o$ (hyperkalemia), a decrease of intracellular adenosine triphosphate (ATP) (hypoxia), and acidosis [27]. From an electrophysiological point of view, these metabolic changes simply produce alterations in the action potential (AP), excitability, conduction velocity (CV), and effective refractive period (ERP), among others, creating a substrate for arrhythmias and fibrillation [26, 27]. In addition, the impact of ischemia in the myocardium is characterized by a high degree of heterogeneity both intramurally and transmurally. In the tissue affected by acute ischemia, two zones can be distinguished: (i) the central ischemic zone corresponding to the core of the tissue suffering from the lack of blood, and (ii) a border zone (BZ) that comprises changes in electrophysiological properties between the healthy and ischemic regions [28, 29]. Proarrhythmic mechanisms of acute ischemia have been extensively investigated, although often in animal models rather than in human ventricles. Studies by Janse et al. [26, 30] in pig and dog hearts highlight the complexity of the proarrhythmic and spatiotemporally dynamic substrate in acute ischemia. Heterogeneity in excitability and repolarization properties across the BZ leads to the establishment of reentry around the ischemic region following ectopic excitation [26, 31]. The same studies also showed intramural reentry in certain cases (highlighting the potential variability in the mechanisms). However, the mechanisms that determine reentry formation and intramural patterns in acute ischemia in the 3D human heart remain unclear, due to the low resolution of intramural recordings.

## 7.2 EQUATIONS THAT GOVERN THE ELECTRICAL ACTIVITY OF THE HEART

The heart shows two types of behavior: electrical and mechanical. All myocardial cells are similar with respect to the mechanical function. However, from an electrical point of view, the cells may be classified into several types. The electric impulses transmitted through the heart are responsible for the rhythmic contraction of the heart muscle/cardiac muscle. When the system works normally, the atria are contracted approximately a sixth of a second before the ventricles, which enables the filling of the ventricles before pumping the blood into the lungs and the peripheral circulation [32]. Another important point of the system is that the ventricles contract synchronously to generate proper blood pumping. Therefore, all cells need to develop an AP in an ordered manner for which the cells must be excited conveniently along the cardiac cycle. To fully understand these phenomena, this chapter describes how the electrical activity in the heart takes place, how it synchronizes, and the mathematical equations that rule them.

### 7.2.1 Governing Equations

This section describes the governing equations of the propagation of the heart electrical activity.

#### *7.2.1.1 Bidomain Model*

The electrical coupling of the cardiomyocytes and the conduction through the ventricles can be mathematically described by a bidomain model [33]. In this model, the cardiac heart tissue is represented by two continuous domains that share the space, that is, the intracellular and extracellular domains coexist spatially. This is opposite to reality because each of them physically takes a fraction of the total volume. In this model, each domain acts as a volume conductor with a different conductivity tensor and different potential, and the ionic currents flow from one domain to another through the cell membrane that acts as a condenser.

The currents in the two domains are given by Ohm's Law:

$$J_i = -\mathbf{M_i}\,\nabla V_i, \tag{7.1}$$

$$J_e = -\mathbf{M_e}\,\nabla V_e, \tag{7.2}$$

where $J_i$ is the intracellular current, $J_e$ is the extracellular current, $\mathbf{M_i}$ and $\mathbf{M_e}$ are the conductivity tensors, and $V_i$ and $V_e$ are the intracellular and extracellular potentials, respectively.

The cell membrane acts as a condenser. Due to its small thickness, the charge stored on one side is compensated immediately on the other side, by which the accumulation of charge at any point is zero, that is:

$$\frac{\partial}{\partial t}(q_i + q_e) = 0, \tag{7.3}$$

where $q_i$ and $q_e$ are the charges in the intracellular and extracellular space, respectively.

In each domain, the flow of current in a point must equal the rate of accumulation plus the ionic current coming out of the point, that is:

$$-\nabla \cdot J_i = \frac{\partial q_i}{\partial t} + \chi J_{ion}, \tag{7.4}$$

$$-\nabla \cdot J_e = \frac{\partial q_e}{\partial t} - \chi J_{ion}, \tag{7.5}$$

where $J_{ion}$ is the current through the membrane. The ionic current is measured by the unit of area of the cellular membrane, whereas the density of charge and the flow of current are measured by unit of volume. The constant $\chi$ represents the cell membrane area-to-volume ratio. On the other hand, the sign of the ionic current is defined as positive when the current leaves the intracellular space and gets into the extracellular.

Introducing Eqs. (7.4), (7.5) in Eq. (7.3), we get the current conservation equation:

$$\nabla \cdot J_i + \nabla \cdot J_e = 0. \tag{7.6}$$

Replacing Eqs. (7.1), (7.2) in Eq. (7.6), we obtained

$$\nabla \cdot (\mathbf{M_i}\,\nabla V_i) + \nabla \cdot (\mathbf{M_e}\,\nabla V_e) = 0. \tag{7.7}$$

The charge of the cell membrane directly depends on the difference of the membrane potential, $V = V_i - V_e$ and the capacitance of the membrane

$$V = \frac{q}{\chi C_m}, \tag{7.8}$$

where $C_m$ is the membrane capacitance and

$$q = \frac{q_i - q_e}{2}. \tag{7.9}$$

Combining Eqs. (7.8), (7.9) and deriving with respect to time, we get

$$\chi C_m \frac{\partial V}{\partial t} = \frac{1}{2}\frac{\partial (q_i - q_e)}{\partial t}.$$

Using Eq. (7.3), we get the relation

$$\frac{\partial q_i}{\partial t} = -\frac{\partial q_e}{\partial t} = \chi C_m \frac{\partial V}{\partial t}.$$

Replacing this latter expression in Eq. (7.4) and using Eq. (7.1), we obtain

$$\nabla \cdot (\mathbf{D_i} \nabla V_i) = C_m \frac{\partial V}{\partial t} + J_{ion}, \tag{7.10}$$

where $\mathrm{D_i} = \mathrm{M_i}/\chi$.

Eqs. (7.7), (7.10) depend on three potentials $V_i$, $V_e$, and $V$. Eliminating $V_i$ from Eqs. (7.7), (7.10), the equations for the bidomain model are obtained

$$\nabla \cdot (\mathbf{D_i} \nabla V) + \nabla \cdot (\mathbf{D_i} \nabla V_e) = C_m \frac{\partial V}{\partial t} + J_{ion}, \tag{7.11}$$

$$\nabla \cdot (\mathbf{D_i}\, \nabla V) + \nabla \cdot ((\mathbf{D_i} + \mathbf{D_e})\, \nabla V_e) = 0. \tag{7.12}$$

Assuming that the heart is surrounded by a nonconductive medium, the normal components of both currents (intracellular and extracellular) are zero at the boundary, by which we have

$$\begin{aligned} \mathbf{n} \cdot J_i &= 0, \\ \mathbf{n} \cdot J_e &= 0, \end{aligned} \tag{7.13}$$

where $\mathbf{n}$ is the outer normal. Using the expression for both currents and eliminating $V_i$, the boundary conditions of the model are obtained

$$\mathbf{n} \cdot (\mathbf{D_i} \nabla V + \mathbf{D_i} \nabla V_e) = 0, \tag{7.14}$$

$$\mathbf{n} \cdot \nabla (\mathbf{D_e}\, \nabla V_e) = 0. \tag{7.15}$$

### *7.2.1.2 Monodomain Model*

As can be observed, the bidomain model represents the electric currents in both the intracellular and extracellular medium. It is represented by a nonlinear parabolic equation coupled with an elliptical equation. Under particular conditions, the bidomain model can be decoupled, allowing us to calculate the transmembrane potential independently of the extracellular potential. Assuming that the conductivity tensors have the same variation in the anisotropy, that is, $\mathbf{D_e} = \lambda \mathbf{D_i}$, where $\lambda$ is a scalar, so $\mathbf{D_e}$ can be eliminated from Eqs. (7.11), (7.12), obtaining

$$\nabla \cdot (\mathbf{D_i} \nabla V) + \nabla \cdot (\mathbf{D_i} \nabla V_e) = C_m \frac{\partial V}{\partial t} + J_{ion}, \tag{7.16}$$

$$\nabla \cdot (\mathbf{D_i}\, \nabla V) + (1+\lambda) \nabla \cdot (\mathbf{D_i}\, \nabla V_e) = 0, \tag{7.17}$$

from Eq. (7.17), we have

$$\nabla \cdot (\mathbf{D_i}\, \nabla V_e) = -\frac{1}{1+\lambda} \nabla \cdot (\mathbf{D_i}\, \nabla V),$$

replacing in Eq. (7.16) and operating, we get the standard formulation for the monodomain model:

$$\nabla \cdot (\mathbf{D}\, \nabla V) = C_m \frac{\partial V}{\partial t} + I_{ion}, \tag{7.18}$$

where $\mathbf{D} = \frac{\lambda}{1+\lambda}\mathbf{D_i}$. With the following boundary condition:

$$\mathbf{n} \cdot \nabla(\mathbf{D}\,\nabla V) = 0. \tag{7.19}$$

The monodomain model takes a numerical and computational complexity that is less than the bidomain. For this reason, the monodomain model is often used to study the propagation of the AP in the heart.

Eq. (7.18) is a parabolic equation describing a reaction-diffusion phenomenon. The part associated with the reaction is determined by the term $J_{ion}$, which is governed by the cellular model. The diffusive (or conductive in this case) part models the propagation of the AP in the tissue.

### 7.2.2 Myocardium Conductance

Within the heart, the fibers are organized transmurally with orientations varying from − 60 degrees (regarding the circumferential axis) to + 60 degrees from the epicardium to the endocardium [34]. The orientation of the muscle fibers in each point of the myocardium can be obtained either by histology, or more recently by using magnetic resonance imaging (MRI), in particular a technique known as diffusion tensor magnetic resonance imaging (DT-MRI) [35]. Born for neuroimaging applications, diffusion tensor imaging (DTI), a special kind of the more general diffusion weighted MRI, is an imaging method that uses the diffusion of water molecules to generate contrast in MR images. Because diffusion of water in tissues is not free, but is affected by the interaction with obstacles such as fibers and heterogeneities in general, water molecule diffusion patterns can be used to identify details about tissue microstructure. DTI, in particular, enables the measurement of the restricted diffusion of water in the myocardium. In DTI, each voxel contains the rate of diffusion and the preferred directions of diffusion. Therefore, assuming that the diffusion is faster along the fiber axis, the eigenvector corresponding to the largest diffusion tensor eigenvector defines the direction of the fiber axis [35]. These are the data that we need to implement in the simulation of the human heart tissue.[1] Even though the heart tissue is truly orthotropic [36], for this work we consider it as transversely isotropic, with the direction of maximum conduction corresponding to the cardiac fiber direction. In the material fiber system, the conductivity tensor is

$$\mathbf{D} = d_o \begin{pmatrix} 1 & 0 & 0 \\ 0 & r & 0 \\ 0 & 0 & r \end{pmatrix}, \tag{7.20}$$

where $d_o$ represents the conductance in the fiber direction and $r \leq 1$ the conductivity ratio between the transversal and longitudinal fibers. In Cartesian coordinates, under conditions of transverse anisotropy, the diffusion tensor can be written as

$$\mathbf{D} = d_o[(1-r)\mathbf{f}\otimes\mathbf{f} + r\mathbf{I}], \tag{7.21}$$

where $\mathbf{f}$ is the fiber orientation, $\mathbf{I}$ is the second-order identity tensor, and $\otimes$ indicates the tensorial product ($(a\otimes b)_{ij} = a_i b_j$). Expressing Eq. (7.21) in components, we obtain

$$\mathbf{D} = d_o\,(1-r)\begin{pmatrix} f_1\,f_1 & f_1\,f_2 & f_1\,f_3 \\ f_2\,f_1 & f_2\,f_2 & f_2\,f_3 \\ f_3\,f_1 & f_3\,f_2 & f_3\,f_3 \end{pmatrix} + d_o\,r\begin{pmatrix} 1 & 0 & 0 \\ 0 & 1 & 0 \\ 0 & 0 & 1 \end{pmatrix}. \tag{7.22}$$

### 7.2.3 Action Potential Models

Hodking and Huxley [37] in 1952 introduced the first mathematical model to reproduce the APs in the cell membrane. Since then, many cardiac cell models, following the formulation established by these researchers, have been developed. AP models can be divided into two main families: (i) phenomenological models and (ii) electrophysiological detailed models.

Phenomenological models macroscopically reproduced the behavior of the cell in terms of the shape and duration of the AP, restitution properties, and CV. Electrophysiological detailed models offer a detailed description of the cellular physiology. They not only include more currents, but also include pumps and exchangers as well as intracellular ion concentration dynamics. Models have been developed for a number of species as well as cell types within the conduction system of the heart, as for example: Stewart et al. [38] for human Purkinjie cells, Maleckar et al. [39] and Nygren

[1] See http://gforge.icm.jhu.edu/gf/project/dtmri_data_sets/.

et al. [40] for human atria, ten Tusscher et al. [8, 41], O'Hara et al. [42], and Carro et al. [43] for human ventricles, Luo and Rudy [44, 45] for guinea pig ventricles, and Shannon et al. [46] for rabbit ventricles, among others.

In this work, we have implemented a number of AP models, namely: Maleckar et al. (MA09) and Nygren et al. (NY98) for atria, ten Tusscher et al. (TP06) and O'Hara et al (OH11) for human ventricles, Stewart et al. (ST99) for Purkinje cells, and Bueno-Orovio et al. (BV08) as a type of phenomenological model. The TP06 model will be extensively used in Chapter 4 to study the acute ischemic heart. In the following, the basic structure of a modern AP model is described in detail.

### 7.2.3.1 *Structure of an Action Potential Model*

As known, it is possible to reproduce the characteristics of AP with simple models. However, an important objective in the modelization of physiological phenomena is to research how the changes in the cell physiology affect the tissue and finally, the studied organ. For this purpose, it is necessary that the models study the cell physiology from the membrane to the ionic channels, which set up the gates for the exchange between the intracellular and extracellular media, including the dynamic mechanism in the cytoplasm.

### 7.2.3.2 *The Cell Membrane*

The cell membrane separates the extracellular medium from cytoplasm. It is formed by a very thin layer of lipid and protein molecules, which are mainly held linked by noncovalent interaction [47]. The lipid molecules are arranged in a continuous double layer of about 5 nm thick; see Fig. 7.1. This bilayer is the basic structure and acts as a relatively impermeable barrier to the passing of most water-soluble molecules.

The protein molecules are dissolved in the lipid bilayer, arranging a connection between the inside and the outside of the cell. The molecules form channels through the membrane through which the ions can pass. Fig. 7.1 schematically shows the structure of the cell membrane with the transport proteins imbibed. Some proteins form pumps and ionic exchangers, needed to maintain the correct concentration of ions in the cell. Both pumps and exchangers have the ability to transport ions in an opposite direction to the gradient generated by the ionic concentration (electric gradient). This process is achieved by using either the gradient of a different ion (exchangers) or consuming chemical energy stored in the form of ATP (pumps). This type of transportation is called active.

Along with the pumps and exchangers, certain proteins form channels in the membrane through which the ions can flow. The flow of ions through these channels is passive and is governed by the concentration gradients and the electric fields. Most of the channels are alternatively selective as to the type of ion that can pass through them. Besides, the channels are capable of opening and closing themselves in response to the changes in the electric field and the ionic concentration. This characteristic is essential in order to propagate the signal in an excitable tissue.

#### 7.2.3.2.1 THE NERNST EQUATION

The ionic concentrations inside and outside the cell differ greatly in cardiomyocytes, that is, a concentration gradient exists for all permeable ions contributing to net ion flux. This concentration gradient induces ions to flow, or diffuse, from regions of high to low concentration. This ion flow produces, as a consequence, the accumulation of ions at the

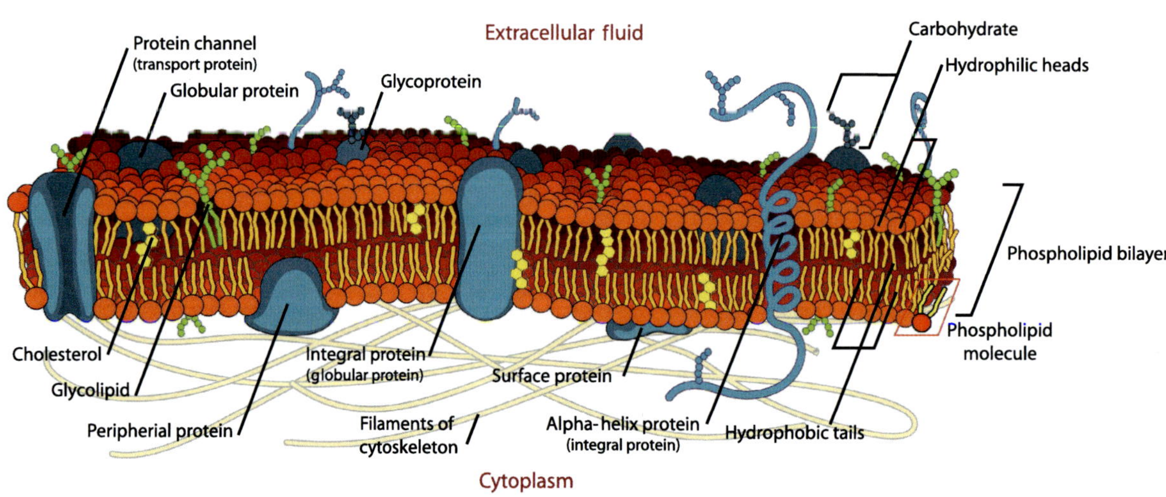


FIG. 7.1 Detail of the cell membrane.

inner and outer membrane surfaces, establishing, therefore, an electric field (a potential difference) within the membrane. Because the ions are charged particles, this electric field exerts forces on the ions crossing the membrane that oppose the diffusional forces established by the difference in ionic concentration. Therefore, to describe membrane ion movements, electric-field forces and diffusional forces should be considered. In this regard, equilibrium is attained when the diffusional force balances the electric field force for all permeable ions.

For a membrane that is permeable to only one type of ion, the electrochemical balance between forces due to the concentration gradient and the potential gradient for a particular ion can be described by the Nernst-Planck-Einstein equation:

$$E_k = \frac{RT}{z_k F} \ln\left(\frac{c_{k,e}}{c_{k,i}}\right), \tag{7.23}$$

where $R$, $F$, and $T$ are the gas constant, the Faraday constant, and the absolute temperature constant, respectively; $E_k$ is the equilibrium voltage across the membrane (Nernst potential) for the $k$th ion; $z_k$ is the valence of the $k$th ion; and $c_{k,\,e}$ and $c_{k,\,i}$ are the extracellular and intracellular concentrations of the $k$th ion, respectively.

#### 7.2.3.2.2 GOLDMAN-HODGKIN-KATZ EQUATION

Assuming that the cell membrane is permeable to a single ion only is not valid. However, it is assumed that when several permeable ions are present, the flux of each is independent of the others (known as the independence principle). According to this principle, and assuming: (i) the membrane is homogeneous and neutral, and (ii) the intracellular and extracellular ion concentrations are uniform and unchanging, the membrane potential is governed by the well-known Goldman-Hodgkin-Katz equation. For $N$ monovalent positive ion species and for $M$ monovalent negative ion species, the potential difference across the membrane is as follows

$$E = -\frac{RT}{F} \log\left(\frac{\sum_{j=1}^{N} P_j[c_j^+]_i + \sum_{j=1}^{M} P_j[c_j^-]_e}{\sum_{j=1}^{N} P_j[c_j^+]_e + \sum_{j=1}^{M} P_j[c_k^-]_i}\right), \tag{7.24}$$

where $[c_j]_i$ and $[c_j]_e$ are the intracellular and extracellular concentrations for the $j$th ion, $P_j$ is the permeability of the intracellular and extracellular concentrations for the $j$th ion, and $E$ is the membrane potential. The permeability for the $j$th ion is defined as

$$P_j = \frac{D_j \beta_j}{h},$$

where $h$ is the thickness of the membrane, $D_j$ is the diffusion coefficient, and $\beta_j$ is the water partition coefficient of the membrane. Both $D_j$ and $\beta_j$ depend on the type of ion and the type of membrane.

### *7.2.3.3 Gates*

Ion channels are specific units located in the cell membrane through which the ions can flow. These channels open and close in response to potential differences or a change in ion concentration. The mechanisms by which these channels open and close are stochastic in nature and may involve complex processes. In the simplest case, the channel, or the channel gate, is considered to be in only two possible states, either open or close. Assigned to these sates, there is a possibility of opening $O$ and a possibility of closure $C$, being stochastic the transition between states. The density of opened channels is $[O]$ and the density of closed channels $[C]$. Furthermore, assume that the density of channels, $[O] + [C]$, is constant. The change between the open state and closed state can be described by a first-order reaction as

$$C \underset{\beta}{\overset{\alpha}{\rightleftarrows}} O, \tag{7.25}$$

where $\alpha$ is the opening rate and $\beta$ is the closing rate of the channels. These rates depend on the membrane potential $V$ in general, even though they could also be modulated by the ion concentration. By the law of mass action, the rate of change from the open state to the closed state is proportional to the concentration of channels in the open state; equally, the rate of change from the closed state to the open state is proportional to the concentration in the closed channels. Therefore, we obtain

$$\frac{d[O]}{dt} = \alpha(V)\,[C] - \beta(V)\,[O].$$

Dividing this equation by the total density [O] + [C] we obtain

$$\frac{dg}{dt} = \alpha(V)(1-g) - \beta(V)g, \tag{7.26}$$

where $g = [O]/([O] + [C])$ is the rate of open channels.

As $\alpha$ and $\beta$ depend on $V$, it is not possible to use a general solution of Eq. (7.26). So Eq. (7.26) can be written as

$$\frac{dg}{dt} = (g_\infty(V) - g)/\tau_g(V), \tag{7.27}$$

with $g_\infty = \alpha/(\alpha+\beta)$ and $\tau_g = 1/(\alpha + \beta)$. Here, $g_\infty$ and $\tau_g$ are constants. The solution of Eq. (7.28) is

$$g(t) = g_\infty + (g_o - g_\infty)\,e^{-t/\tau_g}, \tag{7.28}$$

where $g_o$ is the initial value of $g$, and $g_\infty$ is the value of $g$ in the stable state.

#### *7.2.3.4 Ionic Channels*

The current through an ion channel can be computed using Ohm's Law as the product of the channel conductance times the potential difference between the membrane potential and the equilibrium potential for the specific ion defined by the Nernst potential (Eq. 7.23)

$$I_i = g\,(V - E_i),$$

where $g$ is the permeability (or channel conductance) of the membrane to the ion $i$. Depending on the type of ion, $g$ can be either a constant or a function on the time and on the membrane potential as well as on the ionic concentrations. In general, the conductance of a given channel is given as

$$g = G_{\max} O,$$

where $G_{\max}$ is the maximum channel conductance, and $O$ is the probability that the channel is open.

As explained in the previous section, the cell membrane behaves as a condenser from an electric point of view due to its dielectric characteristics. In addition, the proteins dissolved in the cell membrane form specific units that allow the ionic exchange between the intracellular and extracellular space. Furthermore, some of these units, the ion channels, possess a large specificity to certain ions. Hence, from an electric standpoint, the electric current flowing across the cell membrane during activation can be described using a parallel conductance model, as shown in Fig. 7.2. It consists of a capacitive current plus different currents associated with the different ions under consideration.

In this model, each of these current components is assumed to be independent, that is, each current utilizes its own channel. The modern notation considers a positive current and potential from the inside to the outside.

From the circuit in Fig. 7.2, we obtain the following expression for the current through the membrane

$$J_m = C_m\frac{dV}{dt} + +J_{ex} + J_{pump} + \sum_{j=1}^{n} g_j(V - E_j), \tag{7.29}$$

where $C_m$ is the membrane capacitance, $V$ is the membrane potential, and $g_j$ and $E_j$ are the conductance and the Nernst potential for the ion $j$, respectively. $J_{ex}$ and $J_{pump}$ are currents associated with ion exchangers and ionic pumps (active transport elements) present in the membrane.

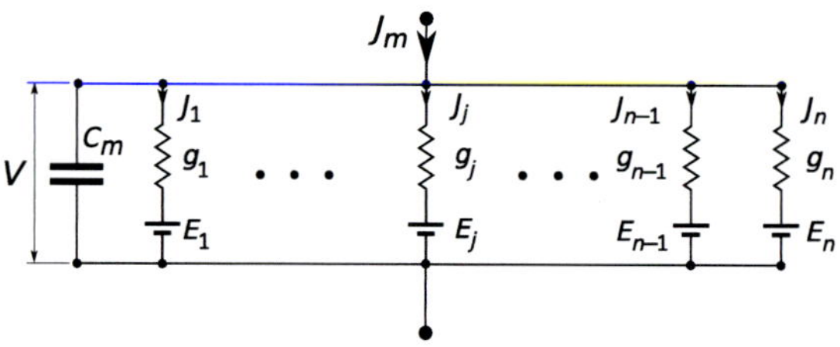


FIG. 7.2 The equivalent circuit of the cell membrane.

#### *7.2.3.5 The Ten Tusscher Action Potential Model*

Modern AP models incorporate many of the features formerly described and their formulation is reduced to Eq. (7.29) in addition to differential equations (7.27) governing the dynamics of the gates and the ionic concentrations in the cytoplasm. The formulation of these models is based on experimental data collected from different animal species. They are, in general, relatively costly from a computational point of view due to the nonlinearity of their equations. Among the most widely used models of AP for human ventricle is the one proposed by ten Tusscher et al. [8, 41] that is shown schematically in Fig. 7.3.

The ten Tusscher model was first introduced in 2004 (TP04) [41]. This first version already specialized the model for the three main types of cardiomyocytes present in the human myocardium (endocardium, midmyocardium, and epicardium) as well as the ionic currents associated with the three main ions (sodium, potassium, and calcium) in addition to intracellular ionic homeostasis and calcium handling. The second version of the model was introduced in 2006 (TP06) [8], where the formulation of the intracellular calcium handling was modified by introducing a subspace to describe the calcium induced calcium release. Fig. 7.3 shows a detailed schematic mode of the TP06 model.

The major innovation of the TP06 model with respect to its predecessor is the modification of the calcium handling and the calcium current through the L-type channels. The improved formulation of calcium handling includes a subspace (SS) between the sarcoplasmic reticulum (SR) and the cell membrane from which the release of calcium into the cytoplasm is performed. Meanwhile, the release of calcium from the SR to SS is carried out through a mechanism controlled by the Ryanodine receptors, whose dynamic is governed by a Markov chain of four states (see Fig. 7.3). However, in the implementation of the TP06, the dimensionality of the Markov chain is reduced to two states in order to reduce the computational cost of the model. With the introduction of the SS, the flow of calcium through the L-type channels injects calcium into the SS, this being inactivated now by the concentration of calcium in the SS $[Ca^{2+}]_{SS}$ instead of $[Ca^{2+}]_i$. In addition, the new formulation includes two gates of inactivation depending on the potential, a slow and a fast one to accommodate additional experimental observations.

In this chapter, the TP06 model has been used to study the electric activity of the human heart in normal and pathological conditions, considering the transmural heterogeneity (different proportions of cells of the epicardium, the endocardium, and the midmyocardium). Its response has also been featured under conditions of ischemia (hyperkalemia, acidosis, and hypoxia). To study the effect of hypoxia, a modification of the $I_{KATP}$ model of Ferrero [48] has been done. Finally, this modified model has been used to study the electrical response of the heart under conditions of acute

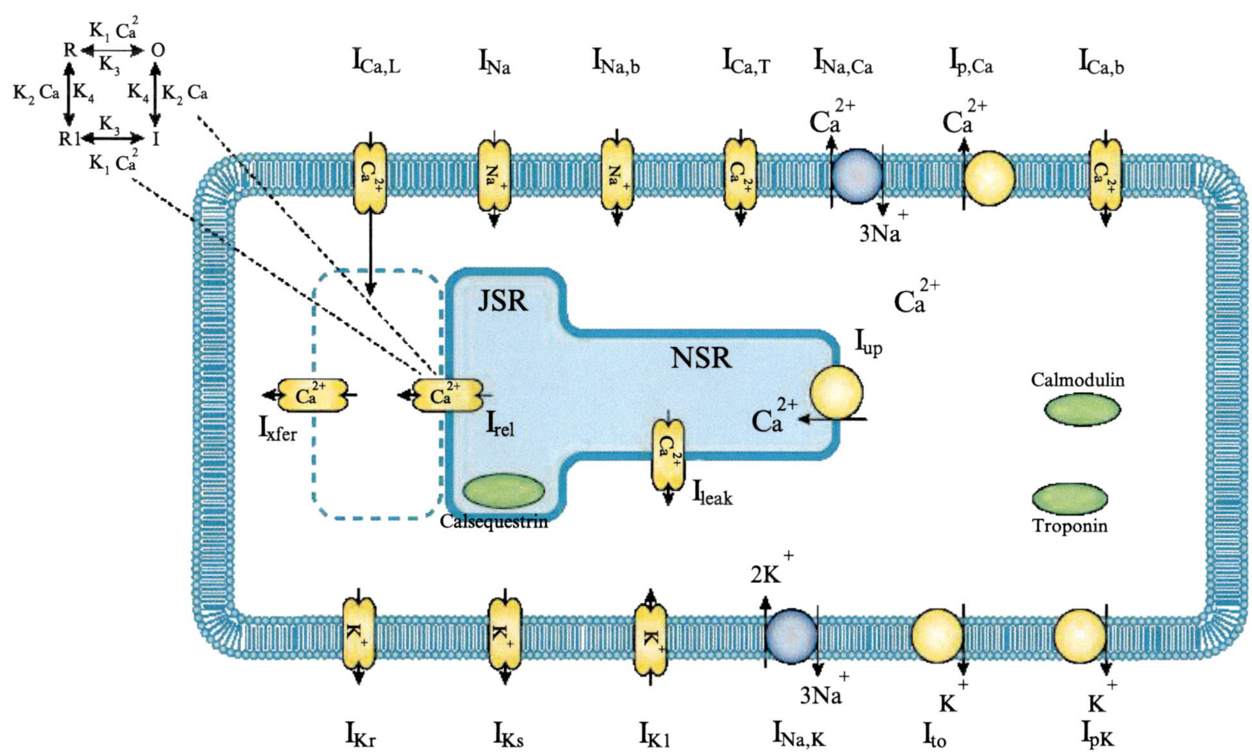


FIG. 7.3 Ten Tusscher model 2006 [8].

regional ischemia (details will be discussed in the following sections). From a computational point of view, the TP06 model has 19 state variables, 14 ionic currents, and requires of a minimum time integration step of 0.02 ms.

### 7.2.4 Numerical Solution of the Electric Activity of the Heart

As discussed before, the monodomain model represents a major simplification of the bidomain model with important advantages for mathematical and computational analysis, which is suitable for studying the electrical behavior of the heart. This section focuses on the numerical solution of the monodomain model described by Eqs. (7.18), (7.19).

From a mathematical and computational view, the problem defined by Eqs. (7.16), (7.17) corresponds to the solution of a linear partial differential equation, which describes the electrical conduction, coupled with a rigid nonlinear system of ODEs describing the transmembrane ion currents, resulting in a problem of nonlinear reaction-diffusion. An efficient way to solve Eqs. (7.16), (7.17) is by application of the splitting technique operators [49]. The decomposition technique of operators has been applied to the monodomain equations [7, 50]. The basic steps are summarized below:

- *Step 1*: Use $V(t)$ as the initial condition for integrating the equation

$$\nabla \cdot (\mathbf{D}\nabla V) = C_m \frac{\partial V}{\partial t} + J_{ion}(V,u), \quad \text{for } t \in [t, t+\Delta t/2]. \tag{7.30}$$

- *Step 2*: Use the result obtained in Step 1 as the initial condition to integrate

$$C_m \frac{\partial V}{\partial t} = -J_{ion}(V,u), \tag{7.31}$$

$$\frac{\partial u}{\partial t} = \mathbf{f}(u,V,t), \quad \text{for } t \in [t, t+\Delta t]. \tag{7.32}$$

- *Step 3*: Use the result obtained in Step 2 as the initial condition for integrating

$$\nabla \cdot (\mathbf{D}\nabla V) = C_m \frac{\partial V}{\partial t} + J_{ion}(V,u), \quad \text{for } t \in [t+\Delta t/2, t+\Delta t]. \tag{7.33}$$

In practice, Steps 1 and 3 can be combined into one, except for the first increment. Therefore, after the initial increase, the algorithm has only two steps, Step I corresponding to the integration of ODEs (Step 2), and Step II corresponding to the integration of the homogeneous parabolic equation (Steps 1 and 3).

- *Step I*: Use $V(t^k)$ as the initial condition to integrate the equation

$$\left.\begin{aligned} C_m \frac{\partial V}{\partial t} &= -J_{ion}(V,\mathbf{u}) - J_{stm}(t), \\ \frac{\partial \mathbf{u}}{\partial t} &= \mathbf{f}(\mathbf{u},V,t), \end{aligned}\right. \quad \text{for } t \in [t^k, t^k+\Delta t]. \tag{7.34}$$

- *Step II*: Use the result obtained in Step I as the initial condition to integrate

$$C_m \frac{\partial V}{\partial t} = \nabla \cdot (\mathbf{D}\nabla V), \quad \text{for } t \in [t^k, t^k+\Delta t]. \tag{7.35}$$

#### 7.2.4.1 *Spatial-Temporal Discretization*

When performing Step II, the computational domain must be discretized in space by a mesh of either finite elements or finite differences to approximate the dependent variables of the problem, $V$ and u, which allows writing Eq. (7.35) as

$$\mathbf{M}\dot{\mathbf{V}} + \mathbf{K}\mathbf{V} = \mathbf{0}, \tag{7.36}$$

where M and K are the mass and stiffness matrices, respectively, obtained by assembling individual element matrices over the entire computational domain.

The most well-known algorithms for integrating in time the semidiscrete system (7.36) are members of the generalized trapezoidal family methods [51]. Let $\mathbf{V}^k$ and $\dot{\mathbf{V}}^k$ denote vectors of the transmembrane potential and its time derivative at each nodal point of the mesh at time $t^k$, respectively, where $k$ is the index of the time step. Then at time $t^{k+1}$ we can write

$$\mathbf{M}\dot{\mathbf{V}}^{k+1} + \mathbf{K}\mathbf{V}^{k+1} = \mathbf{0}, \tag{7.37}$$

$$\mathbf{V}^{k+1} = \mathbf{V}^k + \Delta t \dot{\mathbf{V}}^{k+\theta}, \tag{7.38}$$

$$\dot{\mathbf{V}}^{k+\theta} = (1-\theta)\dot{\mathbf{V}}^k + \theta \dot{\mathbf{V}}^{k+1}, \tag{7.39}$$

where $\theta \in [0, 1]$ is a scalar parameter. Eqs. (7.37)–(7.39) can be combined to obtain an algebraic system of equations to determine $\mathbf{V}^{k+1}$.

When using the operator splitting algorithm for solving the monodomain model, the equations are solved in two steps. First, the electrophysiological cellular model

$$\mathbf{V}^* = \mathbf{V}^k - \Delta t \left( J_{ion}(\mathbf{V}^k, \mathbf{u}) + J_{stm}(t) \right) \tag{7.40}$$

is solved at each mesh point to obtain an intermediate transmembrane potential vector $\mathbf{V}^*$ (Step I). Even though a forward Euler scheme has been used in Eq. (7.40), any other ODE solver can be used to calculate $\mathbf{V}^*$. With this intermediate solution at hand, along with Eqs. (7.38), (7.39) and $\mathbf{M}\dot{\mathbf{V}}^k = -\mathbf{K}\mathbf{V}^k$ from the previous converged time increment, Eq. (7.37) becomes

$$\mathbf{M}\frac{\mathbf{V}^{k+1} - \mathbf{V}^*}{\Delta t} = -\mathbf{K}\left(\theta\, \mathbf{V}^{k+1} + (1-\theta)\, \mathbf{V}^k\right), \tag{7.41}$$

or alternatively

$$\hat{\mathbf{K}}\mathbf{V}^{k+1} = \hat{\mathbf{b}}, \tag{7.42}$$

where $\hat{\mathbf{K}}$ is everything that multiplies onto $\mathbf{V}^{k+1}$, and $\hat{\mathbf{b}}$ contains the other terms in Eq. (7.41). Eq. (7.42) is solved for the entire domain to the obtained $\mathbf{V}^{k+1}$ (Step II).

Hence, the basic algorithm at time $t^{k+1}$ can be summarized, as:

- *Step I*: Use $V^k$ as the initial condition to integrate Eq. (7.40) to obtain $V^*$.
- *Step II*: Use the result obtained in Step I to solve Eq. (7.42) for $V^{k+1}$.

For different values of the parameter $\theta$, different time integration schemes are obtained for integrating the discretized homogeneous parabolic equation system (7.41):

| | |
|---|---|
| $\theta = 0$ | Forward Euler (conditionally stable) |
| $\theta = 0.5$ | Crank-Nicolson scheme (unconditionally stable) |
| $\theta = \frac{2}{3}$ | Galerkin scheme (unconditionally stable) |
| $\theta = 1$ | Backward Euler (unconditionally stable) |

The Crank-Nicolson scheme is second-order accurate in time, whereas the others are first-order accurate in time. However, for $\theta \geq 0.5$, integration schemes are unconditionally stable.

As mentioned before, Step I can be performed using a backward difference approximation in time (implicit integration) or a forward difference approximation in time (explicit integration). Implicit integration requires the solution of a nonlinear system of equations at each point of the mesh, making it computationally costly. However, it ensures the stability of the numerical solution. On the contrary, explicit integration is computationally cheaper but imposes more stringent conditions on the size of the time step in order to avoid numerical instabilities.

### *7.2.4.2 Integration of the Mass Matrix*

For the standard finite element formulation, the elemental mass matrix, $M^e$ of Eq. (7.41), is given by [51]

$$M^e_{ij} = \int_{\Omega_e} N_i N_j d\mathbf{x},$$

where $N_j$ is the shape function of node $j$ of the element $e$ and $C_m = 1$ has been assumed without loss of generality. Remember that $C_m$ is the membrane capacitance, as defined in Eq. (7.29).

When the shape functions N, used to compute $M^e$, are the same used to approximate the potential $V$, the resulting nondiagonal matrix is known as the consistent mass matrix. Using a consistent mass matrix implies that a linear system of equations has to be solved for $V^{k+1}$ when an explicit scheme is used in Eq. (7.41). In order to improve numerical efficiency, the proposed algorithm evaluates $M^e$ using a mass preserving nodal quadrature [52]. Nodal quadrature is based on the use of different base functions to those used to approximate the transmembrane potential, $V$.

This approximation with different shape functions is admissible because it satisfies the finite element criteria of integrability and completeness [52]. In the implementation, we have considered a nodal quadrature to evaluate $M^e$ with $N_i = J_i I$, being $J_i$ the element Jacobian evaluated at node $i$.

## 7.3 VULNERABILITY IN REGIONALLY ISCHEMIC HUMAN HEART: EFFECT OF THE EXTRACELLULAR POTASSIUM CONCENTRATION

Ventricular tachycardia and ventricular fibrillation are two types of cardiac arrhythmias that usually take place during acute ischemia and frequently lead to sudden death. Proarrhythmic mechanisms of acute ischemia have been extensively investigated, although often in animal models rather than in human ventricles. In this work, we investigate how hyperkalemia affects the vulnerability window (VW) to reentry and the reentry patterns in the heterogeneous substrate caused by acute regional ischemia using an anatomically and biophysically detailed human biventricular model.

In recent years, mathematical modeling and computer simulations have been shown to be a useful tool in analyzing electrophysiological phenomena. In particular, one of the major contributions of computer electrophysiology has been the understanding of important relations between electrophysiological parameters [53]. For the ischemic heart, computer models have allowed us to address the role of ischemic abnormalities in cardiac electrophysiological behavior [54]. However, most of these simulations have been restricted to 2D [54, 55] or 3D simulations of the total ischemic heart [56]. Little work has been carried out in modeling the entire heart subjected to acute ischemic conditions [57, 58]. In the work of Weiss et al., they considered heterogeneities caused by ischemia; their characterization of the AP under acute ischemic conditions has mostly relied on the model characterized for guinea pigs [48]. Dutta et al. [58] have investigated how reduced repolarization increases arrhythmic risk in the heterogeneous substrate caused by acute myocardial ischemia. In their work, Dutta et al. [58] developed a human ventricular biophysically detailed model with acute regional ischemia. Even tough macroreentries around the ischemic zone were reported; these reentries self-terminated after three complete circuits of not being able to establish sustained reentry.

In this chapter, we investigate how hyperkalemia affects the VW to reentry and the reentry patterns in the heterogeneous substrate caused by acute regional ischemia using an anatomically and biophysically detailed human biventricular model. The proposed mathematical model is based on the monodomain model [3] for simulating the propagation of the electrical signal of the heart. For the biophysical description of the AP under normal and ischemic conditions, the model proposed by ten Tusscher and Panfilov [8], TP06, is used. In this regard, the model has been modified to account for ischemia by incorporating an ATP-sensitive potassium, $I_{K(ATP)}$, current. By analyzing high spatiotemporal resolution simulation data, we unravel the mechanisms associated with the observed reentrant patterns in acutely ischemic ventricles.

### 7.3.1 Methods

#### 7.3.1.1 *Mathematical Model*

The variation of the transmembrane potential, $V$, in the heart was modeled by means of the monodomain model [3]

$$\nabla \cdot (\mathbf{D}\nabla V) = C_m \frac{\partial V}{\partial t} + J_{ion}(V, \mathbf{w}) + J_{stm}, \tag{7.43}$$

$$\frac{\partial \mathbf{w}}{\partial t} = \mathbf{f}(\mathbf{w}, V, t), \tag{7.44}$$

where $\mathbf{D}$ is the symmetric and positive definite conductivity tensor, $C_m$ the membrane capacitance, $J_{ion}(V, \mathbf{w})$ the transmembrane ionic current, $J_{stm}$ the stimulation current, $\mathbf{w}(\mathbf{w}, V, t)$ is a vector of gating variables and concentrations, $\mathbf{f}$ is a vector valued function, and $t$ refers to time. Both $J_{ion}$ and f depend on the used cellular model. The boundary conditions associated with this model are

$$\mathbf{n} \cdot \nabla(\mathbf{D}\,\nabla V) = 0, \tag{7.45}$$

where n is the outward normal.

The monodomain model represents an important simplification of the more complex bidomain model [3], with important advantages for mathematical analysis and computation. Despite its simplicity, this model is adequate for studying a number of electrophysiologic problems such as ventricular fibrillation or the onset of ischemia in the electric behavior of the heart [8, 54–58].

From a mathematical and computational point of view, the electrophysiology problem is the coupled solution of a linear partial differential equation, describing electric conduction, with a nonlinear stiff system of ODEs describing the cellular ionic currents that lead to a nonlinear reaction-diffusion problem. An efficient way of solving Eqs. (7.43)–(7.45) is by applying the Strang-based operator splitting scheme [49] in combination with a trapezoidal family method for time integrations, in conjunction with the FEM for the spatial discretization [59].

## 7.3.2 Model of Acute Ischemia

Simulation of the ischemic heart requires an accurate description of the organ that incorporates both its muscular structure and heterogeneity (described in the following sections), and an appropriate model of its electrophysiology. This section is dedicated to the characterization of the mathematical model of AP used to perform the numerical simulations of the regional acute ischemic heart. All simulations were performed with a modified version of the ten Tusscher and Panfilov (TP06) model of AP [8].

### 7.3.2.1 *Action Potential Model Under Ischemic Conditions*

One of the most important aspects in simulating the ischemic heart is to incorporate ATP-sensitive potassium current $I_{KATP}$, a dormant depolarization current under physiological conditions that is activated under ischemic conditions [48]. $K_{ATP}$ ion channels have been investigated in different regions of the heart, that is, the atria and ventricle and the sinoatrial (SA) and atrioventricular (AV) nodes, on different species [60–63]. However, very little experimental data regarding the $I_{KATP}$ current for different tissue layers within the ventricle are available. Furukawa et al. [60] have characterized $K_{ATP}$ channels in isolated endocardial and epicardial cells of cats. Experiments by Furukawa et al. suggest that the open probability of $K_{ATP}$ channels reduces with the intracellular ATP concentration, $[ATP]_i$, for both cell types. However, the $[ATP]_i$ concentration responsible for a 50% block of $K_{ATP}$ channels is approximately four times less for endocardial cells than for epicardial cells. Similar observations have been made by Nichols et al. [61] and Venkatesh et al. [62] in the epicardial cells of guinea pigs, and by Light et al. [63] in rabbits.

A modified version of the ten Tusscher cardiac AP model [8] was used in the simulations. The model describes the principal ionic currents through the cardiac cell membrane with a high degree of electrophysiological detail for the three types of cardiac cells. The basic model was completed with the formulation of the ATP-sensitive $K^+$ current ($I_{KATP}$) described by Ferrero et al. [48].

Fig. 7.4 shows the open probability of $K_{ATP}$ channels, $f_{ATP}$, measured for different cell types and different animal species. The figure suggests a better correlation between data corresponding to the same cell type rather than the animal species. This could be interpreted as a low specificity to animal species but a high specificity to cell type. We have adopted this hypothesis to incorporate the $I_{KATP}$ model by Ferrero et al. [48] originally formulated for guinea pigs to the TP06 model for humans. A similar hypothesis has also been adopted by Michailova et al. [64] in developing a heterogeneous model of $I_{KATP}$ for rabbits.

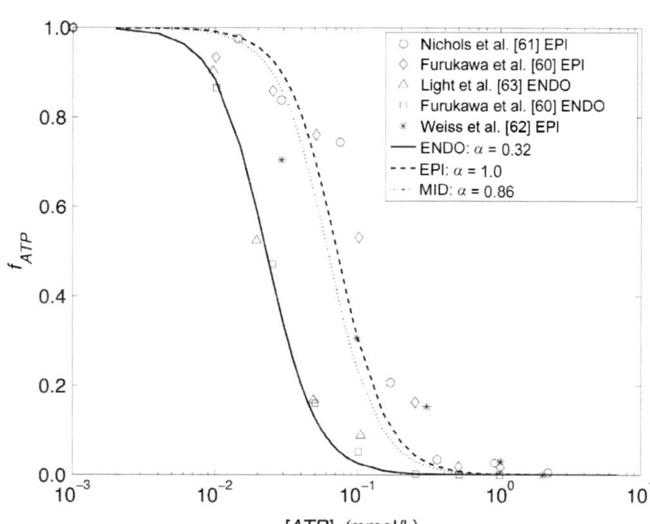


FIG. 7.4 Fraction of open channels for different cell types. *Symbols* represent experimental values for different cell types in different species. *Solid lines* represent the model fit Eq. (7.47).

The procedure followed consisted of modifying the original formulation for the $f_{ATP}$ proposed by Ferrero et al. [48] in order to fit the data from Fig. 7.4 for different cell types. The maximal conductance for the $I_{KATP}$ was modified to adjust the experimental observations on the AP duration (APD) by Furukawa et al. [60]. In other words, this is a 50% reduction in the APD under conditions of hypoxia and hyperkalemia for the epicardium, and a 10% reduction under the same conditions for the endocardium. For the midmyocardium, there are no experimental data available regarding the behavior of $K_{ATP}$ channels. For this case, we have adjusted the value of $f_{ATP}$ such that the same reduction in APD as for epicardial cells was obtained without modifying the maximal conductance of the channel, as proposed by others [64, 65]. In addition, for physiological values of $[ATP]_i$ and $[ADP]_i$, the APD and the resting potential should not be affected by the presence of the $I_{KATP}$ current in the AP model. Hence, the $I_{KATP}$ current has been formulated as [48]

$$I_{KATP} = g_0 \left( \frac{[\mathrm{K}_o^+]}{5.4} \right)^{0.24} f_M f_N f_T f_{ATP} (V - E_K), \tag{7.46}$$

where $g_0$ is the maximum channel conductance in the absence of $Na^+$, $Mg^{2+}$, and ATP; $f_M$, $f_N$, and $f_T$ are correction factors; $f_{ATP}$ is the fraction of opened channels; $V$ is the transmembrane potential; and $E_K$ is the inversion potential of the channel. The maximum channel conductance and the fraction of opened channels, $f_{ATP}$, have been modified with respect to their original formulation

$$f_{ATP} = \frac{1}{1 + ([ATP]_i / K_m)^H}, \tag{7.47}$$

where $[ATP]_i$ is the intracellular concentration of ATP, and $K_m$ (in mmol/L) and $H$ (–) are given as

$$K_m = \alpha(35.8 + 17.9[ADP]_i^{0.256}), \tag{7.48}$$

$$H = 1.3 + 0.74\beta \exp(-0.09[ADP]_i), \tag{7.49}$$

where $[ADP]_i$ is the intracellular concentration of $ADP$ in μmol/L, and $\alpha$ and $\beta$ are fitting parameters that account for the cellular heterogeneity. Parameters $\alpha$ and $\beta$ were identified by fitting experimental data available for different animal models from Fig. 7.4. To adjust $g_0$, 100 stimuli at a basic cycle length (CL) of 1000 ms were applied to an isolated cell. The APD at the last stimulus was measured under physiological and pathological conditions. The variation in APD obtained between the pathological and physiological conditions was used to adjust $g_0$. Table 7.1 shows the adopted $[K^+]_o$, $[ATP]_i$, and $[APD]_i$ values to define physiologic and ischemic conditions for fitting $g_0$.

Table 7.2 summarizes the parameters identified in the fitting process. The table also shows the obtained APD value under physiologic and ischemic conditions defined in Table 7.1, in addition to the APD value obtained when the $I_{KATP}$ current remains inactive. The APD variations between normal and hypoxic conditions given in Table 7.2 are in good agreement with experimental observations by Furukawa et al. [60].

**TABLE 7.1** Extracellular Potassium and Intracellular ADP and ATP Concentrations Defining Physiologic and Ischemic Conditions for the Simulations

| Condition | $[K^+]_o$ (mmol/L) | $[ATP]_i$ (mmol/L) | $[ADP]_i$ (μmol/L) |
|---|---|---|---|
| Normoxia | 5.4 | 6.8 | 15.0 |
| Ischemia | 9.9 | 4.6 | 99.0 |

**TABLE 7.2** Value of the Parameters for the $I_{KATP}$ Current Adapted to the ten Tusscher Model for Different Cell Types, Also Shown Is the APD Under Physiologic and Ischemic Conditions

| | Parameter | | $I_{KATP}$ active | | $I_{KATP}$ inactive |
|---|---|---|---|---|---|
| Cell type | $g_0$ (ms) | $\alpha$ | $APD_{90}^{Nor}$ (ms) | $APD_{90}^{Isq}$ (ms) | $APD_{90}^{Nor}$ (ms) |
| ENDO | 4.5 | 0.32 | 299 | 265 | 299 |
| MID | 4.5 | 0.86 | 393 | 178 | 393 |
| EPI | 4.5 | 1.0 | 298 | 141 | 298 |

#### 7.3.2.2 Heart Model

The geometry of the biventricular heart (the atria have not been considered in the model) and the orientation of the muscle fibers were obtained from DT-MRI from images acquired at Johns Hopkins University [35]. From the segmented image, a regular volumetric mesh was constructed with hexahedral elements and a resolution of 0.4 × 0.4 × 0.4 mm, which gave rise to 1.43 million nodes and 1.29 million hexahedra.

Transmural heterogeneity of electrophysiological behavior across the heart is necessary for normal cardiac function, with excessive heterogeneity contributing to arrhythmogenesis [66]. In this regard, transmural differences in the electrophysiological behavior of the cells were introduced in order to obtain an APD gradient from the endocardium to the epicardium, with the longest APD at the subendocardium [67]. This was achieved by defining a distribution in layers of the three cell types defined in the TP06 models in a proportion of 20% of epicardial cells, 10% of endocardial cells, and the remaining 70% is occupied by M-cells [68]. A recent study based on optical mapping of left-ventricular free wall preparations of human hearts by Glukhov et al. [67] has identified islands of M-cells located at the subendocardium, such as the distribution assumed in this work. These distributions resulted in a positive T wave in the pseudo-ECG, as seen in the bottom panel in Fig. 7.5.

#### 7.3.2.3 Electrophysiological Heterogeneities Under Acute Ischemia

The ischemic region was located in the inferolateral and posterior side of the left ventricle, mimicking the occlusion of the circumflex artery (see Fig. 7.5A). The ischemic region was composed of realistically dimensioned transitional BZs, a normal zone (NZ), and the central zone (CZ) of ischemia in agreement with experimental findings [28] during the early stages of ischemia. In the CZ, $[K^+]_o$ was set at three different values: 7.0, 8.0, and 9.0 in order to study three different time frames during acute ischemia. The inward $Na^+$ and L-type $Ca^{2+}$ currents were scaled by a factor of 0.85 to mimic the effect of acidosis [54, 69], whereas $[ATP]_i$ and $[ADP]_i$ concentrations were set to 5 and 99 mM, respectively [30]. The BZ included a linear variation in electrophysiological properties between the NZ and the ischemic zone (IZ), as shown in experiments [28]. In addition, the model includes a 1.0 mm washed zone (not affected by ischemia) in the endocardium as a result of the interaction between the endocardial tissue and the blood in the ventricular cavities, as suggested by Wilensky et al. [31] (see right panel in Fig. 7.5A). The resulting human ventricular model in acute regional ischemia produced realistic pseudo-ECGs at the six derivations of the standard ECG (see bottom panel in Fig. 7.5), exhibiting the ST elevation in V5–V6 with an acute T wave in V6 and ST depression in V1–V4 consistent with an infarction involving the inferior, lateral, and posterior walls caused by the occlusion of the proximal circumflex artery [70].

#### 7.3.2.4 Stimulation Protocol

This work does not incorporate the specialized conduction system to stimulate the heart. However, Purkinje-like stimulation was simulated by stimulating discrete zones of the endocardium according to the work performed by Durrer et al. [71]. According to this work, endocardial stimulation in the left ventricle starts at three well-defined locations within a time window of 5 ms: (i) a high area on the anterior paraseptal wall just below the attachment of the mitral valve, (ii) a central area on the left surface of the interventricular septum, and (iii) the posterior paraseptal area at about one-third of the distance from the apex to the base. For the right ventricle, we have defined a stimulation area near the insertion of the anterior papillary muscle. In the right ventricle, stimulation starts 5 ms after the onset of the left ventricular potential [71]. Fig. 7.5B shows the location of the four stimulation areas.

The model was preconditioned with endocardial stimulation (S1) consisting of 56 stimulations at a CL of 800 ms (frequency of 1.25 Hz). Following the preconditioning, an extra stimulus, or premature stimulation, (S2) was applied in the subendocardial BZ (see Fig. 7.5C) in agreement with the findings of Janse et al. [30]. The coupling intervals (CI), for example, the time differences between S1 and S2, were varied with a resolution of 1 ms to determine the VW of reentry. In this regard, a depolarization pattern was considered as a reentry if at least two cycles were completed around the ischemic zone. Sustained reentrant patterns were studied for 3 s after the extra stimulus. Reentry patterns and VWs were investigated for different values of extracellular potassium concentration in the CZ.

#### 7.3.2.5 Numerical Simulations

Computations were performed with the GPU-based software TOR [59] using the operator splitting and a semiimplicit scheme with a fixed time step of 0.02 ms. Simulation of 1 s of electric activity took 1.5 h on a GPU Tesla M2090 (6 GB RAM DDR5).

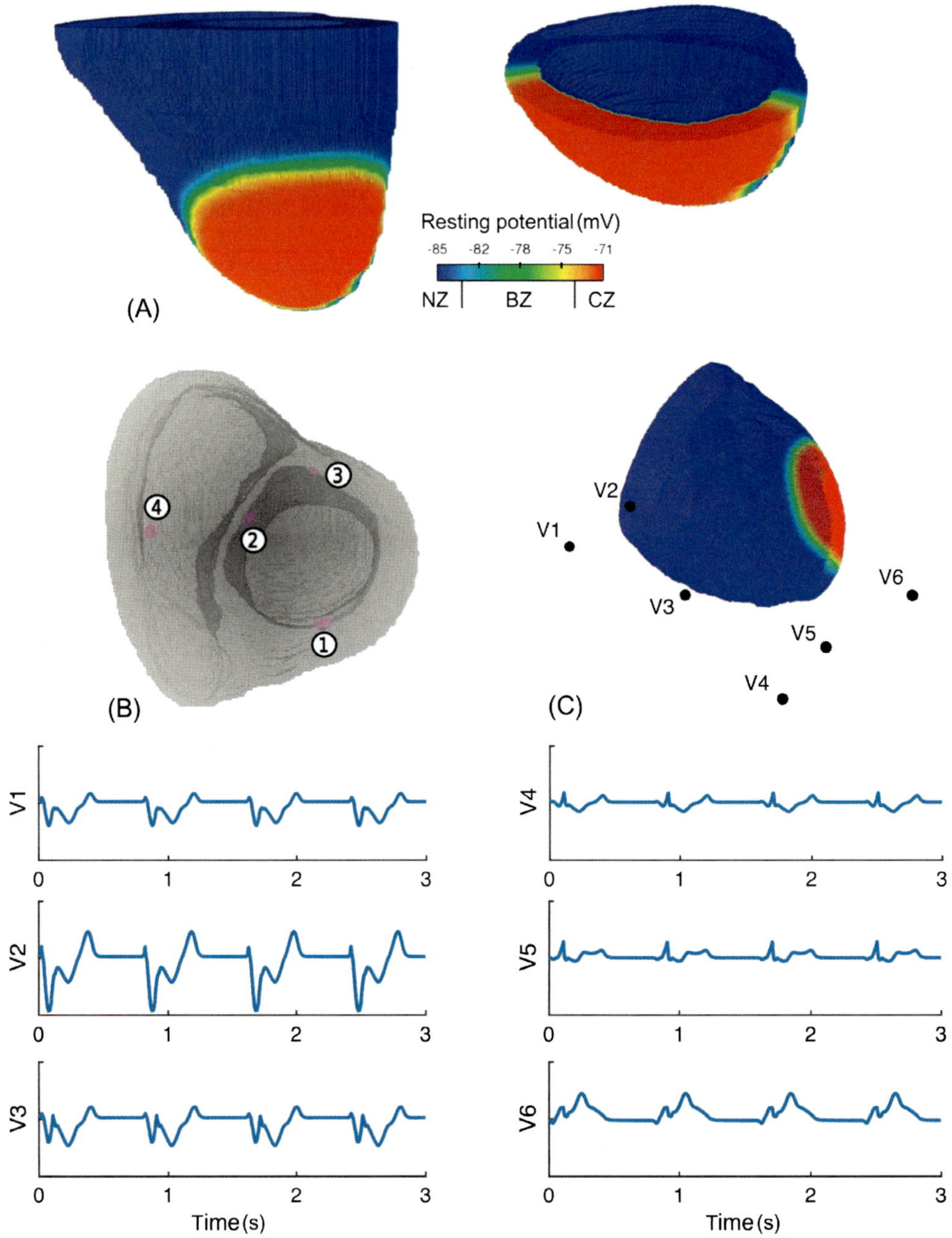


FIG. 7.5 Electrophysiologic heterogeneity in the acute regional ischemic heart. (A) Resting potential in the heterogeneous heart showing higher transmembrane potential in the ischemic tissue (CZ) and the transition through the BZ. The *right panel* shows the details of the washed zone at the endocardium defined in the model as reported in Wilensky et al. [31]. (B) Stimulation sites for the normal SA stimulation according to Durrer et al. [71]. (C) Location of the pseudo-ECG probes corresponding to the six derivations of the standard ECG. A pseudo-ECG is depicted in the *bottom panel*, exhibiting the ST elevation in V5–V6 with an acute T wave in V6 and ST depression in V1–V4 consistent with an infarction involving the inferior, lateral, and posterior walls caused by the occlusion of the proximal circumflex artery. In addition, the positive T wave following the changes in the ST segment is consistent with the inverse relationship between APD and activation times.

### 7.3.3 Results

Results show spatial heterogeneities in the propagated AP, as reported experimentally, throughout the regional ischemic tissue, such as the resting membrane potential (−85.2 mV in NZ and −72.5 mV in the CZ, with potentials varying between these values in the BZ). During a basic beat, activity spreads from three directions into the ischemic area, as shown in Fig. 7.6A. A premature beat was delivered at the point of first activation in the BZ during the basic beat (see point *F* in Fig. 7.6B).

The ectopic beat was delivered at different CI to determine the VW for three levels of extracellular potassium, $[K^+]_o$. Fig. 7.7 shows the VW for different levels of $[K^+]_o$, for an early activation initiating at point *F* in Fig. 7.6.

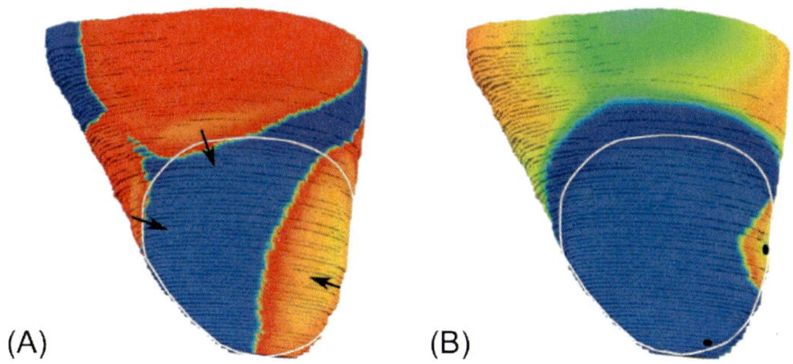


FIG. 7.6 Depolarization pattern through the ischemic tissue (delimited by the *white line*) during basic activation (A), and right after the ectopic beat (B). The ischemic tissue is depolarized by three different fronts coming from the normal activation sites shown in Fig. 7.5. The ectopic beat was delivered at the first point being depolarized within the ischemic tissue (point *F* in the *right panel*).

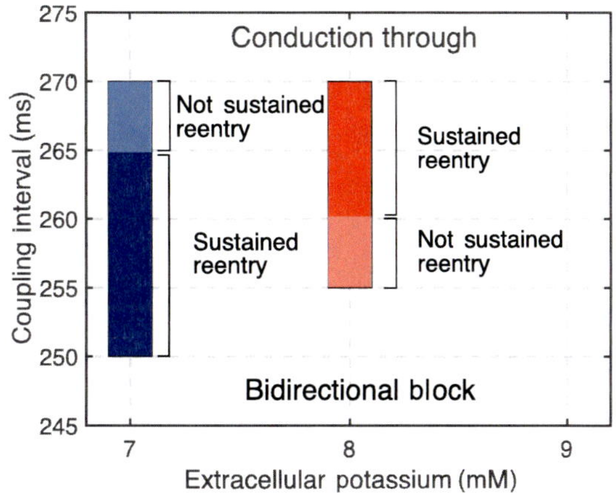


FIG. 7.7 Vulnerability window for different levels of $[K^+]_o$. Early stimulation starts at point *F* in Fig. 7.6.

The extracellular potassium was found to have a significant effect on the VW. For a $[K^+]_o = 7$ mM, the VW was found to be the largest, equal to 20 ms, with a CI between 250 and 270 ms. For CI between 250 and 265 ms, sustained reentry (lasting more than 3 s, about seven to eight beats) was observed, whereas for CIs between 265 and 270 ms, the reentrant activity self-terminated after three reentry circuits (three beats). For $[K^+]_o = 8$ mM, the VW was reduced to 15 ms (CIs between 255 and 270 ms). In this case, sustained reentry was observed for CIs between 255 and 265 ms. On the contrary, for $[K^+]_o = 9.0$ mM no reentrant activity was found. In general, for CIs below the lower bound of the VW, double blocked occurred without generating reentrant activity. On the contrary, for CIs above the upper limit of the VW, conduction through the ischemic zone occurred.

Results in Fig. 7.7 depend on the location of the ectopic stimulation. To explore this fact, the VW was determined for point *F*1 in Fig. 7.6. As in the previous case, no reentrant activity was obtained for $[K^+]_o = 9.0$ mM. For $[K^+]_o = 7.0$ mM the VW was reduced to 10 ms (CI between 265 and 275 ms) with the reentrant activity self-terminating after two beats (800 ms). A similar result was obtained for the case of $[K^+]_o = 8.0$ mM, with reentrant activity observed for CIs between 290 and 310 ms that self-terminated after two reentrant circuits.

The reentrant activity observed for different levels of $[K^+]_o$ and different ectopic sites was associated with ventricular tachycardia in all cases, without degeneration in ventricular fibrillation, at least within the observed window of 3 s after earlier stimulation (see Fig. 7.8). In all cases where sustained reentry was found, the same pattern emerged at the epicardium. After initiating early activity (at point *F* in Fig. 7.6), conduction within the ischemic region slows down (see the frequency of the pseudo-ECG in Fig. 7.8C) and a fragmented wave front with multiple areas of conduction block is observed in the epicardium due to intramural reentry. In successive beats, this pattern evolves into a circus, or double circus, movement within the CZ around an area of diameter about 4.0 cm and a revolution time in the order of 250 ms as shown in Fig. 7.8B for the last recorded beat ($[K^+]_o = 7.0$ mM and CI = 260 ms).

Detailed observation of the activation activity at the endocardium in Fig. 7.8A shows a faster circus movement in the endocardium as compared to the epicardium, with a revolution time on the order of 200 ms. This behavior is due to the presence of the washed zone that allows faster CV. The washed zone was also found to be related to the fragmentation of the wave front at the epicardium because it favored transmural reentry. Fig. 7.9A shows a snapshot of the transmembrane potential at the epicardium and endocardium at $t = 2500$ ms after initiation of earlier stimulation (last recorded beat) for a $[K^+]_o = 7.0$ mM and CI = 260 ms. The figure demonstrates how the double circus closes first

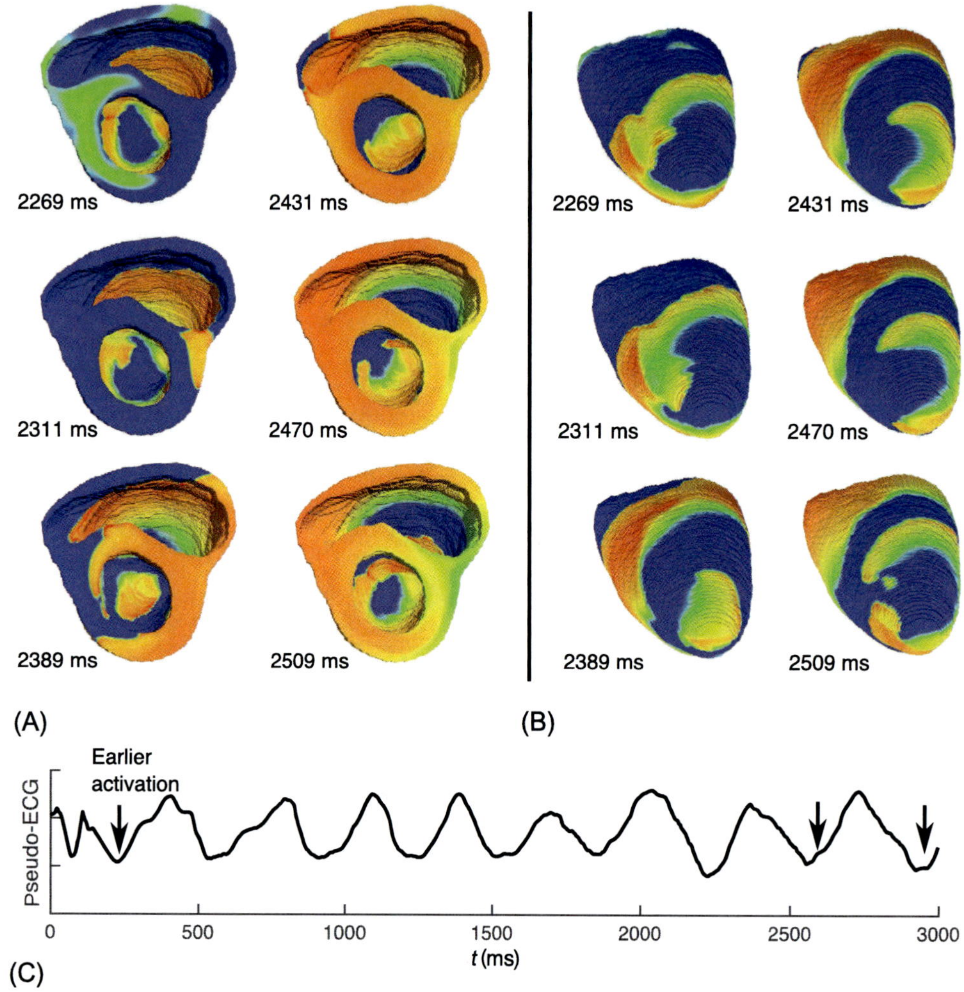


FIG. 7.8 Reentrant patterns at endocardium (A) and epicardium (B) for the last recorded beat for the case $[K^+]_o = 7.0$ mM and CI = 260 ms. Panel C shows the pseudo-ECG corresponding to the derivation V3; the *arrows* indicate the window corresponding to the patterns shown in panels (A) and (B).

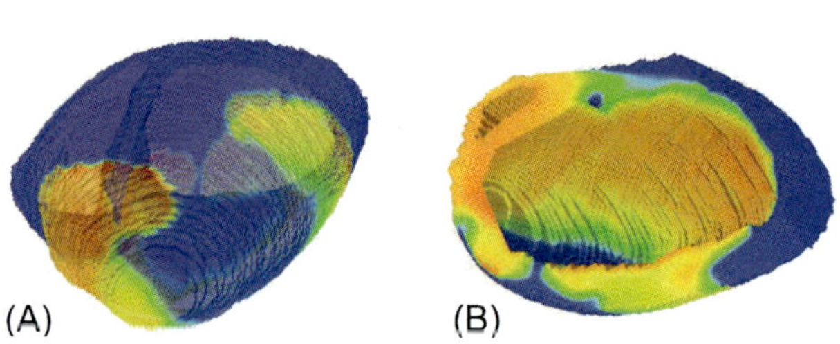


FIG. 7.9 Transmembrane potential at $t = 2500$ ms after initiation of earlier stimulation ($[K^+]_o = 7.0$ mM and CI = 260 ms). (A) Depolarization map at the epicardium and endocardium. (B) Transmural depolarization map.

at the endocardium. The transmural map depicted in Fig. 7.9B shows how the faster endocardial conduction allows intramural conduction, causing this wavefront to emerge at the epicardium at $t = 2509$ ms as shown in Fig. 7.8B.

Even though, basically, one circus movement of relatively large dimensions is responsible for the sustained tachycardia, the pattern of the reentrant wavefront, position, and dimensions changes from beat to beat. Fig. 7.10 shows the changes in the pattern from a single circus to a double circus, a figure of eight, for the case of $[K^+]_o = 7.0$ mM and CI = 260 ms. The reentry was always observed within the ischemic region, and in all cases where sustained reentry was found, the reentrant front was led from the endocardial surface where the pattern was clearly identified for all CIs. Reentry patterns were not altered when the basic stimulation was maintained after delivering the extra stimulus, indicating that tachycardia overrides normal stimulation.

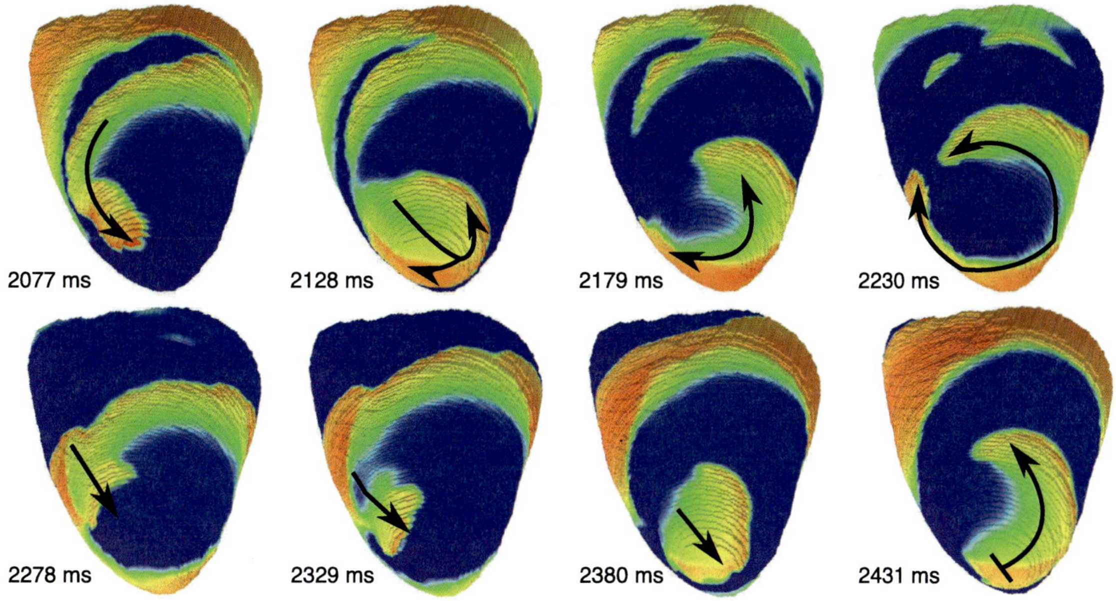


FIG. 7.10 Evolution of the reentry pattern from beat to beat. The sequence indicates an evolution from a single circus to a double circus (figure of eight) to then come back to a single circus pattern ($[K^+]_o = 7.0$ mM and CI = 260 ms).

## 7.3.4 Discussion and Conclusions

We have developed a biventricular human model that combines realistic anatomy and actual muscle fiber orientation obtained from MR-DTI. In addition, the model integrates the biophysical features of membrane kinetics as well as the electrophysiological alterations induced by acute regional ischemia. In this regard, the size of the ischemic area was defined according to previous studies [28]. In despite of not incorporating the specialized conduction system, simulation results are in agreement with experimental studies and clinical observations. Ventricular activation is in agreement with results from Durrer et al. [71]. Pseudo-ECGs at the six derivations of the standard ECG (see bottom panel in Fig. 7.5) are consistent with an infarction caused by the occlusion of the proximal circumflex artery [70].

Extracellular potassium $[K^+]_o$, the element that most influences the ERP in single cell kinetics [48, 54], also has a significant effect on the VW. We found that the size of the VW decreases with $[K^+]_o$. Even though we did not perform simulations for lower values of $[K^+]_o$ than 7 mM, these results are in agreement with the experimental findings from Smith et al. [72] that reported that the peak of arrhythmic events occurs between the first 5–9 min after occlusion, that is, the 1A phase of acute regional ischemia. For this time window, Coronel et al. [28] reported that the accumulation of $[K^+]_o$ in the CZ goes from 6 to 8.5 mM. A possible explanation for this behavior is found in the relationship between the ERP and $[K^+]_o$. As $[K^+]_o$ increases, the ERP also increases, reducing the likelihood of reentry if the size of the ischemic area remains unaltered. In other words, a prolonged ERP in ischemic tissue would require either increased ischemic region size or decreased CV. The size of the ischemic region is defined by the occluded vessel and remains unchanged while local changes in $[K^+]_o$, $[ATP]_i$, and pH occur as demonstrated by many studies [28, 30, 31, 72]. In addition, Smith et al. [72] showed that there is a relatively small increase in tissue resistivity during the first 10 min of ischemia (on the order of 30%–50% of initial values). However, after 15 min of ischemia, a more prominent rise in tissue resistivity begins due to cell-to-cell electrical uncoupling. Electrical uncoupling implies a significant reduction in CV favoring the development of reentry activity.

The macroreentrant patterns of activation obtained in the regionally ischemic human biventricular model were consistent with those reported by Janse et al. [30] in porcine and canine hearts. In all our simulations, reentrant activity was associated with ventricular tachycardia without registering ventricular fibrillation episodes. This is also in agreement with experiments from Smith et al. [72], which reported the onset of ventricular fibrillation to be clustered between 19 and 30 min of ischemia, corresponding to the 1B phase. After stabilization of the reentrant activity, one circuit (single rotor) of a fairly large dimension was basically observed with the activity circling around the CZ. For some beats, double circuits were observed but not sustained, and the reentrant activity continued because of one single reentrant circuit. Patterns from our simulations resemble closely those reported by Janse et al. [30]. For instance, Fig. 8 in [30] shows a single circuit and the eight-shape reentrant pattern also obtained in our simulations (see Fig. 7.10).

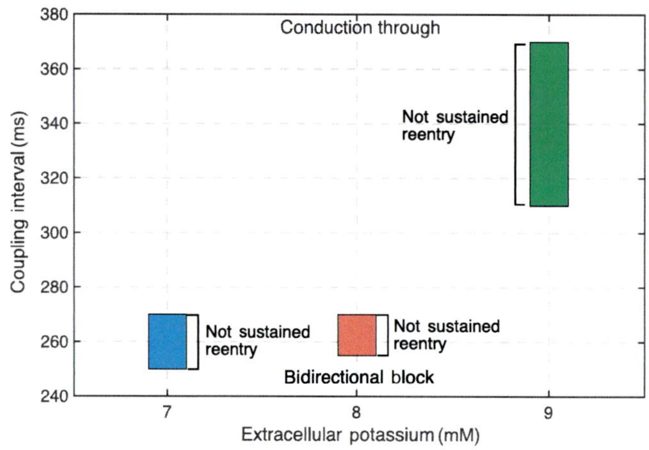


FIG. 7.11 Vulnerability window for different levels of $[K^+]_o$ in a model without the washed-out zone. Early stimulation starts at point *F* in Fig. 7.6.

Our results indicate that reentrant activity may be in part sustained by the presence of a washed zone in the endocardium, as suggested by Wilensky et al. [31]. In order to get a better insight into this, all the simulations were repeated with a model without the washed zone. Fig. 7.11 shows the VW for the model without a washed zone for an ectopic stimulation delivered at point *F* in Fig. 7.6. The most significant finding was that all the reentries self-terminated after two or three complete circuits. While for $[K^+]_o = 7$ mN and $[K^+]_o = 8$ mN the VW was the same but without sustained reentries, for $[K^+]_o = 9$ mM, nonsustained reentrant activity was found for CIs between 310 and 370 ms. The significant difference in the VW between $[K^+]_o = 9$ mM and the other two lower concentrations is associated with the rapid increase in the ERP with $[K^+]_o$. On the contrary, the presence of a VW for $[K^+]_o = 9$ mM in the model without the washed zone is due to the slower CV associated with the ischemic tissue in the endocardium, which allows closing the reentrant circuit that otherwise will find the tissue within the refractory period. These results suggest that the washed zone could act as a proarrhythmic substrate factor helping with establishing sustained ventricular tachycardia.

A number of limitations are associated with this study. The Purkinje system was not present in our model because we were interested in studying the mechanisms associated with reentry rather than in the biophysical mechanisms leading to ectopic excitation. In particular, the study by Janse et al. [30] suggests that earlier activation is most likely due to focal activity localized in the Purkinje fibers close to the ischemic BZ. Future studies, however, are intended to extend our study to evaluate the implication of the Purkinje system in earlier activation during acute regional ischemia and its potential role in contributing to arrhythmia. However, a recent study by Dutta et al. [58] indicates that early activation causing transmural microreentry could be generated by electrotonically triggered EADs at the endocardium. A second important limitation is related to the definition of the ischemic zone itself. We have considered an idealized shape with smooth borders and transitions between the IZ and the NZ. Patient-specific acute regional ischemic areas are expected to have tortuous borders that may contribute to modify the evolution of the reentry patterns and the VW. However, we believe that the general patterns that have emerged in this study will not be greatly modified by the actual shape of the ischemic area as long as the general dimensions of the ischemic area are maintained. In addition, we have only monitored the reentrant activity up to 3 s, during which we have observed perpetuating activity for a restricted vulnerable window. In some cases, that is, CI between 265 and 270 ms for $[K^+]_o = 7.0$ mM, the reentrant activity was spontaneously terminated after completing three reentrant circuits. This type of behavior was also reported in Janse et al. [30], where tachycardia terminated within 30 s after initiation. Additional studies on larger ischemic zones are required in order to determine if the size of the ischemic area may favor the onset of ventricular fibrillation in the ischemic heart. In addition, the location of the ectopic activity is important for both the size of the vulnerable window and the reentrant pattern.

In conclusion, the model predicts the generation of reentry within the ischemic zone due to the heterogeneity in the refractory period between the ischemic affected area and the normal myocardium. The observed patterns obtained with the simulations are in good agreement with experimental studies conducted in porcine and canine hearts subjected to acute regional ischemia. The main results of the simulations can be summarized as follows: (i) As a consequence of the applied extra stimulus that originates an ectopic beat, reentrant activity generated for CIs which range depends on the value of the extracellular potassium concentration $[K^+]_o$; (ii) The reentrant activity, generated due to the extra stimulus initiated as a consequence of the interaction between wavefronts, emerge from the washed zone into the ischemic zone.

## Acknowledgments

The authors thank Dr. Gunnar Seemann at the Karlsruhe Institute of Technology for providing the tetrahedral model of the human atria.

## References

[1] P.J. Hunter, A.J. Pullan, B.H. Smaill, Modeling total heart function, Annu. Rev. Biomed. Eng. 5 (1) (2003) 147–177.
[2] D. Mozaffarian, E.J. Benjamin, A.S. Go, On behalf of the American Heart Association Statistics Committee and Stroke Statistics Subcommittee. Heart disease and stroke statistics 2016 update: a report from the American Heart Association, Circulation 133 (2016) e38–e360.
[3] D. Geselowitz, W. Miller, A bidomain model for anisotropic cardiac muscle, Ann. Biomed. Eng. 11 (3) (1983) 191–206.
[4] C. Henriquez, A brief history of tissue models for cardiac electrophysiology, IEEE Trans. Biomed. Eng. 61 (5) (2014) 1457–1465.
[5] P. Colli-Franzone, L. Pavarino, A parallel solver for reaction-diffusion systems in computational electrocardiology, Math. Models Methods Appl. Sci. 14 (6) (2004) 883–911.
[6] F.H. Fenton, E.M. Cherry, A. Karma, W.J. Rappel, Modeling wave propagation in realistic heart geometries using the phase-field method, Chaos 15 (1) (2005) 013502.
[7] E. Heidenreich, J.M. Ferrero, M. Doblare, J.F. Rodriguez, Adaptive macro finite elements for the numerical solution of monodomain equations in cardiac electrophysiology, Ann. Biomed. Eng. 38 (7) (2010) 2331–2345.
[8] K.H. ten Tusscher, A.V. Panfilov, Alternans and spiral breakup in a human ventricular tissue model, Am. J. Physiol. Heart Circ. Physiol. 291 (3) (2006) 1088–1100.
[9] E.M. Cherry, H.S. Greenside, C.S. Henriquez, Efficient simulation of three-dimensional anisotropic cardiac tissue using an adaptive mesh refinement method, Chaos 13 (3) (2003) 853–865.
[10] P. Colli-Franzone, P. Deuflhard, B. Erdmann, J. Lang, L.F. Pavarino, Adaptivity in space and time for reaction-diffusion systems in electrocardiology, SIAM J. Sci. Comput. 28 (3) (2006) 942–962.
[11] A. Cristoforetti, M. Mase, F. Ravelli, A fully adaptive multiresolution algorithm for atrial arrhythmia simulation on anatomically realistic unstructured meshes, IEEE Trans. Biomed. Eng. 60 (9) (2013) 2585–2593.
[12] M. Bendahmane, R. Bürger, R. Ruiz-Baier, A multiresolution space-time adaptive scheme for the bidomain model in electrocardiology, Numer. Methods Partial Differ. Eq. 26 (6) (2010) 1377–1404.
[13] M.O. Bernabeu, R. Bordas, P. Pathmanathan, J. Pitt-Francis, J. Cooper, A. Garny, D.J. Gavaghan, B. Rodriguez, J.A. Southern, J. P. Whiteley, Chaste: incorporating a novel multi-scale spatial and temporal algorithm into a large-scale open source library, Philos. Trans. R. Soc. A 367 (1895) (2009) 1907–1930.
[14] J.P. Whiteley, Physiology driven adaptivity for the numerical solution of the bidomain equations, Ann. Biomed. Eng. 35 (9) (2007) 1510–1520.
[15] J. Whiteley, An efficient technique for the numerical solution of the bidomain equations, Ann. Biomed. Eng. 36 (8) (2008) 1398–1408.
[16] R. Buchty, V. Heuveline, W. Karl, J.P. Weiss, A survey on hardware-aware and heterogeneous computing on multicore processors and accelerators, Concurr. Comput. Pract. Exp. 24 (7) (2012) 663–675.
[17] J. Nickolls, W. Dallym, The GPU computing era, IEEE Micro 30 (2) (2010) 56–69.
[18] A.R. Sanderson, M.D. Meyer, R.M. Kirby, C.R. Johnson, A framework for exploring numerical solutions of advection-reaction-diffusion equations using a GPU-based approach, Comput. Vis. Sci. 12 (4) (2009) 155–170.
[19] D. Sato, Y. Xie, J.N. Weiss, Z. Qu, A. Garfinkel, A.R. Sanderson, Acceleration of cardiac tissue simulation with graphic processing units, Med. Biol. Eng. Comput. 47 (9) (2009) 1011–1015.
[20] J. Chai, M. Wen, N. Wu, D. Huang, J. Yang, X. Cai, C. Zhang, Q. Yang, Simulating cardiac electrophysiology in the era of GPU-cluster computing, IEICE Trans. Inf. Syst. E96-D (12) (2013) 2587–2595.
[21] E. Bartocci, E.M. Cherry, J. Glimm, R. Grosu, S.A. Smolka, F.H. Fenton, Toward real-time simulation of cardiac dynamics, Proc. of CMSB 2011: the 9th ACM International Conference on Computational Methods in Systems Biology, Paris, France, September 21–23, ACM, 2011, pp. 103–112.
[22] B.M. Rocha, F.O. Campos, R.M. Amorim, G. Plank, R.W.D. Santos, M. Liebmann, G. Haase, Accelerating cardiac excitation spread simulations using graphics processing units, Concurr. Comput. Pract. Exp. 23 (7) (2011) 708–720.
[23] A. Neic, M. Liebmann, E. Hoetzl, L. Mitchell, E.J. Vigmond, G. Haase, G. Plank, Accelerating cardiac bidomain simulations using graphics processing units, IEEE Trans. Biomed. Eng. 59 (8) (2012) 2281–2290.
[24] G. Vigueras, I. Roy, A. Cookson, J. Lee, N. Smith, D. Nordsletten, Toward GPGPU accelerated human electromechanical cardiac simulations, Int. J. Numer. Methods Biomed. Eng. 30 (1) (2014) 117–134.
[25] M. Rubart, D.P. Zipes, Mechanisms of sudden cardiac death, J. Clin. Invest. 115 (9) (2005) 2305–2315.
[26] M.J. Janse, A.G. Kleber, Electrophysiological changes and ventricular arrhythmias in the early phase of regional myocardial ischemia, Circ. Res. 49 (1981) 1069–1081.
[27] E. Carmeliet, Cardiac ionic currents and acute ischemia: from channels to arrhythmias, Physiol. Rev. 79 (3) (1999) 917–1017.
[28] R. Coronel, J.W. Fiolet, F.J. Wilms-Schopman, A.F. Schaapherder, T.A. Johnson, L.S. Gettes, M.J. Janse, Distribution of extracellular potassium and its relation to electrophysiologic changes during acute myocardial ischemia in the isolated perfused porcine heart, Circulation 77 (5) (1988) 1125–1138.
[29] M.J. Janse, J. Cinca, H. Morena, J.W. Fiolet, A.G. Kleber, G.P de Vries, A.E. Becker, D. Durrer, The "border zone" in myocardial ischemia. An electrophysiological, metabolic, and histochemical correlation in the pig heart, Circ. Res. 44 (4) (1979) 576–588.
[30] M.J. Janse, F.J. van Capelle, H. Morsink, A.G. Kleber, F. Wilms-Schopman, R. Cardinal, C.N. d'Alnoncourt, D. Durrer, Flow of "injury" current and patterns of excitation during early ventricular arrhythmias in acute regional myocardial ischemia in isolated porcine and canine hearts. Evidence for two different arrhythmogenic mechanisms, Circ. Res. 47 (2) (1980) 151–165.
[31] R.L. Wilensky, J. Tranum-Jensen, R. Coronel, A.A.M. Wilde, J.W.T. Fiolet, M.J. Janse, The subendocardial border zone during acute ischemia of the rabbit heart: an electrophysiologic, metabolic, and morphologic correlative study, Circulation 74 (1986) 1137–1146.
[32] A.C. Guyton, J. Hall, Textbook of Medical Physiology, thirteenth ed., Saunders-Elsevier, 2015.
[33] L. Tung, A Bi-Domain Model for Describing Ischemic Myocardial Potentials (Ph.D. thesis), Massachusetts Institute of Technology. Department of Electrical Engineering and Computer Science, 1978.

[34] D.D. Streeter, C. Ramon, Muscle pathway geometry in the heart wall, J. Biomech. Eng. 105 (4) (1983) 367–373.

[35] P.A. Helm, A Novel Technique for Quantifying Variability of Cardiac Anatomy Application to the Dyssynchronous Failing Heart (Ph.D. thesis), Johns Hopkins University, 2005.

[36] I. LeGrice, P. Hunter, A. Young, B. Smaill, The architecture of the heart: a data-based model, Philos. Trans. Math. Phys. Eng. Sci. 359 (1783) (2001) 1217–1232.

[37] A.L. Hodgkin, A.F. Huxley, A quantitative description of membrane current and its application to conduction and excitation in nerve, J. Physiol. 117 (4) (1952) 500–544.

[38] P. Stewart, O.V. Aslanidi, D. Noble, P.J. Noble, M.R. Boyett, H. Zhang, Mathematical models of the electrical action potential of Purkinje fibre cells, Philos. Trans. R. Soc. A Math. Phys. Eng. Sci. 367 (1896) (2009) 2225–2255.

[39] M.M. Maleckar, J.L. Greenstein, W.R. Giles, N.A. Trayanova, $K^+$ current changes account for the rate dependence of the action potential in the human atrial myocyte, Am. J. Physiol. Heart C. 297 (2009) 1398–1410.

[40] A. Nygren, C. Fiset, L. Firek, J.W. Clark, D.S. Lindblad, R.B. Clark, W.R. Giles, Mathematical model of an adult human atrial cell: the role of $K^+$ currents in repolarization, Circ. Res. 82 (1) (1998) 63–81.

[41] K.H. ten Tusscher, D. Noble, P.J. Noble, A.V. Panfilov, A model for human ventricular tissue, Am. J. Physiol. Heart Circ. Physiol. 286 (4) (2004) 1573–1589.

[42] T. O'Hara, L. Virág, A. Varró, Y. Rudy, Simulation of the undiseased human cardiac ventricular action potential: model formulation and experimental validation, PLoS Comput. Biol. 7 (5) (2011).

[43] J. Carro, J.F. Rodriguez, P. Laguna, E. Pueyo, A human ventricular cell model for investigation of cardiac arrhythmias under hyperkalaemic conditions, Philos. Trans. R. Soc. A Math. Phys. Eng. Sci. 369 (2011) 4205–4232.

[44] C.H. Luo, Y. Rudy, A dynamic model of the cardiac ventricular action potential. I. Simulations of ionic currents and concentration changes, Circ. Res. 74 (6) (1994) 1071–1096.

[45] C.H. Luo, Y. Rudy, A dynamic model of the cardiac ventricular action potential. II. After depolarizations, triggered activity, and potentiation, Circ. Res. 74 (6) (1994) 1097–1113.

[46] T.R. Shannon, F. Wang, J. Puglisi, C. Weber, D.M. Bers, A mathematical treatment of integrated Ca dynamics within the ventricular myocyte, Biophys. J. 87 (2004) 3351–3371.

[47] B. Alberts, D. Bray, J. Lewis, M. Raff, R. Keith, J.D. Watson, Biologia Molecular de la Celula, Omega S.A., 2002.

[48] J.M.J. Ferrero, J. Saiz, J.M. Ferrero, N.V. Thakor, Simulation of action potentials from metabolically impaired cardiac myocytes: role of ATP-sensitive $K^+$ current, Circ. Res. 79 (2) (1996) 208–221.

[49] G. Strang, On the construction and comparison of difference schemes, SIAM J. Num. Anal. 5 (3) (1968) 506–517.

[50] J. Sundnes, G.T. Lines, A. Tveito, An operator splitting method for solving the bidomain equations coupled to a volume conductor model for the torso, Math. Biosci. 194 (2) (2005) 233–248.

[51] T.J.R. Hughes, The Finite Element Method: Linear Static and Dynamic Finite Element Analysis, Dover Publications, 2000.

[52] O.C. Zienkiewicz, R.L.L. Taylor, J.Z. Zhu, The Finite Element Method: Its Basis and Fundamentals, seventh ed., Butterworth-Heinemann, 2013. vol. 1.

[53] B. Rodriguez, N. Trayanova, D. Noble, Modeling cardiac ischemia, Ann. N. Y. Acad. Sci. 1080 (2006) 395–414.

[54] J.M.J. Ferrero, B. Trenor, B. Rodriguez, J. Saiz, Electrical activity and reentry during acute regional myocardial ischemia: insights from simulations, Int. J. Bifurc. Chaos 13 (2003) 3703–3715.

[55] B. Tice, B. Rodriguez, N. Trayanova, Arrthythmogenicity of transmural heterogeneities in a realistic model of regional ischemia, Heart Rhythm. 2 (5) (2005) S261.

[56] B. Rodriguez, B.M. Tice, J.C. Eason, F. Aguel, N. Trayanova, Cardiac vulnerability to electric shocks during phase 1a of acute global ischemia, Heart Rhythm 1 (6) (2004) 695–703.

[57] D.L. Weiss, M. Ifland, F.B. Sachse, G. Seemann, O. Dossel, Modeling of cardiac ischemia in human myocytes and tissue including spatiotemporal electrophysiological variations, Biomed. Tech. 54 (3) (2009) 107–125.

[58] S. Dutta, A. Mincholé, E. Zacur, A. Quinn, P. Taggart, B. Rodríguez, Early afterdepolarizations promote transmural reentry in ischemic human ventricle with reduced repolarization reserve, Prog. Biophys. Mol. Biol. 120 (2016) 236–248.

[59] A. Mena Tobar, J.M. Ferrero, J.F. Rodríguez, GPU accelerated solver for nonlinear reaction-diffusion systems. Application to the electrophysiology problem, Comput. Phys. Commun. 196 (2015) 280–289.

[60] T. Furukawa, S. Kimura, N. Furukawa, A.L. Bassett, R.J. Myerburg, Role of cardiac ATP-regulated potassium channels in differential responses of endocardial and epicardial cells to ischemia, Circ. Res. 68 (6) (1991) 1693–1702.

[61] C.G. Nichols, C. Ripoll, W.J. Lederer, ATP-sensitive potassium channel modulation of the guinea pig ventricular action potential and contraction, Circ. Res. 68 (1) (1991) 280–287.

[62] J.N. Weiss, N. Venkatesh, S.T. Lamp, ATP-sensitive $K^+$ channels and cellular $K^+$ loss in hypoxic and ischaemic mammalian ventricle, J. Physiol. 447 (1992) 649–673.

[63] P.E. Light, J.M. Cordeiro, R.J. French, Identification and properties of ATP-sensitive potassium channels in myocytes from rabbit Purkinje fibres, Cardiovasc. Res. 44 (2) (1999) 356–369.

[64] A. Michailova, W. Lorentz, A. McCulloch, Modeling transmural heterogeneity of KATP current in rabbit ventricular myocytes, Am. J. Physiol. Cell Physiol. 293 (2) (2007) 542–557.

[65] K. Gima, Y. Rudy, Ionic current basis of electrocardiographic waveforms: a model study, Circ. Res. 90 (8) (2002) 889–896.

[66] B.J. Boukens, M.R. Walton, V.M. Meijborg, R. Coronel, Transmural electrophysiological heterogeneity, the T-wave and ventricular arrhythmias, Prog. Biophys. Mol. Biol. 122 (2016) 202–214.

[67] A.V. Glukhov, V.V. Fedorov, Q. Lou, V.K. Ravikumar, P.W. Kalish, R.B. Schuessler, N. Moazami, I.R. Efimov, Transmural dispersion of depolarization in failing and non-failing human ventricle, Circ. Res. 106 (2010) 981–991.

[68] A. Mincholé, E. Pueyo, J.F. Rodríguez, E. Zacur, M. Doblaré, P. Laguna, Quantification of restitution dispersion from the dynamic changes of the T-wave peak to end, measured at the surface ECG, IEEE Trans. Biomed. Eng. 58 (5) (2011) 1172–1182.

[69] A. Yatani, A.M. Brown, N. Akaike, Effect of extracellular pH on sodium current in isolated, single rat ventricular cells, J. Membr. Biol. 78 (2) (1984) 163–168.
[70] F. Morris, W.J. Brady, ABC of clinical electrocardiography acute myocardial infarction. Part I, BMJ 324 (2002) 831–834.
[71] D. Durrer, R.T. van Dam, G.E. Freud, M.J. Janse, F.L. Meijler, R.C. Arzbaecher, Total excitation of the isolated human heart, Circulation 41 (1970) 899–912.
[72] W.T. Smith, W.F. Fleet, T.A. Johnson, C.L. Engle, W.E. Cascio, The Ib phase of ventricular arrhythmias in ischemic in situ porcine heart is related to changes in cell-to-cell electrical coupling, Circulation 92 (1995) 3051–3060.

## Further Reading

[73] D.G. Allen, P.G. Morris, C.H. Orchard, J.S.A. Pirolo, Nuclear magnetic resonance study of metabolism in the ferret heart during hypoxia and inhibition of glycolysis, J. Physiol. 361 (1) (1985) 185–204.
[74] M.A. Allessie, F.I. Bonke, F.J. Schopman, Circus movement in rabbit atrial muscle as a mechanism of tachycardia. III. The "leading circle" concept: a new model of circus movement in cardiac tissue without the involvement of an anatomical obstacle, Circ. Res. 41 (1977) 9–18.
[75] J. Beaumont, N. Davidenko, J.M. Davidenko, J. Jalife, Spiral waves in two-dimensional models of ventricular muscle: formation of a stationary core, Biophys. J. 75 (1998) 1–14.
[76] N. Bell, M. Garland, Cusp-library: generic parallel algorithms for sparse matrix and graph computations, (2008). Tech. Rep. NVR-2008-008, NVIDIA corporation, December.
[77] A. Bueno-Orovio, E.M. Cherry, F.H. Fenton, Minimal model for human ventricular action potentials in tissue, J. Theor. Biol. 253 (3) (2008) 544–560.
[78] W.C. Cole, C.D. McPherson, D. Sontag, ATP-regulated $K^+$ channels protect the myocardium against ischemia/reperfusion damage, Circ. Res. 69 (3) (1991) 571–581.
[79] R. Coronel, Heterogeneity in extracellular potassium concentration during early myocardial ischaemia and reperfusion: implications for arrhythmogenesis, Cardiovasc. Res. 28 (6) (1994) 770–777.
[80] R. Coronel, F.J.G. Wilms-Schopman, L.R.C. Dekker, M.J. Janse, Heterogeneities in $[K^+]_o$ and TQ potential and the inducibility of ventricular fibrillation during acute regional ischemia in the isolated perfused porcine heart, Circulation 92 (1) (1995) 120–129.
[81] J. Nickolls, I. Buck, M. Garland, K. Skadron, Scalable parallel programming with CUDA, Queue 6 (2) (2008) 40–53.
[82] N.V.I.D.I.A. Corporation, NVIDIA CUDA Programming Guide, June 2008. version 2.0.
[83] D.B. Kirk, W.W. Hwu, Programming Massively Parallel Processors. A Hands-On Approach, Elsevier, 2010.
[84] J.M. Davidenko, R. Salomonsz, A.M. Pertsov, W.T. Baxter, J. Jalife, Effects of pacing on stationary reentrant activity: theoretical and experimental study, Circ. Res. 77 (6) (1995) 1166–1179.
[85] F. Fenton, A. Karma, Vortex dynamics in three-dimensional continuous myocardium with fiber rotation: filament instability and fibrillation, Chaos 8 (1) (1998) 20–47.
[86] S. Filippone, M. Colajanni, Psblas: a library for parallel linear algebra computation on sparse matrices, ACM Trans. Math. Softw. 26 (4) (2000) 527–550.
[87] A. Frank, C.J.C. Choung, R.L. Johnson, A finite-element model of oxygen diffusion in the pulmonary capillaries, J. Appl. Physiol. 82 (1997) 2036–2044.
[88] R.N. Gasser, R.D. Vaughan-Jones, Mechanism of potassium efflux and action potential shortening during ischaemia in isolated mammalian cardiac muscle, J. Physiol. 431 (1) (1990) 713–741.
[89] A.S. Go, D. Mozaffarian, V.L. Roger, et al., Heart disease and stroke statistics 2014 update: a report from the American Heart Association, Circulation 129 (3) (2014) e28–e292.
[90] W.B. Gough, R. Mehra, M. Restivo, R.H. Zeiler, N. El Sherif, Reentrant ventricular arrhythmias in the late myocardial infarction period in the dog. 13. Correlation of activation and refractory maps, Circ. Res. 57 (3) (1985) 432–442.
[91] D.M. Harrild, C.S. Henriquez, A finite volume model of cardiac propagation, Ann. Biomed. Eng. 25 (1997) 315–334.
[92] M.N. Hicks, S.M. Cobbe, Effect of glibenclamide on extracellular potassium accumulation and the electrophysiological changes during myocardial ischaemia in the arterially perfused interventricular septum of rabbit, Cardiovasc. Res. 25 (5) (1991) 407–413.
[93] V. Jacquemet, C.S. Henriquez, Finite volume stiffness matrix for solving anisotropic cardiac propagation in 2-D and 3-D unstructured meshes, IEEE Trans. Biomed. Eng. 52 (8) (2005) 1490–1492.
[94] J.H. Kirkels, C.J. van Echteld, T.J. Ruigrok, Intracellular magnesium during myocardial ischemia and reperfusion: possible consequences for postischemic recovery, J. Mol. Cell Cardiol. 11 (1989) 1209–1218.
[95] A.G. Kleber, Y. Rudy, Basic mechanisms of cardiac impulse propagation and associated arrhythmias, Physiol. Rev. 84 (2) (2004) 431–488.
[96] I. Kodama, A. Wilde, M.J. Janse, D. Durrer, K. Yamada, Combined effects of hypoxia, hyperkalemia and acidosis on membrane action potential and excitability of guinea-pig ventricular muscle, J. Mol. Cell Cardiol. 16 (3) (1984) 247–259.
[97] F.V. Lionetti, GPU Accelerated Cardiac Electrophysiology (Master's thesis), University of San Diego, California, 2010.
[98] C.H. Luo, Y. Rudy, A model of the ventricular cardiac action potential. Depolarization, repolarization, and their interaction, Circ. Res. 68 (6) (1991) 1501–1526.
[99] A. Mahajan, Y. Shiferaw, D. Sato, et al., A rabbit ventricular action potential model replicating cardiac dynamics at rapid heart rates, Biophys. J. 94 (2008) 392–410.
[100] B. Mauroy, Following red blood cells in a pulmonary capillary, ESAIM Proc. 23 (2008) 48–65.
[101] J.G. Murphy, M.A. Lloyd, Mayo Clinic Cardiology, Mayo Clinic Scientific Press and Inform Healthcare USA, Inc., 2007.
[102] G. Karypis, V.M. Kumar, A Software Package for Partitioning Unstructured Graphs, Partitioning Meshes, and Computing Fill-Reducing Orderings of Sparse Matrices, 1998. University of Minnesota, Department of Computer Science/Army HPC Research Center, Minneapolis, MN, version 4.0, September.

[103] J.P. Mills, L. Qie, M. Dao, C.T. Lim, S. Suresh, Nonlinear elastic and viscoelastic deformation of the human red blood cell with optical tweezers, MCB 1 (3) (2004) 169–180.

[104] J.T. Hansen, M.K. Bruce, Nettre's Atlas of Human Physiology, WB Saunders Company, 2002.

[105] S.A. Niederer, E. Kerfoot, A.P. Benson, Verification of cardiac tissue electrophysiology simulators using an N-version benchmark, Philos. Trans. R. Soc. A. 369 (1954) (2011) 4331–4351.

[106] M. Pennacchio, V. Simoncini, Efficient algebraic solution of reaction-diffusion systems for the cardiac excitation process, J. Comput. Appl. Math. 145 (1) (2002) 49–70.

[107] M. Potse, B. Dube, J. Richer, A. Vinet, R. Gulrajani, A comparison of monodomain and bidomain reaction-diffusion models for action potential propagation in the human heart, IEEE Trans. Biomed. Eng. 53 (12) (2006) 2425–2435.

[108] J.L. Puglisi, D.M. Bers, LabHEART: an interactive computer model of rabbit ventricular myocyte ion channels and Ca transport, Am. J. Physiol. Heart C 281 (6) (2001) C2049–C2060.

[109] Z. Qu, J. Kil, F. Xie, A. Garfinkel, J.N. Weiss, Scroll wave dynamics in a three-dimensional cardiac tissue model: roles of restitution, thickness, and fiber rotation, Biophys. J. 78 (6) (2000) 2761–2775.

[110] Z. Qu, H.S. Karagueuzian, A. Garfinkel, J.N. Weiss, Effects of $Na^+$ channel and cell coupling abnormalities on vulnerability to reentry: a simulation study, Am. J. Physiol. Heart Circ. Physiol. 286 (4) (2004) H1310–H1321.

[111] P.L. Rensma, M.A. Allessie, W.J. Lammers, F.I. Bonke, M.J. Schalij, Length of excitation wave and susceptibility to reentrant atrial arrhythmias in normal conscious dogs, Circ. Res. 62 (2) (1988) 395–410.

[112] B.J. Roth, An S1 gradient of refractoriness is not essential for reentry induction by an S2 stimulus, IEEE Trans. Biomed. Eng. 47 (6) (2000) 820–821.

[113] S. Rush, H. Larsen, A practical algorithm for solving dynamic membrane equations, IEEE Trans. Biomed. Eng. 25 (4) (1978) 389–392.

[114] B. Sakmann, G. Trube, Conductance properties of single inwardly rectifying potassium channels in ventricular cells from guinea-pig heart, J. Physiol. 347 (1) (1984) 641–657.

[115] H.I. Saleheen, K.T. Ng, A new three-dimensional finite-difference bidomain formulation for inhomogeneous anisotropic cardiac tissues, IEEE Trans. Biomed. Eng. 45 (1) (1998) 15–25.

[116] J.A. Trangenstein, C. Kim, Operator splitting and adaptive mesh refinement for the Luo-Rudy I model, J. Comput. Phys. 196 (2) (2004) 645–679.

[117] E.J. Vigmond, L.J. Leon, Computationally efficient model for simulating electrical activity in cardiac tissue with fiber rotation, Ann. Biomed. Eng. 27 (2) (1999) 160–170.

[118] C.H. Wang, A.S. Popel, Effect of red blood cell shape on oxygen transport in capillaries, Math. Biosci. 116 (1993) 89110.

[119] A.L. Wit, M.J. Janse, The Ventricular Arrhythmias of Ischemia and Infarction: Electrophysiological Mechanisms, Futura, Mount Kisco, NY, 1993.

[120] G.X. Yan, A.G. Kleber, Changes in extracellular and intracellular pH in ischemic rabbit papillary muscle, Circ. Res. 71 (2) (1992) 460–470.

CHAPTER

# 8

# Towards the Real-Time Modeling of the Heart

*R.R. Rama*,†, S. Skatulla*,†*

*Center for Research in Computational and Applied Mechanics, UCT, Cape Town, South Africa

†Department of Civil Engineering, Computational Continuum Mechanics Research Group, UCT, Cape Town, South Africa

## 8.1 INTRODUCTION

Computational cardiac mechanics have seen a significant amount of research activity during the last decade in an effort to support and supplement clinical and experimental work. In order to simulate the pumping heart, one needs to describe the nonlinear elastic material behavior of the heart muscle tissue [1, 2], the electrophysiology pacing the contraction of the heart muscle [3], the active contraction effect to eject blood into the lungs and systemic circulatory system [2], and the coupling of the blood circulatory system to the heart in terms of varying blood pressure and flow resistance [4] using mathematical and computational models. As a result, those models have become increasingly realistic to the extent that patient-specific heart simulations can provide for qualitative and quantitative predictions of the heart's function in health, injury, and disease leading to advances in diagnostic and therapeutic procedures (see e.g., Baillargeon et al. [5]). However, in those mathematical models stated, complex coupled nonlinear partial and ordinary differential equations are employed and need to be solved using iterative schemes. These computational calculations are extremely time consuming [6]. In Refs. [7–10], it has been found that the required computational resources can vary from 16 to as much as 200 processors for calculation times ranging between 1 and 50 h for only one single heartbeat. Practical medical and research application of heart modeling would usually involve a multitude of simulations for time periods significantly longer than one single heartbeat to study variations in physiological conditions with respect to hemodynamical loads, disease progression, therapeutical measures or medication, etc. This, however, cannot yet be achieved by conventional means of computer modeling. For this reason, the application of such models has been very limited in the medical practice, as they would be usually required to be run on common desktop or laptop machines.

Current advances in computing power do not yet allow for a speed-up in these types of numerical computations. Hence, in the literature, many researchers have explored alternative ways. One popular approach consists of the use of a mass-spring model [11, 12]. As elaborated, for example, in Meier et al. [12], a mass-spring model defines a geometrical mesh in terms of discrete mass points that are interconnected by springs. With the help of the Newtonian law of motion and a time discretization scheme, the deformation of the mesh can be solved for when subjected to external forces. This method is considered simple and computationally efficient [13], and has led to a wide range of applications. For example, Nedel and Thalmann [14] simulated the brachialis muscle of the upper arm. Liu et al. [15] studied a nonbiomechanics related example, namely the hanging of cloth. Using the mass-spring approach, they achieved a calculation frequency ranging from 0.2 to 185 Hz, but with relatively large errors. In Ref. [16], Luboz et al. modeled the deployment of a stent in the femoral artery using an inflated balloon. Their simulations were carried out at a frequency range of 26–53 Hz. Even though high computational speed was achieved using the mass-spring method, Nealen et al. [13] reported that the models had low accuracy, and according to Meier et al. [12], unrealistic behavior, such as delays in deformation propagation and unphysical oscillations. Another common approach for more realistic real-time simulation is through the use of the linear finite element method (FEM) [17, 18]. The latter does not require an update of the stiffness matrix to solve for the mechanical fields such as displacement, stress, and strain. Some examples of its application in the literature are as follows: Cotin et al. [11] proposed an "enhanced linear elastic FEM model" in order to account for the nonlinear behavior

*Advances in Biomechanics and Tissue Regeneration*
https://doi.org/10.1016/B978-0-12-816390-0.00008-X

of the liver. Berkley et al. [19] also employed a linear model for the suturing procedure of biological tissue where they deemed the dynamic effects to be negligible. As such, a calculation frequency of 30 Hz was achieved. Lim and De [20] investigated two different approaches to address the nonlinear material behavior encountered for the surgical tool-tip penetration of a kidney. The key idea consisted of considering nonlinear mechanics in the region where the surgery tool tip is in contact with the organ and treat the rest of the organ in a linear fashion. For the solving procedure, they made use of a mesh-free point collocation-based method and a modified Newton-Raphson scheme, achieving a calculation frequency of 1000 Hz. Lastly, Courtecuisse et al. [21] employed graphical processing units (GPU)-based algorithms of a linear finite element model to emulate the cutting of a liver at a simulation frequency of 29 Hz. Even though linear elastic models provide fast computation times, the solution accuracy in the context of large deformation is usually poor with, for instance, an error rate of up to 30% as encountered in the work of Lim and De [20].

In this chapter, it is proposed to use a reduced order method (ROM) to drastically reduce the computation time of full-cycle heart simulations at high levels of accuracy. Some of the most popular ROM techniques are proper orthogonal decomposition (POD) [22], reduced order basis (ROB) [23], and the proper generalized decomposition (PGD) [24]. The POD and ROB are based on an a posteriori approach, where predefined sets of data are needed to start the computation [22, 23]. Yet, between the two of them, POD has been found to be more computationally efficient [25]. By contrast, the PGD method is based on an a priori procedure and as such does not require predefined sets of data. Even though these ROM techniques have been used across a wide variety of different fields, their application to cardiac modeling is limited. Only POD has been employed so far but targeted specifically toward the electrophysiology of the heart [26, 27]. The applications of POD, ROB, and PGD are mostly geared toward reducing the system of equations, which can consequently be easily solved. However, assembling these equation systems still remains computationally expensive, especially for nonlinear problems.

The POD is well established in solid mechanics, for example, Refs. [28, 29]. For our purpose of achieving real-time modeling, the direct application of POD to, for example, the FEM, does not result in a large enough reduction in calculation time because only the solving of the discrete equation system is accelerated. One particular variant of POD bypasses the setting up of the discrete equation system while exploiting the reduction capabilities of the POD; it is called the proper orthogonal decomposition with interpolation (PODI) method, developed by Ly and Tran [30]. Recent research carried out by Niroomandi et al. and Coelho et al. has found that the PODI approach allows for subsecond calculation times to be achieved and therefore makes high-frequency computation feasible [31, 32]. The PODI method makes use of a collection of datasets of the problem under consideration, describing its mechanics for a range of variations in terms of geometry, material properties, loading conditions, etc. These datasets are then utilized to interpolate the mechanics of a similar problem of the same category where its mechanical behavior is unknown or has not been determined yet. In our case, the moving least square (MLS) approximation [33] has been chosen as the interpolation technique due to its ability in scaling up smoothly to several dimensions when multiparametric simulations are needed. The collection of datasets is stored offline in a database-type format for ease of access and is comprised of full-scale simulation results of the human heart obtained using either the element-free Galerkin method (EFG) or FEM, considering variations in the cardiac tissue characteristics. In practice, it is expected that the used characteristic cardiac parameters will be related to in vivo data. These can be heart anatomy, strain, or hemodynamics data from magnetic resonance imaging (MRI), fiber orientation vectors from diffusion tensor magnetic resonance imaging (DTMRI), or other categorizing parameters such as gender, pathologies, etc. In this sense, a clear advantage of the proposed method is that not only patient-specific heart modeling can be achieved at drastically reduced computation times, but also the accuracy of the computations will be continuously improved by the addition of new datasets over time. A recent paper by Rama et al. [34] addressed this problem and provided an in-depth investigation on the evolution of the computational cost when using PODI. Subsecond calculation times were achieved using a standard desktop computer with only a 1 CPU core and limited memory resources while the calculation errors were kept at minimal levels. Rama et al. applied PODI to heart modeling, which resulted in a calculation frequency of about 550 Hz.

In this chapter, a first application will look at modeling a full heartbeat using PODI [35] and is an extension of the work of Rama et al. [34], where only the diastole filling was investigated. To achieve that, a so-called time standardization method is developed. This procedure is required because the time steps needed for each simulated dataset vary according to the step-size increments or the numerical stability level of the corresponding full-scale simulation. Hence, choosing the solution fields for the PODI calculation necessitates that they are suitably synchronized with each other. This involves the creation of a time line with fixed standardized points applicable to each dataset and an additional temporal PODI calculation to obtain the solution fields at those points.

For our second application, a PODI-based cardiac modeling for geometrical variations in heart anatomy is considered [36]. Every heart shares common anatomical features such as two ventricles and two atria as well as ventricular and aortic valves. Furthermore, a single heartbeat comprises four phases, namely, diastole filling, isovolumetric contraction, ejection, and isovolumetric relaxation where the heart expands, contracts, and twists on itself. In this sense, the heart behavior of different individuals is generally very similar and can be extracted using POD in terms of proper orthogonal modes (POMs). However, despite the similarities, distinct anatomical, physiological, and biomechanical differences remain. Each heart has a different size as well as atrial and ventricular cavity volumes, see, for example, Refs. [37–42]. The elastic material properties of the heart tissue determined from biaxial and triaxial tests by Sommer et al. [43] show clear variations in the stress-strain response, in particular for larger strains. Within the PODI framework, the anatomical differences are of primal importance but difficult to integrate. The problem is linked to the extraction of the POMs. In this work, the PODI database is constructed from a collection of solution fields obtained from simulations carried out for a range of elastic material parameters. For the purpose of projection and interpolation, a subset of these datasets is assembled in a matrix. In order to build the matrix, the data size of all incorporated solution fields and their respective discrete spatial locations must be the same. Due to the anatomical differences of patient-specific hearts, the corresponding discretized computational models will always exhibit a significant variation in nodal numbers and spatial distribution. Consequently, assembling the resulting solution fields into a single matrix will lead to incompatible vectors in the data matrix, effectively preventing the extraction of POMs from it.

Most PODI implementations found in the literature made use of a fixed mesh configuration. Only very recently have variations in geometry and mesh configurations been explored [44–46]. In Amsallem et al. [45], the POMs linked to different discretization layouts were minimized with respect to the POMs of a reference discretization reassembling the geometrical shape of the problem at hand. The method considered POMs extracted from different series of datasets, each computed from different mesh configurations. As such, the resulting POMs were not consistent with each other and an additional step was required to enforce consistency. In González et al. [46], another approach was investigated. The authors embedded different liver geometries on a benchmark cube mesh grid and computed a so-called distance field with respect to the boundary surface of the organ. Following that, they then employed a method called locally linear embedding to find the weight of the different registered liver configurations to reproduce the anatomical model of the liver at hand and to carry out the interpolation. It remains, however, somewhat unclear how the nodes inside the liver geometry were treated. Iuliano and Quagliarella [44] stacked the nodal coordinates in the data matrix such that the extracted POMs reflected the spatial layout of node distribution. This approach only considered an identical number of nodes, mainly focusing on mesh optimization. Generally, however, different mesh discretizations will consist of a unique number of nodes and degrees of freedom. Therefore, for this research to facilitate patient-specific heart modeling, an alternative approach is proposed that will be referred to as the degrees of freedom standardization (DOFS) method.

The DOFS method consists of establishing a set of nodes, the so-called template nodes, that will be common for every dataset in the database. Making use of a three-dimensional (3D) MLS scheme, all solution fields of each involved dataset will be projected onto the set of template nodes and will consequently share the same degrees of freedom and spatial locations. It has to be noted that the examples considered in the DOFS study looked at solution fields that represent only the diastole filling. For the choice of the spatial locations of the template nodes, two different approaches will be explored: a cube grid and a heart-shaped grid. In order to enforce a high degree of accuracy in the interpolation and to ensure that the spatial distribution of the dataset's solution fields is correctly captured by the template nodes of the heart-shaped grid, a nonrigid registration algorithm, the coherent point drift (CPD) method introduced by Myronenko et al. [47], will be used to morph the heart datasets onto the template. The great benefit of this method is that the datasets do not have to possess a geometry of the same size to obtain a good registration [48].

The layout of the chapter is as follows: In Section 8.2, the cardiac mechanics equations along with the model used for the simulation of the heart will be introduced. Section 8.3 deals with the ROM where the POD method will be revisited in Section 8.3.1 and the PODI method will then be elaborated in Section 8.3.2 with a focus on the parametric PODI and the temporal PODI. The time standardization scheme is subsequently outlined in Section 8.4 and the degree of freedom standardization scheme is presented in Section 8.5. Finally, the conclusion of the chapter is given in Section 8.6.

## 8.2 CARDIAC MECHANICS AND MODEL

A full heartbeat cycle consists of four phases: diastolic filling, isovolumetric contraction, ejection, and isovolumetric relaxation. These stages are defined by mathematical models that are introduced in the following.

### 8.2.1 Passive Stress

The heart muscle tissue is adequately modeled by an exponential function coupled with an incompressibility term, as suggested by Usyk and McCulloch [49]. This nonlinear orthotropic hyperelastic strain energy function has been reformulated in terms of the invariants of the Green strain tensor, **E**, by Legner et al. [50] and is used in this research:

$$W=\frac{A}{2}(e^{Q}-1)+A_{\text{comp}}[\det J\ln(\det J)-\det J+1], \tag{8.1}$$

where

$$\begin{aligned} Q &= a_1(\mathrm{tr}(\mathbf{M}_1\mathbf{E}))^2+a_2(\mathrm{tr}(\mathbf{M}_2\mathbf{E}))^2+a_3(\mathrm{tr}(\mathbf{M}_3\mathbf{E}))^2 \\ &\quad +a_4(\mathrm{tr}(\mathbf{M}_1\mathbf{E})^2)+a_5(\mathrm{tr}(\mathbf{M}_5\mathbf{E})^2)+a_6(\mathrm{tr}(\mathbf{M}_3\mathbf{E})^2). \end{aligned} \tag{8.2}$$

$$\begin{aligned} \mathbf{E} &= E_{11}\mathbf{V}_1\otimes\mathbf{V}_1+E_{22}\mathbf{V}_2\otimes\mathbf{V}_2+E_{33}\mathbf{V}_3\otimes\mathbf{V}_3+E_{12}(\mathbf{V}_1\otimes\mathbf{V}_2+\mathbf{V}_2\otimes\mathbf{V}_1) \\ &\quad +E_{13}(\mathbf{V}_1\otimes\mathbf{V}_3+\mathbf{V}_3\otimes\mathbf{V}_1)+E_{23}(\mathbf{V}_2\otimes\mathbf{V}_3+\mathbf{V}_3\otimes\mathbf{V}_2). \end{aligned} \tag{8.3}$$

$$\mathbf{M}_1 = \mathbf{V}_1\otimes\mathbf{V}_1, \quad \mathbf{M}_2=\mathbf{V}_2\otimes\mathbf{V}_2, \quad \mathbf{M}_3=\mathbf{V}_3\otimes\mathbf{V}_3. \tag{8.4}$$

$J=\det\mathbf{F}$ is the Jacobian with **F** denoting the deformation gradient tensor and $A_{\text{comp}}$ is a penalty parameter to control the degree of incompressibility of cardiac tissue. The material constant $A$ is a stress-scaling factor and $a_i$ $(i=1, \ldots, 6)$ are the anisotropy coefficients associated with the local components of **E** corresponding to the preferred material directions, namely fiber axis, $V_1$, sheet axis, $V_2$, and sheet normal axis, $V_3$.

The fiber and the two cross-fiber directions (i.e., sheet and sheet normal) orientation distribution in the heart model are based on the DTMRI data of Rohmer et al. [51], where a human heart was scanned and its fiber angles averaged for different zones of the left ventricle (LV). In order to simplify the procedure of the fiber assignment on the geometry, the basal fiber and sheet angles are assigned throughout the right ventricle (RV) and LV of the heart model. The values retained from the provided experimental data are given in Table 8.1.

The fiber projection algorithm developed by Wong and Kuhl [52] is subsequently employed to obtain the fiber normals on the epicardium and endocardium. Then, with the help of MLS approximations (a brief introduction is provided in Appendix A.1), the 3D distribution of the fiber and cross-fiber directions is generated across the heart walls, as shown in Fig. 8.1.

### 8.2.2 Active Stress

The isovolumetric contraction and relaxation mark the start and end where the myocardium actively contracts. Modeling of the isovolumetric contraction and relaxation phases requires special care to enforce near-constant cavity volume due to the closed heart valves. Here, the approach by Skatulla and Sansour [53] is employed. To emulate the active contraction, the active stress model by Guccione et al. [2] is used where it is assumed that the entire myocardium of the ventricles contracts simultaneously. The active stress model is based on an additive decomposition of the total second Piola-Kirchhoff stress, $\mathbf{S}^{\text{Total}}$, at any point in the myocardium into a passive stress, $\mathbf{S}^{\text{Passive}}$, and an active stress component, $\mathbf{S}^{\text{Active}}$:

$$\mathbf{S}^{\text{Total}}=\mathbf{S}^{\text{Passive}}+\mathbf{S}^{\text{Active}}. \tag{8.5}$$

The passive stress is obtained from the passive strain energy function (Eq. 8.1) and the active stress acts along the fiber direction, $\mathbf{V}_1$, with an active tension magnitude, $T^{\text{active}}$:

$$\mathbf{S}^{\text{active}}=T^{\text{active}}\mathbf{V}_1\otimes\mathbf{V}_1. \tag{8.6}$$

**TABLE 8.1** Fiber and Sheet Angle Along the Epicardium and Endocardium of the Biventricle

| | Angle (degrees) |
|---|---|
| Epicardium fiber, $\theta^{\text{epi}}$ | −57 |
| Endocardium fiber, $\theta^{\text{edo}}$ | 59 |
| Epicardium transverse fiber, $\beta^{\text{epi}}$ | 45 |
| Endocardium transverse fiber, $\beta^{\text{endo}}$ | −10 |

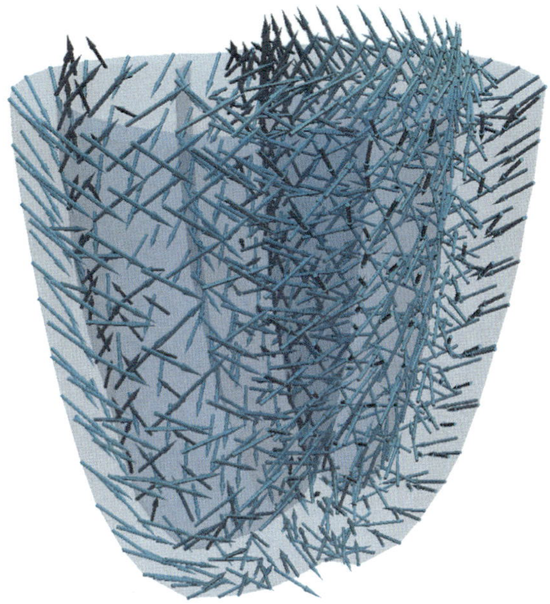

FIG. 8.1 Three-dimensional fiber distribution across the biventricle heart.

Following several studies made on rat cardiac tissue, Guccione et al. isolated the sarcomere length and the calcium ion concentration as the principal features that dictate the contraction mechanism and introduced them via a modified Hill model. The computation of the active contraction force, $T^{active}$, along the fibers of the heart is carried out using the following equation:

$$T^{active} = T_{max} \frac{Ca_0^2}{Ca_0^2 + ECa_{50}^2} C_t, \tag{8.7}$$

where $Ca_0$ is the peak intracellular calcium ion concentration, $ECa_{50}$ the extracellular calcium ion concentration when 50% of peak active contraction force is achieved, and $T_{max}$ is the isometric tension force under maximal activation. $C_t$ is the variable that defines its time transient as

$$C_t = \frac{1}{2}(1 - \cos \omega). \tag{8.8}$$

$\omega$ is a function of time and is defined over three ranges:

$$\omega = \begin{cases} \pi \dfrac{t}{t_0} & \text{when } 0 \leq t < t_0, \\ \pi \dfrac{t - t_0 + t_r}{t_r} & \text{when } t_0 \leq t \leq t_0 + t_r, \\ 0 & \text{when } t_0 + t_r \leq t, \end{cases} \tag{8.9}$$

where the range $0 \leq t < t_0$ represents the time span to linearly reach maximal activation from rest; the range $t_0 \leq t \leq t_0 + t_r$, the relaxation time to reach zero contraction from its peak value; and finally the range $t_0 + t_r \leq t$ when no contraction is present. $t$ is the current time, $t_0$ is the time at which peak tension is reached, and $t_r$ is the time taken for the tension force to dissipate. For the effect of the sarcomere stretch to be accounted for, the relaxation time, $t_r$, is calculated using the actual sarcomere length, $l^{sarcomere}$:

$$t_r = m l^{sarcomere} + b, \tag{8.10}$$

where constant $m$ is the slope of linear relaxation linking relaxation duration to sarcomere length and $b$ its time intercept. The calcium concentration at 50% tension, $ECa_{50}$, in Eq. (8.7) is defined as

$$ECa_{50} = \frac{(Ca_0)_{max}}{\sqrt{e^{B(l^{sarcomere} - l_0^{sarcomere})} - 1}}. \tag{8.11}$$

$(Ca_0)_{max}$ represents the highest intracellular calcium ion concentration, $l_0^{sarcomere}$ is the sarcomere length below which no active tension occurs, and $B$ is a constant governing the shape of the peak isometric tension-length relation.

The active tension force varies throughout the heart, as the actual sarcomere length is computed as a function of the fiber strain, $E_{ff} = \mathbf{E} : \mathbf{V}_1 \otimes \mathbf{V}_1$:

$$l^{sarcomere} = l_R^{sarcomere}\sqrt{2E_{ff}+1}, \tag{8.12}$$

where $l_R^{sarcomere}$ corresponds to the resting sarcomere subjected to zero stress. The active stress model has been successfully applied to hearts belonging to different species such as rats [2], dogs [54], sheeps [10, 55, 56], and humans [57, 58].

### 8.2.3 Windkessel Model

To model the ejection phase, an additional model needs to take into account the behavior and characteristics of the ventricles when connected to the systemic blood circulatory system in order to control the change in pressure, volume, and flow rate. This model also needs to reproduce the duration of the ejection period while incorporating details such as rapid ejection and slow ejection phases. This is facilitated by a so-called Windkessel (WK) model. In the cardiac-related literature, its occurrence is very common. For example, it was used by Creigen et al. [59] to model an artificial heart pump, and by Molino et al. [60] to understand the arterial mechanical characteristics.

In the context of cardiac modeling, many applications can also be found such as in Ottesen and Danielsen [61], where the authors made use of the WK model to simulate an LV with arbitrary heart rate or in the work of Bovendeerd et al. [62], Usyk and McCulloch [49], Sainte-Marie et al. [63], Kerckoffs et al. [64], Sermesant et al. [65], and Kerckhoffs et al. [66] where single ventricular and biventricular heart models successfully reproduced the ejection phase. The WK model is formulated as an ordinary differential equation where three different forms can be distinguished: the two-element WK model, the three-element WK model, and the four-element WK model. As outlined in the works of Paragh et al. [67] and Westerhof et al. [68], the predictive accuracy increases from two- to four-element WK models but less from the three-element WK to the four-element one.

Even though the four-element WK model includes most of the physiological mechanisms of the ejection phase and is very accurate, it has, however, two major drawbacks. First, its formulation consists of a second-order flow rate derivative. Such an order term adds more complexity and instability in a solving scheme. And second, the inductance constant is very difficult to quantify from experimental data [68]. Due to these reasons, the three-element WK model is utilized here. The equation of the three-element WK model is given as follows:

$$\left(1+\frac{R_a}{R_p}\right)I(t) + CR_a\frac{dI(t)}{dt} = \frac{P(t)}{R_p} + C\frac{dP(t)}{dt}, \tag{8.13}$$

where $R_p$ is equivalent to the peripheral resistance, $R_a$ is the flow resistance, and $C$ is the elastic arterial compliance. $P$ is the cavity pressure of the ventricle, which is approximately equal to the aortic pressure [69], while $I$ is the negative rate of change of the ventricular cavity volume.

## 8.3 REDUCED ORDER METHOD

With the cardiac mechanics models relevant for this research introduced, the concept of ROM is now outlined. ROM is a technique commonly used to decrease the complexity of a large system of equations. This is achieved by compressing the whole system to such a point that accuracy is not, in an excessive way, negatively impacted and that the general behavior of the problem, for example, the mechanics, is preserved. As mentioned before, one widely used method classified as an ROM is POD; this will be utilized in the form of the PODI method to achieve real-time modeling of the heart.

### 8.3.1 Proper Orthogonal Decomposition

The POD is a method that can be used to extract features from any dataset consisting of either linear or nonlinear data. In the literature, it is usually found in the form of Kharhunen-Loève decomposition (KLD) [70, 71], singular value decomposition (SVD), or principal component analysis. Even though each of these methods has different derivations, Wu et al. [72] showed the equivalence between those methods and how they can all produce the same solution.

In this chapter, KLD has been chosen to explain the POD. Suppose U is an ensemble set of displacement fields, $u^i$, describing the state of deformation of a body at discrete time steps $t_i$, $i = 1, \ldots, n$, which have been defined over a 1D spatial domain $0 \leq x \leq 1$:

$$\mathbf{U} = \{\mathbf{u}^1, \mathbf{u}^2, \ldots, \mathbf{u}^n\}, \tag{8.14}$$

where for the total number of time steps it holds $n \ll m$, and $m$ is the number of displacement degrees of freedom. Here, it is assumed that **U** is mean centered. That is, the mean vector of the ensemble matrix, v, has been subtracted from each column of the ensemble matrix. As such, **U** will only contain perturbations of the displacement fields with respect to the mean vector. If the displacement field is approximated by a set of basis vectors, $\tilde{\mathbf{\Phi}}$, and coefficients, $\boldsymbol{\alpha}$, through

$$\mathbf{u}^i = \sum_{j=1}^{n} \alpha_j^i \tilde{\mathbf{\Phi}}_j, \tag{8.15}$$

then, the KLD problem is posed as an optimality scheme that requires one to maximize the average projection of U onto $\tilde{\mathbf{\Phi}}$ while being subjected to the constraint of $\tilde{\mathbf{\Phi}}$ being orthonormal:

$$\max_{\tilde{\mathbf{\Phi}}} \langle |(\mathbf{U}, \tilde{\mathbf{\Phi}})|^2 \rangle \quad \text{s.t.} \quad \|\tilde{\mathbf{\Phi}}\|^2 = \mathbf{1}, \tag{8.16}$$

with $(\mathbf{f}, \mathbf{g}) = \int_0^1 \mathbf{f}(x)\mathbf{g}^{\mathrm{T}}(x)dx$, $\langle \cdot \rangle$ being the averaging operation and $\|\mathbf{f}\| = (\mathbf{f}, \mathbf{f})^{12}$. After solving Eq. (8.16) with the help of the Lagrange multiplier method, the following eigenvalue problem was obtained:

$$(\mathbf{R}, \tilde{\mathbf{\Phi}}) = \boldsymbol{\lambda} \tilde{\mathbf{\Phi}} \quad \text{with}\, \mathbf{R} = \frac{1}{n}\mathbf{U}\mathbf{U}^{\mathrm{T}}. \tag{8.17}$$

Solving the earlier produces a set of eigenfunctions that is optimal to the optimization problem stated in Eq. (8.16). R is usually referred to as the kernel and is a positive semidefinite symmetric matrix [73]. It is also alternatively defined as the autocorrelation matrix [74], whose eigenvalues (so-called proper orthogonal values [POVs]) and eigenfunctions (so-called POMs) are represented by $\boldsymbol{\lambda}$ and $\tilde{\mathbf{\Phi}}$, respectively.

If R is computed from a whole set of data $m \times n$, with $m$ being very large, then the resulting matrix would be of size $m \times m$ and extremely large as well. This, therefore, leads to an increase in computational time to find $\boldsymbol{\lambda}$ and $\tilde{\mathbf{\Phi}}$. One method, which is widely used to reduce the system, has been proposed by Sirovich and is commonly known as the snapshot method [75]. If the number of columns (i.e., number of snapshots or time steps, $n$) is smaller than $m$, then Eq. (8.17) can be reduced from an $m$ to an $n$ size system of equations that will, therefore, decrease the size of the autocorrelation matrix to $n \times n$. To do so, Eq. (8.17) is reformulated as follows:

$$(\mathbf{C}, \boldsymbol{\xi}) = \boldsymbol{\lambda}\boldsymbol{\xi} \quad \text{with}\, \mathbf{C} = \frac{1}{n}\mathbf{U}^{\mathrm{T}}\mathbf{U}. \tag{8.18}$$

The eigenvalues of C are the same as those obtained from Eq. (8.17) and can hence be defined to be the POVs. However, the associated eigenvectors are not the POMs. To recover the latter, the following equation is required [76]:

$$\tilde{\mathbf{\Phi}}^i = \frac{1}{\sqrt{n\lambda_i}}\mathbf{U}\boldsymbol{\xi}^i. \tag{8.19}$$

The choice of POMs is essential in minimizing the loss in accuracy of a set of solutions. According to Kerschen et al., the POVs can be used as a guideline to do so [77]. In Ref. [78], Barbic and James showed that for each POM, there is a corresponding POV, which is represented by the eigenvalue obtained from the SVD or KLD method. Usually, the POM represents a modal mode that characterizes, in a mechanical problem, a specific deformation of the domain while the corresponding POV is attributed to the captured energy from that particular deformation [77]. In order to find the total energy discarded when reducing the dataset, Falkiewicz and Cesnik produced the following equation in Ref. [79]:

$$\varepsilon_{\mathrm{rel}} = \frac{\sum_{f=1}^{n-r} \lambda_f}{\sum_{g=1}^{n} \lambda_g} \times 100, \tag{8.20}$$

with $r$ being the number of POMs conserved and $r < n$. In that sense, in order to find a proper balance between selecting a few POMs and conserving the greatest amount of energy, the POD basis, associated with the highest POV, is selected as the former represents the dominant modes of deformation. Interestingly, it was found that with very few modes, one can easily conserve up to 99% of the energy in the system. In Ref. [80], Lin et al. noted that, with their set of data comprising 24 snapshots, only the first two POVs were enough to achieve an energy conservation of 99.9997%, leading us to believe that a 99% energy threshold is enough to achieve a reasonable accuracy.

## 8.3.2 POD With Interpolation

### 8.3.2.1 *Parametric PODI*

For the purpose of real-time modeling, the direct application of POD to a system of equations is not suitable. When applied to the FEM, only the solving part of the overall process is accelerated while the preprocessing time and repetitive equation system assembly time prohibit a drastic drop in calculation time. Hence, in order to circumvent this problem, the PODI method, developed by Ly and Tran [30], has been employed in this work. Their idea starts by first constructing the matrix **U** from an ensemble of displacement vectors, $\{u^1, \ldots, u^p\}$. Each dataset, $u^a$, where $u^a \in \{u^1, \ldots, u^p\}$, corresponds to a set of heart-defining parameters, $\boldsymbol{\theta}^a$, where $\boldsymbol{\theta}^a \in \{\boldsymbol{\theta}^1, \ldots, \boldsymbol{\theta}^p\}$. Here, $\{\boldsymbol{\theta}^1, \ldots, \boldsymbol{\theta}^p\}$ is defined as a sequence of parameter sets, each containing characteristic values of the properties of the respective heart.

From there, the POD calculation is carried out and used to transfer the datasets to a low-dimensional space using the POMs, where an interpolation technique is set up to find the adequate solution. Once this is done, the solution is then projected back to the high-dimensional space. The derivation of the method is given by first expressing U in terms of the basis matrix, $\boldsymbol{\Phi}$, and the coefficient matrix, $\boldsymbol{\Psi}$:

$$\mathbf{U} \approx \boldsymbol{\Phi}\boldsymbol{\Psi}, \tag{8.21}$$

where a single displacement vector, $u^a$, can be defined as:

$$\mathbf{u}^a \approx \psi_1^a \boldsymbol{\Phi}^1 + \psi_2^a \boldsymbol{\Phi}^2 + \cdots + \psi_r^a \boldsymbol{\Phi}^r. \tag{8.22}$$

Now, let us suppose that an unknown displacement field, $\hat{\mathbf{u}}$ (with $\hat{\mathbf{u}} \notin \{\mathbf{u}^1, \ldots, \mathbf{u}^p\}$), which corresponds to the set of parameters $\hat{\theta}$ (where $\boldsymbol{\theta}^1 < \hat{\theta} \leq \boldsymbol{\theta}^p$ and $\hat{\theta} \notin \{\boldsymbol{\theta}^1, \ldots, \boldsymbol{\theta}^p\}$), is to be found using the earlier POD basis. Therefore, Eq. (8.22) can be rewritten as:

$$\hat{\mathbf{u}} \approx \hat{\psi}_1 \boldsymbol{\Phi}^1 + \hat{\psi}_2 \boldsymbol{\Phi}^2 + \cdots + \hat{\psi}_r \boldsymbol{\Phi}^r. \tag{8.23}$$

In Eq. (8.23), $\hat{\mathbf{u}}$ cannot be found as $\hat{\Psi} = (\hat{\psi}_1, \hat{\psi}_2, \ldots, \hat{\psi}_r)$ is unknown and $\hat{\Psi} \notin \{\boldsymbol{\Psi}^1, \ldots, \boldsymbol{\Psi}^p\}$. In this case, Ly and Tran [30] suggested interpolating $\hat{\Psi}$ from matrix $\boldsymbol{\Psi}$ through:

$$\hat{\Psi} = \boldsymbol{\Psi}\mathbf{N}, \tag{8.24}$$

where N are the MLS interpolants that can be derived from

$$\hat{\theta} = \boldsymbol{\Theta}\mathbf{N}, \tag{8.25}$$

as $\hat{\theta}$ and $\boldsymbol{\Theta}$ are both known. $\boldsymbol{\Theta} = (\boldsymbol{\theta}^1, \boldsymbol{\theta}^2, \ldots, \boldsymbol{\theta}^p)$ and is a matrix. Each row of that matrix is defined by specific parameters of a particular set of parameters while the columns are the different sets of parameters. In this chapter, the MLS approximation method [33] was chosen as the interpolation scheme because it can deal with problems of arbitrary dimensionality and different sizes of data points.

### 8.3.2.2 *Temporal PODI*

A temporal PODI calculation is very similar to the parametric one. The difference lies in the definition of the ensemble matrix and the setting up of the interpolants. As discussed in the previous section, the ensemble matrix represents a set of displacement vectors spanning over a parametric space of heart-defining characteristics but corresponding to a particular time step. As such, the interpolation scheme is built up using the parametric space. In this case, the roles of the parametric and temporal space are inverted. Here, the ensemble matrix is defined by a set of displacement vectors spanning over a temporal space and corresponds to a particular parametric point. Hence, the interpolation scheme is constructed using the temporal space. An example of a parametric as compared with a temporal ensemble matrix is given in Fig. 8.2.

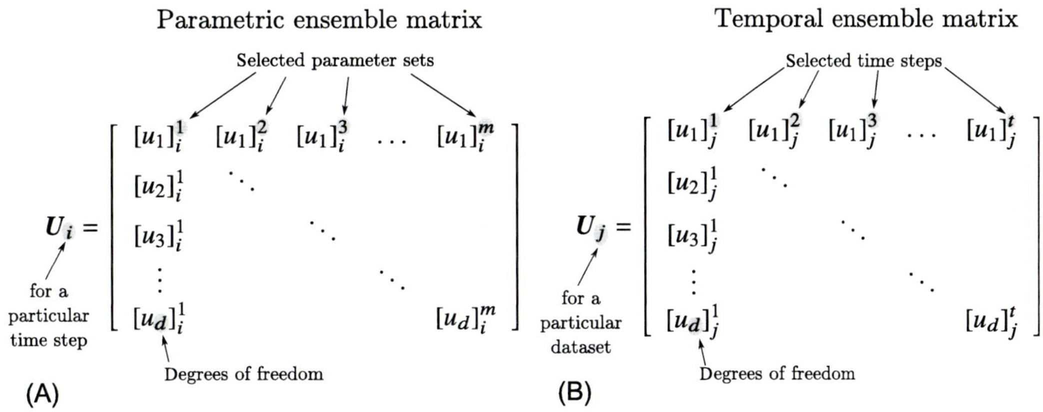


FIG. 8.2 Comparison of (A) a parametric against (B) a temporal PODI ensemble matrix layout.

## 8.4 WHOLE HEART CYCLE MODELING

### 8.4.1 Time Standardization Process

Now with the basics of PODI introduced, we want to elaborate on the time standardization method that is necessary to accomplish full heartbeat computations. All EFG or FEM full-scale simulations carried out during the diastolic filling, isovolumetric contraction, ejection, and isovolumetric relaxation are volume change-driven. That is, a volume increment is prescribed independently to each cavity, and a corresponding displacement field and cavity pressure are determined by solving iteratively the equilibrium and WK equations, respectively. This method leads to robust calculations under multiple loading conditions, as demonstrated in Skatulla and Sansour [53]. For a volume change-driven calculation, especially during the phases of active contraction, the simulation time is a function of the sarcomere length, which in turn is a function of the strains as given in Eqs. (8.10), (8.12). Because for different stress scaling coefficients and fiber directions, the strain state changes at the end of the diastole filling, each simulation starts the isovolumetric contraction phase with a sarcomere length distribution of different magnitude and consequently, different relaxation times. Hence, the simulation timeline of every heart problem varies with the stress scaling coefficients and fiber directions, even though the active stress and WK parameters remain constant. This effect subsequently leads to calculation-step sizes and the number of steps being inconsistent across datasets. As such, at any point along the simulation timeline, two datasets are usually out of phase over the entire heart cycle. That is, for example, one heart could still be at the end of isovolumetric contraction while another one is already in the ejection phase. The variations of the simulation timeline cannot be dealt with using the current implementation of the PODI method. To solve this problem, a time standardization scheme is proposed [35].

The time standardization scheme consists of defining a standardized timeline to which the dataset simulation timeline and its corresponding solutions will be converted. This standardized timeline is a series of equally spaced time points that span a whole heartbeat cycle and are referred to as reference simulation time steps or standardized time steps. The conversion process consists of interpolating the dataset solution fields to the standardized timeline as described in Section 8.3.2.2. Once done, the conventional PODI calculation can proceed in the parametric domain as previously described in Section 8.3.2.1. The PODI results obtained belong to the reference timeline as shown in Fig. 8.3. One major problem that has been encountered during the early implementation stage of this approach is that the last simulation time step of each phase of a heartbeat is not properly captured. This is because as the timeline is discretized by a uniform time interval, it is very unlikely that a standardized time step falls exactly at the change of a heartbeat phase. Very small time intervals can also be considered to minimize this problem, but the PODI calculation time would be negatively impacted. To circumvent this problem, a modified approach is undertaken where the time discretization process is not utilized over a full heartbeat, but instead, across each phase of a heartbeat. To do so, the simulation time step at the start and end of each cardiac phase is identified. The reference simulation time steps are then determined by discretizing the cardiac phase timeline between the start and end points. For each dataset, their cardiac phases are split in an equal number of time points to form a vector of reference simulation time steps, $\mathcal{T}_i^p$. Here, $i$ corresponds to the dataset and $p$ represents the phase of the heartbeat, $p \in$ {DF, IC, EJ, IR} with DF being the diastole filling, IC the isovolumetric contraction, EJ the ejection, and IR the isovolumetric relaxation. Initially, the reference simulation time steps do not have any solution fields. To obtain these solution fields, temporal

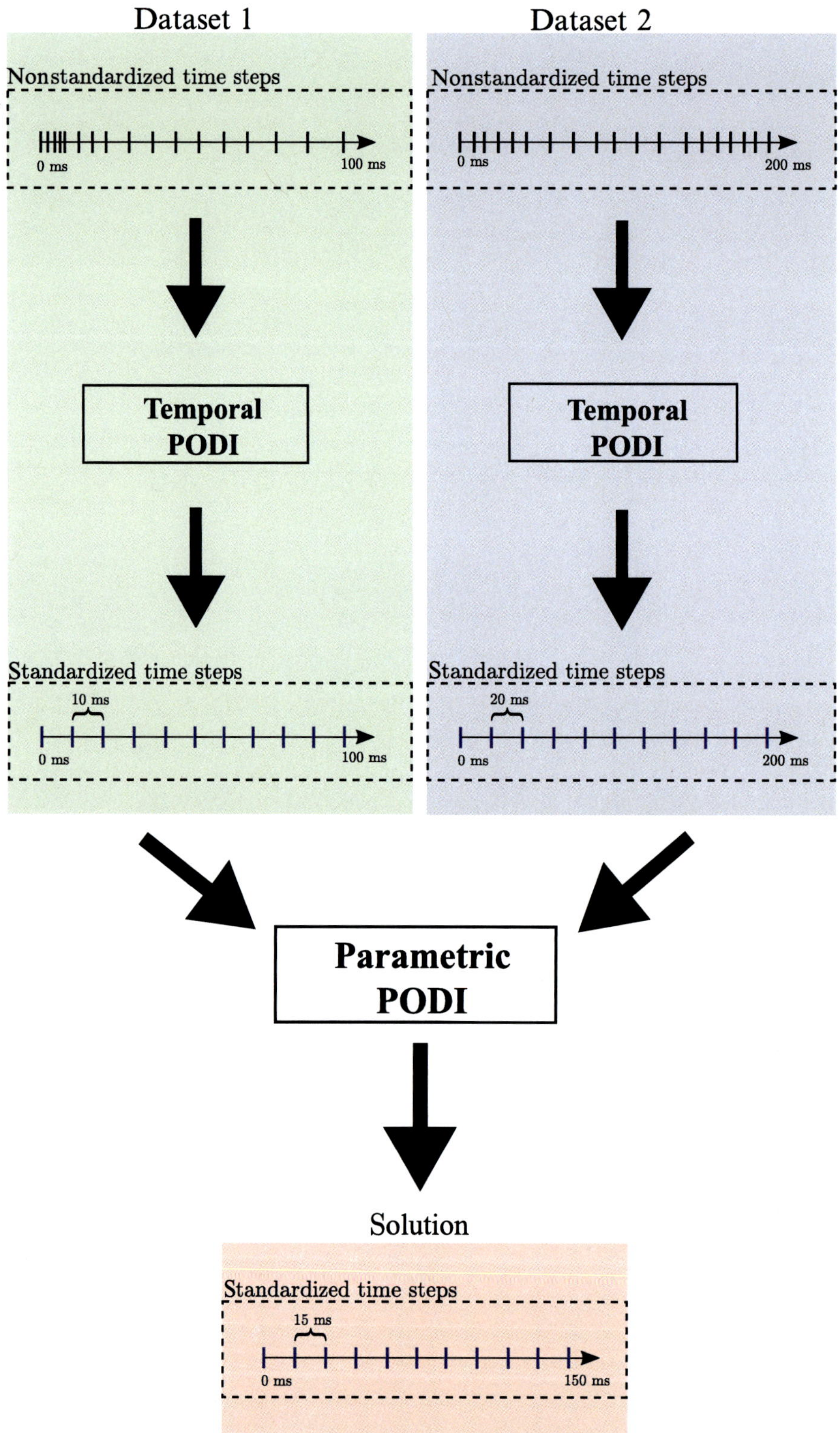


FIG. 8.3 Visual representation of a PODI calculation based on the time standardization process of two selected datasets.

PODI calculations are carried out separately for each dataset. The goal of this set of PODI calculations is to interpolate, in a low-dimensional space, the nonstandardized timeline solution fields onto the standardized one. In that case, an interpolation scheme is built-up based on the actual time steps of the nonstandardized timeline, instead of material parameters, and is therefore referred to as a temporal PODI calculation. This approach has been used in the

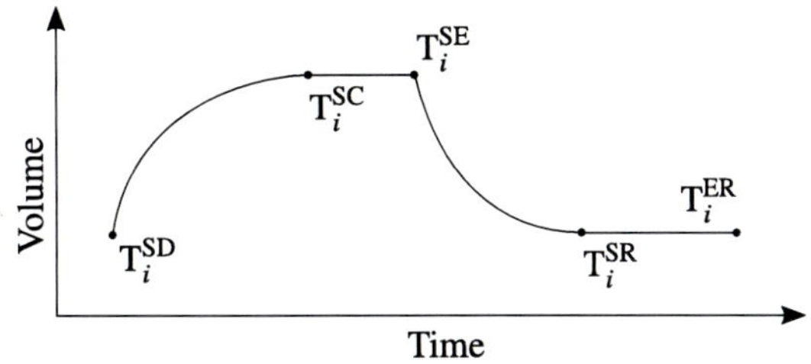


FIG. 8.4 Identification of the phase-change time steps of each phase across a volume-time graph of a dataset, *i*. Note that the graph is not drawn to scale.

literature [74] to obtain a smooth transition of results across a specific timeframe. In the temporal PODI calculations, the supporting time steps are found in the nonstandardized timeline vector, $[\mathbf{T}]_i^p$ and are used to interpolate for $[\mathcal{T}_j]_i^p$, where $\mathcal{T}_j$ is one of the reference time steps. Hence, the MLS interpolants, $\mathbf{N}_{\text{time}}^{\text{MLS}}$, are defined through:

$$[\mathcal{T}_j]_i^p = \mathbf{N}_{\text{time}}^{\text{MLS}} \cdot [\mathbf{T}]_i^p. \tag{8.26}$$

Regarding $\mathbf{U}_j$, the ensemble data matrix of the temporal PODI calculation, it is constructed as described in Section 8.3.2.2. With the data matrix and interpolation scheme set up, the temporal PODI calculation can be carried out for all reference time steps of every selected dataset.

After the datasets' timeline and solution fields have been standardized, the standardized timeline for the PODI problem at hand is now created. The current procedure employed is the interpolation of the starting and ending time steps of each phase, also called phase-change time steps, from the selected datasets. To do so, those phase-change time steps are first compiled in vectors defined by the start of diastole filling, $\mathbf{T}^{\text{SD}}$; the start of isovolumetric contraction, $\mathbf{T}^{\text{SC}}$; the start of ejection, $\mathbf{T}^{\text{SE}}$; the start of isovolumetric relaxation, $\mathbf{T}^{\text{SR}}$; and finally, the end of isovolumetric relaxation, $\mathbf{T}^{\text{ER}}$. The end of diastole filling, the end of isovolumetric contraction and the end of ejection are not considered because they are technically the same as the start of isovolumetric contraction, $\mathbf{T}^{\text{SC}}$; the start of ejection, $\mathbf{T}^{\text{SE}}$; and the start of isovolumetric relaxation, $\mathbf{T}^{\text{SR}}$. For example, the phase-change time steps are first identified for a dataset, *i*, as shown in Fig. 8.4, and then compiled in the phase-change time step vectors as follows

$$[\mathbf{T}^{\text{SD}}]^{\text{T}} = [T_1^{\text{SD}}, \ldots, T_i^{\text{SD}}, \ldots, T_m^{\text{SD}}], \tag{8.27}$$

$$[\mathbf{T}^{\text{SC}}]^{\text{T}} = [T_1^{\text{SC}}, \ldots, T_i^{\text{SC}}, \ldots, T_m^{\text{SC}}], \tag{8.28}$$

$$[\mathbf{T}^{\text{SE}}]^{\text{T}} = [T_1^{\text{SE}}, \ldots, T_i^{\text{SE}}, \ldots, T_m^{\text{SE}}], \tag{8.29}$$

$$[\mathbf{T}^{\text{SR}}]^{\text{T}} = [T_1^{\text{SR}}, \ldots, T_i^{\text{SR}}, \ldots, T_m^{\text{SR}}], \tag{8.30}$$

$$[\mathbf{T}^{\text{ER}}]^{\text{T}} = [T_1^{\text{ER}}, \ldots, T_i^{\text{ER}}, \ldots, T_m^{\text{ER}}], \tag{8.31}$$

where $m$ is the number of selected datasets for the PODI calculation. Once these phase-change time step vectors are compiled, an interpolation scheme is carried out along each phase-change time step vector of the PODI problem at hand. The MLS interpolation scheme is again employed here and the interpolants vector, $\mathbf{N}$, is built up from the selected dataset parameters because, as indicated earlier, the latter is responsible for the evolution of the simulation time steps. The interpolation process is carried out for each end-point vector as follows:

$$\hat{T}^e = \mathbf{N} \cdot \mathbf{T}^e, \tag{8.32}$$

where $e \in \{\text{SD, SC, SE, SR, ER}\}$. Once the phase-change time steps of the problem at hand are obtained, the standardized timeline is determined. The parametric PODI calculation can afterward take place to obtain the solution fields of the time steps, $\hat{\mathbf{T}}^e$, of the problem at hand.

## 8.4.2 PODI Usage and Database Construction

This section means to summarize the general usage of PODI coupled with a database. As such, a step-by-step description of the PODI algorithm is provided. The latter can be split into three main processes: database construction, reduced order calculation, and finally, postprocessing. Each process is discussed here and accompanied by a work flowchart of a complete simulation, as illustrated in Fig. 8.5A, and another chart focusing on the detailed steps of the PODI algorithm, as shown in Fig. 8.5B.

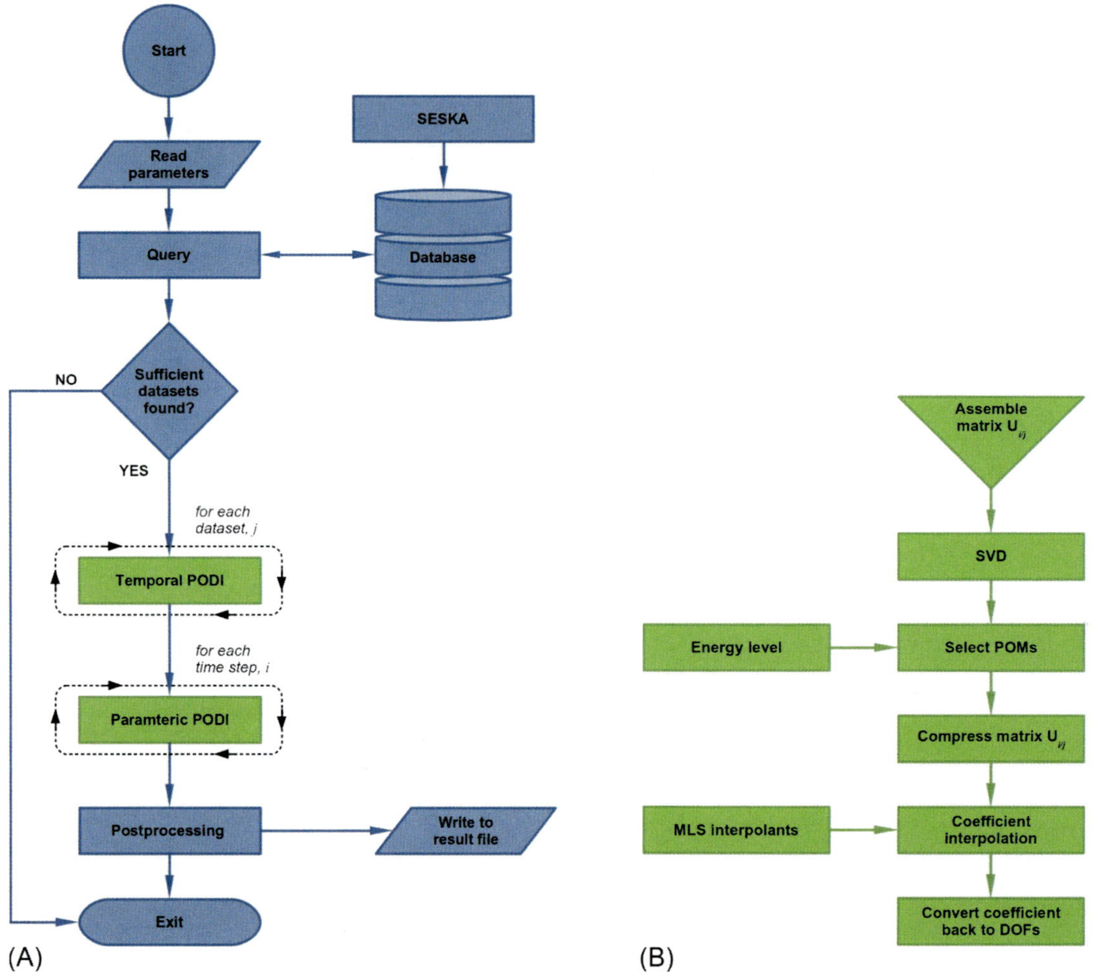


FIG. 8.5 Flowchart of simulation program. (A) Overall process. (B) Paramteric/temporal PODI process for each time-step/dataset, $i/j$.

- *Database construction*: The first step of the database construction procedure is to create all required datasets. An in-house code called SESKA, which is based on the EFG [81] and the FEM, is utilized to generate these datasets that consist of solution fields, such as displacement, stress, and strain, that have been stored offline for each time step of a simulation.

  Regarding the database itself, a very basic database management system is used where all datasets are stored locally on a hard drive, structured into folders. The entry of each dataset and its characterizing parameters referring to the heart's anatomy, hemodynamic loading, pathologies, or any other relevant parameters are then registered in a comma-separated values file. These characterizing parameters are readily available from state-of-the-art medical examination tools, for example, cardiac magnetic resonance (CMR), and need to be provided for the database search.
- *Time standardization*: The datasets closest to the problem at hand are first determined. This is done by feeding the set of parameters of the problem at hand, into a selection algorithm that will query the database and pick the datasets that feature the closest match in terms of problem-defining parameters. These datasets are the "neighboring nodes" in the MLS-based PODI calculation. The timeline of each selected dataset is analyzed and the corresponding start and end points of the four heart phases are determined. Subsequently, the datasets are mapped onto the standardized timeline, separately for each heart phase, via interpolation using the temporal PODI method.
- *Parametric PODI*: In order to obtain the solution of the problem at hand, the parametric PODI calculations are carried out using the previously standardized datasets. From the selected datasets, ensemble matrices are individually built for each solution field, for example, displacement, stress, strain, and time step, $i$. The mean vector v is determined and subtracted from the ensemble in order to obtain the mean centered ensemble matrix, $U_i$. After carrying out the POD analysis of each ensemble matrix $U_i$ and selecting the dominant modes, the matrix is then reduced using $\mathbf{\Phi}^{-1}U_i \approx \mathbf{\Psi}$ according to Eq. (8.21). For the interpolation process in the PODI method, the MLS scheme is then

employed to compute the interpolants and subsequently the PODI coefficients as per Eq. (8.24). In order to recover the full-order solution from the coefficient, Eq. (8.23) is then employed, after which the mean vector v used in calculating the mean centered ensemble matrix $U_i$ is added.

- *Postprocessing and validation*: Once the PODI calculations are completed, the results are saved and viewed by postprocessing software called GiD (CIMNE International Center for Numerical Methods in Engineering). For the validation process, the error in the PODI calculation is then compared with the full-scale simulation solution of the problem at hand using EFG or FEM again, which is assumed to be the exact solution. The error calculation is given as the $\ell_2$ error norm:

$$\varepsilon_{\ell_2}\text{–norm} = \frac{\| \mathbf{U}^{\text{PODI}} - \mathbf{U}^{\text{EFG/FEM}} \|}{\| \mathbf{U}^{\text{EFG/FEM}} \|}. \tag{8.33}$$

### 8.4.3 Numerical Examples

With the cardiac PODI algorithm introduced, its performance is illustrated with two representative examples. The first example consists of an LV with varying preload, that is end-diastolic volume. Subsequently, another more challenging example is considered, which is a biventricle (BV) model with varying end-diastolic and end-isovolumetric contraction pressures.

#### *8.4.3.1 Human Left Ventricle Example*

A patient-specific LV is extracted from MRI images. The geometry is discretized using 2659 tetrahedral elements and 745 nodes. The values of the material constants used in Eq. (8.2) are as given in Table 8.2 while the stress scaling coefficient, $A$, and the compressibility controlling penalty factor, $A_{\text{comp}}$, are fixed to 0.46 and 100 kPa, respectively. The active stress parameters of Eq. (8.7) are listed in Table 8.3. For the three-element WK model, $R_a$ and $R_p$ are calibrated as $1.28 \times 10^7$ Pa s m$^{-3}$ and $1.0 \times 10^8$ Pa s m$^{-3}$, while C is determined as $4.0 \times 10^{-9}$ m$^3$ Pa$^{-1}$. As the Dirichlet boundary condition, the base of the LV is fixed in a direction along the $z$-axis, the LV's long axis, while the endo and epicardium line on the base surface are subjected to an elastic spring force of stiffness $1 \times 10^5$ N m$^{-1}$, emulating the connection of the heart to the major blood vessels. The Neumann boundary condition, which consists of a surface pressure applied to the endocardium surface, is the parameter used to create the PODI database. This is done by varying the magnitude of the surface pressure from 1.0 to 2.0 kPa. Using an increment of 0.1 kPa, a database with a size of 11 datasets is created using the FEM. Across all datasets, the end-isovolumetric contraction pressure is set to 5.5 kPa and the end-isovolumetric relaxation pressure to 0.25 kPa. The energy conservation level is set to 99.999%. The time standardization process discretizes the timeline of each cardiac phase with 200 time points, equally spaced.

Following the creation of the database, the PODI calculation is now undertaken. Six datasets are mobilized. These are the ones associated with an end-diastole pressure of 1.2, 1.3, 1.4, 1.6, 1.7, and 1.8, respectively. The POVs and POMs of the end-diastolic fiber strain data are plotted in Figs. 8.6 and 8.7, respectively. The magnitude of the POVs are, as expected, exponentially distributed across the POMs.

In terms of computation time, the PODI calculation is carried out within 53 s while the full-scale simulation requires 858 s, which represents a speed-up by a factor 16. The calculation is found to be affected by two processes. The first one is reading the datasets into the computer memory, which takes about 14.5 s, while the temporal standardization needs about 23.8 s. The reduced order calculation requires 4.9 s while the postprocessing stage is 9.7 s. The computation times of the PODI, especially the temporal standardization, and the postprocessing stage can still be further reduced if fewer time points were used to discretize the cardiac timeline. Yet, the computational speed of PODI can be

TABLE 8.2 Coefficients of $a_i$, Converted From Usyk et al. [82]

| Parameter | $a_1$ | $a_2$ | $a_3$ | $a_4$ | $a_5$ | $a_6$ |
|---|---|---|---|---|---|---|
| Coefficient | −6.00 | −5.00 | 9.00 | 12.00 | 12.00 | −6.00 |

TABLE 8.3 Active Contraction Parameters of the Human Left Ventricle

| $T_{\max}$ | $\text{Ca}_0$ | $\text{Ca}_0^{\max}$ | $B$ | $l_0$ | $t_0$ | $m$ | $b$ |
|---|---|---|---|---|---|---|---|
| 120 kPa | 4.35 μmol | 4.35 μmol | 4.75 μm$^{-1}$ | 1.58 μm | 220 ms | 1.0489 s μm$^{-1}$ | −1.429 s |

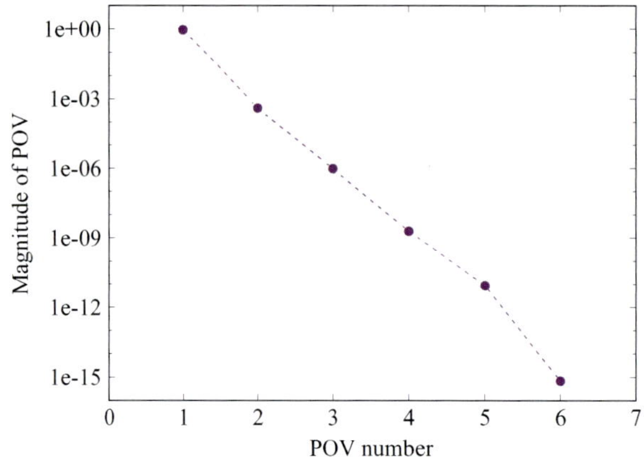


FIG. 8.6 Distribution of the strain POVs at the end of diastolic filling.

considered as encouraging because it takes about 975 Hz to generate the result of a displacement field at a particular step. This calculation frequency includes setting up the interpolants, assembling the $U_i$ matrix, extracting the POVs and POMs, projecting $U_i$ in the low-dimensional space, interpolating the coefficients, and projecting them back in the high-dimensional space.

The next step is to investigate the accuracy of the PODI solution. Before proceeding to the calculation of the error using Eq. (8.33), the exact result, against which the PODI results will be compared to, is defined. This step is required because the full-scale simulation results are based on a different series of time steps, which also have to be standardized. For this standardization procedure, a temporal PODI calculation is again carried out. However, a 100% energy conservation is specified in order to conserve most details of the exact solution during the reduced order interpolation. This process is acknowledged to introduce errors that, however, can be considered to be minimal and negligible. After analyzing the PODI error, the displacement, strain, and stress field solutions are found to have an error of 0.023, 0.022, and 0.027, respectively. These errors can be considered to be low because no visual difference can be observed when comparing the solution plots as given in Figs. 8.8 and 8.9. A similar error magnitude is recorded for the pressure-volume loop of the LV, as the error norm computed from Fig. 8.10 is 0.0279. As far as the end time of each phase is concerned, it is found that they are properly captured by the time standardization process, as shown in Fig. 8.11. The highest error of the end-phase time is recorded at the end of diastole with a $\varepsilon_{\ell_2}$-norm of 0.042 while the lowest one occurs at the end of ejection with a $\varepsilon_{\ell_2}$-norm of 0.007.

### *8.4.3.2 Idealized Biventricle Example*

Due to the unavailability of MRI images to build a BV geometry, an idealized one is constructed. The latter is based on the geometry of Wong and Kuhl [52] using truncated ellipsoids. In this case, the generated BV mesh consists of 550 tetrahedral elements and 193 nodes. The passive and active stress parameters are kept the same as those of the human LV model of the previous example. The three-element WK parameters are different for each ventricle. The LV has been assigned WK parameters of $C = 4.0 \times 10^{-9}$ m$^3$ Pa$^{-1}$, $R_a = 1.0 \times 10^7$ Pa s m$^{-3}$, and $R_p = 1.0 \times 10^8$ Pa s m$^{-3}$ while for the RV, the WK parameters are $C = 1.0 \times 10^{-9}$ m$^3$ Pa$^{-1}$, $R_a = 4 \times 10^7$ Pa s m$^{-3}$, and $R_p = 4.00 \times 10^8$ Pa s m$^{-3}$. The database is built this time by varying the end-diastolic as well as the end-IVC pressure of the LV using the EFG [81]. The pressure range for the end-diastolic pressure is 1–2 kPa with an interval size of 0.25 kPa. For the end-IVC pressure, the range is 4.0–7.0 kPa with an interval of 0.5 kPa. Hence, a total of 35 datasets are created. The right ventricular pressures are determined for each dataset by keeping the LV to RV end-diastole pressure ratio to 1.1:0.95 kPa and the IVC pressure ratio to 5.5:4.65 kPa. It should be noted that the above parameter values and ventricular pressures are not derived from experimental work. Instead, they have been selected based on their effects on cardiac models to produce adequate pressure-volume curves.

Following the PODI calculation, the results are analyzed. The first set of results to be looked at is the performance of the ROM calculation at 99.9% of the energy conserved. Using only one processor, the full EFG simulation takes about 8.5 h while the PODI computation lasts, on average, 27.2 s. This shows that PODI is about 1125 times faster. The PODI calculation time can be broken down into subgroups. The reading of the datasets and time standardization process are the most time-consuming processes because they account for 42% and 41% of the total time, respectively. The post-processing lasts for about 2.7 s, which is 9.84% of the total time. The least time-consuming process is the PODI calculation of all solution fields, as this takes only 1.7 s. If only the time taken for each displacement field for one step is

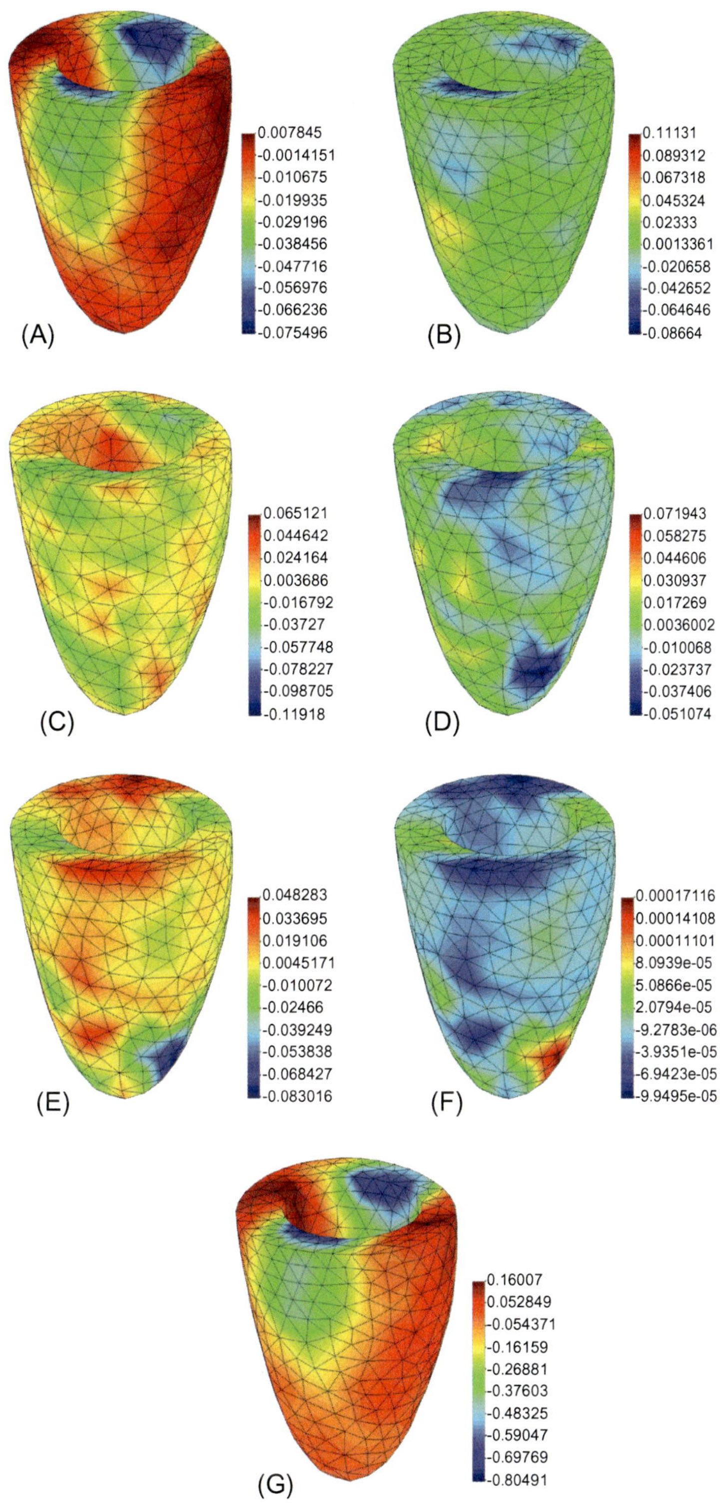


FIG. 8.7 Plot of the fiber strain field (along $\mathbf{V}_1$) of each POM and the corresponding percentage of energy for which it accounts. (A) 99.95%; (B) 0.04%; (C) $1.03 \times 10^{-4}$%; (D) $2.96 \times 10^{-7}$%; (E) $1.27 \times 10^{-8}$%; (F) $2.21 \times 10^{-14}$%; (G) mean.

considered, then it is found that the PODI calculation is running at a frequency of 2700 Hz. The calculation frequency value is higher in this case as compared to the LV example presented in Section 8.4.3.1. This is due to the fact that the BV geometry is discretized with fewer nodes, which consequently leads to a smaller number of degrees of freedom in the parametric, $\mathbf{U}_i$, and temporal, $\mathbf{U}_j$, ensemble matrices.

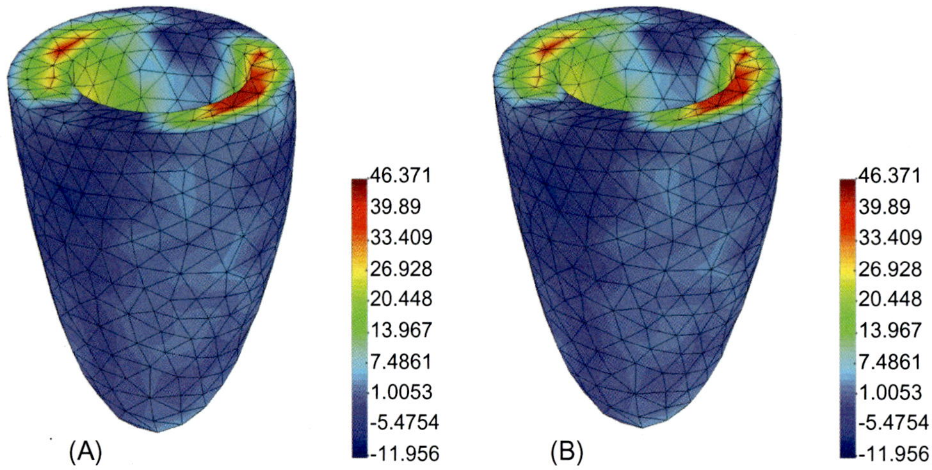


FIG. 8.8 Comparison of fiber stress field (along $\mathbf{V}_1$) of PODI against FEM, at the end of isovolumetric contraction (IVC). (A) PODI; (B) FEM.

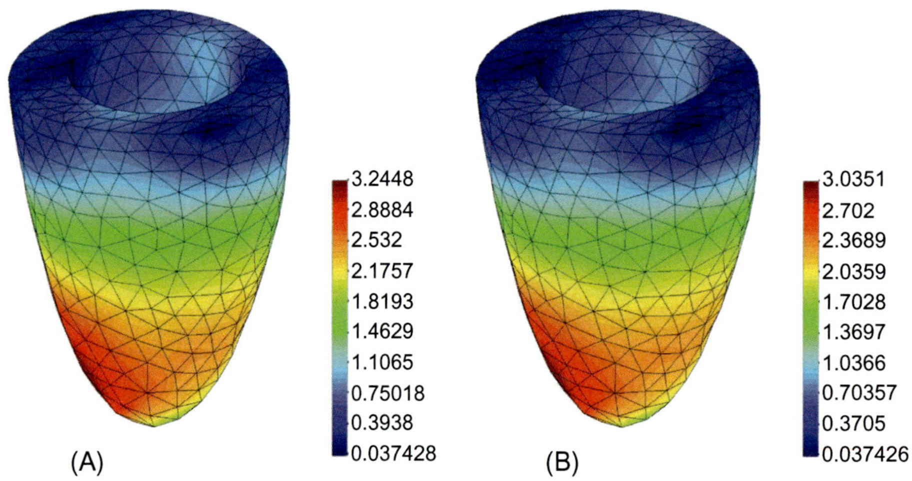


FIG. 8.9 Comparison of displacement field solution of PODI against FEM, at the end of isovolumetric relaxation (IVR). (A) PODI; (B) FEM.

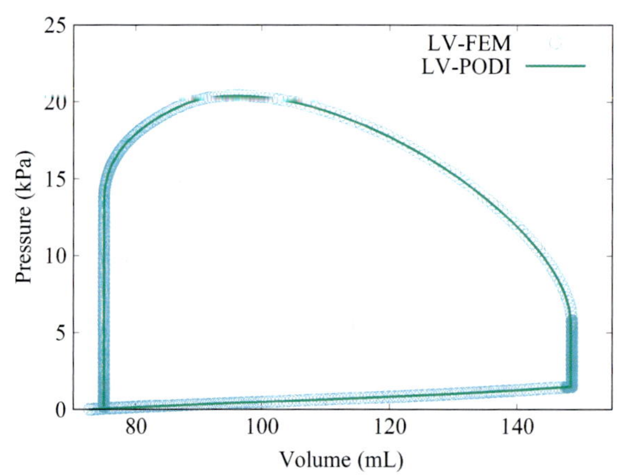


FIG. 8.10 Comparison of the pressure-volume loop generated from PODI against a conventional full-scale simulation using FEM.

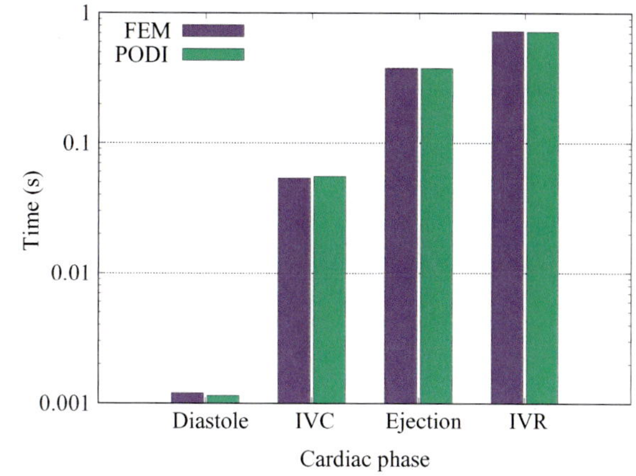


FIG. 8.11 Comparison of end-diastole, end-IVC, end-ejection, and end-IVR time spans generated from PODI against a conventional full-scale simulation using FEM.

After the conventional full-scale simulation results are produced using EFG, the PODI calculation accuracy is now investigated. On average, the error norms of the displacement, stress, and strain fields are 0.024, 0.032, and 0.022, respectively. However, it is observed that those errors are not equally distributed across the different phases of a heartbeat, as shown in Fig. 8.12. The highest error is recorded during the isovolumetric contraction phase. It is about 3.5–4 times higher than during the diastole filling phase. The computation errors of the ejection and isovolumetric relaxation phases are either considerably lower or similar to that of the diastole filling phase. These error values show very little variation as the energy conserved is varied from 70% to 100%. However, for the pressure-volume (PV) relationship of

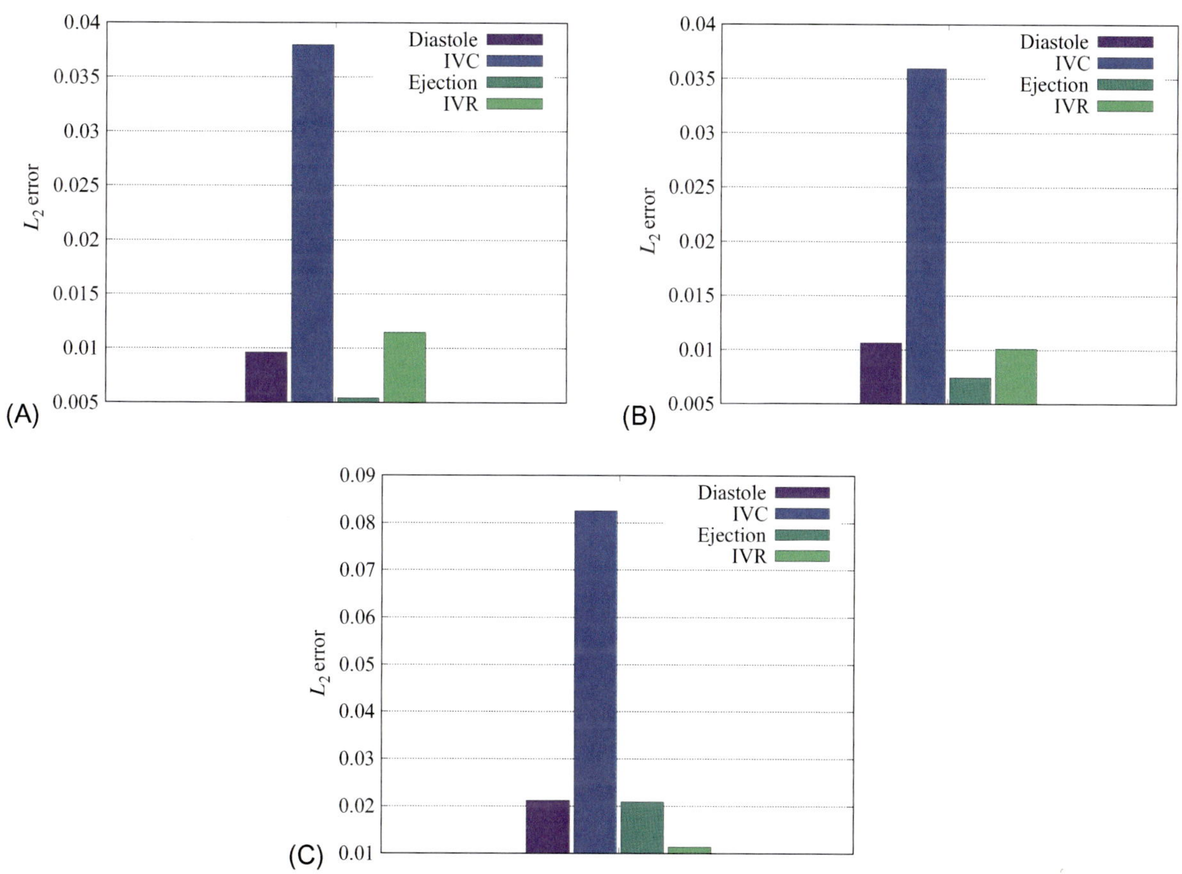


FIG. 8.12 Solution field error across the diastole, isovolumetric contraction (ICV), ejection, and isovolumetric relaxation (IVR). (A) Displacement field; (B) strain field; and (C) stress field.

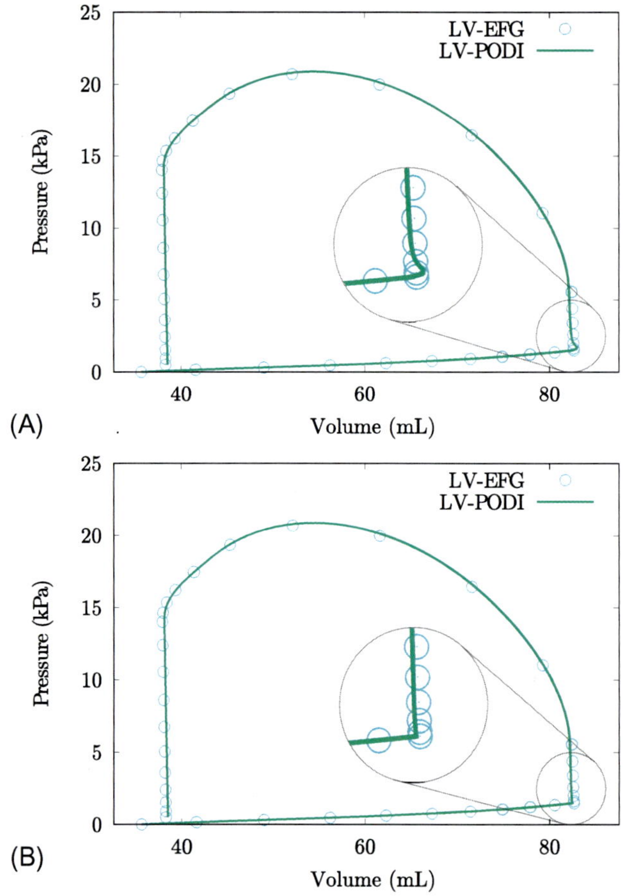


FIG. 8.13 Comparing the left ventricular pressure-volume curves generated by EFG against PODI at increasing energy conservation levels. The magnified region in the plot shows that the increase in POMs conserved provides higher accuracy of the pressure-volume evolution at the start of isovolumetric contraction. (A) Energy conserved: 70%; (B) energy conserved: 99.99%.

the LV, some benefits can be observed. At 70% of the energy conserved, the distribution of the LV PV error across the phases is similar to that of the displacement, stress, and strain fields. The pressure-volume curve at the outset of the IVC shows nonphysical behavior where an expected straight line is not obtained. With the conserved energy increased to 100%, the PODI computation error decreases, which is reflected in Fig. 8.13A and B, respectively, as the kink in the PV curve vanishes. Additionally, as it is expected that the PODI PV-loop of the LV and RV is found in between the PV loops of the mobilized datasets, as shown in Figs. 8.14 and 8.15. The end-diastolic, end-IVC, end-ejection, and end-IVR volumes and pressures recorded are indeed in between those of the PODI selected datasets. Interestingly, it can be

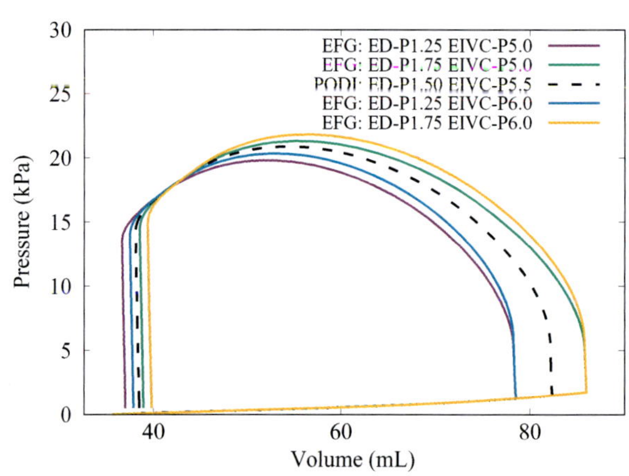


FIG. 8.14 Left ventricular pressure-volume curve of the selected dataset and the PODI solution.

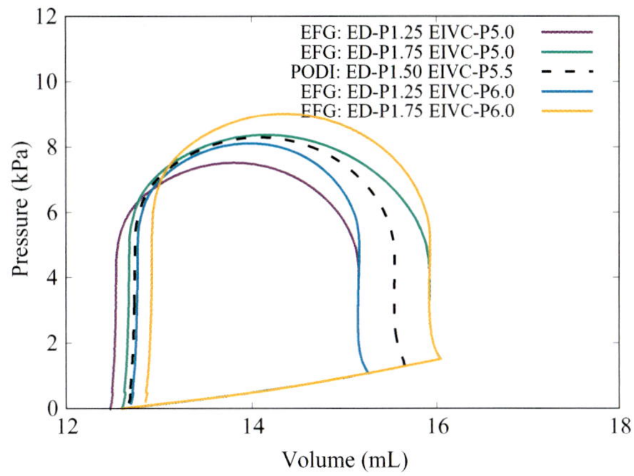


FIG. 8.15 Right ventricular pressure-volume curve of the selected dataset and the PODI solution.

noted that the PODI calculation manages to capture the intersection point of the dataset PV curves for both the LV and the RV, which occurs toward the end of the ejection phase.

The results provided earlier indicate that the most unstable region of the PODI calculation during a full heartbeat simulation occurs across the isovolumetric contraction phase. Studying the left and right ventricular PV loops, given in Figs. 8.14 and 8.15, it can be observed that the IVC phase is the only region where the difference in volume between the closest selected datasets is large. This difference is not as pronounced during the diastole filling, and also to a lesser degree in the ejection and isovolumetric relaxation phases, respectively. Due to the volume difference, the most dominant POMs during IVC and at the start of ejection are less energetic, as shown in Fig. 8.16, where the energy of the first POM drops from 99% to around 94% during the IVC time steps. Other unstable regions can be observed across the heartbeat timeline when looking at the number of POMs conserved, as shown in Fig. 8.17. As expected, an increase in the POMs conserved takes place during the IVC, but that increase also occurs during the start of ejection. However, within the region where the curves of the neighboring datasets meet, the number of POMs conserved decreases. This decrease is due to the fact that the POMs are more energetic because the dataset solution fields are very similar. Also during the isovolumetric relaxation phase, the dataset solution fields are closer to each other, leading to only a few POMs needing to be conserved. However, at the end of the IVR, the required number of POMs conserved increases, as the diastole phase starts again with a larger difference in dataset volumes. Based on these results, it can be concluded that the instabilities in the PODI calculation of a heartbeat arise principally during the IVC and also at the phase transition of IVC to ejection and IVR to diastole. As such, in those regions, it is important to incorporate more solution fields for the PODI calculation. Hence, an approach where the standardized steps are more concentrated in those regions and the PODI calculation has a larger number of selected datasets can be expected to lead to smaller errors.

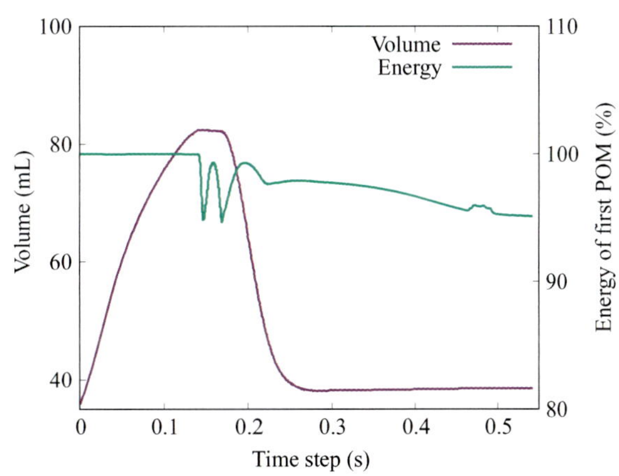


FIG. 8.16 Change of the left ventricular volume and energy of the most dominant mode during one heartbeat.

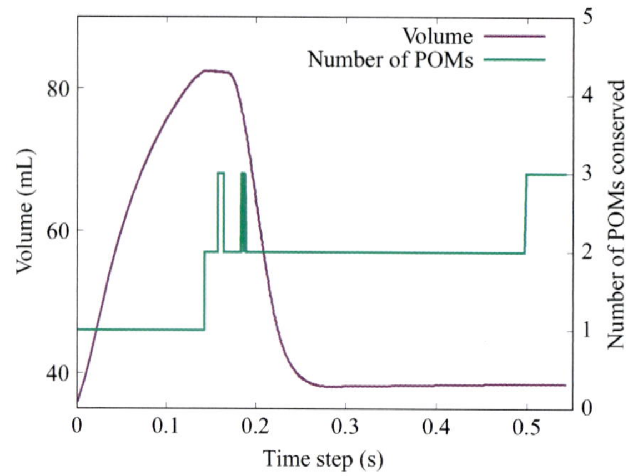


FIG. 8.17 Change of the left ventricular volume and number of POMs conserved during one heartbeat.

## 8.5 PATIENT-SPECIFIC CARDIAC PODI COMPUTATION

### 8.5.1 Degrees of Freedom Standardization Method

In the previous section, all datasets need to have a common geometry and underlying mesh configuration. In the case where different geometries and mesh configurations are used, the collated data matrix $\mathbf{U}_i$ (Eq. 8.21) is ill-set, which is a crucial shortcoming of the PODI approach. Consider as an example where the PODI calculation is carried out from five different datasets with geometries of different dimensions as well as meshes and nodal degrees of freedom: model A = 400 DOFs, model B = 1000 DOFs, model C = 600 DOFs, model D = 2000 DOFs, and model E = 1200 DOFs. Assuming all five models are selected for the PODI computation, then the assembled $\mathbf{U}_i$ matrix will take the following form:

$$\mathbf{U}_i = \begin{bmatrix} u_1^1 & u_1^2 & u_1^3 & u_1^4 & u_1^5 \\ \vdots & \vdots & \vdots & \vdots & \vdots \\ u_{400}^1 & \vdots & \vdots & \vdots & \vdots \\ & \vdots & u_{600}^3 & \vdots & \vdots \\ & u_{1000}^2 & & \vdots & \vdots \\ & & & \vdots & u_{1200}^5 \\ & & & u_{2000}^4 & \end{bmatrix}. \tag{8.34}$$

The type of matrices illustrated in Eq. (8.34) cannot be used for any matrix operations due to incompatible columns due to the missing matrix entries and the row-wise aligned DOFs not being associated with the same spatial locations. In cardiac modeling, such problems are likely to be encountered, especially when moving toward the concept of patient-specific modeling where the heart, for each individual, has distinct anatomical features. Consequently, this research intends to extend the usage of the PODI technique such that realistic heart geometries can be dealt with. It has to be noted that the temporal PODI method does not suffer from the incompatible problem as $U_j$ is built from a single dataset. That is, the total number of rows in each column of $U_j$ corresponds to the total DOFs of the dataset's geometry and is thus constant.

In the literature, two recent papers by Amsallem et al. [45] and González et al. [46] have discussed the problem of extracting POD modes from different mesh configurations. In Amsallem et al., the POMs of different mesh configurations were transformed to the POM, referring to a reference mesh configuration through a minimization process. However, even though the transformed POMs conserved their orthogonal properties, it is not clear if they also conserved their optimality properties as being the most dominant modes, and how the whole transformation procedure impacts the total calculation time. González et al. took a different the approach. The authors embedded different liver geometries in a benchmark cube mesh grid and computed a so-called distant field with respect to the boundary of the organ. Following that, a method called locally linear embedding was employed to find the weightage of each registered liver to reproduce the geometrical shape of the liver at hand and to carry out the interpolation. However, it remains unclear as to how the nodes inside the liver geometry were treated.

In this research, an alternative approach is proposed that is referred to as the DOFS method. This standardization procedure, which takes place during the preprocessing stage, consists of projecting the solution fields onto template nodes [36]. These template nodes are points defined spatially across a template geometry that is the same for every dataset. The projection itself is based on an interpolation scheme that interpolates the solution fields of the given

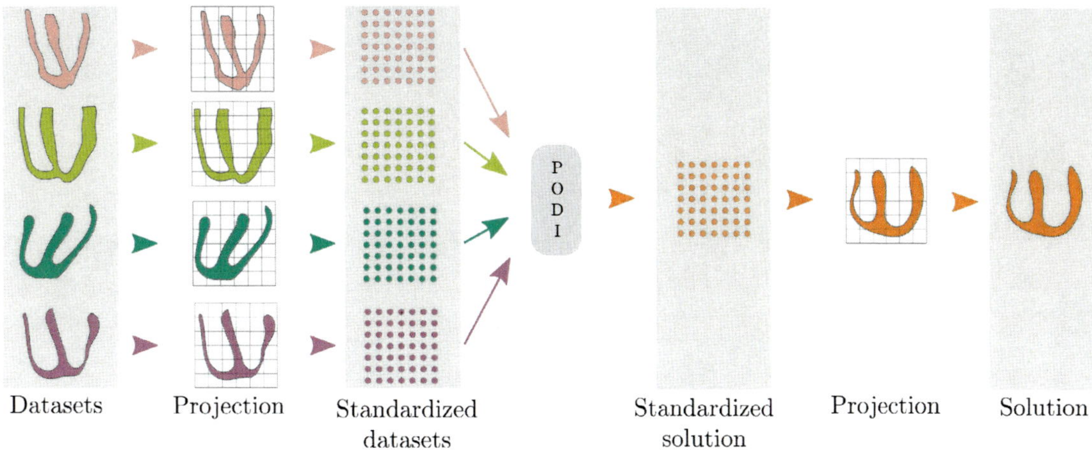


FIG. 8.18 Flowchart of PODI coupled with the degree of freedom standardization method using a cube grid template.

datasets at the template nodes. After the projection process, the solution fields of all datasets share the same degrees of freedom. Consequently, the data matrix, $U_i$, has the same number of column entries aligned to the corresponding DOFs of the template nodes.

In the following, the DOFS method will be explained in full detail. Let us assume that a database of $N$ datasets, each having a different heart geometry with different mesh discretizations, is defined as $\{\boldsymbol{\Omega}^1, \ldots, \boldsymbol{\Omega}^N\}$ with their respective nodal vectors $\{\mathbf{x}^1, \ldots, \mathbf{x}^N\}$. In addition, $\boldsymbol{\Omega}^T$ represents the domain of the template nodes and $\boldsymbol{\Omega}^U$, the geometrical domain for which the solution field is unknown and the PODI calculation will be run for after the standardization has taken place. The corresponding sets of nodes are $\mathbf{x}^T$ and $\mathbf{x}^U$, respectively. As a first step, the given heart geometries $\{\boldsymbol{\Omega}^1, \ldots, \boldsymbol{\Omega}^N\}$ need to be placed in or morphed onto the template geometry. The second step of the standardization procedure is to locate the template nodes, $\mathbf{x}^T$, inside each of those geometries in order to find their relevant neighbors in the corresponding nodal vectors $\{\mathbf{x}^1, \ldots, \mathbf{x}^N\}$. These nodal connectivity lists are required for the MLS approximation scheme to interpolate the solution fields of each dataset at the template nodes. As a result, all datasets are standardized and the $U_i$ matrix is adequately set for the subsequent PODI calculation. A visual representation is shown in Fig. 8.18.

The most challenging aspect of the above-proposed standardization procedure is to locate the template nodes and find their neighbors. A simple strategy would be to identify the neighbors based on their spatial distances from the template nodes. However, this would involve finding distances of all the nodes in each $\boldsymbol{\Omega}^i \in \{\boldsymbol{\Omega}^1, \ldots, \boldsymbol{\Omega}^N\}$ in relation to every node in $\boldsymbol{\Omega}^T$. The computation time for $N$ datasets would be tremendously expensive. Another problem poses the choice of the MLS influence radius, which determines the number of neighboring nodes for each template node and so, the accuracy of the interpolation. The solution to these issues is the point-in-polygon (PIP) algorithm. Different algorithms can be used to solve the PIP problem [83–86]. The method employed in this research is the surface orthogonal method. The latter makes direct use the element-node connectivity list of the mesh where the nodes of the domains $\{\boldsymbol{\Omega}^1, \ldots, \boldsymbol{\Omega}^N\}$, $\boldsymbol{\Omega}^T$, and $\boldsymbol{\Omega}^U$ are the interconnected vertices. The objective of PIP is to find the element in which a node of interest is located. Once identified, the nodes of that element, which is referred to as an element of interest, can then be labeled as neighbors. In this way, only a limited number of elements need to be checked, as the search algorithm can be terminated once a node is found to belong to a specific element. The only circumstance in which the search algorithm can fail is when the node lies on the surface of an element. In this case, the first element in which a node is found is automatically assigned to it. The neighbors are the nodes of the element of interest, and the nodes of the elements connected to the element of interest. This approach also has the benefit that neighboring nodes located across void spaces are automatically excluded.

Two geometrically different configurations are considered as template geometries, namely, a cube and a BV heart model. The advantages and disadvantages of each will be elaborated on in the following.

#### *8.5.1.1 Cube Template Standardization*

The cube template-based standardization is the simpler of two proposed standardization methods. The solution fields are only interpolated at those template nodes of the cube grid that are located inside each dataset's geometry and those outside are disregarded. In order to ensure that all nodes of each dataset take part in the interpolation, the cube grid dimensions must be chosen to be large enough to encase the geometry of each selected dataset. Hence, the preprocessing stage of the PODI calculation involves finding the endmost coordinates along every coordinate direction

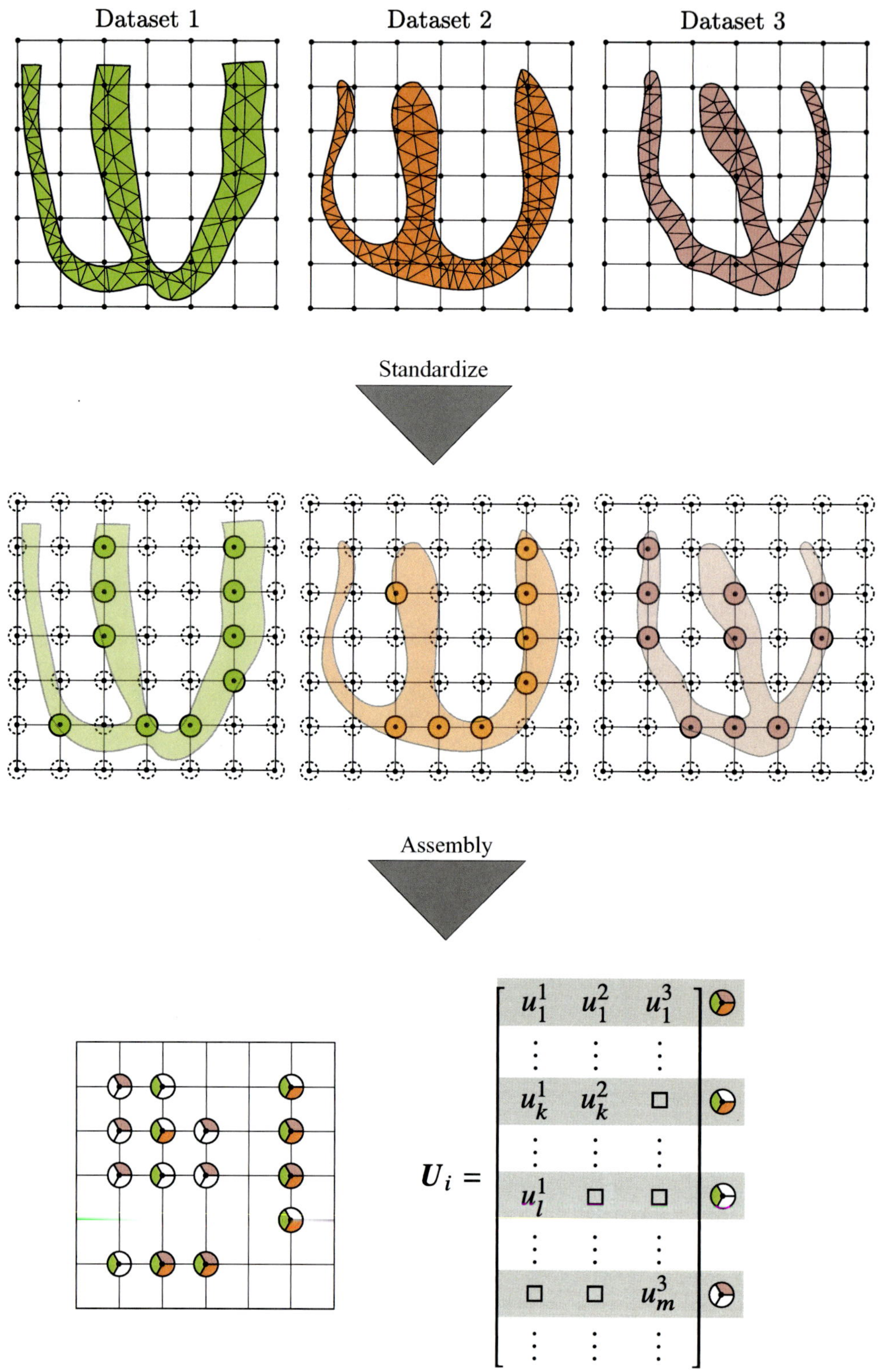


FIG. 8.19 Incompatible **U** entries from cube template registration.

of all heart models to compute the minimum required cube dimensions. Afterward, the cube is discretized using a hexahedral mesh, providing the template node coordinates and their connectivity.

A problem of this method is that the contributions of some dataset surface nodes will not be included. To illustrate this, consider a 2D scenario where the solution fields of three hearts have been standardized using a $7 \times 7$ square grid of template nodes, as depicted in Fig. 8.19. Due to the unique shape of each heart geometry, some template nodes are not found in each geometry and are interpolated for. Accordingly, there are missing column entries in the dataset matrix, $U_i$, which get zero values for lack of a better choice. Although increasing the grid mesh density decreases the number of missing entries relative to the overall number, the cube grid standardization method clearly lacks the flexibility to deal with patient-specific hearts.

#### 8.5.1.2 Heart Template Standardization

A better alternative is to make use of a heart template for the standardization procedure and morph every heart dataset onto the same heart-shaped template. After the morphing process, the dataset nodes are clustered around the template nodes, ensuring that all of them belong to one of the elements of every dataset heart. Accordingly, the dataset matrix, $U_i$, has no missing entries. As a suitable template heart geometry, the statistical average of all selected heart geometries is suggested to minimize the movement of dataset nodes during the morphing process. This optimizes the accuracy of the morphing process and the subsequent interpolation at the template nodes.

Within the field of computer vision, the morphing process is commonly known as registration. It is an important tool for mapping two clouds of points onto each other. Three different techniques are commonly used. The first one is rigid registration, which consists of pure translation; the second is an affine registration that in addition to translation, allows for scaling and rotation. Finally, the third type of registration is called nonrigid registration. Rigid and affine registration methods allow for the preservation of the overall geometric details and affect the whole geometry. In the nonrigid scenario, the geometry can be completely morphed into another one, and the registration can be localized across the geometry. Various examples of heart registration can be found in the literature. In Sermesant et al. [65], a mass center alignment followed by a principal axes-based registration together with an affine registration was employed to deform an idealized BV heart model according to CMR scans. Toussaint et al. [87] made use of a log-diffeomorphic registration to project CMR images of a ventricle onto a perfectly truncated ellipsoid. Lamata et al. [88] introduced a mesh warping technique that required the conversion of the idealized geometry to a binary file, similar to CMR image data before the registration took place. A common aspect of these registration methods is their CMR or binary image requirements. In this research, however, the source geometry and the template geometry are both given in terms of meshes consisting of interconnected nodes. For morphing of the source geometry onto the template one, the CPD method [47] is utilized, which is a nonrigid-based registration method. CPD is a probabilistic method based on the maximum likelihood estimation technique and involves a motion coherence constraint over a velocity field in order to allow for the smooth movement of points from one spatial location to another.

Consider two sets of points, the template points set, $\mathbf{Y}$, and data points set, $\mathbf{X}$, both stored as matrices. The number of rows of these matrices corresponds to the number of points while the number of columns relates to the dimension, $D$, of the points:

$$\begin{aligned} \mathbf{Y} &= (\mathbf{y}_1, \ldots, \mathbf{y}_M)^{\mathrm{T}}, \\ \mathbf{X} &= (\mathbf{x}_1, \ldots, \mathbf{x}_N)^{\mathrm{T}}, \end{aligned}$$

where $M$ is the number of template points and $N$ is the number of data points in their respective set. Then assume the validity of the Gaussian mixture model (GMM) and associate each point in Y with a Gaussian probability density function with the template point as its centroid:

$$p(\mathbf{x}) = \sum_{m=1}^{M+1} P(m) p(\mathbf{x}|m), \tag{8.35}$$

where

$$p(\mathbf{x}|m) = \frac{1}{(2\pi\sigma^2)^{D/2}} e^{-\frac{\|\mathbf{x}-\mathbf{y}_m\|^2}{2\sigma^2}} \tag{8.36}$$

for $m = 1, M$ and

$$p(\mathbf{x}|M+1) = \frac{1}{N}, \tag{8.37}$$

the latter accounting for noise and outliers. Imposing independent and identically distributed data assumptions, $\sigma^2$ denotes equal isotropic covariance, $P(m) = \frac{1}{M}$ equals membership probabilities for every GMM centroid, and $w^{\mathrm{CPD}}$ is a weight constant.

$$p(\mathbf{x}) = w^{\mathrm{CPD}} \frac{1}{N} + (1 - w^{\mathrm{CPD}}) \sum_{m=1}^{M} \frac{1}{M} p(\mathbf{x}|m). \tag{8.38}$$

Next, the GMM centroids are reformulated in terms of a velocity function, $v$, which is used to update the position of Y as:

$$\mathcal{T}(\mathbf{Y},v)=\mathbf{Y}+v(\mathbf{Y}). \tag{8.39}$$

To find an estimate of $v$, a negative log-likelihood function is minimized while imposing independent and identically distributed data assumptions and adding a regularization term, $\phi(v)$:

$$f(v,\sigma^2)=-\sum_{n=1}^{N}\log(p(\mathbf{x}))+\frac{\lambda^{\mathrm{CPD}}}{2}\phi(v), \tag{8.40}$$

with $\lambda^{\mathrm{CPD}}$ being the constant regulating the contribution of the regularization term. In order to find $v$ and $\sigma^2$, the expectation maximization algorithm is employed. The regularization term, $\phi(v)$, is chosen to take the form of:

$$\phi(v)=\int_{\mathbb{R}^D}\frac{|\widetilde{v}(\mathbf{s})|}{\widetilde{G}(\mathbf{s})}, \tag{8.41}$$

where s is the so-called frequency domain variable, $\widetilde{v}$, the Fourier transformation of $v$, and $\widetilde{G}$ is taken to be a Gaussian kernel. A Gaussian kernel is chosen because it is positive definite symmetric and provides a mean to regulate spatial smoothness. More importantly, it also allows the regularization term to be equivalent to the one presented in the motion coherence theory, which involves defining a coherent velocity field across a set of points with no prior data of their motion [89].

The expectation maximization algorithm usually consists of two steps, the E step and the M step. According to Myronenko et al. [47], the M step leads in finding $\mathrm{W}_m^{\mathrm{CPD}}$ from the following expression:

$$(\mathbf{G}+\lambda^{\mathrm{CPD}}\sigma^2(\mathrm{diag}(\mathbf{P})\mathbf{1})^{-1})\mathbf{W}^{\mathrm{CPD}}=\mathrm{diag}(\mathbf{P1})^{-1}\mathbf{PX}-\mathbf{Y}. \tag{8.42}$$

$\mathrm{W}^{\mathrm{CPD}}$, of size $M\times D$, is the list of weight constants. G is known as the Gram matrix with the size of $M\times M$ and whose components are given by

$$G_{ij}=e^{-\frac{1}{2}\left\|\frac{\mathbf{y}_i-\mathbf{y}_j}{\beta^{\mathrm{CPD}}}\right\|^2}, \tag{8.43}$$

where $\beta^{\mathrm{CPD}}$ is a constant and $\|\cdot\|$ is the norm. With Eq. (8.39), Y is updated through $\mathbf{Y}_{\mathrm{update}}=\mathcal{T}(\mathbf{Y},\mathbf{W}^{\mathrm{CPD}})=\mathbf{Y}+\mathbf{GW}^{\mathrm{CPD}}$ while the matrix of posterior probabilities, P, can be evaluated during the E step of the expectation maximization algorithm through the following equation derived from Eq. (8.40):

$$P_{mn}=\frac{\exp^{-\frac{1}{2\sigma^2}\|\mathbf{x}_n-(\mathbf{y}_m+\mathbf{G}(\mathbf{m},\cdot)\mathbf{W}^{\mathrm{CPD}})\|^2}}{\sum_{k=1}^{M}\exp^{-\frac{1}{2\sigma^2}\|\mathbf{x}_n-(\mathbf{y}_k+\mathbf{G}(\mathbf{m},\cdot)\mathbf{W}^{\mathrm{CPD}})\|}+\frac{w^{\mathrm{CPD}}}{1-w^{\mathrm{CPD}}}\frac{(2\pi\sigma^2)^{D/2}M}{N}}, \tag{8.44}$$

where $N_{\mathbf{P}}=\sum_{n=1}^{N}\sum_{m=1}^{M}P^{\mathrm{old}}(m|\mathbf{x}_n)$. Finally, $\sigma^2$ is obtained with the following expression [47]:

$$\sigma^2=\frac{1}{N_{\mathbf{P}}D}(\mathrm{tr}(\mathbf{X}^{\mathrm{T}}\mathrm{diag}(\mathbf{P}^{\mathrm{T}})\mathbf{1X})-2\mathrm{tr}((\mathbf{PX})^{\mathrm{T}}\mathbf{T})+\mathrm{tr}(\mathbf{T}^{\mathrm{T}}\mathrm{diag}(\mathbf{P})\mathbf{1T})). \tag{8.45}$$

To demonstrate CPD registration, consider an idealized 3D human LV created from a half-cut ellipsoid using dimensional data from the literature, as listed in Fig. 8.20.

The LV is then discretized using tetrahedral finite elements to obtain a spatial distribution of points. To represent the template and the data geometry, two different mesh configurations are utilized. The template mesh consists of 974 nodes and 4060 elements while the data mesh has 771 nodes and 3059 elements. Besides the mesh densities, also the heart anatomy of the data mesh is altered in terms of translation, stretch, and rotation. The template mesh and the data mesh are given in Figs. 8.21 and 8.22, respectively.

The CPD algorithm used in this research is based on a MATLAB code written by Myronenko, one of the authors of Ref. [48]. The code can be publicly accessed from: https://sites.google.com/site/myronenko/research/cpd. To start the registration procedure, the MATLAB's code default parameters of $w^{\mathrm{CPD}}=0.7$, $\lambda^{\mathrm{CPD}}=2$, and $\beta^{\mathrm{CPD}}=3$ are assigned. Following the CPD calculation, the results are then analyzed by comparing the deformed data mesh with the original one. In this case, the undeformed data mesh configuration is considered the exact one as its shape is the same as the template mesh. A visual representation of the obtained registered mesh as compared with the template one is shown in Fig. 8.23.

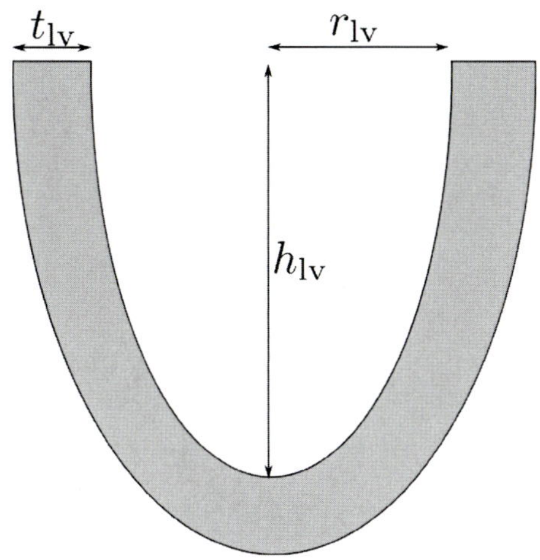


FIG. 8.20 Idealized left ventricle.

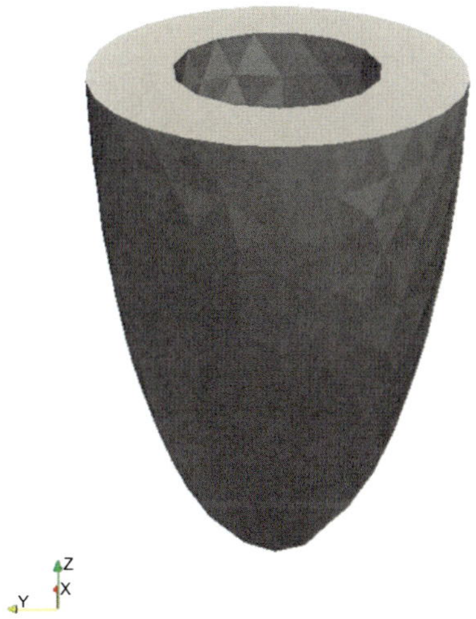


FIG. 8.21 Template mesh.

FIG. 8.22 Deformed data mesh.

FIG. 8.23 Superimposed template mesh (*gray*) and registered data mesh (*green*).

## 8.5.2 Numerical Examples

This section discusses the suitability of the two proposed standardization methods for patient-specific heart modeling. Here, the focus is on the diastolic filling phase where the heart responds passively to the increasing cavity filling pressure. The PODI databases are populated from full-scale simulation results of 58 incremental time steps using the EFG [81], implemented in SESKA [90] and on a desktop computer equipped with an Intel i7 processor (four physical cores clocked at 3.4 GHz) and 8 GB of memory. Given that all datasets have been simulated using the same time steps, the time standardization process was not required.

In heart geometry, an idealized BV model is created that includes important aspects of a real human heart, namely cavity volumes, wall thickness, ventricular diameter, and ventricular depth. A cross-section of the geometry is given in Fig. 8.24, and its dimensions, extracted from the literature, correspond to the end-systole phase of the heart, as presented in Table 8.4.

$h_{LV}^{endo}$ and $h_{RV}^{epi}$ have not been directly obtained from the literature. The former is computed using the diameter-to-depth ratio and the base diameter measure while the latter is the summation of $h_{LV}^{endo}$ and $t_{LV}$. The LV is first created using two halved ellipsoids. On the left side of the LV, the surface is extruded by a distance of $l_{RV}$ to obtain the RV endocardium and $l_{RV} + t_{RV}$ is used for the RV epicardium. The final 3D geometry generated is given in Fig. 8.25.

In terms of boundary conditions, the displacement in the *z*-direction along the surface of the heart's base is fixed to prevent longitudinal rigid body motion. Additionally, an elastic boundary condition with a prescribed stiffness of

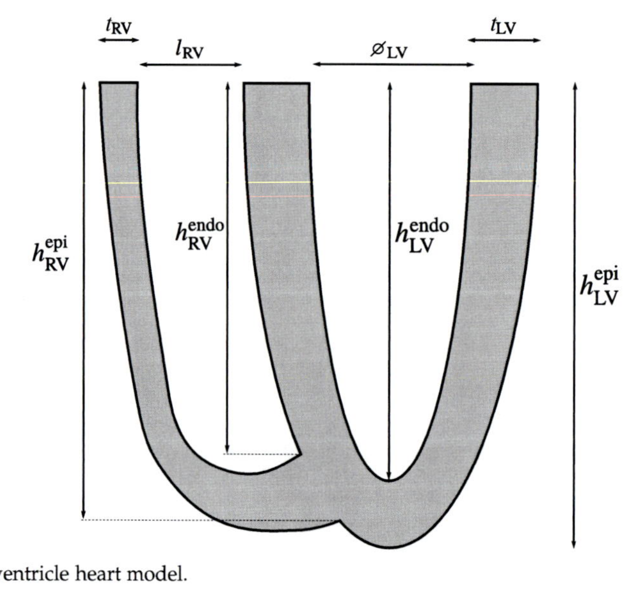


FIG. 8.24 Cross-section of the biventricle heart model.

TABLE 8.4 Dimensions of the Biventricle Model

| | Values | | |
|---|---|---|---|
| | Range | Choice | References |
| **LEFT VENTRICLE** | | | |
| Volume (mL) | 22–48 | 36.98 | [38, 40] |
| Base diameter, $\varnothing_{\mathrm{LV}}$ (cm) | 2.2–4.0 | 3.1 | [41, 91] |
| Wall thickness, $t_{\mathrm{LV}}$ (cm) | 1.09–1.45 | 1.27 | [92] |
| Diameter-depth ratio | 2–3 | 2.37 | [40] |
| Endocardium height, $h_{\mathrm{LV}}^{\mathrm{endo}}$ (cm) | – | 7.35 | – |
| Epicardium height, $h_{\mathrm{LV}}^{\mathrm{epi}}$ (cm) | – | 8.62 | – |
| **RIGHT VENTRICLE** | | | |
| Volume (mL) | 24–62 | 51.21 | [38, 40] |
| Base diameter, $l_{\mathrm{RV}}$ (cm) | 1.2–2.6 | 1.9 | [39] |
| Wall thickness, $t_{\mathrm{RV}}$ (cm) | 3.5–8.5 | 0.63 | [93] |
| Endocardium height, $h_{\mathrm{RV}}^{\mathrm{endo}}$ (cm) | – | 7.00 | – |
| Epicardium height, $h_{\mathrm{RV}}^{\mathrm{epi}}$ (cm) | – | 8.20 | – |

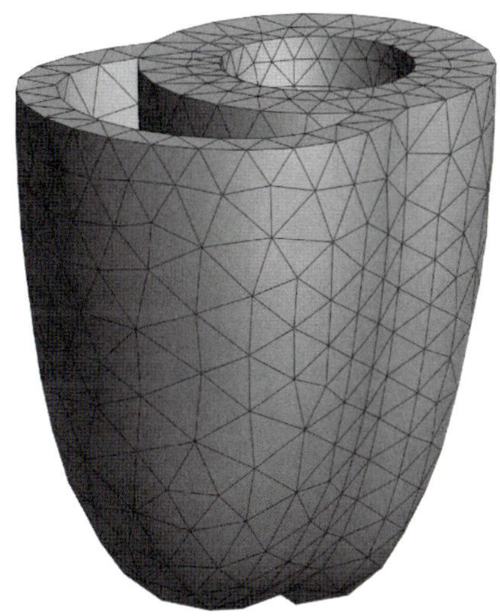

FIG. 8.25 Three-dimensional geometry of a biventricle heart model.

$1 \times 10^{-1}$ kN mm$^{-1}$ is applied along the endocardium and epicardium lines on the base, as shown in Fig. 8.26. This condition allows the thickening and thinning of the base wall as the heart contracts and expands but partially constrains the rotation of the basal myocardium, emulating the presence of the atria and the major blood vessels. Lastly, to model blood filling, a surface pressure is applied to the endocardium wall of each ventricle. The magnitude of those ventricular pressures is given in Table 8.5.

The orthotropic material law (Eq. 8.1) is calibrated according to triaxial test data by Sommer et al. [43], which are the only triaxial experimental data for the human myocardium currently available in the literature. Making use of the standard Levenberg-Marquardt algorithm, the converged values of parameters $A$, $a_1$, $a_2$, $a_3$, $a_4$, $a_5$, and $a_6$, respectively, are listed in Table 8.6. The penalty parameter, $A_{\mathrm{comp}}$, is kept constant at 100 kPa.

To create a PODI database of pseudo patient-specific heart simulation solutions, first, four different tetrahedral mesh configurations of the previously introduced heart model (Fig. 8.25) are considered, as given in Table 8.7. Only BV-1, BV-2, and BV-3 are used for the database, whereas BV-R represents the unknown problem for which a solution is to be computed via the PODI method.

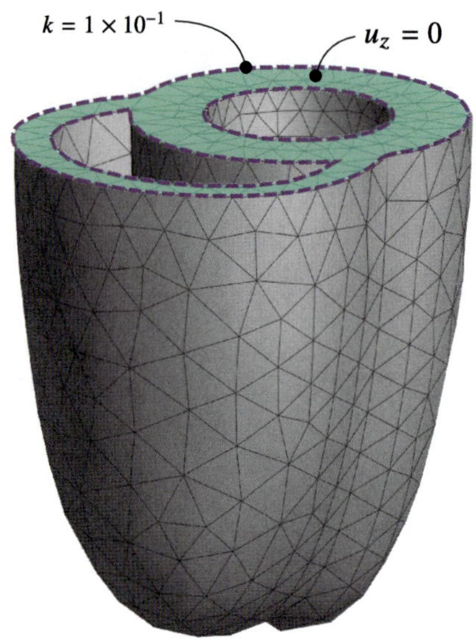


FIG. 8.26 Dirichlet boundary conditions applied to the biventricle heart model.

TABLE 8.5 Ventricular Pressure of the Biventricle Model

| | Pressure (kPa) | References |
|---|---|---|
| Left ventricular pressure (LVP) | 1.5 | [94–96] |
| Right ventricular pressure (RVP) | 1 | [95, 96] |

TABLE 8.6 Passive Material Parameters

| $A$ (kPa) | $a_1$ | $a_2$ | $a_3$ | $a_4$ | $a_5$ | $a_6$ |
|---|---|---|---|---|---|---|
| 0.19 | 12.70 | −8.36 | −8.56 | 11.22 | 14.25 | 9.15 |

TABLE 8.7 BV With Different Mesh Discretizations

| Models | BV-1 | BV-2 | BV-3 | BV-R |
|---|---|---|---|---|
| Nodes | 882 | 902 | 922 | 912 |

Secondly, perturbations are applied to mesh configurations BV-1, BV-2, and BV-3 such that the geometrical shape of the three hearts can be varied. The geometrical perturbation is kept minor as it is expected that, in a realistic scenario, the heart datasets selected from the database would have similar features and characteristics such as age, gender, fitness, diseases, etc. The applied perturbation is a result of two forms of transformation, namely rotation and scaling. A total of nine unique hearts are produced, with one of them illustrated in Fig. 8.27. In order to build a single parametric database, each of these nine heart models is then assigned to a unique stress scaling coefficient value, $A$ (Eq. 8.1), within the range of $0.04\text{kPa} \leq A \leq 0.22\text{kPa}$.

Lastly, the shape of the problem that needs to be solved, BV-R, is also altered as shown in Fig. 8.28 and a stress-scaling factor of $A = 0.13$ is assigned.

Following the dataset standardization, which will be elaborated up on in more detail in Sections 8.5.2.1 and 8.5.2.2, the actual PODI computation considering a targeted energy level of 99% is analog to the approach presented in

FIG. 8.27 Perturbed heart geometry BV-1.

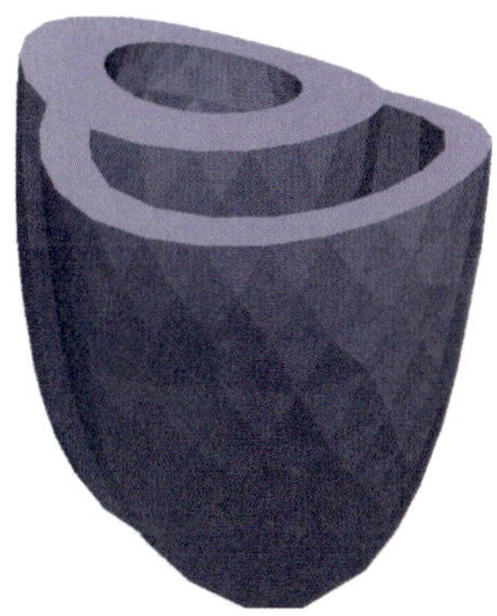

FIG. 8.28 Heart geometry of the problem at hand (BV-R), $A = 0.13$ kPa.

Rama et al. [34]. The PODI results, $U^{PODI}$, were then compared to the full-scale simulation of BV-R, $U^{EFG}_{BV\text{-}R}$, and the $L_2$ norm error, based on Eq. (8.33), was computed as follows:

$$\varepsilon_{\ell_2}^{BV-R}\text{–norm} = \frac{\| \mathbf{U}_i^{PODI} - \mathbf{U}_{i,BV-R}^{EFG} \|}{\| \mathbf{U}_{i,BV-R}^{EFG} \|}. \tag{8.46}$$

### 8.5.2.1 Cube Template Standardization

#### 8.5.2.1.1 COARSE TEMPLATE DISCRETIZATION

Making use of the BV heart model introduced in Section 8.5.2, the application of the cube template standardization process is first investigated. As described in Section 8.5.1.1, the cube grid is constructed in such a way that it encompassed all the heart models, hence ensuring that all data would be captured by the grid. It is discretized with a constant spacing along every coordinate direction, resulting in a total of 294 template nodes, as shown in Fig. 8.29.

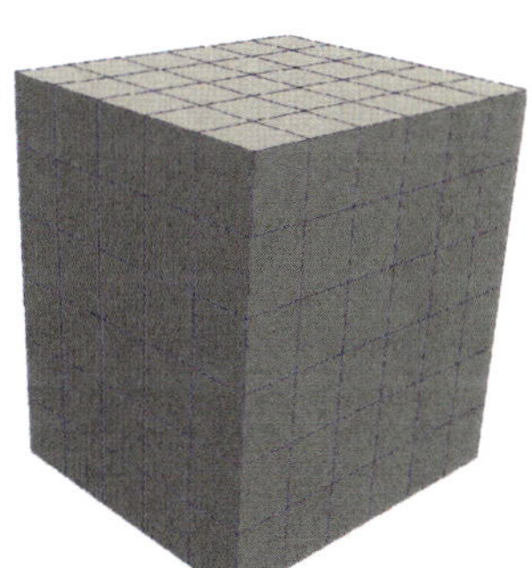

FIG. 8.29 Cube grid with 294 nodes.

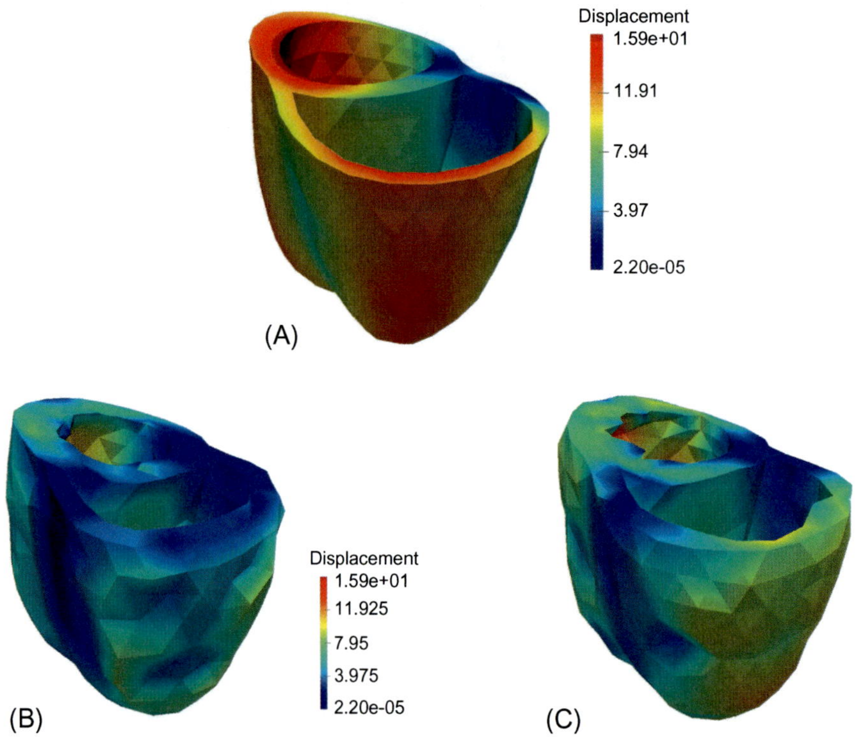


FIG. 8.30 Displacement field solution at end-diastole. (A) EFG. (B) PODI results from cube grid made up of 294 nodes. (C) PODI results from cube grid made up of 990 nodes.

The entire PODI calculation takes on average a total time of about 4 min. This calculation time includes reading the dataset from the database, projecting the results from the hearts to the grid, setting up interpolants, carrying out the PODI calculation, projecting the PODI results from the grid back to the heart of the problem at hand, and postprocessing the results. The costliest operation is projecting the results from the dataset heart geometries to the template grid, and from the grid back to the heart geometry of the problem at hand. This procedure requires about 3.80 min, that is, 95% of the total calculation time. On the other hand, the actual reduced order calculation only needs 0.26 s, where the displacement field computation across all 58 time steps lasts for only 0.029 s. This shows that subsecond calculations can be achieved, as the displacement calculation frequency is about 2000 Hz. This calculation frequency is higher than the one recorded in Rama et al. [34] because the number of template nodes is less here, resulting in smaller dataset matrices, $U_i$.

The $\varepsilon_{\ell_2}^{BV-R}$-norm is found to be 0.46 for the displacement field, 0.89 for the strain field, and 0.98 for the stress field. For the pressure-volume relationship curve, the errors are 0.23 and 0.22 for the LV and the RV, respectively. These errors are higher than those presented in Rama et al. [34] and the final deformed configuration exhibits nonphysical deformations. Solution fields, such as the displacement field shown in Fig. 8.30B, are not smooth, as opposed to the corresponding full-scale simulation result depicted in Fig. 8.30A. The main reason why those nonsmooth solution fields are obtained is that, as explained in Section 8.5.1.1, some of the entries in $U_i$ refer to template nodes, which are located outside the respective model and have been assigned a default zero-value for all solution fields. However, these template nodes need to be included because for other datasets, they are positioned inside the model. Consequently, the projection process leads to solution fields of low magnitude near those nodes, as is clearly visible in Fig. 8.30B.

#### 8.5.2.1.2 REFINED TEMPLATE DISCRETIZATION

Another attempt is made at increasing the number of grid nodes such that the negative effect of the poor surface interpolation can be decreased due to the higher number of grid nodes, located inside the dataset hearts, and raising the PODI solution field's accuracy. To do so, five other grids with different grid node densities are considered, as shown in Table 8.8.

The obtained results show an overall increase in the accuracy of the PODI solutions, as given in Fig. 8.31 and Table 8.9. The displacement field is about 22% more accurate while the LV pressure-volume curve error dropped by 26% as the grid discretization reaches 990 nodes. A cut through the BV models also confirms the drop in error,

TABLE 8.8 Cube Grid With Different Mesh Discretizations

| Cube grid | CG-1 | CG-2 | CG-3 | CG-4 | CG-5 | CG-6 |
|---|---|---|---|---|---|---|
| Nodes | 294 | 336 | 448 | 504 | 720 | 990 |

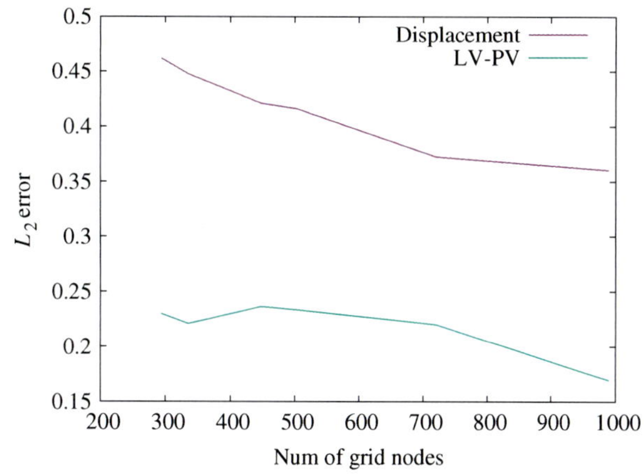


FIG. 8.31 Change in $\varepsilon_{\ell_2}^{\mathrm{BV}-\mathfrak{R}}$-norm as the number of grid nodes increases.

TABLE 8.9 Change in the Error and Calculation Time as the Number of Grid Nodes Increases

| | $L_2$-error | | Calculation time (s) | |
|---|---|---|---|---|
| | **Displacement** | **Visual** | **Total** | **Projection** |
| CG-2 | 0.45 | Deformations | 232 | 219 |
| CG-6 | 0.36 | Deformations | 643 | 628 |

as shown in Fig. 8.32, with the nodes lying inside the wall showing improvement while the lateral wall of the RV exhibits more expansion. However, when the deformed state is more closely investigated, the localized nonphysical deformations are still dominant on the surfaces of the geometry and the magnitude of the displacement field is noticeably lower, as depicted in Fig. 8.30C. The most significant solution improvements can be found at the epicardium surface of the RV.

Regarding the PODI calculation time, it is found to be increasing almost linearly because more grid nodes are present, as shown in Fig. 8.33. This is due to the fact that the PIP algorithm and the template projections require more time.

The three most dominant POMs are found to be needed in order to reach the minimum specified energy limit for the displacement, strain and stress field data. Further investigation of the POVs reveals that the energy of the first, second, and third POMs is about 61.6%, 30.5%, and 7.8%, respectively. The fourth POM is below $1.04 \times 10^{-14}$%. This is in contrast to Rama et al. [34], where the first POM already accounted for about 99% of the energy. The cube standardization method obviously has a scattering effect leading to dataset matrices, $U_i$, with disorganized data due to the

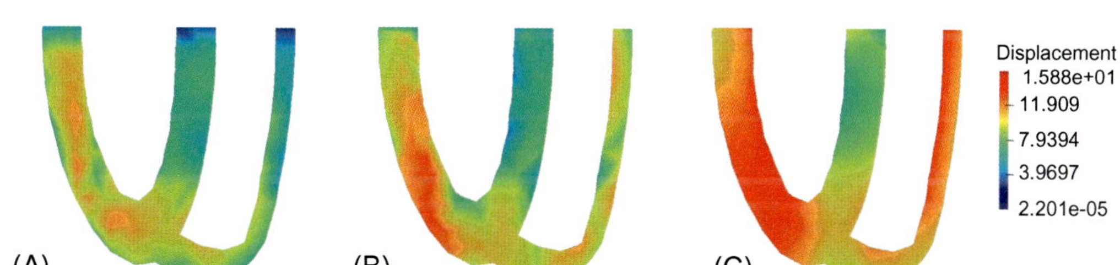


FIG. 8.32 PODI displacement field plot based on the cube template standardization method for different numbers of grid nodes. (A) PODI with 294 grid nodes. (B) PODI with 990 grid nodes. (C) EFG.

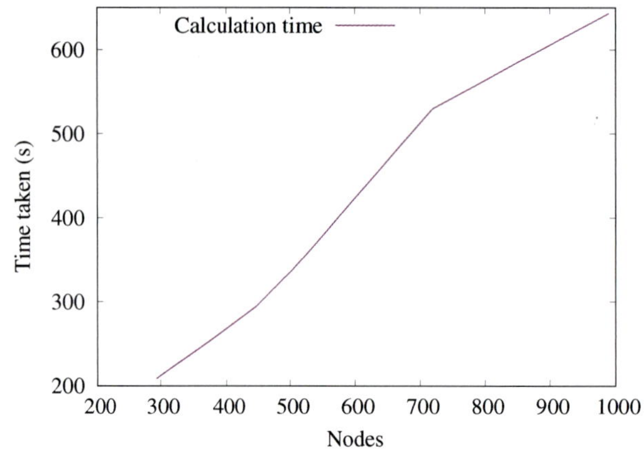


FIG. 8.33 Change in calculation time as the number of grid nodes increases.

introduction of the artificial zero entries. As a result, PODI performs poorly. Consequently, an alternative method where the zero entries are omitted is explored in the next section.

### 8.5.2.2 *Heart Template Standardization*

#### 8.5.2.2.1 COARSE TEMPLATE DISCRETIZATION

An alternative approach to standardize the degrees of freedom is to make use of a heart template coupled with the CPD algorithm. The major advantage of this approach is the elimination of both the incompatible columns and the zero entry problems in the dataset matrix $U_i$. As described in Section 8.5.1.2, the heart template standardization requires two additional steps as compared to the cube template standardization, which involves the registration of every patient-specific heart, including the heart of the problem at hand, onto the template and the reverse projection of the PODI results onto the nodes of the problem at hand. Here, the datasets of the nine heart models and the problem at hand illustrated in Figs. 8.27 and 8.28, respectively, are registered onto the unperturbed BV template heart depicted in Fig. 8.25 with the help of the CPD algorithm with parameters listed in Table 8.10. In this case, the number of nodes of the template heart is 347.

The registration of each heart takes on average 17 s for 285 iterations. One of the mapped geometries is given in Fig. 8.34, where the black mesh grid lines indicate the template geometry. The quality of the registration and resulting meshes is afterward checked to ensure that the elements of the mapped geometries have not substantially degraded. For that, three groups of quantitative results are considered: GA, GB, and GC. GA represents the mesh quality of the original unperturbed meshes, BV1, BV2, and BV3, which are used as a baseline, while GB refers to the meshes perturbed by rotation and translation. Finally, GC represents the quality of the registered meshes. The mesh quality metrics considered are the nodal mean distance (which is computed using a software package called CloudCompare [97], based on Hausdorff's distance measure), the minimum element angle, and the element shape quality. The latter two are determined using the VTK library [98]. The results obtained are compiled as histograms given in Fig. 8.35.

The plot of Fig. 8.35A describes the evolution of the mean distance. A lower value is equivalent to a better registration, as the nodes of the registered mesh are closer to the template mesh. Following the perturbation, the mean distance increased. However, the CPD registration is able to almost restore the mean distance to the GA level, meaning that the meshes recovered their original configurations: BV-1, BV-2, and BV-3. The shape and minimum angle trends are both similar. The perturbation process causes a slight decrease in the mesh quality, which can be considered negligible. Overall, no negative element volumes are recorded. Therefore, the registration can be regarded as successful.

With the registration completed, the PODI simulation is then carried out for the problem with a stress-scaling coefficient of 0.13 kPa and an energy conservation of 99%. The calculation takes in total 5.8 min, including the registration of the four selected datasets and the problem-at-hand geometries on the template heart. As for the cube-shape

TABLE 8.10 CPD Algorithm Parameter Values

| Parameter | $\beta^{CPD}$ | $\lambda^{CPD}$ | $w^{CPD}$ |
|---|---|---|---|
| Value | 2.5 | 0.7 | 0.7 |

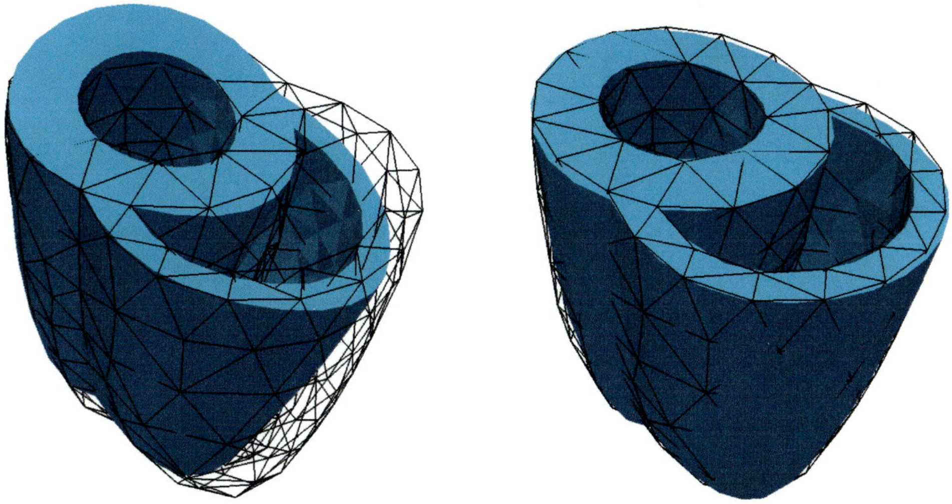

FIG. 8.34 Registration of perturbed BV-1 mesh before registration (*left side*) and after registration (*right side*).

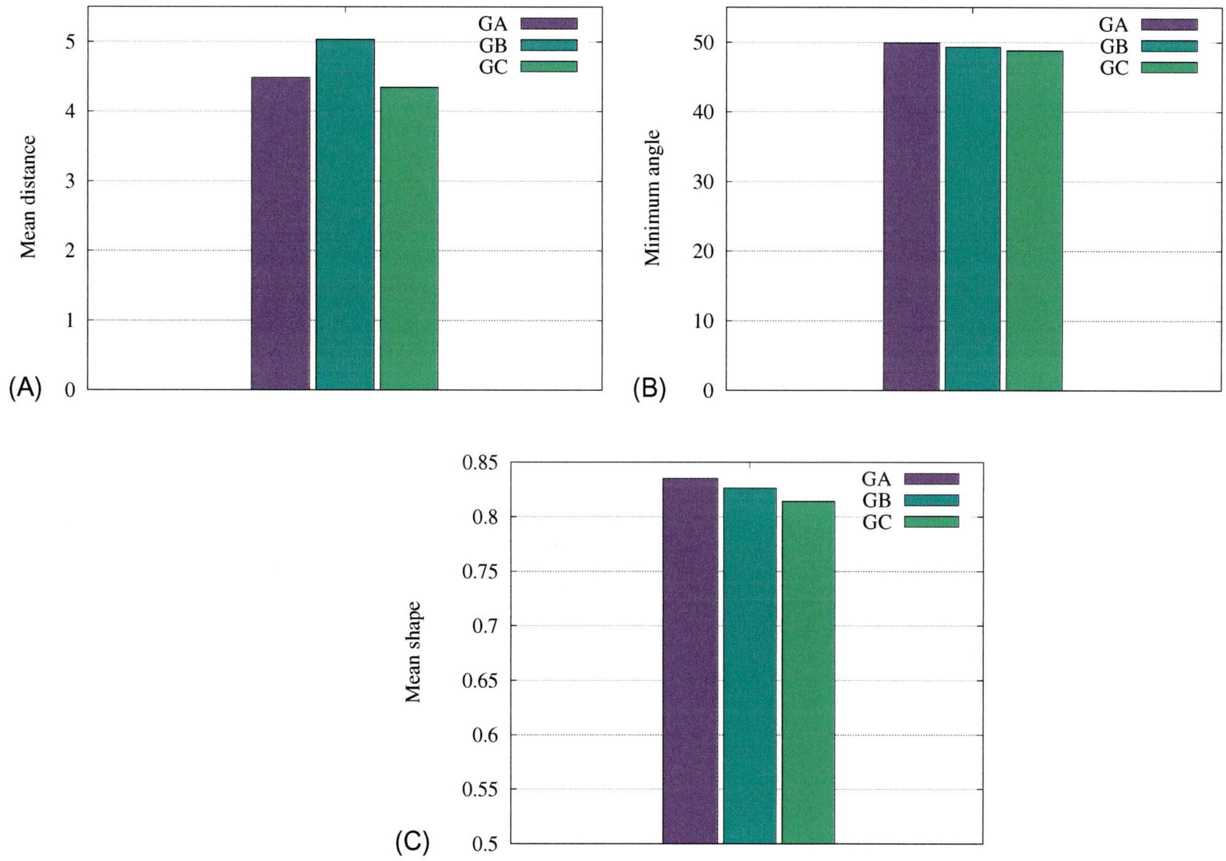


FIG. 8.35 Mesh quality before and after perturbation, and after registration. (A) Mean distance; (B) minimum angle; and (C) mean shape.

template, the projection of the solutions to and from the template is the most expensive operation, as 90% of the calculation time is spent during that process.

In terms of solution accuracy, the $\varepsilon_{\ell_2}^{\mathrm{BV-R}}$-norm obtained for the heart grid template is about 0.22, which represents a 39% improvement as compared to the 990-node cube template. This increase in solution quality and smoothness can also be confirmed when a BV slice is visually compared, first against the corresponding full-scale EFG solution, as given in Fig. 8.36, and second against the cube template solutions, as shown in Fig. 8.32. Also, these results are found to benefit the pressure-volume relationship curves of the LV and RV, as their errors are $3.1 \times 10^{-2}$ and $4.2 \times 10^{-2}$,

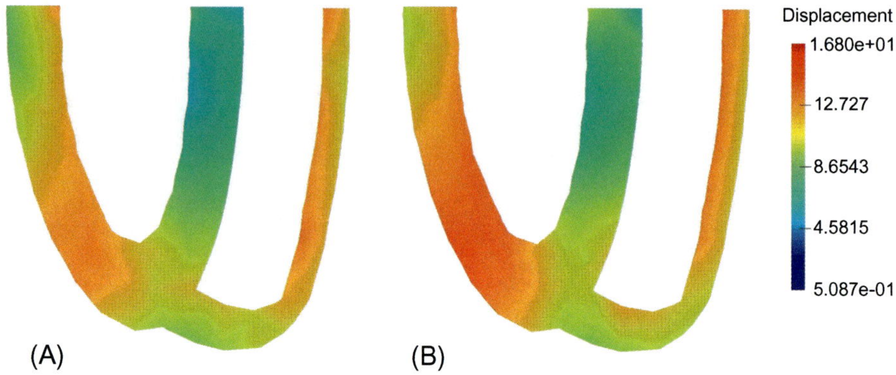


FIG. 8.36 Comparison of the PODI solution obtained from a heart template of 347 nodes against the EFG reference solution. (A) PODI solution; (B) EFG solution.

respectively, which is a drop of 81% and 77% compared to those of the cube template approach. The strain and stress solutions, on the other hand, have a higher error norm, which translates to an increase of 41% and 31%, respectively.

Regarding PODI's calculation characteristics, the three most dominant POMs are conserved in all cases. This is due to the fact that the first POM accounted for only 53.5% while the second and third were 32.7% and 13.7%, respectively. The fourth and last POM is below $1 \times 10^{-13}$%. The energy of the first POM, in this example, is lower than the 62% energy obtained with the cube grid template. This loss in energy seems to be redistributed to the third POM, as the energy of the latter almost doubles. The results presented here show that the POM energies are more scattered for the heart grid template method as opposed to the cube grid template.

Even though the displacement field error norm and the contour plot slice of Fig. 8.36A, show reasonable results, the overall deformation of the BV at the end of the diastole still contains a few localized and irregular mesh distortions, as shown in Fig. 8.37. These irregularities are individual nodes having nonsmooth random movements. This, therefore, leads to a nonsmooth displacement field plot, which is dominant around the RV and especially at the intersection of the left and right ventricular walls. These localized irregularities are supported by the fact that some registration processes, even though being successful globally over the whole heart, visually show unsuccessful mapping around the LV-RV wall connection. One of the main reasons for the unsuccessful mapping is that these locations contained sharp edges, which the CPD algorithm has difficulty in handling properly. Also, this problem can be accentuated, especially when not enough nodes are present around those corners, leading to a lack of information for the registration algorithm to work accurately.

#### 8.5.2.2.2 REFINED TEMPLATE DISCRETIZATION

To solve this problem, the impact of different nodal densities used for the heart grid template on the registration process is now investigated. As such, four additional heart templates are incorporated. The geometry of the templates is kept the same as before, but the discretization is refined continuously. The discretization refinement steps are similar to the cube grid discretization refinements given in Table 8.8, which ensures comparable results. The initial heart

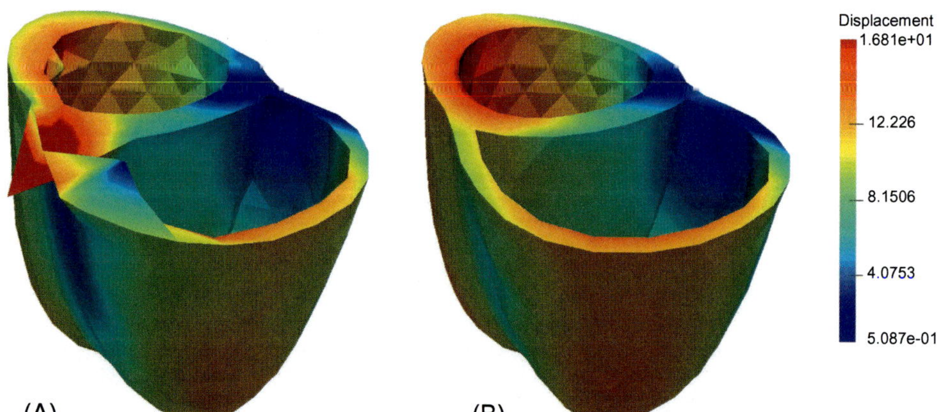


FIG. 8.37 Comparing the EFG reference solution with the PODI solution obtained from a heart template grid with 347 nodes. (A) PODI solution; (B) EFG reference solution.

TABLE 8.11 Mesh Discretization Refinement Levels of the Heart Grid Template

| Heart grid | HG-1 | HG-2 | HG-3 | HG-4 | HG-5 |
|---|---|---|---|---|---|
| Nodes | 347 | 448 | 533 | 714 | 977 |

template is labeled as HG-1 while the refined discretization models are labeled HG-2 to HG-5 where Table 8.11 lists the respective nodal numbers.

Using the new refined heart templates, the performance improvement is separately investigated for the registration of the dataset and the problem-at-hand geometries, and also for the projection of the solution fields to and from the template geometry. Starting with the CPD registration, the mean distance between the nodes, the average minimum element angle, and the element shape are the criteria used to assess the quality of the registration. The compiled results are given in Fig. 8.38.

From Fig. 8.38A, it is found that the mean distance decreases continuously with the increase in the number of template nodes. This effect indicates that the registration has been enhanced, leading to the lowering of the average distance between the template nodes and the registered dataset nodes. The average minimum angle and shape of the registered heart elements also show improvement. As indicated by the curve in Fig. 8.38B and C, the minimum angle and shape increase gradually as more grid nodes are provided, moving closer to the state before the heart mesh was perturbed, shown as dotted lines on their respective plots, being the unperturbed mesh configuration. Based on these results, it can be deduced that providing more nodes to the mesh template promotes better registration with less mesh distortion. This point can be supported by a visual inspection of the registrations given in Fig. 8.39. In all three cases, the registration appears to ameliorate around the LV-RV connection and the RV wall at the basal and apical region.

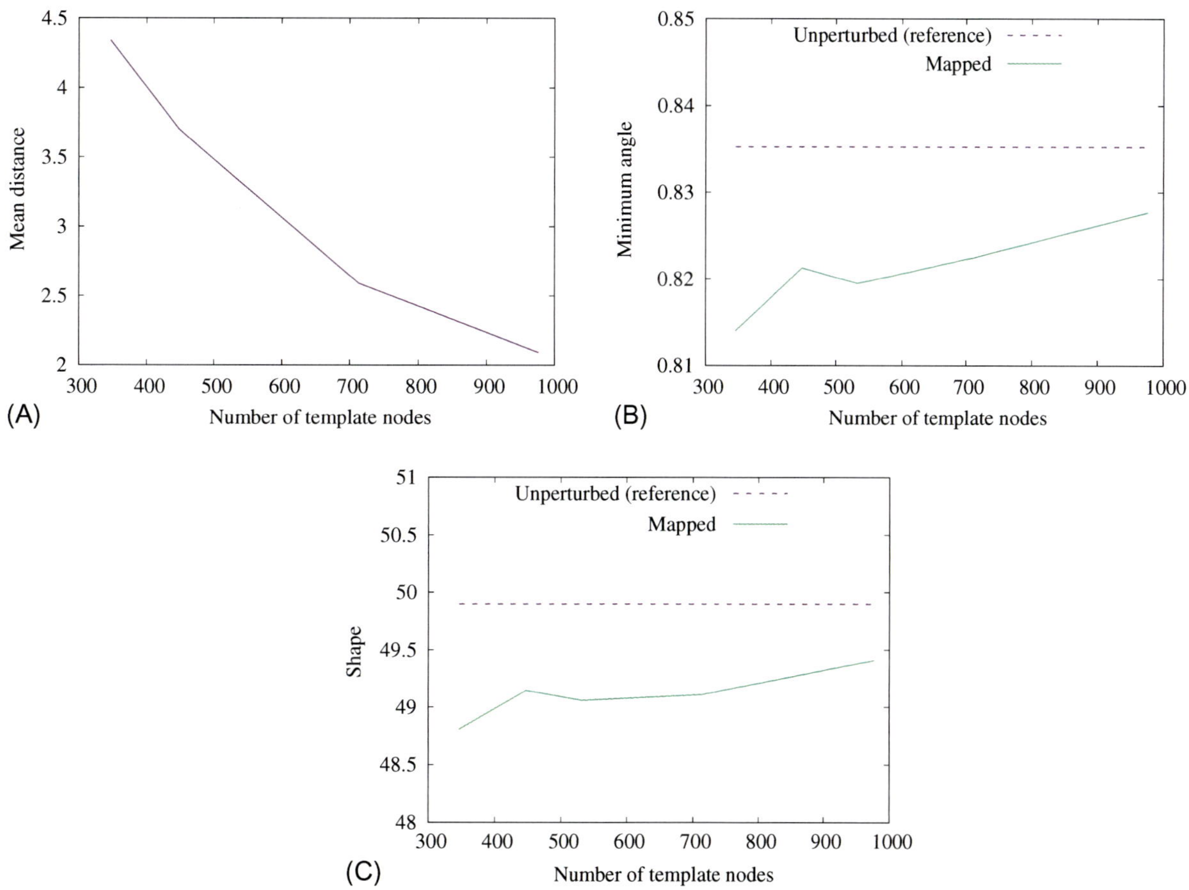


FIG. 8.38 Evolution of the mesh quality measures as the number of template nodes increases. (A) Mean distance; (B) minimum angle; and (C) shape.

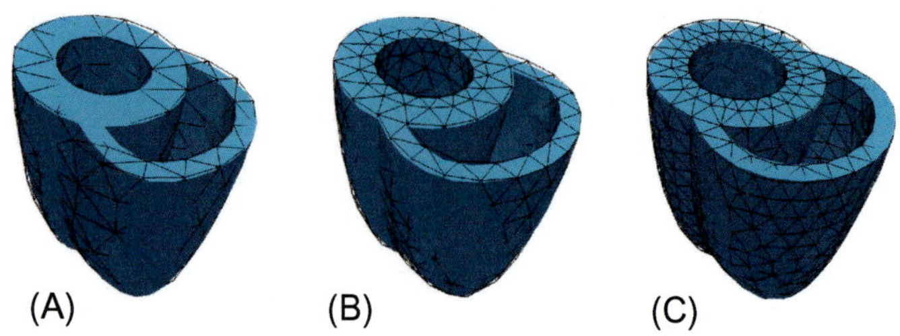


FIG. 8.39 Registration of perturbed BV-1, BV-2, and BV-3 meshes. (A) Model BV-1C registered on HG1. (B) Model BV-1C registered on HG3. (C) Model BV-1C registered on HG5.

Following the registration phase, the PODI calculations are carried out. The displacement field error shows, in Fig. 8.40A and Table 8.12, an improvement of 25% as the number of template grid nodes increases. The LV and RV volumes behave similarly as their error drops by 21% and 81%, respectively, as shown in Fig. 8.40B and C. Finally, with respect to the end-diastole deformation, the final mesh configuration is found to be in better condition. From results obtained with the heart template grid HG-1 to HG-4, irregular mesh distortions are still observed, but with decreasing numbers and severity as the grid node number is increased. Using HG-5, no distortions are visible anymore. The deformed meshes are given in Fig. 8.41. Finally, the energy evolution of the different PODI calculations is analyzed. As before, the three most dominant POMs are conserved to achieve the minimum energy conservation requirement. This is due to the scattered energies across the four POMs. However, the influence of this effect is being limited to the number of grid node increases. Based on Fig. 8.42, it is found that the energy of the first POM changed from 53% to 77%. This rise correlates to the POD method being able to better extract the modes and also isolate the dominant modes more appropriately. This also means that the POD projection of the standardized datasets is carried

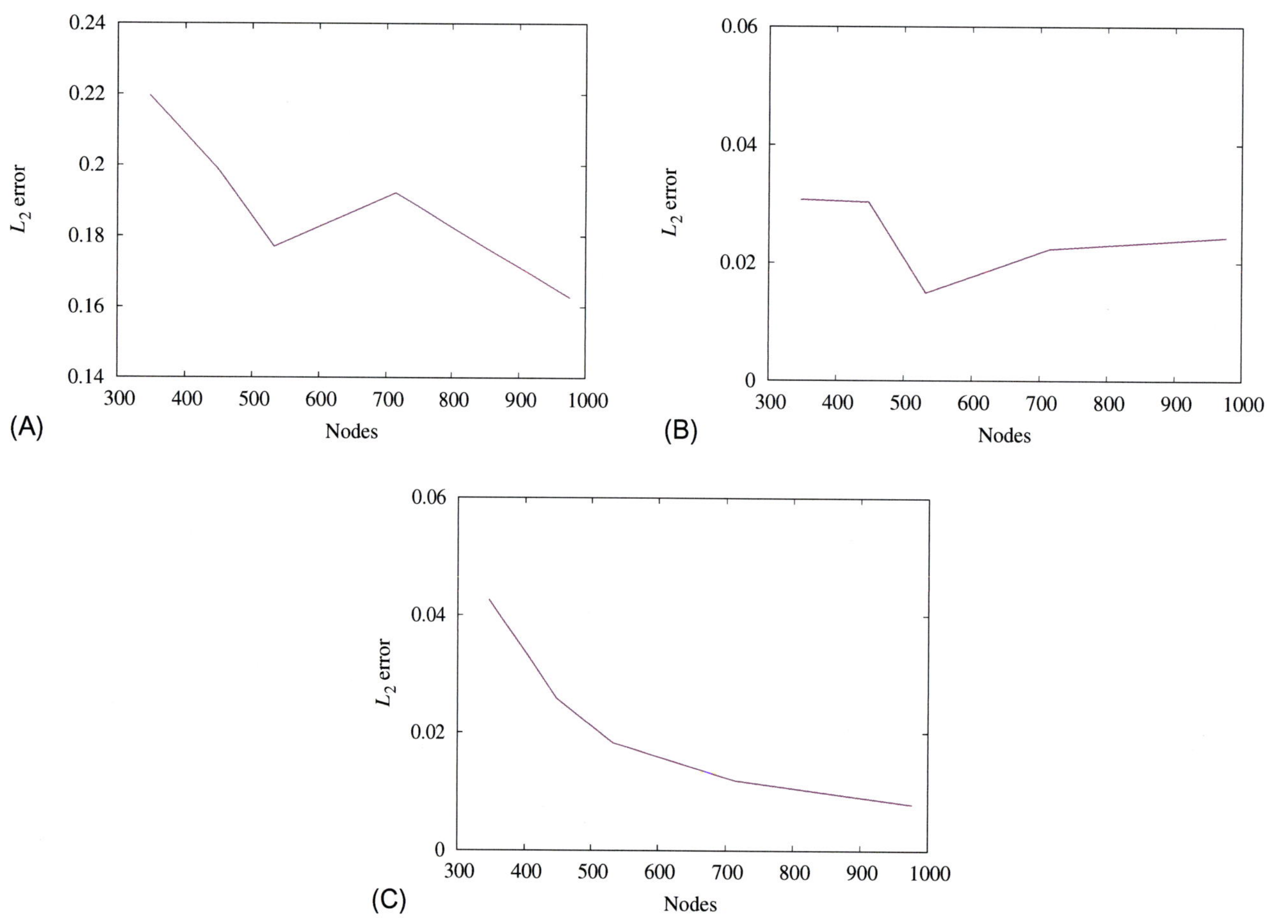


FIG. 8.40 Variation of the $\varepsilon_{\ell_2}^{\mathrm{BV-R}}$-norm as the number of heart template nodes increases. (A) Displacement; (B) left ventricle cavity volume; and (C) right ventricle cavity volume.

TABLE 8.12 Change in the Error and Calculation Time as the Number of Heart Grid Nodes Increases

| | $\varepsilon_{\ell_2}^{BV-R}$-norm | | Calculation time (s) | |
|---|---|---|---|---|
| | **Displacement** | **Visual** | **Total** | **Projection** |
| HG-1 | 0.22 | Deformations | 263 | 235 |
| HG-5 | 0.16 | No deformations | 597 | 564 |

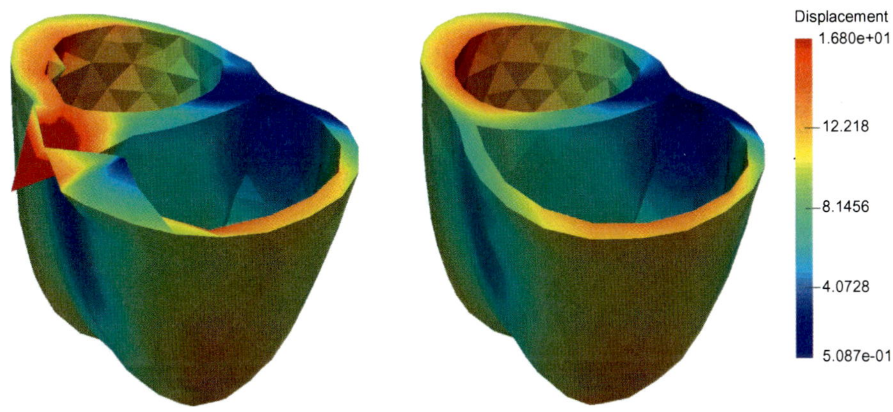


FIG. 8.41 PODI displacement field solutions obtained with heart template grid nodes ranging from 347 to 977. (A) Number of heart grid nodes: 347; (B) number of heart grid nodes: 977.

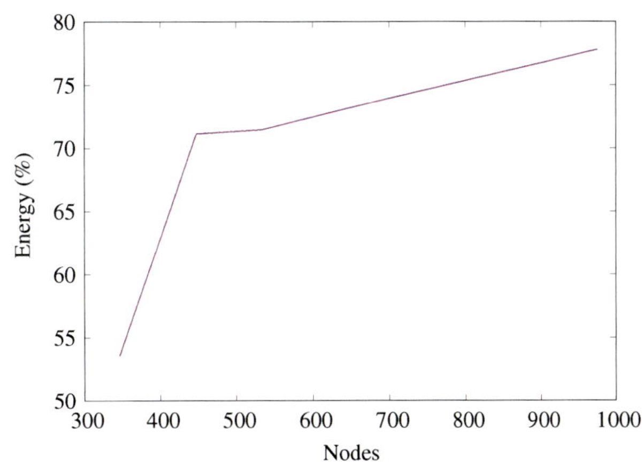


FIG. 8.42 Evolution of the energy of the first POMs as the number of heart template grid nodes is increased.

out more consistently, indicating that a higher amount of solution detail is conserved. Hence, the improvement in registration is directly linked to more accurate PODI solutions.

## 8.6 CONCLUSION

The work presented in this chapter provides an insight on how near real-time modeling of patient-specific hearts can be achieved, taking into account the physiological behavior for an entire heartbeat with the help of the PODI method.

First, in order to achieve the computation of a full heartbeat cycle, the time standardization scheme is proposed to ensure that all datasets used for the parametric PODI calculation are suitably synchronized, as required for the interpolation process. The method means that, for each selected dataset, the respective subsets are identified as belonging to a particular phase of the cardiac cycle such that a standardized timeline can be established and the datasets can be subsequently standardized via temporal PODI calculations. Two examples are considered to demonstrate and study the performance of the developed time standardization scheme. The calculation times of the whole PODI process, including temporal and parametric calculations, are all below 1 min. This represented a speed-up of 15 times or more when compared to their full-scale simulation counterparts, even if the time required for reading the datasets is included, which accounted for more than 25% of the PODI calculation time. The performance could be further improved if the datasets are prestandardized before being stored in the database. The results obtained are found to be very accurate, as the PODI solutions for displacement, stress, and strain fields are within an $\varepsilon_{\ell_2}$ error range of only 0.022–0.032. It is, therefore, demonstrated that this PODI approach is able to capture the heart's behavior for varying hemodynamics in terms of preload and postload.

Second, in order to facilitate patient-specific modeling with arbitrary heart anatomies, two DOFS methods are presented in this work. They are found to solve the problem of incompatible datasets during the assembly of the dataset matrix for the PODI calculation with different levels of accuracy. Using the cube template grid, the solution fields from the selected datasets are standardized by interpolating them at the template nodes. It is found that this operation is, time-wise, costly as opposed to other subprocesses in the PODI calculation. In terms of accuracy, the results obtained are found to be error-prone. That is especially noted when looking at the displacement field where the PODI results show nonphysical behavior. This behavior is identified to be linked to the zero values assigned to nodes during the standardization process in order to account for template nodes lying outside the dataset heart geometry. The second DOFS method considered solves the problems encountered in the cube grid standardization procedure as it no longer introduces the zero values in the dataset matrix. This is achieved by registering the dataset heart geometries to a template heart geometry using the CPD method. As such, all nodes belonging to the dataset hearts are located inside or along the surface of the template heart. The initial results obtained exhibit reduced errors in the displacement fields and pressure-volume relationship curves that decrease by 39% or more. The nonphysical deformation is nevertheless still present but less accentuated. It is subsequently shown that template grid refinement can provide a remedy for this issue by optimizing the CPD registration and completely removing nonphysical deformations. For coarse mesh densities, the cube template required less calculation time than the heart one. But for higher mesh densities, this difference vanishes (see Tables 8.9 and 8.12). In terms of total time, the cube template takes 643 s and the heart template takes 597 s for the finest mesh density. Here, the time required for the template projection alone is 628 s for the cube and 564 s for the heart template. One way of reducing the total calculation time would be to create a database of already standardized datasets instead of standardizing the data during the PODI calculation.

Based on these results, it can be deduced that the presented cardiac PODI framework is suitable to carry out PODI calculations of the whole cardiac cycle using heart geometries having different sizes and mesh discretization, as is the case for patient-specific heart models. Importantly, low calculation times are accomplished at a good solution accuracy.

With those encouraging results, the next steps of this research will be to look at actual patient-specific hearts as obtained from CMR scans and create a database that considers additional characteristics such as gender, age, state of health, fitness, or any other relevant parameters so as to extend its range of applicability.

# APPENDIX

## A.1 Moving Least Square Approximation

As mentioned previously, the PODI method is based on the use of an interpolation technique. In this research, the MLS approximation method [33] is chosen for the PODI calculation as well as the fiber distribution approximation, as it can deal with problems of arbitrary dimensionality and different size of data point sets.

Let us consider a function $f(\boldsymbol{\theta})$ defined over the domain $\mathcal{M}$, which is here not a geometrical domain because each point in $\mathcal{M}$ is not associated with spatial coordinates but with a set of parameters $\boldsymbol{\theta}$ representing the characterizing properties of a problem's solution, for example, stiffness, anisotropy, etc. A possible approximation for $f(\boldsymbol{\theta})$ is given by a polynomial $\mathrm{P}(\boldsymbol{\theta})$ and its nonconstant coefficients $\mathrm{a}(\boldsymbol{\theta})$:

$$f^h(\boldsymbol{\theta}) = \mathbf{P}(\boldsymbol{\theta}) \cdot \mathbf{a}(\boldsymbol{\theta}). \tag{A.1}$$

Now, let domain $\mathcal{M}$ be discretized by a finite number of parameter sets $\{\boldsymbol{\theta}^1, \ldots, \boldsymbol{\theta}^p\}$, the so-called particles scattered in domain $\mathcal{M}$. Each particle is associated with a so-called weight function $\Phi$ of compact support. The size of the support can be individually defined for each particle $\boldsymbol{\theta}^I$, $I = 1, p$ by $\varrho_I$, the so-called influence radius of $\Phi$. The collection of particles with a nonvanishing weight function at point $\boldsymbol{\theta}$ constitutes its support denoted by $\Lambda$. Based on this set of particles, a weighted least square fit in the vicinity of point $\boldsymbol{\theta}$ can be constructed according to

$$J(\mathbf{a}(\boldsymbol{\theta})) := \sum_{I \in \Lambda} [\mathbf{P}(\boldsymbol{\theta}^I \cdot \mathbf{a}(\boldsymbol{\theta}) - f(\boldsymbol{\theta}^I))]^2 \Phi\left(\frac{\boldsymbol{\theta}-\boldsymbol{\theta}^I}{\varrho_I}\right). \tag{A.2}$$

The unknown coefficients $a(\boldsymbol{\theta})$ can be readily determined by minimizing the squared and weighted error between the approximated and actual values of $f(\boldsymbol{\theta}^I)$ at each of the particles $\boldsymbol{\theta}^I \in \Lambda$. That is, minimizing the function $J$ (Eq. A.2), with respect to $a(\boldsymbol{\theta})$. Finally, the coefficients $a(\boldsymbol{\theta})$ are substituted into Eq. (A.1) and the approximation of function $f(\boldsymbol{\theta})$ takes the following form:

$$f^h(\boldsymbol{\theta}) = \mathbf{P}(\boldsymbol{\theta}) \cdot \mathbf{M}^{-1}(\boldsymbol{\theta}) \sum_{I \in \Lambda} \left\{ \mathbf{P}(\boldsymbol{\theta}^I) \Phi\left(\frac{\boldsymbol{\theta}-\boldsymbol{\theta}^I}{\varrho_I}\right) f_I \right\}, \tag{A.3}$$

where $M(\boldsymbol{\theta})$ is the so-called moment matrix of the weight function $\Phi$,

$$\mathbf{M}(\boldsymbol{\theta}) = \sum_{I \in \Lambda} \mathbf{P}(\boldsymbol{\theta}^I) \mathbf{P}(\boldsymbol{\theta}^I) \Phi\left(\frac{\boldsymbol{\theta}-\boldsymbol{\theta}^I}{\varrho_I}\right), \tag{A.4}$$

and $f_I$ are the so-called particle parameters.

Clearly, as the least square fit, Eq. (A.2) only includes a very limited number of sample points, that is, the particles $\boldsymbol{\theta}^I \in \Lambda$, the local character of the approximation $f^h(\boldsymbol{\theta})$ is ensured. With the refinement of the particle distribution the support of the weight functions can be chosen smaller and the approximation $f^h(\boldsymbol{\theta})$ converges for $\varrho_I \to 0$ to the exact function $f(\boldsymbol{\theta})$. The minimum number of supporting particles for any point $\boldsymbol{\theta}$ is dictated by the invertibility requirement of Eq. (A.4), that is, the chosen basis polynomial. The smoothness of the MLS approximation (Eq. A.3) depends on the continuity of the basis polynomial $P \in C^m(\Omega)$ as well as the weight function $\Phi \in C^l(\Omega)$ and it holds $f^h \in C^k$ with $k = \min(l, m)$ [33].

The chosen complete linear basis polynomial in an $n$-dimension is given by

$$\mathbf{P}(\boldsymbol{\theta}) = [1, \theta^1, \ldots, \theta^n], \tag{A.5}$$

and the weight function is based on a cubic spline:

$$w(r) = \begin{cases} \frac{2}{3} - 4r^2 + 4r^3 & \text{for } |r| \leq \frac{1}{2} \\ \frac{4}{3} - 4r + 4r^2 - \frac{4}{3}r^3 & \text{for } |r| \leq \frac{1}{2} \leq 1. \\ 0 & \text{for } |r| > 1 \end{cases} \tag{A.6}$$

For the two-dimensional domain the weight function can be constructed by the following expression:

$$\Phi\left(\frac{\boldsymbol{\theta}-\boldsymbol{\theta}^I}{\varrho_I}\right) = w\left(\frac{\theta_1-\theta_1^I}{\varrho_I}\right), \ldots, w\left(\frac{\theta_2-\theta_2^I}{\varrho_I}\right), \ldots, w\left(\frac{\theta_n-\theta_n^I}{\varrho_I}\right). \tag{A.7}$$

## Acknowledgments

This research has been supported by the Center for High Performance Computing South Africa and the National Research Foundation of South Africa (Grant Numbers 90528, 93111, 104839, and 105858). Opinions expressed and conclusions arrived at are those of the authors and are not necessarily to be attributed to the NRF.

## References

[1] L.L. Demer, F.C.P. Yin, Passive biaxial mechanical properties of isolated canine myocardium, J. Physiol. 339 (1983) 615–630.

[2] J.M. Guccione, L.K. Waldman, A.D. McCulloch, Mechanics of active contraction in cardiac muscle: part II—cylindrical models of the systolic left ventricle, J. Biomech. Eng. 115 (1) (1993) 82–90.

[3] P.C. Franzone, L. Guerri, Spreading of excitation in 3-D models of the anisotropic cardiac tissue. I. Validation of the Eikonal model, Math. Biosci. 113 (2) (1993) 145–209.

[4] T.P. Usyk, I.J. LeGrice, A.D. McCulloch, Computational model of three-dimensional cardiac electromechanics, Comput. Vis. Sci. 4 (4) (2002) 249–257.

[5] B. Baillargeon, N. Rebelo, D.D. Fox, R.L. Taylor, E. Kuhl, The living heart project: a robust and integrative simulator for human heart function, Eur. J. Mech. A Solids 48 (2014) 38–47.

[6] D.B. Davidson, Computational Electromagnetics for RF and Microwave Engineering, second ed., Cambridge University Press, Cambridge, 2010.

[7] P. Lafortune, R. Arís, M. Vázquez, G. Houzeaux, Coupled electromechanical model of the heart: parallel finite element formulation, Int. J. Numer. Methods Biomed. Eng. 28 (2012) 72–86.

[8] S. Niederer, N. Smith, The role of the frank-starling law in the transduction of cellular work to whole organ pump function: a computational modeling analysis, PLoS Comput. Biol. 5 (4) (2009) e1000371, https://doi.org/10.1371/journal.pcbi.1000371.

[9] F.J. Vetter, A.D. McCulloch, Three-dimensional stress and strain in passive rabbit left ventricle: a model study, Ann. Biomed. Eng. 28 (7) (2000) 781–792, https://doi.org/10.1114/1.1289469.

[10] J.C. Walker, M.B. Ratcliffe, P. Zhang, A.W. Wallace, B. Fata, E.W. Hsu, D. Saloner, J.M. Guccione, MRI-based finite-element analysis of left ventricular aneurysm, Am. J. Physiol. Heart Circ. Physiol. 289 (2) (2005) H692–H700, https://doi.org/10.1152/ajpheart.01226.2004.

[11] S. Cotin, H. Delingette, N. Ayache, Real-time elastic deformations of soft tissues for surgery simulation, IEEE Trans. Vis. Comput. Graph. 5 (1) (1999) 62–73, https://doi.org/10.1109/2945.764872.

[12] U. Meier, O. López, C. Monserrat, M.C. Juan, M. Alcañiz, Real-time deformable models for surgery simulation: a survey, Comput. Methods Prog. Biomed. 77 (3) (2005) 183–197, https://doi.org/10.1016/j.cmpb.2004.11.002.

[13] A. Nealen, M. Müller, R. Keiser, E. Boxerman, M. Carlson, Physically based deformable models in computer graphics, Comput. Graph. Forum 25 (4) (2006) 809–836, https://doi.org/10.1111/j.1467-8659.2006.01000.x.

[14] L.P. Nedel, D. Thalmann, Real time muscle deformations using mass-spring systems, Proceedings. Computer Graphics International (Cat. No. 98EX149), IEEE Comput. Soc., 1998, pp. 156–165, https://doi.org/10.1109/CGI.1998.694263.

[15] T. Liu, A.W. Bargteil, J.F. O'Brien, L. Kavan, Fast simulation of mass-spring systems, ACM Trans. Graph. 32 (6) (2013) 1–7, https://doi.org/10.1145/2508363.2508406.

[16] V. Luboz, J. Kyaw-Tun, S. Sen, R. Kneebone, R. Dickinson, R. Kitney, F. Bello, Real-time stent and balloon simulation for stenosis treatment, Vis. Comput. 30 (3) (2013) 1–9, https://doi.org/10.1007/s00371-013-0859-4.

[17] H. Delingette, Toward realistic soft-tissue modeling in medical simulation, Proc. IEEE 86 (3) (1998) 512–523, https://doi.org/10.1109/5.662876.

[18] Z. Lu, V.S. Arikatla, Z. Han, B.F. Allen, S. De, A physics-based algorithm for real-time simulation of electrosurgery procedures in minimally invasive surgery, Int. J. Med. Robot. Comput. Assist. Surg. 10 (4) (2014) 495–504, https://doi.org/10.1002/rcs.1561.

[19] J. Berkley, G. Turkiyyah, D. Berg, M. Ganter, S. Weghorst, Real-time finite element modeling for surgery simulation: an application to virtual suturing, IEEE Trans. Vis. Comput. Graph. 10 (3) (2004) 314–325, https://doi.org/10.1109/TVCG.2004.1272730.

[20] Y.J. Lim, S. De, Real time simulation of nonlinear tissue response in virtual surgery using the point collocation-based method of finite spheres, Comput. Methods Appl. Mech. Eng. 196 (31–32) (2007) 3011–3024, https://doi.org/10.1016/j.cma.2006.05.015.

[21] H. Courtecuisse, H. Jung, J. Allard, C. Duriez, D.Y. Lee, S. Cotin, GPU-based real-time soft tissue deformation with cutting and haptic feedback, Prog. Biophys. Mol. Biol. 103 (2–3) (2010) 159–168, https://doi.org/10.1016/j.pbiomolbio.2010.09.016.

[22] P. Holmes, J.L. Lumley, G. Berkooz, Turbulence, Coherent, Structures, Dynamical Systems and Symmetry, first ed., Cambridge University Press, Cambridge, 1996.

[23] J.S. Hesthaven, G. Rozza, B. Stamm, Certified Reduced Basis Methods for Parametrized Partial Differential Equations, Springer International Publishing, New York, NY, ISBN: 978-3-319-22469-5, 2015.

[24] F. Chinesta, R. Keunings, A. Leygue, The Proper Generalized Decomposition for Advanced Numerical Simulations, Springer International Publishing, Cham, ISBN: 978-3-319-02864-4, 2014, https://doi.org/10.1007/978-3-319-02865-1.

[25] T. Tonn, K. Urban, S. Volkwein, Comparison of the reduced-basis and POD a posteriori error estimators for an elliptic linear-quadratic optimal control problem, Math. Comput. Model. Dyn. Syst. 17 (4) (2011) 355–369, https://doi.org/10.1080/13873954.2011.547678.

[26] M. Boulakia, E. Schenone, J.-F. Gerbeau, Reduced-order modeling for cardiac electrophysiology. Application to parameter identification, Int. J. Numer. Methods Biomed. Eng. 28 (6–7) (2012) 727–744.

[27] C. Corrado, J. Lassoued, M. Mahjoub, N. Zemzemi, Stability analysis of the POD reduced order method for solving the bidomain model in cardiac electrophysiology, Math. Biosci. 272 (2016) 81–91.

[28] S. Han, B. Feeny, Application of proper orthogonal decomposition to structural vibration analysis, Mech. Syst. Signal Process. 17 (5) (2003) 989–1001, https://doi.org/10.1006/mssp.2002.1570.

[29] I.T. Georgiou, J. Sansour, Analyzing the finite element dynamics of nonlinear in-plane rods by the method of proper orthogonal decomposition, in: S. Idelsohn, E. Onate, E. Dvorkin (Eds.), Computational Mechanics, New Trends and Applications, CIMNE, Barcelona, Spain, 1998.

[30] H.V. Ly, H.T. Tran, Modeling and control of physical processes using proper orthogonal decomposition, in: H.T. Banks, K.L. Bowers, J. Lund (Eds.), Computation and Control VI Proceedings of the Sixth Bozeman Conference, vol. 33, 2001, pp. 223–236, https://doi.org/10.1016/S0895-7177(00)00240-5.

[31] R.F. Coelho, P. Breitkopf, C. Knopf-Lenoir, P. Villon, Bi-level model reduction for coupled problems. Application to a 3D wing, Struct. Multidiscip. Optim. 39 (4) (2009) 401–418.

[32] S. Niroomandi, I. Alfaro, E. Cueto, F. Chinesta, Real-time deformable models of non-linear tissues by model reduction techniques, Comput. Methods Programs Biomed. 91 (3) (2008) 223–231, https://doi.org/10.1016/j.cmpb.2008.04.008.

[33] P. Lancaster, K. Salkauskas, Surfaces generated by moving least squares methods, Math. Comput. 37 (1981) 141–158, https://doi.org/10.1090/S0025-5718-1981-0616367-1.

[34] R.R. Rama, S. Skatulla, C. Sansour, Real-time modelling of diastolic filling of the heart using the proper orthogonal decomposition with interpolation, Int. J. Solids Struct. (2016) 1–12, https://doi.org/10.1016/j.ijsolstr.2016.04.003.

[35] R.R. Rama, S. Skatulla, Towards real-time modelling of passive and active behaviour of the human heart using PODI-based model reduction, Comput. Struct. (2018), https://doi.org/10.1016/j.compstruc.2018.01.002.

[36] R.R. Rama, S. Skatulla, Towards real-time cardiac mechanics modelling with patient-specific heart anatomies, Comput. Methods Appl. Mech. Eng. 328 (2018) 47–74, https://doi.org/10.1016/j.cma.2017.08.015.

[37] F. Haddad, S.A. Hunt, D.N. Rosenthal, D.J. Murphy, Right ventricular function in cardiovascular disease, part I: anatomy, physiology, aging, and functional assessment of the right ventricle, Circulation 117 (11) (2008) 1436–1448, https://doi.org/10.1161/CIRCULATIONAHA.107.653576.

[38] L.E. Hudsmith, S.E. Petersen, J.M. Francis, M.D. Robson, S. Neubauer, Normal human left and right ventricular and left atrial dimensions using steady state free precession magnetic resonance imaging, J. Cardiovasc. Magn. Reson. 7 (5) (2009) 775–782, https://doi.org/10.1080/10976640500295516.

[39] J. Kjaergaard, C.L. Petersen, A. Kjaer, B.K. Schaadt, J.K. Oh, C. Hassager, Evaluation of right ventricular volume and function by 2D and 3D echocardiography compared to MRI, Eur. J. Echocardiogr. 7 (6) (2006) 430–438, https://doi.org/10.1016/j.euje.2005.10.009.

[40] J. Sandstede, C. Lipke, M. Beer, S. Hofmann, T. Pabst, W. Kenn, S. Neubauer, D. Hahn, Age- and gender-specific differences in left and right ventricular cardiac function and mass determined by cine magnetic resonance imaging, Eur. Radiol. 10 (3) (2000) 438–442, https://doi.org/10.1007/s003300050072.

[41] P. Stolzmann, H. Scheffel, P.T. Trindade, A.R. Plass, L. Husmann, S. Leschka, M. Genoni, B. Marincek, P.A. Kaufmann, H. Alkadhi, Left ventricular and left atrial dimensions and volumes: comparison between dual-source CT and echocardiography, Investig. Radiol. 43 (5) (2008) 284–289, https://doi.org/10.1097/RLI.0b013e3181626853.

[42] L.G. Rudski, W.W. Lai, J. Afilalo, L. Hua, M.D. Handschumacher, K. Chandrasekaran, S.D. Solomon, E.K. Louie, N.B. Schiller, Guidelines for the echocardiographic assessment of the right heart in adults: a report from the American Society of Echocardiography, J. Am. Soc. Echocardiogr. 23 (7) (2010) 685–713, https://doi.org/10.1016/j.echo.2010.05.010. quiz 786–788.

[43] G. Sommer, A.J. Schriefl, M. Andrä, M. Sacherer, C. Viertler, H. Wolinski, G.A. Holzapfel, Biomechanical properties and microstructure of human ventricular myocardium, Acta Biomater. (2015), https://doi.org/10.1016/j.actbio.2015.06.031.

[44] E. Iuliano, D. Quagliarella, Proper orthogonal decomposition, surrogate modelling and evolutionary optimization in aerodynamic design, Comput. Fluids 84 (2013) 327–350, https://doi.org/10.1016/j.compfluid.2013.06.007.

[45] D. Amsallem, R. Tezaur, C. Farhat, Real-time solution of computational problems using databases of parametric linear reduced-order models with arbitrary underlying meshes, http://arxiv.org/abs/1506.07153, 2015.

[46] D. González, E. Cueto, F. Chinesta, Computational patient avatars for surgery planning, Ann. Biomed. Eng. 44 (1) (2016) 35–45, https://doi.org/10.1007/s10439-015-1362-z.

[47] A. Myronenko, A. Myronenko, X. Song, X. Song, M.Á. Carreira-Perpiñán, M.Á. Carreira-Perpiñán, Non-rigid point set registration: coherent point drift, in: B. Schölkopf, J. Platt, T. Hoffman (Eds.), Advances in Neural Information Processing Systems, vol. 19, MIT Press, Cambridge, MA, 2007, pp. 1009–1016. ISBN: 0162-8828, https://doi.org/10.1109/TPAMI.2010.46.

[48] A. Myronenko, X. Song, Point set registration: coherent point drift, IEEE Trans. Pattern Anal. Mach. Intell. 32 (12) (2010) 2262–2275, https://doi.org/10.1109/TPAMI.2010.46.

[49] T.P. Usyk, A.D. McCulloch, Computational method for soft tissue biomechanics, Biomech. Soft Tissue Cardiovasc. Syst. 441 (2002) 273–342, https://doi.org/10.1007/978-3-7091-2736-0_7.

[50] D. Legner, S. Skatulla, J.M. Bewu, R.R. Rama, B.D. Reddy, C. Sansour, N.H. Davies, T. Franz, Studying the influence of hydrogel injections into the infarcted left ventricle using the element-free Galerkin method, Int. J. Numer. Methods Biomed. Eng. 30 (2014) 416–429, https://doi.org/10.1002/cnm.2610.

[51] D. Rohmer, A. Sitek, G.T. Gullberg, Reconstruction and visualization of fiber and laminar structure in the normal human heart from ex vivo diffusion tensor magnetic resonance imaging (DTMRI) data, Investig. Radiol. 42 (11) (2007) 777–789, https://doi.org/10.1097/RLI.0b013e3181238330.

[52] J. Wong, E. Kuhl, Generating fibre orientation maps in human heart models using Poisson interpolation, Comput. Methods Biomech. Biomed. Eng. 17 (11) (2014) 1217–1226, https://doi.org/10.1080/10255842.2012.739167.

[53] S. Skatulla, C. Sansour, On a path-following method for non-linear solid mechanics with applications to structural and cardiac mechanics subject to arbitrary loading scenarios, Int. J. Solids Struct. 96 (2016) 181–191.

[54] J.M. Guccione, K.D. Costa, A.D. McCulloch, Finite element stress analysis of left ventricular mechanics in the beating dog heart, J. Biomech. 28 (10) (1995) 1167–1177.

[55] J.C. Walker, M.B. Ratcliffe, P. Zhang, A.W. Wallace, E.W. Hsu, D.A. Saloner, J.M. Guccione, Magnetic resonance imaging-based finite element stress analysis after linear repair of left ventricular aneurysm, J. Thorac. Cardiovasc. Surg. 135 (5) (2008) 1094–1102.

[56] J.F. Wenk, P. Eslami, Z. Zhang, C. Xu, E. Kuhl, J.H. Gorman III., J.D. Robb, M.B. Ratcliffe, R.C. Gorman, J.M. Guccione, A novel method for quantifying the in-vivo mechanical effect of material injected into a myocardial infarction, Ann. Thorac. Surg. 92 (3) (2011) 935–941.

[57] L.C. Lee, L. Ge, Z. Zhang, M. Pease, S.D. Nikolic, R. Mishra, M.B. Ratcliffe, J.M. Guccione, Patient-specific finite element modeling of the cardiokinetix parachute device: effects on left ventricular wall stress and function, Med. Biol. Eng. Comput. 52 (6) (2014) 557–566.

[58] M. Genet, L.C. Lee, R. Nguyen, H. Haraldsson, G. Acevedo-Bolton, Z. Zhang, L. Ge, K. Ordovas, S. Kozerke, J.M. Guccione, Distribution of normal human left ventricular myofiber stress at end diastole and end systole: a target for in silico design of heart failure treatments, J. Appl. Physiol. 117 (2) (2014) 142–152.

[59] V. Creigen, S.V. Mourik, V. Rottsch, M. Vellekoop, Modeling a heart pump, in: Proceedings of the 58th European Study Group Mathematics With Industry, 2007, pp. 7–26.

[60] P. Molino, C. Cerutti, C. Julien, G. Cuisinaud, M.-P. Gustin, C. Paultre, Beat-to-beat estimation of Windkessel model parameters in conscious rats, Am. J. Physiol. Heart Circ. Physiol. 274 (1) (1998) H171–H177.

[61] J.T. Ottesen, M. Danielsen, Modeling ventricular contraction with heart rate changes, J. Theor. Biol. 222 (3) (2003) 337–346.

[62] P.H.M. Bovendeerd, T. Arts, J.M. Huyghe, D.H. Van Campen, R.S. Reneman, Dependence of local left ventricular wall mechanics on myocardial fiber orientation: a model study, J. Biomech. 25 (10) (1992) 1129–1140.

[63] J. Sainte-Marie, D. Chapelle, R. Cimrman, M. Sorine, Modeling and estimation of the cardiac electromechanical activity, Comput. Struct. 84 (2006) 1743–1759.

[64] R.C.P. Kerckoffs, P.H.M. Bovendeerd, J.C.S. Kotte, F.W. Prinzen, K. Smits, T. Arts, Homogeneity of cardiac contraction despite physiological asynchrony of depolarization: a model study, Ann. Biomed. Eng. 31 (5) (2003) 536–547.

[65] M. Sermesant, P. Moireau, O. Camara, J. Sainte-Marie, R. Andriantsimiavona, R. Cimrman, D.L.G. Hill, D. Chapelle, R. Razavi, Cardiac function estimation from MRI using a heart model and data assimilation: advances and difficulties, Med. Image Anal. 10 (4) (2006) 642–656, https://doi.org/10.1016/j.media.2006.04.002.

[66] R.C.P. Kerckhoffs, M.L. Neal, Q. Gu, J.B. Bassingthwaighte, J.H. Omens, A.D. McCulloch, Coupling of a 3D finite element model of cardiac ventricular mechanics to lumped systems models of the systemic and pulmonic circulation, Ann. Biomed. Eng. 35 (1) (2007) 1–18.

[67] S. Parragh, B. Hametner, S. Wassertheurer, Influence of an asymptotic pressure level on the Windkessel models of the arterial system, IFAC-PapersOnLine 48 (1) (2015) 17–22.

[68] N. Westerhof, J.-W. Lankhaar, B.E. Westerhof, The arterial Windkessel, Med. Biol. Eng. Comput. 47 (2) (2009) 131–141.

[69] A.M. Katz, Physiology of the Heart, fifth ed., Lippincott Williams & Wilkins, Philadelphia, PA, 2010.

[70] M. Loève, Function aleatoire de second ordre, Comptes Rendus Acad. Sci., Paris, 1945, p. 220.

[71] K. Karhunen, Zur Spektraltheorie Stochastischer Prozesse, Ann. Acad. Sci. Fennicae A1 (1946) 34.

[72] C.G. Wu, Y.C. Liang, W.Z. Lin, H.P. Lee, S.P. Lim, A note on equivalence of proper orthogonal decomposition methods, J. Sound Vibr. 265 (2003) 1103–1110, https://doi.org/10.1016/S0022-460X(03)00032-4.

[73] B.-T. Tan, K. Willcox, M. Damodaran, Applications of properorthogonal decomposition for inviscid transonic aerodynamics, Singapore-MIT Alliance, Tech. Rep. (2003). Available from, http://dspace.mit.edu/bitstream/handle/1721.1/3694/HPCES002.pdf?sequence=2.

[74] P. Druault, P. Guibert, F. Alizon, Use of proper orthogonal decomposition for time interpolation from PIV data: application to the cycle-to-cycle variation analysis of in-cylinder engine flows, Exp. Fluids 39 (2005) 1009–1023, https://doi.org/10.1007/s00348-005-0035-3.

[75] L. Sirovich, Turbulence and the dynamics of coherent structures, Appl. Math. XLV (3) (1987) 561–582.

[76] H. Li, Z. Luo, J. Chen, Numerical simulation based on POD for two-dimensional solute transport problems, Appl. Math. Model. 35 (5) (2011) 2489–2498, https://doi.org/10.1016/j.apm.2010.11.064.

[77] G. Kerschen, J.-C. Golinval, A.F. Vakakis, L.A. Bergman, The method of proper orthogonal decomposition for dynamical characterisation and order reduction of mechanical systems: an overview, Nonlinear Dyn. 41 (1–3) (2005) 147–169, https://doi.org/10.1007/s11071-005-2803-2.

[78] J. Barbic, D. James, Real-time subspace integration for St. Venant-Kirchhoff deformable models, ACM Trans. Graph. 24 (3) (2005) 982–990, https://doi.org/10.1145/1073204.1073300.

[79] N.J. Falkiewicz, C.E.S. Cesnik, Proper orthogonal decomposition for reduced-order thermal solution in hypersonic aerothermielastic simulation, AIAA J. 49 (5) (2011) 994–1009, https://doi.org/10.2514/1.J050701.

[80] W.Z. Lin, Y.J. Zhang, E.P. Li, Proper orthogonal decomposition in the generation of reduced order models for interconnects, EEE Trans. Adv. Packag. 31 (3) (2008) 627, https://doi.org/10.1109/TADVP.2008.927820.

[81] J. Dolbow, T. Belytschko, An introduction to programming the meshless element F reeGalerkin method, Arch. Comput. Methods Eng. 5 (3) (1998) 207–241, https://doi.org/10.1007/BF02897874.

[82] T.P. Usyk, R. Mazhari, A.D. McCulloch, Effect of laminar orthotropic myofiber architecture on regional stress and strain in the canine left ventricle, J. Elast. 61 (2000) 143–164, https://doi.org/10.1023/A:1010883920374.

[83] K. Hormann, A. Agathos, The point in polygon problem for arbitrary polygons, Comput. Geom. Theory Appl. 20 (3) (2001) 131–144, https://doi.org/10.1016/S0925-7721(01)00012-8.

[84] M. Chen, T. Townsend, Efficient and consistent algorithms for determining the containment of points in polygons and polyhedra, in: G. Marechal (Ed.), Eurographics 87, Elsevier Science Publishers B.V. (North-Holland), Amsterdam, 1987, pp. 423–437. 10.1.1.173.1916.

[85] D.G. Alciatore, R. Miranda, A winding number and point-in-polygon algorithm, Technical report, Colorado State University, 1995.

[86] P.S. Heckbert, Graphics Gems IV, Academic Press, Boston MA, 1994.

[87] N. Toussaint, C.T. Stoeck, T. Schaeffter, S. Kozerke, M. Sermesant, P.G. Batchelor, In vivo human cardiac fibre architecture estimation using shape-based diffusion tensor processing, Med. Image Anal. 17 (8) (2013) 1243–1255, https://doi.org/10.1016/j.media.2013.02.008.

[88] P. Lamata, M. Sinclair, E. Kerfoot, A. Lee, A. Crozier, B. Blazevic, S. Land, A.J. Lewandowski, D. Barber, S. Niederer, N. Smith, An automatic service for the personalization of ventricular cardiac meshes, J. R. Soc. Interface 11 (91) (2014) 20131023, https://doi.org/10.1098/rsif.2013.1023.

[89] A.L. Yuille, N.M. Grzywacz, A mathematical analysis of the motion coherence theory, Int. J. Comput. Vis. 3 (2) (1989) 155–175, https://doi.org/10.1007/BF00126430.

[90] SESKA, Computational Continuum Mechanics Research Group, University of Cape Town, South Africa, (2017). http://www.ccm.uct.ac.za/.

[91] L.A. Simmons, A.G. Gillin, R.W. Jeremy, Structural and functional changes in left ventricle during normotensive and preeclamptic pregnancy, Am. J. Physiol. 283 (4) (2002) H1627–H1633, https://doi.org/10.1152/ajpheart.00966.2001.

[92] S.F. Yiu, M. Enriquez-Sarano, C. Tribouilloy, J.B. Seward, A.J. Tajik, Determinants of the degree of functional mitral regurgitation in patients with systolic left ventricular dysfunction: a quantitative clinical study, Circulation 102 (12) (2000) 1400–1406, https://doi.org/10.1161/01.CIR.102.12.1400.

[93] M. Oikawa, Y. Kagaya, H. Otani, M. Sakuma, J. Demachi, J. Suzuki, T. Takahashi, J. Nawata, T. Ido, J. Watanabe, K. Shirato, Increased [18F] fluorodeoxyglucose accumulation in right ventricular free wall in patients with pulmonary hypertension and the effect of epoprostenol, J. Am. Coll. Cardiol. 45 (11) (2005) 1849–1855, https://doi.org/10.1016/j.jacc.2005.02.065.

[94] W.A. Goetz, E. Lansac, H.-S. Lim, P.A. Weber, C.M.G. Duran, Left ventricular endocardial longitudinal and transverse changes during isovolumic contraction and relaxation: a challenge, Am. J. Physiol. Heart Circ. Physiol. 289 (1) (2005) H196–H201, https://doi.org/10.1152/ajpheart.00867.2004.

[95] C.R. Greyson, Pathophysiology of right ventricular failure, Crit. Care Med. 36 (1 suppl) (2008) S57–S65, https://doi.org/10.1097/01.CCM.0000296265.52518.70.

[96] V.B. Ho, G.P. Reddy, Cardiovascular Imaging, first ed., Elsevier-Health Sciences Division, Philadelphia, PA, 2010.

[97] CloudCompare, CloudCompare (version 2.6). (2016). http://www.cloudcompare.org/.

[98] W. Schroeder, K. Martin, B. Lorensen, The Visualization Toolkit, ISBN: 978-1-930934-19-1, 2006.

CHAPTER

# 9

# Computational Musculoskeletal Biomechanics of the Knee Joint

*Hafedh Marouane, Aboulfazl Shirazi-Adl, Masoud Sharifi*

Division of Applied Mechanics, Department of Mechanical Engineering, Polytechnique, Montréal, QC, Canada

## 9.1 INTRODUCTION

Due to large relative movements and a distal location in the body, human knee joints experience loads and movements of substantial magnitude during various occupational, recreational, and regular daily living activities [1]. This demanding mechanical environment exposes knee joints to a host of painful deformities, injuries, and degenerations involving patellofemoral (PF) and/or tibiofemoral (TF) structures. With osteoarthritis (OA) as a painful and debilitating disease (present in ~10% of the general population and >70% of those over age 65) that affects the knee more than any other weight-bearing joint in the human body, total knee replacements approaching 1 million per year in the United States alone, anterior cruciate ligament (ACL) damage as the most common sports injury with ~100k new incidents per year in the United States and ~50% reconstruction rate, >500k per year arthroscopic partial meniscectomy in the United States, and finally also millions of corrective/preventive osteotomy surgeries and biological repairs (tissue engineering), the knee joint is in the spotlight in immediate need for more effective preventive and treatment programs [2–7]. The situation is alarming due both to the dramatic increase in these interventions, especially in younger and more active age groups that expect to remain active even after surgery, and to the ever-growing portion of the population with obesity and aging that are common OA risk factors.

An improved in-depth understanding of the biomechanics of the knee joint is therefore necessary for more efficient design and management of preventive and treatment programs of these injuries. Due to inherent challenges in experimental studies (in vivo and ex vivo) and the associated limitations, invasiveness, and burdens in time, effort, and cost, computational approaches have long been recognized as reliable, important, and complementary methods in various areas of biomechanics and biomedical engineering. The primary advantage of these numerical tools lies in robust control over boundary conditions, loading, geometry, and material properties allowing for the sensitivity and statistical analyses in output measures as input parameters vary. Moreover, temporal and spatial variations in internal forces, contact stresses/areas/centers, and tissue stresses/strains are invaluable outputs that are difficult, if not impossible, to quantify in experimental investigations. In response, several finite element (FE) models with different degrees of precision and refinement have been developed [8–14]. Likewise, various musculoskeletal (MS) models of the lower extremity have been constructed with the objective to further existing understanding of the knee joint functional biomechanics in normal and disturbed conditions under more physiological load and movement conditions [15–24]. While former FE models provide valuable detailed information (i.e., tissue stresses and strains) in joint constituent materials, latter MS models offer crucial results on activation patterns in musculature and resulting global joint loads in complex physiological activities such as gait [25–27].

Following a simplified two-dimensional planar model of the knee by Yamaguchi et al. [28] that accounted for the kinematics of both TF and PF joints in the sagittal plane, numerous models at different complexities have been introduced toward more realistic and accurate simulations and results. Such developments have often involved complex three-dimensional finite element models. In MS simulations, however, no such detailed knee joints are usually used. Instead, the most common knee joint description is an idealized planar one that constrains the knee motion and kinetics



Advances in Biomechanics and Tissue Regeneration
https://doi.org/10.1016/B978-0-12-816390-0.00009-1

into a single sagittal plane. In fact, joints are commonly simplified with constrained kinematics. In most MS models the human knee joint is either described as an idealized frictionless spherical joint with rotations only [29] or a simple planar joint [16]. OpenSim, developed in the 1990s, is a popular interactive software for the development and analysis of human MS systems [16]. This publically available software is regularly used by many researchers [30, 31]. Another more recently developed commercially available MS package is the AnyBody Modeling System (AnyBody Technology A/S, Aalborg, Denmark) that is also used for the estimation of muscle and joint contact forces [32–34].

Our recent lower extremity hybrid kinematics-driven MS model investigations, by explicit incorporation of a validated detailed FE model of the entire knee joint, offer improved estimations both at the global (muscle and joint forces) and local knee joint (tissue stresses and strains, contact pressure, and center) levels [15, 19, 20, 22, 23, 35–37] by full consideration of the active–passive synergy (though currently only at the knee joint). This chapter starts with a short description of the joint passive tissues and our knee joint FE and lower extremity MS models followed by some sample results and validation of model predictions.

## 9.2 METHODS

### 9.2.1 Passive Tissues

#### *9.2.1.1 Cartilage*

As a fluid-saturated nonhomogeneous composite tissue, articular cartilage provides smooth articulation at joint surfaces, absorbs impacts, and redistributes applied stresses to the underlying bone. Articular cartilage is made mainly of collagen (type II) (50%–73% dry weight), proteoglycans (15%–30% dry weight), and water (58%–78% weight). The composition and structure of articular cartilage change with depth, from the joint surface to the anchorage at bone [38–45]. At the superficial zone, collagen fibrils are horizontally oriented parallel to the articular surface, whereas they become rather random in the transitional zone and finally turn perpendicular to the subchondral bone–cartilage interface in the deep zone [40, 46, 47]. Collagen content is highest in the superficial and deep zones of articular cartilage and lowest in the middle zone [41]. In tandem with proteoglycan content, the equilibrium modulus of the nonfibrillar solid matrix increases downward along with the depth from the articular surface to the subchondral junction [41].

This layered structure along with the nonlinearity and tension-compression differences in the collagen fibril properties results in a highly nonlinear, nonhomogeneous composite fibrous tissue [48–52]. The tissue is saturated with water, the flow of which (mobile portion) gives rise to the time-dependent viscoelastic behavior leading to common stress relaxation (drop in stress under constant strain) and creep (increase in deformation under constant loads) effects. The water content, as in other similar fibrous tissues such as intervertebral discs and menisci, varies with time depending on the tissue composition, external load history, and osmolality of surrounding media and as such plays a crucial role in the load bearing especially in the transient (short-term) periods where fluid pressurization plays a crucial mechanical role in supporting applied loads in nondegenerate conditions [53].

#### *9.2.1.2 Ligaments*

Similar to other biological soft tissues, the ligaments of the knee are made of a ground substance reinforced with collagen fibers (mainly type I) [54]. Four major ligaments (ACL, anterior cruciate ligament; PCL, posterior cruciate ligament; LCL, lateral collateral ligament; and MCL, medial collateral ligament), among others, stiffen and control the relative movements of TF joint. ACL and PCL can each be separated into two distinct bundles: anteromedial (am) and posterolateral (pl) in ACL and anterolateral (al) and posteromedial (pm) in PCL. These bundles experience different patterns of length changes during active/passive knee flexion [55, 56]. Cadaver studies have confirmed the primary roles of ACL-am in high flexion and ACL-pl at near extension angles during passive knee flexion or under anterior tibial forces [57–59]. Depending on loading conditions, one or more of these ligaments act as primary restraints in resisting applied force movements and enhancing joint stability. Though modulated by joint flexion angle, MCL and LCL are the primary constraints in adduction-abduction rotations, whereas AP translations and internal-external rotations are resisted primarily by cruciate ligaments [59, 60].

#### *9.2.1.3 Meniscus*

Meniscus is a semilunar shaped fibrocartilaginous tissue with extremities inserted into the intercondylar eminence at the proximal tibial plateau. It is composed mainly of a dense network of collagen fibrils (mainly type I), proteoglycans, and water. Similar to cartilage, its fluid content gives rise to tissue time-dependent response. While the collagen fibrils at the top, bottom, and peripheral surfaces show no major preferred orientations, they are nevertheless circumferentially oriented in the bulk of the tissue in between these surfaces [61]. In addition, radial tie fibers are present that increase the tensile resistance of the meniscus [62]. Similar to the articular cartilage, the structural inhomogeneity and anisotropy of menisci dominate its tensile behavior [63].

Primary meniscal functions include the uniform transfer of the joint load across the cartilage [64], shock absorption [63, 65], augmentation of the stiffness—stability—of the joint [66, 67], assistance in joint lubrication and articulation, prevention of hyperextension, and protection of the joint [68, 69]. Pressure measurements have shown that 45%–70% of the applied compression force is transmitted through the menisci [63, 68, 70, 71]. In joints after total meniscectomy, contact stresses could double with a 50%–70% reduction in contact areas [72]. A 10% reduction in meniscal contact area secondary to the partial meniscectomy produces ~65% increase in peak contact stresses [72], leading likely to early development of OA [73–75].

The mechanical properties of menisci have been extensively studied under compressive, tensile, and shear load conditions [63, 76–78]. Tensile properties, as in articular cartilage, vary with tissue depth and from a direction to another, depending on the spatial organization of collagen fibrils within the tissue [77].

### 9.2.2 Knee Joint Passive Finite Element (FE) Model

Numerous FE models with different degrees of complexity and accuracy have been developed to study the knee joint biomechanics under various loads and movements [79]. The first comprehensive FE model of the TF joint is that of Bendjaballah et al. [11] with the model reconstructed from CT images and direct measurements of a cadaver knee. Menisci were simulated as a nonlinear nonhomogeneous composite of an isotropic bulk reinforced by collagen fibrils with strain-dependent nonlinear material properties [80], ligaments as nonlinear elements with initial strains in different bundles, and articular cartilage layers as a simple isotropic homogeneous elastic material. Each bony structure was taken as a rigid body represented by a primary reference node. The material properties were derived from the data available in the literature [11, 81, 82]. Each meniscus matrix was stiffened by a higher modulus of 15 MPa at both ends (~5 mm length), which was inserted into the tibial eminence to simulate its horns [83]. Articulations at the cartilage–cartilage (i.e., uncovered areas) and cartilage–meniscus (i.e., covered areas) were simulated as large displacement frictionless contact [83, 84].

In later refinements and developments of this model, the articular cartilage layers at both TF and PF joints were also represented as a depth-dependent nonhomogeneous nonlinear composite of incompressible bulk matrices reinforced by collagen fibril networks [85–87]. In superficial zones of all cartilage layers and bounding surfaces of menisci, membrane elements were used to represent homogeneous in-plane distribution of fibrils with random orientations [78]. Despite such isotropic distribution, however, a direction-dependent response prevails due to the strain dependency in fibril material properties and anisotropy in strain field. The collagen fibril properties (types I and II) were taken nonlinear based on earlier studies [88]. In the transitional zone with random fibrils (i.e., no dominant orientations), continuum brick elements that take the principal strain directions as the material principal axes represent collagen fibrils. In the deep zone, however, vertical fibrils were modeled with vertical membrane elements similar to horizontal superficial ones except in offering resistance only in local fibril directions [86]. In the bulk region of each meniscus in between peripheral surfaces, collagen fibrils that are dominant in the circumferential direction were represented by membrane elements with local material principal axes defined initially in orthogonal circumferential and radial directions.

Thickness of membrane elements in different regions of cartilage and menisci was computed based on fibril volume fraction in each zone. In the cartilage, the equivalent collagen fibril content in the superficial zone was estimated based on reported tissue properties in tension [89–91] and type II collagen stress-strain curve [85]. A total volume fraction of 15% was estimated in the superficial zone in agreement with the mean value of 14% reported for its wet weight.

In accordance with earlier investigations [92, 93], the transient response of water-saturated articular cartilage and meniscus under higher strain rates could be computed either by a biphasic approach or equivalently by an incompressible elastic analysis using bulk equilibrium moduli. Examining this equivalency at various Poisson's ratios using our earlier nonhomogeneous axisymmetric model of cartilage [86], indentation results at 20% strain applied in 0.5 s demonstrated a significant sensitivity in transient reaction force [86, 87]. Nearly incompressible Poisson's ratios in the range of 0.4999–0.5 could yield results identical to those computed with biphasic simulations. Since loading cycles of daily activities like walking and running last for only a fraction of second, an incompressible elastic model can hence alternatively be employed with no loss of accuracy to compute the transient response. In this case, the hydrostatic pressure in the elastic model represents the transient pore pressure in its equivalent poroelastic model. An equivalent compressible elastic material can also be employed in which case greater equilibrium moduli should be used depending on the Poisson's ratio considered.

Nonlinear spring elements were employed to model various ligaments of the TF and PF joints. MCL wrapped around the proximal medial bony edge of the tibia with peripheral attachments to the medial meniscus [11, 81]. Each ligament was simulated by a number of uniaxial elements with different cross-sectional areas (based on the literature) and initial strains (based on the literature and comparison of results with available measurements) [82, 94–96]. Knee FE model, including both TF and PF joints and associated soft tissues (but not bony structures), is shown in Fig. 9.1.

### 9.2.3 Lower Extremity Musculoskeletal (MS) Model

Resolution of kinetic redundancy toward the estimation of unknown muscle forces in MS models of the human body during various activities remains a formidable challenge. Inverse dynamics is the common method of choice when compared with the forward dynamic simulations where activity in muscles are prescribed, say based on measured activation via limited surface electromyography (EMG), and then continuously updated with constraints on some kinematics trajectories, measured contact forces, joint moments, and/or objective functions [30]. In the former, joint moments are initially evaluated (inverse dynamics) by equations of motion using measured joint kinematics, external loads, and body anthropometric characteristics. The redundant muscle forces are subsequently estimated, either using an optimization [21, 25, 97–101] or an EMG-driven [17, 18, 102–105] method. There are also hybrid EMG-assisted optimization (EMGAO) versions of these two approaches [106, 107]. In the optimization-based methods, muscle forces are estimated by minimizing a single or multiple objective functions, such as the sum of muscle forces, system margin of stability, or muscle activations to different powers [15, 20, 21, 25, 108–110].

The predicted muscle forces are generally validated qualitatively by comparison of estimated muscle activation levels with normalized recorded EMG under the same activity [20, 110] or the predicted contact forces with data from patients with instrumented knee implants [97, 101]. Predictions have been found sensitive to many factors, such as recorded muscle activation patterns [97], muscle weighting [99, 101], and musculotendon properties and lever arms. In a study on the sensitivity of muscle force estimations to changes in musculotendon properties, Redl et al. [111] found that changes in the muscle fiber length and tendon rest length of vasti were most critical to model force estimates in normal gait. EMG-driven models often use Hill-type muscle models (accounting for activation, fiber length/velocity, and pennation angle) when estimating muscle forces [17, 31]. Muscle gains are evaluated in a manner to match joint moments evaluated based on inverse dynamics [17, 18, 27, 102] or to minimize the error between predicted and measured (via instrumented implants) joint contact forces [103]. While EMG-driven approaches are biological in using recorded individual's muscle activity with inter- and intrasubject volitional variations, they remain susceptible to major assumptions and shortcomings associated with the limited available surface EMG, location on large and deep

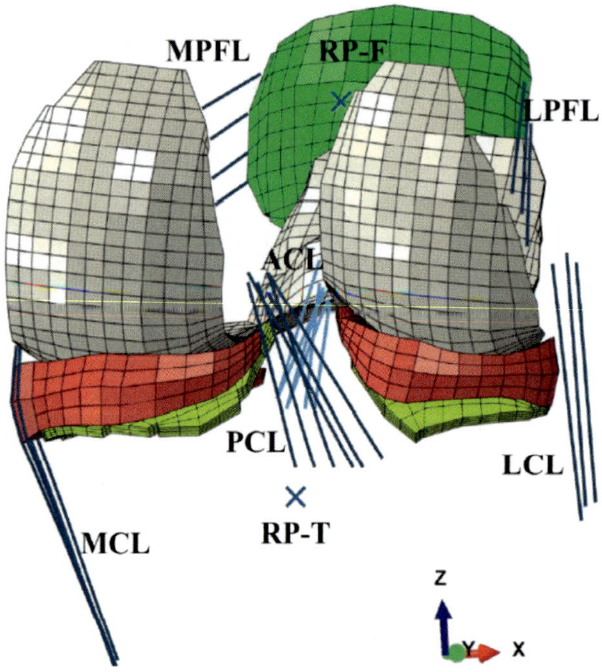


FIG. 9.1 Posterior view of the knee joint FE model showing articular cartilage layers, menisci, and major ligaments. Rigid bony structures (i.e., femur, tibia, and patella) are represented by their reference primary nodes and not shown.

muscles, cross-talk considerations, complexity in force-EMG relation, signal processing and normalization, and introduction of gain factors [103, 112].

Using a hybrid MS model of the lower extremity with our detailed validated FE model of the knee joint, we resolve the redundancy and estimate muscle forces using a kinematics-driven optimization approach [15, 20, 22, 35, 36]. In this way, not only the passive properties of the knee joint are accurately represented, but also both joint moments (from inverse dynamics and gait data) and joint kinematics (from gait data) are used to drive the model resulting in a synergistic passive-active simulation. By considering the measured kinematics, this kinematics-driven MS model can also fall into the category of a biological approach as joint kinematics and human posture are controlled in static and transient movements by the central nervous system.

## 9.3 EQUILIBRIUM APPLICATIONS: BOUNDARY CONDITIONS AND LOADING

In knee joint biomechanics, under both active and passive conditions, proper consideration of joint loading and boundary conditions are of prime importance as they both substantially affect results and subsequent comparisons and validations. Here as follows are some sample applications of our FE model dealing with boundary conditions and loading.

Careful selection of stable and fully unconstrained boundary conditions in experimental and model studies of any complex articulation such as the knee joint is crucial. On the one hand, the constraints should be sufficient to avoid instability (hypermobility), while on the other hand, they should not be excessive to artificially and inadvertently overconstraint and stiffen the physiological response. In addition, if the intention is to represent and compare with an existing study, being in vivo, in vitro, or in silico, the model should replicate as closely as possible the corresponding boundary conditions. For example, to investigate the passive TF response in full extension under 1000 N, Bendjaballah et al. [11] applied the force on the primary (reference) node of the femur with the tibia completely fixed. For a stable and unconstrained response, the femoral flexion-extension (F-E) and adduction-abduction (add-abd) rotations were also fixed while leaving internal-external (I-E) rotations and all three translations at the femur free. Additional constraint on I-E rotations was subsequently found to have a significant effect on knee biomechanics especially in the meniscectomized case; the knee became stiffer in the axial direction, the ratio of medial/lateral contact load increased, and the coupled displacements decreased as coupled I-E rotations were fixed [11]. In a similar study, but with the refined and improved version of the model under up to 2000 N compression at full extension, Shirazi and Shirazi-Adl [87] used equivalent, but reversed, boundary conditions with the femur fixed and tibia left free except in add-abd and F-E rotations. They also found a much stiffer response as tibial coupled I-E rotations were also constrained. Foregoing boundary conditions in the knee joint under axial compression force assure unconstrained motions with no interference with the natural biomechanical roles of menisci, ligaments, and articular surfaces. Besides and equally important, artifact loads are eliminated, and the response is no more dependent on the location where the large axial compression force is applied on the femur or on the tibia.

Due to the joint instability and artifact moments caused by large compression forces, in vitro biomechanical investigations of the knee joint face the dilemma of how (joint constraints) and where (anterior-posterior, A-P and medial-lateral, M-L, locations) to apply compression forces of physiological magnitudes [71, 113]. To circumvent these difficulties under compression forces, in vitro studies impose additional constraints on rotations in sagittal and frontal planes [114, 115]. A novel approach was proposed in our recent study [37] where the compression load was applied at the joint mechanical balance point (MBP) identified as a point at which the applied compression does not cause any coupled rotations in sagittal and frontal planes (Fig. 9.2). Analyses were carried out at different TF flexion angles (0, 15, 30, and 45 degrees), while the tibia was fully free (even in the F-E and add-abd rotations unlike earlier studies presented in the foregoing paragraph) and the femur fully fixed at the desired flexion angle. For a robust approach, to determine the unique location of the joint MBP under a specific axial compression force, tibial coupled rotations were initially constrained, and the associated required reaction moments were estimated. Subsequently, the location of the compression force was shifted in A-P and M-L directions (by the ratio of these required moments divided by the applied compression force) to eliminate these undesired (artifact) moments [37]. The computed MBP location varied with the joint flexion angle and compression force magnitude (Fig. 9.2).

To study the passive tibiofemoral joint at full extension under add-abd moments of up to 15 Nm [116], the femur was left free except in F-E rotations, whereas the tibia was completely fixed. The joint laxity in this loading case is found to be relatively unaffected by additional restraint on the femoral I-E rotations. The joint is found much stiffer under adduction moment. Using a refined version of the model, Marouane et al. [37] carried out similar analyses under up to 20 Nm moments at various flexion angles and compression preloads (up to 1800 N) applied at associated MBP location.

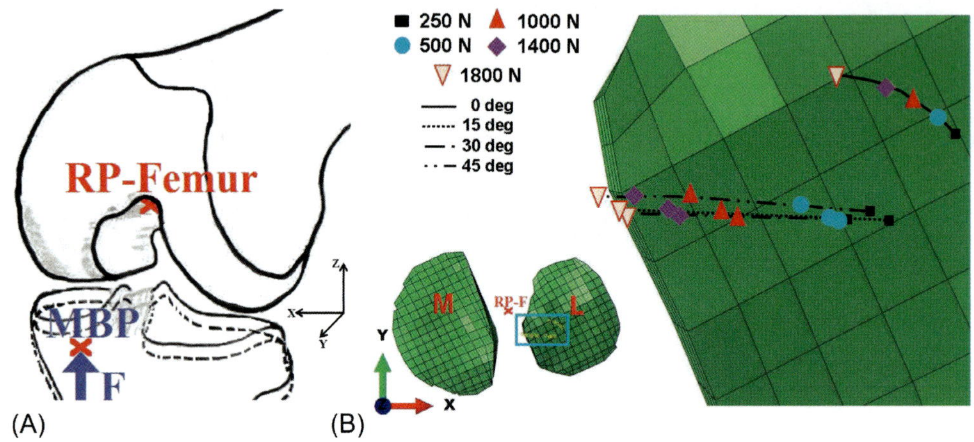


FIG. 9.2 (A) Mechanical balance point (MBP) approach; when applied at these locations, the compression preloads do not cause any coupled sagittal and frontal rotations in the fully unconstrained tibia. (B) Shifts in the location of the MBP projected on the tibial plateau as a function of the joint flexion angle and compression preload.

Compression preload substantially increased the joint moment bearing capacities and instantaneous angular rigidities in both frontal (add-abd) and sagittal (F-E) planes. The add-abd laxities diminished with compression preloads despite concomitant substantial reductions in collateral ligament forces. It was concluded that the augmented passive moment resistance under larger compression preloads expected in daily activities such as gait should not be overlooked in the knee joint MS models. A noticeable increase in the knee passive moment resistance diminishes the portion of the net moment to be resisted by the musculature resulting in smaller muscle forces and hence lower joint loads.

The knee joint passive response was also studied at full extension under a femoral posterior drawer force of up to 200 N acting alone or combined with a 1500 N compression preload [117]. For an unconstrained response, while avoiding the adverse effect of load positioning on joint kinematics [118], the femur was left free in three translations while fixed in all three rotations. A posterior drawer force of 200 N was applied onto the femur with and without a 1500 N compression preload. For the tibia, on the other hand, I-E and add-abd rotations were left free while constraining F-E rotations and three translations. An equivalent reversed unconstrained set of boundary conditions was also examined, that is, free rotations on the femur while loading the tibia under free translations, a different boundary condition that yielded almost the same results.

To apply quadriceps and hamstring muscle forces while simulating an active joint loading condition, measurement studies restrain the tibial A-P translation at a point away from the joint to counterbalance the moment of muscle forces while preserving the joint flexion angle at a desired level [71, 119–123]. This constraint, however, apart from generating extensor/flexor moments, introduces artifact tibial A-P shear forces, the magnitude of which depends on the distal location of restraint (i.e., lever arm) and the joint moment (i.e., muscle forces) [14]. The likely effect of such constrain on the joint response in flexion-extension (0–90 degrees) under isolated and combined hamstrings (205.5 N) and quadriceps (411 N) activation was investigated [14]. The femur was completely fixed, while the tibia and patella were left free in all directions. The effect of tibial restraint at two locations (20 cm or 30 cm distal to the joint level) on results was studied and compared with the reference boundary condition of the tibia constrained by pure moments (Fig. 9.3). The tibial restraint by a force markedly influenced, depending on the joint moment and restraining lever arm, the tibial A-P translation, TF contact forces, and forces in cruciate ligaments (Fig. 9.3). The restraining forces, especially when placed closer to the joint (i.e., smaller lever arm), reduced forces in cruciate ligaments at critical joint angles: in ACL at near full extension and in PCL at larger flexion angles. Ahmed et al. [71] reported the effect of this restraint lever arm on results and estimated that any moment arm >40 cm had no discernible effect on the pressure distribution on the patellar cartilage. Others reported that ACL force and tibial anterior translation significantly increased as the resisting force shifted distally from the joint (i.e., larger lever arm) [124, 125].

The importance of proper loading was also demonstrated when simulating the closed kinetic chain squat exercises at different flexion angles [126]. The effect of two loading configurations generating identical joint moments was considered: a realistic vertical reaction force at foot and an idealized pure sagittal moment similar to that used by Cohen et al. [127]. In corroboration of earlier studies [14] described earlier, the pure moment loading markedly influenced TF and PF contact forces/areas and forces in cruciate ligaments.

Knee morphological aspects have extensively been investigated in search for factors that could play a role in the risk of ACL rupture in both sexes and in higher prevalence of noncontact ACL ruptures in female athletes. Many image-based studies of patients with noncontact ACL rupture versus the control subjects have identified the higher posterior tibial slope (PTS) as a risk factor [128, 129]. To simulate the effect of changes in PTS on tibial translation and ACL force

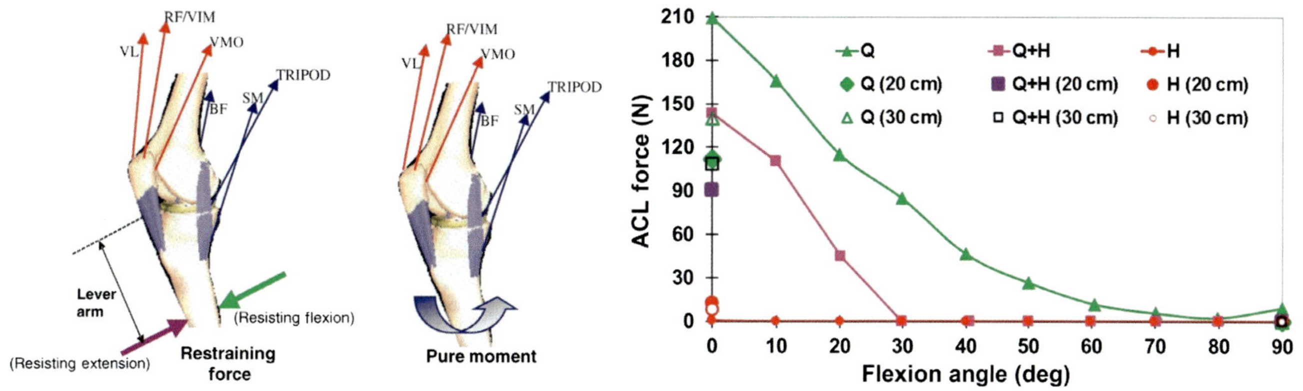


FIG. 9.3 Schematic representations of muscle forces and loads in two distinct loading conditions *(left)* and computed forces in ACL for cases with quadriceps activation alone (Q), hamstrings activation alone (H) and coactivation in both (Q+H) under pure moment (0–90 degrees) and restraining forces (only at 0- and 90-degree flexion angles *(right)* [14].

under compression [36] and during simulated gait [19], the initial medial and lateral PTSs were altered both by rigidly rotating tibial cartilage layers around local lateral-medial axes at the center of the respective tibial articulations. In this manner, minimal changes were made in tibial articular geometries. Ligament footprints were not altered; their lengths and orientations remained unchanged. These changes in PTS hence affected only the tibial slope with minimal effects on the geometry of the articular cartilage layers and overlying menisci. In accordance with image-based studies, results demonstrated that steeper PTS is a major risk factor in markedly increasing anterior tibial translation (ATT), ACL force, and hence its vulnerability to injury.

MS modeling of the lower extremity is promising to improve the current understanding of the knee joint function and injuries and associated prevention and treatment programs. Due to the complexity, numerous assumptions are often made when estimating muscle forces and joint contact loads. The knee is commonly idealized as a planar (2-D) joint with its motion constrained to remain in the sagittal plane [16, 30, 97, 130–133], neglecting thus both displacements and equilibrium equations in remaining planes. With muscle forces predicted, the static equilibrium in the frontal plane is consequently considered to estimate TF compartmental loads neglecting the knee joint passive resistance and assuming medial and lateral contact centers [27, 100, 103]. To evaluate the effects of such assumptions, our hybrid MS model of the lower extremity incorporating our detailed validated 3-D knee FE model was used to simulate the stance phase of gait [15, 20, 22]. To drive the musculoskeletal model, kinetics (hip/knee/ankle joint moments), as well as kinematics (hip/knee/ankle joint rotations) data, were taken from the mean of asymptomatic subjects collected in gait [134, 135]. Ground reaction force magnitudes were based on measurements of Hunt et al. [136]. The consistency of these two datasets at various periods of gait was assured by applying the latter forces on the foot at locations that generated the knee joint moments reported in the former studies. Substantial unbalanced knee joint moments reaching, at 25% stance, 30 Nm in abduction and 12 Nm in internal direction were found neglected in the 2-D model when estimating muscle forces. The model with an idealized planar 2-D knee joint substantially diminished muscle forces, ACL force, and TF contact forces/stresses when compared with the realistic 3-D model, see Fig. 9.4.

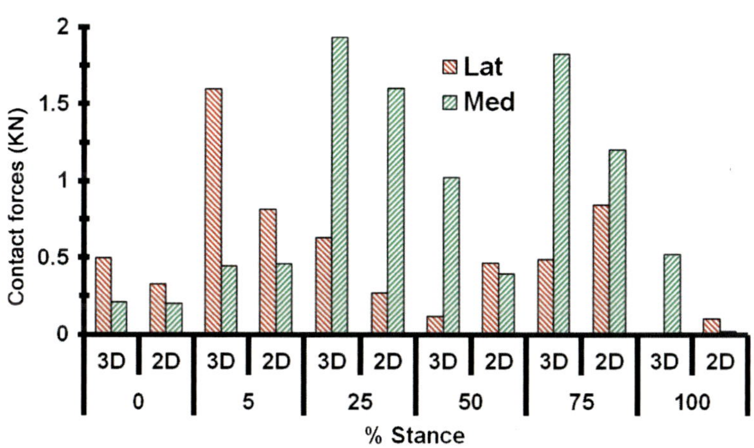


FIG. 9.4 Computed tibiofemoral compartmental contact forces on the medial and lateral plateaus in 3-D and 2-D models. In the 2-D model the knee joint out-of-sagittal plane rotations and moment equilibrium equations, which are both considered in 3-D model based on gait data, are totally neglected [23].

## 9.4 JOINT STABILITY ANALYSES

Stability of a mechanical or biological structure is defined as its ability to withstand small perturbations without hypermobility or excessive motions causing damage to the system. As a musculoskeletal system, knee joint stability is maintained by an intricate interplay between active musculature and passive tissues. Knee instability usually manifests itself by giving way, excessive laxity, and pain. So, knee joint stability assessment and evaluation of parameters affecting it are crucial in injury prevention and treatment managements. In clinical context, knee joint stability is usually evaluated by its laxity under external physiological loads and disturbances. There are a few tests to evaluate the knee stability for different injuries; for example, Lachman and pivot shift tests are performed to detect anterior cruciate ligament injuries [137, 138].

The first stability analysis of the human spine by the minimization of its potential energy was performed by Bergmark [139]. We have implemented the same approach to evaluate for the first time the stability of the knee joint in gait [140]. To do so, muscle forces are initially calculated in the equilibrium phase of the study as described in preceding paragraphs. At the final deformed configuration, all muscles are then replaced by uniaxial elements with a force-dependent axial stiffness [132]. Muscle stiffness $k=q\frac{F}{L}$ is taken linearly proportional to the computed total (passive and active) muscle force ($F$) and inversely proportional to its current length ($L$). $q$ is a constant dimensionless coefficient taken the same for all muscles [106, 139, 141]. After replacing muscles with springs, perturbation and buckling analyses, under a unit load at and along GRF and at the deformed loaded configuration, are performed. Stability analyses are iteratively performed for different values of $q$ to evaluate the minimum (critical) $q$ below which the system ceases to be stable [140]. A lower critical $q$ suggests a more stable system so that, at the extreme, the critical $q=0$ indicates that the passive system is stable under given loads and does not need any additional stiffness from musculature for stability although muscle forces are very likely needed to maintain equilibrium.

Short of a nonlinear postbuckling analysis, linear buckling and perturbation analyses are two common stability tools to evaluate the stability (divergence type) of a structure under given static-dynamic loading conditions. Buckling analysis at a given $q$ predicts the reserve additional load (buckling load) that the system can support before exhibiting instability; at the neutral stable condition, this buckling load approaches zero. The perturbation analysis, on the other hand, quantifies the system response under a small perturbation that tends to infinity as the system becomes unstable. The critical $q$ in the intact knee joint at heel strike was found at about 14. In this case, as $q$ approaches the critical $q$, displacement in the perturbation analysis substantially increases, whereas the lowest buckling (or reserve) load drops to nil (Fig. 9.5).

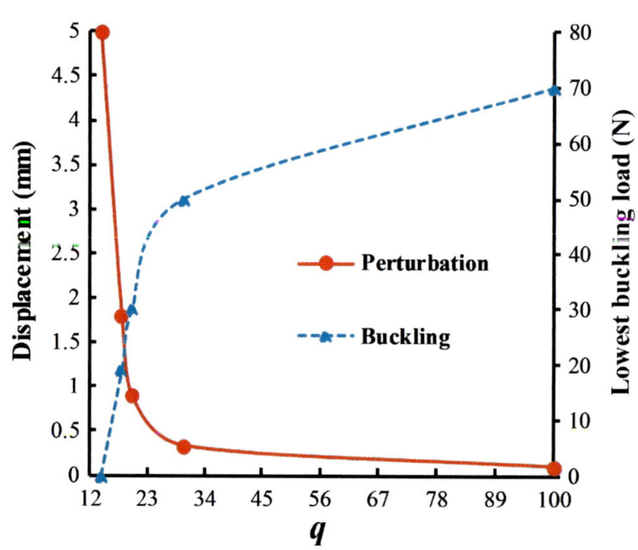


FIG. 9.5 Displacement at unit perturbation load along the ground reaction force *(on the left side)* and lowest buckling force (*Fcr*) *(on the right side)* at different values of $q$ (muscle stiffness coefficient) in linear perturbation and buckling analyses performed at the deformed loaded configuration of the intact knee joint at the initiation of contact (heel strike) in gait.

## 9.5 VALIDATION

Computational models, if accurate and properly validated, are powerful and reliable tools in advancing our understanding of joint function in normal and perturbed conditions and in prevention and treatment programs [142–144]. Our model of the knee joint has constantly and extensively been validated at different stages of its development over the last 20 years [11] by comparison of its predictions with available reported measurements under various loading and boundary conditions, both passive (e.g., axial compression load [11, 36, 37, 87], F-E [84], add-abd [37, 116], A-P drawer [81, 83, 117, 145] load-displacement conditions) and active (e.g., F-E [14, 82, 94], open-closed kinetic chain exercises [126, 146, 147] and during gait [15, 20, 22, 23, 35, 36]). Overall, satisfactory agreements have been found and are presented here for some conditions.

Results of the TF model on the contact forces/pressures/areas under axial compression forces up to 1800 N have been compared with available measurements (Table 9.1 [37]) demonstrating good agreements. Despite differences in femoral constraints and loadings, in vitro studies of the TF joint at full extension under axial compression [68, 70, 114, 148, 149] report similar nonlinear stiffening in the axial direction.

Under a small compression preload of 10 N, the TF FE model computed tangent add-abd angular stiffnesses of 5.38, 1.29, 1.13, and 0.83 Nm/deg in frontal plane at 0-, 15-, 30-, and 45-degree joint flexion angles, respectively [37]. The presence of compression preload (up to 1800 N) applied at MBP substantially increased foregoing TF stiffnesses at all flexion angles [37], see Fig. 9.6. With no joint compression, Markolf et al. [161] measured comparable in vitro stiffness values (mean $\pm$ standard deviation) of $11.0 \pm 7.5$, $1.6 \pm 1.2$, $1.1 \pm 0.8$ and $0.8 \pm 0.9$ Nm/deg at 0-, 10-, 20-, and 45-degree flexion angles, respectively. Moreover, Creaby et al. [162] measured, in vivo in seated asymptomatic participants with relaxed knee at 20-degree flexion, the midrange angular stiffness of $1.62 \pm 0.68$ Nm/deg, whereas Bendjaballah et al. [116] and Marouane et al. [37] computed passive stiffnesses in the frontal plane of ~4.5 Nm/deg at full extension and 1.46 Nm/deg at 15-degree flexion, respectively.

Computed cruciate coupling and screw-home mechanism in passive TF joint during F-E were found in agreement with the literature [84]. Screw-home motion of internal tibial rotation during joint flexion and external rotation during extension was computed. Prediction of ~16-degree internal tibial rotation at 90-degree femoral flexion fell in the range of reported values of 14–36 [163, 164], $6.5 \pm 4$ [161], 14.5 [165], $19.2 \pm 4$ [166], and $20 \pm 4$ degrees [167] at 90-degree flexion angle. In passive unconstrained F-E of TF joint, ACL force diminished from 65 N at 10-degree hyperextension to 31 N at 90-degree flexion. This change is due to a shift in load resistance from ACL-pl bundle at hyperextension to ACL-am bundle in flexion angles that corroborates with measurements [56]. The PCL resistance, on the other hand, initiated at flexion angles beyond 20–30 degrees and reached 35 N (by PCL-al bundle only) at 90-degree flexion. The foregoing predictions agree with others reporting increased force/strain in the PCL after 40–50 degrees of flexion [168] and increased strain in ACL anteromedial bundles with flexion after 40 degrees of flexion [169].

In the passive TF joint at full extension under single and combined femoral posterior drawer and axial compression forces, computed primary femoral laxity of 3.6 mm under 100-N posterior femoral force alone [117] compared well with the mean anterior tibial laxities of 2.8–6.7 mm reported under 100-N anterior tibial force at full extension [118, 170, 171]. The posterior translation of the femur relative to the tibia under pure compression, which is due to the posterior slope of the tibia [171] agreed with reported values [170, 171]. The relative magnitude and increases with compression preload of femoral translation in drawer were also in agreement with measurements [170–173]. The estimated force of 170 N in ACL matched measured values of 75–162 N reported under ~100 N drawer at full extension [174, 175]. Addition of axial compression preload further increased ACL force under drawer from ~1.6 times, in agreement with 150% in measurements [175], to more than twofold that of the applied drawer force. Prediction of the marked increase in ACL force in the presence of compression preload [36, 37, 87] corroborates well with earlier findings [171, 173, 176–179] but contradicts others suggesting that compression preload in drawer protects the ACL from damage [180].

Our estimated muscle forces during the stance phase of gait [15, 20] were also compared with recorded muscle activities (EMG). Predicted activation levels in quadriceps, hamstrings, and gastrocnemii muscles were found in very satisfactory agreement with values in the literature [27, 102, 181–183] and followed the same relative trends as in measured EMG activities [134, 135]. The computed hamstring forces peaked right after the HS at 5% period [15, 20]. Reported normalized EMG activities in the superficial lateral and medial hamstrings [134] are also the highest right at the HS and decrease thereafter. To qualitatively validate our predictions with normalized EMG measurements collected on the same asymptomatic subjects whose kinematics and kinetics were used to drive our MS model [134], the computed muscle forces were normalized to their maximum forces estimated at 0.6 MPa times muscle physiological cross-section (Fig. 9.7). Coefficients of determination, $R^2$, are computed higher than 0.7 in all muscles. At the

**TABLE 9.1** Comparison of Contact Area/Pressure at Different Compressions and Flexion Angles in the Passive TF Model

| Tibial contact area ($mm^2$) | | | 0 N | 400 N | 500 N | 800 N | 1000 N | 1200 N | 1400 N | 1500 N | 1800 N |
|---|---|---|---|---|---|---|---|---|---|---|---|
| Refined model [37] | **0°** | L/M/**T** | 168/132/**300** | – | 484/208/**692** | 544/237/**782** | 570/260/**830** | 580/269/**849** | 586/300/**836** | – | – |
| | **15°** | L/M | – | 344/199 | 380/232 | 423/301 | 460/342 | 486/374 | 505/404 | – | – |
| | **30°** | L/M | – | 363/170 | 382/196 | 440/264 | 466/335 | 475/366 | 491/382 | – | – |
| | **45°** | L/M | – | 376/151 | 396/188 | 426/265 | 446/281 | 458/324 | 472/343 | – | – |
| Less refined model [37] | **0°** | L/M/**T** | 169/100/**270** | – | 485/258/**743** | 526/327/**853** | 574/327/**902** | 585/351/**937** | 604/388/**992** | 604/378/**982** | 651/453/**1104** |
| Shirazi et al. [87] | **0°** | **T** | **355** | – | **893** | – | **1083** | – | – | **1214** | **1253** |
| Poh et al. [150] | **0°** | L/M/**T** | | | | | 403 (±120)/374 (±87)/**777 (±89)** @ 1800 N | | | | |
| Seitz et al. [151] | **0°** | L/M | – | – | 323 ± 160/313 ± 119 @ 500 N | | | 376 ± 161/392 ± 108 @ 1200 N | | | - |
| | **30°** | | | | 302 ± 128/348 ± 164 @ 500 N | | | 364 ± 148/410 ± 92 @ 1200 N | | | |
| Marzo and Gurske-DePerio [152] | **0°** | L/M | | | | | 571 ± 80/594 ± 59 @ 1800 N | | | | |
| Paci et al. [153] | **0°** | M | – | – | – | – | 327 | – | – | – | – |
| | **15°** | | | | | | 336 | | | | |
| | **30°** | | | | | | 300 | | | | |
| | **45°** | | | | | | 277 | | | | |
| Brown and Shaw [154] | **0°** | **T** | – | – | **1225 ± 180** | – | **1250 ± 100** | – | – | **1340 ± 100** | – |
| Ahmed and Burke [71] | **0°** | **T** | – | **1200** @ 445 N | – | **1650** @ 890 N | – | – | **1800** @ 1335 N | – | **2000** @ 1779 N |
| Huang et al. [155] | **0°** | L | – | 238 ± 135 | – | – | – | 300 ± 100 | – | – | – |
| | **15°** | | | 280 ± 120 | | | | 300 ± 95 | | | |
| | **30°** | | | 280 ± 130 | | | | 330 ± 130 | | | |
| | **45°** | | | 300 ± 120 | | | | 310 ± 120 | | | |
| Paletta et al. [156] | **0°** | L | – | – | – | – | – | – | – | – | 407 |
| | **15°** | | | | | | | | | | 363 |
| Lee et al. [115] | **0°** | M | – | – | – | – | – | – | – | – | 533 ± 48 |
| | **30°** | | | | | | | | | | 477 ± 84 |
| Fukubayashi and Kurosawa [157] | **0°** | L/M/**T** | – | | 420 ± 60/530 ± 150/**960 ± 170** @ 500 N | | 510 ± 70/640 ± 180/**1150 ± 200** @ 1000 N | | | – | – |
| Kurosawa et al. [114] | **0°** | **T** | – | – | **1130 ± 250** | – | **1300 ± 300** | – | – | **1410 ± 320** | – |
| Krause et al. [148] | **0°** | **T** | – | – | – | – | **2084** | – | – | – | – |
| Walker et al. [149, 158] | **0°** | **T** | **330** | – | – | – | – | – | – | **1514** | – |

| Mean contact pressure (MPa) | | | 200 N | 400 N | 500 N | 800 N | 1000 N | 1200 N | 1400 N | 1500 N | 1800 N |
|---|---|---|---|---|---|---|---|---|---|---|---|
| Refined model [37] | **0°** | L/M/**T** | – | – | 0.92/0.62/**0.78** | 1.29/0.81/**1.08** | 1.53/0.91/**1.26** | 1.78/1.05/**1.46** | 2.06/1.12/**1.64** | – | – |
| | **15°** | L | | 0.99 | 1.09 | – | – | 1.75 | 1.93 | | |
| | **30°** | L | | 0.99 | 1.14 | 1.46 | 1.65 | 1.86 | 2.05 | – | – |
| | **45°** | L | | 1.00 | 1.13 | 1.53 | 1.74 | 1.96 | 2.16 | – | – |
| | | | | | | | | | | – | – |
| Less refined model [37] | **0°** | L/M/**T** | – | – | 0.87/0.48/**0.7** | 1.26/0.58 **0.95** | 1.43/0.73/**1.11** | 1.67/0.82/**1.28** | 1.86/0.89/**1.41** | 1,98/0.99/**1.52** | 2,16/1.03/**1.61** |
| Poh et al. [150] | **0°** | L/M/**T** | | | | | 1.93 (±0.6)/3.61 (±0.72)/**2.73 (±0.49)** @ 1800 N | | | | |
| Shirazi et al. [87] | **0°** | **T** | – | **0.58** | **0.65** | – | **1.0** | – | – | **1.29** | – |
| Haut Donahue et al. [159, 160] | **0°** | L/M | – | – | – | 0.94/0.72 | – | 1,59/1.36 | – | – | – |
| Kurosawa et al. [114] | **0°** | **T** | – | – | **0.47 ± 0.12** | – | **0.8 ± 0.2** | – | – | **1.1 ± 0.28** | – |
| Krause et al. [148] | **0°** | **T** | – | – | – | – | **0.48 ± 0.08** | – | – | – | – |
| Huang et al. [155] | **0°** | L | – | 1.45 ± 0.95 | – | – | – | 2.6 ± 0.8 | – | – | – |
| | **15°** | | | 1.32 ± 0.76 | | | | 2.86 ± 1.32 | | | |
| | **30°** | | | 1 ± 0.5 | | | | 2.7 ± 1 | | | |
| | **45°** | | | 0.85 ± 0.42 | | | | 2.72 ± 1.2 | | | |

*L*, lateral; *M*, medial; *T*, total.

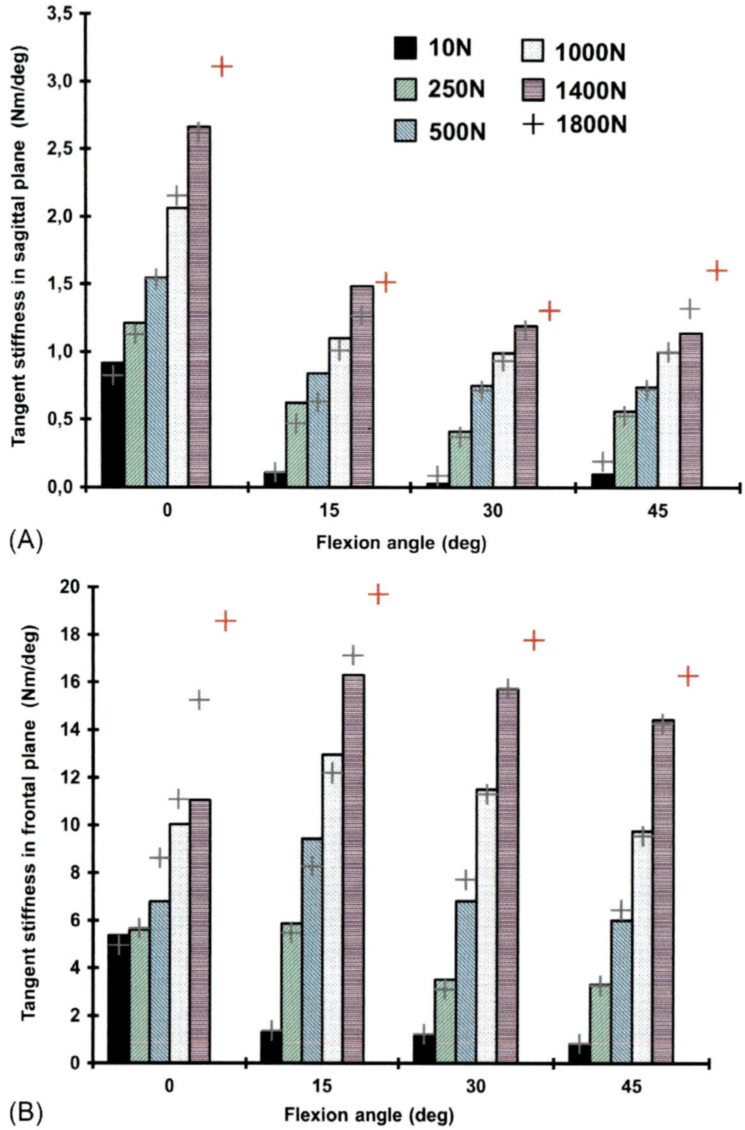


FIG. 9.6 Instantaneous (tangent) F-E (A) and add-abd (B) angular rigidities of the passive tibiofemoral joint as a function of the compression preload and joint flexion angle (+: results by a less refined version of the same model) [37].

initiation of contact (HS at 0% stance), estimated values are overall smaller than measurements that could partly be due to the absence of coactivity in the model employed here. Moreover, concerns on collected surface EMG measurements in larger and deeper muscles, cross-talk effects, processing and normalization of signals, and finally force-EMG relations should be taken into account whenever comparing force estimations with collected EMG data.

Apart from the early periods of stance (0% and 5%) and due mainly to add-abd rotations, the medial plateau carried a larger portion of contact forces than did the lateral plateau [15, 20, 22], which agrees with earlier estimates [104, 184]. In accordance with our predictions, others also reported small forces or none at all on the lateral compartment at the toe-off period at the end of stance [132, 185]. In vivo measurements of knee contact forces provide a valuable opportunity for the global validation of MS models by direct comparison between predicted/measured contact forces that could also indirectly validate estimated muscle forces and algorithm used. Some studies have measured in vivo knee contact forces of patients by instrumented knee implants during gait on treadmill [186–188] and over ground [1, 189–191]. Maximum total contact force ranged from 1.8 to 3.0 BW, typically remaining between 2.0 and 2.5 BW. For the most part, MS model studies overestimate tibial contact forces during gait with maximum total contact forces ranging from 1.8 to 8.1 BW with most estimates falling in the range of 3.0–3.5 BW [15, 19, 20, 22, 25, 110, 132, 181, 182, 192]. With our MS model during the stance phase of gait in asymptomatic subjects, the maximum contact forces (up to 4 BW) were found at 25% and 75% stance periods that are larger than 2.5–3 BW measured in vivo in patients with instrumented knee implants [1].

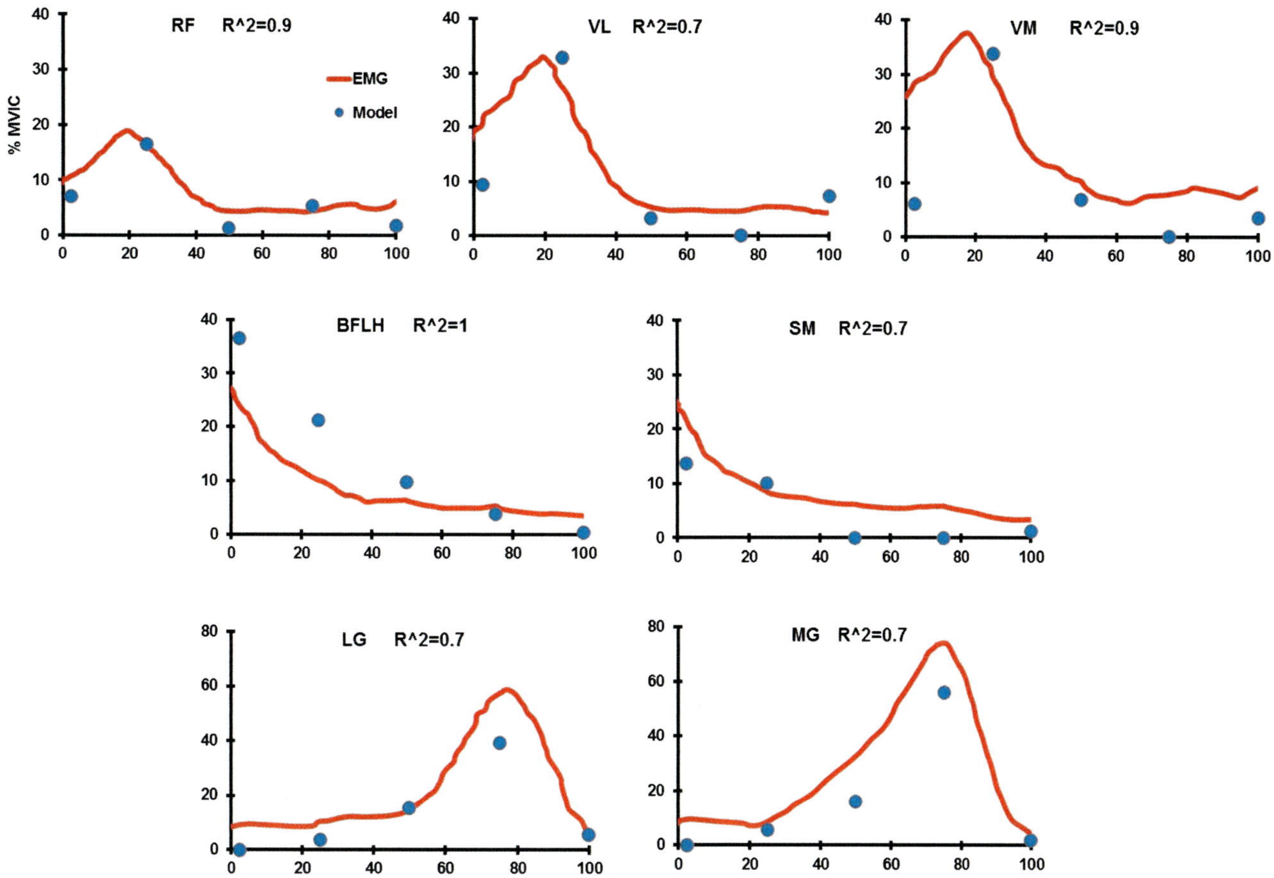


FIG. 9.7 Ensemble averaged normalized surface EMG [127] and estimated muscle forces normalized to their isometric maximal values (taking the maximum muscle stress at 0.6 MPa) during the stance phase for various knee muscles. For uniformity, the estimated values at 0% and 5% periods of stance are averaged to present a single data point at HS. Coefficients of determination as a measure on the goodness of fit are also listed at each period.

## 9.6 FUTURE DIRECTIONS

In our MS model of the lower extremity, only the knee joint was represented in details. On the other hand, hip and ankle joints were modeled as frictionless spherical joints. To improve the accuracy of estimations, it is preferable to incorporate also these articulations with accurate passive properties as much as possible. In addition, due to the expected variations in morphology, musculature, and material properties, subject-specific models taking account of individual characteristics are recommended to compute personalized response [193]. The resulting predictions will then help establish more personalized protocols for effective injury prevention, treatment, and postoperative rehabilitation. Toward this goal, statistical modeling and sensitivity analyses will be of great help.

With the rising incidence of medial knee joint OA, knee adduction moment (KAM) is commonly introduced as a surrogate measure of load on the medial plateau and hence as a marker where its reduction is the main focus of various interventions (e.g., orthoses, shoe insoles, gait modification, and osteotomy) that aim to prevent the development and progression of OA. However, some recent in vivo studies using instrumented implants have questioned such direct relationship and qualified the correlation between KAM and the medial compartment load as poor to average [194]. Similarly, questions have been raised on the association between pain/symptoms and reduction in KAM when wearing wedged insoles [195]. The internal load distribution is however influenced mainly by changes in the knee adduction rotation (rather than in KAM) as demonstrated in our earlier simulations at mid-stance phase of gait [35]. Confirmation of this finding during the entire stance phase of gait is, however, essential to back up and generalize such conclusion. Knee adduction rotation and moment should hence be altered within reported measurements one at a time at each stance period [196].

Quantification of the stability margin (reserve) of the intact/injured/reconstructed human knee joints is crucial in performance evaluation, injury prevention, implant design, and treatment managements. Joint hypermobility (instability) has been associated with pain, implant failure, injury, and OA [197]. Dynamic stability of the knee joint in daily activities is maintained by a delicate interplay between the passive tissues and active musculature (voluntary and reflex). Future developments should hence quantify the mechanical stability of the human knee joint in daily activities such as gait. The role of muscle antagonistic coactivity at different levels and modifications in gait kinematics-kinetics should be investigated with focus on both the joint stability margin and the muscle/contact/ligament forces. Such works could also shed light on the mechanisms at play when investigating the distinct behaviors of copers and non-copers following an ACL rupture.

## Acknowledgments

The current work was supported by a grant from the Natural Sciences and Engineering Research Council of Canada (RGPIN5596).

## References

[1] I. Kutzner, B. Heinlein, F. Graichen, A. Bender, A. Rohlmann, A. Halder, A. Beier, G. Bergmann, Loading of the knee joint during activities of daily living measured in vivo in five subjects, J. Biomech. 43 (2010) 2164–2173.
[2] F. Guilak, Biomechanical factors in osteoarthritis, Best Pract. Res. Clin. Rheumatol. 25 (2011) 815–823.
[3] L. Murphy, T.A. Schwartz, C.G. Helmick, J.B. Renner, G. Tudor, G. Koch, A. Dragomir, W.D. Kalsbeek, G. Luta, J.M. Jordan, Lifetime risk of symptomatic knee osteoarthritis, Arthritis Care Res. 59 (2008) 1207–1213.
[4] J.N. Katz, B.E. Earp, A.H. Gomoll, Surgical management of osteoarthritis, Arthritis Care Res. 62 (2010) 1220–1228.
[5] S. Kim, Changes in surgical loads and economic burden of hip and knee replacements in the US: 1997–2004, Arthritis Care Res. 59 (2008) 481–488.
[6] E. Losina, T.S. Thornhill, B.N. Rome, J. Wright, J.N. Katz, The dramatic increase in total knee replacement utilization rates in the United States cannot be fully explained by growth in population size and the obesity epidemic, J. Bone Joint Surg. Am. 94 (2012) 201–207.
[7] H. Jones, P.C. Rocha, Prevention in ACL injuries, in: Sports Injuries, Springer, 2012, pp. 33–42.
[8] N. Yang, P. Canavan, H. Nayeb-Hashemi, B. Najafi, A. Vaziri, Protocol for constructing subject-specific biomechanical models of knee joint, Comput. Methods Biomech. Biomed. Eng. 13 (2010) 589–603.
[9] M. Kazemi, L. Li, A viscoelastic poromechanical model of the knee joint in large compression, Med. Eng. Phys. 36 (2014) 998–1006.
[10] E. Pena, B. Calvo, M. Martinez, M. Doblare, A three-dimensional finite element analysis of the combined behavior of ligaments and menisci in the healthy human knee joint, J. Biomech. 39 (2006) 1686–1701.
[11] M.Z. Bendjaballah, A. Shirazi-Adl, D. Zukor, Biomechanics of the human knee joint in compression: reconstruction, mesh generation and finite element analysis, Knee 2 (1995) 69–79.
[12] K. Halonen, M. Mononen, J. Töyräs, H. Kröger, A. Joukainen, R. Korhonen, Optimal graft stiffness and pre-strain restore normal joint motion and cartilage responses in ACL reconstructed knee, J. Biomech. (2016).
[13] V.B. Shim, T.F. Besier, D.G. Lloyd, K. Mithraratne, J.F. Fernandez, The influence and biomechanical role of cartilage split line pattern on tibio-femoral cartilage stress distribution during the stance phase of gait, Biomech. Model. Mechanobiol. 15 (2016) 195–204.
[14] W. Mesfar, A. Shirazi-Adl, Knee joint mechanics under quadriceps-hamstrings muscle forces are influenced by tibial restraint, Clin. Biomech. 21 (2006) 841–848.
[15] M. Adouni, A. Shirazi-Adl, Evaluation of knee joint muscle forces and tissue stresses-strains during gait in severe OA versus normal subjects, J. Orthop. Res. 32 (2014) 69–78.
[16] S.L. Delp, J.P. Loan, M.G. Hoy, F.E. Zajac, E.L. Topp, J.M. Rosen, An interactive graphics-based model of the lower extremity to study orthopaedic surgical procedures, IEEE Trans. Biomed. Eng. 37 (1990) 757–767.
[17] D.G. Lloyd, T.F. Besier, An EMG-driven musculoskeletal model to estimate muscle forces and knee joint moments in vivo, J. Biomech. 36 (2003) 765–776.
[18] K. Manal, T.S. Buchanan, An electromyogram-driven musculoskeletal model of the knee to predict in vivo joint contact forces during normal and novel gait patterns, J. Biomech. Eng. 135 (2013) 021014.
[19] H. Marouane, A. Shirazi-Adl, J. Hashemi, Quantification of the role of tibial posterior slope in knee joint mechanics and ACL force in simulated gait, J. Biomech. (2015).
[20] M. Adouni, A. Shirazi-Adl, R. Shirazi, Computational biodynamics of human knee joint in gait: From muscle forces to cartilage stresses, J. Biomech. (2012).
[21] Z.F. Lerner, D.J. Haight, M.S. DeMers, W.J. Board, R.C. Browning, The effects of walking speed on tibiofemoral loading estimated via musculoskeletal modeling, J. Appl. Biomech. 30 (2014) 197.
[22] H. Marouane, A. Shirazi-Adl, M. Adouni, Alterations in knee contact forces and centers in stance phase of gait: a detailed lower extremity musculoskeletal model, J. Biomech. 49 (2016) 185–192.
[23] H. Marouane, A. Shirazi-Adl, M. Adouni, 3D active-passive response of human knee joint in gait is markedly altered when simulated as a planar 2D joint, Biomech. Model. Mechanobiol. (2016) 1–11.
[24] N.H. Yang, H. Nayeb-Hashemi, P.K. Canavan, A. Vaziri, Effect of frontal plane tibiofemoral angle on the stress and strain at the knee cartilage during the stance phase of gait, J. Orthop. Res. 28 (2010) 1539–1547.
[25] H.J. Kim, J.W. Fernandez, M. Akbarshahi, J.P. Walter, B.J. Fregly, M.G. Pandy, Evaluation of predicted knee-joint muscle forces during gait using an instrumented knee implant, J. Orthop. Res. 27 (2009) 1326–1331.
[26] W.R. Taylor, M.O. Heller, G. Bergmann, G.N. Duda, Tibio-femoral loading during human gait and stair climbing, J. Orthop. Res. 22 (2004) 625–632.
[27] C.R. Winby, D.G. Lloyd, T.F. Besier, T.B. Kirk, Muscle and external load contribution to knee joint contact loads during normal gait, J. Biomech. 42 (2009) 2294–2300.

[28] G.T. Yamaguchi, F.E. Zajac, A planar model of the knee joint to characterize the knee extensor mechanism, J. Biomech. 22 (1989) 1–10.
[29] M.K. Horsman, H. Koopman, H. Veeger, F. van der Helm, The Twente Lower Extremity Model: a comparison of maximal isometric moment with the literature, in: The Twente Lower Extremity Model, 2007, p. 65.
[30] D.G. Thelen, F.C. Anderson, S.L. Delp, Generating dynamic simulations of movement using computed muscle control, J. Biomech. 36 (2003) 321–328.
[31] E.M. Arnold, S.R. Ward, R.L. Lieber, S.L. Delp, A model of the lower limb for analysis of human movement, Ann. Biomed. Eng. 38 (2010) 269–279.
[32] D.W. Wagner, M.P. Reed, J. Rasmussen, Assessing the importance of motion dynamics for ergonomic analysis of manual materials handling tasks using the AnyBody Modeling System, in: SAE Technical Paper, 2007.
[33] M. Damsgaard, J. Rasmussen, S.T. Christensen, E. Surma, M. de Zee, Analysis of musculoskeletal systems in the AnyBody Modeling System, Simul. Model. Pract. Theory 14 (2006) 1100–1111.
[34] M. De Zee, L. Hansen, C. Wong, J. Rasmussen, E.B. Simonsen, A generic detailed rigid-body lumbar spine model, J. Biomech. 40 (2007) 1219–1227.
[35] M. Adouni, A. Shirazi-Adl, Partitioning of knee joint internal forces in gait is dictated by the knee adduction angle and not by the knee adduction moment, J. Biomech. 47 (2014) 1696–1703.
[36] H. Marouane, A. Shirazi-Adl, M. Adouni, J. Hashemi, Steeper posterior tibial slope markedly increases ACL force in both active gait and passive knee joint under compression, J. Biomech. 47 (2014) 1353–1359.
[37] H. Marouane, A. Shirazi-Adl, M. Adouni, Knee joint passive stiffness and moment in sagittal and frontal planes markedly increase with compression, Comput. Method Biomech. Biomed. Eng. 18 (2015) 339–350.
[38] E. Koay, K. Athanasiou, Articular cartilage biomechanics, mechanobiology, and tissue engineering, in: Biomechanical Systems Technology: Volume 3: Muscular Skeletal Systems, 2009, pp. 1–37.
[39] I. Clarke, Articular cartilage: a review and scanning electron microscope study. II. The territorial fibrillar architecture, J. Anat. 118 (1974) 261.
[40] M. Kääb, I. Ap Gwynn, H. Nötzli, Collagen fibre arrangement in the tibial plateau articular cartilage of man and other mammalian species, J. Anat. 193 (1998) 23–34.
[41] V.C. Mow, X.E. Guo, Mechano-electrochemical properties of articular cartilage: their inhomogeneities and anisotropies, Annu. Rev. Biomed. Eng. 4 (2002) 175–209.
[42] V.C. Mow, M.H. Holmes, W. Michael Lai, Fluid transport and mechanical properties of articular cartilage: a review, J. Biomech. 17 (1984) 377–394.
[43] A. Ratcliffe, P.R. Fryer, T.E. Hardingham, The distribution of aggregating proteoglycans in articular cartilage: comparison of quantitative immunoelectron microscopy with radioimmunoassay and biochemical analysis, J. Histochem. Cytochem. 32 (1984) 193.
[44] D. Shepherd, B. Seedhom, Thickness of human articular cartilage in joints of the lower limb, Ann. Rheum. Dis. 58 (1999) 27–34.
[45] W. Wilson, J. Huyghe, C. Van Donkelaar, Depth-dependent compressive equilibrium properties of articular cartilage explained by its composition, Biomech. Model. Mechanobiol. 6 (2007) 43–53.
[46] R. Minns, F. Steven, The collagen fibril organization in human articular cartilage, J. Anat. 123 (1977) 437.
[47] N. Broom, D. Marra, Ultrastructural evidence for fibril-to-fibril associations in articular cartilage and their functional implication, J. Anat. 146 (1986) 185.
[48] L. Li, J. Soulhat, M. Buschmann, A. Shirazi-Adl, Nonlinear analysis of cartilage in unconfined ramp compression using a fibril reinforced poroelastic model, Clin. Biomech. 14 (1999) 673–682.
[49] J. Soulhat, M. Buschmann, A. Shirazi-Adl, A fibril-network-reinforced biphasic model of cartilage in unconfined compression, J. Biomech. Eng. 121 (1999) 340–347.
[50] L. Li, M. Buschmann, A. Shirazi-Adl, The asymmetry of transient response in compression versus release for cartilage in unconfined compression, J. Biomech. Eng. 123 (2001) 519–522.
[51] C.-Y. Huang, A. Stankiewicz, G.A. Ateshian, V.C. Mow, Anisotropy, inhomogeneity, and tension–compression nonlinearity of human glenohumeral cartilage in finite deformation, J. Biomech. 38 (2005) 799–809.
[52] M.H. Doweidar, M. Doblaré, Finite element modeling and simulation of the multiphysic behavior of articular cartilage, in: Numerical Methods and Advanced Simulation in Biomechanics and Biological Processes, Elsevier, 2018, pp. 37–53.
[53] L. Li, M. Buschmann, A. Shirazi-Adl, A fibril reinforced nonhomogeneous poroelastic model for articular cartilage: inhomogeneous response in unconfined compression, J. Biomech. 33 (2000) 1533–1541.
[54] V. Duthon, C. Barea, S. Abrassart, J. Fasel, D. Fritschy, J. Ménétrey, Anatomy of the anterior cruciate ligament, Knee Surg. Sports Traumatol. Arthrosc. 14 (2006) 204–213.
[55] J. Hollis, S. Takai, D. Adams, S. Horibe, S.-Y. Woo, The effects of knee motion and external loading on the length of the anterior cruciate ligament (ACL): a kinematic study, J. Biomech. Eng. 113 (1991) 208–214.
[56] S.L. Woo, R.J. Fox, M. Sakane, G.A. Livesay, T.W. Rudy, F.H. Fu, Biomechanics of the ACL: measurements of in situ force in the ACL and knee kinematics, Knee 5 (1998) 267–288.
[57] A. Amis, G. Dawkins, Functional anatomy of the anterior cruciate ligament. Fibre bundle actions related to ligament replacements and injuries, Bone Joint J. 73 (1991) 260–267.
[58] M.T. Gabriel, E.K. Wong, S.L.Y. Woo, M. Yagi, R.E. Debski, Distribution of in situ forces in the anterior cruciate ligament in response to rotatory loads, J. Orthop. Res. 22 (2004) 85–89.
[59] M. Sakane, R.J. Fox, S.L.Y.W. Glen, A. Livesay, G. Li, F.H. Fu, In situ forces in the anterior cruciate ligament and its bundles in response to anterior tibial loads, J. Orthop. Res. 15 (1997) 285–293.
[60] K.L. Markolf, D.M. Burchfield, M.M. Shapiro, M.F. Shepard, G.A.M. Finerman, J.L. Slauterbeck, Combined knee loading states that generate high anterior cruciate ligament forces, J. Orthop. Res. 13 (1995) 930–935.
[61] R. Aspden, A model for the function and failure of the meniscus, Eng. Med. 14 (1985) 119–122.
[62] D. Skaggs, W. Warden, V. Mow, Radial tie fibers influence the tensile properties of the bovine medial meniscus, J. Orthop. Res. 12 (1994) 176–185.

[63] D.C. Fithian, M.A. Kelly, V.C. Mow, Material properties and structure-function relationships in the menisci, Clin. Orthop. Relat. Res. (1990) 19.
[64] I.M. Levy, P. Torzilli, R. Warren, The effect of medial meniscectomy on anterior-posterior motion of the knee, J. Bone Joint Surg. Am. 64 (1982) 883–888.
[65] A.S. Voloshin, J. Wosk, Shock absorption of meniscectomized and painful knees: a comparative in vivo study, J. Biomed. Eng. 5 (1983) 157–161.
[66] E.L. Radin, F. de Lamotte, P. Maquet, Role of the Menisci in the Distribution of Stress in the Knee, Clin. Orthop. Relat. Res. 185 (1984) 290–294.
[67] I.M. Levy, P. Torzilli, J. Gould, R. Warren, The effect of lateral meniscectomy on motion of the knee, J. Bone Joint Surg. Am. 71 (1989) 401–406.
[68] B. Seedhom, Transmission of the load in the knee joint with special reference to the role of the menisci part I: anatomy, analysis and apparatus, Eng. Med. 8 (1979) 207–219.
[69] S. Kawamura, K. Lotito, S.A. Rodeo, Biomechanics and healing response of the meniscus, Oper. Tech. Sports Med. 11 (2003) 68–76.
[70] N. Shrive, J. O'connor, J. Goodfellow, Load-bearing in the knee joint, Clin. Orthop. Relat. Res. 131 (1978) 279–287.
[71] A. Ahmed, D. Burke, In-vitro measurement of static pressure distribution in synovial joints – Part I: Tibial surface of the knee, J. Biomech. Eng. 105 (1983) 216.
[72] M.E. Baratz, F.H. Fu, R. Mengato, Meniscal tears: the effect of meniscectomy and of repair on intraarticular contact areas and stress in the human knee A preliminary report, Am. J. Sports Med. 14 (1986) 270–275.
[73] N. Caplan, D.F. Kader, Knee joint changes after meniscectomy, in: Classic Papers in Orthopaedics, Springer, 2014, pp. 173–175.
[74] T. Fairbank, Knee joint changes after meniscectomy, J. Bone Joint Surg. (Br.) 30 (1948) 664–670.
[75] R.E. Jones, E.C. Smith, J. Reisch, Effects of medial meniscectomy in patients older than forty years, J. Bone Joint Surg. Am. 60 (1978) 783–786.
[76] V.C. Mow, A. Ratcliffe, Structure and function of articular cartilage and meniscus, Basic Orthop. Biomech. 2 (1997) 113–177.
[77] M. Tissakht, A. Ahmed, Tensile stress-strain characteristics of the human meniscal material, J. Biomech. 28 (1995) 411–422.
[78] W. Zhu, K.Y. Chern, V.C. Mow, Anisotropic viscoelastic shear properties of bovine meniscus, Clin. Orthop. Relat. Res. (1994) 34.
[79] A.E. Peters, R. Akhtar, E.J. Comerford, K.T. Bates, Tissue material properties and computational modelling of the human tibiofemoral joint: a critical review, PeerJ 6 (2018) e4298.
[80] A. Shiraz-Adl, Strain in fibers of a lumbar disc: analysis of the role of lifting in producing disc prolapse, Spine 14 (1989) 96–103.
[81] K. Moglo, A. Shirazi-Adl, Biomechanics of passive knee joint in drawer: load transmission in intact and ACL-deficient joints, Knee 10 (2003) 265–276.
[82] W. Mesfar, A. Shirazi-Adl, Biomechanics of the knee joint in flexion under various quadriceps forces, Knee 12 (2005) 424–434.
[83] K. Moglo, A. Shirazi-Adl, On the coupling between anterior and posterior cruciate ligaments, and knee joint response under anterior femoral drawer in flexion: a finite element study, Clin. Biomech. 18 (2003) 751–759.
[84] K. Moglo, A. Shirazi-Adl, Cruciate coupling and screw-home mechanism in passive knee joint during extension–flexion, J. Biomech. 38 (2005) 1075–1083.
[85] R. Shirazi, A. Shirazi-Adl, Analysis of articular cartilage as a composite using nonlinear membrane elements for collagen fibrils, Med. Eng. Phys. 27 (2005) 827–835.
[86] R. Shirazi, A. Shirazi-Adl, Deep vertical collagen fibrils play a significant role in mechanics of articular cartilage, J. Orthop. Res. 26 (2008) 608–615.
[87] R. Shirazi, A. Shirazi-Adl, M. Hurtig, Role of cartilage collagen fibrils networks in knee joint biomechanics under compression, J. Biomech. 41 (2008) 3340–3348.
[88] S.A. Shirazi-Adl, S.C. SHRIVASTAVA, A.M. AHMED, Stress analysis of the lumbar disc-body unit in compression a three-dimensional nonlinear finite element study, Spine 9 (1984) 120–134.
[89] G. Kempson, M. Freeman, S. Swanson, Tensile properties of articular cartilage, 1968.
[90] G. Kempson, H. Muir, C. Pollard, M. Tuke, The tensile properties of the cartilage of human femoral condyles related to the content of collagen and glycosaminoglycans, Biochim. Biophys. Acta (BBA) Gen. Subjects 297 (1973) 456–472.
[91] S.-Y. Woo, W. Akeson, G. Jemmott, Measurements of nonhomogeneous, directional mechanical properties of articular cartilage in tension, J. Biomech. 9 (1976) 785–791.
[92] T.D. Brown, R.J. Singerman, Experimental determination of the linear biphasic constitutive coefficients of human fetal proximal femoral chondroepiphysis, J. Biomech. 19 (1986) 597–605.
[93] G.A. Ateshian, B.J. Ellis, J.A. Weiss, Equivalence between short-time biphasic and incompressible elastic material responses, J. Biomech. Eng. 129 (2007) 405–412.
[94] W. Mesfar, A. Shirazi-Adl, Biomechanics of changes in ACL and PCL material properties or prestrains in flexion under muscle force-implications in ligament reconstruction, Comput. Method Biomech. Biomed. Eng. 9 (2006) 201–209.
[95] K. Moglo, A. Shirazi-Adl, On the coupling between anterior and posterior cruciate ligaments, and knee joint response under anterior femoral drawer in flexion: a finite element study, Clin. Biomech. 18 (2003) 751–759.
[96] A. Shirazi-Adl, K. Moglo, Effect of changes in cruciate ligaments pretensions on knee joint laxity and ligament forces, Comput. Methods Biomech. Biomed. Eng. 8 (2005) 17–24.
[97] M.S. DeMers, S. Pal, S.L. Delp, Changes in tibiofemoral forces due to variations in muscle activity during walking, J. Orthop. Res. 32 (2014) 769–776.
[98] B.A. Knarr, J.S. Higginson, Practical approach to subject-specific estimation of knee joint contact force, J. Biomech. (2015).
[99] Z.F. Lerner, M.S. DeMers, S.L. Delp, R.C. Browning, How tibiofemoral alignment and contact locations affect predictions of medial and lateral tibiofemoral contact forces, J. Biomech. 48 (2015) 644–650.
[100] R.H. Miller, A.Y. Esterson, J.K. Shim, Joint contact forces when minimizing the external knee adduction moment by gait modification: a computer simulation study, Knee (2015).
[101] K.M. Steele, M.S. DeMers, M.H. Schwartz, S.L. Delp, Compressive tibiofemoral force during crouch gait, Gait Posture 35 (2012) 556–560.
[102] T.F. Besier, M. Fredericson, G.E. Gold, G.S. Beaupré, S.L. Delp, Knee muscle forces during walking and running in patellofemoral pain patients and pain-free controls, J. Biomech. 42 (2009) 898–905.
[103] P. Gerus, M. Sartori, T.F. Besier, B.J. Fregly, S.L. Delp, S.A. Banks, M.G. Pandy, D.D. D'Lima, D.G. Lloyd, Subject-specific knee joint geometry improves predictions of medial tibiofemoral contact forces, J. Biomech. 46 (2013) 2778–2786.

[104] D. Kumar, K.S. Rudolph, K.T. Manal, EMG-driven modeling approach to muscle force and joint load estimations: case study in knee osteoarthritis, J. Orthop. Res. 30 (2012) 377–383.
[105] M. Sartori, M. Reggiani, D. Farina, D.G. Lloyd, EMG-driven forward-dynamic estimation of muscle force and joint moment about multiple degrees of freedom in the human lower extremity, PLoS One 7 (2012) e52618.
[106] J. Cholewicki, S.M. McGill, EMG assisted optimization: a hybrid approach for estimating muscle forces in an indeterminate biomechanical model, J. Biomech. 27 (1994) 1287–1289.
[107] D. Gagnon, N. Arjmand, A. Plamondon, A. Shirazi-Adl, C. Larivière, An improved multi-joint EMG-assisted optimization approach to estimate joint and muscle forces in a musculoskeletal model of the lumbar spine, J. Biomech. 44 (2011) 1521–1529.
[108] N. Arjmand, A. Shirazi-Adl, Sensitivity of kinematics-based model predictions to optimization criteria in static lifting tasks, Med. Eng. Phys. 28 (2006) 504–514.
[109] H. Mokhtarzadeh, L. Perraton, L. Fok, M.A. Muñoz, R. Clark, P. Pivonka, A.L. Bryant, A comparison of optimisation methods and knee joint degrees of freedom on muscle force predictions during single-leg hop landings, J. Biomech. 47 (2014) 2863–2868.
[110] M. Adouni, A. Shirazi-Adl, Consideration of equilibrium equations at the hip joint alongside those at the knee and ankle joints has mixed effects on knee joint response during gait, J. Biomech. 46 (2013) 619–624.
[111] C. Redl, M. Gfoehler, M.G. Pandy, Sensitivity of muscle force estimates to variations in muscle–tendon properties, Hum. Mov. Sci. 26 (2007) 306–319.
[112] N. Arjmand, D. Gagnon, A. Plamondon, A. Shirazi-Adl, C. Lariviere, A comparative study of two trunk biomechanical models under symmetric and asymmetric loadings, J. Biomech. 43 (2010) 485–491.
[113] K.L. Markolf, W.L. Bargar, S.C. Shoemaker, H.C. Amstutz, The role of joint load in knee stability, J. Bone Joint Surg. Am. 63 (1981) 570.
[114] H. Kurosawa, T. Fukubayashi, H. Nakajima, Load-bearing mode of the knee joint: physical behavior of the knee joint with or without menisci, Clin. Orthop. Relat. Res. (1980) 283.
[115] S.J. Lee, K.J. Aadalen, P. Malaviya, E.P. Lorenz, J.K. Hayden, J. Farr, R.W. Kang, B.J. Cole, Tibiofemoral contact mechanics after serial medial meniscectomies in the human cadaveric knee, Am. J. Sports Med. 34 (2006) 1334–1344.
[116] M. Bendjaballah, A. Shirazi-Adl, D. Zukor, Finite element analysis of human knee joint in varus-valgus, Clin. Biomech. 12 (1997) 139–148.
[117] R. Shirazi, A. Shirazi-Adl, Analysis of partial meniscectomy and ACL reconstruction in knee joint biomechanics under a combined loading, Clin. Biomech. 24 (2009) 755–761.
[118] T.W. Rudy, M. Sakane, R.E. Debski, S.L. Woo, The effect of the point of application of anterior tibial loads on human knee kinematics, J. Biomech. 33 (2000) 1147–1152.
[119] H.-U. Buff, L.C. Jones, D.S. Hungerford, Experimental determination of forces transmitted through the patello-femoral joint, J. Biomech. 21 (1988) 17–23.
[120] T.Q. Lee, S.H. Anzel, K.A. Bennett, D. Pang, W.C. Kim, The influence of fixed rotational deformities of the femur on the patellofemoral contact pressures in human cadaver knees, Clin. Orthop. Relat. Res. 302 (1994) 69–74.
[121] L. Draganich, T. Andriacchi, G. Andersson, Interaction between intrinsic knee mechanics and the knee extensor mechanism, J. Orthop. Res. 5 (1987) 539–547.
[122] W. Senavongse, F. Farahmand, J. Jones, H. Andersen, Quantitative measurement of patellofemoral joint stability: force-displacement behavior of the human patella in vitro, J. Orthop. Res. 21 (2003) 780.
[123] F. Farahmand, M.N. Tahmasbi, A. Amis, The contribution of the medial retinaculum and quadriceps muscles to patellar lateral stability—an in-vitro study, Knee 11 (2004) 89–94.
[124] K.A. Jurist, J.C. Otis, Anteroposterior tibiofemoral displacements during isometric extension efforts. The roles of external load and knee flexion angle, Am. J. Sports Med. 13 (1985) 254–258.
[125] M.G. Pandy, K.B. Shelburne, Dependence of cruciate-ligament loading on muscle forces and external load, J. Biomech. 30 (1997) 1015–1024.
[126] M. Adouni, A. Shirazi-Adl, Knee joint biomechanics in closed-kinetic-chain exercises, Comput. Method Biomech. Biomed. Eng. 12 (2009) 661–670.
[127] Z.A. Cohen, H. Roglic, R.P. Grelsamer, J.H. Henry, W.N. Levine, V.C. Mow, G.A. Ateshian, Patellofemoral stresses during open and closed kinetic chain exercises. An analysis using computer simulation, Am. J. Sports Med. 29 (2001) 480–487.
[128] J. Hashemi, N. Chandrashekar, H. Mansouri, B. Gill, J.R. Slauterbeck, R.C. Schutt, E. Dabezies, B.D. Beynnon, Shallow medial tibial plateau and steep medial and lateral tibial slopes new risk factors for anterior cruciate ligament injuries, Am. J. Sports Med. 38 (2010) 54–62.
[129] B. Sonnery-Cottet, P. Archbold, T. Cucurulo, J.-M. Fayard, J. Bortolletto, M. Thaunat, T. Prost, P. Chambat, The influence of the tibial slope and the size of the intercondylar notch on rupture of the anterior cruciate ligament, J. Bone Joint Surg. (Br.) 93 (2011) 1475–1478.
[130] K. Shelburne, M. Pandy, Determinants of cruciate-ligament loading during rehabilitation exercise, Clin. Biomech. 13 (1998) 403–413.
[131] K.B. Shelburne, M.G. Pandy, F.C. Anderson, M.R. Torry, Pattern of anterior cruciate ligament force in normal walking, J. Biomech. 37 (2004) 797–805.
[132] K.B. Shelburne, M.R. Torry, M.G. Pandy, Contributions of muscles, ligaments, and the ground-reaction force to tibiofemoral joint loading during normal gait, J. Orthop. Res. 24 (2006) 1983–1990.
[133] B.J. Fregly, T.F. Besier, D.G. Lloyd, S.L. Delp, S.A. Banks, M.G. Pandy, D.D. D'Lima, Grand challenge competition to predict in vivo knee loads, J. Orthop. Res. 30 (2012) 503–513.
[134] J.L. Astephen, Biomechanical Factors in the Progression of Knee Osteoarthritis, School of Biomedical Engineering, Dalhousie University, Halifax, 2007.
[135] J.L. Astephen, K.J. Deluzio, G.E. Caldwell, M.J. Dunbar, Biomechanical changes at the hip, knee, and ankle joints during gait are associated with knee osteoarthritis severity, J. Orthop. Res. 26 (2008) 332–341.
[136] A.E. Hunt, R. M. Smith, M. Torode, A.-M. Keenan, Inter-segment foot motion and ground reaction forces over the stance phase of walking, Clin. Biomech. 16 (2001) 592–600.
[137] H. Galway, D. MacIntosh, The lateral pivot shift: a symptom and sign of anterior cruciate ligament insufficiency, Clin. Orthop. Relat. Res. 147 (1980) 45–50.
[138] J.S. Torg, W. Conrad, V. Kalen, Clinical I diagnosis of anterior cruciate ligament instability in the athlete, Am. J. Sports Med. 4 (1976) 84–93.

[139] A. Bergmark, Stability of the lumbar spine: a study in mechanical engineering, Acta Orthop. Scand. 60 (1989) 1–54.

[140] M. Sharifi, A. Shirazi-Adl, H. Marouane, Computational stability of human knee joint at early stance in gait: effects of muscle coactivity and anterior cruciate ligament deficiency, J. Biomech. (2017).

[141] J. Crisco Iii, M.M. Panjabi, The intersegmental and multisegmental muscles of the lumbar spine: a biomechanical model comparing lateral stabilizing potential, Spine 16 (1991) 793–799.

[142] H.B. Henninger, S.P. Reese, A.E. Anderson, J.A. Weiss, Validation of computational models in biomechanics, Proc. Inst. Mech. Eng. H J. Eng. Med. 224 (2010) 801–812.

[143] A.C. Jones, R.K. Wilcox, Finite element analysis of the spine: towards a framework of verification, validation and sensitivity analysis, Med. Eng. Phys. 30 (2008) 1287–1304.

[144] M. Viceconti, S. Olsen, L.-P. Nolte, K. Burton, Extracting clinically relevant data from finite element simulations, Clin. Biomech. 20 (2005) 451–454.

[145] M.Z. Bendjaballah, A. Shirazi-Adl, D. Zukor, Biomechanical response of the passive human knee joint under anterior-posterior forces, Clin. Biomech. 13 (1998) 625–633.

[146] W. Mesfar, A. Shirazi-Adl, Knee joint biomechanics in open-kinetic-chain flexion exercises, Clin. Biomech. Kidlington 23 (2008) 477–482.

[147] W. Mesfar, A. Shirazi-Adl, Computational biomechanics of knee joint in open kinetic chain extension exercises, Comput. Method Biomech. Biomed. Eng. 11 (2008) 55–61.

[148] W. Krause, M. Pope, R. Johnson, D. Wilder, Mechanical changes in the knee after meniscectomy, J. Bone Joint Surg. Am. 58 (1976) 599.

[149] P.S. Walker, M.J. Erkman, The role of the menisci in force transmission across the knee, Clin. Orthop. Relat. Res. 109 (1975) 184.

[150] S.Y. Poh, K.S.A. Yew, P.L.K. Wong, S.B.J. Koh, S.L. Chia, S. Fook-Chong, T.S. Howe, Role of the anterior intermeniscal ligament in tibiofemoral contact mechanics during axial joint loading, Knee (2011).

[151] A.M. Seitz, A. Lubomierski, B. Friemert, A. Ignatius, L. Dürselen, Effect of partial meniscectomy at the medial posterior horn on tibiofemoral contact mechanics and meniscal hoop strains in human knees, J. Orthop. Res. 30 (2012) 934–942.

[152] J.M. Marzo, J. Gurske-DePerio, Effects of medial meniscus posterior horn avulsion and repair on tibiofemoral contact area and peak contact pressure with clinical implications, Am. J. Sports Med. 37 (2009) 124–129.

[153] J.M. Paci, M.G. Scuderi, F.W. Werner, L.G. Sutton, P.F. Rosenbaum, J.P. Cannizzaro, Knee medial compartment contact pressure increases with release of the type I anterior intermeniscal ligament, Am. J. Sports Med. 37 (2009) 1412–1416.

[154] T.D. Brown, D.T. Shaw, In vitro contact stress distribution on the femoral condyles, J. Orthop. Res. 2 (1984) 190–199.

[155] A. Huang, M. Hull, S.M. Howell, The level of compressive load affects conclusions from statistical analyses to determine whether a lateral meniscal autograft restores tibial contact pressure to normal: a study in human cadaveric knees, J. Orthop. Res. 21 (2003) 459–464.

[156] G.A. Paletta, T. Manning, E. Snell, R. Parker, J. Bergfeld, The effect of allograft meniscal replacement on intraarticular contact area and pressures in the human knee a biomechanical study, Am. J. Sports Med. 25 (1997) 692–698.

[157] T. Fukubayashi, H. Kurosawa, The contact area and pressure distribution pattern of the knee: a study of normal and osteoarthrotic knee joints, Acta Orthop. 51 (1980) 871–879.

[158] P. Walker, J. Hajek, The load-bearing area in the knee joint, J. Biomech. 5 (1972) 581–589.

[159] T.L. Haut Donahue, M. Hull, A finite element model of the human knee joint for the study of tibio-femoral contact, J. Biomech. Eng. 124 (2002) 273.

[160] T.L. Haut Donahue, M. Hull, M.M. Rashid, C.R. Jacobs, How the stiffness of meniscal attachments and meniscal material properties affect tibio-femoral contact pressure computed using a validated finite element model of the human knee joint, J. Biomech. 36 (2003) 19–34.

[161] K. Markolf, J. Mensch, H. Amstutz, Stiffness and laxity of the knee – the contributions of the supporting structures. A quantitative in vitro study, J. Bone Joint Surg. Am. 58 (1976) 583.

[162] M.W. Creaby, T.V. Wrigley, B.W. Lim, K.A. Bowles, B.R. Metcalf, R.S. Hinman, K.L. Bennell, Varus–valgus laxity and passive stiffness in medial knee osteoarthritis, Arthritis Care Res. 62 (2010) 1237–1243.

[163] D. Wilson, J. Feikes, J. O'connor, Ligaments and articular contact guide passive knee flexion, J. Biomech. 31 (1998) 1127–1136.

[164] D.R. Wilson, J. Feikes, A. Zavatsky, J. O'connor, The components of passive knee movement are coupled to flexion angle, J. Biomech. 33 (2000) 465–473.

[165] S.C. Shoemaker, K.L. Markolf, Effects of joint load on the stiffness and laxity of ligament-deficient knees, J. Bone Joint Surg. Am. 67 (1985) 136–146.

[166] H. Kurosawa, P. Walker, S. Abe, A. Garg, T. Hunter, Geometry and motion of the knee for implant and orthotic design, J. Biomech. 18 (1985) 487–499.

[167] Y.-F. Hsich, L. Draganich, Knee kinematics and ligament lengths during physiologic levels of isometric quadriceps loads, Knee 4 (1997) 145–154.

[168] B. Beynnon, J. Yu, D. Huston, B. Fleming, R. Johnson, L. Haugh, M.H. Pope, A sagittal plane model of the knee and cruciate ligaments with application of a sensitivity analysis, J. Biomech. Eng. 118 (1996) 227–239.

[169] J. Bach, M. Hull, H. Patterson, Direct measurement of strain in the posterolateral bundle of the anterior cruciate ligament, J. Biomech. 30 (1997) 281–283.

[170] K.L. Markolf, W.L. Bargar, S.C. Shoemaker, H.C. Amstutz, The role of joint load in knee stability, J. Bone Joint Surg. Am. 63 (1981) 570–585.

[171] P.A. Torzilli, X. Deng, R.F. Warren, The effect of joint-compressive load and quadriceps muscle force on knee motion in the intact and anterior cruciate ligament-sectioned knee, Am. J. Sports Med. 22 (1994) 105–112.

[172] H.H. Hsieh, P.S. Walker, Stabilizing mechanisms of the loaded and unloaded knee joint, J. Bone Joint Surg. Am. 58 (1976) 87–93.

[173] G. Li, T.W. Rudy, C. Allen, M. Sakane, S.L. Woo, Effect of combined axial compressive and anterior tibial loads on in situ forces in the anterior cruciate ligament: a porcine study, J. Orthop. Res. 16 (1998) 122–127.

[174] S. Takai, S.L. Woo, G.A. Livesay, D.J. Adams, F.H. Fu, Determination of the in situ loads on the human anterior cruciate ligament, J. Orthop. Res. 11 (1993) 686–695.

[175] K.L. Markolf, D.M. Burchfield, M.M. Shapiro, M.F. Shepard, G.A. Finerman, J.L. Slauterbeck, Combined knee loading states that generate high anterior cruciate ligament forces, J. Orthop. Res. 13 (1995) 930–935.

[176] B.C. Fleming, P.A. Renstrom, B.D. Beynnon, B. Engstrom, G.D. Peura, G.J. Badger, R.J. Johnson, The effect of weightbearing and external loading on anterior cruciate ligament strain, J. Biomech. 34 (2001) 163–170.
[177] E.G. Meyer, R.C. Haut, Excessive compression of the human tibio-femoral joint causes ACL rupture, J. Biomech. 38 (2005) 2311–2316.
[178] E.G. Meyer, T.G. Baumer, J.M. Slade, W.E. Smith, R.C. Haut, Tibiofemoral contact pressures and osteochondral microtrauma during anterior cruciate ligament rupture due to excessive compressive loading and internal torque of the human knee, Am. J. Sports Med. 36 (2008) 1966–1977.
[179] S.J. Wall, D.M. Rose, E.G. Sutter, S.M. Belkoff, B.P. Boden, The role of axial compressive and quadriceps forces in noncontact anterior cruciate ligament injury a cadaveric study, Am. J. Sports Med. 40 (2012) 568–573.
[180] K.L. Markolf, J.F. Gorek, J.M. Kabo, M.S. Shapiro, Direct measurement of resultant forces in the anterior cruciate ligament. An in vitro study performed with a new experimental technique, J. Bone Joint Surg. Am. 72 (1990) 557–567.
[181] K. Shelburne, M. Torry, M. Pandy, Muscle, ligament, and joint-contact forces at the knee during walking, Med. Sci. Sports Exerc. 37 (2005) 1948–1956.
[182] Y.-C. Lin, J.P. Walter, S.A. Banks, M.G. Pandy, B.J. Fregly, Simultaneous prediction of muscle and contact forces in the knee during gait, J. Biomech. 43 (2010) 945–952.
[183] R. Neptune, F. Zajac, S. Kautz, Muscle force redistributes segmental power for body progression during walking, Gait Posture 19 (2004) 194–205.
[184] D. Kumar, K.T. Manal, K.S. Rudolph, Knee joint loading during gait in healthy controls and individuals with knee osteoarthritis, Osteoarthr. Cartil. 21 (2013) 298–305.
[185] T.M. Guess, G. Thiagarajan, M. Kia, M. Mishra, A subject specific multibody model of the knee with menisci, Med. Eng. Phys. 32 (2010) 505–515.
[186] D.D. D'Lima, S. Patil, N. Steklov, J.E. Slamin, C.W. Colwell Jr., The Chitranjan Ranawat Award: in vivo knee forces after total knee arthroplasty, Clin. Orthop. Relat. Res. 440 (2005) 45–49.
[187] D. Zhao, S.A. Banks, D.D. D'Lima, C.W. Colwell, B.J. Fregly, In vivo medial and lateral tibial loads during dynamic and high flexion activities, J. Orthop. Res. 25 (2007) 593–602.
[188] D.D. D'Lima, N. Steklov, S. Patil, C.W. Colwell, The Mark Coventry Award: in vivo knee forces during recreation and exercise after knee arthroplasty, Clin. Orthop. Relat. Res. 466 (2008) 2605–2611.
[189] D.D. D'Lima, S. Patil, N. Steklov, J.E. Slamin, C.W. Colwell, Tibial forces measured in vivo after total knee arthroplasty, J. Arthroplast. 21 (2006) 255–262.
[190] I. Kutzner, S. Küther, B. Heinlein, J. Dymke, A. Bender, A.M. Halder, G. Bergmann, The effect of valgus braces on medial compartment load of the knee joint–in vivo load measurements in three subjects, J. Biomech. 44 (2011) 1354–1360.
[191] S. Taylor, P. Walker, J. Perry, S. Cannon, R. Woledge, The forces in the distal femur and the knee during walking and other activities measured by telemetry, J. Arthroplast. 13 (1998) 428–437.
[192] P.F. Catalfamo, G. Aguiar, J. Curi, A. Braidot, Anterior cruciate ligament injury: compensation during gait using hamstring muscle activity, Open Biomed. Eng. J. 4 (2010).
[193] A. Kłodowski, M.E. Mononen, J.P. Kulmala, A. Valkeapää, R.K. Korhonen, J. Avela, I. Kiviranta, J.S. Jurvelin, A. Mikkola, Merge of motion analysis, multibody dynamics and finite element method for the subject-specific analysis of cartilage loading patterns during gait: differences between rotation and moment-driven models of human knee joint, Multibody Syst. Dyn. 37 (2016) 271–290.
[194] J.P. Walter, D.D. D'Lima, C.W. Colwell, B.J. Fregly, Decreased knee adduction moment does not guarantee decreased medial contact force during gait, J. Orthop. Res. 28 (2010) 1348–1354.
[195] R.K. Jones, G.J. Chapman, L. Forsythe, M.J. Parkes, D.T. Felson, The relationship between reductions in knee loading and immediate pain response whilst wearing lateral wedged insoles in knee osteoarthritis, J. Orthop. Res. 32 (2014) 1147–1154.
[196] D.A. Wilson, J.L.A. Wilson, G. Richardson, M.J. Dunbar, Changes in the functional flexion axis of the knee before and after total knee arthroplasty using a navigation system, J. Arthroplast. 29 (2014) 1388–1393.
[197] M.R. Mulvey, G.J. Macfarlane, M. Beasley, D.P. Symmons, K. Lovell, P. Keeley, S. Woby, J. McBeth, Modest association of joint hypermobility with disabling and limiting musculoskeletal pain: results from a large-scale general population–based survey, Arthritis Care Res. 65 (2013) 1325–1333.

CHAPTER

# 10

# Determination of the Anisotropic Mechanical Properties of Bone Tissue Using a Homogenization Technique Combined With Meshless Methods

*Marco Marques**,†, *Jorge Belinha**,‡, *António F. Oliveira*§, *Renato M. Natal Jorge**,†

***Institute of Science and Innovation in Mechanical and Industrial Engineering (INEGI), Porto, Portugal †Faculty of Engineering of University of Porto (FEUP), Porto, Portugal ‡School of Engineering, Polytechnic of Porto (ISEP), Porto, Portugal §ICBAS—Abel Salazar Institute of Biomedical Sciences, Porto, Portugal**

## 10.1 INTRODUCTION

Bones are the main integrant of the skeletal system by which the body supports, protects, and moves itself as well as stores and produces blood cells. Bone is a complex structure that consists of two different tissue types: the cortical bone, a thin and stiff outer layer, and the trabecular bone, which is more flexible and has a foam-like inner structure [1–3]. Bone biomechanics are based on the idea that bone provides a high load-carrying capacity and that bone tissue is structurally optimized for this mechanical function [4, 5]. Considering this purpose, bone has a mechanism, bone remodeling, that allows its microstructural integrity to be continuously maintained. This process can lead to bone removal via osteoclasts or bone regeneration via osteoblasts [6–11]. It also occurs in other biological processes such as growth, reinforcement, and resorption. Bone remodeling has been continuously studied, resulting in the development of semiempirical mathematical descriptions. These models simulate and predict experimental results using, for example, computer science methodologies such as finite element methods (FEM). The efforts to understand the bone remodeling phenomenon have led to a continuous development of semiempirical mathematical descriptions, but also to better comprehension of the bone structure. It was observed that bone has different functional requirements at different scales. This is the reason why some authors start to classify bone as a hierarchical multiscale material, with different structural levels from the macroscale (whole bone) to the subnanoscale (hydroxyapatite crystals, constituent of the inorganic phase of bone) [12–17]. Because bone has different functional requirements at different scales, it was necessary to investigate the mechanical properties of its distinct components and the structural relationships across different scales [18, 19]. The evolution of the hierarchical bone classification also led to the evolution of models that allow studying bone biological and mechanical processes by incorporating a multiscale approach.

Mechanoregulatory models are defined by laws that only consider the influence of mechanical factors in bone remodeling. In bioregulatory models, only the biochemical factors are considered while in the mechanobioregulatory models, both mechanical and biochemical factors are considered. Despite being more representative in comparison with mechanoregulatory and bioregulatory models, mechanobioregulatory models are more complicated due to the higher number of assumptions and restrictions involved in their formulation. The mechanoregulatory was the first to appear, in 1960 by Pauwels [20]. In 1999, Adam and coworkers [21] created one of the first bioregulatory modes to characterize bone behavior. Later, the mechanobioregulatory models appear as a combination of the

*Advances in Biomechanics and Tissue Regeneration*
**https://doi.org/10.1016/B978-0-12-816390-0.00010-8**

mechanoregulatory with the bioregulatory. The first mathematical model related to mechanoregulation of bone remodeling that described Wolff's law was developed in 1960 by Pauwels [20]; it was later applied in 1965 [22]. Wolff's law, developed by anatomist and surgeon Julius Wolff, states that bone adapts itself to applied loads. Wolff reported that the directions of the applied external loads directly influence the direction of the trabecular bone by changing the trabecular bone's physical disposition and distribution. Today, it is generally accepted that bone remodeling is mainly caused by the transient nature of its strain/stress fields, induced by the external loads applied in its physical boundary. From 1960 until now, many other models were created by using novel ideas or by modifying/enhancing existent models [23–39]. In 2012, Belinha et al. [40] developed a material law that permits correlating the bone apparent density with the bone level of stress. Using this new material law, a biomechanical remodeling model was developed as an adaptation of Carter's models for predicting bone density distribution that assumes that bone structure is a gradually self-optimizing anisotropic biological material that maximizes its own structural stiffness [27, 29, 30, 41]. Peyroteo et al. [42, 43] developed another model, considering the Belinha et al. [40] material law as part of a mechanoregulation model. Among many bioregulatory models [21, 44–48], one of the most known bioregulatory models is Komarova's model [45]. Komarova's model describes the population dynamics of bone cells accordingly with the number of osteoclasts and osteoblasts at a single basic multicellular unit (BMU). The development of these mechanoregulatory and bioregulatory models led to the development of mechanobioregulatory modes, being one of the first versions developed by Lacroix and coworkers in 2002. Using the models developed by Prendergast et al. [34] and Huiskes et al. [49], this first model considered cellular processes together with mechanical factors by incorporating the random walk of mesenchymal stem cells [35, 36]. One other numerical model was developed by Mousavi and Doweidar [50] that allows studying mesenchymal stem cell differentiation to osteoblasts as well as osteoblast proliferation due to mechanical stimulations. The latest developed mechanobiological remodeling model was developed in 2016, where it was included in the hormonal regulation and biochemical coupling of bone cell populations, the mechanical adaptation of the tissue, and factors that influence the microstructure on bone turnover rate [51].

One key factor in all these models is the characterization of the bone mechanical properties. The first models considered bone as an isotropic material, a simplistic approach to the behavior of trabecular bone that disregards the importance of orientation in the remodeling process [31, 41, 52, 53]. Later, models started to consider material density and orientation with bone anisotropic mechanical properties, taking into account the trabecular architecture features [38, 54–56]. More recently, some works have started to characterize bone mechanical properties using the fabric tensor concept [57–62]. The fabric tensor is a symmetric second-rank tensor that characterizes the arrangement of a multiphase material, encoding the orientation and anisotropy of the material.

Numerical methods combined with computer science are widely used in a variety of areas, from civil engineering to mechanical engineering to chemistry to biomechanics. These methodologies allow studying and analyzing, in silicio, the behavior of materials and structures, being first used in the biomechanics area in 1972 by Huiskes and coworkers [63] to evaluate stresses in human bones. Today, FEM is one of the most popular discrete numerical methods [64] while other methods such as meshless have started to appear. Meshless methods evolved from the first developed meshless method dated from 1977, where Gingold et al. [65] proposed smoothed-particle hydrodynamics, to more recent methods such as the natural neighbor radial point interpolation method (NNRPIM) [66] and the natural radial element method (NREM) [67].

The main difference between meshless methods and the FEM is the methodology to discretize the problem domain. Meshless methods, in opposition to FEM, do not use elements to establish nodal/element connectivity, and so discretize the problem domain using an unstructured node set that can be distributed regularly or irregularly. Because of this, meshless methods can have advantages such us the capability to discretize high complex problem domains using information gathered directly from medical images, a feature of high importance in the biomechanics field.

In meshless methods, the nodal discretization is constructed just by using the spatial coordinates of the nodes, allowing us to define individually the material properties of each node. The concept of influence domain, equivalent to elements in FEM, defines how each node interacts with its neighbor by using geometrical and mathematical constructions. Only meshless methods that use nodal-dependent constructions of the integration mesh are called truly meshless methods because they allow directly defining the spatial position and the integration weight of all integration points only using the spatial positions of the nodes. The untrue meshless methods use a nodal-independent background integration mesh to establish the system of equations from the integro-differential equations ruling the physical phenomenon under study [1]. Meshless methods are used in many different fields of mechanics, such us laminates [68–78], two-dimensional (2D) and three-dimensional (3D) linear elasticity [79], plate and shell bending problems [80], composites [81, 82], fractures [83–85], etc. Meshless is also widely used in biomechanics. They are used, for example, to study the behavior of bone response to the insertion of implants [86–89], to analyze the behavior of soft tissue under stress [90, 91], to evaluate the behavior of the inner ear [92, 93], and for bone remodeling computational research [42, 43, 86, 87, 94–102].

Some of the models and methodologies presented in this chapter analyze structures like bone, which has an underlying microstructure. The behavior of the structure at its microscale significantly influences the behavior of the structure observed at the macroscale. Finding the relation across scales will allow developing multiscale models capable of predicting the behavior of the macroscale using the microscale, and vice versa. Homogenization techniques allow homogenizing the mechanical properties of the heterogeneous material under study, thereby allowing substituting this material with an equivalent homogeneous material. This homogenization can be integrated into multiscale methods allowing us to define, for example, the mechanical properties of a highly complex microstructure such as trabecular bone, and replacing this microstructure by a simple structure with equivalent mechanical proprieties. This simplification allows relating the multiscales and simplifying the problem complexity, solving it at the macroscale [103].

The main objective of this chapter is to show how it is possible to combine a new homogenization technique (applied to the trabecular bone microscale) with meshless methods, aiming to achieve a low-cost and efficient multiscale technique.

## 10.2 HOMOGENIZATION TECHNIQUE

In this section, the homogenization technique that allows expeditiously defining the homogenized mechanical properties of trabecular bone at its microscale is fully described. This technique allows defining the mechanical properties of a trabecular bone representative volume element (RVE). In this technique, images generated by micro-CT are used, allowing us to acquire information on trabecular bone morphology. In this homogenization technique, in order to define the mechanical properties of the trabecular bone RVE, the fabric tensor concept and a bone tissue phenomenological law were used. Fabric tensors can be obtained by two different methodologies: mechanical-based or morphologic-based. In the case of the morphologic-based methodology, the information of the interface between the phases of the material is used to obtain the orientation distribution function (ODF). Micro-CT images provide information about the changes of the phase of trabecular bone that is required to define the ODF data, and so define the fabric tensor. This process is further explained in this section. First, aiming to define an RVE from a 2D micro-CT, the images must be segmented. Thus, the obtained RVE describes the local trabecular bone microscale morphology and the information regarding the changes of phase of the RVE is recorded. The image segmentation creates binarized information, ones and zeros, that can identify what is bone, ones, and what is void space, zeros. These binarizations can be obtained using image-processing methodologies, such as Ostu's method [104]. Fig. 10.1C presents an example of the image type that is acquired using a micro-CT to be used in this methodology. These binary images, $I_s$, are employed in the methodology developed by Whitehouse [105] to define the fabric tensor. This methodology is considered the gold standard in this kind of application because there exist a large number of works sustaining its appropriateness to predict the mechanical properties of trabecular bone [60, 62, 105–107]. When the ODF data is acquired by this method, disposed on a polar plot, and fitted into an ellipse, it is possible to obtain parameters that can be correlated with the material orientation, allowing us to further define the trabecular bone mechanical properties.

### 10.2.1 Fabric Tensor Morphologic-Based Method

In this methodology, the number of interceptions between a parallel family line in direction $\iota$, and the interface between both phases of the material is counted, $\mathrm{Int}(\iota)$. The length of the parallel lines family, $h$, is also considered. The parameters $h$ and $\iota$ define the ODF, which in this case is called mean interception length (MIL) (Eq. 10.1).

$$\mathrm{MIL}(\iota) = \frac{h}{\mathrm{Int}(\iota)} \tag{10.1}$$

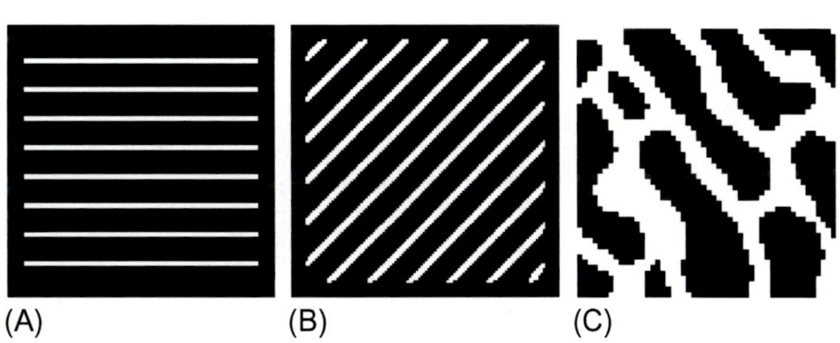

FIG. 10.1 In this figure, the $I_s$ used in this work are presented. (A) Benchmark image 1; (B) benchmark image 2; (C) trabecular bone.

The dimensional information of $I_s$ is used to define the size of an image containing the family of parallel lines with $\iota = 0$ degrees. Fig. 10.2A represents an example of an image containing a family of parallel lines $I_\iota$, in this case, with $\iota = 0$ degrees. Counting the interceptions of those parallel lines with changes of phase of RVE, represented by the boundaries of the $I_s$, it is possible to obtain the orientation-dependent feature.

Rotating the family of parallel line images with $\iota$ between 0 and 180 degrees using a defined angle increment, and then counting the interception of the family of parallel lines with $I_s$, it is possible to obtain the ODF of the $I_s$. The generated data for $\iota$ between [180,360] degrees is a [0,180] degrees data repetition because the orientation-dependent feature is not influenced by the direction. For example, Fig. 10.2 represents the rotation of $I_\iota$ between 0 and 180 degrees with a $\iota$ increment of 45 degrees.

To better understand how the ODF data are acquired using $I_s$ and $I_\iota$, Figs. 10.3–10.5 are presented. Each image makes reference to one of the images presented in Fig. 10.1, with $\iota$ between 0 and 180 degrees, using an increment of 45 degrees. In each one of the images, five pixel colors—black, blue, cyan, red, and pink—are presented, as a result of the combination of multiple image information. The blue pixels represent the corresponding white pixels from the $I_s$ image. The black pixels represent the background of the $I_s$ image. This information is constant in each set of images

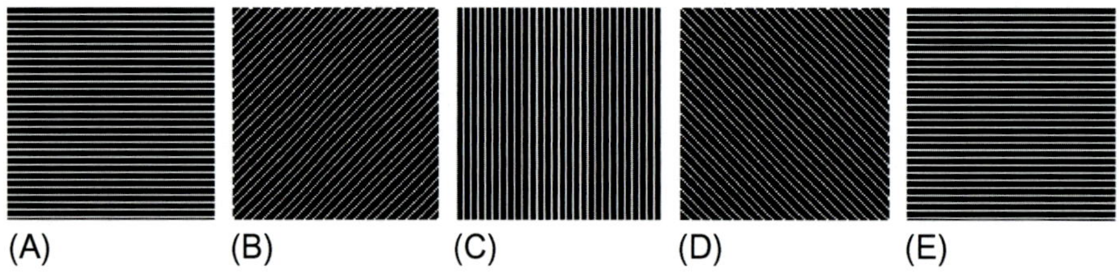


FIG. 10.2 Images of parallel line rotation $I_\iota$ with an angle increment of 45 degrees, within the interval of $\iota = [0, 180\,\text{degrees}]$. (A) $I_{pl}$ 0 degrees; (B) $I_{pl}$ 45 degrees; (C) $I_{pl}$ 90 degrees; (D) $I_{pl}$ 135 degrees; (E) $I_{pl}$ 180 degrees.

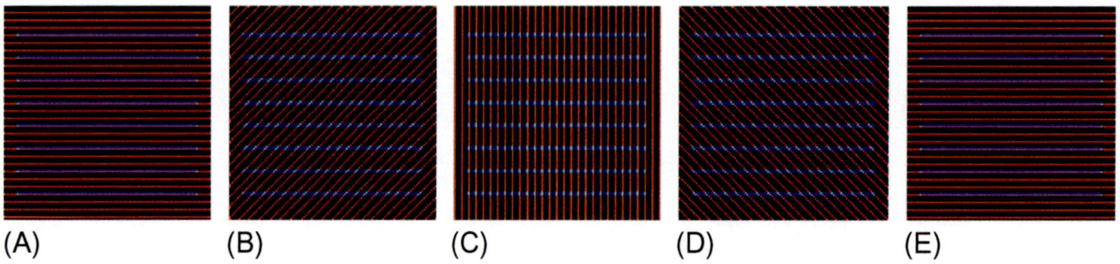


FIG. 10.3 Grid lines rotation interceptions of Fig. 10.1A. (A) 0 degrees; (B) 45 degrees; (C) 90 degrees; (D) 135 degrees; (E) 180 degrees.

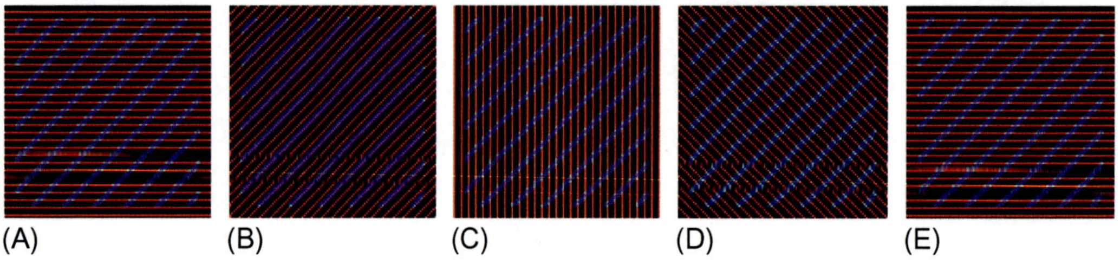


FIG. 10.4 Grid lines rotation interceptions of Fig. 10.1B. (A) 0 degrees; (B) 45 degrees; (C) 90 degrees; (D) 135 degrees; (E) 180 degrees.

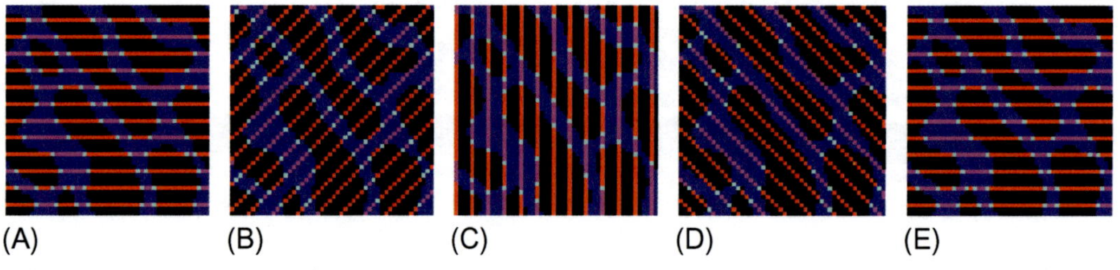


FIG. 10.5 Grid lines rotation interceptions of Fig. 10.1C. (A) 0 degrees; (B) 45 degrees; (C) 90 degrees; (D) 135 degrees; (E) 180 degrees.

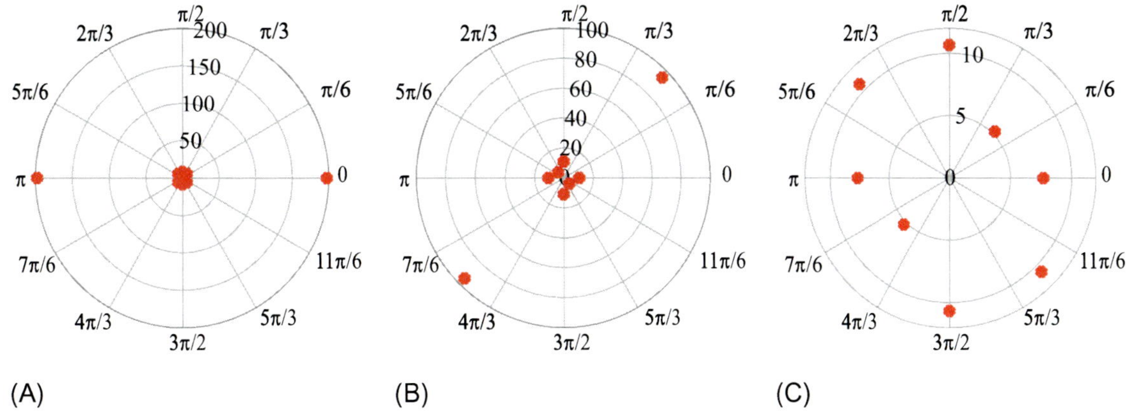


FIG. 10.6 Polar plot of ODF data, the *red points*, from Fig. 10.1A–C, respectively.

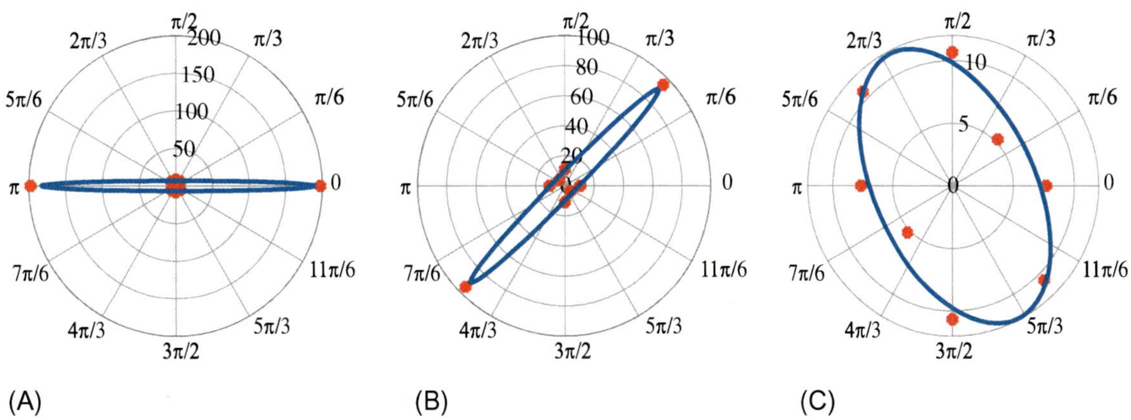


FIG. 10.7 Polar plot of ODF data and corresponding fitted ellipse from Fig. 10.1A–C, respectively. (A) $\theta = 0$ degrees; (B) $\theta = 45$ degrees; (C) $\theta = 117$ degrees.

with an origin in the same $I_s$ image. The only pixels that change in these images are the pixels in red, cyan, and pink. The red pixels represent the $I_t$. The creation of this $I_t$ results in five different images, as can be observed in each set of images (Figs. 10.3–10.5). The union of the pink pixels with the cyan pixels represents the intersection of $I_s$ with each of the $I_t$ images. The methodology to acquire the ODF data only needs the information of the material phase change, and for this reason, only the cyan pixels are used to obtain the ODF data. Counting the number of cyan pixels that result from combining $I_s$ with each $I_t$, and considering the length of the parallel family lines (Eq. 10.1), the resulting ODF data is plotted in Fig. 10.6.

Fitting an ellipse into this data, it is possible to obtain the material orientation of the trabecular RVE. In Fig. 10.7, the fitted ellipses correspond to the ODF data present in Fig. 10.6. Considering Wolff's law, it is understandable that this ellipse is aligned with the RVE preferential trabecular directions, as it has the result of the functional requirements of the trabecular bone. From the fitted ellipse, it is possible to obtain the ellipse minor axis length, $\beta_{\min}$, major axis length, $\beta_{\max}$, and $\theta$, the angle of the ellipse major axis with the polar plot horizontal axis, which for the case of Fig. 10.1C, represents the preferential trabecular direction. For the case of Fig. 10.1A and B their $\theta$ angle represent the preferential direction of the created benchmark images, with $\theta = 0$ and $\theta = 45$, respectively.

## 10.2.2 Phenomenological Material Law Method

Merging the information obtained using the fabric tensor concept and the phenomenological material law, it is possible to define the homogenized orthotropic mechanical properties of an RVE. To use the phenomenological material law defined by Belinha and coworkers [40], it was required to define the average apparent density, $\rho_{app}$. This was achieved by using the binary image $I_s$ information as represented in Eq. (10.2), with $\alpha_w$ being the number of white pixels and $\alpha_b$ the number of black pixels.

$$\rho_{\text{app}} = \left(\frac{\alpha_w}{\alpha_b}\right)\rho_{\text{app}}^{\text{cortical}} \tag{10.2}$$

Using the $\rho_{\text{app}}$, it is possible to define the axial Young's modulus, $E_{\text{axial}}$. To define the transverse elastic modulus, $E_{\text{trasnv}}$, the relation between the ellipse minor axis length, $\beta_{\text{min}}$, the major axis length, $\beta_{\text{max}}$, and the axial elastic modulus, $E_{\text{axial}}$, can be used, as Eq. (10.3) shows.

$$E_{\text{transv}} = \frac{(\|\beta_{\text{min}}\| E_{\text{axial}})}{\|\beta_{\text{max}}\|} \tag{10.3}$$

Poisson's coefficient, $\nu$, can be calculated according to the mixture theory using the relation between white and black pixels, as represented in Eq. (10.4).

$$\nu = \frac{0.0(\alpha_b) + 0.3(\alpha_w)}{\alpha_t} \tag{10.4}$$

with $\alpha_t$ being the total number of pixels of the binary image $I_s$. The shear modulus, $G$, can be expeditiously calculated using Eq. (10.5).

$$G = \frac{E_{\text{axial}}}{2(1+\nu)} \tag{10.5}$$

Using the homogenized material properties ($E_{\text{axial}}$, $E_{\text{trasnv}}$, $\nu$, and $G$), it is possible to define the constitutive matrix $\boldsymbol{c}_{ox'y'}$ for the $ox'y'$ local coordinate system. Transforming $\boldsymbol{c}_{ox'y'}$ with the transformation rotation matrix $\boldsymbol{T}$, it is possible to define the material constitutive matrix in the global axis.

## 10.3 VALIDATION

To validate this homogenization technique, some tests are performed, which allowed understanding the behavior of the methodology used to acquire the fabric tensor. Thus, three numerical studies were tested, one concerning the influence of the size of the RVE, another related to scale analysis, and a third concerning the rotation effect of the RVE in the acquisition of the $\theta$ ellipse parameter.

### 10.3.1 Scale Study

In order to understand the influence of the size of the RVE, three distinct models were constructed based on Fig. 10.1A–C. The models presented in Fig. 10.8a and d are benchmark fabricated unitary binary images with a well-defined material orientation, 0 and 45 degrees, respectively.

Alternatively, it was also used a realistic trabecular model RVE, represented in Fig. 10.8g and obtained from Fig. 10.1C, a micro-CT binary image, was also used. The three models were repeated $r_n \times r_n$, being $r_n = 1, 2, \ldots, 10$. Applying the homogenization technique to all the RVE and corresponding repetitions and comparing the results between each element of the constitute matrix (Fig. 10.9), it is perceptible that the scale of the RVE does not affect significantly the acquisition of the mechanical properties. The small changes visible in Fig. 10.9C occur because the unitary image, Fig. 10.8g, is not periodic, which means that the repetition of the image results in the creation of new changes of phase.

### 10.3.2 Rotation Study

To understand the influence of the rotation in this methodology, the already presented RVEs were rotated with respect to their initial position using an increment of 20 degrees in the interval between [0,180 degrees]. For the cases presented in this chapter (Fig. 10.1), the average difference between the obtained material orientation and the expected angle was of 5.83 degrees. This difference can be explained by the changes occurring on the source image upon the rotation process, as can be observed in Fig. 10.10, where the red circle marks the changes in the same region in different rotated images. In Fig. 10.11, it is perceptible that by applying a rotation to the image, the principal direction of the fitted ellipse, $\theta$, reflects the applied rotation.

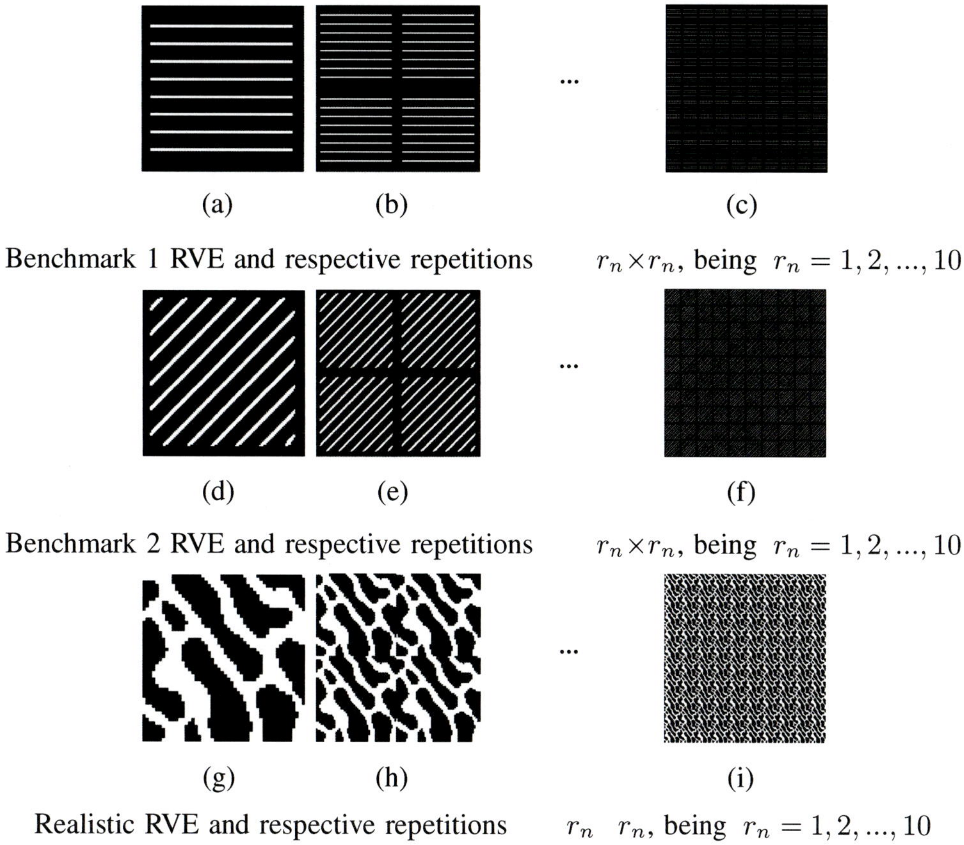


FIG. 10.8 Model set used to validate the behavior of the methodology used to define the fabric tensor.

### 10.3.3 Structural Application

This homogenization methodology is intended to be used as an improvement of existing methodologies, usually used in highly heterogeneous problem domains. Thus, it was necessary to compare the structural response of a heterogeneous material domain with the structural response of the corresponding homogenized material domain, whose mechanical properties were obtained using the proposed methodology. That is, to concede the different problem domains used in this study to be equivalent, the mechanical properties of the homogenous domain had to be defined using the information of the heterogeneous models problem domain. This allows defining equivalent models despite the different levels of heterogeneousness.

The heterogeneous RVE was defined using a heterogeneous domain (Fig. 10.1C). This problem domain (Fig. 10.12B) is complex and is formed by two different materials, the trabecular bone and the void space. The homogeneous RVE (Fig. 10.12A) was defined by a homogeneous domain, discretized by a set of uniformly distributed nodes and integration points, with the same homogenized material properties acquired using the described methodology. The RVEs were constructed with an $L \times L$ dimension. To define the problem, a displacement of $0.1 \times L$ at the nodes of the top layer, $y = L$, was imposed. The nodes at $x = 0$ and $x = L$ were constrained on the $Ox$ direction, $\bar{u} = 0$, and the nodes at $y = 0$ and $y = L$ were constrained on the $Oy$ direction, $\bar{v} = 0$. Two different numerical approaches, the FEM and the NNRPIM, were used the compare the mechanical behavior of these RVEs. The integration mesh constructed within the FEM is fundamentally different from the integration mesh constructed with the NNRPIM formulation, resulting in very different positions of the integration points. Thus, in order to compare the stress field obtained with the two different RVEs, the concept of von Misses homogenized stress, $\sigma^h_{eff}$, is used. The $\sigma^h_{eff}$, defined by Eq. (10.6), allows combining the stress field in one scalar value, facilitating the comparison of different models. In this equation, $n_Q$ represents the number of integration points, which discretize the problem domain, that do not belong to the vicinity of the domain boundary, typically 2%, avoiding the inaccurate stress concentrations that appear near the domain boundary.

$$\sigma^h_{eff} = \frac{1}{n_Q} \sum_{i=1}^{n_Q} \sigma(\boldsymbol{x}_i)_{eff} \tag{10.6}$$

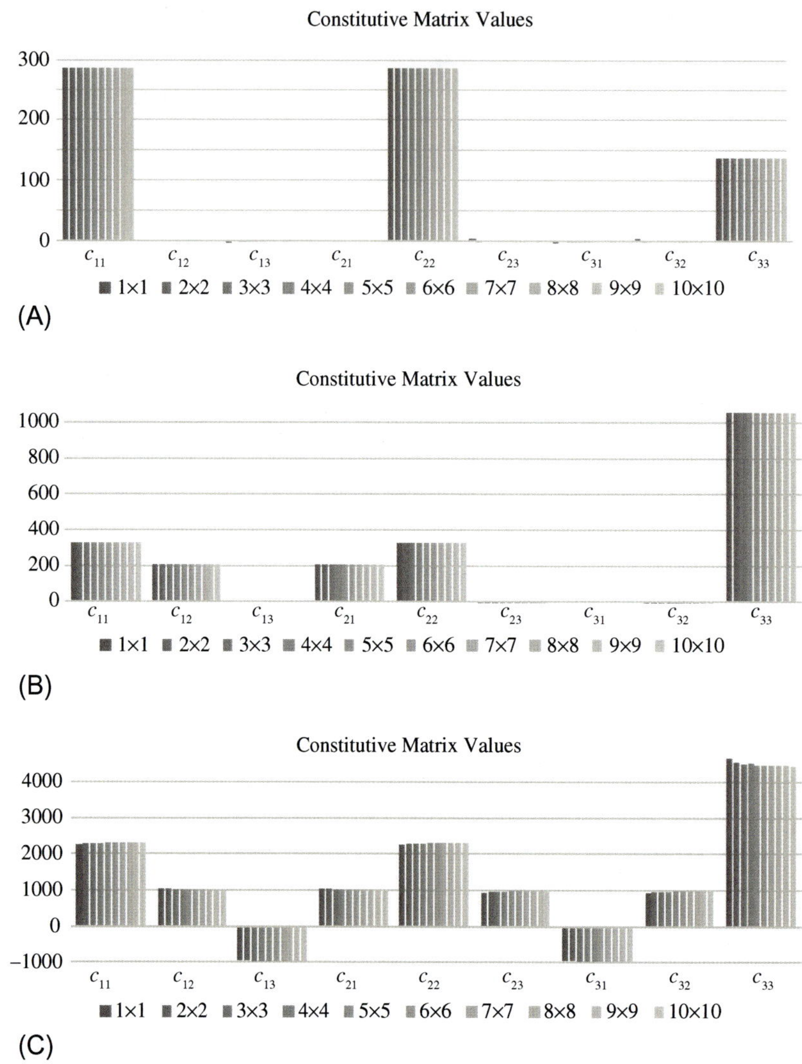


FIG. 10.9 Constitutive matrix values obtained using a unitary RVE and up to 10 × 10 repetitions. (A) Results for the benchmark 1 RVE with principal direction of 0 degrees. (B) Results for the benchmark 2 RVE with principal direction of 45 degrees. (C) Results for the trabecular RVE.

FIG. 10.10 Image rotation process and inherent morphologic change. (A) Original image; (B) 20 degrees rotation; (C) 40 degrees rotation; (D) 60 degrees rotation; (E) 80 degrees rotation.

The acquisition of the $\sigma_{eff}^{h}$, obtained for each analyzed RVE, using both FEM and NNRPIM methodologies (Fig. 10.13) provides the necessary data to validate the developed methodology. It is perceptible by this figure that by increasing the level of detail and the size of the heterogeneous RVE, the value of the homogenized stress decreases. Thus, when the analysis uses a heterogeneous model following a 4 × 4 repetition, the obtained homogenized stress is very close with the homogenized stress obtained with the homogeneous RVE, indicating that the presented homogenization technique is capable of accurately obtaining the homogenized orthotropic material properties of a trabecular patch. Comparing the FEM with the NNRPIM meshless method, despite the equivalent results in the homogeneous RVE, the results of heterogeneous RVEs are not so close. This difference in the results can be explained by the locking effects that occur in the FEM.

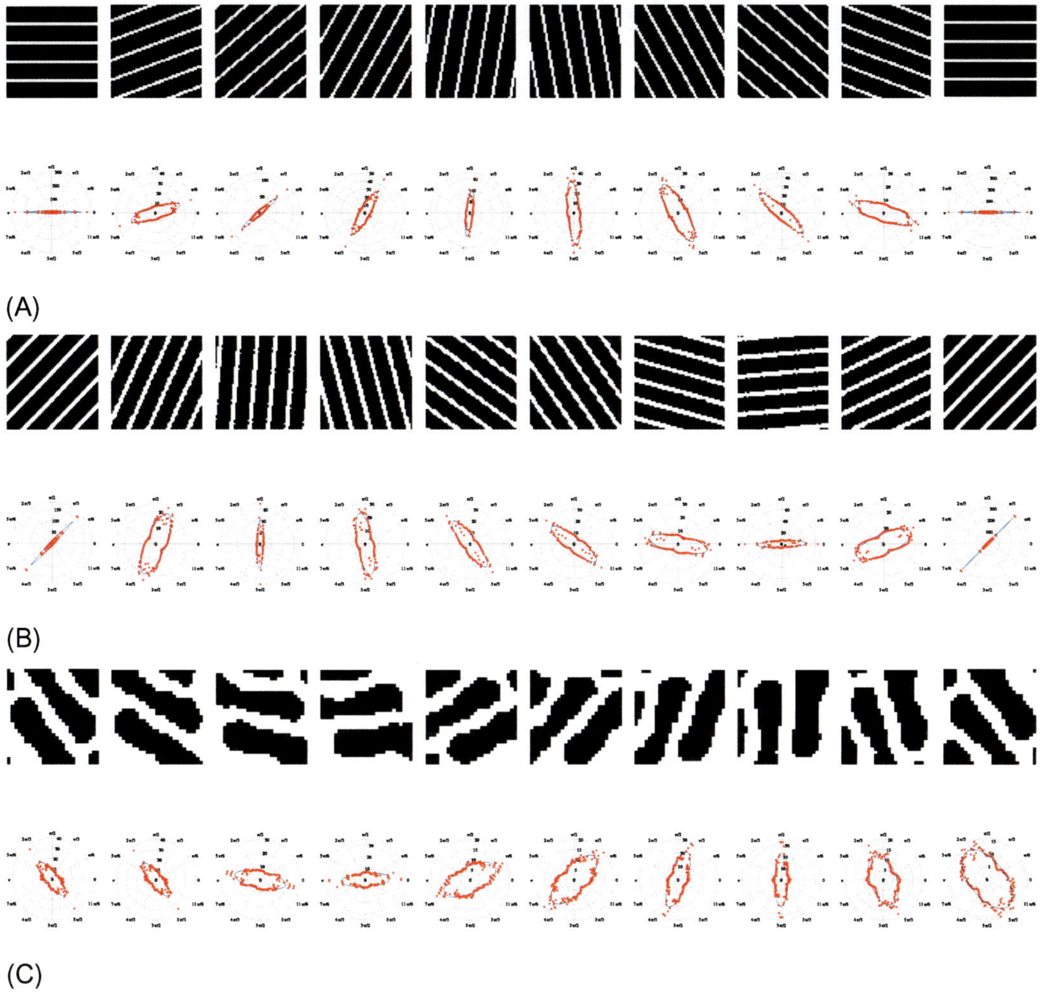

FIG. 10.11 Set of images with respective ODF polar plot, and respective rotations with an increment angle of 20 degrees, between the interval $[0, 180\,\text{degrees}]$. (A) Benchmark 1 RVE rotation and ODF polar plots. (B) Benchmark 2 RVE rotation and ODF polar plots. (C) Trabecular RVE rotation and ODF polar plots.

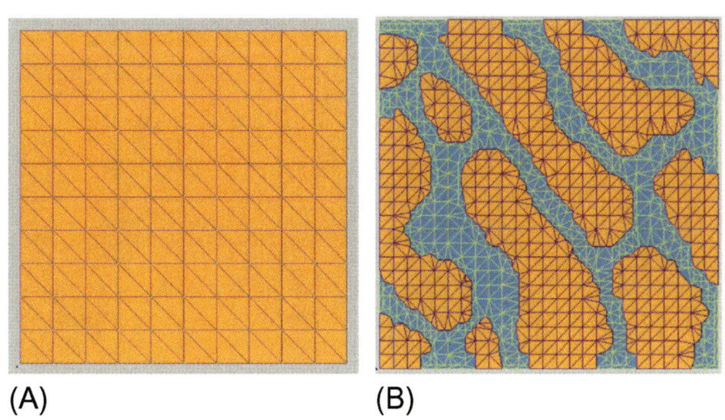

FIG. 10.12 (A) Discretized homogeneous RVE (11 × 11 nodes uniformly distributed). (B) Example of a discretized heterogeneous RVE created using micro-CT image information.

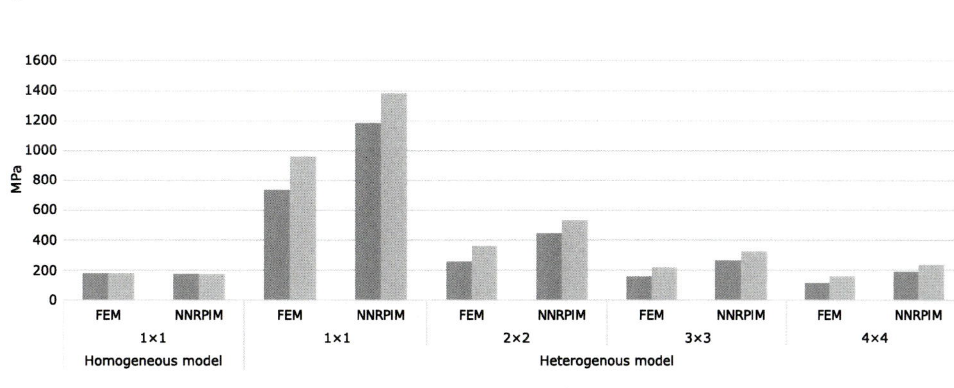


FIG. 10.13 Homogenized von Mises effective stress $\sigma_{eff}^{h}$ obtained with the FEM and the NNRPIM.

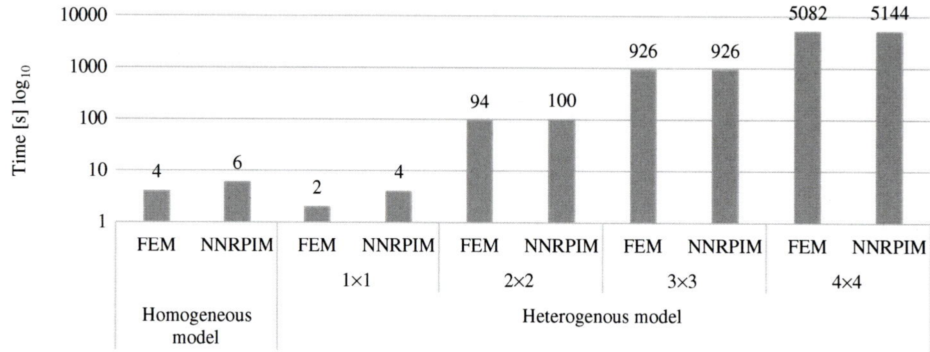


FIG. 10.14 Computational cost (in seconds) of each analysis.

Each analysis has its own computational cost. In Fig. 10.14, the time lapse of each structural analysis is shown. Observing the computational cost of each analysis, it is possible to understand that the analysis of the homogenized RVE is much faster than heterogeneous RVEs. Generally, the multiscale techniques use highly discretized RVEs, with a high computational cost associated. As this example shows, the proposed homogenization methodology is capable of reducing the cost of the multiscale analysis, enabling more demanding simulations.

## 10.4 CONCLUSIONS

The presented methodology allows defining the mechanical properties of a micro-CT RVE patch without any a prior knowledge. Using a morphologic methodology to acquire the ODF data from the micro-CT RVE, and combining this data with a phenomenological material law, a methodology was defined that allows defining a homogeneous material that is equivalent to a heterogeneous material. The defined material mechanical properties are directly related with the trabecular anisotropy encoded in the fabric tensor and with the material law developed by Belinha et al. [40]. It was demonstrated that the methodology is stable and provides good results, even when considering different RVE scales and different material principal directions. It was also shown that the NNRPIM is capable of producing accurate and smooth microscale variable fields (at the RVE scale), which allows obtaining accurate final homogenized variable fields. Multiscale techniques usually use highly discretized RVEs. The homogenization technique here proposed showed that, when combined with meshless methods, it is capable of reducing the cost of analyzing highly heterogeneous domains. Thus, using this methodology in multiscale analyses will allow simulating more complex problems with lower costs. Also, as bone is a hierarchal material, this methodology might be a powerful tool to understand the remodeling process, using a multiscale approach, where the mechanical properties of trabecular bone can be defined at the microscale, considering the trabeculae architecture.

## Acknowledgments

The authors acknowledge the funding provided by Ministério da Ciência, Tecnologia e Ensino Superior—Fundação para a Ciência e a Tecnologia (Portugal), under Grants SFRH/BD/110047/2015, and by project funding MIT-EXPL/ISF/0084/2017. Additionally, the authors gratefully acknowledge the funding of Project NORTE-01-0145-FEDER-000022—SciTech—Science and Technology for Competitive and Sustainable Industries, cofinanced by Programa Operacional Regional do Norte (NORTE2020), through Fundo Europeu de Desenvolvimento Regional (FEDER).

## References

[1] J. Belinha, Meshless Methods in Biomechanics: Bone Tissue Remodelling Analysis, Lecture Notes in Computational Vision and Biomechanics, vol. 2, Springer Netherlands, Dordrecht, 2014, p. 328, https://doi.org/10.1007/978-94-007-4174-4.

[2] C. VanPutte, J. Regan, A. Russo, R. Seeley, T. Stephens, P. Tate, Seeley's Anatomy & Physiology, tenth ed., vol. 2, McGraw-Hill Companies, New York, 2003, pp. 8–9.

[3] Q.-H. Qin, Mechanics of Cellular Bone Remodeling, CRC Press, Taylor & Francis Group, Boca Raton, 2012.

[4] D.R. Carter, M.C. Van Der Meulen, G.S. Beaupré, Mechanical factors in bone growth and development, Bone 18 (1 suppl) (1996) 5S–10S, https://doi.org/10.1016/8756-3282(95)00373-8.

[5] V.C. Mow, A. Ratcliffe, A.R. Poole, Cartilage and diarthrodial joints as paradigms for hierarchical materials and structures, Biomaterials 13 (2) (1992) 67–97.

[6] L.F. Bonewald, The amazing osteocyte, J. Bone Miner. Res. 26 (2) (2011) 229–238.

[7] J.A. Buckwalter, M.J. Glimcher, R.R. Cooper, R. Recker, Bone biology. II: formation, form, modeling, remodeling, and regulation of cell function, Instr. Course Lect. 45 (1996) 387–399.
[8] R.B. Martin, D.B. Burr, N.A. Sharkey, D.P. Fyhrie, Skeletal Tissue Mechanics, Springer, New York, NY, 2015. https://doi.org/10.1007/978-1-4939-3002-9.
[9] E. Ozcivici, Y.K. Luu, B. Adler, Y.-X. Qin, J. Rubin, S. Judex, C.T. Rubin, Mechanical signals as anabolic agents in bone, Nat. Rev. Rheumatol. 6 (1) (2010) 50–59.
[10] A.G. Robling, P.J. Niziolek, L.A. Baldridge, K.W. Condon, M.R. Allen, I. Alam, S.M. Mantila, J. Gluhak-Heinrich, T.M. Bellido, S.E. Harris, C.H. Turner, Mechanical stimulation of bone in vivo reduces osteocyte expression of SOST/sclerostin, J. Biol. Chem. 283 (9) (2008) 5866–5875.
[11] S. Theoleyre, Y. Wittrant, S.K. Tat, Y. Fortun, F. Redini, D. Heymann, The molecular triad OPG/RANK/RANKL: involvement in the orchestration of pathophysiological bone remodeling, Cytokine Growth Factor Rev. 15 (6) (2004) 457–475.
[12] E. Lucchinetti, Composite models of bone properties, in: Bone Mechanics Handbook, CRC Press, 2001, pp. 12-1–12-19.
[13] E. Hamed, I. Jasiuk, Elastic modeling of bone at nanostructural level, Mater. Sci. Eng. R Rep. 73 (3–4) (2012) 27–49.
[14] A. Barkaoui, B. Tlili, A. Vercher-Martínez, R. Hambli, A multiscale modelling of bone ultrastructure elastic proprieties using finite elements simulation and neural network method, Comput. Methods Prog. Biomed. 134 (2016) 69–78.
[15] E. Hamed, Y. Lee, I. Jasiuk, Multiscale modeling of elastic properties of cortical bone, Acta Mech. 213 (1–2) (2010) 131–154.
[16] I. Jasiuk, Micromechanics of bone modeled as a composite material, in: Micromechanics and Nanomechanics of Composite Solids, Springer International Publishing, Cham, 2018, pp. 281–306, https://doi.org/10.1007/978-3-319-52794-9_10.
[17] R.O. Ritchie, M.J. Buehler, P. Hansma, Plasticity and toughness in bone, Phys. Today 62 (6) (2009) 41–47.
[18] W.J. Landis, The strength of a calcified tissue depends in part on the molecular structure and organization of its constituent mineral crystals in their organic matrix, Bone 16 (5) (1995) 533–544.
[19] S. Weiner, W. Traub, Bone structure: from angstroms to microns, FASEB J. 6 (3) (1992) 879–885.
[20] F. Pauwels, Eine neue theorie über den Einfluβ mechanischer reize auf die differenzierung der stützgewebe, Z. Anat. Entwicklungsgesch. 121 (6) (1960) 478–515.
[21] J.A. Adam, A simplified model of wound healing (with particular reference to the critical size defect), Math. Comput. Model. 30 (5–6) (1999) 23–32.
[22] F. Pauwels, Gesammelte Abhandlungen zur funktionellen Anatomie des—Friedrich Pauwels—Springer, Springer, 1965, p. 543. Available from: http://www.springer.com/us/book/9783642868429.
[23] H.M. Frost, The Laws of Bone Structure, Thomas, Springfield, IL, 1964.
[24] D.H. Hegedus, S.C. Cowin, Bone remodeling II: small strain adaptive elasticity, J. Elast. 6 (4) (1976) 337–352.
[25] S.C. Cowin, R.R. Nachlinger, Bone remodeling III: uniqueness and stability in adaptive elasticity theory, J. Elast. 8 (3) (1978) 285–295.
[26] S.C. Cowin, A.M. Sadegh, G.M. Luo, An evolutionary Wolff's law for trabecular architecture, J. Biomech. Eng. 114 (1) (1992) 129–136.
[27] D.R. Carter, T.E. Orr, D.P. Fyhrie, Relationships between loading history and femoral cancellous bone architecture, J. Biomech. 22 (3) (1989) 231–244.
[28] D.R. Carter, Mechanical loading history and skeletal biology, J. Biomech. 20 (11–12) (1987) 1095–1109.
[29] D.P. Fyhrie, D.R. Carter, A unifying principle relating stress to trabecular bone morphology, J. Orthop. Res. 4 (3) (1986) 304–317.
[30] R.T. Whalen, D.R. Carter, C.R. Steele, Influence of physical activity on the regulation of bone density, J. Biomech. 21 (10) (1988) 825–837.
[31] R. Huiskes, H. Weinans, H.J. Grootenboer, M. Dalstra, B. Fudala, T.J. Slooff, Adaptive bone-remodeling theory applied to prosthetic-design analysis, J. Biomech. 20 (11–12) (1987) 1135–1150.
[32] H.E. Pettermann, T.J. Reiter, F.G. Rammerstorfer, Computational simulation of internal bone remodeling, Arch. Comput. Methods Eng. 4 (4) (1997) 295–323.
[33] T.J. Reiter, F.G. Rammerstorfer, H.J. Bohm, Numerical Algorithm for the Simulation of Bone Remodeling, in: American Society of Mechanical Engineers, vol. 17, Bioengineering Division (Publication) BED, 1990, pp. 181–184. Available from: http://www.scopus.com/inward/record.url?eid=2-s2.0-0025555125&partnerID=tZOtx3y1.
[34] P.J. Prendergast, R. Huiskes, K. Søballe, Biophysical stimuli on cells during tissue differentiation at implant interfaces, J. Biomech. 30 (6) (1997) 539–548.
[35] D. Lacroix, P.J. Prendergast, G. Li, D. Marsh, Biomechanical model to simulate tissue differentiation and bone regeneration: application to fracture healing, Med. Biol. Eng. Comput. 40 (1) (2002) 14–21.
[36] D. Lacroix, P.J. Prendergast, A mechano-regulation model for tissue differentiation during fracture healing: analysis of gap size and loading, J. Biomech. 35 (9) (2002) 1163–1171.
[37] L.M. McNamara, P.J. Prendergast, Bone remodelling algorithms incorporating both strain and microdamage stimuli, J. Biomech. 40 (6) (2007) 1381–1391.
[38] M. Doblaré, J.M. García, Anisotropic bone remodelling model based on a continuum damage-repair theory, J. Biomech. 35 (1) (2002) 1–17.
[39] B.M. Mulvihill, P.J. Prendergast, Mechanobiological regulation of the remodelling cycle in trabecular bone and possible biomechanical pathways for osteoporosis, Clin. Biomech. 25 (5) (2010) 491–498.
[40] J. Belinha, R.M. Natal Jorge, L.M.J.S. Dinis, A meshless microscale bone tissue trabecular remodelling analysis considering a new anisotropic bone tissue material law, Comput. Methods Biomech. Biomed. Eng. 5842 (2012) 1–15.
[41] D.R. Carter, D.P. Fyhrie, R.T. Whalen, Trabecular bone density and loading history: regulation of connective tissue biology by mechanical energy, J. Biomech. 20 (8) (1987) 785–794.
[42] M. Peyroteo, J. Belinha, L. Dinis, R. Natal Jorge, The mechanologic bone tissue remodeling analysis, in: Numerical Methods and Advanced Simulation in Biomechanics and Biological Processes, 2018, pp. 303–323, https://doi.org/10.1016/b978-0-12-811718-7.00017-4.
[43] M.M.A. Peyroteo, J. Belinha, S. Vinga, L.M.J.S. Dinis, R.M. Natal Jorge, Mechanical bone remodelling: comparative study of distinct numerical approaches, Eng. Anal. Bound. Elem. (2018), https://doi.org/10.1016/j.enganabound.2018.01.011.
[44] A. Bailón-Plaza, M.C.H. Van Der Meulen, A mathematical framework to study the effects of growth factor influences on fracture healing, J. Theor. Biol. 212 (2) (2001) 191–209.
[45] S.V. Komarova, R.J. Smith, S.J. Dixon, S.M. Sims, L.M. Wahl, Mathematical model predicts a critical role for osteoclast autocrine regulation in the control of bone remodeling, Bone 33 (2) (2003) 206–215.

[46] V. Lemaire, F.L. Tobin, L.D. Greller, C.R. Cho, L.J. Suva, Modeling the interactions between osteoblast and osteoclast activities in bone remodeling, J. Theor. Biol. 229 (3) (2004) 293–309.

[47] P. Pivonka, J. Zimak, D.W. Smith, B.S. Gardiner, C.R. Dunstan, N.A. Sims, T.J. Martin, G.R. Mundy, T. John Martin, G.R. Mundy, Model structure and control of bone remodeling: a theoretical study, Bone 43 (2) (2008) 249–263.

[48] M.D. Ryser, N. Nigam, S.V. Komarova, Mathematical modeling of spatio-temporal dynamics of a single bone multicellular unit, J. Bone Miner. Res. 24 (5) (2009) 860–870.

[49] R. Huiskes, W.D.V. Driel, P.J. Prendergast, K. Søballe, A biomechanical regulatory model for periprosthetic fibrous-tissue differentiation, J. Mater. Sci. Mater. Med. 8 (12) (1997) 785–788.

[50] S.J. Mousavi, M.H. Doweidar, Role of mechanical cues in cell differentiation and proliferation: a 3D numerical model, PLoS ONE 10 (5) (2015) e0124529.

[51] C. Lerebours, P.R. Buenzli, S. Scheiner, P. Pivonka, A multiscale mechanobiological model of bone remodelling predicts site-specific bone loss in the femur during osteoporosis and mechanical disuse, Biomech. Model. Mechanobiol. 15 (1) (2016) 43–67.

[52] R.T. Hart, D.T. Davy, K.G. Heiple, A computational method for stress analysis of adaptive elastic materials with a view toward applications in strain-induced bone remodeling, J. Biomech. Eng. 106 (4) (1984) 342.

[53] G.S. Beaupré, T.E. Orr, D.R. Carter, An approach for time-dependent bone modeling and remodeling-theoretical development, J. Orthop. Res. 8 (5) (1990) 651–661.

[54] C.R. Jacobs, J.C. Simo, G.S. Beaupre, D.R. Carter, Adaptive bone remodeling incorporating simultaneous density and anisotropy considerations, J. Biomech. 30 (6) (1997) 603–613.

[55] P. Fernandes, J.M. Guedes, H. Rodrigues, Topology optimization of three-dimensional linear elastic structures with a constraint on "perimeter", Comput. Struct. 73 (6) (1999) 583–594.

[56] M. Doblaré, J.M. García, Application of an anisotropic bone-remodelling model based on a damage-repair theory to the analysis of the proximal femur before and after total hip replacement, J. Biomech. 34 (9) (2001) 1157–1170.

[57] J. Hazrati Marangalou, K. Ito, B. van Rietbergen, A novel approach to estimate trabecular bone anisotropy from stress tensors, Biomech. Model. Mechanobiol. 14 (1) (2015) 39–48.

[58] R. Moreno, Ö. Smedby, D.H. Pahr, Prediction of apparent trabecular bone stiffness through fourth-order fabric tensors, Biomech. Model. Mechanobiol. 15 (4) (2016) 831–844.

[59] J. Kabel, B. van Rietbergen, A. Odgaard, R. Huiskes, Constitutive relationships of fabric, density, and elastic properties in cancellous bone architecture, Bone 25 (4) (1999) 481–486.

[60] S.C. Cowin, S.B. Doty, Tissue Mechanics, Springer Science, New York, NY, ISBN 0-387-36825-6, 2007.

[61] J.C. Rice, S.C. Cowin, J.A. Bowman, On the dependence of the elasticity and strength of cancellous bone on apparent density, J. Biomech. 21 (2) (1988) 155–168.

[62] P.K. Zysset, A review of morphology-elasticity relationships in human trabecular bone: theories and experiments, J. Biomech. 36 (10) (2003) 1469–1485.

[63] R. Huiskes, E.Y.S. Chao, A survey of finite element analysis in orthopedic biomechanics: the first decade, J. Biomech. 16 (6) (1983) 385–409.

[64] O.C. Zienkiewicz, R.L. Taylor, The Finite Element Method, fourth ed., McGraw-Hill, London, 1994.

[65] R.A. Gingold, J.J. Monaghan, Smooth particle hydrodynamics: theory and application to non-spherical stars, Mon. Notices R. Astron. Soc. 181 (1977) 375–389.

[66] L.M.J.S. Dinis, R.M. Natal Jorge, J. Belinha, Analysis of 3D solids using the natural neighbour radial point interpolation method, Comput. Methods Appl. Mech. Eng. 196 (13–16) (2007) 2009–2028.

[67] J. Belinha, L.M.J.S. Dinis, R.M. Natal Jorge, The natural radial element method, Int. J. Numer. Methods Eng. 93 (12) (2013) 1286–1313.

[68] J. Belinha, L.M.J.S. Dinis, Elasto-plastic analysis of plates by the element free Galerkin method, Eng. Comput. 23 (5) (2006) 525–551.

[69] J. Belinha, L.M.J.S. Dinis, Nonlinear analysis of plates and laminates using the element free Galerkin method, Compos. Struct. 78 (3) (2007) 337–350.

[70] L.M.J.S. Dinis, R.M. Natal Jorge, J. Belinha, Analysis of plates and laminates using the natural neighbour radial point interpolation method, Eng. Anal. Bound. Elem. 32 (3) (2008) 267–279.

[71] J. Belinha, L.M.J.S. Dinis, R.M. Natal Jorge, The natural neighbour radial point interpolation method: dynamic applications, Eng. Comput. 26 (8) (2009) 911–949.

[72] L.M.J.S. Dinis, R.M. Natal Jorge, J. Belinha, An unconstrained third-order plate theory applied to functionally graded plates using a meshless method, Mech. Adv. Mater. Struct. 17 (2) (2010) 108–133.

[73] L.M.J.S. Dinis, R.M. Natal Jorge, J. Belinha, Static and dynamic analysis of laminated plates based on an unconstrained third order theory and using a radial point interpolator meshless method, Comput. Struct. (19) 1771–1784, 10.1016/j.compstruc.2010.10.015.

[74] J. Belinha, L.M.J.S. Dinis, R.M. Natal Jorge, Analysis of thick plates by the natural radial element method, Int. J. Mech. Sci. 76 (2013) 33–48.

[75] S. Moreira, J. Belinha, L.M.J.S. Dinis, R.M. Natal Jorge, Análise de vigas laminadas utilizando o natural neighbour radial point interpolation method, Rev. Internac. Metod. Numer. Calc. Disen. Ingr. 30 (2) (2014) 108–120.

[76] B.V. Farahani, J.M. Berardo, R. Drgas, J.M.A. César de Sá, A.J.M. Ferreira, J. Belinha, The axisymmetric analysis of circular plates using the radial point interpolation method, Int. J. Comput. Methods Eng. Sci. Mech. 16 (6) (2015) 336–353.

[77] B.V. Farahani, J. Berardo, J. Belinha, A.J.M. Ferreira, P.J. Tavares, P.M.G.P. Moreira, On the optimal shape parameters of distinct versions of RBF meshless methods for the bending analysis of plates, Eng. Anal. Bound. Elem. 84 (2017) 77–86.

[78] B.V. Farahani, J. Berardo, J. Belinha, A.J.M. Ferreira, P.J. Tavares, P. Moreira, An optimized RBF analysis of an isotropic Mindlin plate in bending, Proc. Struct. Integr. 5 (2017) 584–591.

[79] L.M.J.S. Dinis, R.M. Natal Jorge, J. Belinha, Analysis of 3D solids using the natural neighbour radial point interpolation method, Comput. Methods Appl. Mech. Eng. 196 (13–16) (2007) 2009–2028.

[80] L.M.J.S. Dinis, R.M. Natal Jorge, J. Belinha, A 3D shell-like approach using a natural neighbour meshless method: isotropic and orthotropic thin structures, Compos. Struct. 92 (5) (2010) 1132–1142.

[81] J. Belinha, L.M.J.S. Dinis, R.M. Natal Jorge, Composite laminated plate analysis using the natural radial element method, Compos. Struct. 103 (2013) 50–67.

[82] D.E.S. Rodrigues, J. Belinha, F.M.A. Pires, L.M.J.S. Dinis, R.M. Natal Jorge, Homogenization technique for heterogeneous composite materials using meshless methods, Eng. Anal. Bound. Elem. (2018), https://doi.org/10.1016/j.enganabound.2017.12.012.

[83] J.M.C. Azevedo, J. Belinha, L.M.J.S. Dinis, R.M. Natal Jorge, Crack path prediction using the natural neighbour radial point interpolation method, Eng. Anal. Bound. Elem. 59 (2015) 144–158.

[84] J. Belinha, J.M.C. Azevedo, L.M.J.S. Dinis, R.M. Natal Jorge, The natural neighbor radial point interpolation method extended to the crack growth simulation, Int. J. Appl. Mech. 8 (1) (2016) 1650006.

[85] B.V. Farahani, R. Amaral, J. Belinha, P.J. Tavares, P. Moreira, A GTN failure analysis of an AA6061-T6 Bi-failure specimen, Proc. Struct. Integr. 5 (2017) 981–988.

[86] J. Belinha, L.M.J.S. Dinis, R.M. Natal Jorge, The analysis of the bone remodelling around femoral stems: a meshless approach, Math. Comput. Simul. 121 (2016) 64–94.

[87] J. Belinha, L.M.J.S. Dinis, R.M. Natal Jorge, The mandible remodeling induced by dental implants: a meshless approach, J. Mech. Med. Biol. 15 (4) (2015) 1550059.

[88] H.M.S. Duarte, J.R. Andrade, L.M.J.S. Dinis, R.M. Natal Jorge, J. Belinha, Numerical analysis of dental implants using a new advanced discretization technique, Mech. Adv. Mater. Struct. 23 (4) (2015) 467–479.

[89] C.S.S. Tavares, J. Belinha, L.M.J.S. Dinis, R.M. Natal Jorge, The elasto-plastic response of the bone tissue due to the insertion of dental implants, in: Procedia Engineering, vol. 110, 2015, pp. 37–44, https://doi.org/10.1016/j.proeng.2015.07.007.

[90] M.H. Doweidar, B. Calvo, I. Alfaro, P. Groenenboom, M. Doblaré, A comparison of implicit and explicit natural element methods in large strains problems: application to soft biological tissues modeling, Comput. Methods Appl. Mech. Eng. 199 (25–28) (2010) 1691–1700.

[91] M. Marques, J. Belinha, L.M.J.S. Dinis, R. Natal Jorge, A brain impact stress analysis using advanced discretization meshless techniques, Proc. Inst. Mech. Eng. H 232 (3) (2018) 257–270.

[92] C.F. Santos, J. Belinha, F. Gentil, M. Parente, B. Areias, R.N. Jorge, Biomechanical study of the vestibular system of the inner ear using a numerical method, Proc. IUTAM 24 (2017) 30–37.

[93] C.F. Santos, J. Belinha, F. Gentil, M. Parente, R.N. Jorge, The free vibrations analysis of the cupula in the inner ear using a natural neighbor meshless method, Eng. Anal. Bound. Elem. (2018), https://doi.org/10.1016/j.enganabound.2018.01.002.

[94] M. Doblaré, E. Cueto, B. Calvo, M.A. Martínez, J.M. Garcia, J. Cego nino, On the employ of meshless methods in biomechanics, Comput. Methods Appl. Mech. Eng. 194 (6–8) (2005) 801–821.

[95] J.M. García, M. Doblaré, E. Cueto, Simulation of bone internal remodeling by means of the $\alpha$-shape-based natural element method, in: European Congress on Computational Methods in Applied Sciences and Engineering (ECCOMAS 2000), 2000, pp. 11–14.

[96] K.M. Liew, H.Y. Wu, T.Y. Ng, Meshless method for modeling of human proximal femur: treatment of nonconvex boundaries and stress analysis, Comput. Mech. 28 (5) (2002) 390–400.

[97] J.D. Lee, Y. Chen, X. Zeng, A. Eskandarian, M. Oskard, Modeling and simulation of osteoporosis and fracture of trabecular bone by meshless method, Int. J. Eng. Sci. 45 (2–8) (2007) 329–338.

[98] F. Taddei, M. Pani, L. Zovatto, E. Tonti, M. Viceconti, A new meshless approach for subject-specific strain prediction in long bones: evaluation of accuracy, Clin. Biomech. 23 (9) (2008) 1192–1199.

[99] F. Buti, D. Cacciagrano, F. Corradini, E. Merelli, L. Tesei, M. Pani, Bone remodelling in BioShape, Electron. Notes Theor. Comput. Sci. 268 (C) (2010) 17–29.

[100] J. Belinha, R.M. Natal Jorge, L.M.J.S. Dinis, Bone tissue remodelling analysis considering a radial point interpolator meshless method, Eng. Anal. Bound. Elem. 36 (11) (2012) 1660–1670.

[101] J. Belinha, R.M. Natal Jorge, L.M.J.S. Dinis, A meshless microscale bone tissue trabecular remodelling analysis considering a new anisotropic bone tissue material law, Comput. Methods Biomech. Biomed. Eng. 5842 (2012) 1–15, https://doi.org/10.1080/10255842.2012.654783.

[102] S.F. Moreira, J. Belinha, L.M.J.S. Dinis, R.M. Natal Jorge, A global numerical analysis of the "central incisor/local maxillary bone" system using a meshless method, MCB Mol. Cell. Biomech. 11 (3) (2014) 151–184.

[103] V.P.H.U. Nguyen, M. Stroeven, L.J. Sluys, Multiscale continuous and discontinuous modeling of heterogeneous materials: a review on recent developments, J. Multiscale Model. 03 (4) (2011) 229–270.

[104] N. Otsu, A threshold selection method from gray-level histograms, IEEE Trans. Syst. Man Cybern. 9 (1) (1979) 62–66.

[105] W.J. Whitehouse, The quantitative morphology of anisotropic trabecular bone, J. Microsc. 101 (2) (1974) 153–168.

[106] K. Mizuno, M. Matsukawa, T. Otani, M. Takada, I. Mano, T. Tsujimoto, Effects of structural anisotropy of cancellous bone on speed of ultrasonic fast waves in the bovine femur, IEEE Trans. Ultrason. Ferroelectr. Freq. Control 55 (7) (2008) 1480–1487.

[107] A. Odgaard, Three-dimensional methods for quantification of cancellous bone architecture, Bone 20 (4) (1997) 315–328.

CHAPTER

# 11

# Analysis of the Biomechanical Behavior of Osteosynthesis Based on Intramedullary Nails in Femur Fractures

*Sergio Gabarre*, Jorge Albareda†,‡,§, Luis Gracia¶,‖, Sergio Puértolas¶,‖, Elena Ibarz¶,‖, Antonio Herrera‡,§,‖*

***Vlaams Instituut voor Biotechnologie, Leuven, Belgium †Department of Orthopaedic Surgery and Traumatology, Lozano Blesa University Hospital, Zaragoza, Spain ‡Aragón Health Research Institute, Zaragoza, Spain §Department of Surgery, University of Zaragoza, Zaragoza, Spain ¶Department of Mechanical Engineering, University of Zaragoza, Zaragoza, Spain ‖Aragón Institute for Engineering Research, Zaragoza, Spain**

## 11.1 INTRODUCTION

Femoral shaft fractures are among the most severe injuries of the skeleton. In particular, these fractures are the most serious of the long bones of the body, characterized by high morbidity and mortality [1, 2], and are frequently associated with significant complications and sequelae. They represent around 13% of total skeleton fractures [3].

For this reason, it is necessary that they are treated because of their complexity, seeking the most appropriate method depending on the characteristics and location of the fracture (Fig. 11.1), as well as the patient. Although there is no universally accepted classification, these fractures have been classified according to their location by Wiss et al. [4]

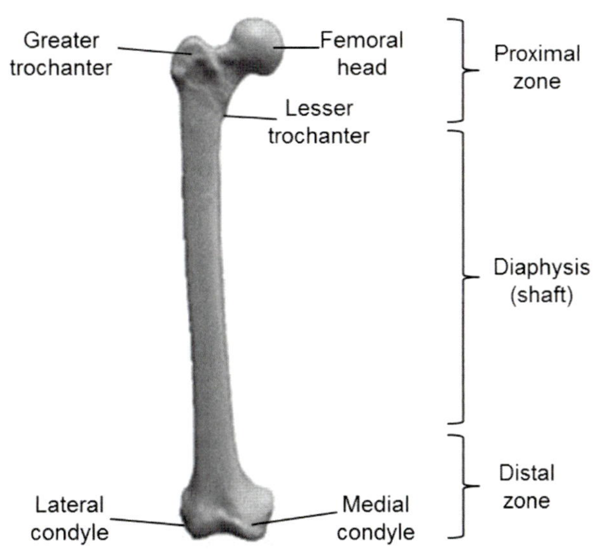


FIG. 11.1 Femoral zones.



Advances in Biomechanics and Tissue Regeneration
https://doi.org/10.1016/B978-0-12-816390-0.00011-X

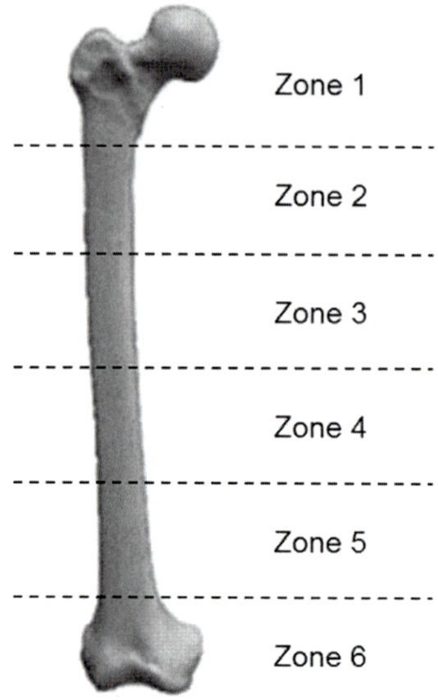


FIG. 11.2 Location of femoral fractures according to Wiss' classification.

(Fig. 11.2). In addition, diaphyseal fractures, by their degree of comminution, have been classified by Winquist and Hansen [5] (Fig. 11.3), type IV being the most difficult to treat and with increased complications and sequelae because of their instability.

The treatment of these fractures, always surgical, since the 1980s has been done by using intramedullary nails [6], which have many designs. These nails have revolutionized the treatment of diaphyseal femoral fractures, increasing their indications to the totality of fractures between zones 2 and 5 of Wiss et al. [4], regardless of the type of fracture, and presenting high values of consolidation with complications and sequelae [7, 8]. Since then in a bid to improve results, changes in nail design, morphology, material manufacture, screw configuration, and surgical approach have been introduced. Currently, there are stainless steel or titanium nails; slotted instead of grooved nails; hollow or solid nails to give greater rigidity; and oblique, transverse, spiral blade screws, etc. These developments have increased the stability and degree of fixation of the screws in the osteoporotic bone, and with anterograde or retrograde surgical approaches have increased their indications to more distal fractures. There are also reamed or unreamed nails to minimize vascular injury. Reamed nails decrease the risk of pulmonary embolism caused by the increase in intramedullary pressure produced during reaming [9], although this point continues to be controversial because some authors do not find a significant difference in pulmonary embolism between reamed and unreamed nails [10]. That is to say, there are multiple therapeutic possibilities but there is no consensus on the indication of each type of nail or surgical approach to the different types of fractures.

Regarding the use of clinical findings as an aid in decision making, most are satisfactory, but a definitive conclusion has not been reached as to therapeutic indication.

The use of reamed or unreamed nails is a persistent discussion [7]. In a meta-analysis performed in [11], there is scientific evidence of the best results of reamed nails against unreamed nails in terms of resurgeries, consolidation delays, and pseudarthrosis. Similar results with both types of nails and techniques as to breakage of the implant and the production of distress and respiratory failure [11] have been found. However, complications have been found in the use of reamed nails, especially in polytraumatized patients with lung injury [7], and because of this unreamed nails have been designed in an attempt to decrease the respiratory impact of reamed nails. However, this theoretical beneficial effect of unreamed nails in polytraumatized patients with respiratory involvement is not clinically proven [11]. There is some controversy because a number of authors have achieved excellent results with few complications using unreamed nails [12].

With respect to retrograde nails, Papadokostakis et al. [13], in a meta-analysis that studied the results of treatment with a retrograde nail in distal and diaphyseal fractures, found that this type of nail is a treatment option for distal fractures, but not for diaphyseal fractures because it produces high rates of pain in knees and a greater number of pseudarthroses and resurgeries than when using anterograde nails. These higher rates of failure in retrograde nails are due to the use of unreamed nails of small diameter, smaller than the diameter of the femoral medullary canal. At present, comparative results with anterograde nails are being discussed. Thus retrograde nails are preferable in patients with difficult access to the greater trochanter, such as obese and pregnant patients, and in patients with ipsilateral fracture of the tibia, which is tactically advisable when treating both fractures with a unique approach [7].

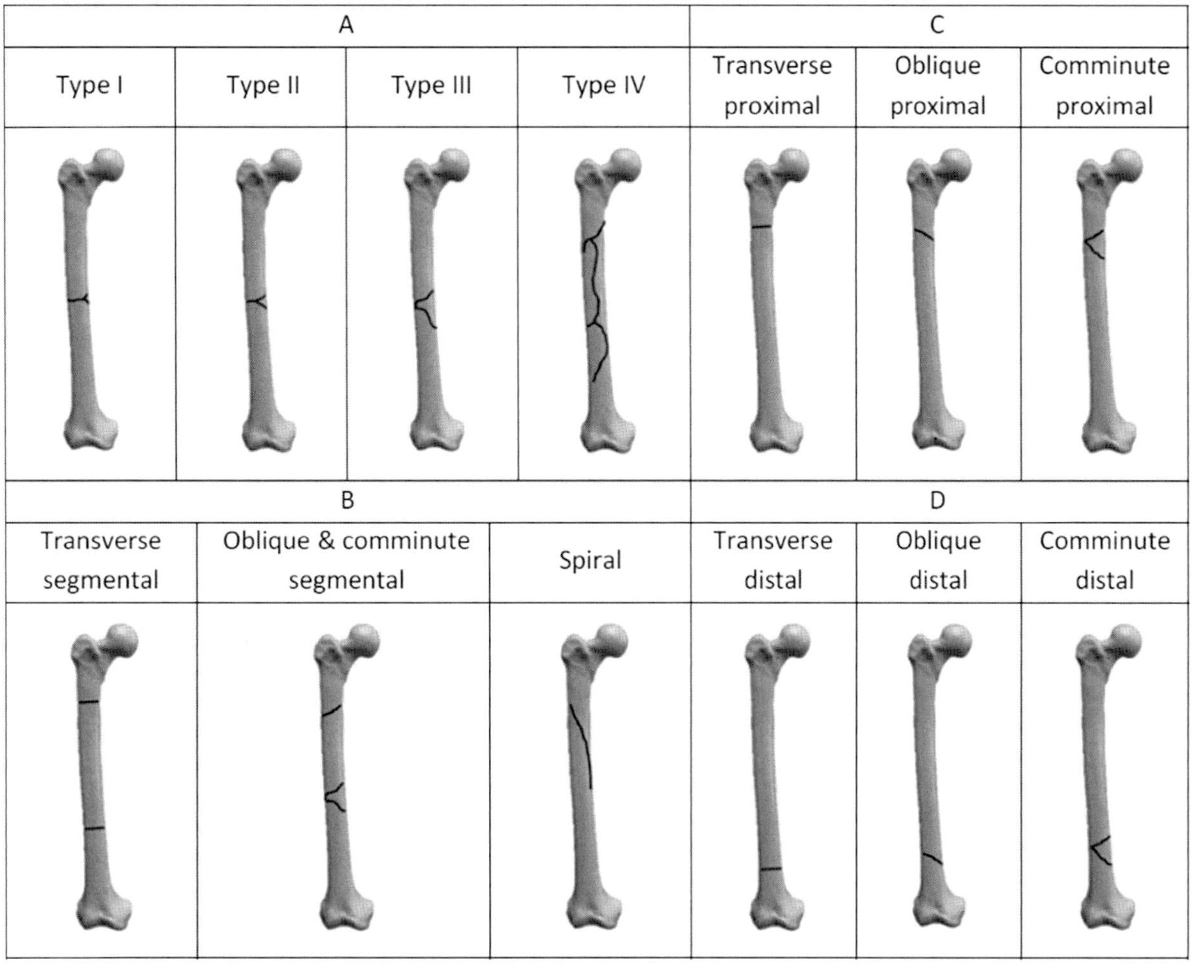


FIG. 11.3 Femoral fractures according to Winquist and Hansen's classification.

Despite the high number of designs, techniques, and materials, the intramedullary anterograde reamed nail with a static combination of screws continues to be the reference treatment of fractures of femurs located between the 2 and 5 Wiss zones [4, 14, 15], depending on the success of the treatment, the characteristics of the fracture, the body habits of the patient, the associated lesions, and the experience of the surgeon using this technique [7].

Experimentation with artificial bones or corpses in the lab, or with experimental animals, trying to study the biomechanical behavior of various types of osteosyntheses in different fractures, is an important aid to clinical practice for determining the therapeutic indication appropriate for every type of fracture. In this field, there are many studies of all kinds of variables in terms of models of nails, techniques, and types of fracture, but the results are inconclusive and sometimes divergent, particularly regarding their application to human clinical practice, which must be corroborated with clinical studies. Research in experimental animals presents application difficulties due to anatomical differences and load conditions, including complex application studies on cadaver bone or plastic anatomical models [16].

Due to differences between experimentation in vivo and in vitro, finite element (FE) models have emerged as a powerful tool that simulates different biological systems in both physiological and pathological conditions, although there are few articles studying the behavior of intramedullary nails in femoral bone. With regard to fractures of the femur and biomechanical behavior of the different osteosynthesis techniques, experimental works studying multiple variables have been developed [17]. The location and type of fracture is a factor of utmost importance. The most studied fractures have been supracondylar or distal femoral fractures because of their greater complexity, greater number of complications, and multiple treatment options. Traditionally, their treatment has been based on plates associated with dynamic screws, screws, or monoblock sheets; however, due to the lateral location of the plate, frequent medial collapses in unstable fractures are produced [18]. For this reason, retrograde intramedullary nails have been designed specifically to treat this type of fracture in an attempt to improve the biomechanical behavior of the fracture and implant, and minimize surgical aggression; however, this type of fracture is still subject to breakage of the implants at the level of the holes of the unplaced screws [19]. Nevertheless, in biomechanical experimental studies using artificial bones, the classic and traditional anterograde nails introduced in a retrograde manner have shown better results in terms of stability of this distal fracture of the femur compared to those specifically designed as retrograde nails for this type of fracture [20].

Despite the new designs, anterograde nails continue to be used with good clinical results in these complex fractures, although some nails have increased the number of distal screws and the possibility of placing them on different planes of space to increase the stability of the fracture [21]. With respect to its diaphyseal location, Montanini et al. [22] on an FE model of the diaphyseal fracture treated with intramedullary nails found that immediately after the fracture, the loading should not exceed the critical tensions of breakage of the implant into the hole in the proximal screw. These stresses in the nail disappear once the fracture is consolidated, which has been clinically confirmed [23–25].

Comparing only the biomechanical behavior of the intramedullary nails and osteosynthesis plates in fractures at the diaphyseal location using FE models, plates can allow greater stability to the focus of the fracture, while nails suffer increased deformities in the monopodal support. The increase in the diameter of the nail is critical because an increase of 2mm decreases the deformation of the nails by 40% [26] and therefore increases the chances of failure. This work confirms the results published by Heiney et al. [27] in the case of unstable distal fractures of the femur, where comparing nails with plates showed that there was a greater chance of failure of the implant in nails than in plates.

A debated and unsolved point is nail locking. On some models, the proximal screw is only one oblique from the greater to the lesser trochanter, while in other models there are two proximal transverse screws. Placement depends on the type of fracture, the type of screw, its diameter, stresses caused by screws and brittle points in the nail holes of unplaced screws, its proximity to the fracture focus, placement plans, etc.; these points remain clinically and biomechanically unclarified. In terms of the diameter of the screws, there must be a compromise between a maximum value, which does not exceed 50% of the diameter of the nail to ensure its resistance [28], and a minimum value, which ensures their resistance to the loads and stresses to which they are subjected. The smaller is the diameter of the nail, smaller should be the diameter of the screws. This is why some models of unreamed nails have increased their diameter to allow larger diameter screws.

The proximity of the distal screws to the fracture focus is an important point. The closer the screws are to fracture focus increases the stresses and forces supported by screws, while if they are further away from the focus of fracture, rotational stability improves and their chances of failure are lower [29]. They must always be positioned perpendicular to the axis of the nail with at least two screws, except in transverse fractures without comminution in which a unique screw may be sufficient.

Wahnert et al. [30] compared the stability achieved with different types of distal screws (screws and coiled sheets) and with a nail plate using an artificial model of osteoporotic femoral distal fracture subjected to rotational and axial loads. The conclusion is that in these fractures and against rotational stresses, distal locking of the nail with four screws with different angles is greater regarding the stability granted to the fracture than other types of locking systems with two screws lateral to medial or using a coiled sheet, and is similar to the stability granted with the nail plate. Locking with four distal screws obtained the best biomechanical results in terms of joint stability against rotational and axial stresses, and results on distal bolts have been confirmed in clinical studies [21]. Chen et al. [19], in a biomechanical study using FEs and artificial bones, explored the stresses in the screws and the rigidity of osteosynthesis in a retrograde nail in the treatment of distal femoral fractures. They came to the conclusion that distal screws are more important with respect to the stability of the fracture than proximal screws. A screw placed next to a fracture increases the rigidity of the mounting in oblique fractures, although this increase is not transcendent in transverse fractures and an unplaced screw determines an increase in stresses in the whole nail by 70%, facilitating their failure by breakage of the implant.

Nail material has also been studied. Pérez et al. [31] examined the biomechanical behavior of nails of stainless steel and titanium using an FE model in femoral fractures in children. The model is not applicable to fractures in adults or to the behavior of the intramedullary nails, but the conclusion is that titanium behaves best, since stainless steel creates stress-shielding areas in the bone that increase the risk of refracture once implants are removed. However, Kaiser et al. [32] obtained different conclusions in terms of nail material. In this case, they compared intramedullary nails of steel or titanium in the stabilization of diaphyseal femoral fractures using artificial bones, and concluded that steel allows greater stiffness to the mounting and that titanium must only be used in cases of allergy to metals or in cases where future scans by magnetic resonance imaging are needed. The behavior of different materials for intramedullary nails in the treatment of diaphyseal fractures of the distal femur in adult has not yet been studied.

In conclusion, primitive stainless steel anterograde intramedullary nails with a static combination of screws are nowadays the reference treatment for femoral fractures from zones 2 to 5 of Wiss; however, there has been no clear demonstration of the superiority of other nails and techniques specifically designed for the treatment of certain fractures. Broad discussions on the material to be used (stainless steel or titanium), on the route of entry of the nails (anterograde or retrograde), reamed or unreamed, and the placement and types of distal screws depending on the type of fracture to treat are still ongoing.

## 11.2 METHODOLOGY OF SIMULATION

As already indicated, treatment of fractures of the femur between the 2 and 5 zones of Wiss is performed with the intramedullary nail with different combinations of screws. There are multiple designs of nail, steel or titanium, reamed and unreamed, solid or hollow, anterograde or retrograde, and various types of locking system. The use of one or other depends on the type and location of the fracture, the characteristics of the patient, and the experience of the surgeon. There is no scientific, clinical, or experimental evidence that demonstrates the best results and indications for each type of nail in each type of fracture.

In view of the difficulties experienced with in vitro testing or in experiments with living subjects, FE simulation models have been developed to carry out research on biomechanical systems with high reproducibility, versatility, and limited cost. These models allow the study to be repeated as many times as desired, being a nonaggressive investigation of modified starting conditions (loads, material properties, etc.). However, work continues on the achievement of increasingly realistic models that allow placement of the generated results and predictions into a clinical setting. To that end, it is necessary to use meshes suitable for every particular problem, regarding both type of element and size. This is necessary to perform a sensitivity analysis to determine the optimal features or, alternatively, the minimum mesh necessary to achieve the required accuracy [33].

Thus through the development of a computational model based on the FE method, it is possible to study the biomechanical behavior of the same nail made of two materials (steel or titanium), introduced with or without reaming, anterograde or retrograde, with different types of locking systems in different types of stable and unstable fractures of the femur. This is done to find the best indication and therapeutic technique for each type of femoral fracture from the subtrochanteric to the supracondylar region.

For this purpose, the methodology consists of the development of an FE model of a femur, on which will be simulated various types of fractures in the subtrochanteric, diaphyseal, and supracondylar areas, stabilized through various assemblies and materials of intramedullary nails. The mechanical strength of the nail against bending and compressive loads is also studied to determine its maximum strength. Subsequently, a comparative analysis of the different types of fixation in fractures is developed to verify what is the optimal solution in each of the analyzed cases. The biomechanical results are evaluated in correlation with observed clinical results. Thus it is possible to observe biomechanical needs for each type of fracture to find the optimum combination of variables (type of nail, locking system, material, surgical approach, etc.) ideal for their treatment.

To develop an FE model, the first phase is to generate the geometry of the different parts that define the model. One of the most significant aspects of biomechanical systems is their geometric complexity, which greatly complicates the generation of accurate simulation models. Thus the use of scanners together with three-dimensional (3D) images obtained by computed tomography (CT) generate geometric models that combine high accuracy in the external form with an excellent definition of internal interfaces. The method requires not only appropriate software tools, capable of processing images, but also compatibility with the programs used later to generate the FE model [33].

Development of the model of a healthy femur is crucial to perfect the whole process of simulation, and to obtain reliable results. A 3D FE model of the femur from a 55-year-old male donor was developed. To obtain a faithful geometry, the bone was scanned using a 3D Roland Picza laser scanner (Fig. 11.4).

This device offers two sweep modes: a plane-based and a rotary sweep with a scanning resolution of 0.2 mm. The scanner provides a first approximation of the outer geometry represented by a cloud of points. By means of its own scanning software (Roland Dr. Picza 3) [34], general and rough cleaning operations were performed and eventually treated afterwards in Pixform software [35]. The scanned file was subjected to a specific cleaning protocol to pull the 3D image out of scan noise (deleting spikes, cleaning abnormal and nonmanifold surfaces, fixing bad normals, etc.). Those initial rough geometries with noisy faces and screws attached to a support to fix firmly the bone while scanning are shown in Fig. 11.5.

Local operations could also be carried out as smoothing and bridging gaps, because a fully closed geometry was required to continue with the process. Fig. 11.6 displays the final geometry of the femur after the cleaning and geometry treatment process was accomplished.

Eventually, when a closed geometry was obtained, it was finally wrapped by an ensemble of Bézier parametric surfaces as is exhibited in Fig. 11.7. The order and number of these surfaces were varied until a proper fidelity was reached. Analysis tools were provided in this software to check deviation from the original geometry of the surfaces generated. Afterward, they were exported to I-DEAS 11 NX Series software [36]. First, each volume was checked for suitability for meshing. If not, problematic surfaces were identified and the possible source problem was referred back to Pixform and Rhinoceros software [37].

FIG. 11.4 Roland LPX-250 3D laser scanner.

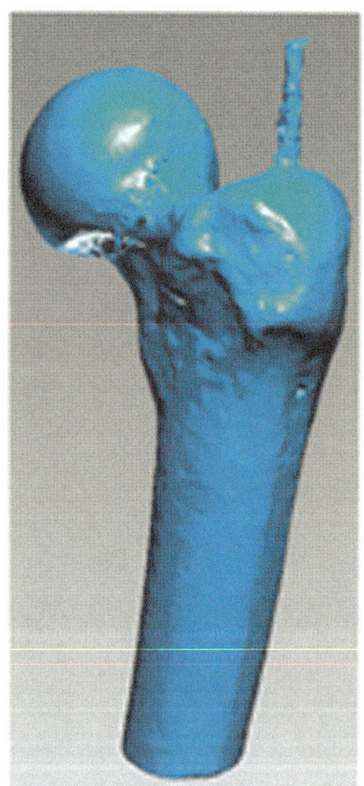

FIG. 11.5 Initial geometries directly obtained from scanning.

Because the 3D scanner provided only the outer geometry, CT images of the femur were needed to quantify mineral bone density and eventually assign its corresponding Young's modulus. A CT image treatment was performed with Mimics software (Belgium) [38]. A CT scan (512 × 512 acquisition matrix, field of view = 240 mm, slice thickness = 0.5 mm in plane resolution) was obtained using a Toshiba Aquilion 64 scanner (Toshiba Medical Systems Zoetermeer, Netherlands). Stacks of images from each bone were processed using Mimics. A threshold of 700 Hounsfield units was chosen to start cleaning the stack of images for the bone. This threshold served to establish an approximate border between cortical and cancellous bone.

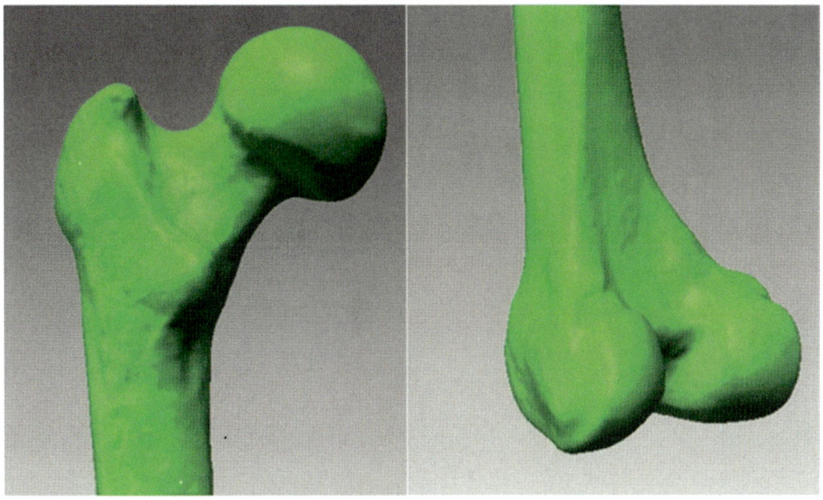

FIG. 11.6 Smoothed and treated geometry of the femur.

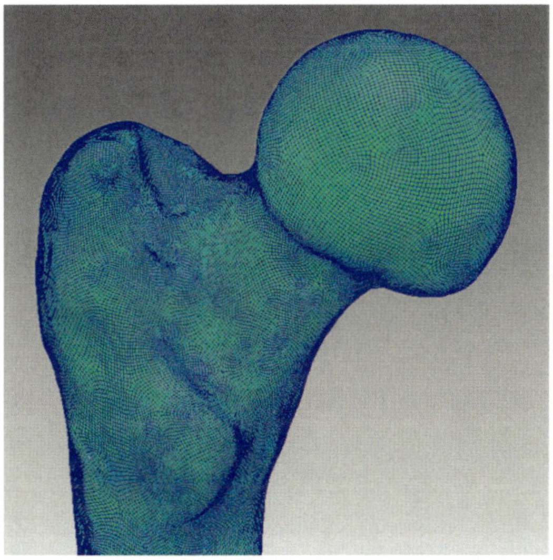

FIG. 11.7 Surface ensemble wrapping previously treated scanned geometry for the proximal femur.

Masks, 3D objects, and 3D polylines were generated during an iterative process until a smoothed and properly cleaned geometry was obtained without any artifacts, spikes, or geometric irregularities. After this, the whole set of polylines generated was exported to I-DEAS. The initial set of polylines was reduced to smaller equally spaced groups depending on its distance: 4 mm, 3 mm, 2 mm, and 1 mm. The minimal distance was placed at the femoral head and distal femur part. These areas are of major interest because they are the region where boundary conditions are imposed. Fig. 11.8 shows a superposition of the 3D volumes of cortical bone with the corresponding CT image at a certain level (available with clipping in Mimics).

The next step of the process was the proper alignment and orientation of the initial 3D geometry with the one exported from Mimics. Homolog polylines from the scanned anatomical model were generated (see red lines in Fig. 11.9). These polylines were obtained by cutting the bony geometry by auxiliary planes at the same previous Z levels. The aim was to create a connection between both geometries. Each geometry level or bone cut was treated by means of a novel algorithm to assign the corresponding density values between both geometries. The connection between corresponding levels was made to determine the start/end homolog points for each level represented in Fig. 11.9 by the pairs of black arrows on the right side of the image.

The intramedullary nail Stryker S2 model (Stryker, Mahwah, NJ, USA) was used for the study, with a length of 380 mm, a wall thickness of 2 mm, and an outer diameter of 13 mm. The corresponding locking screws have an outer diameter of 5 mm. The geometrical model for the nail and the screws was generated by means of the NX I-DEAS program (Figs. 11.10–11.12).

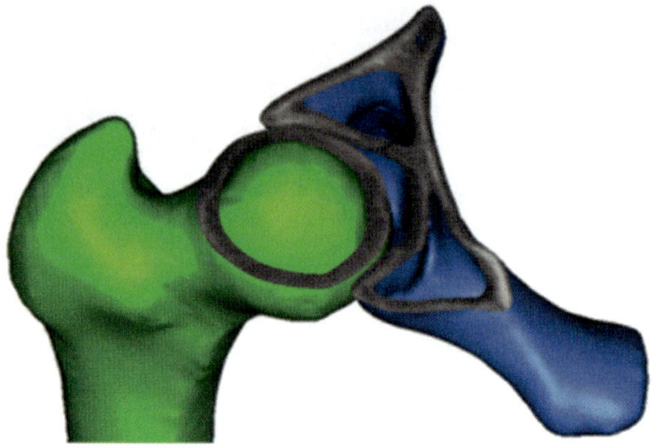

FIG. 11.8 Cross-section of femur volume belonging to cortical bone with the CT image corresponding to this level.

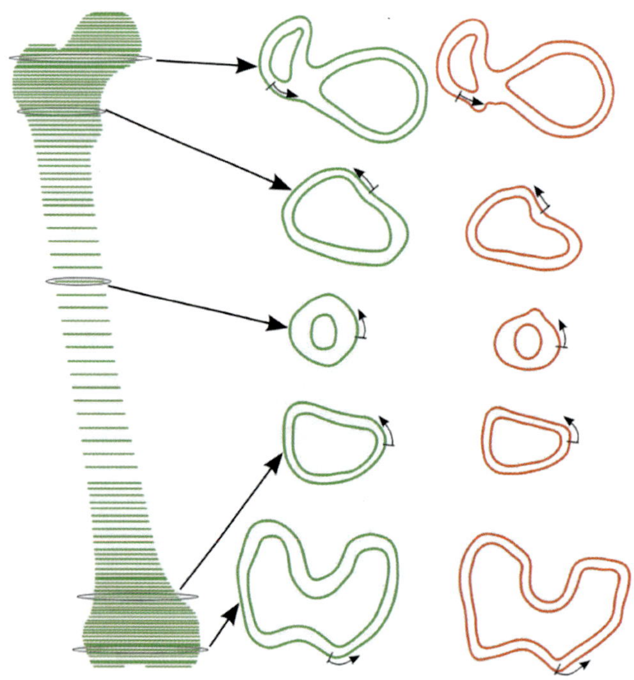

FIG. 11.9 Assignment scheme between Mimics *(left)* polylines and bone cut *(right)* polylines.

On the geometrical model of the healthy femur, it is necessary to make the appropriate modifications to simulate different fractures, according to the classification of comminution of Winquist and Hansen, considering different locations (subtrochanteric, diaphyseal, and supracondylar fractures). For this purpose, pairs of outdated uneven surfaces around the desired fracture gap were generated (Fig. 11.13). This process was carried out in NX I-DEAS.

After obtaining the geometry of the fractured femoral bone, the intramedullary nail, and the screws, the intramedullary nail with the corresponding screws was positioned in the femur using NX I-DEAS software in the same way as one would carry out a real surgery. This assembly of the computer-aided design model was performed under the supervision of a surgeon.

After defining the geometry, the mesh can be generated. Bone, nail, and screws were modeled with linear tetrahedra with a reference size of 1.5 mm, using NX I-DEAS. Two details of the final mesh of bone are shown in Fig. 11.14.

Afterwards, an interpolation technique was adopted to assign the property to every mesh element of each bone located in between consecutive splines. Each tetrahedron was reduced to its barycenter, projecting it to each plane between where it was situated. Once all these projections were done, density assignment to every projection was calculated by means of the developed algorithm in FORTRAN [39]. The stiffness was assigned to every bone element depending on the previous density assignment.

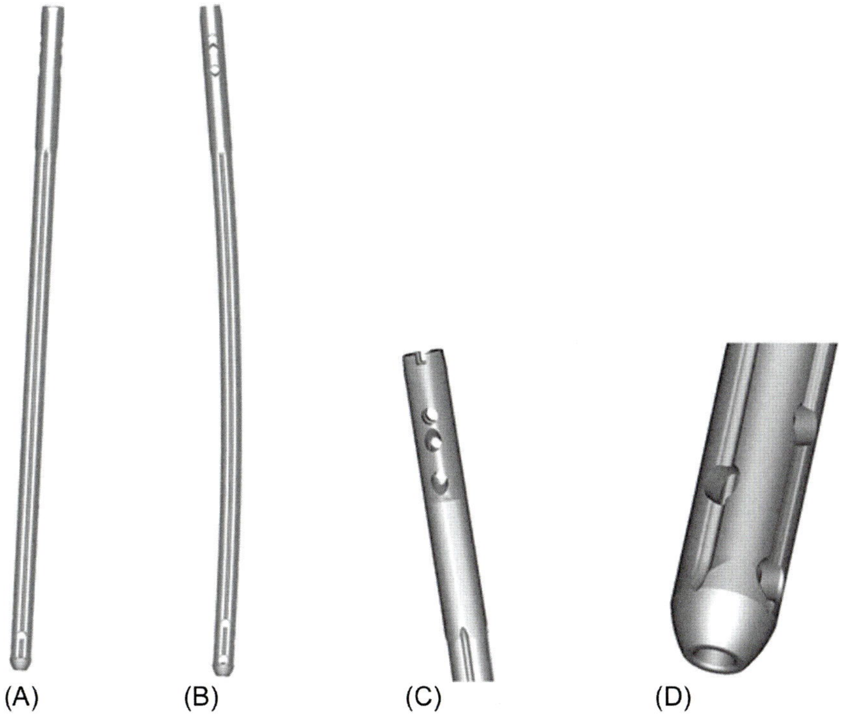


FIG. 11.10 Geometrical model of the intramedullary nail: (A) front view of antegrade nail; (B) sagittal view of antegrade nail; (C) detail in perspective of the head of the nail; (D) detail in perspective of the tip of the nail.

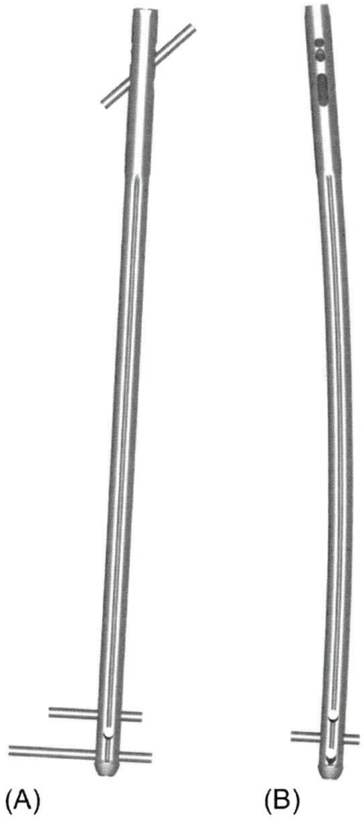


FIG. 11.11 Geometrical model of the set nail and screws: (A) front view; (B) sagittal view.

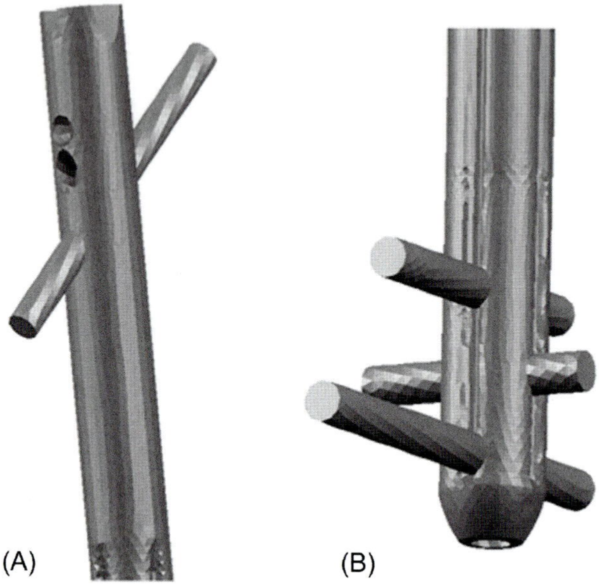


FIG. 11.12 Detail of screws in the intramedullary nail: (A) head; (B) tip of the nail.

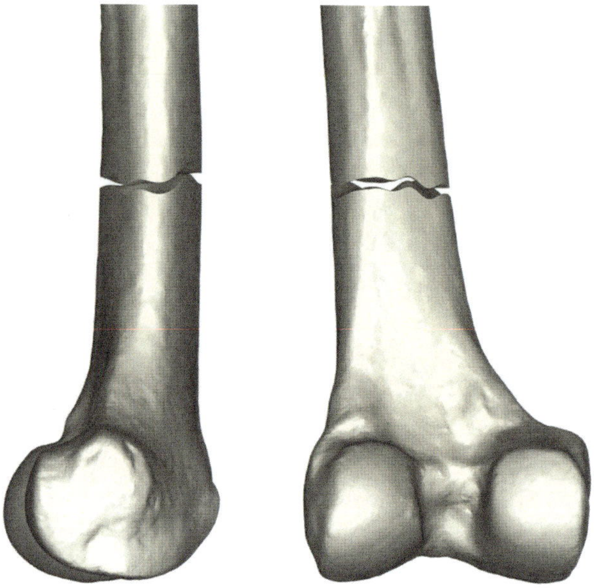

FIG. 11.13 Detail of surfaces around the fracture gap.

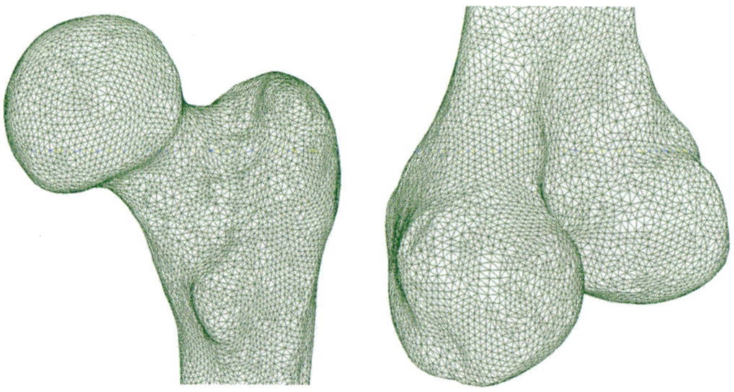

FIG. 11.14 Final mesh of the femur.

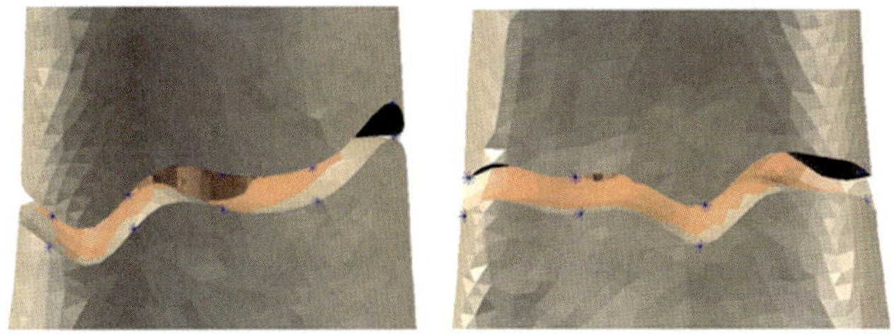

FIG. 11.15 Perspective details of the fracture generated along with the homologous points to measure the micromovement (marked with *dots*).

Subsequently, in the fracture site, pairs of homologous points were determined (Fig. 11.15). These points, selected from the mesh nodes located opposite to each other, will be used to measure the micromovements and identifying trends in forming bone callus to the imposed loads. The model of the fractured femur was meshed according to the previously described conditions.

In the same way, the intramedullary nail and screws were meshed again using NX I-DEAS software. The FE model of the intramedullary nail is shown in Fig. 11.16.

A linear tetrahedral was used to mesh the complete osteosynthesis model. The final FE model is shown in Fig. 11.17.

To guarantee the accuracy of the FE results, a sensitivity analysis was performed to determine the minimal mesh size required for an accurate simulation. For this purpose, a mesh refinement was performed to achieve a convergence toward a minimum of the potential energy, with a tolerance of 1% between consecutive meshes. As an example, the statistics corresponding to one of the FE models are presented in Table 11.1.

In the FE simulation, the appropriate characterization of the mechanical behavior of the different materials, which is usually very complex, is essential. Once the inner interface between the cortical and trabecular bone was determined in the way explained before, material properties were assigned to the FE model in NX I-DEAS. They were assumed linear elastic isotropic properties for the bone, with variable values related to the processed CT images [40]. The metallic nail was made of 316 LVM steel or Ti-6L-4V and the metallic screws were made of 316 LVM steel, both assumed to be linear elastic isotropic. Table 11.2 summarizes the mechanical properties values used in different materials.

Concerning the load conditions, a load case associated with an accidental support of the leg at early postoperative stage has been considered. This load was quantified to be about 25% of the maximum gait load. According to Orthoload's database (Fig. 11.18), the hip reaction force and abductor force, referring to 45% of the gait, corresponded to the maximum and most representative load [42]. Forces generated by the abductor muscles were applied to the proximal area of the greater trochanter, in agreement with most classic authors' opinions [43, 44] (Fig. 11.19). Fully constrained boundary conditions (Fig. 11.20) were applied at the distal part of each femur (at the condyles).

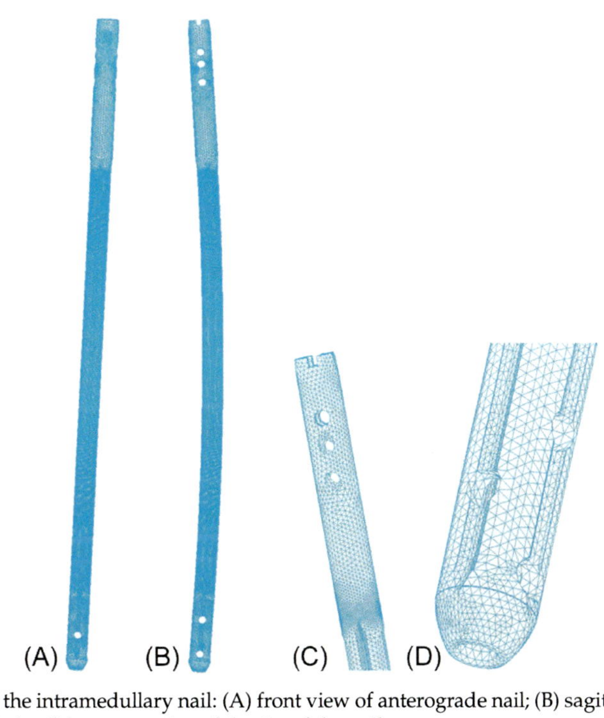


FIG. 11.16 Finite element model of the intramedullary nail: (A) front view of anterograde nail; (B) sagittal view of anterograde nail; (C) detail in perspective of the head of the nail; (D) detail in perspective of the tip of the nail.

FIG. 11.17 Cross-section of the femur joint, intramedullary nail, and screws.

TABLE 11.1 Mesh Statistics of the Finite Element Models

| Part of model | Element type | Number of elements |
|---|---|---|
| Femoral bone | 4-node linear tetrahedron | 427,939 |
| Cortical bone | 4-node linear tetrahedron | 216,169 |
| Trabecular bone | 4-node linear tetrahedron | 211,770 |
| Intramedullary nail | 4-node linear tetrahedron | 574,547 |
| Screw #1 | 4-node linear tetrahedron | 2417 |
| Screw #2 | 4-node linear tetrahedron | 1454 |
| Screw #3 | 4-node linear tetrahedron | 1111 |
| Screw #4 | 4-node linear tetrahedron | 2818 |
| Total | | 1,007,869 |

TABLE 11.2 Material Properties

| Elastic isotropic | Young's modulus (MPa) | Poisson coefficient |
|---|---|---|
| Cortical bone[a] [41] | 20,000 | 0.3 |
| Trabecular bone[a] [41] | 959 | 0.12 |
| 316 LVM steel[b] | 192,360 | 0.3 |
| Ti-6L-4V[b] | 113,760 | 0.34 |

[a] *Values supplied by the manufacturer.*
[b] *Value corresponding to the maximum bone density. Elemental values were assigned according to the explained algorithm.*

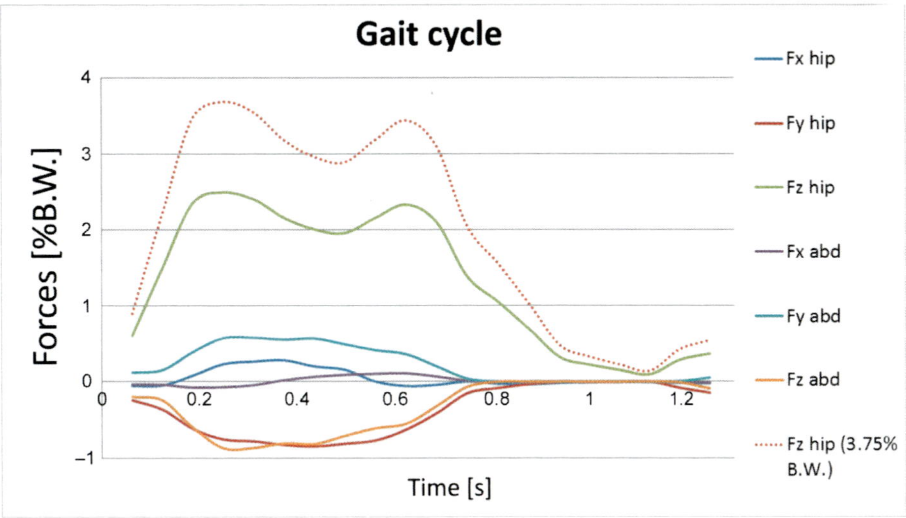


FIG. 11.18 Simulated gait cycle introducing hip reaction forces and abductor (abd) muscle forces. The maximal hip reaction force was 2.54% body weight (BW). The *red dotted line* is the $F_z$ hip reaction force for the scaled 3.75% BW gait cycle.

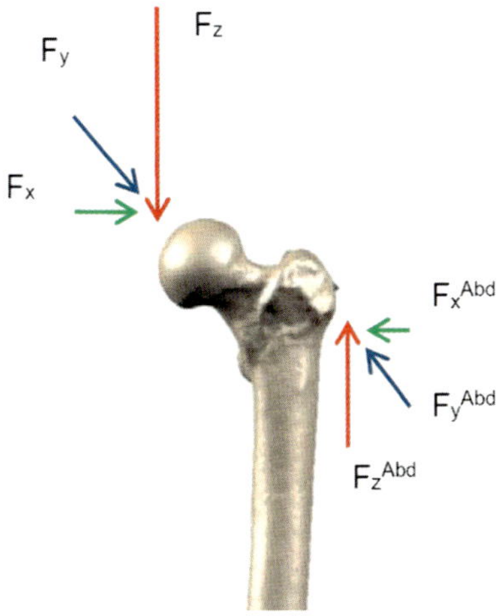


FIG. 11.19 Scheme of the load conditions applied to the model. *Abd*, abductor.

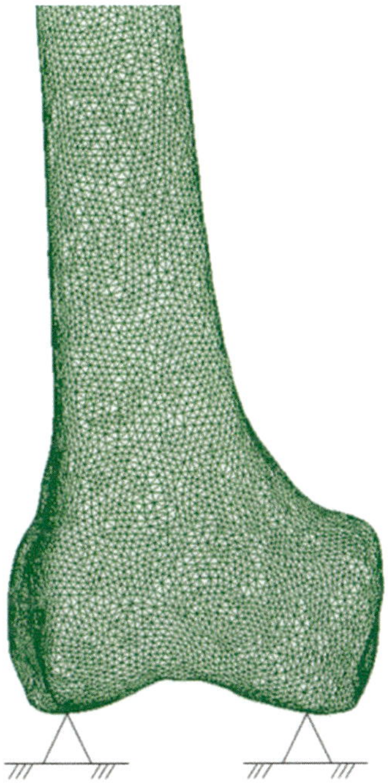

FIG. 11.20 Boundary conditions applied: femoral condyles constrained.

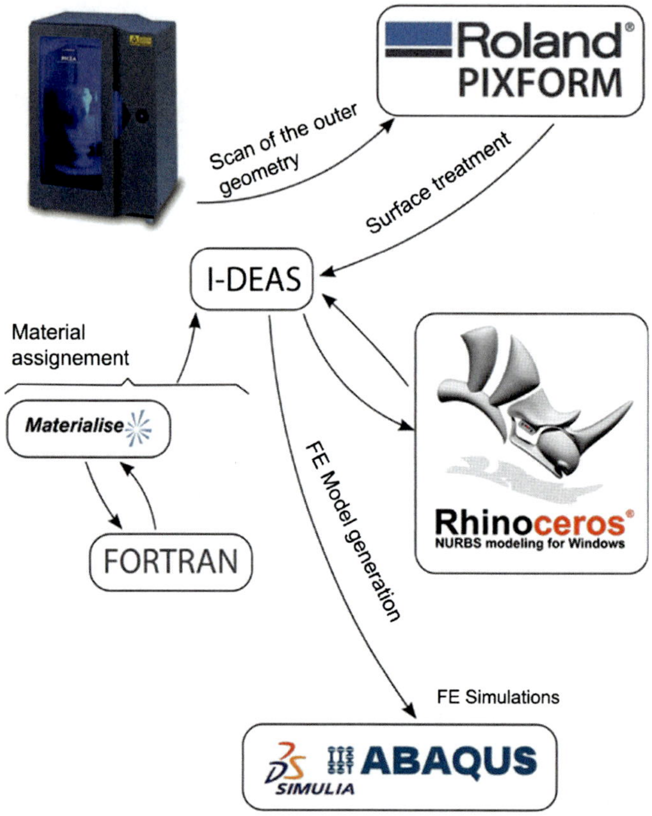


FIG. 11.21 Illustration of workflow followed through the acquisition of the geometry, finite element model generation, material assignment, and eventually simulations developed.

A key issue in the FE models is the interaction between the different constitutive elements of the biomechanical system, especially when it results in essential conditions affecting the behavior to be analyzed. In this way, the biomechanical behavior of this kind of osteosynthesis depends basically on the conditions of contact between the intramedullary nail and bone, so that the correct simulation of the interaction conditions determines the reliability of the model. The study was focused on the immediately postoperative stage. Thus the interaction at the fracture site did not take into account any biological healing process. Contact interaction was assumed between the outer surface of the nail and the inner cortex of the medullary canal of the femur. Tied interaction between screws and cortical bone was considered, whereas contact between screws and femoral nail was simulated. The selected friction values of bone/nail and nail/screws were 0.1 and 0.15, respectively, in accordance with the literature [45–47]. Of interest, other similar studies modeled bone/nail interaction as frictionless [22, 48].

The Abaqus 6.11 program [49] was employed for the calculations and postprocessing the results of the previously generated models in NX I-DEAS software.

To summarize, a schematic workflow is depicted in Fig. 11.21, exhibiting the overall software used to generate FE models and perform different simulations.

## 11.3 TYPES OF FRACTURES AND OSTEOSYNTHESIS

For the treatment of fractures to femurs between the 2 and 5 Wiss zones, an intramedullary nail with corresponding screws is used. However, the characteristics of this indication are not unique, since there are multiple combinations according to the geometric design of the nail, its material, the type of surgical approach, locking system, etc. Thus the use of one or another system depends on the type and location of the fracture, the characteristics of the patient, and the experience of the surgeon.

To help surgeons choose the best osteosynthesis in each case, two different studies were performed (A and B). The "A" study consisted of an analysis of the biomechanical behavior of a single system of osteosynthesis for different types of fracture (in terms of type and position along the femur). On the other hand, the "B" study corresponded to an analysis of the biomechanical behavior of different systems of osteosynthesis (with different locking systems) for different types of fracture with the same location in the femur (distal).

All the considered fractures were modeled as transverse by means of an irregular surface developed to represent a closer geometry to the actual fracture. The effect of the gap size was unclear in the literature. So, the majority of the reviewed in vivo studies referred to a gap size ranging from 0.6 to 6 mm [40, 50], whereas in FE simulation articles it ranged from 0.7 to 10 mm [47, 51].

Thus for the "A" study, three different fracture gaps were studied: 0.5 mm (considered as a noncomminuted fracture), 3 mm (the most referenced value found in the literature, representing a mid-value), and 20 mm as an example of a comminuted fracture. In addition to this, three localizations of the fracture were studied: proximal, medial, and distal for each gap size. Only one combination of screws was studied: one oblique placed proximally and two transversely at the distal part. On the other hand, the purpose of the "B" study was to investigate the optimal screw combination and gap size for a single distal fracture location, considering the same three gap sizes: 0.5, 3, and 20 mm, respectively. Thus four combinations of locking screws were considered: one oblique proximal screw combined with four configurations of the three distal ones, two lateral-medial and one anteroposterior. Table 11.3 summarizes the list of FE models simulated for the "A" and "B" studies (9 and 12 FE models, respectively). These models will be duplicated, since each one of these cases is carried out considering the two studied materials (stainless steel and titanium) of the nail.

Finally, to validate the conclusions obtained from simulations, a clinical follow-up was carried out for both studies, approved by the Ethics Committee of the Institute of Health Sciences of Aragón (protocol number CI PI15/0214). Thus for the "A" study, a sample of 55 patients, 24 males and 31 females, with a mean age of 52.5 years was obtained, all of them treated with femoral nail Stryker S2. Localizations of fractures were 32 in the right femur and 33 in the left femur. On the other hand, for the "B" study, a sample of 15 patients, 6 males and 9 females, with a mean age of 53.2 years was obtained, all of them treated with the same nail as in the "A" study. Localizations of fractures were 10 in the right femur and 5 in the left femur. The grade of comminution was measured in both cases according to the scale of Winquist and Hansen [5]. The distribution of cases corresponding to fracture localization and fracture grade are included, for the "A" and "B" studies, in Table 11.4.

**TABLE 11.3** Different Configurations Considered in the Finite Element Simulation

| FE model | Proximal screws | Distal screws | Fracture location | Gap size | | Screw configuration |
|---|---|---|---|---|---|---|
| A-01 | Oblique (#1) | 2L/M (#2, #3) | Proximal | 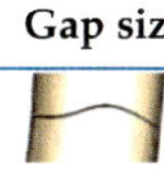 | 0.5 mm |   |
| A-02 | Oblique (#1) | 2L/M (#2, #3) | Proximal |  | 3 mm | |
| A-03 | Oblique (#1) | 2L/M (#2, #3) | Proximal |  | 20 mm | |
| A-04 | Oblique (#1) | 2L/M (#2, #3) | Medial | 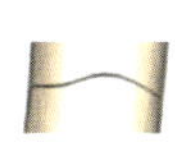 | 0.5 mm | |
| A-05 | Oblique (#1) | 2L/M (#2, #3) | Medial |  | 3 mm | |
| A-06 | Oblique (#1) | 2L/M (#2, #3) | Medial |  | 20 mm | |
| A-07 | Oblique (#1) | 2L/M (#2, #3) | Distal | 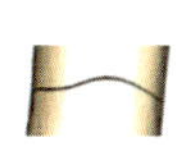 | 0.5 mm | |
| A-08 | Oblique (#1) | 2L/M (#2, #3) | Distal |  | 3 mm | |

*Continued*

**TABLE 11.3** Different Configurations Considered in the Finite Element Simulation—cont'd

| FE model | Proximal screws | Distal screws | Fracture location | Gap size | | Screw configuration |
|---|---|---|---|---|---|---|
| A-09 | Oblique (#1) | 2L/M (#2, #3) | Distal |  | 20mm | |
| B-01 | Oblique (#1) | 2L/M+1 A/P screws (#2, #3, #4) | Distal | 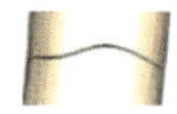 | 0.5mm |  |
| B-02 | Oblique (#1) | 2L/M+1 A/P screws (#2, #3, #4) | Distal |  | 3mm | |
| B-03 | Oblique (#1) | 2L/M+1 A/P screws (#2, #3, #4) | Distal |  | 20mm | |
| B-04 | Oblique (#1) | 1L/M+1 A/P screws (#2, #3) | Distal | 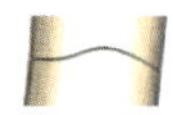 | 0.5mm | |
| B-05 | Oblique (#1) | 1L/M+1 A/P screws (#2, #3) | Distal |  | 3mm | |
| B-06 | Oblique (#1) | 1L/M+1 A/P screws (#2, #3) | Distal |  | 20mm | |
| B-07 | Oblique (#1) | 1L/M+1 A/P screws (#3, #4) | Distal | 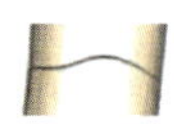 | 0.5mm | |
| B-08 | Oblique (#1) | 1L/M+1 A/P screws (#3, #4) | Distal |  | 3mm | |
| B-09 | Oblique (#1) | 1L/M+1 A/P screws (#3, #4) | Distal | 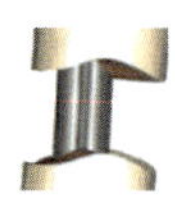 | 20mm | |
| B-10 | Oblique (#1) | 2L/M screws (#2, #4) | Distal | 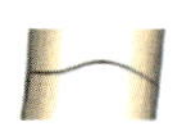 | 0.5mm | |
| B-11 | Oblique (#1) | 2L/M screws (#2, #4) | Distal |  | 3mm | |
| B-12 | Oblique (#1) | 2L/M screws (#2, #4) | Distal |  | 20mm | |

*A/P*, anteroposterior; *L/M*, lateral-medial.

## 11.4 RESULTS

The FE simulations allowed the mobility results for the different cases analyzed to be obtained. Fig. 11.22 shows, for the "A" study, the deformed shape amplified (×25) and the vertical displacement maps corresponding to non-comminuted fractures (gap size 0.5mm), mid-value gap (gap size 3mm), and comminuted (gap size 20mm). In Fig. 11.23, the same results can be observed for all four combinations of screws and steel nail corresponding to the "B" study.

TABLE 11.4 Statistics for the Clinical Follow-Up

| Study | Wiss zone | Cases | Comminution grade | Cases |
|---|---|---|---|---|
| A | 2 | 7 | None | 29 |
| | 3 | 11 | 1 | 9 |
| | 4 | 22 | 2 | 9 |
| | 5 | 15 | 3 | 1 |
| | | | 4 | 7 |
| | Total | 55 | Total | 55 |
| B | 5 | 9 | None | 9 |
| | 5 | 5 | 2 | 5 |
| | 5 | 1 | 4 | 1 |
| | Total | 15 | Total | 15 |

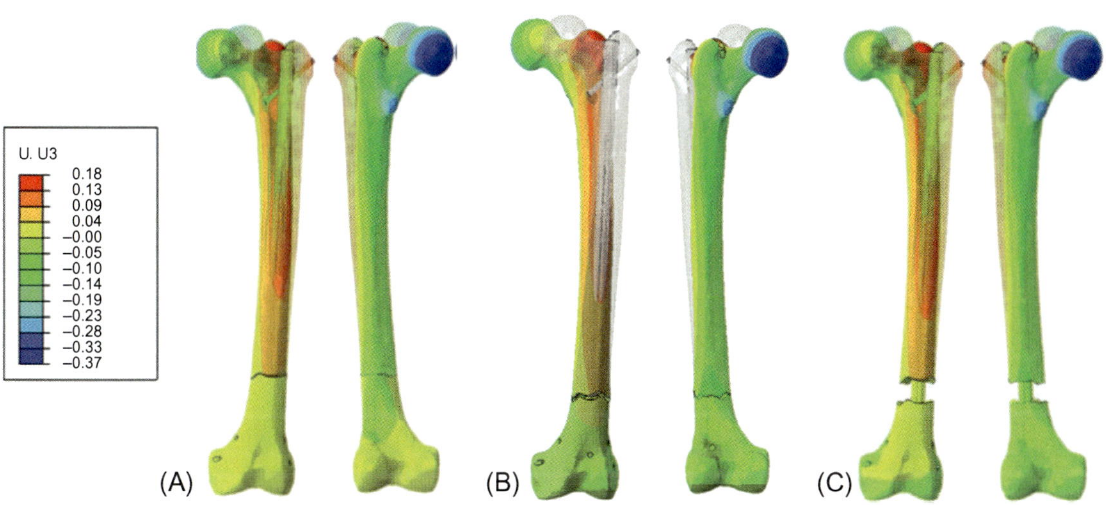


FIG. 11.22 Deformed shape (×25) and vertical displacement maps, for the "A" study, corresponding to distal fractures: (A) noncomminuted (gap size 0.5 mm); (B) mid-value gap (gap size 3 mm); (C) comminuted (gap size 20 mm).

The study of micromotions at the fracture site was measured as the relative motion between pairs of homologous points defined from opposed nodes depicted in Fig. 11.15. The maximum amplitude of micromotion between homologous points at the fracture site for steel and titanium nails is reported in Fig. 11.24 for the "A" study and in Fig. 11.25 for the "B" study, respectively.

Thus Fig. 11.24A shows that the most rigid behavior belongs to the distal fracture (40.69–66.43 μm), followed by the medial one (51.96–73.39 μm), and proximal one (60.29–90.29 μm). Micromotion amplitude follows the same growing tendency with the increase in gap size for all three fracture locations. For the titanium nail, Fig. 11.24B shows the same tendency at the three fracture locations observed previously: micromotions at distal fracture ranges from 62.02 to 123.71 μm, followed by the medial one (ranging from 75.88 to 139.80 μm), and finally the proximal one (varying from 93.07 to 140.83 μm). If the ratio of the amplitudes between both materials is calculated, a pitchfork of 1.46–2.00 is obtained, which is located within the range of Young's modulus ratio for both materials (1.69).

On the other hand, in Fig. 11.25 it can be observed that the most rigid behavior of both nail materials corresponds to the fourth interlocking system: 40.69 μm (gap size of 0.5 mm) and 48.33 μm (gap size of 3 mm), whereas the first one (three distal screws) shows the best stability in terms of micromotions for the biggest gap size of 20 mm: 63.50 μm. The second and third screw combinations exhibit a similar behavior when the nail material is changed to titanium among the three gap sizes.

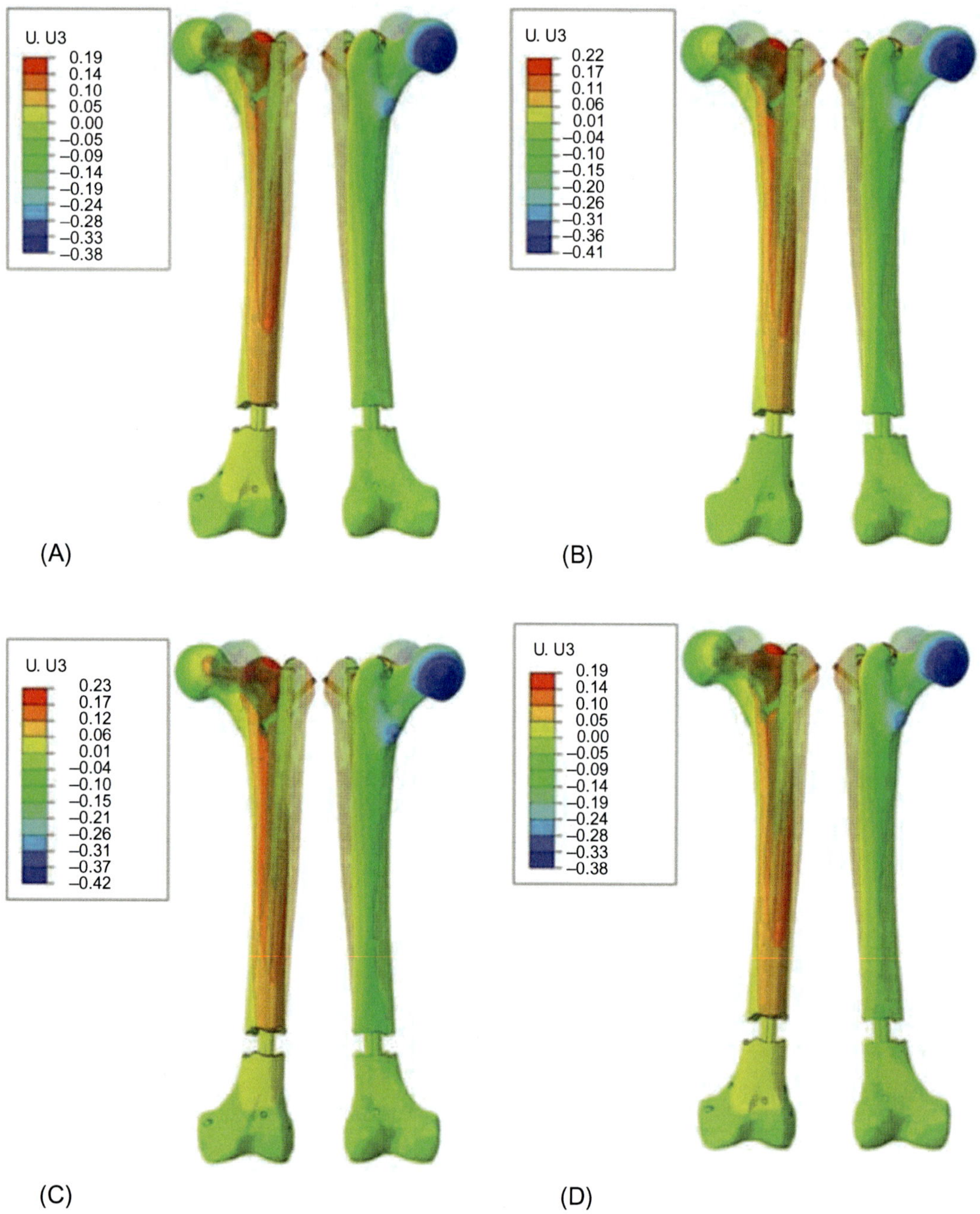


FIG. 11.23 Deformed shape (×25) and vertical displacement maps, for the "B" study, corresponding to a distal fracture: (A) 1st interlocking system; (B) 2nd interlocking system; (C) 3rd interlocking system; (D) 4th interlocking system.

With respect to the evaluation of global stability by measuring the displacement at the head of the nail (insertion point at the trochanter), Figs. 11.26 and 11.27 show the results obtained for the "A" and "B" studies for both intramedullary materials. In this way, when evaluating global stability, in the "A" study, the trend is reversed with respect to the amplitude of axial micromotion. In this case, the proximal fracture is the most rigid, followed by medial fracture and distal fracture. This result is obtained because when the physiological loads at the head of the femur are applied, the intramedullary nail blocks the global movement of the femoral head "sooner" for the proximal fracture than for the distal one. According to gap size influence, there is a marked increase in the interfragmentary movement as well as global stability when the gap increases. Thus for the steel nail, values range from 1.33 mm (proximal fracture, 0.5 mm

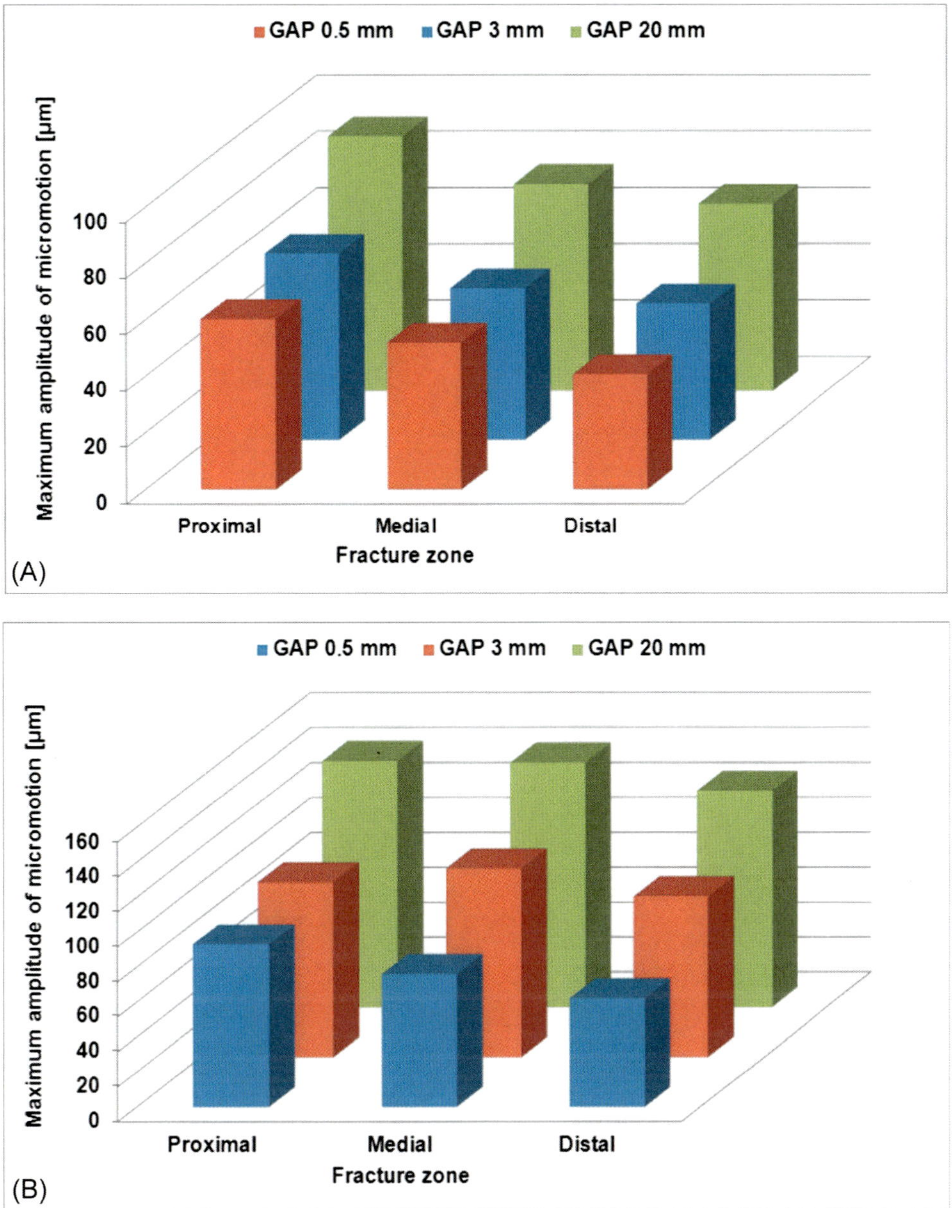


FIG. 11.24 Amplitude of axial micromotion (μm), for the "A" study, corresponding to different nail materials: (A) steel intramedullary nail; (B) titanium intramedullary nail.

gap) to 2.01 mm (distal fracture, 20 mm gap), whereas the titanium nail yields a higher rate of global movement: 1.62 mm (proximal fracture, 0.5 mm gap) to 3.14 mm (distal fracture, 20 mm gap). By calculating the ratio of the global movement between both materials a pitchfork of 1.22–1.56 is obtained.

However, in the "B" study, global stability of each fixation system follows similar tendencies as the aforementioned amplitude of micromotion for the steel nail and titanium nail. The global movement at the top of the nail was measured yielding the most rigid behavior for the fourth interlocking system: 1.75–2.01 mm for the steel nail, whereas for the titanium nail, the first screw combination showed the smallest motion for the first interlocking system: 2.81 and 2.80 mm (3 mm and 20 mm gap size, respectively). For the smallest gap size, the fourth interlocking system was again the most stable in terms of global movement (2.36 mm). Analogously to the analyzed micromotions, the second and third fixation systems yield similar results for both materials in the two gaps associated with comminuted fractures.

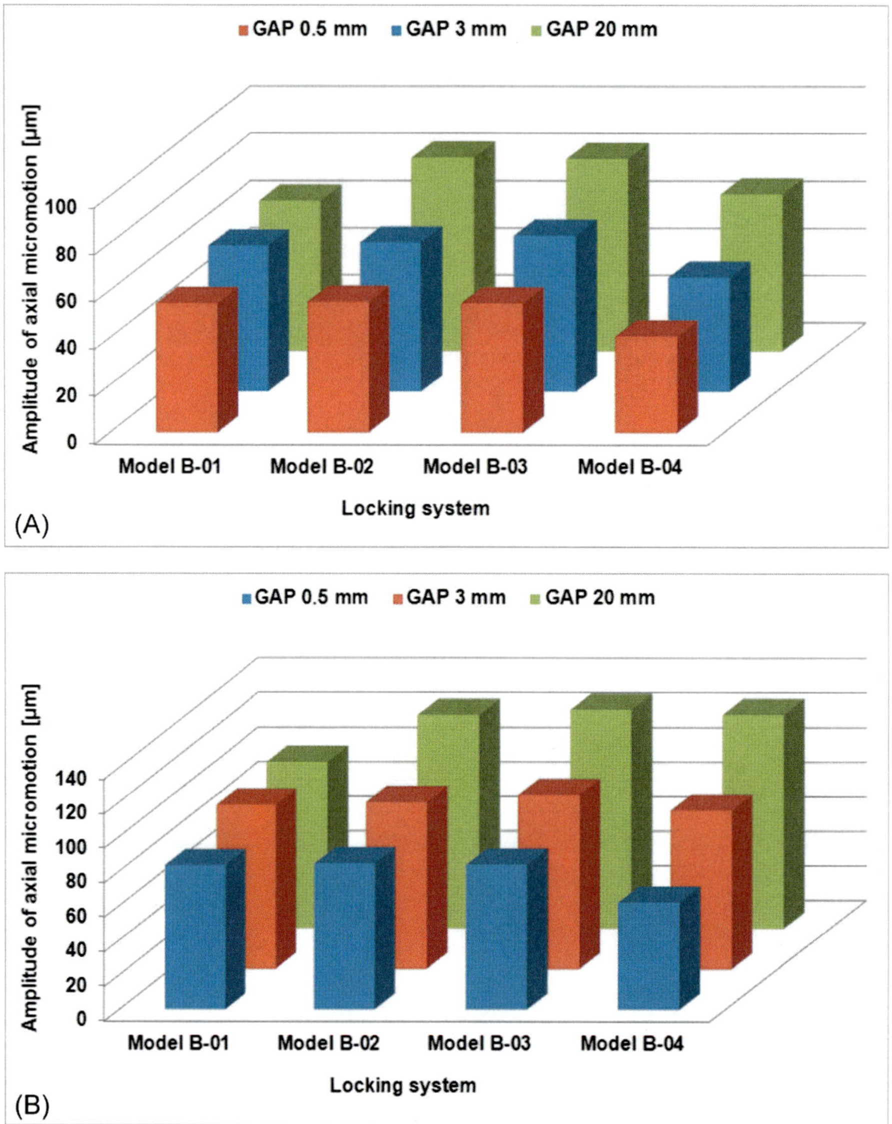


FIG. 11.25 Amplitude of axial micromotion (μm), for the "B" study, corresponding to different nail materials: (A) steel intramedullary nail; (B) titanium intramedullary nail.

Concerning intramedullary nails, Figs. 11.28 and 11.29 show the von Mises stress maps in the nail corresponding to proximal and distal fractures, respectively, for a gap of 3.0 mm. As can be seen, the maximum stress values in the nail are located at the position corresponding to the site of fracture (i.e., near the top of the nail for proximal fracture and near the bottom of the nail for distal fracture), due to the bending effect produced on the nail connecting the two parts of the fractured femur. Moreover, a high stress concentration appears in the screw hole nearest to the fracture site. In any case, due to the low level of load applied, the values do not affect the material yielding stress, so the nail strength is not compromised.

Finally, for the screws, Fig. 11.30 shows the von Mises stress maps for a proximal fracture (gap size 3 mm). The figure shows a high stress concentration in the upper screw, very near to the fracture site, while the lower screws are quite discharged. In the same way, Fig. 11.31 shows the von Mises stress maps for a distal fracture (gap size 3 mm). In this case, the biomechanical behavior is completely different, showing similar stress values in both upper and lower screws. Although the stresses tend to concentrate in the lower screws, the presence of two screws allows for a better

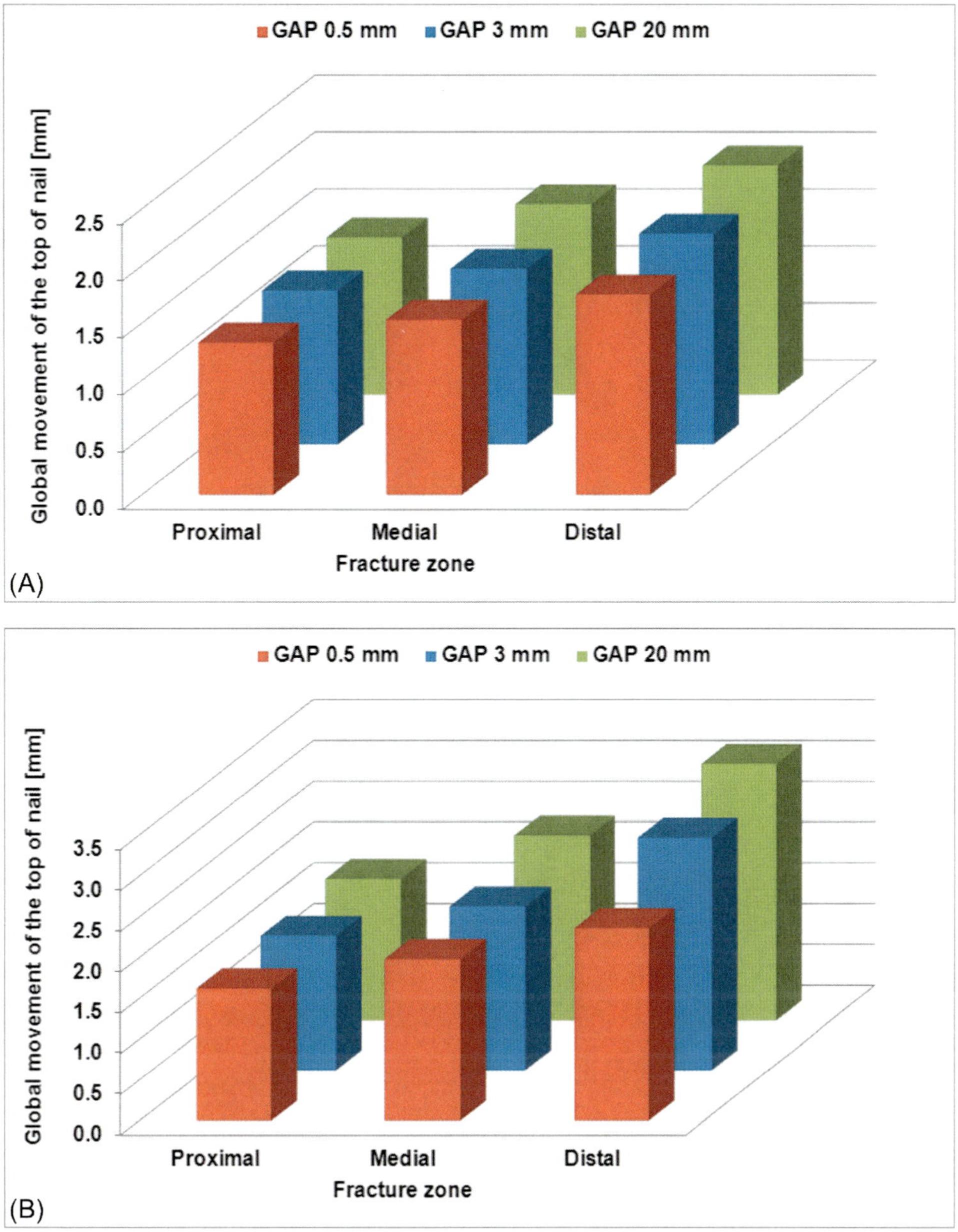


FIG. 11.26 Global movement of the top of the nail (mm), for the "A" study, corresponding to different nail materials: (A) steel intramedullary nail; (B) titanium intramedullary nail.

load transmission between nail and bone, generating a moment that balances the bending effect appearing in that zone.

With respect to the clinical follow-up, Table 11.5 shows the mean time of the fracture consolidation, except for Grade 3 in the "A" study, since it was not finally considered because only one case was assessed, and Grades 2 and 3 in the "B" study, since there are no cases. Thus, for both studies, it can be observed that the healing time increases with higher comminution grade.

In view of Table 11.5, the clinical results are in accordance with the FE simulations results, obtaining a longer healing period for fractures with worse stability.

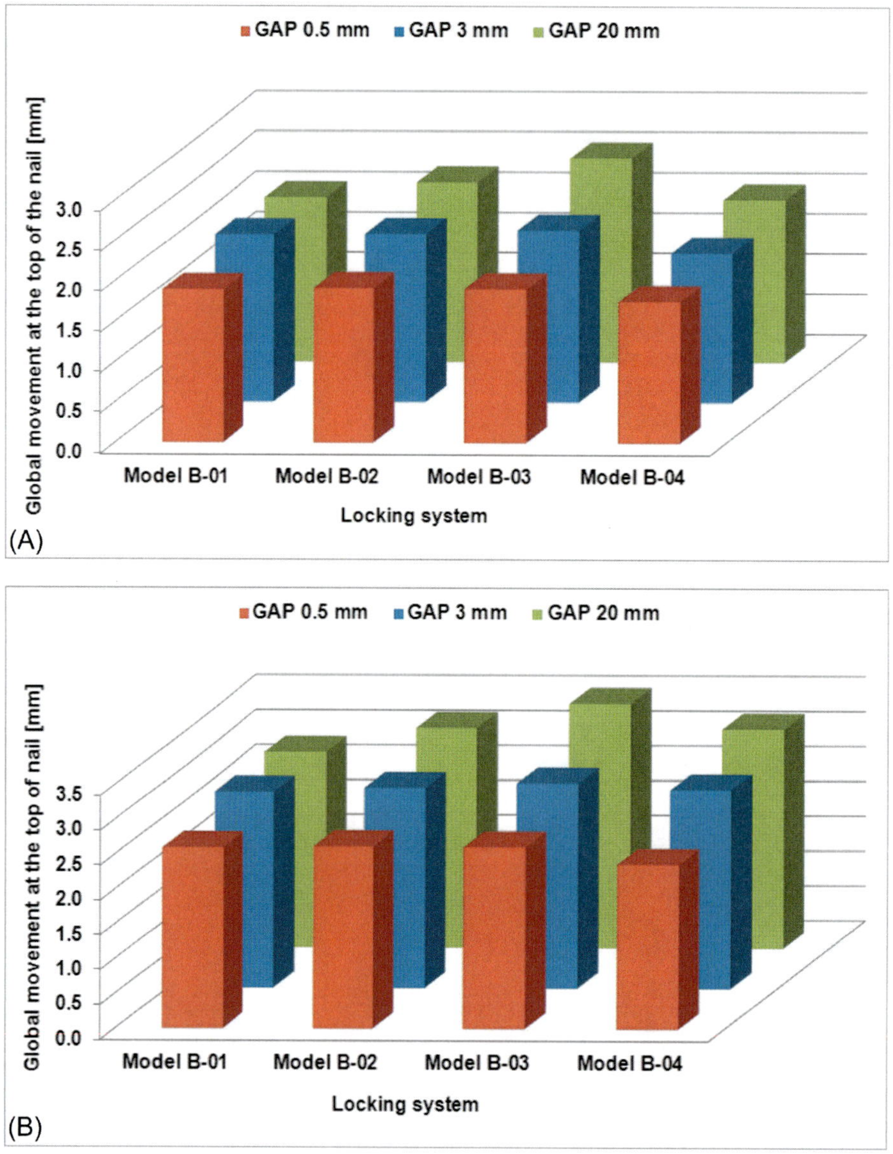


FIG. 11.27 Global movement of the top of the nail (mm), for the "B" study, corresponding to different nail materials: (A) steel intramedullary nail; (B) titanium intramedullary nail.

## 11.5 CONCLUSIONS

Different FE models have been developed, on the one hand, to analyze various types of fractures in the subtrochanteric and diaphyseal supracondylar area with several gap sizes, stabilized with a single combination of screws for the intramedullary nail, and, on the other hand, to characterize the stability of different interlocking systems and identify the optimal one for every type of fracture in the distal location. In addition, the mechanical strength of the nail against bending and compression efforts was studied comparing two nail materials: stainless steel and titanium alloy.

The results of the FE simulations were compared with a set of clinical cases included in the clinical follow-up. In this way, the following conclusions were obtained:

- A good agreement between clinical results and the simulated fractures in terms of gap size was found. Noncomminuted fractures have a minimum mean consolidation time (4.1 months), which coincides with appropriate mobility at the fracture site obtained in the FE simulations, whereas comminuted fractures have a higher mean consolidation period (7.1 months), corresponding to excessive mobility at the fracture site obtained by means of FE simulations. The healing time rises as the comminution grade increases.

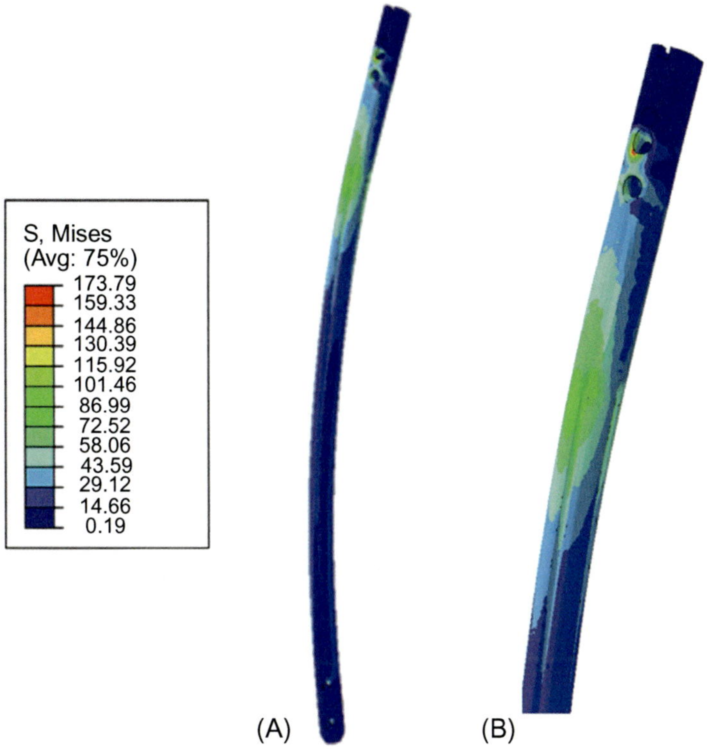


FIG. 11.28 Von Mises stress maps in the nail for proximal fracture (gap size 3mm): (A) whole nail; (B) detail corresponding to the fracture site.

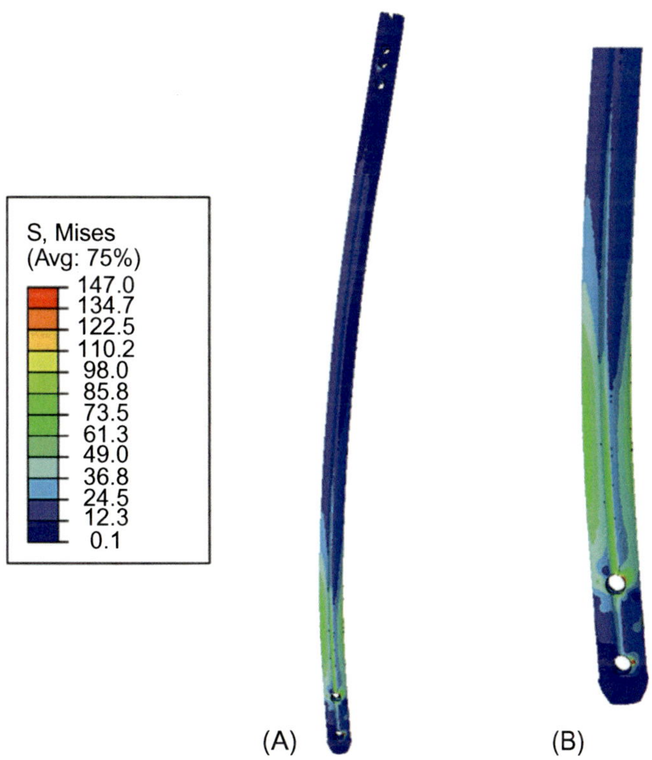


FIG. 11.29 von Mises stress maps in the nail for distal fracture (gap size 3mm): (A) whole nail; (B) detail corresponding to the fracture site.

- Regarding the best nail material, the mobility rate with the titanium nail was higher than with the steel nail. So, the steel nail confers a stiffer fixation system, which is better for osteosynthesis. In particular, the obtained results between both nail materials (stainless steel and titanium alloy) show a higher mobility when using titanium nails, which produce a higher rate of strains at the fracture site, amplitude of micromotions, and bigger global movements compared to stainless steel nails.

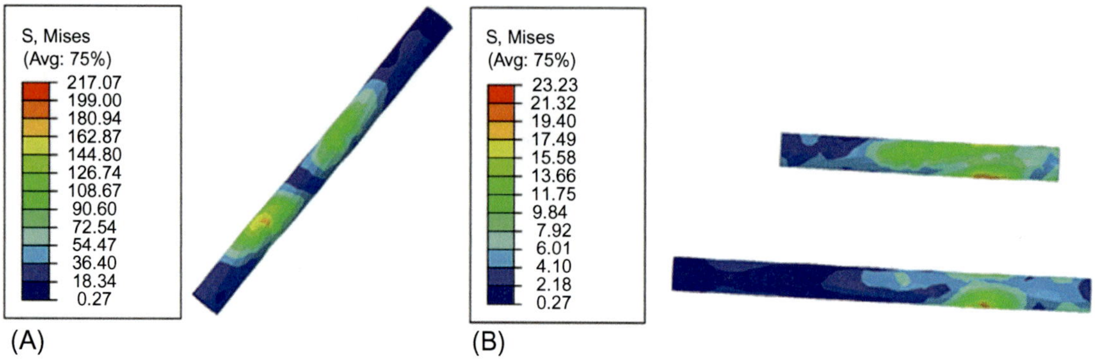


FIG. 11.30 Von Mises stress maps in the screws, proximal fracture (gap size 3mm): (A) upper screw; (B) lower screws.

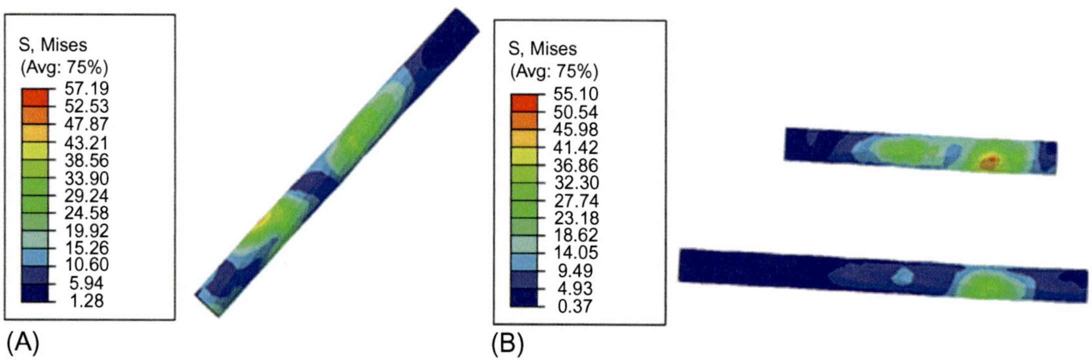


FIG. 11.31 Von Mises stress maps in the screws, distal fracture (gap size 3mm): (A) upper screw; (B) lower screws.

TABLE 11.5 Time Consolidation for the Clinical Follow-Up (Months)

| Study | Comminution grade | Mean time of the consolidation |
|---|---|---|
| A | Noncomminutes | 4.1 |
| | 1 | 4.9 |
| | 2 | 6.2 |
| | 4 | 7.1 |
| B | Noncomminutes | 4.8 |
| | 1 | 5.2 |
| | 2 | 5.2 |

- Among the studied combinations of distal screws, the one with two distal screws medial-lateral provided the best results in terms of stability at the fracture site and global movement at the top of the nail along the three fracture gap sizes. This tendency is because the locking effect is maximized when the distance between the distal screws is increased. This parameter is limited by the proximity to the fracture site and the distance to the femoral condyles.

In conclusion, an anterograde locked nail is particularly useful in the treatment of a wide range of supracondylar fractures with proximal extension into the femoral diaphysis, which confirms that this technique is nowadays the reference surgical treatment for this kind of fracture.

## Acknowledgments

This research has been partially financed by the Fundacion Mutua Madrileña (Research Projects: AP162632016) and by the Government of Spain: Ministry of Economy and Competitiveness (Research Project: DPI2016-77745-R).

## References

[1] M. Braten, T. Terjesen, I. Rossvoll, Femoral shaft fractures treated by intramedullary nailing. A follow-up study focusing on problems related to the method, Injury 26 (1995) 379–383.

[2] P. Wolinsky, N. Tejwani, J.H. Richmond, K.J. Koval, K. Egol, D.J. Stephen, Controversies in intramedullary nailing of femoral shaft fractures, Instr. Course Lect. 51 (2002) 291–303.

[3] C.M. Court-Brown, B. Caesar, Epidemiology of adult fractures: a review, Injury 37 (2006) 691–697.

[4] D.A. Wiss, C.H. Fleming, J.M. Matta, D. Clark, Conminuted and rotationally unstable fracture of the femur treated with interlocking nail, Clin. Orthop. 212 (1986) 35–47.

[5] R.A. Winquist, S.T. Hansen, Conminuted fractures of the femoral shaft treated by intramedullary nailing, Orthop. Clin. North Am. 11 (1980) 633–648.

[6] R.J. Brumback, J.P. Reilly, A. Poka, R.P. Lakatos, G.H. Bathon, A.R. Burgess, Intramedullary nailing of femoral shaft fractures. Part I: Decision manking errors with interlocking fixation, J. Bone Joint Surg. 70A (1988) 1441–1451.

[7] W.M. Ricci, B. Gallagher, G.J. Haidukewych, Intramedullary nailing of femoral shaft fractures: current concepts, J. Am. Acad. Orthop. Surg. 17 (2009) 296–305.

[8] P. Tornetta, D. Tiburzi, Reamed versus nonreamed anterograde femoral nailing, J. Orthop. Trauma 14 (1) (2000) 15–19.

[9] D. Heim, U. Schiegel, S.M. Perren, Intramedullary pressure in reamed and undreamed nailing of the femur and tibia: an in vitro study in intact human bones, Injury 24 (Suppl 3) (1993) S56–S63.

[10] P.R. Wolinsky, D. Banit, R.E. Parker, Y. Shyr, J.R. Snapper, E.J. Rutherford, K.D. Johnson, Reamed intramedullary femoral nailing after induction of an "ARDS-like" state in sheep: effect on clinically applicable markers of pulmonary function, J. Orthop. Trauma 12 (3) (1998) 169–175.

[11] X. Duan, T. Li, A.Q. Mohammmed, Z. Xiang, Reamed intramedullary nailing versus undreamed intramedullary nailing for shaft fracture of femur. A systematic literature review, Arch. Orthop. Trauma Surg. 131 (2011) 1445–1452.

[12] D. Choudary, B. Kanthimathi, A prospective comparative study of reamed vs. Unreamed nailing in fractures shaft of tibia, Malays. Orthop. J. 6 (3) (2012) 21–26.

[13] G. Papadokostakis, C. Papakostidis, R. Dimitriou, P.V. Giannoudis, The role and efficacy of retrograding nailing for the treatment of diaphyseal and distal femoral fractures: a systematic review of the literature, Injury 36 (2005) 813–822.

[14] K.S. Shih, C.C. Hsu, T.P. Hsu, A biomechanical investigation of the effects of static fixation and dynamization after interlocking femoral nailing: a finite element study, J. Trauma 72 (2012) 46–53.

[15] T.M. Tupis, G.T. Altman, D.T. Altman, H.A. Cook, M.C. Miller, Femoral bone strains durig antegrade nailing: a comparison of two entry points with identical nails using finite element analysis, Clin. Biomech. 27 (2012) 354–359.

[16] M.M. Kaiser, G. Zachert, R. Wendlandt, M. Rapp, M. Eggert, C. Stratmann, L.M. Wessel, A.P. Schulz, B.J. Kienast, Biomechanical analysis of a synthetic femoral spiral fracture model: do end caps improve retrograde flexible intramedullary nail fixation? J. Orthop. Surg. Res. 6 (2011) 46.

[17] R.J. Eveleigh, A review of biomechanical studies of intramedullary nails, Med. Eng. Phys. (5) (1995) 323–331.

[18] R.W. Sanders, M. Swiontkowski, H. Rosen, D. Helfet, Complex fractures and malunion of the distal femur: results of treatment with double plate, J. Bone Joint Surg. 73A (1991) 341–346.

[19] S.H. Chen, T.H. Yu, C.H. Chang, Y.H. Lu, Biomechanical analysis of retrograde intramedullary nail fixation in distal femoral fractures, Knee 15 (2008) 384–389.

[20] C.C. Wu, C.L. Tai, Retrograde nailing of a femoral supracondyle, Orthopedics 35 (2012) 491–496.

[21] S.G. Kulkarni, A. Varsheya, G.S. Kulkarni, M.G. Kulkarni, V.S. Kulkarni, R.M. Kulkarni, Antegrade interloking nailing for distal femoral fractures, J. Orthop. Surg. 20 (2012) 48–54.

[22] R. Montanini, V. Filardi, In vitro biomechanical evaluation of antegrade femoral nailing at early and late postoperative stages, Med. Eng. Phys. 32 (2010) 889–897.

[23] E. Schneider, M.C. Michel, M. Genge, K. Zuber, R. Ganz, S.M. Perren, Loads acting in an intramedullary nail during fracture healing in the human femur, J. Biomech. 34 (7) (2001) 849–857.

[24] J.L. Franklin, R.A. Winquist, S.K. Bernischke, S.T. Hansen, ST, broken intramedullary nails, J. Bone Joint Surg. 70A (1988) 1463–1471.

[25] M. Sánchez, J. Albareda, J.M. Cardona, J. Fuentelsaz, J.J. Fernandez, F. Seral, Biomechanical study of the Grosse-Kempf femoral nail, Int. Orthop. 21 (1997) 115–118.

[26] M. Charles-Harris, D. Lacroix, I. Proubasta, J.A. Planell, Intramedullary nails vs osteosynthesis plates for femoral fracture stabilization: a finite element analysis, J. Appl. Biomater. Biomech. 3 (3) (2005) 157–167.

[27] J.P. Heiney, M.D. Barnett, G.A. Vrabe, A.J. Schoenfeld, A. Baji, G.O. Njus, Distal femoral fixation: a biomechanical comparison of trigen retrograde intramedullary (im) nail, dynamic condylar screw (DCS), and locking compression plate (LCP) condylar plate, J. Trauma 66 (2009) 443–449.

[28] M.R. Bong, F.J. Kummer, J.K. Koval, K.A. Egol, Intramedullary nailing of the lower extremity: biomechanical and biology, J. Am. Acad. Orthop. Surg. 15 (2007) 97–106.

[29] S.B. Antekeier, R.L.J. Burden, M.J. Voor, C.S. Roberts, Mechanical study of the safe distance between distal femoral fracture site and distal locking screw in antegrade intramedullary nailing, J. Orthop. Trauma 19 (2005) 693–697.

[30] D. Wähnert, K.L. Hoffmeier, G. von Oldenburg, R. Fröber, G.O. Hofmann, T. Mückley, Internal fixation of type-C distal femoral fractures in osteoporotic bone, J. Bone Joint Surg. 92 (2010) 1442–1452.

[31] A. Pérez, A. Mahar, C. Negus, P. Newton, T. Impelluso, A computational evaluation of the effect of intramedullary nail material properties on the stabilization of simulated femoral shaft fractures, Med. Eng. Phys. 30 (2008) 755–760.

[32] M.M. Kaiser, L.M. Wessel, G. Zachert, C. Stratmann, R. Eggert, N. Gros, M. Schulze-Hessing, B. Kienast, M. Rapp, Biomechanical analysis of a synthetic femur spiral fracture model: influence of diferent materials on the stiffness in flexible intramedullary nailing, Clin. Biomech. 26 (2011) 592–597.

[33] A. Herrera, L. Gracia, E. Ibarz, J.J. Panisello, J. Cegoñino, J. Mateo, J. Rodríguez-Vela, S. Puértolas, Finite element simulation. Applications in orthopaedic surgery and traumatology, in: D. Moratal (Ed.), Finite Element Analysis, IntechOpen, Rijeka, Croatia, 2010, pp. 121–150.

[34] Roland DG Corporation, Dr PICZA 3, User Manual, http://support.rolanddga.com/docs/documents/departments/technical%20services/manuals%20and%20guides/drpicz3e.pdf, 2001. Accessed 28 June 2018.
[35] Roland DG Corporation, Pixform Pro II Software, http://support.rolanddga.com/docs/Documents/departments/Technical%20Services/Manuals%20and%20Guides/RU_PixformProII.pdf, 2008. Accessed 28 June 2018.
[36] Siemens, I-deas® 11 NX Series PLM software, http://www.plm.automation.siemens.com/, 2013. Accessed 28 June 2018.
[37] Rhinoceros® software, https://www.rhino3d.com/es/, 2018. Accessed 28 June 2018.
[38] Materialise Mimics software, https://www.materialise.com/es/medical/software/mimics, 2018. Accessed 28 June 2018.
[39] Fortran software, http://www.fortran.com/the-fortran-company-homepage/fortran-tools-libraries-and-application-software/, 2018. Accessed 28 June 2018.
[40] L.E. Claes, H.J. Wilke, P. Augat, S. Rubenacker, K.J. Margevicius, Effect of dynamization on gap healing of diaphyseal fractures under external fixation, Clin. Biomech. 10 (1995) 227–234.
[41] A. Herrera, J.J. Panisello, E. Ibarz, J. Cegonino, J.A. Puertolas, L. Gracia, Long-term study of bone remodelling after femoral stem: a comparison between DEXA and finite element simulation, J. Biomech. 40 (2007) 3615–3625.
[42] Loading of orthopaedic implants, OrthoLoad, 2018. https://orthoload.com/. Accessed 28 June 2018.
[43] H. Weinans, R. Huiskes, H.J. Grootenboer, Effects of fit and bonding characteristics of femoral stems on adaptative bone remodeling, J. Biomech. Eng. 116 (4) (1994) 393–400.
[44] J. Kerner, R. Huiskes, G.H. van Lenthe, H. Weinans, B. van Rietbergen, C.A. Engh, A.A. Amis, Correlation between pre-operative periposthetic bone density and post-operative bone loss in THA can be explained by strain-adaptative remodeling, J. Biomech. 32 (1999) 695–703.
[45] J.A. Grant, N.E. Bishop, N. Gotzen, C. Sprecher, M. Honl, M.M. Morlock, Artificial composite bone as a model of human trabecular bone: the implant-bone interface, J. Biomech. 40 (2007) 1158–1164.
[46] S. Eberle, C. Gerber, G. von Oldenburg, S. Hungerer, P. Augat, Type of hip fracture determines load share in intramedullary osteosynthesis, Clin. Orthop. Rel. Res. 467 (2009) 1972–1980.
[47] S.H. Chen, M.C. Chiang, C.H. Hung, S.C. Lin, H.W. Chang, Finite element comparison of retrograde intramedullary nailing and locking plate fixation with/without an intramedullary allograft for distal femur fracture following total knee arthroplasty, Knee 21 (2014) 224–231.
[48] S. Samiezadeh, P. Tavakkoli Avval, Z. Fawaz, H. Bougherara, Biomechanical assessment of composite versus metallic intramedullary nailing system in femoral shaft fractures: a finite element study, Clin. Biomech. 29 (2014) 803–810.
[49] Abaqus software, Dassault Systèmes, https://www.3ds.com/es/productos-y-servicios/simulia/productos/abaqus/, 2018. Accessed 28 June 2018.
[50] T. Yamaji, K. Ando, S. Wolf, P. Augat, L. Claes, The effect of micromovement on callus formation, J. Orthop. Sci. 6 (2001) 571–575.
[51] P. Augat, J. Burger, S. Schorlemmer, T. Henke, M. Peraus, L. Claes, Shear movement at the fracture site delays healing in a diaphyseal fracture model, J. Orthop. Res. 21 (2003) 1011–1017.

CHAPTER

# 12

# Biomechanical Study in the Calcaneus Bone After an Autologous Bone Harvest

*Javier Bayod López*[*], Ricardo Becerro de Bengoa Vallejo*[†], and Marta E. Losa Iglesias*[‡]

[*]Group Applied Mechanics and Bioengineering, School of Engineering and Architecture, University of Saragossa, Zaragoza, Spain [†]Department of Nursing, School of Nursing, Physiotherapy and Podiatry, Complutense University, Madrid, Spain [‡]Faculty of Health Sciences, Rey Juan Carlos University, Madrid, Spain

## 12.1 INTRODUCTION

The extraction of bone stock for autologous graft is a technique performed routinely in foot and ankle surgery [1–3]. Autologous bone grafts in the foot are performed in procedures such as distraction osteotomies, revision surgeries, nonunion fractures, and joint fusion [3, 4]. The iliac crest is a common donor site, but it is associated with complication rates from 10% to 49% [3, 5, 6], including potential visceral injury, chronic pain, numbness, bruising, infection, and delayed healing [7–14]. Other donor sites, including the proximal and distal tibia, fibular, distal radius, and greater trochanter, have fewer relative complications, but generally less bone can be harvested [3, 4].

In foot and ankle surgery, the calcaneus provides an optimal site to procure an autologous bone graft. Using ankle-block anesthesia and small incisions, the bone can be successfully harvested and surgically implanted using the same operative field [7]. Importantly, only minor complications are associated with calcaneus bone harvest. A 2-year outcome study of 17 patients who had undergone foot surgery and calcaneal bone harvest reported minor incisional symptoms in three patients. Medial heel pain was reported in five cases, including three ascribed to plantar fasciitis, and one with pain prior to the surgical procedure caused by a clubfoot deformity [7]. This study demonstrated that the calcaneus is a safe option for donor grafts in foot and ankle surgical procedures, rather than the proximal tibia, distal tibia, or iliac crest that are traditionally used [7].

Heel fracture represents one significant complication that can occur with calcaneal bone harvest. One study evaluated the clinical outcomes of 19 patients who had foot surgery using autogenous, tricortical bone grafts harvested from the calcaneus [4]. Allogenic cubes were used in 1 cohort (9 patients) to fill the defect, and the remaining 10 patients received no tissue replacement. After 6 months, two patients succumbed to a heel fracture. One fracture, in a patient from the first cohort, was inferior to the graft site, and a patient from the second cohort experienced a fracture posterior to the graft site [4]. Unfortunately, it was unclear if the fractures resulted from the size of the graft, emphasizing the importance of future investigations to establish the maximum graft size for safe extraction.

Based on this the objective of the current study was to determine the effects on the mechanical properties of the foot due to progressive calcaneus bone removal. To address this, a three-dimensional (3-D) finite element (FE) model was developed. With this approach, we evaluated six conditions of principal stress. One variable was the intact foot, and the remaining five conditions included a model where a piece of bone of variable depth (maximum 7.5 mm) had been removed. Because of potential increases in mechanical stress in the calcaneus secondary to contraction of the Achilles tendon, we also evaluated mechanical properties of the foot with increasing traction forces assigned to the Achilles.

*Advances in Biomechanics and Tissue Regeneration*
https://doi.org/10.1016/B978-0-12-816390-0.00012-1

## 12.2 METHODS

The methodology used was a 3-D FE model [15] (Fig. 12.1) that was created based on computed tomography (CT) images obtained from a healthy male volunteer (36 years old, 169-cm height, and 69-kg weight) with no foot pain or deformities. Twenty-eight foot bones were incorporated into the 3-D model: talus, calcaneus, cuboid, navicular, three wedges, five metatarsals, five proximal phalanges, four middle phalanges, five phalanges, and two sesamoids. This model also included the following ligaments: posterior talocalcaneal, calcaneus, navicular, tarsometatarsal, intermetatarsal, Lisfranc, calcaneal cuboid–calcaneus–navicular, plantar, and the plantar fascia.

Tetrahedral elements that were 1-mm long comprised the mesh [15]. The connections between the different bones were modeled by cartilaginous joints. The bone was modeled differentiating between cortical and cancellous bone (Fig. 12.2).

Both entities were considered elastic and isotropic. The mechanical properties of cortical bone were 17,000 MPa with a Poisson's ratio of 0.3. Cancellous bone properties were 700 MPa and a Poisson's ratio of 0.3 [16, 17]. To model the cartilage, we employed an isotropic elastic material that exhibited Young's modulus of 10 MPa and a 0.4 Poisson's ratio [18]. To properly model the ligaments, we utilized a set of 483 dimensional, incompressible elements comprised of two distinguishing groups. Rigid structures, such as the plantar fascia and plantar ligaments, were depicted with a cross-sectional area 290.7 $mm^2$ and a 350-MPa Young's modulus and a Poisson's ratio of 0.3 [19]. The remaining, less rigid ligaments were modeled with Young's modulus of 260 MPa, Poisson's ratio of 0.3, and 18.4 $mm^2$ cross-sectional area [19]. The final model, consisting of 137,718 nodes and 735,062 elements, was then imported and assembled in the Abaqus (v. 6.14,1, Dassault Systèmes, Vélizy-Villacoublay, France).

Our investigation focused on the stance phase of gait. Displacement boundaries were established between the metatarsal bones and sesamoids and the lower nodes of the calcaneus. The foot carried a total load defined as 300 N oriented 10° from vertical, including the body-weight surface load [18]. Additionally, a concentrated force (150 N) was applied to the calcaneus at the posterior aspect to mimic the force of the Achilles tendon resulting from contraction of the triceps surae muscle group [20].

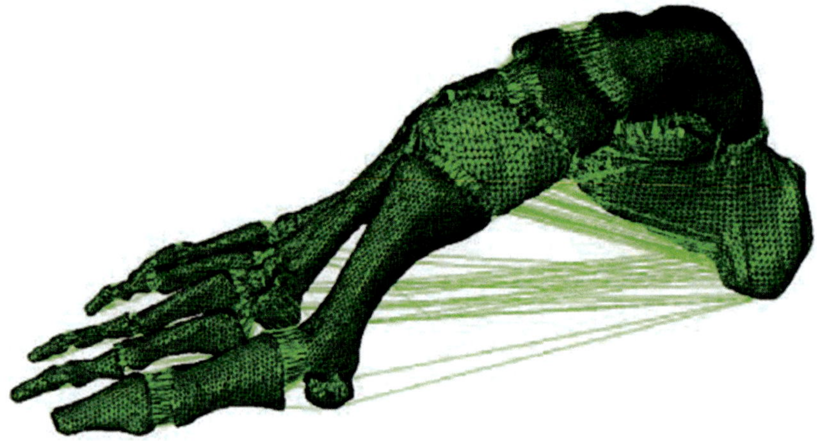

FIG. 12.1 Finite element model.

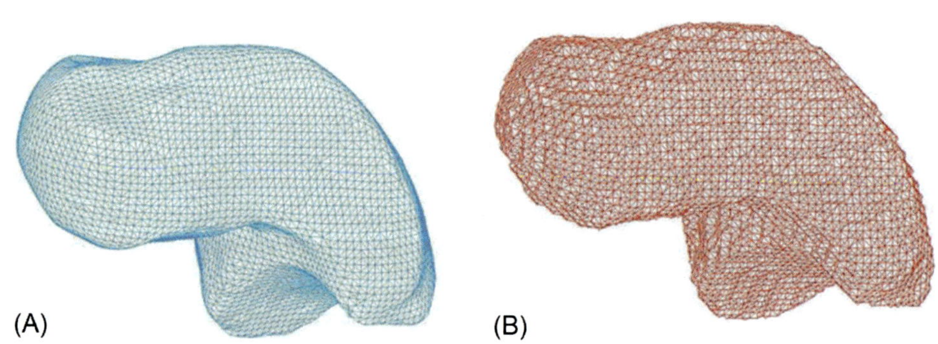


FIG. 12.2 Distinction between cortical (A) and cancellous (B) bones in the finite element mesh.

To determine the effect of calcaneal bone excision in a sequential manner, calculations were performed using a model of an intact foot as a reference point. Sequential elimination of calcaneus mesh elements from the model mimicked surgical extraction until a region of 24 × 7 × 7.5 mm (1260 mm$^3$) had been removed. The initial excision began at the top of the dorsal posterior tuberosity of the calcaneus, creating the first layer in a block with dimensions of 24 × 7 × 1.5 mm (252 mm$^3$). Subsequent excisions removed layers approximately 1.5 mm deep, creating blocks of 24 × 7 × 3 mm (504 mm$^3$), 24 × 7 × 4.5 mm (756 mm$^3$), 24 × 7 × 6 mm (1008 mm$^3$), and 24 × 7 × 7.5 mm (1260 mm$^3$).

Additionally, in the intact foot model and the model in which 7.5 mm of bone had been extracted, loads of 600 N and 750 N (2 and 2.5 times walking) were applied to the talus to evaluate the impact of gait variations.

Finally, we evaluated the effect of Achilles tendon traction on the calcaneus during gait.

The typical traction while walking is 150 N. Consistent with evaluating loads on the talus at 2 and 2.5 by walking, we evaluated Achilles tendon loads of 300 and 375 N with a constant value of 300 N maintained against the talus.

## 12.3 RESULTS

### 12.3.1 Displacements Varying the Talus Load and Constant Achilles Tendon Load Based on the Amount of Bone Extraction

There was no change in either anterior/posterior (AP) or medial/lateral (ML) translation of calcaneus with increasing loads on the talus and a constant Achilles tendon load based on the amount of bone extraction (Table 12.1). There was, however, an increase in translation across conditions as the load on the talus increased.

Maximum principal stresses (tension) were concentrated at the posterior aspect of the calcaneus (Table 12.2; Fig. 12.3).

TABLE 12.1 Displacements as a Function of Load Variation on the Talus and Depth of Bone Removal. Load on the Achilles Tendon Remains Constant

| | | Depth of bone excision | | | | | |
|---|---|---|---|---|---|---|---|
| **Talus load** | **Displacement (mm)** | **Intact** | **24×7×1.5** | **24×7×3** | **24×7×4.5** | **24×7×6** | **24×7×7.5** |
| 300 N | AP | 1.445 | 1.446 | 1.452 | 1.461 | 1.467 | 1.469 |
| | ML | 0.543 | 0.543 | 0.541 | 0.537 | 0.536 | 0.535 |
| 600 N | AP | 2.348 | 2.349 | 2.362 | 2.376 | 2.386 | 2.389 |
| | ML | 0.891 | 0.891 | 0.891 | 0.894 | 0.895 | 0.895 |
| 750 N | AP | 2.797 | 2.798 | 2.814 | 2.831 | 2.842 | 2.895 |
| | ML | 1.07 | 1.071 | 1.075 | 1.078 | 1.08 | 1.081 |

These data were previously published in Bayod et al. [24].

TABLE 12.2 Principal Stress as a Function of Load Variation on the Talus and Depth of Bone Removal. Load on the Achilles Tendon Remains Constant

| | | Depth of bone excision | | | | | |
|---|---|---|---|---|---|---|---|
| **Talus load** | **Principal stress (MPa)** | **Intact** | **24×7×1.5** | **24×7×3** | **24×7×4.5** | **24×7×6** | **24×7×7.5** |
| 300 N | Tension | 23.22 | 23.22 | 31.02 | 33.95 | 34.23 | 34.27 |
| | Compression | −35.27 | −35.27 | −35.57 | −36.18 | −36.4 | −36.48 |
| 600 N | Tension | 51.14 | 51.14 | 51.17 | 51.12 | 51.09 | 51.08 |
| | Compression | −93.88 | −93.9 | −94.25 | −95.09 | −95.39 | −95.49 |
| 750 N | Tension | 65.3 | 65.31 | 65.33 | 65.25 | 65.2 | 65.19 |
| | Compression | −123.6 | −123.6 | −124 | −125 | −125.3 | −125.54 |

These data were previously published in Bayod et al. [24].

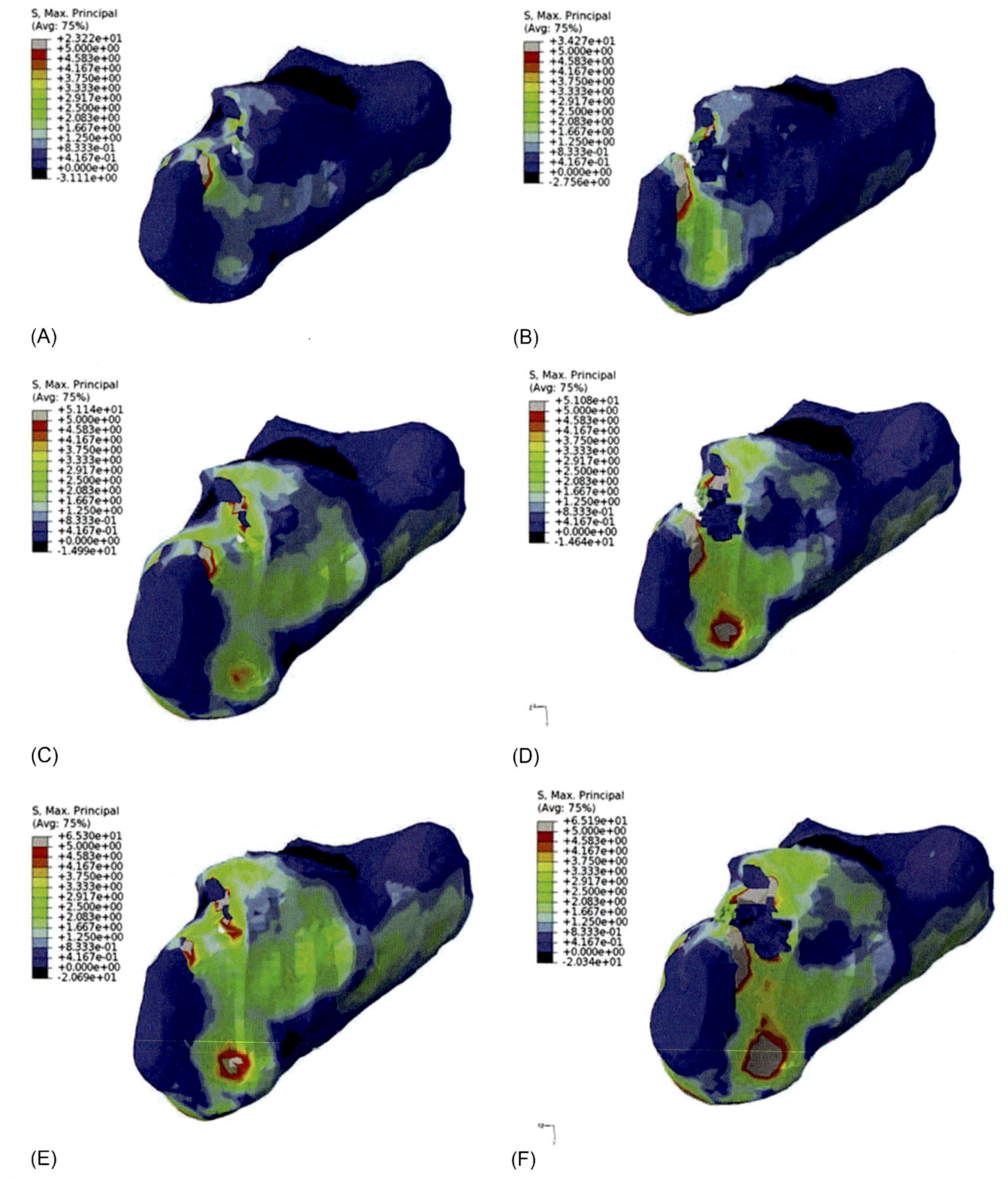


FIG. 12.3 Maximum principal stress in an intact foot with loads of 300 (A), 600 (C), and 750 N (E), respectively, and with 7.5 mm bone excision and loads of 300 (B), 600 (D), and 750 N (F), respectively.

However, as the volume of bone removal increased and the load on the talus increased in both the intact model and models with bone removal, we observed that maximum principle stresses were concentrated around the region of the extracted bone (Fig. 12.3).

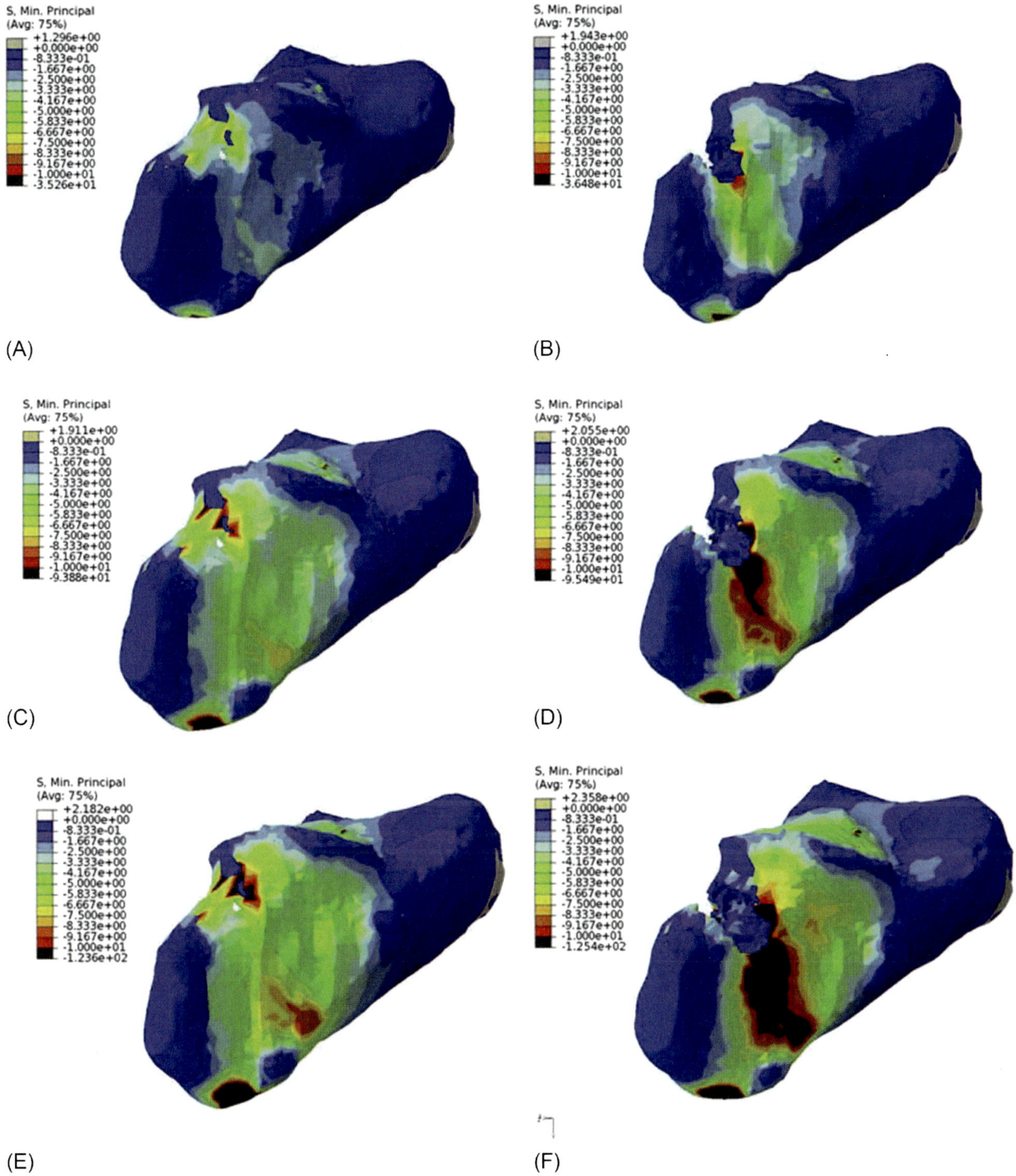


FIG. 12.4 Minimum principal stress in an intact foot with loads of 300 (A), 600 (C), and 750 N (E), respectively, and with 7.5 mm bone excision and loads of 300 (B), 600 (D), and 750 N (F), respectively.

Minimum principal stress (compression) was concentrated in a different way at the calcaneus zone (Table 12.2; Fig. 12.4). When the load on the talus and the volume of bone extraction increased, compression stress in the healthy model and the simulated model with removal of bone material were very different. In the intact model the compression stress was localized from the dorsoposterior aspect of the calcaneus and the support zone (Fig. 12.4). In the extracted bone model, the compression stress was localized in the "central area" of the calcaneus called the "neutral area" (Fig. 12.4).

### 12.3.2 Displacements Varying Achilles Tendon Load Based on the Amount of Bone Extraction

There was no change in either AP or ML translation in calcaneus with increasing Achilles tendon load based on the amount of bone extraction (Table 12.3). There was, however, an increase in translation across conditions as the load on the Achilles tendon increased.

Maximum principle stresses (tension) were concentrated at the posterior aspect of the calcaneus (Table 12.3; Fig. 12.4). However, as the volume of bone removal increased and the load on the calcaneus increased in both the intact model and models with bone removal, we observed that maximum principal stresses were concentrated around the region of the extracted bone (Fig. 12.3).

Minimum principal stress (compression) was concentrated at the support zone (Table 12.2; Fig. 12.4). When the load on the talus and the volume of bone extraction increased, compression stress in the healthy model and the simulated model with removal of bone material extended from the edge of the bone extracted site to the bottom and sides of the calcaneus, near the support zone planting (Fig. 12.4).

Principal stresses varying by the Achilles tendon load (Table 12.4; Figs. 12.5 and 12.6).

## 12.4 DISCUSSION

We used a 3-D FE model to create sequential simulations of calcaneus bone removal to mimic a graft harvest. The size of the maximum donor site was 1.30 $cm^3$. Based on our results, we suggest that a calcaneal bone harvest should not exceed a volume of $2.4 \times 0.7 \times 0.75$ cm.

Because we were interested in evaluating the mechanical properties of the calcaneus during functional tasks, we focused on the stance phase of gait, when the heel is on the ground. We also evaluated experimental conditions that mimicked daily tasks and had the potential to increase calcaneal stress, including conditions with an increase in force

TABLE 12.3 Displacements as a Function of Achilles Tendon Load and Depth of Bone Removal. Load on the Talus Remains Constant

| | | Depth of bone excision | | | | | |
|---|---|---|---|---|---|---|---|
| **Talus load** | **Displacement (mm)** | **Intact** | **24×7×1.5** | **24×7×3** | **24×7×4.5** | **24×7×6** | **24×7×7.5** |
| 300N | AP | 1.445 | 1.446 | 1.452 | 1.461 | 1.467 | 1.469 |
| | ML | 0.543 | 0.543 | 0.541 | 0.537 | 0.536 | 0.535 |
| 600N | AP | 1.984 | 1.985 | 1.992 | 2 | 2.013 | 2.016 |
| | ML | 0.738 | 0.738 | 0.735 | 0.73 | 0.728 | 0.727 |
| 750N | AP | 2.254 | 2.254 | 2.261 | 2.276 | 2.286 | 2.289 |
| | ML | 0.836 | 0.836 | 0.832 | 0.827 | 0.874 | 0.824 |

These data were previously published in Bayod et al. [24].

TABLE 12.4 Principal Stress as a Function of Achilles Tendon Load and Depth of Bone Removal. Load on the Talus Remains Constant

| | | Depth of bone excision | | | | | |
|---|---|---|---|---|---|---|---|
| **Achilles tendon load** | **Principal stress (MPa)** | **Intact** | **24×7×1.5** | **24×7×3** | **24×7×4.5** | **24×7×6** | **24×7×7.5** |
| 150N | Tension | 23.22 | 23.22 | 31.02 | 33.95 | 34.23 | 34.27 |
| | Compression | −35.26 | −35.27 | −35.57 | −36.18 | −36.4 | −36.48 |
| 300N | Tension | 33.33 | 37.23 | 62.73 | 68.37 | 68.86 | 68.95 |
| | Compression | −19.23 | −18.98 | −31.95 | −20.54 | −20.54 | −20.65 |
| 375N | Tension | 42.16 | 46.98 | 78.58 | 85.58 | 86.18 | 86.28 |
| | Compression | −23.74 | −19.77 | −36.93 | −42.63 | −19.96 | |
| | | | | | | −19.34 | |

These data were previously published in Bayod et al. [24].

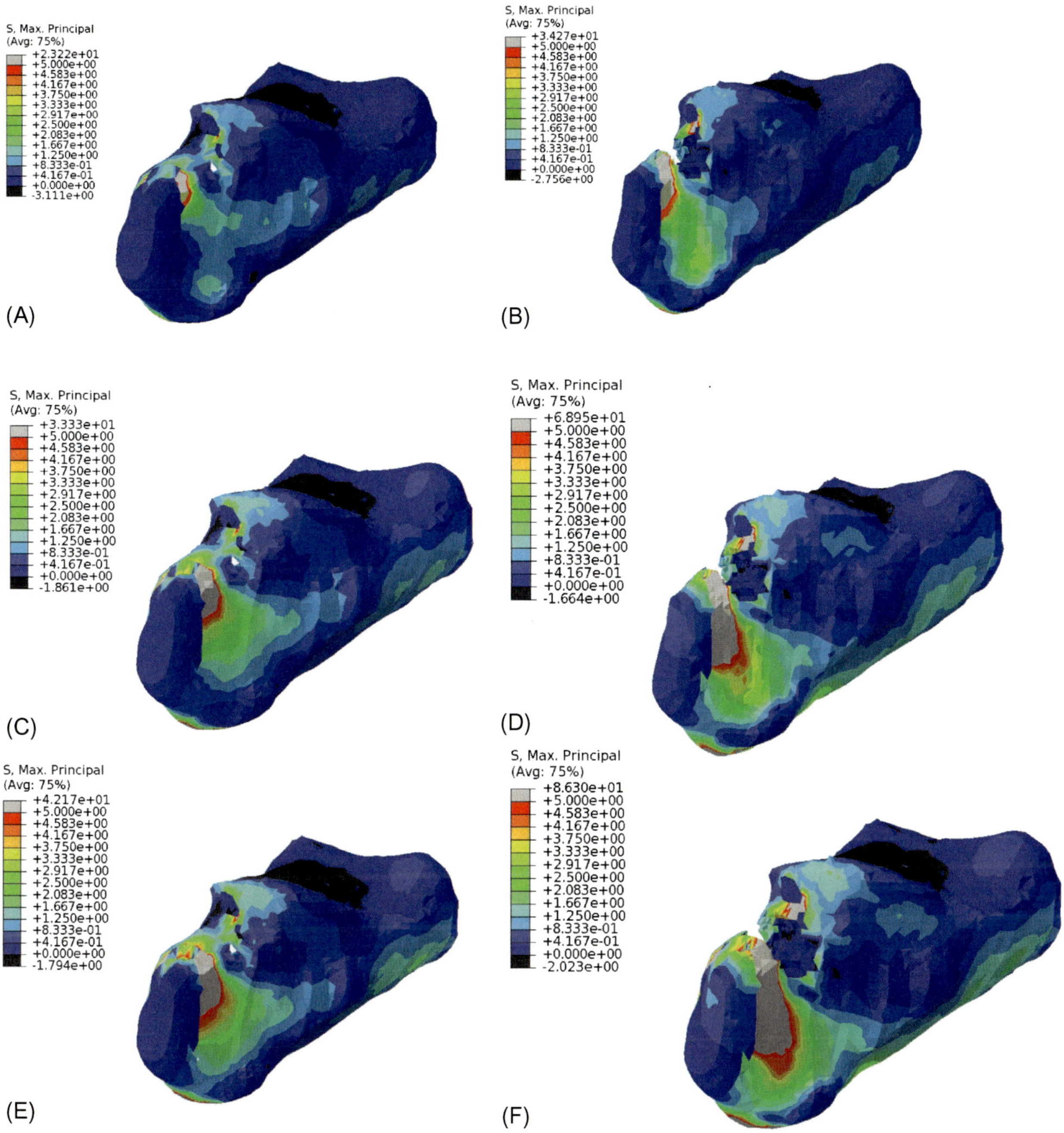


FIG. 12.5 Maximum principal stress in an intact foot with loads of 150 (A), 300 (C), and 375 N (E), respectively, on the Achilles tendon and with 7.5 mm bone excision and loads of 150 N (B), 300 N (D), and 375 N (F), respectively, on the Achilles tendon.

application to the talus and to the calcaneus through traction applied by the Achilles tendon. The results from these simulations indicate that as the volume of the bone extracted from the calcaneus increases, there is a redistribution of stresses that differs significantly from an intact foot. This redistribution is further magnified with increasing loads. Even though an increase in the volume of bone harvest did not significantly affect the maximum stress that we identified, in cases where the calcaneus was vulnerable to injury, stresses did increase.

This stress redistribution in the calcaneus may create an opportunity for a fracture risk.

The calcaneus consists of a peripheral cortical layer of compact tissue that coats a framework of spongy tissue. This lamina exists in an orientation to effectively fulfill the functional requirements of the foot [21]. Thus its structure is

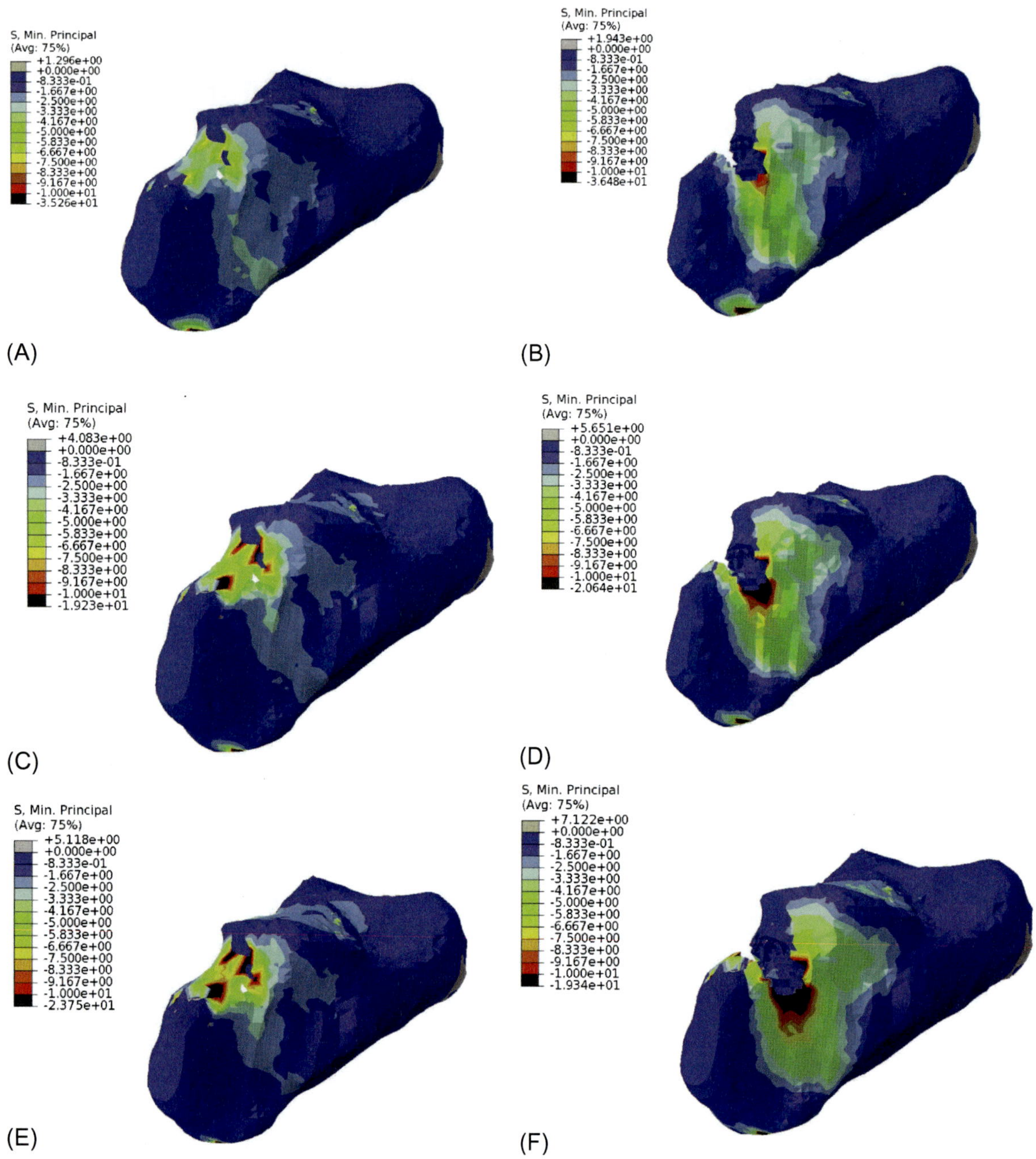


FIG. 12.6 Minimum principal stress in an intact foot with loads of 150 (A), 300 (C), and 375 N (E), respectively, on the Achilles tendon and with 7.5 mm bone excision and loads of 150 N (B), 300 N (D), and 375 N (F), respectively, on the Achilles tendon.

related both to bone strength and load transmission. However, the bone distribution is not homogenous. An area of sparse (or absent) mineralization exists in the anterior portion of the calcaneus. This region is termed the "neutral triangle" or Ward's triangle [22, 23]. One study measured the cortical thickness of these internal calcaneal trabecular arrays in 14 dry, frozen specimens [23]. Those results concluded that the primary and secondary fracture lines, those often associated with calcaneus fractures, correlated with the trabecular patterns and often initiating in the neutral triangle [23]. Regions lacking trabeculae, or parallel to organized trabeculae, were the weakest plane of stress resistance.

In the current study, compressive and tensile stress was redistributed to the borders of the site of graft harvest. Additionally, the posterior aspect of the calcaneus was a source of stress during loaded conditions. In instances of maximum bone extraction, greater stress values were recorded compared with the intact foot. Furthermore, we noted greater compressive stresses upon maximum bone extraction in the anterior calcaneus that corresponded to the neutral triangle. Thus the qualitative load redistribution is clinically significant even when stress values are equivalent to the intact foot in models that have experienced bone harvest.

Our study provides important results to guide the clinical management of a postoperative patient after they have received a calcaneal bone harvest. After simulated increasing bone removal, tensile stress also dramatically increased when Achilles tendon tension was included in the simulation. Furthermore, forces surrounding the donor site were redistributed. These findings support the notion of an increased risk of calcaneal fracture when including traction exerted by the Achilles tendon after graft harvest. Feeney et al. [4] described the utilization of a below-the-knee cast worn for 4–10 weeks when autologous calcaneal bone was harvested for use in foot surgery. This was followed by partial weight bearing for 2–4 weeks while wearing a lace up training style shoe. Feeney et al. [4] reported one of the cases of calcaneal fracture in their investigation was sustained at the 8-week mark when first weight bearing without a cast. The authors of that study suggested that to avoid future fractures, a longer period of weight bearing may be helpful. However, we advocate that ankle stabilization, such as in a cast or posterior splint positioned in some degree of plantarflexion would be useful to minimize the Achilles tendon traction and the fracture risk.

## 12.5 CONCLUSION

Using the calcaneus for autologous bone harvest in surgical procedures involving the foot is associated with many advantages and few adverse outcomes. One of the more serious comorbidities of this procedure is fracture of the calcaneus. Traditionally, the size of harvested bone is determined at the discretion of the physician performing the surgery. Results from our study provide important guidelines for optimal maximum bone extraction for this type of procedure and effective immobilization positioning in postoperative management.

## Acknowledgments

The authors gratefully acknowledge the support of the Ministry of Economy and Competitiveness of the government of Spain through the project DPI2016-77016-R.

## References

[1] S.A. Alter, L. Licovski, Bone grafting for reconstructive osteotomies of the foot, J. Foot Ankle Surg. 35 (1996) 418–427.
[2] K.T. Mahan, H.J. Hillstrom, Bone grafting in foot and ankle surgery. A review of 300 cases, J. Am. Podiatr. Med. Assoc. 88 (1998) 109–118, https://doi.org/10.7547/87507315-88-3-109.
[3] S.M. Raikin, K. Brislin, Local bone graft harvested from the distal tibia or calcaneus for surgery of the foot and ankle, Foot Ankle Int. 26 (2005) 449–453.
[4] S. Feeney, S. Rees, M. Tagoe, Tricortical calcaneal bone graft and management of the donor site, J. Foot Ankle Surg. 46 (2007) 80–85.
[5] S.D. Schulhofer, L.M. Oloff, Iliac crest donor site morbidity in foot and ankle surgery, J. Foot Ankle Surg. 36 (1997) 155–158. discussion 161.
[6] E.M. Younger, M.W. Chapman, Morbidity at bone graft donor sites, J. Orthop. Trauma 3 (1989) 192–195.
[7] K.R. Biddinger, G.A. Komenda, L.C. Schon, M.S. Myerson, A new modified technique for harvest of calcaneal bone grafts in surgery on the foot and ankle, Foot Ankle Int. 19 (1998) 322–326.
[8] J.K. DeOrio, D.C. Farber, Morbidity associated with anterior iliac crest bone grafting in foot and ankle surgery, Foot Ankle Int. 26 (2005) 147–151.
[9] C. Hierholzer, D. Sama, J.B. Toro, M. Peterson, D.L. Helfet, Plate fixation of ununited humeral shaft fractures: effect of type of bone graft on healing, J. Bone Joint Surg. Am. 88 (2006) 1442–1447.
[10] M.H. Hofbauer, R.J. Delmonte, M.L. Scripps, Autogenous bone grafting, J. Foot Ankle Surg. 35 (1996) 386–390.
[11] K.T. Mahan, Calcaneal donor bone grafts, J. Am. Podiatr. Med. Assoc. 84 (1994) 1–9, https://doi.org/10.7547/87507315-84-1-1.
[12] R.W. Mendicino, E. Leonheart, P. Shromoff, Techniques for harvesting autogenous bone graft of the lower extremity, J. Foot Ankle Surg. 35 (1996) 428–435.
[13] N. Nigro, D. Grace, Radiographic evaluation of bone grafts, J. Foot Ankle Surg. 35 (1996) 378–385.
[14] B.N. Summers, S.M. Eisenstein, Donor site pain from the ilium. A complication of lumbar spine fusion, J. Bone Joint Surg Br. 71 (1989) 677–680.
[15] J.M. Garcia-Aznar, J. Bayod, A. Rosas, R. Larrainzar, R. Garcia-Bogalo, M. Doblare, L.F. Llanos, Load transfer mechanism for different metatarsal geometries: a finite element study, J. Biomech. Eng. 131 (2009), 021011. https://doi.org/10.1115/1.3005174.
[16] G.N. Duda, F. Mandruzzato, M. Heller, J. Goldhahn, R. Moser, M. Hehli, L. Claes, N.P. Haas, Mechanical boundary conditions of fracture healing: borderline indications in the treatment of unreamed tibial nailing, J. Biomech. 34 (2001) 639–650.
[17] M.J. Gomez-Benito, P. Fornells, J.M. Garcia-Aznar, B. Seral, F. Seral-Innigo, M. Doblare, Computational comparison of reamed versus unreamed intramedullary tibial nails, J. Orthop. Res. 25 (2007) 191–200, https://doi.org/10.1002/jor.20308.
[18] A. Gefen, Stress analysis of the standing foot following surgical plantar fascia release, J. Biomech. 35 (2002) 629–637.

[19] J.T. Cheung, M. Zhang, A.K. Leung, Y.B. Fan, Three-dimensional finite element analysis of the foot during standing – a material sensitivity study, J. Biomech. 38 (2005) 1045–1054.
[20] A. Simkin, Structural analysis of the human foot in standing posture, Tel Aviv University, Tel Aviv, Israel, 1982.
[21] F.J.F. Camacho, P.M. Martinez, R.R. Torres, A.C. Garcia, L.G. Pellico, Densitometric analysis of the human calcaneus, J. Anat. 189 (1996) 205–209.
[22] M. Harty, Anatomic considerations in injuries of the calcaneus, Orthop. Clin. North Am. 4 (1973) 179–183.
[23] F.F. Sabry, N.A. Ebraheim, J.N. Mehalik, A.T. Rezcallah, Internal architecture of the calcaneus: implications for calcaneus fractures, Foot Ankle Int 21 (2000) 114–118.
[24] J. Bayod, R. Becerro-de-Bengoa-Vallejo, M.E. Losa-Iglesias, M. Doblaré, Mechanical stress redistribution in the calcaneus after autologous bone harvesting, J. Biomech. 45 (7) (2012 Apr 30) 1219–1226.

# PART II

# MECHANOBIOLOGY AND TISSUE REGENERATION

CHAPTER

# 13

# Multidimensional Biomechanics Approaches Though Electrically and Magnetically Active Microenvironments

*S. Ribeiro*[*,†], *C. Garcia-Astrain*[‡], *M.M. Fernandes*[*,¶], *S. Lanceros-Mendez*[‡,§], *and C. Ribeiro*[*,¶]

**[*]Center/Department of Physics, University of Minho, Braga, Portugal [†]Centre of Molecular and Environmental Biology (CBMA), Universidade do Minho, Braga, Portugal [‡]BCMaterials, Basque Center for Materials, Applications and Nanostructures, UPV/EHU Science Park, Leioa, Spain [§]IKERBASQUE, Basque Foundation for Science, Bilbao, Spain [¶]CEB—Centre of Biological Engineering, University of Minho, Braga, Portugal**

## 13.1 RELEVANCE OF ELECTRIC AND MECHANICAL CLUES FOR TISSUE ENGINEERING

Tissue engineering approaches usually involve a biocompatible material in combination with stem cells and different stimuli to repair tissues or organs. Cell adhesion is influenced by several parameters such as the surface chemistry of the scaffold and its surface charge or topography. To induce stem cell differentiation to the desired lineage, stem cells require extracellular stimuli, such as chemical (growth factors) and physical clues (i.e., mechanical stimulation). Physical signals are particularly relevant, as cell development is influenced by these stimuli and cell activity can be modulated in in vitro models that mimic the body microenvironment. Moreover, cell adhesion, proliferation, and differentiation can be also regulated by using active scaffolds that provide the appropriate environment for specific cell responses. The effect of external stimuli over cell attachment is also a key point, since focal adhesions are the predominant mechanism by which cells mechanically connect to and apply traction forces on the extracellular matrix (ECM) [1]. Although the cellular response to electrical stimuli remains still unknown, some regulatory membrane proteins and enzymes are sensitive to electric fields. When a cell attaches to a surface, it receives information from the environment by means of ion channels and receptors present in the membrane and starts developing focal adhesions [2].

The relevance of electrical phenomena in the human body was realized in the 18th century by von Haller and later by Galvani and Volta who demonstrated the dependence of muscles and nerve cells on electricity [3]. Major functions of cells, such as metabolism and growth, are influenced by electrical processes at different stages. Cells maintain a difference in potential and modulate it when necessary, can switch current on and off, and vary current flow or store charge [4]. Among the different clues determining tissue functionality, electrical and electromechanical ones are essential for tissues such as bone or muscle [5]. Further, movement and migration can be guided by electric fields in a variety of different cell types such as corneal, epidermal, and epithelial cells [6–9]. Moreover, these electric fields can also modulate the phenotypes of vascular endothelial cells, regenerate nerve fibers, and influence ligament healing [10–12].

*Advances in Biomechanics and Tissue Regeneration*
https://doi.org/10.1016/B978-0-12-816390-0.00013-3

### 13.1.1 Bone

Yasuda et al. were the first to report the piezoelectric effect in dry bones [13, 14]. This effect can be also found in other tissues such as tendon, ligaments, cartilage, skin, dentin, collagen, deoxyribonucleic acids (DNA), and cell membranes [5, 15–17].

Bones consists of a solid matrix and a fluid phase containing blood and extracellular fluid. The solid phase is composed by a crystalline mineral phase (calcium, phosphate, and carbonate), an amorphous mineral phase, collagen fibers, and a ground substance [3]. Thus bone consists mainly of three phases: collagen (which is piezoelectric), extracellular minerals, and pores. Typically a mechanical stress induces a polarization variation, and the application of an electric field produces the converse effect, a change in the material geometry or strain. The electroactive properties stem from the crystalline nature of collagen, and the displacement of hydrogen bonds in these crystals. Apparently, when the fiber is hydrated, the crystalline structure of collagen also changes, and the bound water favors a change in the crystal symmetry, reducing the piezoelectric properties [18, 19]. The piezoelectric response of human bone was quantified by means of a piezoresponse force microscope resulting to be 7–8 $pCN^{-1}$ [20].

Bone tissue shows a large potential to repair and regenerate itself through complex feedback mechanisms, where electromechanical processes are essential due to its piezoelectric nature. The first report by Yasuda et al. was later verified after the observation of electrical variations and electric potential regeneration when the bone is mechanically stressed [14, 21]. As a consequence of these mechanical stresses, electrical signals are produced and stimulate bone growth and remodeling [22]. Osteocytes, which play an important role in the structural regeneration of bone and bone mechanotransduction, seem to be the responsible for bone growth under piezoelectric signals [23, 24]. Then, these cells communicate with other cells, such as osteoblasts and osteoclasts for bone regeneration.

The enhancement and stimulation of osteogenic activities after the application of electrical stimulation has been demonstrated. Thus osteoblasts are affected by electromechanical signals, and, due to the piezoelectric nature of bone, the mechanical stimuli are converted into electrical [25, 26]. Moreover, under dynamic mechanical conditions, the growth and differentiation of osteoblasts in a piezoelectric material can be enhanced.

### 13.1.2 Collagen and Other Piezoelectric Tissues

Other tissues containing collagen, such as tendons and ligaments, also display piezoelectricity and, thus, undergo an electrical potential variation when a mechanical stress is applied [27, 28]. The piezoelectricity of dry tendons has been measured resulting in a decrease of the piezoelectric coefficient as hydration increases [19, 29–31]. Other soft tissues, such as skin, callus, cartilage, and tendons seem to be related with the orientation of the protein fibers [15, 32]. Fibrous molecules such as collagen, keratin, fibrin, elastin, or cellulose, present in connective tissues, show also piezoelectric properties. Electrical polarization variations were also verified in hair [33] when subjected to stress and in DNA [27]. Pineal gland tissues also contain noncentrosymmetric material, which is also piezoelectric [34].

### 13.1.3 Cardiac Tissue

The myocardium is a highly organized structure with unique electrical and mechanical properties. Thus cardiomyocyte growth and maturation seem to be influenced by mechanical loading and electric fields. Some heart characteristics such as size and performance can be guided by these stimuli. The myocardium is formed by fibers with a multilayered helical architecture essential for heart contraction for which mechanical stretching and electrical current stimulation are essential for the development of the tissue. Apparently, when mechanical and electrical stimuli are combined, they are able to promote contractility, calcium handling, protein expression, or cell proliferation [35, 36]. Combined mechanical and electrical stimulations proved to influence also recellularization, cardiomyocyte differentiation, and tissue remodeling in 2–4 days [37]. Electric fields also increase the mitochondrial content, influencing the conduction velocity, cell orientation, and contractile force. On the other hand, mechanical stimuli induce chemical and electrical responses in cardiomyocytes. The use of mechanical stretches between 10% and 15% proved to improve cardiac tissue structure and force development [38].

### 13.1.4 Nerve Tissues

Neurons transmit electrochemical signals across the nervous system. Thus they are affected by electrical stimuli [4]. The information is sent by axons, and their activity is associated to electrical variations. If an electric current goes through the neuron's membrane, it results in an action potential. Under an appropriate stimulus the cell membrane goes to depolarization from resting state, followed by repolarization to the resting state, reversing its polarity for a short period of time. In this way, by changing the electric potential of a nerve cell, an action potential can be produced. Electric fields ranging from 0.1 to 10 V/cm have been shown to influence the direction of neurite growth and increase the neurite initiation [39]. Moreover, these electric fields can also influence the rate and orientation of neurite outgrowth in vitro [11].

In this way, electroactive materials and, particularly, piezoelectric ones show a huge potential for tissue engineering applications. These materials vary their surface polarization under mechanical loading and influence cell morphology, adhesion, proliferation, and differentiation [40], mimicking the stimuli present under many physiological conditions. When compared with nonpiezoelectric controls, a higher cell growth and differentiation are observed for piezoelectric materials [11, 41]. Moreover the use of these active scaffolds has proved to promote neurite extension and neuronal differentiation [42]. The use of piezoelectric materials, with mechanically induced variations of surface charge, leads to enhanced osteogenic differentiation of human adipose stem cells [43]. Positively charged surfaces have been also reported to promote higher adhesion [44] and spreading [43, 45].

Moreover, for tissues such as bone, which are subjected to mechanoelectrical solicitations during movement, tissue reparation when immobilized is more complex as the natural stimuli are absent [46]. Thus the use of smart materials able to induce mechanical or electrical stimulation to the tissue during damage repair or in cell cultures within bioreactors shows great potential as an alternative to the already existing tissue engineering strategies.

## 13.2 PRINCIPLES FOR ELECTRIC AND MECHANICAL CLUES

The tissue development, repair, and/or regeneration of cells and organs and also the cell behavior and function require multiple physiological clues, not only the (bio)chemical ones but also the physical signals, which may be electrical and mechanical [47, 48].

With respect to the mechanical forces, cells are constantly exposed to them (depending on the tissue, they can be shear, compressive, or tensile forces) [49], which in turn exert forces to their environment, modulating their behavior (cell migration, proliferation, and differentiation) [47, 50]. Also, they play a key role in diverse cellular processes, ranging from proliferation to transcription to organogenesis [51], due, for example, to the focal adhesion complexes and the internal remodeling of the cytoskeletal architecture, to the signalization of the second messenger (such as the intracellular $Ca^{2+}$), or to the gene expression changes. So, cells are constantly monitoring the extracellular parameters of the surrounding microenvironment to respond to these changes appropriately and modify their behavior, through a process termed mechanotransduction [52]. Basically, each cellular process must begin with mechanotransduction, which is the conversion of the mechanical forces into biochemical or electrical signals that will remodel the cells and tissues at the structural and functional levels [53].

The microenvironment found is dependent on the tissue and can be highly distinct. For example, skeletal muscle cells are found embedded in a 3-D tissue undergoing mechanical stretching and compression, while endothelial cells are found in a 2-D interface in contact with the fluid, exposed also not only to stretching and compression during pulsatile blood flow but also to fluid shear stress [51, 54]. In this way, for the design of the mechanical platforms, it must be taken into consideration the mechanobiological niche of each tissue.

The bioelectric fields are generated by specific ion channels and pumps within cell membranes, which guide the development and regeneration of many tissues, such as cartilage, nervous system, and vascular endothelial cells [55, 56].

The knowledge that cells and tissues are able to receive from physical stimuli to translate them into biochemical and biological responses has been paving the way for the development of novel smart materials to be applied in regenerative medicine [57–60].

One such example is the development of electrically and magnetically active materials/scaffolds for tissue engineering purposes. These materials are able to efficiently induce cell seeding, growth, and differentiation, taking advantage of the mechanotransduction properties of the cells [4]. Living cells show many properties that are typical of electrical systems. They generate electromotive force, use varying resistances in series or in parallel, regulate the potential differences whenever needed, switch on and off, control and rectify current flow, and store charge. Across the plasma membrane, in the presence of electrical voltage, the environment inside the cell remains more negative than outside [4]. Due to all these properties the application of small electric fields on different cells such as corneal [6, 7] and epithelial [9] has been described to guide them to move and migrate in culture. Electric fields have also been reported to modulate phenotypes of vascular endothelial cells [10] and to enhance nerve fiber outgrowth in vitro [11] and have been widely used in bone and cartilage regeneration [12, 56, 61].

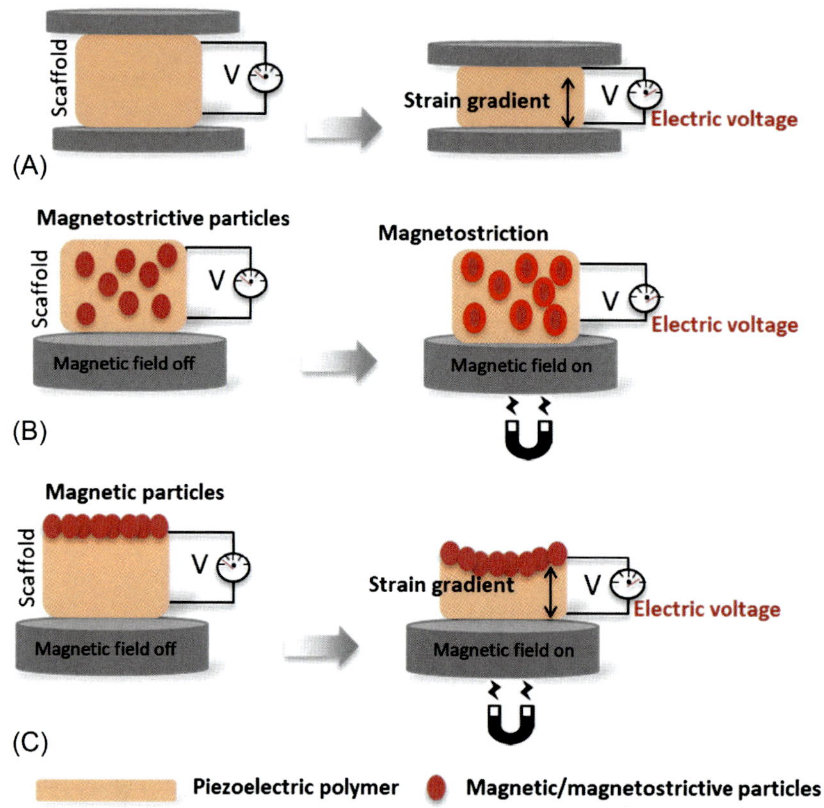


FIG. 13.1 Schematic representation of the (A) mechanoelectric properties of a scaffold upon the application of mechanical stimuli. (B) and (C) magnetoelectric properties of scaffolds upon the application of magnetic stimulus.

Therefore materials able to induce a surface electrical charge are able to stimulate the growth of cells and tissues in culture constituting a suitable approach for tissue engineering applications. Examples of such materials are the materials possessing mechanoelectric, magnetic, and magnetoelectric properties. Mechanoelectric materials are mostly constituted by piezoelectric materials that respond to mechanical stimuli, inducing an electrical potential variation (Fig. 13.1A). Magnetic and magnetoelectric materials used for tissue engineering applications are mainly composites comprising magnetic or magnetostrictive particles and a piezoelectric polymer. Due to their magnetic component, they sense a magnetic field that induces a mechanical stimulation on the scaffold due to the incorporated magnetic or magnetostrictive properties, which further induce an electrical polarization variation due to the piezoelectric phase present in the same scaffold (Fig. 13.1B and C).

## 13.3 ELECTRIC AND ELECTROMECHANICAL CLUES

New advances in tissue engineering have been carried out based on the application of different kind of stimuli (electric and/or mechanical) to influence cell response and fate [62]. Table 13.1 summarizes relevant experimental works where electrical and/or mechanical stimuli were used.

Some studies rely on the application of electrical or mechanical stimuli without the use of active polymers (such as conductive or piezoelectric). However, in the previous years, active polymers have been used for different tissue engineering areas (such as bone, muscle, and nerve) to induce these stimuli more naturally, resembling and taking advantage of the presence of electrical or mechanical signals within the body. For example, neurite outgrowth was significantly improved when conductive polymer was electrically stimulated (Fig. 13.2). Also, there are tissues that improve differentiation with the combination of electrical and mechanical stimuli, as in the case of the muscle.

Piezoelectric polymers such as poly(vinylidene fluoride) (PVDF) and vinylidene fluoride (VDF) copolymers are the most widely researched polymers to develop scaffolds with mechanoelectric properties [63]. These polymers possess

TABLE 13.1 Cells, Applied Stimuli (Electrical and/or Mechanical) and Material Used for Specific Representative Biomedical Applications

| | | Stimuli | | | |
|---|---|---|---|---|---|
| **Tissue** | **Cells** | **Electric** | **Mechanical** | **Material** | **Reference** |
| Cartilage | Mesenchymal stem cell MSC chondrogenic | | 1 Hz with 10% deformation | Poly(vinylidene fluoride-trifluoroethylene) (PVDF-TrFE) and polycaprolactone (PCL) fibers | [69] |
| | Adipose-derived stem cells (ADSCs) | 20 min application of 1 kHz, 20 mV/cm | | – | [70] |
| Cardiac | Human cardiac adipose tissue-derived progenitor cells (cardiac ATDPCs) | 2 ms pulses of 50 mV/cm at 1 Hz | 10% stretching | Fibrin hydrogel | [71] |
| | Human embryonic stem cells (hESC, line H13) | 1 V/mm at 1 Hz (1 and 90 s) | | – | [72] |
| | Cardiomyocites of neonatal rat cardiac cells | Biphasic rectangular pulses, 1 ms, 1 Hz, 3 V/cm | 5% stretch with 50% duty cycle at 1 Hz | Fibrin gel | [73] |
| | Cardiomyocytes | biphasic pulses, 5 V/cm, 0.2 Hz, 1 ms | | Collagen/fibrin 3-D matrices | [74] |
| | Cardiac adipose tissue-derived progenitor cells (cardiac ATDPCs) | Monophasic square-wave pulses, 2 ms, 50 mV/cm, 1 Hz | | Silicone-patterned surface | [75] |
| Skin | Human cutaneous fibroblasts | 0.05 V/mm (DC) | | Polypyrrole/poly(L-lactic acid) (PPy/PLLA) membranes | [76] |
| | Human skin fibroblasts | 100 mV/mm (DC) | | PPy/PLLA membranes | [77] |
| | Human dermal fibroblasts | 100 mV (DC) | | PPy-PDLLA membranes | [78] |
| Muscle | Satellite cells | 2 Hz | | In vivo | [79] |
| | Vascular smooth muscle cells | 50 μA (AC) 0.05, 5, and 500 Hz | | PPy substrates | [80] |
| | Skeletal: the murine-derived muscle cell line C2C12 | 20 V, 50 ms pulse, 1 Hz | | Polydimethylsiloxane (PDMS) micropatterned substrates | [81] |
| | Muscle precursor cells (MPC) | 70 mV/cm, 33mHz | | Micropatterned PLLA membranes | [82] |
| | Mouse C2C12 myoblast cells | 5 V, 10 ms, 1 Hz | | Graphene oxide/polyacrylamide (GO/PAAM) composite hydrogels | [83] |
| | C2C12 | Intermittent EF cycles 10 min separated by 20 min for 3 h | | PCL aligned topography | [84] |
| Nerve | Rat PC-12 | 0.1 V, 10 uA | | PPy film as anode | [85] |
| | Adipose-derived stem cells toward the neuronal lineage (C17.2) | 1.5 mA | | – | [86] |
| | Mouse neuronal cerebellum stem cells | 100 mV (DC) | | Polyaniline (PANI) with PCL/gelatin nanofibers | [87] |
| | NIH-3T3 fibroblasts | 0–200 m (DC) | | PANI/poly(L-lactide-co-ε-caprolactone) (PLCL) nanofibers | [88] |

*Continued*

TABLE 13.1 Cells, Applied Stimuli (Electrical and/or Mechanical) and Material Used for Specific Representative Biomedical Applications—cont'd

| Tissue | Cells | Stimuli | | Material | Reference |
|---|---|---|---|---|---|
| | | Electric | Mechanical | | |
| | Rat neuronal phaeochromocytoma (PC12) | 100 mV (DC)200 | | PPy film | [85] |
| | PC12 | $10\,mV\,cm^{-1}$ | | Poly(lactic-*co*-glycolic acid) (PLGA) films coated with PPy | [89] |
| | PC12 | 0, 2, 8 and 20 μA/mm (DC) | | PDLLA/CL membrane coated with PPy | [90] |
| | Dorsal root ganglia (DRG) | 10 V (DC) | | PCL/PPy nanofibers | [91] |
| | Cochlear neural explants | Biphasic pulses, 100 μs, $\pm 1\,mA/cm^2$, 250 Hz | | PPy film | [92] |
| | PC12 | 0.1 V, 1 Hz | | Copolymer of hydroxyl-capped poly(L-lactide) (PLA) and carboxyl-capped aniline pentamer (AP) film | [93] |
| | Schwann cells | 100 mV (DC) | | PPy/chitosan membrane | [94] |
| | Human umbilical cord mesenchymal stem cells (huMSCs) | 100 mV/mm (DC) | | PPy/PLA nanofibers | [95] |
| | Mousse neuroblastoma cells (Nb2a) | 2–3 mV, 1200 Hz | | Poled PVDF fibers | [11] |
| | PC12 | 40 mV | | PLA/PPy fibers | [96] |
| | PC12 | 100 mV | | PPy films | [97] |
| | PC12 | 10 mV/cm | | PPy-PLGA electrospun meshes | [98] |
| | PC12 | | Ultrasound stimulation, 20 W, 40 kHz | Stable glycol-chitosan-boron nitride nanotubes (BNNT) | [99] |
| | Exogenous human neural progenitor cells | +1 V to −1 V square wave 1 kHz | | PPy scaffold | [100] |
| | Rat spinal cord neurons | | Vibration base, 50 Hz | PVDF film | [101] |
| | SH-SY5Y neuron-like cells | | Ultrasounds, 0.1, 0.2, 0.4 and $0.8\,W/cm^2$ | Barium titanate nanoparticles | [102] |
| | SH-SY5Y neuron-like cells | | Ultrasounds, 5 s twice a day | PVDF-TrFE/barium titanate film | [103] |
| Bone | Rat bone marrow-derived mesenchymal stem cells (MSCs) | 0.2, 0.4, and 0.7 V/min for 60 min | | 3-D collagen fibers | [104] |
| | Human adipose stem cells (hASCs) | | 1 Hz with maximum amplitude of ≈1 mm | β-PVDF "poled +" film | [43] |
| | MC3T3-E1 preosteoblasts | | Frequency of 1 Hz | β-PVDF "poled −" film | [41] |
| | Osteoblasts | 10 mA, 10 Hz (AC) for 6 h/day | | PLLA/carbon nanotube (CNT) nonporous substrates | [105] |
| | Osteoblasts | 100 μA (DC) | | PLA/CNTs nanofiber | [106] |
| | Goat marrow stromal cells (GMCs) | | A lab rotator (model DSR 2800V, Digisystem Laboratory Instruments) | β-PVDF film | [107] |

TABLE 13.1 Cells, Applied Stimuli (Electrical and/or Mechanical) and Material Used for Specific Representative Biomedical Applications—cont'd

| Tissue | Cells | Stimuli | | Material | Reference |
|---|---|---|---|---|---|
| | | Electric | Mechanical | | |
| | Rat bone marrow stromal cells | 0.35 V/cm for 4 h | | PPy films | [108] |
| | human adipose-derived mesenchymal stem cells | 200 μA (DC) for 4 h/day | | PPY/PCL scaffold | [109] |
| | BMSC and MC3T3-E1 | 500 mV, 1 kHz | | Self-doped sulfonated polyaniline-based interdigitated electrodes | [110] |
| | SaOS-2 osteoblast-like cell culture | ~20–60 mV | Twice a day for 10 s | P(VDF-TrFE)/BNNTs film | [111] |
| Wound healing | NIH 3T3 cells (mouse embryo fibroblasts) | | Intermittent deformation of 8% at 0.5 Hz for 24 h | Polyurethane/PVDF fibers | [112] |
| | Skin fibroblasts | 100 mV/mm, 10 s stimulation within a period of 1200 s or 300 s stimulation within a period of 600 s | | PPy on the surface of polyethylene terephthalate (PET) substrates | [113] |
| Blood vessels | Human umbilical vein endothelial cells | 400 mV/cm 30 min/day | | PANI-coated PCL fibers | [114] |

high electroactive properties, including piezoelectric, pyroelectric, and ferroelectric properties [64]. Materials composed by these polymers develop voltage when a mechanical stress is applied, thus promoting the adhesion and proliferation of different types of cells [5].

PVDF, in particular, is a semicrystalline biocompatible polymer possessing high mechanical strength, thermal stability, chemical resistance, and hydrophobic properties [41, 45, 65, 66]. It is biocompatible, being, therefore, a promising material for biomedical applications, which showed to influence cellular response when both the phase and polarization of the material were evaluated [65, 67]. Thus the polarization of PVDF influenced the adsorption of fibronectin, being, therefore, an important factor to consider in further studies. The influence of polymer surface charge on MC3T3-E1 preosteoblasts cultivated under static and dynamic conditions was also studied [41], and it was concluded that positively charged PVDF films promote higher osteoblast adhesion and proliferation, which further increased under dynamic stimulation. The application of this polymer is not only confined to academic research. Due to its stability, strength, and biocompatibility, PVDF has been approved by the FDA, and it has been used in surgical mesh form for human implants and surgery [68].

## 13.4 MAGNETIC, MAGNETOMECHANIC, AND MAGNETOELECTRIC MATERIALS

Another strategy able to induce the mechanotransduction effect on cells is the application of magnetic stimuli on magnetic responsive materials. The use of magnetic materials in biomedicine has been widely explored; many applications are based on the possibility of preparing nanoscaled magnets. These materials possess important properties that have allowed applications in different medical areas including neurology [115], ophthalmology [116], dentistry [117], and cardiology [118]. In fact the size of these magnetic nanomaterials has been playing a key role, imparting unique properties to the material. Specific interactions with cells, viruses, and proteins, which

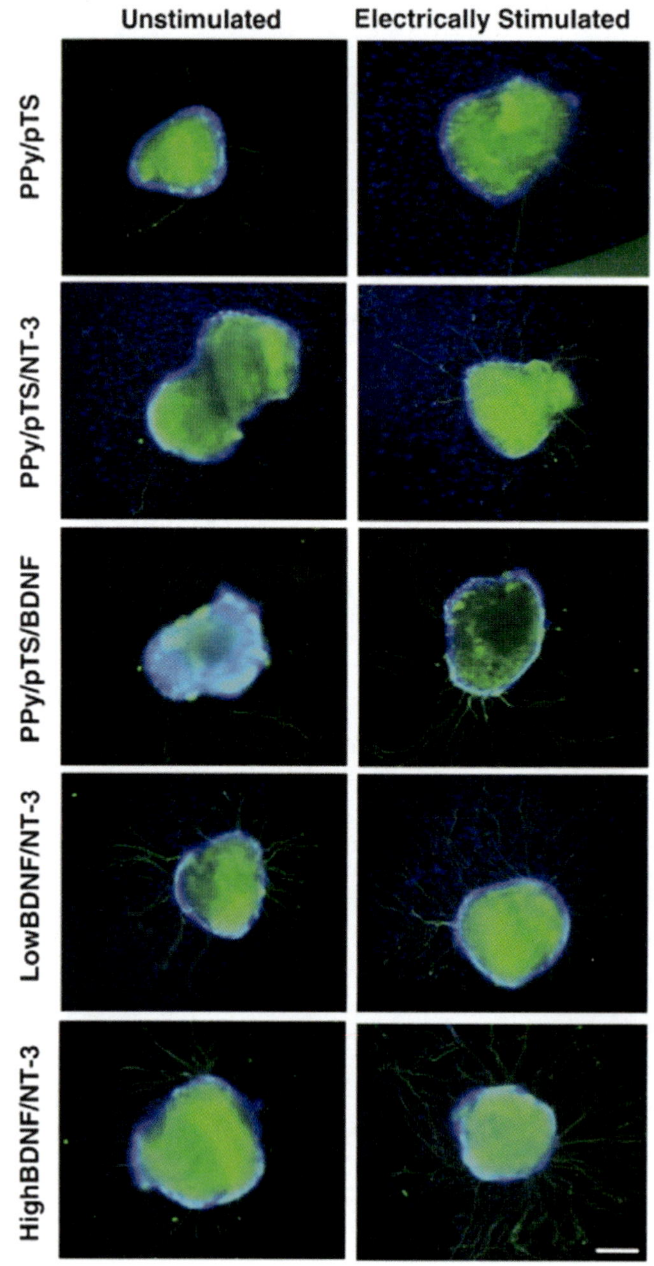


FIG. 13.2 Representative images of cochlear neural explants grown on polypyrrole/sodium salt polymers with and without neurotrophin. Neurites were visualized by immunocytochemistry with a neurofilament-200 primary antibody and a fluorescent secondary antibody *(green)*. Cell nuclei are labeled with DAPI *(blue)*. *(Reproduced with permission from R. Ravichandran, S. Sundarrajan, J.R. Venugopal, S. Mukherjee, S. Ramakrishna, Applications of conducting polymers and their issues in biomedical engineering, J. Royal Soc. Interf. 7 (2010) S559–S579.)*

ultimately induce cellular growth or death, and the possibility of entering the body and reach spaces that are inaccessible by other materials [119, 120] are just few examples of the important properties these materials hold. Moreover, at sizes below 20nm, magnetic nanoparticles exhibit superparamagnetic behavior, and no remanent magnetization is observed when the magnetic field is removed, making them suitable for in vivo applications since it prevents aggregation and enables to easily redisperse rapidly after withdrawing the magnetic field [119]. Besides the nanoparticles the use of magnetic liposomes has been increasingly investigated in the field of biomedicine due to their resemblance with the cellular membrane that allows better interaction with cells. Moreover, both nanoparticles and liposomes are easily functionalized with other materials, allowing enhanced interactions and specific binding to the targeted biological entities, which ultimately imparts colloidal stability and biocompatibility [121]. Also, one of the advantages of using magnetic nanomaterials in biomedicine is the fact that they are easily traceable and localized inside the body through the action of a magnetic field, thus using minimally invasive methods [122].

Magnetic nanomaterials may be divided, according to their morphology, in magnetic micro-/nanoparticles, which may be categorized into two different structures: pure metals (Co, Fe, Ni, and Ti) and metal oxides (iron oxides $Fe_2O_3$ or $Fe_3O_4$ and ferrites such as $BaFe_{12}O_{19}$ and $CoFe_2O_4$) and magnetic nanocomposites [123, 124]. Magnetic nanoparticles have been commonly used in a variety of cells and tissues for tissue engineering applications [125]. Also, they have been playing an important role in cell separation and immunoassays, drug targeting and delivery, gene delivery and transfection [126], and magnetic resonance imaging (MRI) contrast agents [127, 128]. These particles vary in size, surface chemistry, magnetic properties, and bulk chemistry and often consist of a magnetic iron oxide core coated with a biocompatible polymer. This allows their functionalization enabling them to attach binding molecules such as antibodies, peptides, and other functional groups [129]. Magnetic nanoparticles may be attached to functional sites, such as the cell membrane and/or on internal cellular components, thus acting as transducers of magnetic fields and enabling the noninvasive control of various cellular functions [57]. Early findings have shown that by functionalizing magnetic microparticles with ligands able to attach on different receptors on the cell surface, it was possible to study the mechanical linkage between the cell membrane receptor and the cytoskeletal network [130]. Since then, basic scientific research has been conducted on a variety of cell types to evaluate the mechanotransduction phenomenon assisted by these particles and respective coatings [131–137].

These types of magnetic nanoparticles may be incorporated as a filler to obtain composites with magnetoelectric properties. The obtained magnetic nanocomposites are a very interesting material since it allows to remotely mechanically and electrically stimulate tissues from outside of the human body [138, 139] and for specific cell cultures in bioreactors [140]. The possibility to remotely control tissue stimulation without the need of patient movement is certainly an innovative approach. As previously mentioned and depicted in Fig. 13.1B and C, the mechanical deformation induced by a magnetic field due to the magnetostriction or magnetic properties of one of the components of the composite results in an electrical polarization variation due to the piezoelectric effect of the other phase, allowing large magnetoelectric effects at room temperature [141, 142]. Thus the magnetic actuation ability of the magnetoelectric composite allows the mechanical and electrical stimuli of neighboring cells [143].

Two main types of polymer-based magnetoelectric composites can be found in the literature: laminated composites and particulate micro- and nanocomposites [143]. Examples of that are the P(VDF-TrFE)/Metglas 2605SA1 laminates that possess high magnetoelectric response ($383\,V/cm\,Oe^{-1}$) [144] and P(VDF-TrFE)/$CoFe_2O_4$ nanocomposites that show lower magnetoelectric response ($42\,mV/cm\,Oe^{-1}$) but present higher flexibility, simple fabrication, easy shaping, the possibilities of miniaturization, and the absence of degradation at the piezoelectric/magnetostrictive interface [145]. This is not the only advantages of nanocomposites; they also allow the development of geometries suitable for tissue engineering, namely, spheres [141] and fiber mats [146], which may be tuned for specific tissue engineering approaches.

Therefore these composites, more precisely magnetoelectric biomaterials, have begun to be used as novel approach for tissue engineering applications and have already demonstrated to promote the cell proliferation of preosteoblast cells by the application of a varying magnetic field [46, 147]. With these kinds of biomaterials, it is possible to provide mechanical or mechanoelectric stimulus to deliver higher cell proliferation (Fig. 13.3).

Static or alternating magnetic fields applied in some clinical studies have already been reported, inducing the integration and regeneration of tissues into ceramics [148]. Collagen magnetic scaffolds, which were fixed with external magnets in vivo, were also proven to induce controlled regeneration of bone cells [149, 150].

Therefore tissue engineering approaches using mechanoelectric, magnetic, and magnetoelectric composites hold great promise, since it actively responds to biomimetic stimuli that control processes of cell regeneration and homeostasis.

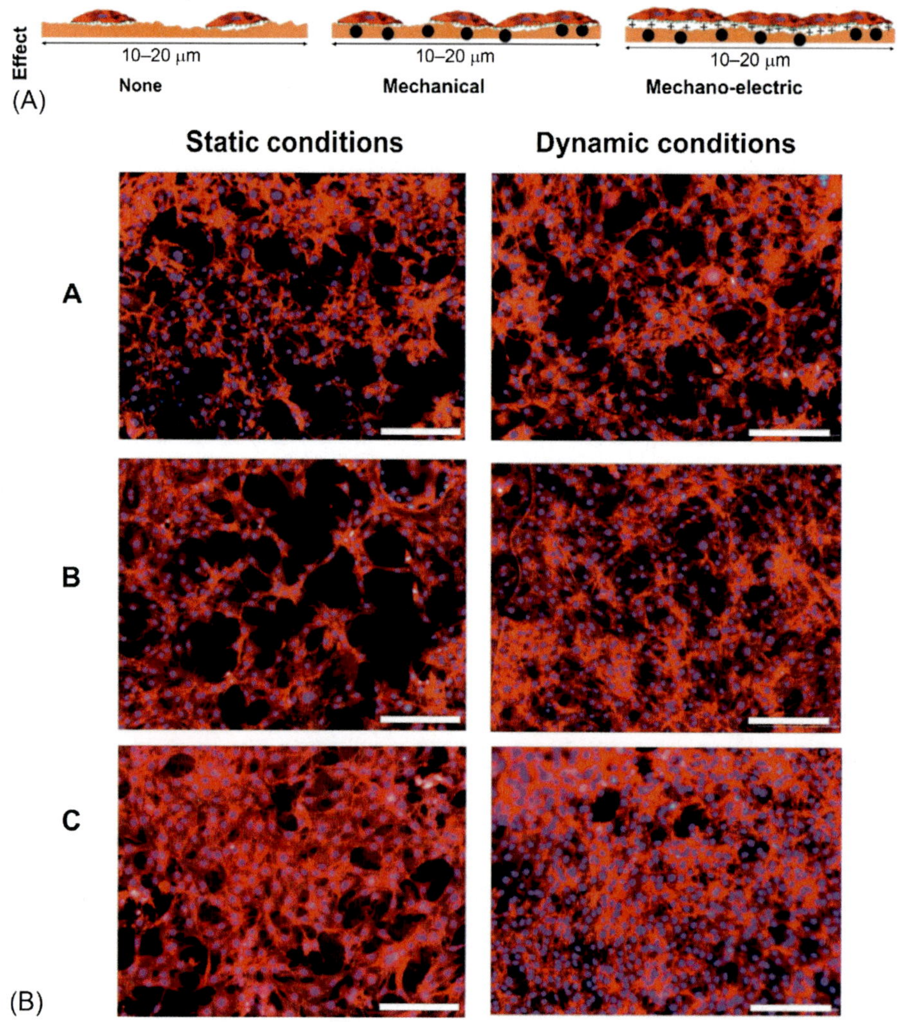


FIG. 13.3 (A) Representative images of the different stimuli provided by the scaffolds and (B) representative images of preosteoblast culture after 72h on PVDF-TrFE nonpoled (**A**, none stimuli), PVDF-TrFE/TD nonpoled (**B**, mechanical stimuli), and PVDF-TrFE/TD "poled +" (**C**, mechano-electric stimuli) scaffolds with static and dynamic conditions. Scale bar = 200 μm. *(Reproduced with permission from C. Ribeiro, V. Correia, P. Martins, F.M. Gama, S. Lanceros-Mendez, Proving the suitability of magnetoelectric stimuli for tissue engineering applications, Colloids Surf. B 140 (2016) 430–436.)*

## 13.5 CONCLUSIONS

It was demonstrated that the application of electrical and/or mechanical stimuli can improve the regeneration success of different cell/tissues, mimicking in vivo electromechanical microenvironments and allowing the development of novel tissue engineering strategies. A large variety of materials is already available, and several proof-of-concept investigations indicate the suitability of this approach. Nevertheless, further investigation is needed to perfectly match the suitable microenvironment for different cell types, including the right combination of biophysical and biochemical stimuli for proper tissue regeneration.

## Acknowledgments

The authors thank the Fundação para a Ciência e Tecnologia (FCT) for the financial support under framework of the Strategic Funding UID/FIS/04650/2013, project PTDC/EEI-SII/5582/2014, and project POCI-01-0145-FEDER-028237. SR, MMF, and CR also thank the FCT for the grants SFRH/BD/111478/2015, SFRH/BPD/121464/2016, and SFRH/BPD/90870/2012, respectively. Finally the authors acknowledge funding by the Spanish Ministry of Economy and Competitiveness (MINECO) through the project MAT2016-76039-C4-3-R (AEI/FEDER, UE) and from the Basque Government Industry Department under the ELKARTEK and HAZITEK program.

# References


[1] K. Burridge, C. Guilluy, Focal adhesions, stress fibers and mechanical tension, Exp. Cell Res. 343 (2016) 14–20.

[2] C.Y. Ning, Z.N. Zhou, G.X. Tan, Y. Zhu, C.B. Mao, Electroactive polymers for tissue regeneration: developments and perspectives, Prog. Polym. Sci. 81 (2018) 144–162.

[3] J.J. Telega, R. Wojnar, Piezoelectric effects in biological tissues, J. Theor. Appl. Mech. 40 (3) (2002).

[4] L. Ghasemi-Mobarakeh, M.P. Prabhakaran, M. Morshed, M.H. Nasr-Esfahani, H. Baharvand, S. Kiani, S.S. Al-Deyab, S. Ramakrishna, Application of conductive polymers, scaffolds and electrical stimulation for nerve tissue engineering, J. Tissue Eng. Regen. Med. 5 (2011) e17–e35.

[5] C. Ribeiro, V. Sencadas, D.M. Correia, S. Lanceros-Méndez, Piezoelectric polymers as biomaterials for tissue engineering applications, Colloids Surf. B: Biointerfaces 136 (2015) 46–55.

[6] G.L. Sulik, H.K. Soong, P.C.T. Chang, W.C. Parkinson, S.G. Elner, V.M. Elner, Effects of steady electric fields on human retinal pigment epithelial cell orientation and migration in culture, Acta Ophthalmol. 70 (1992) 115–122.

[7] M. Zhao, A. Agius-Fernandez, J.V. Forrester, C.D. McCaig, Orientation and directed migration of cultured corneal epithelial cells in small electric fields are serum dependent, J. Cell Sci. 109 (1996) 1405–1414.

[8] J. Pu, C.D. McCaig, L. Cao, Z. Zhao, J.E. Segall, M. Zhao, EGF receptor signalling is essential for electric-field-directed migration of breast cancer cells, J. Cell Sci. 120 (2007) 3395–3403.

[9] E. Wang, M. Zhao, J.V. Forrester, C.D. McCaig, Bi-directional migration of lens epithelial cells in a physiological electrical field, Exp. Eye Res. 76 (2003) 29–37.

[10] A. Bouaziz, A. Richert, A. Caprani, Vascular endothelial cell responses to different electrically charged poly(vinylidene fluoride) supports under static and oscillating flow conditions, Biomaterials 18 (1997) 107–112.

[11] R.F. Valentini, T.G. Vargo, J.A. Gardella Jr., P. Aebischer, Electrically charged polymeric substrates enhance nerve fibre outgrowth in vitro, Biomaterials 13 (1992) 183–190.

[12] P.-h.G. Chao, H.H. Lu, C.T. Hung, S.B. Nicoll, J.C. Bulinski, Effects of applied DC electric field on ligament fibroblast migration and wound healing, Connect. Tissue Res. 48 (2007) 188–197.

[13] I. Yasuda, K. Noguchi, T. Sata, Dynamic callus and electric callus, J. Bone Joint Surg. 37 (1955) 1292–1293.

[14] E. Fukada, I. Yasuda, On the piezoelectric effect of bone, J. Phys. Soc. Jpn. 12 (1957) 1158–1162.

[15] M.H. Shamos, L.S. Lavine, Piezoelectricity as a fundamental property of biological tissues, Nature 213 (1967) 267–269.

[16] E. Fukada, Electricl phenomena in biorheology, Biorheology 19 (1982) 15–27.

[17] H. Athenstaedt, Pyroelectric and piezoelectric behaviour of human dental hard tissues, Arch. Oral Biol. 16 (1971) 495–501.

[18] J.C. Anderson, C. Eriksson, Electrical properies of wet collagen, Nature 218 (1968) 166–168.

[19] J.C. Anderson, C. Eriksson, Piezoelectric properties of dry and wet bone, Nature 227 (1970) 491–492.

[20] C. Halperin, S. Mutchnik, A. Agronin, M. Molotskii, P. Urenski, M. Salai, G. Rosenman, Piezoelectric effect in human bones studied in nanometer scale, Nano Lett. 4 (2004) 1253–1256.

[21] C.A.L. Bassett, R.O. Becker, Generation of electric potentials by bone in response to mechanical stress, Science 137 (1962) 1063–1064.

[22] H.M. Frost, Wolff law and bones structural adaptations to mechanical usage – an overview for clinician, Angle Orthod. 64 (1994) 175–188.

[23] A.A. Marino, R.O. Becker, Piezoelectric effect and growth control in bone, Nature 228 (1970) 473–474.

[24] S. Baiotto, M. Zidi, Theoretical and numerical study of a bone remodeling model: the effect of osteocyte cells distribution, Biomech. Model. Mechanobiol. 3 (2004) 6–16.

[25] K. Anselme, Osteoblast adhesion on biomaterials, Biomaterials 21 (2000) 667–681.

[26] B. Miara, E. Rohan, M. Zidi, B. Labat, Piezomaterials for bone regeneration design – homogenization approach, J. Mech. Phys. Solids 53 (2005) 2529–2556.

[27] E. Fukada, History and recent progress in piezoelectric polymers, IEEE Trans. Ultrason. Ferroelectr. Freq. Control 47 (2000) 1277–1290.

[28] C.R. West, A.E. Bowden, Using tendon inherent electric properties to consistently track induced mechanical strain, Ann. Biomed. Eng. 40 (2012) 1568–1574.

[29] E. Fukada, I. Yasuda, Piezoelectric effects in collagen, Jpn. J. Appl. Phys. 3 (1964) 117–118.

[30] A.A. Marino, R.O. Becker, Piezoelectricity in hydrated frozen bone and tendon, Nature 253 (1975) 627–628.

[31] D. Gross, W.S. Williams, Streaming potential and the electromechanical response of physiologically-moist bone, J. Biomech. 15 (1982) 277–295.

[32] L.S. Lavine, I. Lustrin, M.H. Shamos, R.A. Rinaldi, A.R. Liboff, Electric enhancement of bone healing, Science 175 (1972) 1118–1121.

[33] A.J.P. Martin, Tribo-electricity in wool and hair, Proc. Phys. Soc. 53 (1941) 186–189.

[34] S.B. Lang, A.A. Marino, G. Berkovic, M. Fowler, K.D. Abreo, Piezoelectricity in the human pineal gland, Bioelectrochem. Bioenerg. 41 (1996) 191–195.

[35] J.-L. Ruan, N.L. Tulloch, M.V. Razumova, M. Saiget, V. Muskheli, L. Pabon, H. Reinecke, M. Regnier, C.E. Murry, Mechanical stress conditioning and electrical stimulation promote contractility and force maturation of induced pluripotent stem cell-derived human cardiac tissue, Circulation 134 (2016) 1557–1567.

[36] D. Kai, M.P. Prabhakaran, G. Jin, S. Ramakrishna, Biocompatibility evaluation of electrically conductive nanofibrous scaffolds for cardiac tissue engineering, J. Mater. Chem. B 1 (2013) 2305–2314.

[37] B. Wang, G. Wang, F. To, J.R. Butler, A. Claude, R.M. McLaughlin, L.N. Williams, A.L. de Jongh Curry, J. Liao, Myocardial scaffold-based cardiac tissue engineering: application of coordinated mechanical and electrical stimulations, Langmuir 29 (2013) 11109–11117.

[38] H. Cao, B.J. Kang, C.A. Lee, K.K. Shung, T.K. Hsiai, Electrical and mechanical strategies to enable cardiac repair and regeneration, IEEE Rev. Biomed. Eng. 8 (2015) 114–124.

[39] G.N. Li, D. Hoffman-Kim, Tissue-engineered platforms of axon guidance, Tissue Eng. B Rev. 14 (2008) 33–51.

[40] N. Weber, Y.S. Lee, S. Shanmugasundaram, M. Jaffe, T.L. Arinzeh, Characterization and in vitro cytocompatibility of piezoelectric electrospun scaffolds, Acta Biomater. 6 (2010) 3550–3556.

[41] C. Ribeiro, S. Moreira, V. Correia, V. Sencadas, J.G. Rocha, F.M. Gama, J.L. Gómez Ribelles, S. Lanceros-Méndez, Enhanced proliferation of pre-osteoblastic cells by dynamic piezoelectric stimulation, RSC Adv. 2 (2012) 11504–11509.

[42] Y.-S. Lee, T.L. Arinzeh, The influence of piezoelectric scaffolds on neural differentiation of human neural stem/progenitor cells, Tissue Eng. A 18 (2012) 2063–2072.
[43] C. Ribeiro, J. Pärssinen, V. Sencadas, V. Correia, S. Miettinen, V.P. Hytönen, S. Lanceros-Méndez, Dynamic piezoelectric stimulation enhances osteogenic differentiation of human adipose stem cells, J. Biomed. Mater. Res. Part A 103 (2015) 2172–2175.
[44] G.B. Schneider, A. English, M. Abraham, R. Zaharias, C. Stanford, J. Keller, The effect of hydrogel charge density on cell attachment, Biomaterials 25 (2004) 3023–3028.
[45] P. Martins, A.C. Lopes, S. Lanceros-Mendez, Electroactive phases of poly(vinylidene fluoride): determination, processing and applications, Prog. Polym. Sci. 39 (2014) 683–706.
[46] C. Ribeiro, V. Correia, P. Martins, F.M. Gama, S. Lanceros-Mendez, Proving the suitability of magnetoelectric stimuli for tissue engineering applications, Colloids Surf. B: Biointerfaces 140 (2016) 430–436.
[47] J.L. Hoon, M.H. Tan, C.G. Koh, The regulation of cellular responses to mechanical cues by rho gtpases, Cell 5 (2016).
[48] C. Brás-Pereira, E. Moreno, Mechanical cell competition, Curr. Opin. Cell Biol. 51 (2018) 15–21.
[49] T.J. Kirby, J. Lammerding, Emerging views of the nucleus as a cellular mechanosensor, Nat. Cell Biol. 20 (2018) 373–381.
[50] V.M. Weaver, Cell and tissue mechanics: the new cell biology frontier, Mol. Biol. Cell 28 (2017) 1815–1818.
[51] N.V. Bukoreshtliev, K. Haase, A.E. Pelling, Mechanical cues in cellular signalling and communication, Cell Tissue Res. 352 (2013) 77–94.
[52] J.M. Barnes, L. Przybyla, V.M. Weaver, Tissue mechanics regulate brain development, homeostasis and disease, J. Cell Sci. 130 (2017) 71–82.
[53] M.L. McCain, K.K. Parker, Mechanotransduction: the role of mechanical stress, myocyte shape, and cytoskeletal architecture on cardiac function, Pflugers Archiv Eur. J. Physiol. 462 (2011) 89–104.
[54] A. Trumbull, G. Subramanian, E. Yildirim-Ayan, Mechanoresponsive musculoskeletal tissue differentiation of adipose-derived stem cells, Biomed. Eng. Online 15 (2016) 43.
[55] M. Levin, Bioelectric mechanisms in regeneration: unique aspects and future perspectives, Semin. Cell Dev. Biol. 20 (2009) 543–556.
[56] X. Yuan, D.E. Arkonac, P.H.G. Chao, G. Vunjak-Novakovic, Electrical stimulation enhances cell migration and integrative repair in the meniscus, Sci. Rep. 4 (2015).
[57] S.H. Cartmell, J. Dobson, The use of magnetic particles in tissue engineering, in: Nanotechnologies for the Life Sciences, Wiley, 2011.
[58] C.M. Bidan, M. Fratzl, A. Coullomb, P. Moreau, A.H. Lombard, I. Wang, M. Balland, T. Boudou, N.M. Dempsey, T. Devillers, A. Dupont, Magneto-active substrates for local mechanical stimulation of living cells, Sci. Rep. 8 (2018) 1464.
[59] S.J. Mousavi, M.H. Doweidar, Encapsulated piezoelectric nanoparticle–hydrogel smart material to remotely regulate cell differentiation and proliferation: a finite element model, Comput. Mech. (2018) 1–19.
[60] S.J. Mousavi, M.H. Doweidar, Numerical modeling of cell differentiation and proliferation in force-induced substrates via encapsulated magnetic nanoparticles, Comput. Methods Prog. Biomed. 130 (2016) 106–117.
[61] B. Baker, R.O. Becker, J. Spadaro, A study of electrochemical enhancement of articular cartilage repair, Clin. Orthop. Relat. Res. 102 (1974) 251–267.
[62] R. Balint, N.J. Cassidy, S.H. Cartmell, Electrical stimulation: a novel tool for tissue engineering, Tissue Eng. B: Rev. 19 (2013) 48–57.
[63] J. Serrado Nunes, A. Wu, J. Gomes, V. Sencadas, P.M. Vilarinho, S. Lanceros-Méndez, Relationship between the microstructure and the microscopic piezoelectric response of the α- and β-phases of poly(vinylidene fluoride), Appl. Phys. A 95 (2009) 875–880.
[64] J.-C. Dubois, Ferroelectric polymers: Chemistry, physics, and applications. Edited by Hari Singh Nalwa, Marcel Dekker, New York 1995, XII, 895 pp., hardcover, $225.00, ISBN 0-8247-9468-0, Adv. Mater. 8 (1996) 542.
[65] C. Ribeiro, J.A. Panadero, V. Sencadas, S. Lanceros-Méndez, M.N. Tamaño, D. Moratal, M. Salmerón-Sánchez, J.L. Gómez Ribelles, Fibronectin adsorption and cell response on electroactive poly(vinylidene fluoride) films, Biomed. Mater. 7 (2012) 035004–035013.
[66] H. Lee, B. Bhushan, Nanotribology of polyvinylidene difluoride (PVDF) in the presence of electric field, J. Colloid Interface Sci. 360 (2011) 777–784.
[67] R. Costa, C. Ribeiro, A.C. Lopes, P. Martins, V. Sencadas, R. Soares, S. Lanceros-Mendez, Osteoblast, fibroblast and in vivo biological response to poly(vinylidene fluoride) based composite materials, J. Mater. Sci. Mater. Med. 24 (2012) 395–403.
[68] T. Yildirim, Stimuli-responsive polymeric nanoparticles biomedical applications, in: Chemisch-Geowissenschaftliche Fakultät, Friedrich-Schiller-Universität Jena, 2017, p. 324.
[69] S.M. Damaraju, Y. Shen, E. Elele, B. Khusid, A. Eshghinejad, J. Li, M. Jaffe, T.L. Arinzeh, Three-dimensional piezoelectric fibrous scaffolds selectively promote mesenchymal stem cell differentiation, Biomaterials 149 (2017) 51–62.
[70] M. Mardani, S. Roshankhah, B. Hashemibeni, M. Salahshoor, E. Naghsh, E. Esfandiari, Induction of chondrogenic differentiation of human adipose-derived stem cells by low frequency electric field, Adv. Biomed. Res. 5 (2016) 97.
[71] A. Llucià-Valldeperas, C. Soler-Botija, C. Gálvez-Montón, S. Roura, C. Prat-Vidal, I. Perea-Gil, B. Sanchez, R. Bragos, G. Vunjak-Novakovic, A. Bayes-Genis, Electromechanical conditioning of adult progenitor cells improves recovery of cardiac function after myocardial infarction, Stem Cells Transl. Med. 6 (2017) 970–981.
[72] E. Serena, E. Figallo, N. Tandon, C. Cannizzaro, S. Gerecht, N. Elvassore, G. Vunjak-Novakovic, Electrical stimulation of human embryonic stem cells: cardiac differentiation and the generation of reactive oxygen species, Exp. Cell Res. 315 (2009) 3611–3619.
[73] K.Y. Morgan, L.D. Black, Mimicking isovolumic contraction with combined electromechanical stimulation improves the development of engineered cardiac constructs, Tissue Eng. Part A 20 (2014) 1654–1667.
[74] T. Boudou, W.R. Legant, A. Mu, M.A. Borochin, N. Thavandiran, M. Radisic, P.W. Zandstra, J.A. Epstein, K.B. Margulies, C. S. Chen, A microfabricated platform to measure and manipulate the mechanics of engineered cardiac microtissues, Tissue Eng. Part A 18 (2012) 910–919.
[75] A. Llucià-Valldeperas, B. Sanchez, C. Soler-Botija, C. Gálvez-Montón, C. Prat-Vidal, S. Roura, J. Rosell-Ferrer, R. Bragos, A. Bayes-Genis, Electrical stimulation of cardiac adipose tissue-derived progenitor cells modulates cell phenotype and genetic machinery, J. Tissue Eng. Regen. Med. 9 (2015) E76–E83.
[76] G. Shi, Z. Zhang, M. Rouabhia, The regulation of cell functions electrically using biodegradable polypyrrole-polylactide conductors, Biomaterials 29 (2008) 3792–3798.
[77] G. Shi, M. Rouabhia, S. Meng, Z. Zhang, Electrical stimulation enhances viability of human cutaneous fibroblasts on conductive biodegradable substrates, J. Biomed. Mater. Res. Part A 84 (2008) 1026–1037.

[78] G. Shi, M. Rouabhia, Z. Wang, L.H. Dao, Z. Zhang, A novel electrically conductive and biodegradable composite made of polypyrrole nanoparticles and polylactide, Biomaterials 25 (2004) 2477–2488.

[79] Q. Wan, S.S. Yeung, K.K. Cheung, S.W. Au, W.W. Lam, Y.H. Li, Z.Q. Dai, E.W. Yeung, Optimizing electrical stimulation for promoting satellite cell proliferation in muscle disuse atrophy, Am. J. Phys. Med. Rehabil. 95 (2016) 28–38.

[80] A.S. Rowlands, J.J. Cooper-White, Directing phenotype of vascular smooth muscle cells using electrically stimulated conducting polymer, Biomaterials 29 (2008) 4510–4520.

[81] P. Bajaj, B. Reddy, L. Millet, C. Wei, P. Zorlutuna, G. Bao, R. Bashir, Patterning the differentiation of C2C12 skeletal myoblasts, Integr. Biol. 3 (2011) 897–909.

[82] M. Flaibani, L. Boldrin, E. Cimetta, M. Piccoli, P. De Coppi, N. Elvassore, Muscle differentiation and myotubes alignment is influenced by micropatterned surfaces and exogenous electrical stimulation, Tissue Eng. A 15 (2009) 2447–2457.

[83] H. Jo, M. Sim, S. Kim, S. Yang, Y. Yoo, J.H. Park, T.H. Yoon, M.G. Kim, J.Y. Lee, Electrically conductive graphene/polyacrylamide hydrogels produced by mild chemical reduction for enhanced myoblast growth and differentiation, Acta Biomater. 48 (2017) 100–109.

[84] U.H. Ko, S. Park, H. Bang, M. Kim, H. Shin, J.H. Shin, Promotion of myogenic maturation by timely application of electric field along the topographical alignment, Tissue Eng. Part A 24 (2018) 752–760.

[85] C.E. Schmidt, V.R. Shastri, J.P. Vacanti, R. Langer, Stimulation of neurite outgrowth using an electrically conducting polymer, Proc. Natl. Acad. Sci. U. S. A. 94 (1997) 8948–8953.

[86] L. Jaatinen, S. Salemi, S. Miettinen, J. Hyttinen, D. Eberli, The combination of electric current and copper promotes neuronal differentiation of adipose-derived stem cells, Ann. Biomed. Eng. 43 (2015) 1014–1023.

[87] L. Ghasemi-Mobarakeh, M.P. Prabhakaran, M. Morshed, M.H. Nasr-Esfahani, S. Ramakrishna, Electrical stimulation of nerve cells using conductive nanofibrous scaffolds for nerve tissue engineering, Tissue Eng. Part A 15 (2009) 3605–3619.

[88] S.I. Jeong, I.D. Jun, M.J. Choi, Y.C. Nho, Y.M. Lee, H. Shin, Development of electroactive and elastic nanofibers that contain polyaniline and poly(L-lactide-co-ε-caprolactone) for the control of cell adhesion, Macromol. Biosci. 8 (2008) 627–637.

[89] J.Y. Lee, J.-W. Lee, C.E. Schmidt, Neuroactive conducting scaffolds: nerve growth factor conjugation on active ester-functionalized polypyrrole, J. R. Soc. Interface 6 (2009) 801–810.

[90] Z. Zhang, M. Rouabhia, Z. Wang, C. Roberge, G. Shi, P. Roche, J. Li, L.H. Dao, Electrically conductive biodegradable polymer composite for nerve regeneration: electricity-stimulated neurite outgrowth and axon regeneration, Artif. Organs 31 (2007) 13–22.

[91] J. Xie, M.R. MacEwcm, S.M. Willerth, X. Li, D.W. Moran, S.E. Sakiyama-Elbert, Y. Xia, Conductive core-sheath nanofibers and their potential application in neural tissue engineering, Adv. Funct. Mater. 19 (2009) 2312–2318.

[92] B.C. Thompson, R.T. Richardson, S.E. Moulton, A.J. Evans, S. O'Leary, G.M. Clark, G.G. Wallace, Conducting polymers, dual neurotrophins and pulsed electrical stimulation – dramatic effects on neurite outgrowth, J. Control. Release 141 (2010) 161–167.

[93] L. Huang, X. Zhuang, J. Hu, L. Lang, P. Zhang, Y. Wang, X. Chen, Y. Wei, X. Jing, Synthesis of biodegradable and electroactive multiblock polylactide and aniline pentamer copolymer for tissue engineering applications, Biomacromolecules 9 (2008) 850–858.

[94] J. Huang, X. Hu, L. Lu, Z. Ye, Q. Zhang, Z. Luo, Electrical regulation of Schwann cells using conductive polypyrrole/chitosan polymers, J. Biomed. Mater. Res. Part A 93 (2010) 164–174.

[95] J. Zhou, L. Cheng, X. Sun, X. Wang, S. Jin, J. Li, Q. Wu, Neurogenic differentiation of human umbilical cord mesenchymal stem cells on aligned electrospun polypyrrole/polylactide composite nanofibers with electrical stimulation, Front. Mater. Sci. 10 (2016) 260–269.

[96] L. Tian, M.P. Prabhakaran, J. Hu, M. Chen, F. Besenbacher, S. Ramakrishna, Synergistic effect of topography, surface chemistry and conductivity of the electrospun nanofibrous scaffold on cellular response of PC12 cells, Colloids Surf. B: Biointerfaces 145 (2016) 420–429.

[97] N. Gomez, C.E. Schmidt, Nerve growth factor-immobilized polypyrrole: Bioactive electrically conducting polymer for enhanced neurite extension, J. Biomed. Mater. Res. Part A 81 (2007) 135–149.

[98] J.Y. Lee, C.A. Bashur, A.S. Goldstein, C.E. Schmidt, Polypyrrole-coated electrospun PLGA nanofibers for neural tissue applications, Biomaterials 30 (2009) 4325–4335.

[99] G. Ciofani, S. Danti, D. D'Alessandro, L. Ricotti, S. Moscato, G. Bertoni, A. Falqui, S. Berrettini, M. Petrini, V. Mattoli, A. Menciassi, Enhancement of neurite outgrowth in neuronal-like cells following boron nitride nanotube-mediated stimulation, ACS Nano 4 (2010) 6267–6277.

[100] P.M. George, T.M. Bliss, T. Hua, A. Lee, B. Oh, A. Levinson, S. Mehta, G. Sun, G.K. Steinberg, Electrical preconditioning of stem cells with a conductive polymer scaffold enhances stroke recovery, Biomaterials 142 (2017) 31–40.

[101] N. Royo-Gascon, M. Wininger, J.I. Scheinbeim, B.L. Firestein, W. Craelius, Piezoelectric substrates promote neurite growth in rat spinal cord neurons, Ann. Biomed. Eng. 41 (2013) 112–122.

[102] A. Marino, S. Arai, Y. Hou, E. Sinibaldi, M. Pellegrino, Y.T. Chang, B. Mazzolai, V. Mattoli, M. Suzuki, G. Ciofani, Piezoelectric nanoparticle-assisted wireless neuronal stimulation, ACS Nano 9 (2015) 7678–7689.

[103] G.G. Genchi, L. Ceseracciu, A. Marino, M. Labardi, S. Marras, F. Pignatelli, L. Bruschini, V. Mattoli, G. Ciofani, P(VDF-TrFE)/$BaTiO_3$ nanoparticle composite films mediate piezoelectric stimulation and promote differentiation of SH-SY5Y neuroblastoma cells, Adv. Healthc. Mater. 5 (2016) 1808–1820.

[104] S. Sun, I. Titushkin, M. Cho, Regulation of mesenchymal stem cell adhesion and orientation in 3D collagen scaffold by electrical stimulus, Bioelectrochemistry 69 (2006) 133–141.

[105] R. Bizios, K.R. Ullmann, P.R. Supronowicz, P.M. Ajayan, D.W. Metzger, B.P. Arulanandam, Novel current-conducting composite substrates for exposing osteoblasts to alternating current stimulation, J. Biomed. Mater. Res. 59 (2002) 499–506.

[106] S. Shao, S. Zhou, L. Li, J. Li, C. Luo, J. Wang, X. Li, J. Weng, Osteoblast function on electrically conductive electrospun PLA/MWCNTs nanofibers, Biomaterials 32 (2011) 2821–2833.

[107] M.T. Rodrigues, M.E. Gomes, J.F. Mano, R.L. Reis, β-PVDF membranes induce cellular proliferation and differentiation in static and dynamic conditions, Mater. Sci. Forum (2008) 72–76.

[108] W.W. Hu, Y.T. Hsu, Y.C. Cheng, C. Li, R.C. Ruaan, C.C. Chien, C.A. Chung, C.W. Tsao, Electrical stimulation to promote osteogenesis using conductive polypyrrole films, Mater. Sci. Eng. C 37 (2014) 28–36.

[109] J. Zhang, M. Li, E.T. Kang, K.G. Neoh, Electrical stimulation of adipose-derived mesenchymal stem cells in conductive scaffolds and the roles of voltage-gated ion channels, Acta Biomater. 32 (2016) 46–56.

[110] Y. Min, Y. Liu, Y. Poojari, J.C. Wu, B.E. Hildreth, T.J. Rosol, A.J. Epstein, Self-doped polyaniline-based interdigitated electrodes for electrical stimulation of osteoblast cell lines, Synth. Met. 198 (2014) 308–313.

[111] G.G. Genchi, E. Sinibaldi, L. Ceseracciu, M. Labardi, A. Marino, S. Marras, G. De Simoni, V. Mattoli, G. Ciofani, Ultrasound-activated piezoelectric P(VDF-TrFE)/boron nitride nanotube composite films promote differentiation of human SaOS-2 osteoblast-like cells, Nanomedicine (2017).

[112] H.F. Guo, Z.S. Li, S.W. Dong, W.J. Chen, L. Deng, Y.F. Wang, D.J. Ying, Piezoelectric PU/PVDF electrospun scaffolds for wound healing applications, Colloids Surf. B: Biointerfaces 96 (2012) 29–36.

[113] Y. Wang, M. Rouabhia, Z. Zhang, Pulsed electrical stimulation benefits wound healing by activating skin fibroblasts through the TGFβ1/ERK/NF-κB axis, Biochim. Biophys. Acta Gen. Subj. 1860 (2016) 1551–1559.

[114] Y. Li, X. Li, R. Zhao, C. Wang, F. Qiu, B. Sun, H. Ji, J. Qiu, C. Wang, Enhanced adhesion and proliferation of human umbilical vein endothelial cells on conductive PANI-PCL fiber scaffold by electrical stimulation, Mater. Sci. Eng. C 72 (2017) 106–112.

[115] M.A. Howard 3rd, M.S. Grady, R.C. Ritter, G.T. Gillies, W.C. Broaddus, R.G. Dacey, Magnetic neurosurgery, Stereotact. Funct. Neurosurg. 66 (1996) 102–107.

[116] O. Ergeneman, C. Bergeles, M.P. Kummer, J.J. Abbott, B.J. Nelson, Wireless intraocular microrobots: opportunities and challenges, in: J. Rosen, B. Hannaford, R.M. Satava (Eds.), Surgical Robotics: Systems Applications and Visions, Springer US, Boston, MA, 2011, pp. 271–311.

[117] X.Z. Qu, M.Y. Wang, H.S. Ong, C.P. Zhang, Post-operative hemimaxillectomy rehabilitation using prostheses supported by zygoma implants and remaining natural teeth, Clinics 71 (2016) 575–579.

[118] A. Da Costa, J.B. Guichard, C. Roméyer-Bouchard, A. Gerbay, K. Isaaz, Robotic magnetic navigation for ablation of human arrhythmias, Med. Devices (Auckland, N.Z.) 9 (2016) 331–339.

[119] M. Faraji, Y. Yamini, M. Rezaee, Magnetic nanoparticles: synthesis, stabilization, functionalization, characterization, and applications, J. Iran. Chem. Soc. 7 (2010) 1–37.

[120] S.F. Medeiros, A.M. Santos, H. Fessi, A. Elaissari, Stimuli-responsive magnetic particles for biomedical applications, Int. J. Pharm. 403 (2011) 139–161.

[121] S. Gil, J.F. Mano, Magnetic composite biomaterials for tissue engineering, Biomater. Sci. 2 (2014) 812–818.

[122] J.M.D. Teresa, C. Marquina, D. Serrate, R. Fernandez-Pacheco, L. Morellon, P.A. Algarabel, M.R. Ibarra, From magnetoelectronic to biomedical applications based on the nanoscale properties of advanced magnetic materials, Int. J. Nanotechnol. 2 (2005) 3–22.

[123] J. Kudr, Y. Haddad, L. Richtera, Z. Heger, M. Cernak, V. Adam, O. Zitka, Magnetic nanoparticles: from design and synthesis to real world applications, Nano 7 (2017) 243.

[124] C.V. Fernandes, F. António, R. Clarisse, B.-L. Manuel, M. Pedro, L.-M. Senentxu, Advances in magnetic nanoparticles for biomedical applications, Adv. Healthc. Mater. 7 (2018) 1700845.

[125] A. Ito, H. Honda, Magnetic nanoparticles for tissue engineering, in: Nanotechnologies for the Life Sciences, Wiley-VCH Verlag GmbH & Co. KGaA, 2007.

[126] S.C. McBain, H.H.P. Yiu, J. Dobson, Magnetic nanoparticles for gene and drug delivery, Int. J. Nanomedicine 3 (2008) 169–180.

[127] J. Dobson, Gene therapy progress and prospects: magnetic nanoparticle-based gene delivery, Gene Ther. 13 (2006) 283–287.

[128] R. Weissleder, G. Elizondo, J. Wittenberg, C.A. Rabito, H.H. Bengele, L. Josephson, Ultrasmall superparamagnetic iron oxide: characterization of a new class of contrast agents for MR imaging, Radiology 175 (1990) 489–493.

[129] T. Neuberger, B. Schöpf, H. Hofmann, M. Hofmann, B. von Rechenberg, Superparamagnetic nanoparticles for biomedical applications: possibilities and limitations of a new drug delivery system, J. Magn. Magn. Mater. 293 (2005) 483–496.

[130] N. Wang, J.P. Butler, D.E. Ingber, Mechanotransduction across the cell surface and through the cytoskeleton, Science 260 (1993) 1124–1127.

[131] A.R. Bausch, F. Ziemann, A.A. Boulbitch, K. Jacobson, E. Sackmann, Local measurements of viscoelastic parameters of adherent cell surfaces by magnetic bead microrheometry, Biophys. J. 75 (1998) 2038–2049.

[132] A.R. Bausch, W. Möller, E. Sackmann, Measurement of local viscoelasticity and forces in living cells by magnetic tweezers, Biophys. J. 76 (1999) 573–579.

[133] M. D'Addario, P.D. Arora, R.P. Ellen, C.A.G. McCulloch, Regulation of tension-induced mechanotranscriptional signals by the microtubule network in fibroblasts, J. Biol. Chem. 278 (2003) 53090–53097.

[134] M. Glogauer, J. Ferrier, C.A. McCulloch, Magnetic fields applied to collagen-coated ferric oxide beads induce stretch-activated Ca2+ flux in fibroblasts, Am. J. Phys. Cell Phys. 269 (1995) C1093–C1104.

[135] M. Glogauer, P. Arora, G. Yao, I. Sokholov, J. Ferrier, C.A. McCulloch, Calcium ions and tyrosine phosphorylation interact coordinately with actin to regulate cytoprotective responses to stretching, J. Cell Sci. 110 (1997) 11–21.

[136] M. Glogauer, J. Ferrier, A new method for application of force to cells via ferric oxide beads, Pflugers Arch. 435 (1997) 320–327.

[137] H. Pommerenke, C. Schmidt, F. Dürr, B. Nebe, F. Lüthen, P. Müller, J. Rychly, The mode of mechanical integrin stressing controls intracellular signaling in osteoblasts, J. Bone Miner. Res. 17 (2002) 603–611.

[138] J. Dobson, Remote control of cellular behaviour with magnetic nanoparticles, Nat. Nanotechnol. 3 (2008) 139–143.

[139] R. Guduru, S. Khizroev, Magnetic field-controlled release of paclitaxel drug from functionalized magnetoelectric nanoparticles, Part. Part. Syst. Charact. 31 (2014) 605–611.

[140] C. Ribeiro, V. Correia, P. Martins, F.M. Gama, S. Lanceros-Mendez, Proving the suitability of magnetoelectric stimuli for tissue engineering applications, Colloids Surf. B: Biointerfaces 140 (2016) 430–436.

[141] R. Gonçalves, P. Martins, D.M. Correia, V. Sencadas, J.L. Vilas, L.M. León, G. Botelho, S. Lanceros-Méndez, Development of magnetoelectric $CoFe_2O_4$ /poly(vinylidene fluoride) microspheres, RSC Adv. 5 (2015) 35852–35857.

[142] Y. Li, Z. Wang, J. Yao, T. Yang, Z. Wang, J.-M. Hu, C. Chen, R. Sun, Z. Tian, J. Li, L.-Q. Chen, D. Viehland, Magnetoelectric quasi-(0-3) nanocomposite heterostructures, Nat. Commun. 6 (2015) 6680.

[143] P. Martins, S. Lanceros-Méndez, Polymer-based magnetoelectric materials, Adv. Funct. Mater. 23 (2013) 3371–3385.

[144] J. Jin, S.-G. Lu, C. Chanthad, Q. Zhang, M.A. Haque, Q. Wang, Multiferroic polymer composites with greatly enhanced magnetoelectric effect under a low magnetic bias, Adv. Mater. 23 (2011) 3853–3858.

[145] P. Martins, R. Gonçalves, S. Lanceros-Mendez, A. Lasheras, J. Gutiérrez, J.M. Barandiarán, Effect of filler dispersion and dispersion method on the piezoelectric and magnetoelectric response of CoFe2O4/P(VDF-TrFE) nanocomposites, Appl. Surf. Sci. 313 (2014) 215–219.

[146] R. Gonçalves, P. Martins, X. Moya, M. Ghidini, V. Sencadas, G. Botelho, N.D. Mathur, S. Lanceros-Mendez, Magnetoelectric $CoFe_2O_4$/polyvinylidene fluoride electrospun nanofibres, Nanoscale 7 (2015) 8058–8061.

[147] C. Ribeiro, D.M. Correia, S. Ribeiro, M.M. Fernandes, S. Lanceros-Mendez, Piezo-and magnetoelectric polymers as biomaterials for novel tissue engineering strategies, MRS Adv. (2018) 1671–1676.

[148] P.A. Glazer, M.R. Heilmann, J.C. Lotz, D.S. Bradford, Use of electromagnetic fields in a spinal fusion. A rabbit model, Spine 22 (1997) 2351–2356.

[149] S. Bakbak, R. Kayacan, O. Akkuş, Effect of collagen fiber orientation on mechanical properties of cortical bone, J. Biomech. 44 (2011) e11–e21.

[150] S. Panseri, A. Russo, M. Sartori, G. Giavaresi, M. Sandri, M. Fini, M.C. Maltarello, T. Shelyakova, A. Ortolani, A. Visani, V. Dediu, A. Tampieri, M. Marcacci, Modifying bone scaffold architecture *in vivo* with permanent magnets to facilitate fixation of magnetic scaffolds, Bone 56 (2013) 432–439.

CHAPTER

# 14

# Using 3-D Printing and Bioprinting Technologies for Personalized Implants

*Julien Barthes**,†, Edwin-Joffrey Courtial‡, Esteban Brenet*, Celine Blandine Muller*,†, Helena Knopf-Marques*,†,§, Christophe Marquette‡, Nihal Engin Vrana*,†*

*INSERM UMR 1121, 11 rue Humann, Strasbourg, France †Protip Medical, 8 Place de l'Hôpital, Strasbourg, France ‡3dFAB Université Lyon 1—CNRS 5246 ICBMS, Lyon, France §Université de Strasbourg, Faculté de Chirurgie Dentaire, Fédération de Médecine Translationnelle de Strasbourg, Fédération de Recherche Matériaux et Nanosciences Grand Est, Strasbourg, France

## 14.1 INTRODUCTION

The last two centuries have seen a steady increase in average life expectancy all around the world, particularly due to the advances such as antibiotics, vaccines, availability of better healthcare, and improved hygiene. However, the aging societies pose another health risk, chronic diseases. Some of these diseases can be handled by pharmaceutical means and rehabilitation. But in some cases, the damage to a given tissue/organ (the knees, pancreas, kidney, etc.) is so extensive that there is a need for either a device that can take over the function of the tissue/organ (such as femur implants or dialysis systems) or completely replacement of the malfunctioning organ (transplantation). As there is a persistent donor, shortage for transplants and the current implants are not potential remedies for several diseases; over the last 40 years, a new field tissue engineering and regenerative medicine (TERM) has grown with the promise of providing tissues/organs on demand without any potential risk of rejection or disease transmission using cells and materials in configurations that can take over tissue function and can be fully or partially integrated with the host.

The advances in the TERM field resulted in the availability of many artificial tissues (particularly skin, cartilage, and bone) and organs (such as the gall bladder), which has been applied successfully in clinical settings [1]. However, the most success was generally obtained where the tissue structure is mostly isotropic, the cell types are either limited or organized in a spatially distinct manner, and the function is structural [2]. The more complex organs with mechanically active parts where high degree of innervation and synchronization is required or organs with complex biochemical functions have not been strong points of regenerative medicine. One of the roadblocks in this aspect is the precise control of cellular spatial distribution and microscale material properties. The recent answer to these challenges has been the development of 3-D printing systems that can simultaneously handle the cellular component either alone or together with biomaterials. The simultaneous printing of cells and materials or cell aggregates is called bioprinting. Bioprinting is a well-known technology that assembles the cells and natural or synthetic cell matrices [3], by integrating living materials, motion control, computer-aided design software, and biomaterials together to achieve highly accurate biomimetic constructs [3, 4]. In this chapter, we will first provide the available bioprinting methods together with specific examples related to the development of artificial tissues/organ with bioprinting. Then, we will continue with 3-D printing of implantable devices that does not contain cells for structural support where anatomical dispersity of patients necessitates personalization and use of 3-D printing methods is picking up the pace.


https://doi.org/10.1016/B978-0-12-816390-0.00014-5

## 14.2 BIOPRINTING

Three-dimensional printing is a manufacturing process in which materials are assembled in a predetermined manner by a process under computer control to create complex 3-D objects. The initial use of 3-D printing in the medical fields was for patient-specific anatomical models for reconstructive surgeries. This was followed by patient-specific implant designs such as structures for replacing cranium parts, as such replacements are generally the result of damage that is unique to each patient [5]. Another example of its use is in the manufacture of vascularized tissue, bioprinting [6].

Three-dimensional bioprinting is an important tool for creating complex tissues; it enables a breakthrough in the field of tissue engineering. With this tool, it is possible to manufacture 3-D structures with specific (multi)materials that have a good compatibility (both biologically and anatomically) with the human body (e.g., hydrogels) via a model in predefined dimensions [7]. Once printed, the material will be placed in an in vitro culture environment (prevascularization in vitro) to transform it into tissue that can subsequently be used for transplants or for partial replacement of tissues (cardiac patches, ligaments, cartilage or bone segments, etc.). The success of artificial tissues and their integration is directly related to their ability to be vascularized. Therefore, the structuring of hydrogels or other materials to allow a better vascularization is very important, and the bioprinting technique is particularly appropriate for that [8]. The other potential option is to use the host body as a bioreactor for the maturation of the tissue in vivo (in situ tissue engineering), but this is not suitable for all cases.

Recently, Kolesky et al. reported an important work describing thick, vascularized human tissues with programmable cellular heterogeneity that are capable of long-term (more than 6 weeks) perfusion on-chip fabricated by multimaterial 3-D bioprinting [7]. With their results of in situ development of human mesenchymal stem cells (hMSCs) within tissues containing a pervasive, perfusable, endothelialized vascular network, they demonstrated that the 3-D tissue microenvironments enable the exploration of emergent biological phenomena. Moreover, their 3-D tissue manufacturing platform opens new avenues of fabrication and investigation of human tissues for both ex vivo and in vivo applications [7]. Bioprinted tissue constructs have also been widely used for disease modeling [9]. The fabrication of disease models using the bioprinted technique is recently commercialized, especially for cancer drug screening [3].

### 14.2.1 Bioprinting Techniques

Bioprinting technology is an important fabrication methodology for producing cell-laden scaffolds, cells in aggregated forms to obtain tissues and organs. In this section, we provide a brief overview of the bioprinting techniques used for developing artificial tissue constructs for transplantation. Up to date, bioprinting techniques can be mainly divided into three categories, namely, laser-, inkjet-, or extrusion-based bioprinting.

**In laser-based bioprinting**, a laser pulse guides an individual cell from a source to a substrate, as shown in Fig. 14.1A. The cells, suspended in a solution, are transferred from a donor slide to a collector slide by the laser pulse. The transfer happens due to the bubble formation by the laser, which forces cells to move toward the collector substrate. This technique allow precise deposition of cells in relatively small 3-D structures [10, 11]. However, it is limited for construction of milliscale and beyond structures. Researchers have demonstrated the feasibility of cell deposition by using this bioprinting technique. One example of this outstanding tool for the generation of multicellular 3-D constructs mimicking tissue functions was the study realized by Koch et al. They demonstrated the 3-D arrangement of viable cells by laser-assisted bioprinting as multicellular grafts analogous to native archetype and the formation of tissue by these cells. For that, fibroblasts and keratinocytes were embedded in collagen and printed in 3-D as a demonstration for skin tissue [12].

**The inkjet-based bioprinting** is very similar to inkjet printers used in offices and with personal computers. However, in this case, the ink is a "bioink" made of hydrogel and cells that is printed in the form of droplets, Fig. 14.1B [13, 14]. Inkjet bioprinters are relatively cheap and can work under ambient conditions without strict requirements. It has been commonly used for creation of blood vessel-like channels capable of transporting oxygen and nutrients. Christensen et al. presented a work in which they used inkjet bioprinting technique for the fabrication of vessel-like channels with bifurcations printed in sodium alginate only and mouse fibroblast containing alginate bioinks. The postprinting fibroblast cell viability of printed cellular tubes was found to be above 90% even after a 24 h incubation, considering the control effect [15].

**In extrusion-based bioprinting**, a deposition system (based on air pressure or piston-based extrusion) dispenses the bioink under the control of a computer, resulting in a precise deposition of encapsulated cells in cylindrical filaments, Fig. 14.1C. This technique is based on a combination of fluid dispensing system (pneumatic or mechanical) and

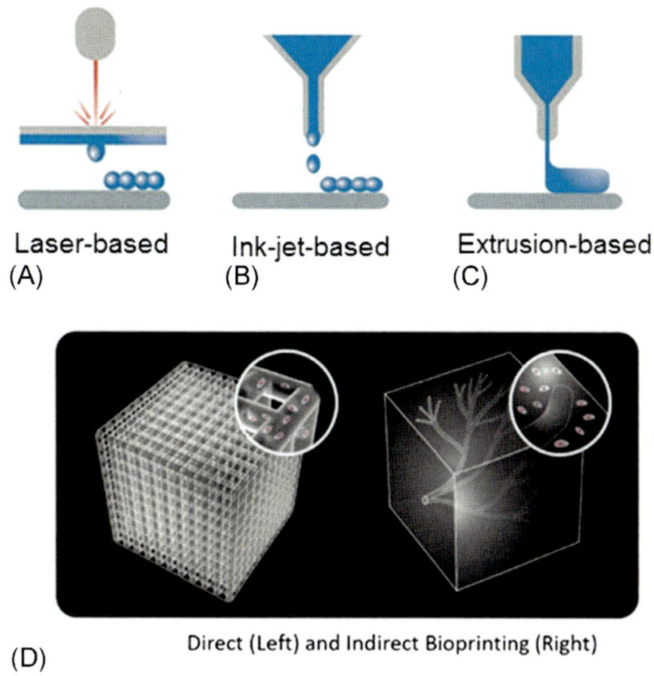


FIG. 14.1 Representative drawings of the three most common bioprinting methods: (A) laser-based, (B) inkjet-based, and (C) extrusion-based bioprinting. (D) Representative direct and indirect bioprinted structures. In direct bioprinting, cell containing base material is printed according to the designed 3-D structure; in indirect bioprinting, a sacrificial level was removed to enable perfusable areas for providing nutrient and gas transfer for the encapsulated cells in the bulk structure. *(Reprinted with permission from N. Nagarajan, et al., Enabling personalized implant and controllable biosystem development through 3D printing, Biotechnol. Adv. (2018).)*

an automated system for extrusion and writing [16]. In the last couple of years, several researchers have tried to use a fugitive bioink in extrusion-based bioprinting to create vascular channels [17]. The fugitive bioink is removed afterward by thermally induced reverse cross-linking leaving a network behind [18]. It is well known that this technique is very convenient for producing 3-D cell-laden structures. For instance, Bertassoni et al. used extrusion-based bioprinting for the fabrication of microchannel networks within cell-laden GelMA hydrogels as a model platform. They demonstrated that the fabricated microchannels resulted in improved mass transport, viability, and differentiation of cells in cell-laden GelMA hydrogels [19]. Table 14.1 shows a comparison of bioprinting techniques in which the resolution, commonly used materials, gelation speed, advantages, and disadvantages of each technique are presented [10].

There are two bioprinting approaches that explore engineering vascular networks within the engineered tissue constructs through indirect and direct bioprinting, Fig. 14.1D. In the indirect approach, a negative supportive mold is created initially, which is then used to cast the desired polymer scaffold through a suitable drying method. Frequently, freeze-drying approach is used as it causes less shrinkage and can reproduce the designs accurately. In the case of direct approach, the scaffolds are produced directly from the model material, through processes such as extrusion printing [3].

TABLE 14.1 Comparison Among Laser-, Inkjet-, and Extrusion-Based Bioprinting Techniques [10]

| | Laser-based | Inkjet-based | Extrusion-based |
|---|---|---|---|
| Resolution | High | Medium | Medium-low |
| Materials | Cells in media | Liquids, hydrogels | Hydrogels, cell aggregates |
| Gelation speed | High | High | Medium |
| Advantages | High accuracy, single-cell manipulation, high-viscosity material | Affordable, versatile | Multiple compositions, good mechanical properties |
| Disadvantages | Relatively harsh conditions for cells, low scalability, low viscosity prevents buildup in 3-D | Low viscosity prevents buildup in 3-D, low strength | Shear stress on nozzle tip wall can negatively affect the cells, limited number of biomaterials can be used |

In the indirect bioprinting approach, first, a negative supportive mold is created, Fig. 14.1D. This approach is widely used for the fabrication of vasculatures in which a sacrificial ink is easily removed, where this fugitive ink was used to print tubular networks within the construct [3]. One example is a recent research developed by Kang et al. that they describe a system for deposition of cell-laden hydrogels together with synthetic biodegradable polymers in integrated patterns and anchored on sacrificial hydrogels [20]. The obtained cell-laden hydrogel is important to protect cell viability and to promote growth and expansion; at the same time, the adjacent sacrificial scaffolding (Pluronic F127) was used to provide the initial structural and architectural integrity.

The direct bioprinting approach can also be used as an alternative for organ fabrication. Bertassoni et al. [21] used the direct bioprinting to precisely deposit cells and cell-laden materials with the objective of generating controlled tissue architecture [22]. Their work shows a strategy for bioprinting of photolabile cell-laden methacrylated gelatin (GelMA) hydrogels in which encapsulated hepatocyte cells preserved high cell viability for at least 8 days.

## 14.3 MATERIALS

Depending on the printing technique, the composition of the materials used in 3-D bioprinting processes will differ. It is, therefore, necessary to define the differences of the materials used according to the different printing techniques to determine the improvement that each compound brings to the print quality [23]. Those materials are known as "bioinks," which is used as a term making reference to original conventional inkjet printing inks, means the bioprintable materials used in 3-D bioprinting processes in which cells are deposited in a spatially controlled pattern to fabricate living tissues and organs. In this section, properties of materials suitable as bioinks, particularly hydrogel-forming materials, used in laser-, inkjet- and extrusion-based bioprinting will be described.

In tissue engineering, hydrogels are classified as naturally derived hydrogels (based on agarose, alginate, collagen, chitosan, fibrin, gelatin, hyaluronic acid, etc.) and synthetically derived hydrogels (such as Pluronic, Matrigel, polyethylene glycol (PEG), methacrylated gelatin, polydimethylsiloxane (PDMS), etc.). Table 14.2 summarizes some hydrogels used as bioink for 3-D printing with their key points.

### 14.3.1 Natural Hydrogels

**Agarose** is a natural polysaccharide, usually extracted from certain red seaweed species. It is a linear polymer with a molecular weight of about 120 kDa. It undergoes gradual gelation at low temperature and liquefies at the temperatures ranging from 20°C to 70°C [23]. Agarose shows some limitations for 3-D printing in function of low cell adhesion and spreading on and in it. Nevertheless, it can be used as a mold material for 3-D culture of cell aggregates [28]. Agarose has been used in extrusion-based [29], inkjet-based [30], and laser-based bioprinting [24].

TABLE 14.2 Key Points of Natural and Synthetic Hydrogels Used as Bioink for 3-D Printing

| Hydrogel | Material | Key points |
|---|---|---|
| Natural | Agarose | Positive: viscoelastic nature, rapid gelation mechanism<br>Negative: nondegradable, low cell adhesion [24] |
| | Alginate | Positive: high biocompatibility, various choice of cross-linking<br>Negative: rapid degradation, not highly cell adhesive [25] |
| | Collagen | Positive: natural dominant component of connective tissues, high level of mimicking of native ECM environment.<br>Negative: cells deposited in collagen are not homogeneously distributed, low mechanical properties and instability [23] |
| | Gelatin | Positive: highly available, easy-to-obtain material while being highly biocompatible<br>Negative: poor bioprintability and stability in physiological conditions [17] |
| Synthetic | Pluronic | Positive: temperature-induced gelation makes it ideal for creating perfusable channels<br>Negative: poor solubility, required 4°C for solubilization |
| | Methacrylated gelatin | Positive: suitable biological properties and tunable physical characteristics [26]<br>Negative: UV light and photoinitiator requirements can have negative effects on cells |
| | PEG | Positive: printable in all types of bioprinting [27]<br>Negative: highly hydrophilic and not ideal for cell remodeling |

**Alginate** is a polysaccharide distributed generally in the cell walls of brown algae. It is a popular hydrogel for bioprinting processes due to its high biocompatibility and various choices of cross-linking. It undergoes ionic cross-linking in the presence of calcium chloride or sulfate ($CaCl_2$ or $CaSO_4$, respectively) [31]. Alginate is a very good candidate for extrusion-based [32, 33] and laser-based bioprinting [34, 35]. However, it lacks cellular adhesion signals.

**Collagen** is the main structural protein in the extracellular space in the various connective tissues in animal bodies. It consists of polypeptide chains self-assembled to form triple helices of elongated fibrils. There are 28 types of collagen, which are categorized as fibrillar (such as collagen type I) and nonfibrillar (such as Collagen type IV). Collagen type I is fibrillar and the most common type of collagen in many tissues. It is obtained from natural sources (e.g., tails of rats, calf hide, or pork skin) [36]. Collagen matrix enables cell adhesion, attachment, and growth due to abundant integrin-binding domains. However, there are some limitations of the use of collagen type I in the 3-D bioprinting in function of the slow gelation rate, which can cause nonhomogeneity of cell distribution in collagen [23]. Another inconvenience is the tendency of collagen to gel with small fluctuations in temperature. Nevertheless, it has been used as a bioink for extrusion-based [37] and inkjet-based [38] bioprinting.

**Gelatin**, a thermoreversible natural polymer, is a mixture of peptides and proteins produced by partial hydrolysis of collagen extracted from the skin, bones, and connective tissues. It is a good candidate for bioink used in extrusion-based printing as it is highly accessible and cheap and has good biocompatibility. However, gelatin is hardly bioprinted in its native form because of its poor mechanical properties. To improve its bioprintability and stability in physiological conditions, several chemicals (such as cross-linking with glutaraldehyde) and physical cues (temperature-induced gelation) have been used [17, 39].

## 14.3.2 Synthetic Hydrogels

**Pluronic®F-127** is a trademark for poloxamers that can be defined as nonionic triblock copolymers composed of a central hydrophobic chain of polyoxypropylene (poly(propylene oxide)) flanked by two hydrophilic chains of polyoxyethylene (poly(ethylene oxide)). Poloxamers have temperature dependents gelation and self-assembly properties. A particularly useful property of poloxamers is that at low temperatures, high-concentration poloxamer solutions are in liquid, whereas at higher temperatures, they form reversible gels [40, 41]. The reversible properties of Pluronic are very useful for creating perfusable channels within bulky cell-laden constructs. At room temperature or at higher temperatures, it is in solid form, so it can be surrounded by a second type of hydrogel and then placed at 4°C to liquefy it. Pluronic is used in extrusion-based bioprinting [42].

**Gelatin methacrylated (GelMA)** was developed in response to some limitations presented by natural hydrogels, such as extensive contraction, poor mechanical properties, and rapid degradation and stability at body temperature [26, 43]. The photo-cross-linkable methacrylated gelatin hydrogels are synthesized by adding methacrylate groups to the amine-containing side groups of gelatin [44]. Chen et al. demonstrated ECFC-driven vascular morphogenesis in GelMA hydrogels for vascular tissue engineering [45]. The polymerization of GelMA hydrogels can be achieved in 15 s of UV light exposure in the presence of a photoinitiator. With this rapid polymerization, it would be a critical feature to avoid hydrogel dissemination at the implantation site (for potential in situ bioprinting applications) [46]. GelMA hydrogels have been bioprinted by inkjet-based [47] and extrusion-based [21] techniques to manufacture cell-laden constructs with high cell viability.

**Poly ethylene glycol (PEG)** is a linear polyether compound with many applications from industrial manufacturing to medicine. It can be conjugated with many biomolecules as proteins, enzymes, and lipids [23]. A research developed by Benoit et al. showed that cells encapsulated in PEG survive even without the addition of biological constituents although they are unable to remodel the hydrogel [48]. The advantage PEG is its mechanical properties that can be manipulated by changing its chemistry. For that, the addition of diacrylate and methacrylate groups is beneficial for improving the mechanical property of the resulting photo-cross-linkable hydrogels. However, those additives require photo-cross-linking by exposure to UV light, which can dramatically reduce cell viability. PEG hydrogels can be printable in all types of bioprinting: extrusion-based, inkjet-based, and laser-based [27].

Beyond the replacement of tissues, generally, another common medical need is structural support to the tissues in the form of implants. Some common examples of such structures are tracheal, esophageal, and vascular stents and dental, orthopedic, and breast implants. Although the industrialized implants are currently in use, the anatomical and structural variations between the patients result in implants and medical devices that are not perfect fits, which result in problems such as mechanical mismatch-related complications, pain, excessive inflammation, pseudotumors, and granuloma formation. Thus, there is a great need for personalization of such implants. In soft tissues, one of the most widely used materials for such implants is polydimethylsiloxane. In the next section, we will demonstrate its use in otorhinolaryngology setting as an example of 3-D printing in medical implants where the presence of cells is not required.

**Polydimethylsiloxane (PDMS)** belongs to a group of polymeric organosilicon compounds. It has been widely used to fabricate microfluidic systems due to its rheological properties, and it is a low-cost material [49]. Additionally, PDMS elastomer is used in a wide range of biomaterial applications including cell culture substrates, flexible electronics, and medical devices [50]. Ozbolat et al. have recently demonstrated that the 3-D printed PDMS samples possess improved mechanical and cell adhesion properties compared with traditionally manufactured samples using casting process. Three-dimensional printing of PDMS not only enables the generation of 3-D models of tissues and organs but also brings a new concept in surface engineering for cell adhesion studies [51].

## 14.4 3D PRINTING OF PERSONALIZED SILICONE IMPLANT

The development of personalized implant has become a necessity for some applications that require high degree of anatomical conformation. To overcome these problems, 3-D printing is one of the processes that will enable the manufacture of custom implants. The combination between the computer tomography (CT) scan from patient and the computer-aided design (CAD) can be employed to generate the personalized implants. The resulting implant can be then produced with high degree of fidelity without requiring expensive molds. As explained in the previous paragraph, many materials have been used to develop 3-D printed medical implants, and most of them are water-soluble materials. Another kind of polymers, elastomers, and specifically silicone has received a great interest in the last few years for developing 3-D printed personalized implants. Even if silicone has been widely used to develop soft implants, most of these implants are produced using injection molding process that limit the application for the manufacturing of custom-made implants [52]. This can be explained by the difficulty of 3-D printing of silicone, as the technique has limitations in handling viscous liquids just before the curing limit. In this part, we will focus our discussion on the printing of silicone-based materials to develop personalized implants.

### 14.4.1 Soft 3-D Implant Printing: Example of Silicone

Elastomers have been widely used for biomedical applications especially in tissue engineering and to develop medical devices mainly due to their mechanical properties (highly elastic with low Young's modulus) close to natural tissues. Indeed, elastomers also referred as « Rubbers » are a special class of polymers that are very elastic and composed of a cross-linked network with a glass transition below room temperature, which allow them to bend and flex within the body temperature. Despite of their mechanical properties that can mimic tissues, elastomers are easy to sterilize and easy to process using molding or liquid injection molding. Among elastomers, the main medical grades that have been successfully produced and commercialized are of silicone-, polyurethane-, and polyvinyl chloride-based materials [53].

In this class of materials, silicone is the most widely used for biomedical applications. Silicone, also known as « PDMS: polydimethylsiloxane » (Fig. 14.2), has been used in the medical field for more than 70 years. These materials represent a good candidate for implantable devices thanks to its low reactivity and relatively low immune response. Its strong Si—O—Si (siloxane) backbone and the presence of methyl $CH_3$ confer high chemical stability and outstanding flexibility (high tear strength and high elasticity). Moreover, silicone-based materials are biocompatible and bio-inert and have high stability in rather abrasive physiological environment [54, 55].

Below are the main properties that contributed to the success of silicone-based materials in the medical field:

- Thermal stability
- Chemical stability
- Electrical insulation
- High gas permeability
- Mechanical properties close to natural tissues (highly stretchable and high compliance with soft tissue)

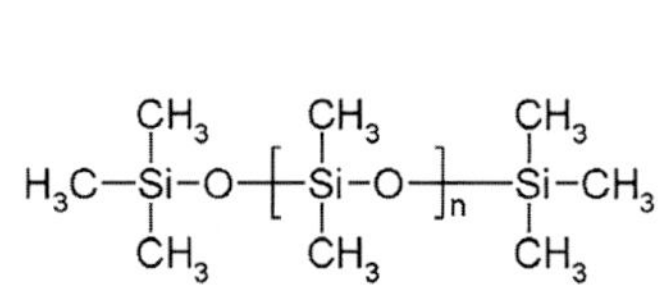

FIG. 14.2 Chemical formula of PDMS.

In the medical field, silicone implants have been mainly used for the following applications [56–62]:

- Development of artificial organs (urethra, trachea, etc.)
- Restoration of organ or tissue functions (orthopedic implants, cardiac valves, cochlear implants, hydrocephalic shunt, intralaryngeal implants, etc.)
- Reconstructive and aesthetic surgeries (breast, cheek, scrotum, orbital implant, and ORL implants)
- Development of disposable medical devices (contact lenses, catheters, etc.)

### 14.4.2 Silicone ORL Implant and the Need of Personalization

In this part, we will focus our discussion on the development of soft tissue facing implants in the otorhinolaryngology field and more precisely in three specifics areas: the larynx, trachea, and bronchus. With these examples, we aim to demonstrate the need, in some specific situations, to develop personalized implants that perfectly fit a given patient's anatomy. For this purpose, we will show how 3-D printing can facilitate the manufacture of personalized implants.

In this area, one of the most common implants used are the stents. These devices are used to maintain the diameter of lumen either in the trachea and larynx to prevent stenosis related to different pathologies to block the organ, that is, respiration. These implants are directly implanted in the larynx or the trachea. Different kinds of stents can be used for these purposes depending on the position of the stenosis: tracheal stents, larynx stents, or a combination of tracheal and larynx stents. These stents are designed to be implanted for a limited period of time and are supposed to be removed at the end. So, one of the specifications for these implants is the need of bio-inertness (no integration with the surrounding tissues) meaning that cells and tissues should not adhere on the implant so that the removal of the implant will be possible. Moreover, the implant should be noncytotoxic, transparent, and compatible with X-ray imaging techniques. Silicone materials fulfill all these specifications such that they are available in medical grade, transparent, visible to x-ray, and bio-inert, and its mechanical properties are similar to tissues that explain why most of the implants developed for such applications are made of silicone.

We will now give a brief explanation on the anatomy of this area and the related pathologies.

The laryngotracheobronchial tract constitutes the only "aerial tube" that allow the vital function of breathing (Fig. 14.3).

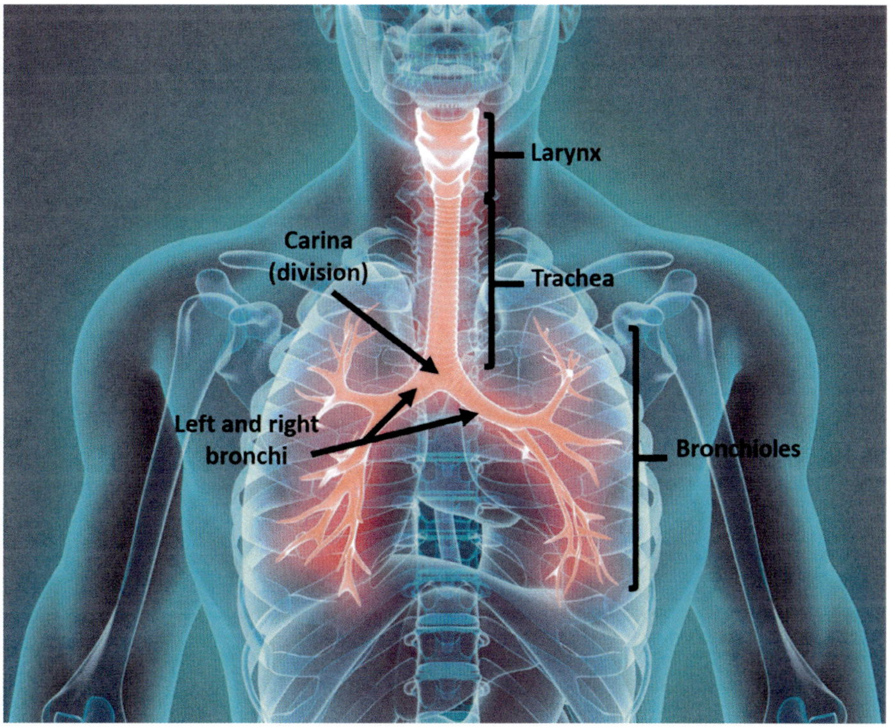


FIG. 14.3 Laryngotracheobronchial tract.

Besides the ventilation, the larynx is responsible for two other essential functions: swallowing and phonation. It is located at the intersection between the digestive and aerial ways. Any injury to the integrity of this part can quickly put the patient's life at risk. If the diameter of this single respiratory tract is reduced, the consequence on respiratory function can be dramatic and even lethal. Moreover, in this area, patient can also suffer from swallowing disorders that can cause the passage of the food bolus directly in the airways, which cause inhalation pneumopathies. If not treated correctly, these inhalation pneumopathies can be fatal.

#### 14.4.2.1 *Different Types of Stenosis of the Respiratory Tract*

**a.** *Laryngeal stenosis*

Laryngeal stenosis corresponds to a narrowing of the diameter of the larynx. They are most often acquired in adults and congenital in children [63]. A laryngeal dyspnea is a difficulty in inspiration that can quickly affect patient respiratory capacity leading to respiratory distress. Moreover, depending on the cause of the stenosis, the mobility of the larynx may be impaired, preventing it from performing its role of protective sphincter for the lower respiratory tract. It is therefore quickly necessary to perform a tracheotomy (establishment of an alternative route for airflow by introducing an incision in the trachea secured with a cannula) to shunt the obstacle and allow normal breathing and protection of the lungs. Despite the progress of anesthesia and intensive care equipment, the vast majority of laryngeal stenosis is related to translaryngeal tracheal intubation at any age [64].

Many other causes may be responsible for laryngeal stenosis:

- Posttracheostomy injuries are rare and the consequence of an incorrect technique [65]
- External laryngeal trauma [66]
- Laryngotracheal burns by inhalation of smoke
- Not only sequelae of benign or malignant laryngeal tumors but also surgical treatment of these tumors by partial laryngectomies
- Radiotherapy, which can lead to major laryngeal stenosis on larynx [67]
- Inflammatory systemic diseases responsible for obstructive laryngeal granulations

**b.** *Tracheobronchial stenosis*

We can distinguish benign stenosis from malignant stenosis.

Benign stenosis due to postintubation represents more than 90% of the cases, Fig. 14.4B–E. There are tracheal lesions that lead to stenosis [68] Lesions can remain microscopic without sequelae. In the most unfavorable cases, these lesions can lead to severe stenosis within a few hours. Causes increasing the development of these lesions are multifactorial and will not be detailed here. Malignant stenosis is rare. The most common histologies are cystic adenoid carcinomas, squamous cell carcinomas, sarcomas, and lymphomas, Fig. 14.4F–G.

Finally, bronchial stenosis is in majority in connection with bronchopulmonary malignant tumors like bronchocellular carcinomas. Benign tumoral etiologies such as papillomatosis or inflammatory granulomatous type are also found.

**c.** *The contrast tomodensitometry (CT)*

CT is considered as the gold standard in the characterization of stenosis: its caliber, its location, its length, but also for research purposes and its etiological assessment [69]. It also allows the realization of 3-D reconstructions that allow excellent visual representation of the stenosis (Fig. 14.5). This "virtual endoscopy" technique makes it possible to explore stenosis that is too tight to allow an endoscope to proceed, in particular, to ensure the examination of structures (whether they are healthy or not) below the stenosis [70].

#### 14.4.2.2 *Management of Stenosis: Development of Silicone Soft Implants*

The management of these diseases is extremely complex and to this day poorly standardized. The only way to definitively cure the patient is to respect the stenotic area and perform an end-to-end anastomosis of the extremities. Nevertheless, these procedures, such as tracheal resection/anastomosis, remain difficult surgeries, which are performed only in the context of a limited stenosis, in young, otherwise healthy patients. Whether at the laryngeal, tracheal, or bronchial level, the majority of the treatments are done endoscopically, either in a palliative context for tracheobronchial cancers or because of the too large extent of the stenosis at the tracheal level (greater than 50% of the tracheal length). To avoid resection, several implants that can be used to ensure the continuity of the airways have been developed in form of stents or plugs. In this section, we will present some silicone implants that have been developed to treat

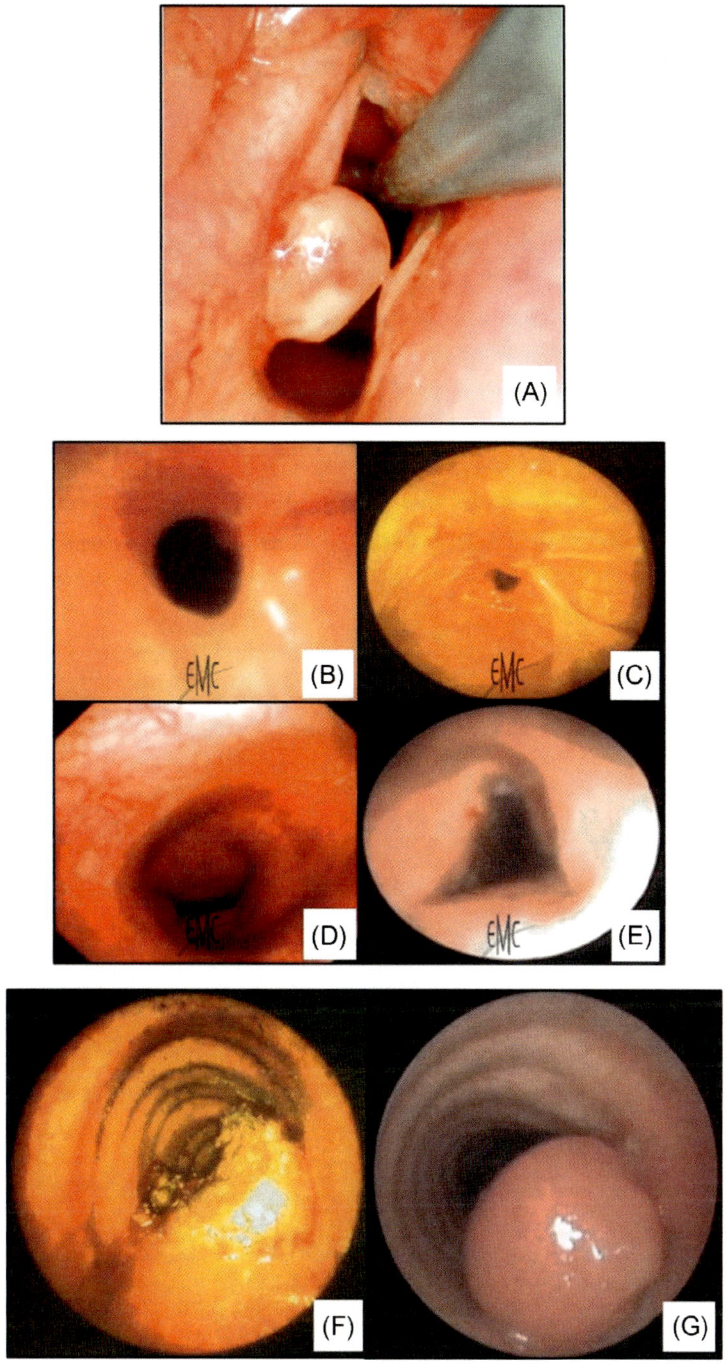


FIG. 14.4 (A) Inflammatory granuloma of the vocal fold in laryngoscopy. Endoscopic views of the three types of benign tracheal stenosis: (B) simple stenosis in diaphragm. (C) and (D) are complex stenoses. (E) Posttracheostomy stenosis in "B." Tumor stenosis: (F) squamous cell carcinoma of the trachea. (G) Adenoid cystic carcinoma of the trachea. Pictures (A), (F), and (G) were kindly provided by the University of Reims. *(Reprinted with permission from C.A. Righini, et al., Sténoses trachéales de l'adulte. EMC, Oto-Rhino-Laryngol. 10 (1) (2014) 1–15 (Article 20-760-A-10).)*

these diseases. These implants are manufactured using injection molding technique, and we will demonstrate the limitations of this technique from the point of view of implant personalization.

**a.** *Laryngeal level*

There are generally two types of laryngeal prostheses used to treat these diseases, Fig. 14.6A–C. The most established implant is the Montgomery "T" tube. It is used for multidilated laryngeal stenosis and for postoperative calibration. The anterior horizontal part allows breathing during the pose in the same way as a tracheotomy tube.

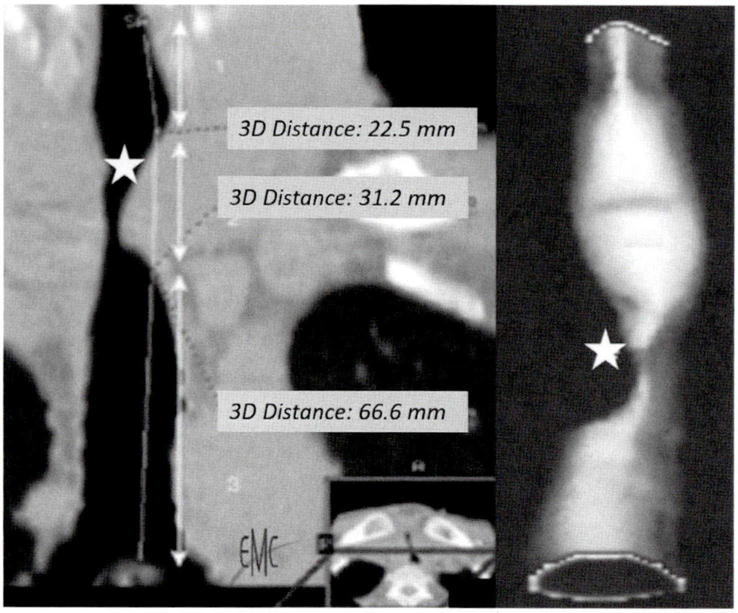


FIG. 14.5 CT appearance of tracheal stenosis in frontal section with 3-D reconstruction, in the right, giving a good representation of the level and importance of the stenosis *(white star)*. Identification of suprastenotic (1), stenotic (2), and substenotic (3) areas. *(Reprinted with permission from C.A. Righini, et al., Sténoses trachéales de l'adulte. EMC, Oto-Rhino-Laryngol. 10 (1) (2014) 1–15 (Article 20-760-A-10).)*

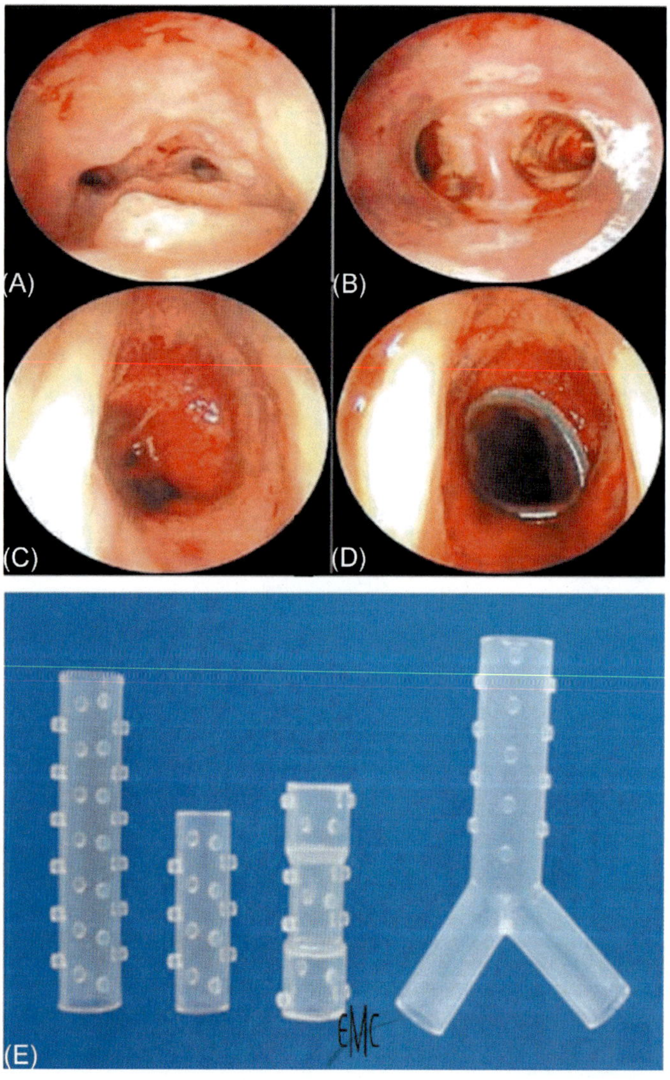


FIG. 14.6 Endoscopic management of stenoses: (A) stenosis of the carina and both bronchi stem, (B) endoscopic postdilation aspect of this same stenosis, (C) Nonoperable malignant tumor stenosis of the trachea, (D) palliative endoscopic treatment using a silicone prosthesis allowing instantaneous repermeabilization of the respiratory tract. (E) Different types of silicone prostheses: from *left* to *right*: large tracheal prosthesis, small tracheal prosthesis, diabolo tracheal prosthesis, "Y" prosthesis for carina's stenosis [68]. Pictures (D), (E), (F), and, (G) were kindly provided by the University of Reims.

The most recent implant design that can be used for restenosis patients (but aimed more generally at patients with swallowing disorders) is the NewBreez implant developed by the company PROTiP Medical, which fits between the vocal cords and recreates a true laryngeal sphincter [71, 72].

**b.** *Tracheobronchial level*

Even incomplete, the endoscopic management always has an immediate effectiveness, often with immediately evident outcomes. For complex stenosis, the placement of an endotracheal prosthesis after dilatation at the first endoscopy is justified by the risk of partial inefficiency of the simple dilation and the very high risk of rapid recurrence, Fig. 14.6A–D [73]. The advantage of the silicone-based materials for the development of these soft implants is its inherent flexibility characteristics, bio-inertness, being not traumatic for the tracheal mucosa, and low immune response, Fig. 14.6E.

#### *14.4.2.3 Complications Related to Standard Prostheses*

As all these implants will be inserted in the laryngotracheobronchial tract, a geometry close to patient anatomy is required. For most of the cases, all the available standard sizes developed by the manufacturer is sufficient. Nevertheless, depending on patient medical background (cancer, multiple surgeries, other diseases), larynx, trachea, or bronchus tracts can be deformed and then the standard stents or intralaryngeal implant can no longer be used and then a specific custom implant needs to be designed.

If the implant does not fit correctly patient anatomy, the implant can migrate in the airway tract or just be inefficient due to the lack of anatomical conformity and that can lead to severe medical complications such as the following:

- Excessive inflammation
- Pulmonary aspiration
- Infection
- Airway obstruction with asphyxia
- Mucosal necrosis (excessive pressure exerted by the implant on the tracheal wall)

Most of these medical complications will require the removal and the replacement of the prosthesis leading to emergency hospitalization. These complications impact directly patient quality of life and increase the cost of medical care.

To overcome this problem, personalized 3-D printed implants can be an interesting solution because it provides an implant with a geometry that fits patient anatomy, which can decrease dramatically the risk of complications.

In an European study of 263 patients treated with prosthesis for tracheal stenosis, the percentage of migration and obstruction was respectively 18.6 and 5.7% [74].

### 14.4.3 Benefits of 3D Printing

As mentioned earlier, the benefits of 3-D printing technique to develop medical implants will be at different levels:

- For customization and personalization
- Cost efficiency
- Time efficiency

First, 3-D printing technology allows the design of custom-made medical implants that can really fit to patient anatomy, and this will increase the efficiency of the implant, prevent implant related complications due to the anatomical and mechanical misfits, and overall increase the patient quality of life after implantation. For example, for patients having anatomical defects that prevent the use of standard implants prepared using injection molding technique such as deviation of the tracheal or laryngeal tracts, bronchus malformation, or cardiac malformation, this technique is able to produce implant with complex geometries that will take into the consideration the specific features of the patient anatomy for offering conformity.

Another important aspect is the cost efficiency. This technique will allow producing cheaper medical implants but currently only when it concerns small series. Other conventional techniques for the moment are less expensive for large-scale production. Techniques such as injection molding are more economic for the production of big series of implants because once a mold is produced, it can be used for the production of thousands of similar implants. Nevertheless, this manufacturing technique does not allow to develop personalized implant that fits patient anatomy, and a size range must be fixed for a given implant as the limiting step of developing molds will only allow a specific number of implant size. For a personalized injection molded implant, a new mold needs to be designed and produced for each implant, which is due to the current costs related to mold design and production, not economically feasible.

The last aspect is the time efficiency of 3-D printing. With this technique, a personalized implant can be printed in less than few hours that is faster than any other manufacturing techniques, due to the decreasing number of steps from the medical imaging to implant production. Although for standard implants the impact of this fact is less evident for certain implantations, the implants need to be implanted in a short time (few days), and so with injection molding technique, it will not be possible to produce a personalized implant in such a short notice since the production of a new mold can take more than a month [75, 76].

### 14.4.4 Different Steps to Print Personalized Medical Implant

One of the main advantages of 3-D printing is its ability to generate complex geometry that fits perfectly patient anatomy, which enables the manufacturing of custom-made implants. To be able to print these implants, first, a digital 3-D object is designed using CAD software (such as SolidWorks or AutoCAD) and then saved as a printable STL file (stereolithography files). Then, this CAD model will be further sliced into layers using a slicer software to generate the G-Code and, finally, 3-D printed. This 3-D object will be generated through the translation of x-ray, MRI, or CT scans. CT scan is generally the most common technique used to generate the STL files since it has high spatial resolution and a wide range of applications and it is the reference technique before most of the surgical interventions [77].

Obtaining a 3-D printed personalized implant from CT scan involves multiple steps, Fig. 14.7 [78, 79, 80]:

- **Image acquisition** (X-ray, MRI, and CT scans): images are acquired with the suitable device, and then, they are exported into digital imaging and communication in medicine (DICOM) format. DICOM is a format that has been created to enable the storage and the transmission of medical images coming from different imaging machines and different manufacturers, and it is widely adopted by hospitals.
- **Image segmentation:** This process will enable the conversion of 3-D anatomical information acquired with X-ray, MRI, or CT scan to 3-D digital model that represents the specific patient anatomy. This step will convert 3-D volumetric data into 2-D planar data. The area of interest will be segmented by applying a threshold and 2-D projections (axial, sagittal, and coronal) will be obtained to recreate the exact geometry pertaining to the anatomical region of interest.
- **3-D reconstruction**: 3-D digital model is reconstructed using the mask obtained after segmentation and save as STL file.
- **3-D printing:** 3-D model is sliced into 2-D layers using a 3-D slicing software to generate the machine code that will be used by the 3-D printer to print the implant.

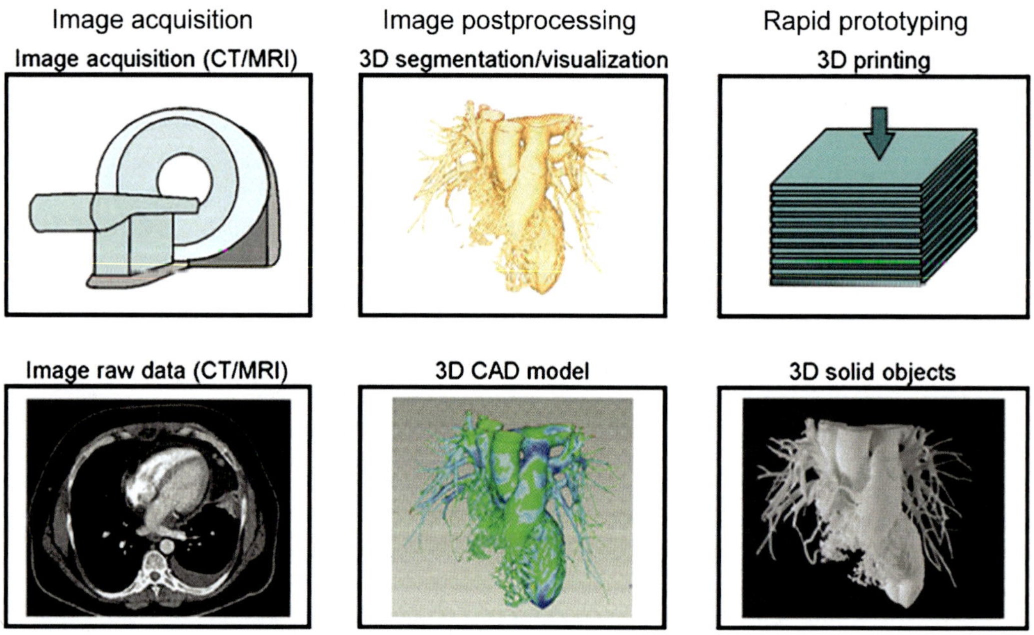


FIG. 14.7 Different steps to generate and 3-D print personalized implant. *(Reprinted with permission from F. Rengier, et al., 3D printing based on imaging data: review of medical applications, Int. J. Comput. Assist. Radiol. Surg. 5 (4) (2010) 335–341.)*

## 14.4.5 3-D Printing of Silicone for Healthcare

### *14.4.5.1 Technology and Challenge*

The selection of 3-D printing technologies to process elastomer materials depends on the chemical reactivity and rheological properties of the material. Two major technologies are used to print silicone materials: UV light technology [81] and liquid deposition modeling (LDM) [82], using UV and thermal curing, respectively. Nevertheless, in healthcare applications, the use of UV-cured silicone rubber can present toxicity due to the presence of unreacted photoinitiator leakage that has strong cytotoxicity [83]. Therefore, the use of thermally cured silicone rubber together with LDM is the most advantageous approach, provided that the rheological properties are adapted.

The LDM 3-D printing technology uses rheological properties of the material to flow it through a nozzle (predeposition) and maintain the shape of the 3-D object (postdeposition) up to the end of the cross-linking reaction. In the predeposition step, the flow of silicone through a nozzle is easier if the formulation presents a shear thinning effect and an adequate yield stress value regarding the extrusion system (mechanic, pneumatic, or screw-based). In the postdeposition part, the thixotropic behavior must be minimized to ensure the layer stability when the next layer will be deposited. If the thixotropic time is low enough, the yield stress character will be recovered, and the 3-D object shape fidelity will be ensured. Thus, the value of yield stress is related to the capacity of the material to build simple or complex geometries as shown in the next section.

The progression of rheological properties of thermally cured silicone is related to the chemical composition and the kinetics of the phase transition (from liquid to solid). Two kinds of chemical reactions can be used: polycondensation within monocomponent silicone and polyaddition within bicomponent silicone.

### *14.4.5.2 Mono-Component Silicone*

RTV-1 silicone rubbers are monocomponent products that are free flowing or paste- like in consistency. They react with atmospheric moisture to form flexible rubbers (RTV-1 = room temperature vulcanizing, 1-component) [84]. By virtue of their outstanding properties, these silicone rubbers are ideal for many sealing, bonding, and coating applications.

During the manufacturing process, terminal OH groups of the polysiloxane react with the cross-linking agent, generating curable products. The reaction itself takes place on exposure to atmospheric moisture and is accompanied by the liberation of hydrolysis products. This reaction, which is also referred to as vulcanization, starts with the formation of a solid skin at the surface of the rubber and continues gradually toward the inside.

In 3-D printing, these materials are really interesting since they can attain phase transition within tens of minutes. In these conditions, first layers of silicone are quickly cured and acquire strong mechanical properties that then help to keep the shape of the 3-D object. However, release of volatile or soluble parts of these materials often occurs that are toxic and hinder their use for healthcare applications.

### *14.4.5.3 Bi-Component Silicone*

For RTV-2 or LSR silicone, the chemical reaction consists of an addition curing reaction leading to the binding of Si-H groups to vinyl groups. Salts or platinum, palladium, or rhodium complexes may serve as catalysts [84]. If platinum-olefin complexes are used, curing will take place at room temperature. Platinum complexes containing nitrogen are used to trigger addition reaction at elevated temperatures (e.g., Pt complexes with pyridine, benzonitrile, or benzotriazole).

In 3-D printing, the kinetic of hydrosilylation can be managed with the use of inhibitors to keep the flow properties of silicone constant throughout the whole additive manufacturing process. However, contrary to RTV-1, a specific attention must be given to yield stress character when using bicomponent silicone in 3-D printing. Indeed, as the progression of rheological properties is slow to maintain the flow properties during printing, bicomponent silicone with low yield stress value cannot be printed (Fig. 14.8) into complex geometries presenting overhang structures, important mass/area ratio or bridges. In this case, silicone formulation has to be adapted with the addition of yield stress modulating agents such as polyethylene glycol (PEG), which reacts with silica (contained in silicone) to form a more stable macromolecular network [85].

## 14.4.6 Rheological Properties of Printable Silicone

### *14.4.6.1 Rheological Testing and Parameters*

The determination of silicone rheological behavior for 3-D printing can be performed using stress- or strain-controlled rheometers. With respect to the aforementioned rheological properties pertinent to the 3-D printing process,

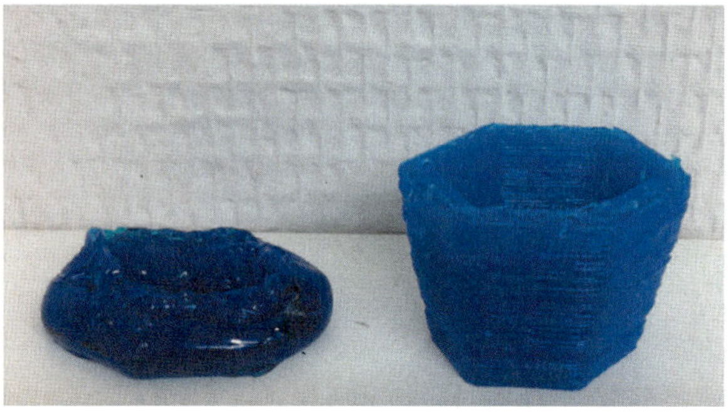

FIG. 14.8 Silicone 3-D printing with LSR *(left)* versus LSR+PEG *(right)*. Pictures were kindly provided by 3dFAB, University Lyon 1.

three specific tests can be performed to ensure the printability of silicone: shear thinning effect, yield stress character, and thixotropic behavior (Fig. 14.10).

Shear thinning is a characteristic phenomenon of some non-Newtonian fluids in which the fluid viscosity decreases with increasing shear stress (the opposite of shear thickening). In 3-D printing, a shear thinning effect is required to decrease the viscosity of silicone while flowing through the nozzle, that is, where wall shear rate is the higher. In this case, the consequent wall shear stress will be sufficiently low to flow highly viscous silicone through nozzle. Experimentally, the shear thinning effect can be documented through the access to stress rate test in flow mode.

Yield stress character describes the capacity of materials to keep its shape under a predetermined pressure. It is the most important rheological property in 3-D printing that impacts predeposition and postdeposition behavior. If the yield stress character of the material is too low, the flowing through nozzle is easy, but maintaining 3-D printed silicone complex shape is impossible. If the yield stress character of the material is too high, the 3-D printed silicone complex shape will be easy to maintain, but the flowing through nozzle will be highly challenging. Therefore, an adequate yield stress value must be found to flow material regarding deposition system and to keep the shape of 3-D object. The yield stress can be measured through stress-controlled rheometer using stress ramp test in flow mode.

As a typical example, LSR and PEG mixing in different ratio can be used to control silicone yield stress. A large range of yield stresses allow users to print 3-D objects, but the value of yield stress is related to the complexity of geometry (Fig. 14.9). A human ear is a complex geometry with high overhang and mass/area ratio: a yield stress around 1500 Pa will be then required.

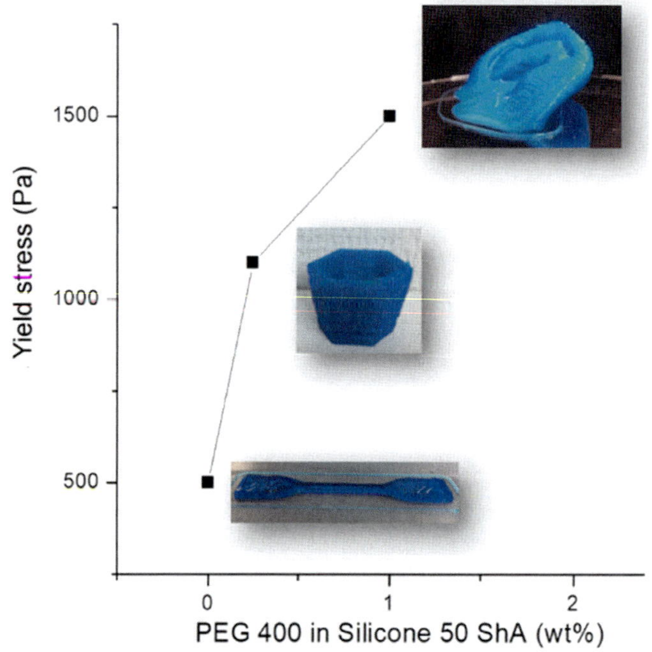


FIG. 14.9 Accessible 3-D printed objects for different values of yield stress with silicone and PEG mix. Graphics were kindly provided by 3dFAB, University Lyon 1.

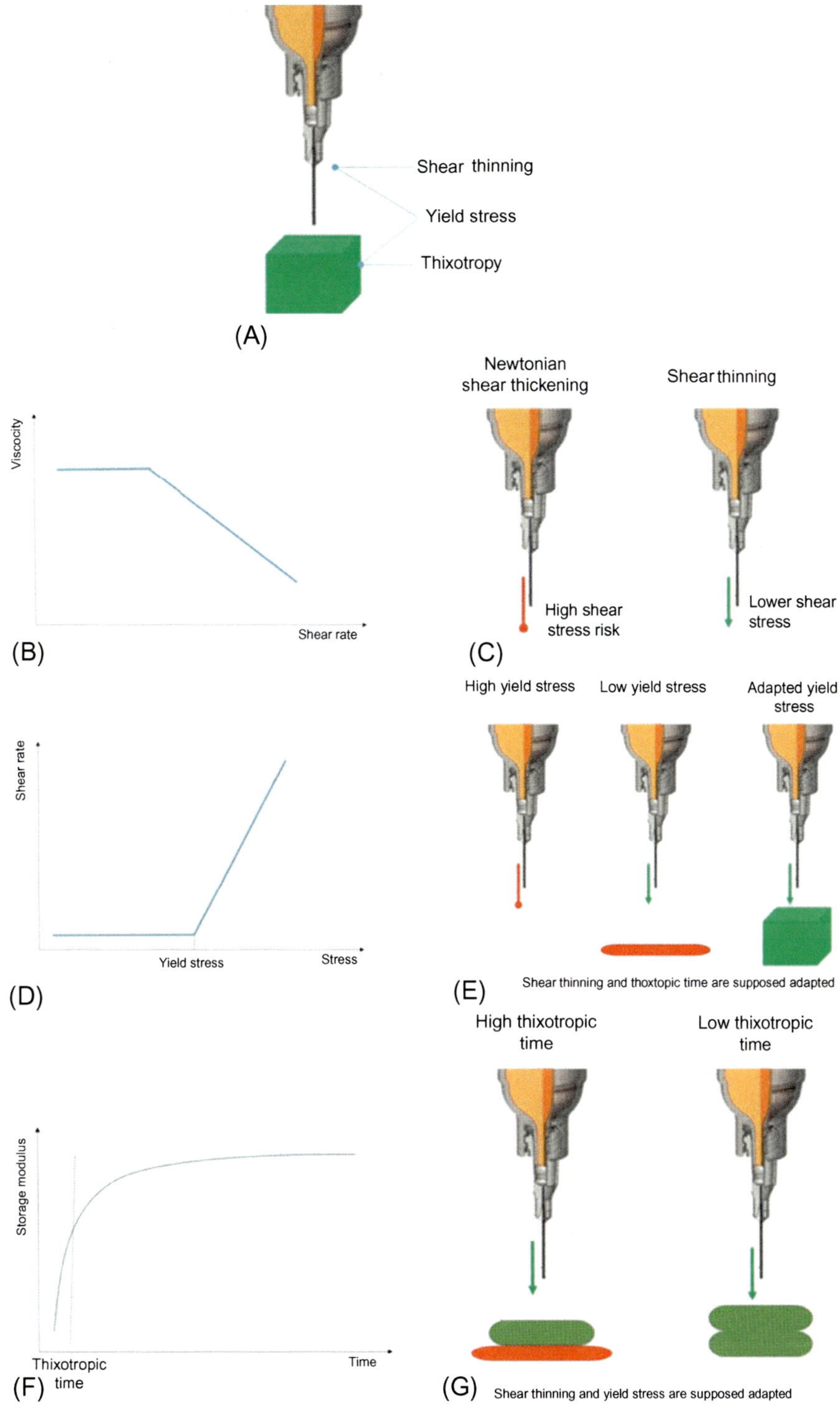


FIG. 14.10 (A) Rheological properties of interest and localization. (B) Shear thinning behavior. (C) Consequence of shear thinning behavior on 3-D printing. (D) Yield stress behavior. (E) Consequence of yield stress behavior on 3-D printing. (F) Thixotropic behavior. (G) Consequence of thixotropic behavior on 3-D printing. Graphics were kindly provided by 3dFAB, University Lyon 1.

Thixotropy is a time-dependent shear thinning property. It can be used to describe the restructuration time of material after a destructuration step. In 3-D printing, the destructuration occurs when the material flows through nozzle and the restructuration begins just after material extrusion. If the used material presents a high thixotropy (long thixotropy time), the recovering of the rheological properties will be slow, and the shape of the printed layer will not be maintained when the next layer will be deposited. The thixotropic behavior can be measured with two-step transient tests: stress growth in flow mode and time sweep test in oscillatory mode.

Furthermore, another rheological testing can be performed to ensure the stability of silicone throughout additive manufacturing process using time sweep test in oscillatory mode.

## 14.5 CONCLUSION

One of the current challenges in the biomedical field is to incorporate the patient-specific conditions into the treatment options. Personalization of implantable devices is one of the aspects of this general problem, as personalization ensures anatomical and biomechanical conformity that can result in evasion of significant complications. The advances in 3-D printing have enabled the production of such implants; however, the constraints of the printing process need to be carefully assessed, and the rheological properties of the base material have to be adjusted accordingly to achieve high-fidelity and mechanically robust 3-D printed structures. The rheological evolution of 3-D printed, remodelable structures containing cellular components is the next frontier in this area, and the control of spatiotemporal changes in mechanical properties must be carefully designed for management of the risks related to in vivo implantation.

### Acknowledgments

This work has received funding from FUI FASSIL. This project has received funding from the European Union's Horizon 2020 research and innovation program under grant agreement no 760921 (PANBioRA).

### Conflict of Interest Statement

J.B., C.B.M., and N.E.V. are full-time employees of Protip Medical. N.E.V. is stockholder of Protip Medical. The presentation of the PROTiP Medical products in the chapter was not done for publicity purposes, and their inclusion is purely based on our R&D activities in 3-D printed silicone-based implants.

### References

[1] M. Lutolf, J. Hubbell, Synthetic biomaterials as instructive extracellular microenvironments for morphogenesis in tissue engineering, Nat. Biotechnol. 23 (1) (2005) 47.
[2] S.V. Murphy, A. Atala, Organ engineering–combining stem cells, biomaterials, and bioreactors to produce bioengineered organs for transplantation, BioEssays 35 (3) (2013) 163–172.
[3] N. Nagarajan, et al., Enabling personalized implant and controllable biosystem development through 3D printing, Biotechnol. Adv. (2018).
[4] R. Suntornnond, et al., A highly printable and biocompatible hydrogel composite for direct printing of soft and perfusable vasculature-like structures, Sci. Rep. 7 (1) (2017) 16902.
[5] B.L. Eppley, A.M. Sadove, Computer-generated patient models for reconstruction of cranial and facial deformities, J. Craniofac. Surg. 9 (6) (1998) 548–556.
[6] A. Hasan, et al., Microfluidic techniques for development of 3D vascularized tissue, Biomaterials 35 (26) (2014) 7308–7325.
[7] D.B. Kolesky, et al., Three-dimensional bioprinting of thick vascularized tissues, Proc. Natl. Acad. Sci. 113 (12) (2016) 3179–3184.
[8] J.S. Miller, et al., Rapid casting of patterned vascular networks for perfusable engineered three-dimensional tissues, Nat. Mater. 11 (9) (2012) 768.
[9] W. Peng, D. Unutmaz, I.T. Ozbolat, Bioprinting towards physiologically relevant tissue models for pharmaceutics, Trends Biotechnol. 34 (9) (2016) 722–732.
[10] A.B. Dababneh, I.T. Ozbolat, Bioprinting technology: a current state-of-the-art review, J. Manuf. Sci. Eng. 136 (6) (2014) 061016.
[11] W. Wang, G. Li, Y. Huang, Modeling of bubble expansion-induced cell mechanical profile in laser-assisted cell direct writing, J. Manuf. Sci. Eng. 131 (5) (2009) 051013.
[12] L. Koch, et al., Skin tissue generation by laser cell printing, Biotechnol. Bioeng. 109 (7) (2012) 1855–1863.
[13] T. Boland, et al., Application of inkjet printing to tissue engineering, Biotechnol. J. Healthc. Nutr. Technol. 1 (9) (2006) 910–917.
[14] B. Derby, Bioprinting: inkjet printing proteins and hybrid cell-containing materials and structures, J. Mater. Chem. 18 (47) (2008) 5717–5721.
[15] K. Christensen, et al., Freeform inkjet printing of cellular structures with bifurcations, Biotechnol. Bioeng. 112 (5) (2015) 1047–1055.
[16] V. Mironov, Printing Technology to Produce Living Tissue, Taylor & Francis, 2003.
[17] I.T. Ozbolat, M. Hospodiuk, Current advances and future perspectives in extrusion-based bioprinting, Biomaterials 76 (2016) 321–343.
[18] D.B. Kolesky, et al., Bioprinting: 3D bioprinting of vascularized, heterogeneous cell-laden tissue constructs, Adv. Mater. 26 (19) (2014) 2966.
[19] L.E. Bertassoni, et al., Hydrogel bioprinted microchannel networks for vascularization of tissue engineering constructs, Lab Chip 14 (13) (2014) 2202–2211.
[20] H.-W. Kang, et al., A 3D bioprinting system to produce human-scale tissue constructs with structural integrity, Nat. Biotechnol. 34 (3) (2016) 312.
[21] L.E. Bertassoni, et al., Direct-write bioprinting of cell-laden methacrylated gelatin hydrogels, Biofabrication 6 (2) (2014) 024105.
[22] B. Derby, Printing and prototyping of tissues and scaffolds, Science 338 (6109) (2012) 921–926.
[23] M. Hospodiuk, et al., The bioink: a comprehensive review on bioprintable materials, Biotechnol. Adv. 35 (2) (2017) 217–239.
[24] L. Koch, et al., Laser printing of skin cells and human stem cells, Tissue Eng. Part C: Methods 16 (5) (2009) 847–854.
[25] I.T. Ozbolat, W. Peng, V. Ozbolat, Application areas of 3D bioprinting, Drug Discov. Today 21 (8) (2016) 1257–1271.
[26] K. Yue, et al., Synthesis, properties, and biomedical applications of gelatin methacryloyl (GelMA) hydrogels, Biomaterials 73 (2015) 254–271.

[27] G. Gao, et al., Inkjet-bioprinted acrylated peptides and PEG hydrogel with human mesenchymal stem cells promote robust bone and cartilage formation with minimal printhead clogging, Biotechnol. J. 10 (10) (2015) 1568–1577.
[28] C.M. Livoti, J.R. Morgan, Self-assembly and tissue fusion of toroid-shaped minimal building units, Tissue Eng. A 16 (6) (2010) 2051–2061.
[29] D.F. Duarte Campos, et al., The stiffness and structure of three-dimensional printed hydrogels direct the differentiation of mesenchymal stromal cells toward adipogenic and osteogenic lineages, Tissue Eng. A 21 (3–4) (2014) 740–756.
[30] T. Xu, et al., Inkjet printing of viable mammalian cells, Biomaterials 26 (1) (2005) 93–99.
[31] K.K. Lai, R. Renneberg, W.C. Mak, High efficiency single-step biomaterial-based microparticle fabrication via template-directed supramolecular coordination chemistry, Green Chem. 18 (6) (2016) 1715–1723.
[32] I.T. Ozbolat, H. Chen, Y. Yu, Development of 'Multi-arm Bioprinter' for hybrid biofabrication of tissue engineering constructs, Robot. Comput. Integr. Manuf. 30 (3) (2014) 295–304.
[33] Y. Zhang, Y. Yu, I.T. Ozbolat, Direct bioprinting of vessel-like tubular microfluidic channels, J. Nanotechnol. Eng. Med. 4 (2) (2013) 020902.
[34] H. Gudapati, et al., Alginate gelation-induced cell death during laser-assisted cell printing, Biofabrication 6 (3) (2014) 035022.
[35] B. Guillotin, et al., Laser assisted bioprinting of engineered tissue with high cell density and microscale organization, Biomaterials 31 (28) (2010) 7250–7256.
[36] P. Fratzl, Collagen: structure and mechanics, an introduction, in: Collagen, Springer, 2008, pp. 1–13.
[37] C.M. Smith, et al., Three-dimensional bioassembly tool for generating viable tissue-engineered constructs, Tissue Eng. 10 (9–10) (2004) 1566–1576.
[38] A. Skardal, et al., Bioprinted amniotic fluid-derived stem cells accelerate healing of large skin wounds, Stem Cells Transl. Med. 1 (11) (2012) 792–802.
[39] X. Wang, et al., Generation of three-dimensional hepatocyte/gelatin structures with rapid prototyping system, Tissue Eng. 12 (1) (2006) 83–90.
[40] M. Mesa, et al., Morphology and porosity characteristics control of SBA-16 mesoporous silica. Effect of the triblock surfactant Pluronic F127 degradation during the synthesis, Solid State Sci. 7 (8) (2005) 990–997.
[41] S. Wang, et al., Receptor-mediated drug delivery systems targeting to glioma, Nanomaterials 6 (1) (2015) 3.
[42] W. Wu, A. DeConinck, J.A. Lewis, Omnidirectional printing of 3D microvascular networks, Adv. Mater. 23 (24) (2011) H178–H183.
[43] D. Hanjaya-Putra, et al., Controlled activation of morphogenesis to generate a functional human microvasculature in a synthetic matrix, Blood 118 (2011) 804–815.
[44] J.W. Nichol, et al., Cell-laden microengineered gelatin methacrylate hydrogels, Biomaterials 31 (21) (2010) 5536–5544.
[45] Y.C. Chen, et al., Functional human vascular network generated in photocrosslinkable gelatin methacrylate hydrogels, Adv. Funct. Mater. 22 (10) (2012) 2027–2039.
[46] R.-Z. Lin, et al., Transdermal regulation of vascular network bioengineering using a photopolymerizable methacrylated gelatin hydrogel, Biomaterials 34 (28) (2013) 6785–6796.
[47] E. Hoch, et al., Chemical tailoring of gelatin to adjust its chemical and physical properties for functional bioprinting, J. Mater. Chem. B 1 (41) (2013) 5675–5685.
[48] D.S. Benoit, A.R. Durney, K.S. Anseth, Manipulations in hydrogel degradation behavior enhance osteoblast function and mineralized tissue formation, Tissue Eng. 12 (6) (2006) 1663–1673.
[49] B.-H. Jo, et al., Three-dimensional micro-channel fabrication in polydimethylsiloxane (PDMS) elastomer, J. Microelectromech. Syst. 9 (1) (2000) 76–81.
[50] T.J. Hinton, et al., 3D printing PDMS elastomer in a hydrophilic support bath via freeform reversible embedding, ACS Biomater. Sci. Eng. 2 (10) (2016) 1781–1786.
[51] V. Ozbolat, et al., 3D printing of PDMS improves its mechanical and cell adhesion properties, ACS Biomater. Sci. Eng. 4 (2) (2018) 682–693.
[52] C.S. O'Bryan, et al., Self-assembled micro-organogels for 3D printing silicone structures, Sci. Adv. 3 (5) (2017) e1602800.
[53] J. Black, G. Hastings, Handbook of Biomaterials Properties, Chapman and Hall, London, 1998.
[54] R. Yoda, Elastomers for biomedical applications, J. Biomater. Sci. Polym. Ed. 9 (6) (1998) 561–626.
[55] J.E. Puskas, Y. Chen, Biomedical application of commercial polymers and novel polyisobutylene-based thermoplastic elastomers for soft tissue replacement, Biomacromolecules 5 (4) (2004) 1141–1154.
[56] R.R. De Nicola, Permanent artificial (silicone) urethra, J. Urol. 63 (1) (1950) 168–172.
[57] W.E. Neville, P.J. Bolanowski, H. Soltanzadeh, Prosthetic reconstruction of the trachea and carina, J. Thorac. Cardiovasc. Surg. 72 (4) (1976) 525–538.
[58] S.V. Gohil, et al., Chapter 8: Polymers and composites for orthopedic applications, in: S. Bose, A. Bandyopadhyay (Eds.), Materials for Bone Disorders, Academic Press, 2017, pp. 349–403.
[59] V.L. Gott, D.E. Alejo, D.E. Cameron, Mechanical heart valves: 50 years of evolution, Ann. Thorac. Surg. 76 (6) (2003) S2230–S2239.
[60] R.J. Thomas, A model for cochlear implant electrode insertion and force evaluation: results with a new electrode design and insertion technique, Laryngoscope 115 (8) (2005) 1325–1339.
[61] S. Kalousdian, M.S. Karlan, M.A. Williams, Silicone elastomer cerebrospinal fluid shunt systems, Neurosurgery 42 (4) (1998) 887–892.
[62] J. Curtis, A. Colas, Chapter II.5.18: Medical applications of silicones, in: B.D. Ratner, et al. (Eds.), Biomaterials Science, third ed., Academic Press, 2013, pp. 1106–1116.
[63] O. Cuisnier, et al., Surgical and/or endoscopic treatment of acquired tracheal stenosis in adult patients, Ann. Otolaryngol. Chir. Cervicofac. 121 (1) (2004) 3–13.
[64] E.A. Weymuller, Laryngeal injury from prolonged endotracheal intubation, Laryngoscope 98 (1988) 1–15.
[65] J.L. Stauffer, D.E. Olson, T.L. Petty, Complications and consequences of endotracheal intubation and tracheotomy. A prospective study of 150 critically ill adult patients, Am. J. Med. 70 (1) (1981) 65–76.
[66] H.A. Gaissert, R.H. Lofgren, H.C. Grillo, Upper airway compromise after inhalation injury. Complex strictures of the larynx and trachea and their management, Ann. Surg. 218 (5) (1993) 672–678.
[67] K.M. Greven, et al., Distinguishing tumor recurrence from irradiation sequelae with positron emission tomography in patients treated for larynx cancer, Int. J. Radiat. Oncol. Biol. Phys. 29 (4) (1994) 841–845.

[68] C.A. Righini, et al., Sténoses trachéales de l'adulte. EMC, Oto-Rhino-Laryngol. 10 (1) (2014) 1–15 (Article 20-760-A-10).
[69] P.M. Boiselle, K.S. Lee, A. Ernst, Multidetector CT of the central airways, J. Thorac. Imaging (2005) 20–23.
[70] G.R. Ferretti, et al., Benign abnormalities and carcinoid tumors of the central airways: diagnostic impact of CT bronchography, AJR Am. J. Roentgenol. 174 (5) (2000) 1307–1313.
[71] V. Bourinet, et al., Experience with transcordal silicone stents in adult laryngotracheal stenosis: a bicentric retrospective study, Respir. Int. Rev. Thorac. Dis. 95 (6) (2018) 441–448.
[72] T. Raguin, et al., Method for dealing with severe aspiration using a new concept of intralaryngeal prosthesis: a case report, Head Neck 38 (10) (2016) E2504–E2507.
[73] J.M. Vergnon, et al., Efficacy of tracheal and bronchial stent placement on respiratory functional tests, Chest 107 (3) (1995) 741–746.
[74] M.C. Dumon, et al., Silicone tracheobronchial endoprosthesis, Rev. Mal. Respir. 16 (4 Pt 2) (1999) 641–651.
[75] C.L. Ventola, Medical applications for 3D printing: current and projected uses, Pharmacy and Therapeutics 39 (10) (2014) 704–711.
[76] C. Scott, Johnson & Johnson Looks Toward a Future of Personalized Medicine Through 3D Printing, Available from: https://3dprint.com/190785/johnson-and-johnson-medicine/, 2017.
[77] K. Tappa, U. Jammalamadaka, Novel biomaterials used in medical 3D printing techniques, J. Funct. Biomater. 9 (1) (2018) 17.
[78] F. Rengier, et al., 3D printing based on imaging data: review of medical applications, Int. J. Comput. Assist. Radiol. Surg. 5 (4) (2010) 335–341.
[79] D. Mitsouras, et al., Medical 3D printing for the radiologist, RadioGraphics 35 (7) (2015) 1965–1988.
[80] M. Vuķicevic, et al., Cardiac 3D printing and its future directions, JACC Cardiovasc. Imaging 10 (2) (2017) 171–184.
[81] Home: ACEO® 3D Silicone Printing. [Online], Available: https://www.aceo3d.com/. Accessed 30 November 2017.
[82] Home - ViscoTec Pumpen- u. Dosiertechnik GmbH. [Online], Available: https://www.viscotec.de/fr/. Accessed 11 July 2018.
[83] C.G. Williams, et al., Biomaterials 26 (11) (2005) 1211–1218.
[84] EUROPE, S, Chemical reactions on the finished silicone, Available from: http://www.silicones.eu/science-research/chemistry/chemical-reactions-on-the-finished-Silicone.
[85] J.-N. Paquien, et al., Rheological studies of fumed silica–polydimethylsiloxane suspensions, Colloids Surf. A Physicochem. Eng. Asp. 260 (1) (2005) 165–172.

## Further Reading

[86] Y. Thibout, J.-M. Vergnon, H. Dutau, Traitements endoscopiques du cancer bronchique, EMC-Pneumologie 10 (1) (2013) 1–10 (Article 6-000-M-10).

CHAPTER

# 15

# Computational Simulation of Cell Behavior for Tissue Regeneration

*S.Jamaleddin Mousavi**,†,‡, *Mohamed H. Doweidar**,†,‡

*Mechanical Engineering Department, School of Engineering and Architecture (EINA), University of Zaragoza, Zaragoza, Spain †Aragón Institute of Engineering Research (I3A), University of Zaragoza, Zaragoza, Spain ‡Biomedical Research Networking Center in Bioengineering, Biomaterials and Nanomedicine (CIBER-BBN), Zaragoza, Spain

## 15.1 INTRODUCTION

In this chapter, a numerical discrete model is represented to consider the role of cell migration in different processes such as cell differentiation, cell proliferation, and cell morphology. Cell migration is essential for normal tissue development and morphogenesis of the human body and organ systems. Therefore, over the last few decades, the investigation of cell behavior in the presence of different stimulating cues has become a hotbed for researchers. It is well known that cell migration regulates numerous physiological processes, such as morphogenesis [1–6], tissue development [7, 8], cell differentiation and proliferation [9–11], and pathological processes such as wound healing [12, 13] and tumor metastasis [14, 15].

In the case of tissue development, the tissue should be generated in a correct geometry with a proper cell type. Abnormal cell migration may lead to uncontrolled states such as invasion and the metastasis of cancer. In such cases, cells may migrate in individual routines or in groups of cells as tightly associated epithelial sheets or clusters (e.g., Drosophila border cells and zebrafish lateral line primordium), or they may possess a mesenchymal character such as during gastrulation and neural crest migration [1]. Wound healing is another coordinated multicell response programmed through a defined timetable in which each phase prepares the wound for subsequent phases that are required for reestablishing the tissue. One of the most important stages in this timetable is cell migration, by which fibroblasts migrate in the direction of the wound to improve the matrix structure and to modify the wound contraction [16]. Stimuli that regulate cell behavior may change the rate of cell migration toward the wound to speed up wound healing [17]. Additionally, a change of cell morphology is another significant parameter in wound healing during which cells lengthen themselves during their migration toward wound locations [12] to cover the wound through changing their shape to fill all intercellular gaps [13]. Further, malignant cells invade healthy tissues during metastasis in response to different conditions. For instance, neoplastic cells follow this process to come into lymphatic and blood vessels to spread into the circulatory system, causing metastatic development in distant organs [18].

Stem cells have the potential to proliferate or differentiate into different cell phenotypes. Different signaling such as physicochemical factors, including particular mechanical mechanisms, can control both processes. However, the control of stem cell lineage specification by mechanical cues is less understood. However, certain key themes have been experimentally proven. For instance, stem cells can experience any alterations in the stiffness of their surrounding microenvironment and consequently differentiate to a specific lineage specification. Besides, external mechanical forces exerted on the stem cells can control stem cell differentiation to a specific cell phenotype. In addition, the differentiation process can affect the mechanical properties of the cells and their specific subcellular components. The combination of these three fundamental concepts allows introducing a new theory for the behavior of stem cells [19, 20].

*Advances in Biomechanics and Tissue Regeneration*
https://doi.org/10.1016/B978-0-12-816390-0.00015-7

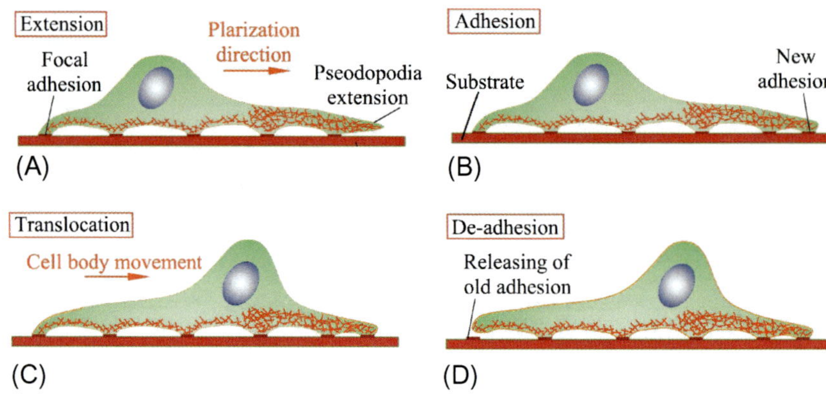


FIG. 15.1 The cyclic four main steps of cell migration. (A) Extension, (B) development of new adhesion, (C) translocation of cell body, and (D) deadhesion.

As shown in Fig. 15.1, cell migration includes a number of orchestrated and cyclic processes, including the extension of pseudopodia, the formation of new adhesions, the translocation of the cell body, and the release of old adhesions [21–24]. During this process, the cell adheres to its extracellular matrix (ECM) by means of different forces, including the concurrent traction forces (acto-myosin forces) and random protrusion force (generated by the active polymerization of the actin network) [21, 24]. The effect of the cell shape on the balance of traction forces and its effect on cell behavior is less understood. However, experiments show that it depends on the orientational distribution and the number of stress fibers within the cell. All these parameters, in turn, depend on the magnitude and symmetry characteristics of the ECM stiffness [25] and the cell internal deformation [26, 27].

A general pattern of cellular differentiation is also a challenging topic in tissue repair. Nevertheless, it is experimentally well known that cell differentiation and proliferation can be prompted by a mechanosensing process and cell ECM interaction during cell migration [28–30]. To demonstrate this hypothesis, the first attempt was made by Engler et al. [28], demonstrating that, on a two-dimensional matrix, the stiffness (mechanotaxis) can guide the human mesenchymal cell (MSC) fate. In such a way, when cells are cultured on soft ECMs mimicking the elasticity of brain tissue (a stiffness of 0.1–1 kPa), they differentiate into neuronal precursors; on matrices with intermediate stiffness mimicking muscle (a stiffness of 8–17 kPa), they induce myogenic commitment while on relatively rigid matrices such as collagenous bone (a stiffness of 25–40 kPa), they differentiate to osteoblasts. Similar results were reported by Huebsch et al. [30] within a three-dimensional (3D) hydrogel synthetic ECM.

It is well known that, in addition to mechanotaxis (durotaxis) [24, 31, 32], the cell behavior can be actively controlled by other stimuli such as chemotaxis [33–37], thermotaxis [38, 39], and/or electrotaxis [40–42]. To understand it comprehensively, we need to discover the role of the above-mentioned cues. Many experimental works [33, 35, 43, 44] address that cells migrate directionally along even a shallow gradient of chemical substances such as growth factors or attracting agents [33, 35, 36, 43]. Chemoattraction is thought to play a crucial role in guiding cells in many immunobiological processes such as reaching leukocytes to the infection locations. However, the mechanisms by which a cell transduces a chemotactic cue into a certain movement still remain elusive [34]. Besides, in in vivo, thermotaxis may be considered as a complementary signal to chemotaxis because each mechanism is active in a specific region where the other is ineffective [45]. For instance, trophoblasts invade the endometrium, the inner membrane of a uterus, by means of thermotaxis. These cells subjected to oxygen and thermal gradients do not migrate in response to the oxygen gradient (a chemotactic signal) but they migrate in response to thermal gradients less than 1°C toward the warmer locations [39]. Furthermore, recent in vitro studies have demonstrated that, when stationary cells are exposed to direct current electric fields (dcEFs), they effectively migrate toward cathode or anode poles, depending on the cell type [40, 46–50]. For instance, epithelia generate a steady voltage across themselves, driving an electric current in the wounded sites [42, 47]. In the rat cornea injury, an electric current of about 10 $\mu A/cm^2$ is measured. Besides, the skin of a fingertip wound is able to create a lateral electric field (EF) in the range of 40–200 mV/mm [42]. In the last few years, it has been shown that the calcium ion, $Ca^{2+}$, is involved in the electrical-field-induced cell response [49, 51–56].

Cells migrate either individually or in a population of cells. Many experimental works have widely studied single-cell migration [43, 57]. Nevertheless, collective cell migration is vastly dominant in many cell types such as those related to tumor cells [23] and many physiological and pathological processes such as tissue remodeling [58] and wound healing [59]. The tendency of cell-cell attraction during collective guidance is recognized. However, similar behavior of cell migration has been observed for fibroblasts, which are less-cohesive cell types [48, 60]. Cell migration can be considered collective when two or more cells make contact and maintain their cell-cell connections during migration, at least occasionally [21]. In the presence of other cells, the cell-cell attraction may affect cell-ECM adhesion and may facilitate cell-cell contacts. Hence, the dynamics of collective cell migration result in complex changes in multicellular tissue structures. Disparate single cell migration, collective cell migration serves to keep the tissue intact during remodeling. The cell-cell interactions and cell motility coordination during multicell migration can be studied from two perspectives. First, how do the cells affect each other? To what radius does a cell transmit force and communicate for transmitting information? Second, how can cell-cell interactions affect their individual and collective behavior? Understanding these perspectives can help to answer many questions such as Does collective cell migration speed up or delay cell movement? How do cell slugs affect each other? Some of these questions may be answered via experimental works, but to profoundly answer these questions and much more, numerical studies are inevitable.

Cell migration on two-dimensional (2D) surfaces takes place during the reepithelialization of wounds, the scanning of leukocytes along the inner blood vessel wall, or inner epithelial surfaces [21]. Although 2D studies have enhanced our insights into many contexts such as the basic mechanisms by which cells migrate, interact with the ECM, and change their speed or direction, they may sometimes impose an artificial apical-based cell polarity that may not exist through 3D in vivo processes. For example, the studies of Hakkinen et al. [3] showed that cell morphology strongly depends on the ECM dimensionality because the cells tend to be less elongated and more spread on 2D matrices than in 3D matrices (Fig. 15.2). This is attributed to the number of integrins and receptors, which are associated with cell-ECM interaction. Limited integrins and receptors of the cell can participate in 2D cell-ECM adhesion. In contrast, the capability of the cells to move in a 3D ECM not only depends on the viscosity and stiffness of the ECM, but also on the density of the fibers. Therefore, to fully understand the underlying mechanisms by which cells migrate in vivo, it is necessary to study the cell migration in 3D environments, too. This explains why experimental studies in 3D matrices have begun to grow gradually in the last few years [31, 61].

Numerical modeling of cell migration has a relatively short history. For the first time, as early as 1970, Keller et al. [62, 63] developed a model using partial differential equations to study the biochemical regulation of bacterial movement. A decade later, in the 1980s, a distinguished field was developed for research in which numerical models were proposed for the movement of isolated individual cells. A key early work in this direction was presented by Oster [64, 65]. Several works have been successively developed by different researchers [66–68]. Keller's highly effective equations became the basis of phenomenological models ranging from slime mold slugs [69] to tumor angiogenesis [70] and wound healing [71].

Numerical models, which consider cell behavior, have recently increased. For instance, several numerical models have been recently developed to study the behavior of cell populations [37, 72], cell morphology [36, 37, 73–75], and cell fate [76, 77]. Each model has a kind of limitation and there is no comprehensive model to simultaneously consider cell behavior in a multisignaling environment, collective cell migration, and cell shape changes. Although computational models such as [36, 78] concurrently deal with mechanotactic and chemotactic signals, they miss the thermotactic and electrotactic influence. Some numerical models only focus on the migration of a specific cell type [79], consider the 2D cell shape using a hybrid cellular Potts model [80], or study 2D cell-cell interaction with a defined cell configuration [79, 81, 82]. Each model is developed using a specific method. For example, there are energy-based [75] and coarse-grained [73, 74] models studying ECM rigidity influences on cell morphology and migration as well as continuum mechanics models studying the chemotactic effect on cell migration and cell shape change [78]. In most

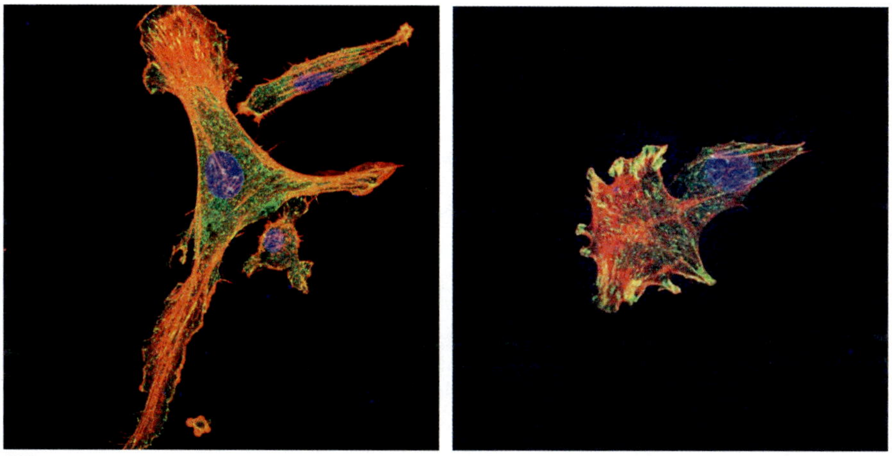

(A) (B)

FIG. 15.2 Cell elongation and morphology on a 2D (A) and within a 3D (B) ECM with identical stiffness.

of these models, the actual forces acting on the cell body are not considered [79, 81–83]. However, the main limitation in many of them is that they have ignored the cell mechanosensing process that is an essential feature in cell-ECM interaction [36, 79, 82]. In addition, the above-mentioned 2D models, which simulate cell shape change, do not study cell configuration in a free mode. Instead, they restrict the cell shape to a rigid ellipse by which the cell shape change is represented by an alteration of the aspect ratio of the ellipse (the ratio of the major axis to the minor axis). A typical rigid mode of cell configuration is also assumed in some 3D models [84] in which a 3D epithelial cell interacts with a plane ECM. The main limitation of this model is that the cell can have different rigid shapes such as a hexagonal prism, columnar, cuboidal, and/or squamous. Vermolen et al. [37] presented a 3D phenomenological numerical model to investigate the effect of chemotactic cues on cell morphology. Although in their model the cell can take irregular shapes, their model is formulated in such a way that the cell velocity fundamentally depends on the chemical gradient, which is not sufficiently precise according to many experimental works [24, 31, 85]. They also assumed that the cell volume changes when the cell sends out pseudopods. This is an unrealistic assumption since recent experimental investigations [86–88] have demonstrated that the overall cellular volume remains constant as the cell shape changes. Han et al. [89] investigated the spatiotemporal dynamics of cell migration using a biochemical-mechanical contractility model that incorporates the traction forces developed by the cell during cell migration in 2D ECMs. Unfortunately, their model does not include cell shape changes during cell migration. In their model, the formation of a new adhesion regulates a reactivation of stress fiber assembly within the cell and predicts the spatial distribution of traction forces.

Besides the concerns discussed earlier, signaling mechanisms by which the microenvironment stiffness controls cell differentiation and proliferation are not computationally considered at the cell level. Several mechanobiological macrolevel models have been established to describe cell lineage specification during bone fracture healing [76, 77, 90–95]. All these numerical models are able to predict the general patterns of cellular differentiation due to external mechanical stimuli in the macrolevel. A 2D model presented by Stops et al. [77] considers cell differentiation and proliferation in a collagen-glycosaminoglycan scaffold subjected to mechanical strain and perfusive fluid flow. Their findings indicate that specific combinations of scaffold strains and inlet fluid flows define the specific cell fate. Besides, the 2D model developed by Kang et al. [76] to simulate bone fracture healing is formulated based on the density of each cell phenotype. It is assumed that the cell differentiation and proliferation can be modulated according to the magnitude and frequency of mechanical stimuli. According to their numerical results, the bone healing process can be improved when the magnitude and the frequency of the mechanical stimuli are employed as control factors of cell proliferation. To our knowledge, there is no numerical models to consider cell differentiation and/or proliferation based on the mechanosensing process during cell migration.

In addition to the experimental research, theoretical studies and numerical models also deliver deep insights into cell behavior. Experimental investigations of cell behavior are relatively time consuming and expensive. Nevertheless, the numerical models can save experimental resources and time. These models not only associate the experimental results to the first principles but also describe the behavior and sensitivity of the systems as a function of each parameter. Furthermore, mathematical models are helpful in identifying the key parameters that play a main role in defining the overall behavior of the system. Hence, they can lead to new and more effective in vitro experiments.

Consequently, the main aim of the present chapter is to cover different aspects of cell behavior such as cell migration, differentiation, proliferation, shape changes, and cell-cell interactions in a 3D multisignaling ECM. For this purpose, we first modeled the mechanosensing process of a single cell. Next, the model will be extended to include individual cell migration, cell-cell interaction, and collective cell migration due to pure mechanotaxis. Afterward, the effect of different signals will be added to consider the behavior of individual cells and populations of cells in a 3D multisignaling matrix. To consider these processes, for the sake of simplification, initially a constant spherical cell morphology is assumed. Subsequently, the model will be extended to study the cell morphological changes due to mechanotaxis, thermotaxis, chemotaxis, and electrotaxis. Finally, the basic model of cell migration will be employed to include cell differentiation and proliferation because of the mechanosensing process.

## 15.2 METHODOLOGY

Cell migration is similar to many other physical events. It depends on the equilibrium of effective forces acting on the cell body. Traction force, which is generated due to cell internal deformation, is transmitted through the integrins at the focal adhesions to the ECM [13, 24]. This force is the only directional force, in the absence of other external cues, that guides the cell movement together with a random protrusion force. In the presence of thermotactic and chemotactic signals, the direction of this force is modified according to the pointed direction by those stimuli. In the presence of electrotaxis, the cell is exposed to an independent electrostatic force. Therefore, in a multisignaling environment, all these effective forces on the cell body should be in equilibrium with the opposing drag force. To predict the cell response to multiple signals received from the cell environment, we developed a 3D computational model using a discrete finite element methodology considering the cell as a group of finite elements. It is considered that ECM and the cell constitute the working domain. The working domain is divided into a number of subdomains (depending on the cell number) and each subdomain is considered to represent a cell. The cell behavior is modeled using two different strategies:

1. Constant spherical cell morphology to investigate cell migration, differentiation, and proliferation in the presence of different stimuli.
2. Variable cell morphology to study the cell morphology in the presence of different stimuli.

This allows us to predict the cell behavior and response when it is surrounded by different microenvironmental physical characteristics.

### 15.2.1 Mechanotaxis

The presented model can be employed to simulate adherent cells that are cultured on 2D or within 3D ECM. Cells have a special internal structure that is prepared to detect the stiffness of the matrix in which they reside. For instance, fibroblasts preferentially move toward stiffer ECMs [26, 96]. This phenomenon is known as mechanotaxis by which a cell moves directionally toward stiffer regions within its ECM [97]. In the mechanosensing step, the cell senses its ECM by exerting a sensing force to diagnose its microenvironment and to obtain information about its ECM rigidity. When the cell determines its surrounding mechanical conditions, it starts to pull itself toward the stiffer and/or more fixed region. The active apparatuses cellular elements regulating the cell mechanosensing process are the actin filaments and the Myosin II machinery and the passive mechanical ones of the cell body are the microtubules and the cell membrane (see Fig. 15.3A) [26, 32, 72, 97, 98]. The cytoplasmic cytoskeletal (CSK) is linked with the external ECM through focal adhesions and transmembrane integrins that are assumed to be totally rigid in the present model. This scheme agrees with the tensegrity hypothesis [96] because the deformation of external ECM is balanced by tensile forces generated in the actin CSK. External forces are also considered to be another possible cause of the deformation of the ECM and cell. During cell migration, certain mechanical forces such as the traction force, the drag force, and the protrusion random force will affect the cell [24, 97, 99].

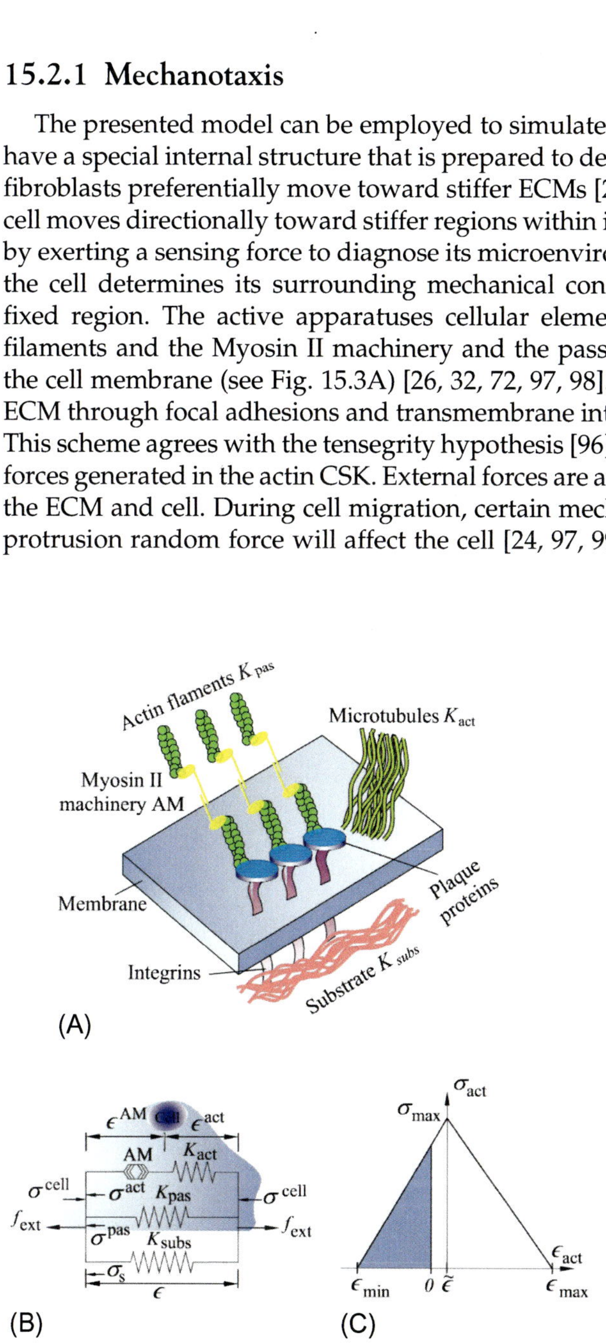


FIG. 15.3 Mechanosensing model of an adherent cell. (A) Schematic diagram of the relevant mechanical constituents of a cell [26]. (B) Cell mechanical model. $K_{act}$, $K_{pas}$, and $K_{subs}$ denote the stiffness modulus of actin filaments, the passive components of the cell, and the ECM, respectively. $f_{ext}$ stands for external forces applied to the cell or the ECM. (C) Dependence of the active, $\sigma^{act}$, and passive, $\sigma^{pas}$, contractile stresses on the deformation of the cell, $\epsilon$. $\epsilon_{act}$ is the deformation of the active contractile elements. $\sigma_{max}$ stands for the maximum contractile stress exerted by the actin-myosin machinery, and $\epsilon_{min}$ and $\epsilon_{max}$ are the corresponding shortening and lengthening strains of the contractile elements with respect to the unloaded length at which active stress becomes zero. $\sigma_s$ is the stress of the ECM [26].

#### 15.2.1.1 Traction Force

Following Mousavi et al. [100], the one-dimensional (1D) model represented in Fig. 15.3B can be particularized to calculate the cell internal stresses within 3D ECMs. Assuming that the contractile forces exerted by a cell are isotropic [26], the change of the length of each element is then interpreted as its corresponding volumetric strain [101]. Therefore, the effective stress transmitted by each cell node located on the cell-ECM interface can be defined by

$$\sigma_i^{\text{cell}} = \sigma_i^{\text{act}} + \sigma_i^{\text{pas}} \tag{15.1}$$

where $\sigma_i^{\text{act}}$ and $\sigma_i^{\text{pas}}$ are the contractile stresses transmitted through the actin bundles via the $i$th node on the cell-ECM interface. The myosin II machinery generates the first internally while the second is produced by the passive resistance of the cell, basically by the CSK microtubules and the membrane. In Fig. 15.3B, $\sigma^{\text{cell}}$ can also be interpreted as the average cell stress that bears the submembrane plaque in agreement with the integrin-mediated mechanosensing hypothesis [102]. In addition, $\epsilon$ denotes the local volumetric strain. In the model herein, the local strain is computed from the deformation of the cell external nodes along the direction of the traction force exerted at the corresponding node. $\epsilon^{\text{act}}$ stands for the deformation of the active contractile element. This deformation relates to the fact that the real physical change of the overlap between the myosin and actin filaments occurs when active forces are applied. Finally, $\epsilon_a$ represents the deformation of the actin bundles that encourages the transmitted active forces.

Simplifying the cell-ECM structure under moderate cell and ECM strains, we approximate the unidimensional constitutive behavior of the cell by a simple linear-elastic spring [26]. Therefore, for the calculation of the contractile stress, $\sigma^{\text{act}}$, as a function of the contractile element deformation, a simple piecewise linear constitutive model has been used (see Fig. 15.3C). If $\epsilon_{\min} \leq \epsilon \leq \epsilon_{\max}$, the active stress, $\sigma^{\text{act}}$, affects cell total stress, $\sigma^{\text{cell}}$, else it is 0 and $\sigma^{\text{cell}}$ is equal to $\sigma^{\text{pas}}$. The stress experienced by the passive contractile elements is directly proportional to the stiffness of the passive elements and cell deformation as

$$\sigma^{\text{pas}} = K_{\text{pas}}\epsilon^{\text{cell}} \tag{15.2}$$

Therefore, by substitution of Eq. (15.2) in Eq. (15.1) and considering the stress curve of the active elements in Fig. 15.3C, at each point of the cell surface, the cell stress transmitted to the ECM can be calculated as

$$\sigma_i^{\text{cell}} = \begin{cases} K_{\text{pas}}\epsilon_i^{\text{cell}} & \epsilon_i^{\text{cell}} < \epsilon_{\min} \text{ or } \epsilon_i^{\text{cell}} > \epsilon_{\max} \\ K_{\text{pas}}\epsilon_i^{\text{cell}} + \dfrac{K_{\text{act}}\sigma_{\max}(\epsilon_{\min} - \epsilon_i^{\text{cell}})}{K_{\text{act}}\epsilon_{\min} - \sigma_{\max}} & \epsilon_{\min} \leq \epsilon_i^{\text{cell}} \leq \tilde{\epsilon} \\ K_{\text{pas}}\epsilon_i^{\text{cell}} + \dfrac{K_{\text{act}}\sigma_{\max}(\epsilon_{\max} - \epsilon_i^{\text{cell}})}{K_{\text{act}}\epsilon_{\max} - \sigma_{\max}} & \tilde{\epsilon} \leq \epsilon_i^{\text{cell}} \leq \epsilon_{\max} \end{cases} \tag{15.3}$$

where $\tilde{\epsilon} = \sigma_{\max}/K_{\text{act}}$. The characteristic spring constant can be identified with the volumetric stiffness modulus of the microtubules, $K_{\text{pas}}$, myosin II, $K_{\text{act}}$, and ECM, $K_{\text{subs}}$.

Physically, during the cell movement, the contraction of the actin-myosin apparatus drives forward the translocation of the cell body and causes traction forces on the ECM [24, 99]. Actually, the movement of cells within a complex embryo or organism is guided by a complex interplay between chemical and physical signals such as ECM stiffness [24, 31], boundary conditions, and generated forces due to cell-cell and cell-ECM interactions [24, 99]. Anyway, the simplified model described earlier is here employed to predict cell migration as a function of cell internal deformation.

The prominent aspect of the presented approach for cell modeling is that the cell can have any morphology, if there is an interest to consider a variable morphology, and can be represented by any number of finite elements. For this purpose, an algorithm has been used to track the key parameters required for cell migration at each time step, considering important processes for cell migration such as the asymmetry of the cell and traction force generation. Moreover, several aspects associated with the ECM such as stiffness, boundary conditions, and their effects on the direction of cell movement can be considered.

It is considered that a cell first exerts sensing forces to sense its ECM. These forces act at each finite element node of the membrane toward the cell centroid. The cell deformation due to these sensing forces is shown by the dashed lines in Fig. 15.4. Therefore, the cell strain at each finite element node of the cell surface (membrane) in the direction of the corresponding sensing force can be written as

$$\epsilon_i^{\text{cell}} = \frac{\text{n}_i\text{N}_i}{\text{CN}_i} \tag{15.4}$$

where $\text{N}_i$ is a point on the surface of the undeformed cell (solid line), $\text{n}_i$ is the same point on the surface of the deformed cell (dashed line), and C is the cell center. The net traction force of a cell is the result of the local traction forces exerted by the cell at its front and rear, which can be calculated as [27]

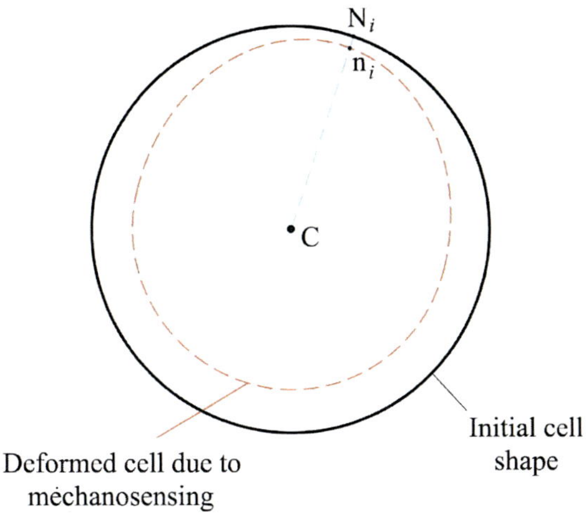


FIG. 15.4 Schematic illustration of the cell deformation resulting from the sensing force in the cell mechanosensing stage. $N_i$ is a point on the surface of the undeformed cell (*solid line*), $n_i$ is the same point on the surface of the deformed cell (*dashed line*), and C is the cell center.

$$\mathbf{F}_i^{\text{trac}} = \sigma_i^{\text{cell}} Skn_{\text{r}}\psi\mathbf{e}_i \tag{15.5}$$

where $S$ stands for a proportionality model parameter with units of area. $k$ is the binding constant for the integrins at the front and rear of the cell to the ligands in the ECM and $n_{\text{r}}$ is the total number of available receptors at the front and rear of the cell. Finally, $\psi$ represents the concentration of the ligands at the leading edge of the cell in the ECM [27]. Therefore, The resultant traction force, $\mathbf{F}_{\text{trac}}^{\text{net}}$, can be calculated by the summation of the traction forces applied at each node as

$$\mathbf{F}_{\text{trac}}^{\text{net}} = \sum_{i=1}^{n} \mathbf{F}_i^{\text{trac}} \tag{15.6}$$

where $n$ is the number of the nodes located on the cell surface.

In a multicellular system, cells deform when they contact each other to occupy all the ECM [72, 100]. Therefore, when cell migration is considered with a constant cell shape, to avoid interference of two cells it is assumed that

$$\| \mathbf{r}_j - \mathbf{r}_i \| \geq 2r \tag{15.7}$$

where $\mathbf{x}_i$ and $\mathbf{x}_j$ are the position vector of each cell centroid (Fig. 15.5).

During the mechanosensing process, when two or more cells come into contact with each other, their in-common nodal points on the cell surface are not able to send out any pseudopods to sense the ECM [100, 103, 104]. Therefore, these cells do not exert any sensing force at those nodes until they are separated again. However, in these nodes, the

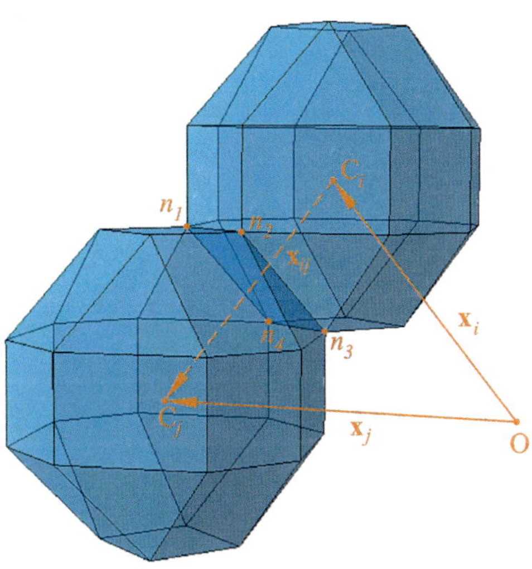


FIG. 15.5 Two in-contact cells with four shared nodes, $n_1$, $n_2$, $n_3$, and $n_4$.

corresponding deformation and consequently the nodal traction forces are not zero due to deformation compatibility. This means that all the cell nodes, including the nodes that are in contact ($n_1$, $n_2$, $n_3$, and $n_4$ in Fig. 15.5), deform together [100, 105].

#### 15.2.1.2 Protrusion Force

During cell migration, cells send out local protrusions to probe their environment by exerting a protrusion force. This force is generated by actin polymerization and must be distinguished from the cytoskeletal contractile force [27]. It arises from cell-matrix attachments at the new sites of lamellipodia and filopodia development, which have a stochastic nature during cell migration [106]. This causes cells to move along a directed random path toward the effective signals. The direction and magnitude of the protrusion force are chosen randomly at each time step. It is remarkable that the order of the protrusion force magnitude is the same as that of the traction force, but with lower amplitude [27, 100, 107–109]. Therefore, we randomly estimate it as

$$\mathbf{F}_{\text{prot}} = \kappa F_{\text{net}}^{\text{trac}} \mathbf{e}_{\text{rand}} \tag{15.8}$$

where $\mathbf{e}_{\text{rand}}$ represents a random unit vector and $F_{\text{net}}^{\text{trac}}$ is the modulus of the net traction force while $0 \leq \kappa < 1$ is a random number [100].

#### 15.2.1.3 Drag Force

By contrast, the drag force comes from the viscous resistance to the cell movement. In a Maxwell solid, the needed force for deforming the ECM depends on the deformation rate and, accordingly, the velocity. The main objective here is to imply a velocity-dependent opposing force to associate the viscoelastic character of the cell surrounding the ECM. To define a velocity-dependent opposing force associated with the linear viscoelastic character of the ECM surrounding the cell, we have assumed a simplification that the ECM is a viscoelastic medium [27]. At a microscale, the viscous resistance dominates while the inertial resistance of a viscous fluid is small enough to be ignored. Stokes [110] described the drag force of a sphere at the limit of negligible convection as

$$F_{\text{D}}^{\text{s}} = \beta v \tag{15.9}$$

where $v$ is the relative velocity and $\beta$ is often referred to as the Stokes' drag regime for a small spherical object moving slowly through a viscous fluid expressed as

$$\beta = 6\pi r \eta(E_{\text{sub}}) \tag{15.10}$$

where $r$ is the object radius and $\eta(E_{\text{sub}})$ is the effective medium viscosity. In an ECM with a stiffness gradient, we assume that it is linearly proportional to the ECM stiffness, $E_{\text{sub}}$, at each point. Therefore, it can be calculated as

$$\eta(E_{\text{sub}}) = \eta_{\text{min}} + \lambda E_{\text{sub}} \tag{15.11}$$

where $\eta_{\text{min}}$ is the minimum viscosity of the ECM corresponding to minimum stiffness and $\lambda$ is the proportionality coefficient. The viscosity coefficient may eventually be saturated with ECM rigidity; however, this saturation occurs outside the ECM rigidity range suitable for the cell phenotypes considered here [111].

The drag of nonspherical solid particles will depend on the degree of nonsphericity as well as their orientation to the flow (the drag will generally be anisotropic with respect to direction). When cell morphology changes during migration, Eq. (15.9) will not deliver an accurate representation of the drag force. Its definition in the case of irregular particles is further complicated by the randomness of the shapes and dynamics. However, a review of experimental studies of the mean drag of irregularly shaped particles suggests that it is reasonable and appropriate to use a shape factor, $f_{\text{shape}}$, to moderate the Stokes expression as [112, 113]

$$F_{\text{drag}} = f_{\text{shape}} F_{\text{D}}^{\text{s}} \tag{15.12}$$

Nevertheless, it is expected that only approximate and probabilistic predictions are possible for highly irregular particles. A wide variety of shape-characterizing parameters has been suggested for irregular particles, the most common and most successful being the Corey Shape Factor (CSF). This shape factor employs three lengths of a particle in mutually perpendicular directions, being representative of cell surface area changes [112]; the cell's longest dimension, $l_{\text{max}}$, the shortest dimension, $l_{\text{min}}$, and the intermediate or medium dimension, $l_{\text{med}}$. Herein, we take advantage of a convenient version of CSF used by Loth [112] to estimate the shape factor as

$$f_{\text{shape}} = \left(\frac{l_{\max} l_{\text{med}}}{l_{\min}^2}\right)^{0.09} \tag{15.13}$$

This is nearly identical to the expression proposed by Dressel [114], except that he used a 0.1 exponent. Other shape factors have also been suggested to characterize the shape irregularity, but the impact of the max-med-min area factor tends to be the strongest [112, 115]. It is notable that this shape factor in the case of a spherical cell shape delivers one.

### 15.2.2 Chemotaxis and Thermotaxis

In the presence of chemotaxis or thermotaxis, the cell polarization direction depends not only on the mechanotactic signal, but also on those additional stimuli. We assume that the presence of these cues changes neither the properties of a typical cell nor its surrounding ECM. Because the exerted traction forces by the cell depend on the cell type and the mechanical properties of the ECM, the presence of these cues does not affect the magnitude of the net traction force but changes its direction. The realignment of the net traction force in the presence of chemical and thermal cues changes based on the direction of chemotaxis and thermotaxis gradients. Therefore, it is assumed that a part of the net traction force is guided by mechanotaxis by a unit vector of

$$\mathbf{e}_{\text{mech}} = \frac{\mathbf{F}_{\text{net}}^{\text{trac}}}{\| \mathbf{F}_{\text{net}}^{\text{trac}} \|} \tag{15.14}$$

while the rest is guided by chemotaxis, $\mathbf{e}_{\text{ch}}$, and/or thermotaxis, $\mathbf{e}_{\text{th}}$, cues, respectively, as

$$\mathbf{e}_{\text{ch}} = \frac{\nabla[C]}{\| \nabla[C] \|} \tag{15.15}$$

$$\mathbf{e}_{\text{th}} = \frac{\nabla T}{\| \nabla T \|} \tag{15.16}$$

where $\nabla$ is the gradient operator and [C] and $T$ are the chemoattractant concentration and the temperature, respectively. Consequently, based on each cue, the effective direction on the cell and the effective force due to mechanotaxis, chemotaxis, and thermotaxis is calculated as

$$\mathbf{F}_{\text{eff}} = F_{\text{net}}^{\text{trac}}(\mu_{\text{mech}}\mathbf{e}_{\text{mech}} + \mu_{\text{ch}}\mathbf{e}_{\text{ch}} + \mu_{\text{th}}\mathbf{e}_{\text{th}}); \quad \mu_{\text{mech}} + \mu_{\text{ch}} + \mu_{\text{th}} = 1 \tag{15.17}$$

where $\mu_{\text{mech}}$, $\mu_{\text{ch}}$, and $\mu_{\text{th}}$ are the efficient factors of mechanotaxis, chemotaxis, and thermotaxis cues, respectively.

### 15.2.3 Electrotaxis

Based on the experimental observations, the electrotactic response of the cell is controlled by an intracellular influx of $Ca^{2+}$ [47, 48]. However, it is still a controversial question. For instance, $Ca^{2+}$ dependence of electrotaxis has been observed in neural crest cells, embryo mouse fibroblasts, and fish and human keratocytes [40, 47, 49, 60, 116]. By contrast, $Ca^{2+}$ independent electrotaxis has been reported in mouse fibroblasts [51]. Therefore, the precise mechanism behind the $Ca^{2+}$ role in electrotaxis is not well known.

A simple cell in rest has a negative membrane potential [47]. If this cell is exposed to a dcEF, the side of its plasma membrane that faces the cathode depolarizes while the other side that faces the anode hyperpolarizes [40, 47, 60]. In the case of a cell with negligible voltage-gated conductance, the hyperpolarized membrane side attracts $Ca^{2+}$ by passive electrochemical diffusion. This side of the cell, then, contracts and propels the cell toward the cathode. This process continues until the voltage-gated $Ca^{2+}$ channels (VGCCs) near the cathodal side open (depolarized) to allow $Ca^{2+}$ influx (see Fig. 15.6). Consequently, the intracellular $Ca^{2+}$ level rises on both the anodal and cathodal sides of the cell. The resultant of the generated force, in this case, would depend on the balance between the opposing magnetic contractile forces [47]. That is why some cells reorient toward the cathode, such as embryo fibroblasts [49], human keratinocytes [48, 117], fish epidermal cells [116], and human retinal pigment epithelial cells [118], while some others do so toward the anode, such as human granulocytes [119] and metastatic human breast cancer cells [120]. To define the effective electrical force in the presence of an electrotactic cue, let us assume a cell located in a uniform EF. As a consequence, it is ionized and has acquired a charge. The experienced force by this individual charged cell can be obtained by [121]

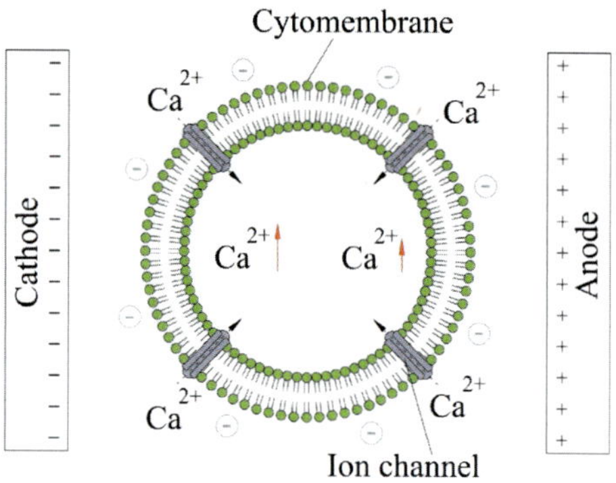


FIG. 15.6 A typical cell within a dcEFs. A simple cell in the resting state has a negative membrane potential [47]. By exposing it to a negligible voltage-gated conductance to a dcEF, due to passive electrochemical diffusion, it attracts $Ca^{2+}$ from its hyperpolarized membrane near the anode. Consequently, this side of the cell contracts, propelling the cell toward the cathode. Therefore, voltage-gated $Ca^{2+}$ channels (VGCCs) near the cathode (depolarized side) open and a $Ca^{2+}$ influx occurs. In such a cell, the intracellular $Ca^{2+}$ level rises on both sides. The direction of cell movement, then, depends on the difference of the opposing magnetic contractile forces that are generated by the cathode and anode [47].

$$\mathbf{F}^{\mathrm{EF}} = E\Omega S\mathbf{e}_{\mathrm{EF}} \tag{15.18}$$

where $E$ stands for uniform EF and $\Omega$ denotes the surface charge density of the cell. $\mathbf{e}_{\mathrm{EF}}$ is a unit vector in the direction of the EF toward the cathode or anode, depending on the cell type. The time course of the translocation response during exposing a cell to a dcEF demonstrates that the cell velocity versus translocation varies depending on the dcEF strength. Experiments of Nishimura et al. [48] on human keratinocytes indicate that the net migration velocity is maximal when the dcEF strength is about 100 mV/mm while it decreases by reducing the dcEF strength. They reported that increasing the dcEF strength to 400 mV/mm does not change the net migration velocity of the cell. As previously cited, it is thought that the $Ca^{2+}$ influx plays a role in this process [47, 48, 50, 122–124]. In other words, increasing the concentration of intracellular $Ca^{2+}$ correlates with the magnitude of the imposed dcEF. Therefore, the cell surface charge is directly proportional to the imposed dcEF strength [47, 48]. Consequently, we assume a linear relationship between the cell surface charge and the applied dcEF strength as

$$\Omega = \begin{cases} \dfrac{\Omega_{\mathrm{satur}}}{E_{\mathrm{satur}}} E & E \le E_{\mathrm{satur}} \\ \Omega_{\mathrm{satur}} & E > E_{\mathrm{satur}} \end{cases} \tag{15.19}$$

where $\Omega_{\mathrm{satur}}$ is the saturation value of the surface charge and $E_{\mathrm{satur}}$ is the maximum dcEF strength that causes $Ca^{2+}$ influx.

Besides the electrical force exerted by dcEF to each cell in the ECM, cells experience a cell-cell electrostatic force because of cell charge. Thus, the generated force between $i$th and $j$th cells, $\mathbf{F}_{ij}^{EF}$, can be expressed as

$$\mathbf{F}_{ij}^{\mathrm{EF}} = \frac{k_{\mathrm{e}}}{\epsilon_{\mathrm{r}}} \left( \frac{\Omega S}{\| \mathbf{r}_{ij} \|} \right)^2 \mathbf{e}_{ij} \tag{15.20}$$

where $\mathbf{r}_{ij}$ stands for the vector passing from the centroids of two cells, as shown in Fig. 15.5, and $k_{\mathrm{e}}$ is the coulomb's constant in a vacuum. $\epsilon_{\mathrm{r}}$ is the dielectric constant (relative permittivity) of the medium and, finally, $\mathbf{e}_{ij}$ is the direction of the generated force between the two cells, which can be in repelling or absorbing directions depending on the two cell types.

$$\mathbf{e}_{ij} = \pm \frac{\mathbf{r}_{ij}}{\| \mathbf{r}_{ij} \|} \tag{15.21}$$

Assuming $n_{\mathrm{c}}$ cells within the ECM, the resultant electrostatic force exerted by a cell population on the $i$th cell can be calculated as

$$\mathbf{F}_{ip}^{\mathrm{EF}} = \sum_{j=1}^{n_c - 1} \mathbf{F}_{ij}^{\mathrm{EF}} \tag{15.22}$$

Consequently, the total electrostatic force on a cell in the presence of a dcEF and other cells can be calculated as

$$\mathbf{F}_{\mathrm{tot}}^{\mathrm{EF}} = \mathbf{F}^{\mathrm{EF}} + \mathbf{F}_{ip}^{\mathrm{EF}} \tag{15.23}$$

### 15.2.4 Force Equilibrium

In this model, the main forces from different sources, given by Eqs. (15.8), (15.17), (15.23), have been considered to define the cell migration process. Therefore, assuming that the contribution of the cell inertia is negligible compared to the other forces due to the microscale of the problem, force equilibrium yields opposing drag force as

$$\mathbf{F}_{\mathrm{drag}} + \mathbf{F}_{\mathrm{eff}} + \mathbf{F}_{\mathrm{prot}} + \mathbf{F}_{\mathrm{tot}}^{\mathrm{EF}} = 0 \tag{15.24}$$

The instantaneous velocity of the cell may therefore be written as

$$v = \frac{\| \mathbf{F}_{\mathrm{drag}} \|}{6\pi r \eta} \tag{15.25}$$

with the net polarization direction

$$\mathbf{e}_{\mathrm{pol}} = \frac{\mathbf{F}_{\mathrm{drag}}}{\| \mathbf{F}_{\mathrm{drag}} \|} \tag{15.26}$$

Finally, the incremental translocation vector of an individual cell over a certain small time increment, $\tau$, is calculated as

$$\mathbf{d} = v\tau\mathbf{e}_{\mathrm{pol}} \tag{15.27}$$

### 15.2.5 Discretization of the Cell and ECM Domains

As previously discussed, cell migration is composed of several coordinated cyclic processes. Guided by the aforementioned experimental observations [125], it is coupled with the cell traction forces. Therefore, only the dominant modes of cell morphological changes are considered by the cell body retraction at the rear and extension at the front. Referring to Fig. 15.7, we represent a working domain by $\Lambda \subset R^3$ . Hence, considering $\mathbf{X}$ the global coordinates, the initial domain of the cell can be described by

$$\Omega' = \{\mathbf{x}'(\mathbf{X}') | \mathbf{x}'(\mathbf{X}') \in \Lambda : \forall \| \mathbf{x}' \| \leq r\} \tag{15.28}$$

where $\mathbf{X}'$ denotes the cell local coordinates located in the cell centroid. Accordingly, the points located on the cell surface (membrane) can be represented by $\partial\Omega'$. Thus, the ECM domain can be defined as

$$\Omega = \{\mathbf{x}(\mathbf{X}) | \mathbf{x}(\mathbf{X}) \in \Lambda, \ \ \mathbf{x}(\mathbf{X}) \notin \Omega'\} \tag{15.29}$$

During cell migration, both ECM, $\Omega$, and cell, $\Omega'$, domains change such that $\Omega' \cup \Omega = \Lambda$ and $\Omega' \cap \Omega = \emptyset$.

To correctly incorporate $k$, $\psi$, and $n_r$ parameters in the cell front and rear (see Eq. 15.5, it is necessary to identify the cell front and back during cell translocation. To this end, let us assume $\chi$ is a plane that passes by the cell centroid O and

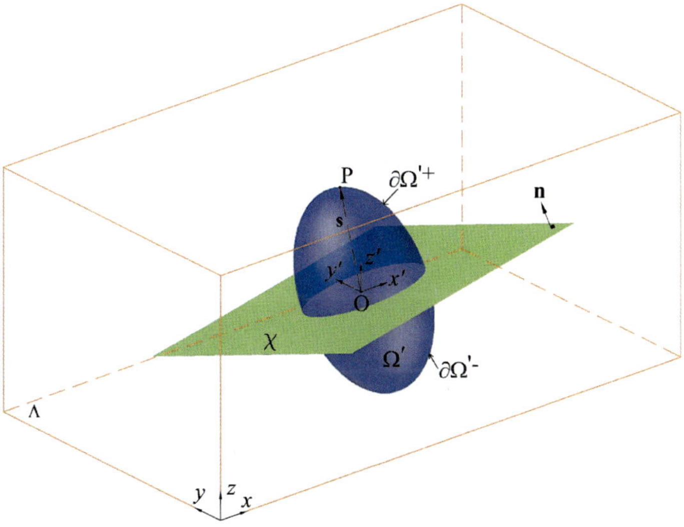


FIG. 15.7 Schematic demonstration of the cell extension and retraction in the front and rear, respectively, at each time step. $\Lambda \subset R^3$, $\Omega$, and $\Omega'$ represent the 3D working space, ECM, and the cell domains, respectively. $\mathbf{X}$ stands for the global coordinates and $\mathbf{X}'$ represents the local cell coordinates located in the cell centroid, O. $\chi$ is a plane passing by the cell centroid with unit normal vector $\mathbf{n}$ in the direction the cell polarization direction, $\mathbf{e}_{\mathrm{pol}}$. P denotes a finite element node located on the cell membrane, $\partial\Omega$. $\partial\Omega'^{+}$ and $\partial\Omega'^{-}$ are the finite element nodes located on the front and rear of the cell membrane, respectively.

has a normal unit vector $\mathbf{n}$ in the direction of $\mathbf{e}_{\mathrm{pol}}$, and $\mathbf{s}(\mathbf{X}')$ is a position vector of an arbitrary node located on $\partial\Omega'$ (see Fig. 15.7). Hence, the projection of $\mathbf{s}$ in the direction of $\mathbf{n}$ can be defined by

$$\delta = \mathbf{n} \cdot \mathbf{s} \tag{15.30}$$

Thereby, nodes with negative $\delta$ are located on the rear cell surface $\partial\Omega'^{-}$ while nodes with positive $\delta$ belong to the front cell $\partial\Omega'^{+}$, satisfying $\partial\Omega' = \partial\Omega'^{+} \cup \partial\Omega'^{-}$.

## 15.2.6 Cell Migration, Considering Constant Cell Shape

When we are interested in some cell behavior such as cell migration, differentiation, proliferation, and/or apoptosis, it is more efficient to consider a constant cell shape during cell activities. Therefore, at each time step, the cell displacement is calculated by Eq. (15.27) and the new cell center coordinates are calculated. When the cell displacement is high enough (i.e., if it is greater than the length of one element of the working domain), the cell centroid is moved to the center of the next neighbor element (1 of the 24 existing neighboring elements). Otherwise, the cell will keep its position until the accumulation of the calculated displacement is equal to or more than one element size.

## 15.2.7 Cell Migration, Considering Cell Shape Change and Remodeling

During cell migration, the cell extends a protrusion from the membrane vertex whose position vector is approximately in the direction of cell polarization. On the contrary, it retracts the trailing end from the surface vertex whose position vector is totally in the opposite direction of cell polarization. Thus, the maximum value of $\delta$ (Eq. 15.30) delivers the surface node located on $\partial\Omega'^{+}$ from which the cell must be extended while the minimum value of $\delta$ represents the surface node located on $\partial\Omega'^{-}$ from which the cell must be retracted. Assume $e^{\mathrm{ex}} \in \Omega$ is the finite element that the surface node with the maximum δ belongs to it, and $e^{\mathrm{re}} \in \Omega'$ is the finite element that the surface node with the minimum value of $\delta$ belongs to it. To update the cell shape changes during migration, simply, $e^{\mathrm{re}}$ is moved from the $\Omega'$ domain to the $\Omega$ domain; in contrast, $e^{\mathrm{ex}}$ is eliminated from the $\Omega$ domain and is included in the $\Omega'$ domain [126, 127].

Note that, based on [128], the cell is prevented from obtaining an infinitely thin shape during migration, considering that the cell can extend approximately 10% of its whole volume as pseudopodia.

It is interesting to evaluate the cell elongation, $\epsilon_{\mathrm{elong}}$, and the cell morphological index (CMI) during its migration. Therefore, they can be estimated by

$$\epsilon_{\mathrm{elong}} = 1 - \frac{\sqrt{l_{\min} l_{\mathrm{med}}}}{l_{\max}} \tag{15.31}$$

$$\mathrm{CMI}(t) = \frac{S(t)}{S_{\mathrm{in}}} \tag{15.32}$$

where $S_{\mathrm{in}}$ denotes the initial area of the cell membrane (spherical cell shape).

## 15.2.8 MSC Differentiation and Apoptosis

Many experimental studies reported that sensing of the mechanical properties of the ECM is essential for different cell types to undergo differentiation, proliferation, and apoptosis [85, 129–131]. For instance, a determined deformation experienced by an MSC stimulates its differentiation into a specific cell type, which is due to the differences in their tissue of origin [91, 132, 133]. Besides, the magnitude and duration of the mechanical signal sensed by the cell and the degree of preconditioning are also considered important factors. Experimental works demonstrate that a typical cell may undergo apoptosis when it senses a high deformation [129, 131]. For instance, Kearney et al. [131] demonstrated that specific tensile strain applied on MSCs can trigger cell apoptosis, mediated by the cell internal strain. In vitro experiments showed that mechanical injury beyond 20% of the tensile cyclic strain causes significant neuronal cell death [129] and a 10% tensile cyclic strain imposed continuously in a 2D MSC culture leads to a considerable apoptosis.

Guided by well-known experimental works [85, 129–131], it is here assumed that an MSC, $m$, is able to differentiate into a certain cell type $i$, $i \in \{s, c, l\}$, where $s$, $c$, and $l$ represent osteoblast, chondrocyte, and neuroblast lineage specifications [134, 135]. Mechanoregulation of cell differentiation is presented by cell internal deformation (the mechanical signal) in the cell polarization direction. Therefore, the deformation of each node located on the cell surface in its polarization direction can be obtained by

$$\gamma_i = \mathbf{e}_{\text{pol}} : \boldsymbol{\epsilon}_i : \mathbf{e}_{\text{pol}}^T \tag{15.33}$$

where $\boldsymbol{\epsilon}_i$ stands for the strain tensor of the $i$th node on the cell surface in the mechanosensing phase. Consequently, the cell internal deformation can be calculated as

$$\gamma(\mathbf{x},t) = \sum_{i=1}^{n} \gamma_i \tag{15.34}$$

where $\mathbf{x}$ is the position vector of the cell centroid at the time $t$. $\gamma_l \leq \gamma(\mathbf{x}, t) \leq \gamma_u$ varies spatially and temporally during cell migration [76, 91, 136] where $\gamma_l$ and $\gamma_u$ are the lower and upper bounds of cell internal deformations leading to cell differentiation, respectively.

Experimental observations [137–139] show that cell differentiation is both a mechanobiological and a time-dependent process. This means that cells must be present at a certain level of maturity to be active in the differentiation or proliferation stages [76, 137–139]. This maturation period can, in turn, be regulated by the mechanical signals received by a cell and depends on the cell type and its ECM. The stronger mechanical signals (less internal deformation) can decrease the cell maturation time. However, cells require a minimum time period to enter into the differentiation or proliferation phase after their culture [138]. Consequently, we here assume a linear relationship between the mechanical signal sensed by cultured cells and their maturation time as [140]

$$\tau_{\text{mat}}(\gamma,t) = \tau_{\text{min}} + \tau_{\text{p}}\gamma(\mathbf{x},t) \tag{15.35}$$

where $\tau_{\text{min}}$ is the minimum time that a typical cell needs to go into the differentiate or proliferate phase and $\tau_{\text{p}}$ is a time proportionality. Therefore, in addition to cell type, each cell must be also represented by a maturation index (MI) described by

$$\text{MI} = \begin{cases} \dfrac{t}{\tau_{\text{mat}}} & t \leq \tau_{\text{mat}} \\ 1 & t > \tau_{\text{mat}} \end{cases} \tag{15.36}$$

MI $< 1$ implies that a typical MSC is unable to differentiation or proliferation even if the mechanical cue is proper. MI $= 1$ implies that a cell $j \in \{m, s, c, l\}$ is completely mature and active for differentiation or proliferation if the mechanical conditions are appropriate. It is here assumed that the evolution of cell MI is an irreversible process, meaning that during cell migration it cannot be reduced except when the cell phenotype changes due to differentiation or one mother cell proliferate into two daughter cells. Considering these conditions, the process of MSC differentiation and apoptosis related to mechanical signals and maturation can be represented by [76, 140]

$$\text{Cell phenotype} = \begin{cases} s & \gamma_l < \gamma \leq \gamma_s \ \&\ \text{MI} = 1 \\ c & \gamma_s < \gamma \leq \gamma_c \ \&\ \text{MI} = 1 \\ l & \gamma_c < \gamma \leq \gamma_u \ \&\ \text{MI} = 1 \\ \text{Apoptosis} & \gamma > \gamma_{\text{apop}} \\ \text{No differentiation} & \text{Otherwise} \end{cases} \tag{15.37}$$

It is noteworthy that small cyclic strains that may cause fatigue apoptosis of a typical cell are not considered here.

### 15.2.9 Cell Proliferation

Cell proliferation is a process that results in an increase in cell population. During this process, two daughter cells are formed from a single mother cell. It follows four determined stages, including the first growth phase, the synthesis phase, the second growth phase, and the mitosis phase [11, 141]. During the first growth phase (G1 phase), a huge content of biological materials is synthesized by the cell. When it has completed the cell synthesis phase (S phase), the cell starts to replicate its DNA. When the S phase is completed, the cell undergoes the second growth phase (G2 phase), which leads to the mitosis phase (M phase). Subsequently, the cell chromosomes are reorganized and a mother cell division produces two daughter cells. This instant is critical because some cells may temporarily enter into the quiescence stage (G0 phase) and stop proliferation [11, 141].

The objective here is to model the cell proliferation process consistent with the aforementioned biological stages, assuming that there are enough oxygen or nutrient resources for the cultured cells. Hence, here, the dominant stages of the cell division cycle are reduced into two main steps, meaning that during the G1, S, and G2 phases, the cell undergoes maturation and growth. Once cell maturation is achieved, based on the strength of the mechanical signal sensed by the cell, it enters into the M phase and forms two nonmature daughter cells. Consequently, in the present model, each cell is in the quiescence phase unless it delivers two daughter cells. In other words, a cell is either under maturation or in the proliferation stage. This process can be represented by

$$\text{Cell proliferation} = \begin{cases} \text{1 Mother cell} \rightarrow \text{2 Daughter cells} & \gamma \leq \gamma_i^{\text{prof}} \\ & \text{and MI} = 1 \\ \text{No cell division} & \text{Otherwise} \end{cases} \tag{15.38}$$

where $i \in \{m, s, c, l\}$ and $\gamma_i^{\text{prof}} < \gamma_u$ is the mechanical signal defining the proliferation limit of the $i$th cell [76]. After division of a mother cell, the position of two daughter cells is assumed as

$$\begin{aligned} \mathbf{x}_{\text{daut}}^{(1)} &= \mathbf{x}_{\text{moth}} \\ \mathbf{x}_{\text{daut}}^{(2)} &= \mathbf{x}_{\text{moth}} + 2r\mathbf{e}_{\text{rand}} \end{aligned} \tag{15.39}$$

where "moth" and "daut" subscripts represent mother and daughter cells, respectively.

## 15.3 NUMERICAL IMPLEMENTATION AND APPLICATIONS

The described model is implemented in the commercial software ABAQUS coupled with a user-defined element subroutine (UEL) [142]. The parameters of the model are listed in Table 15.1. An ECM is mesh using 16,000 3D hexahedral elements and with 18,081 nodes. The initial cell radius is assumed to be 15 μm with a total number of 24 nodes on its surface. In the following sections, we will illustrate some interesting applications of the model.

### 15.3.1 Effect of ECM Depth on Cell Mechanosensing and Migration

In this application, we study the cell migration due to the effect of the ECM depth when a sloped surface is constrained while the rest of the boundaries are free, as shown in Fig. 15.8. The elastic modulus of the ECM is assumed to be 100 kPa and the cell is initially located at the maximum depth of the ECM. The results show that, as shown in Fig. 15.8A, the cell tends to migrate on the ECM surface in a random directional trajectory toward locations with lower depth because, during the cell mechanosensing process, the cell senses less internal deformation in the lower depth direction, which is more rigid due to a constrained sloped surface [24, 100]. Fig. 15.8B shows the deformations generated in the ECM due to the sensing forces.

### 15.3.2 Cell Behavior Within a Multisignaling ECM

To consider the effect of different signals on cell behavior, we assume that there is a stiffness gradient in the $x$-direction within the ECM (from 10 kPa at $x = 0$ to 100 kPa at $x = 800$ in Fig. 15.9). Afterward, we add other cues such as thermal gradient ($T = 35°\text{C}$ at $x = 0$ and $T = 39°\text{C}$ at $x = 800\,\mu\text{m}$), chemical gradient ($C = 10^{-4}$ M at $x = 800$), and dcEF ($E = 100\,\text{mV/mm}$ with cathode located at $x = 800\,\mu\text{m}$) in the ECM to evaluate the effect of different stimuli on the cell behavior compared with pure mechanotaxis. Next, two cases of multicell migration and cellular morphology changes will be studied.

#### *15.3.2.1 Multicell Migration Within a Multisignaling ECM*

Experimental observations show that cell populations migrate directionally toward stiffer regions of the ECM in the presence of a stiffness gradient (mechanotaxis) [24, 31], toward warmer sites in the presence of a thermal gradient (thermotaxis) [146, 147], toward higher concentrations of nutrients when there is a chemotactic attraction [148], and/or toward the cathode or anode (depending on cell type) in the presence of dcEF (electrotaxis) [47, 48, 50, 122–124].

To apply the present model, initially the cells are distributed in the first quarter of the ECM. The results show that, first, the cells tend to migrate toward each other, even when they are located in the stiffer regions, stimulated by the stretched regions between cells. However, the final destination of the cells is dominated by the migration along the existent gradients or toward the cathode, regardless of the primary distribution of the cells (see Fig. 15.9). In the case of pure stiffness gradient (Fig. 15.9A), despite the maximum stiffness being located at the end of the ECM, the cells do

TABLE 15.1 ECM and Cell Parameters

| Symbol | Description | Value | References |
|---|---|---|---|
| $\eta_{\text{min}}$ | ECM viscosity | 1000 Pa s | [27, 143] |
| $\lambda$ | Proportionality coefficient | 0.4 μm min | [111] |
| $K_{\text{pas}}$ | Stiffness of microtubules | 2.8 kPa | [144] |
| $K_{\text{act}}$ | Stiffness of myosin II | 2 kPa | [144] |
| $\epsilon\text{max}$ | Maximum strain of the cell | 0.9 | [100, 109] |
| $\epsilon\text{min}$ | Minimum strain of the cell | −0.9 | [100, 109] |
| $\sigma\text{max}$ | Maximum contractile stress exerted by actin-myosin machinery | 0.1 kPa | [4, 97] |
| $k$ | Binding constant at rear and front of the cell | $10^8\ \text{mol}^{-1}$ | [27] |
| $n_r$ | Number of available receptors at the back and front of the cell | $10^5$ | [27] |
| $\psi$ | Concentration of the ligands at rear and front of the cell | $10^{-5}$ mol | [27] |
| $\Omega_{\text{satur}}$ | Saturation value of surface charge density | $10^{-4}\ \text{C/m}^2$ | [46] |
| $E_{\text{satur}}$ | Maximum dcEF causing $Ca^{2+}$ influx | 100 mV/mm | [47, 48] |
| $k_e$ | Coulomb's constant | $9 \times 10^9\ \text{m}^2\ \text{C}^{-2}$ | – |
| $\epsilon_r$ | Dielectric constant | $10^7$ | [145] |
| $\tau_{\text{min}}$ | Minimum time needed for cell proliferation | 4 days | [76, 138] |
| $\tau_p$ | Time proportionality | 200 days | [76, 138] |
| $\gamma_l$ | Lower bound of cell internal deformation leading to osteoblast differentiation | 0.005 | [76] |
| $\gamma_s$ | Upper bound of cell internal deformation leading to osteoblast differentiation | 0.04 | [76] |
| $\gamma_c$ | Upper bound of cell internal deformation leading to chondrocyte differentiation | 0.1 | [76] |
| $\gamma_u$ | Upper bound of cell internal deformation leading to neuroblast differentiation | 0.5 | Estimated |
| $\gamma_{\text{apop}}$ | Cell internal deformation leading to cell apoptosis | 1 | [76] |
| $\gamma_i^{\text{prof}}$ | Limit of cell proliferation | 0.2 | [76] |

not achieve this limit due to the free boundary surface at $x = 800$, where they "feel" higher deformation. In this case, the cell aggregation will be located at $x = 598 \pm 10$ μm. Depending on the stimulus added to the ECM, combined with the stiffness gradient, the center of cell aggregation can displace further toward the limit surface. For example, in the presence of a thermal gradient, cells migrate toward warmer sites and aggregate around $x = 641 \pm 10$μm (Fig. 15.9B) while in the presence of a chemical gradient, cells displace more toward the higher concentration of the nutrients and accumulate around $x = 688 \pm 5$μm (Fig. 15.9C). However, the strongest signal is the electrotaxis, which translates the cell aggregation center to $x = 736 \pm 6$μm (Fig. 15.9D). Aggregation of the cell population due to mechanotaxis, thermotaxis, chemotaxis, and electrotaxis is consistent with the experimental observations [31, 39, 43, 149]. Furthermore, many experiments have demonstrated the dominant effect of dcEF on multicell and individual cell migration [41, 42, 47, 124].

Moreover, the obtained results show that cell-cell interaction may delay cell migration [121]. Fig. 15.10 shows the cell net traction force during cell migration versus average cell translocation in the $x$-direction. In all the cases, in the first interval the average cell net traction force decreases. This is because, when cells are accumulated, only the outer cells can move while the cells trapped inside the aggregation cannot send out protrusions and remain immobile. Once the cells come into contact with each other, they cannot exert more sensing forces in the common nodes so that their net traction force decreases. Therefore, the average cell net traction force reduces until the cells move in the direction of the existing gradient or toward the cathode. Each small cell aggregation behaves like a single cell. Small cell slugs start to migrate along the gradient direction or toward the cathode, causing a reduction of their interspaces that in turn leads to an increase of cell net traction force. This enables the slugs to sense each other and migrate toward each other to form a bigger cell aggregation. Consequently, in the last interval of multicell migration, the average cell net traction force again decreases (see Fig. 15.10).

FIG. 15.8 Migration of a cell located at the maximum depth of the ECM with constrained sloped surface and constant stiffness of 100 kPa. (A) Trajectory of the migration of the cell due to the change of the ECM depth. (B) Deformation of the ECM in the $x$-direction due to the mechanosensing process.

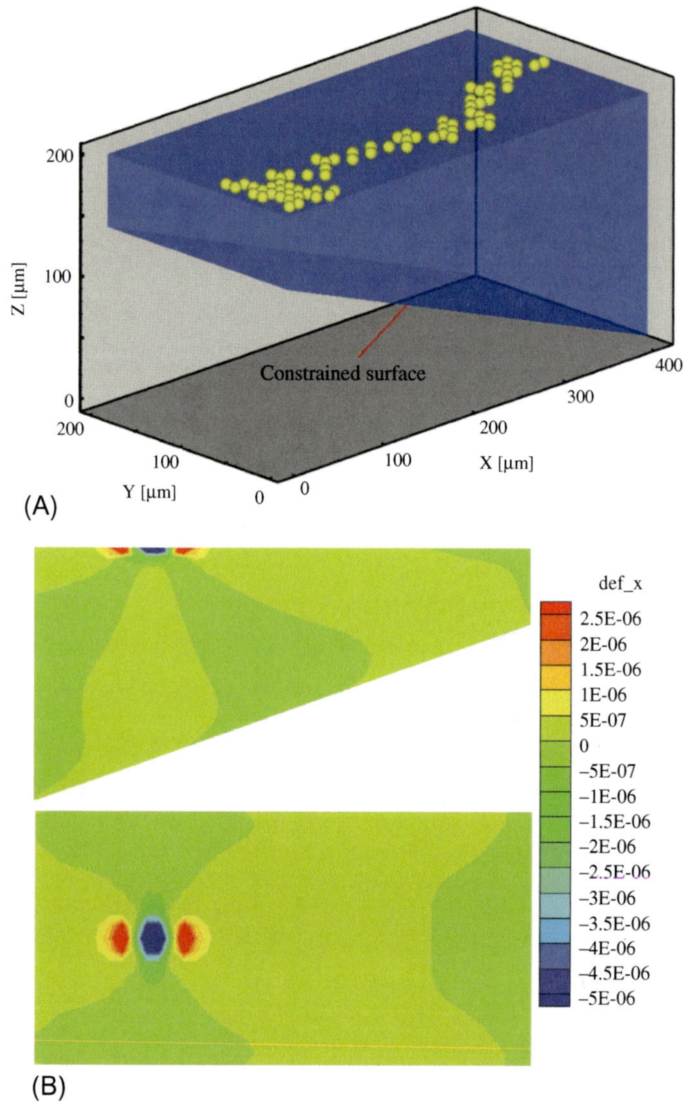


#### 15.3.2.2 *Single Cell Morphology Within a Multisignaling ECM*

Experimental observations show that, within 3D ECMs, cells actively migrate and elongate their body in the direction of higher stiffness [31, 61, 150], thermal [39, 45], and/or chemical [35, 151] gradients and/or toward the cathode [49] or anode [119, 152] (depending on the cell type) in the presence of dcEF.

In the beginning, for all cases considered here, it is assumed that the cell has a spherical morphology and starts to move from one corner of the ECM with minimum stiffness. The cell configuration in the intermediate and last steps is presented in Fig. 15.11 for pure mechanotaxis and combined with thermotaxis, chemotaxis, or electrotaxis. The results show that, independent of the initial position of the cell, it tends to migrate in the direction of the present gradients and/or the cathode/anode, during which it becomes gradually elongated in the direction of the corresponding cue. The cell maximum elongation occurs in the intermediate region of the ECM in all the cases because it is far from the unconstrained boundary surfaces [105]. When the cell reaches the free boundary at $x = 400\mu m$, the cell elongation diminishes (see Fig. 15.11). In the case of pure mechanotaxis, despite the maximum elastic modulus at $x = 400\mu m$, the cell does not reach this plane, but it moves around an imaginary equilibrium plane (IEP) located at $x = 351 \pm 5$ μm (Fig. 15.11A). However, the cell may extend random protrusions in different directions as well as to the end of the ECM. The combination of mechanotaxis with other cues can elongate the cell more and displace it further toward the corresponding cue. For example, along with the thermal gradient, the IEP displaces toward the warmer region to locate at $x = 359 \pm 3$ μm (Fig. 15.11B). In the presence of the chemical gradient and dcEF, the IEP displaces toward the

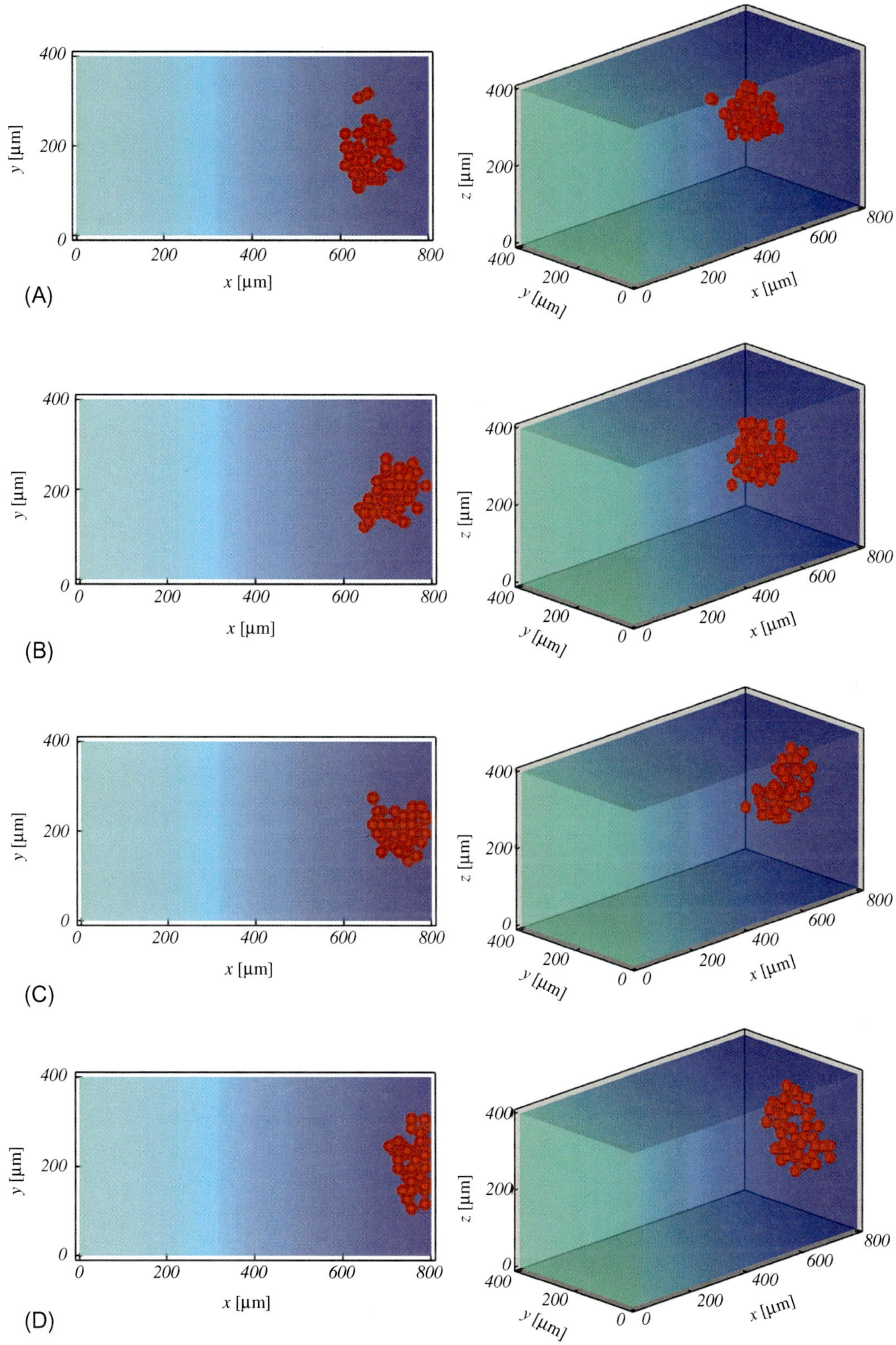


FIG. 15.9 Multicell migration in the presence of (A) pure stiffness gradient and along with (B) thermal gradient, (C) chemical gradient, or (D) dcEF.

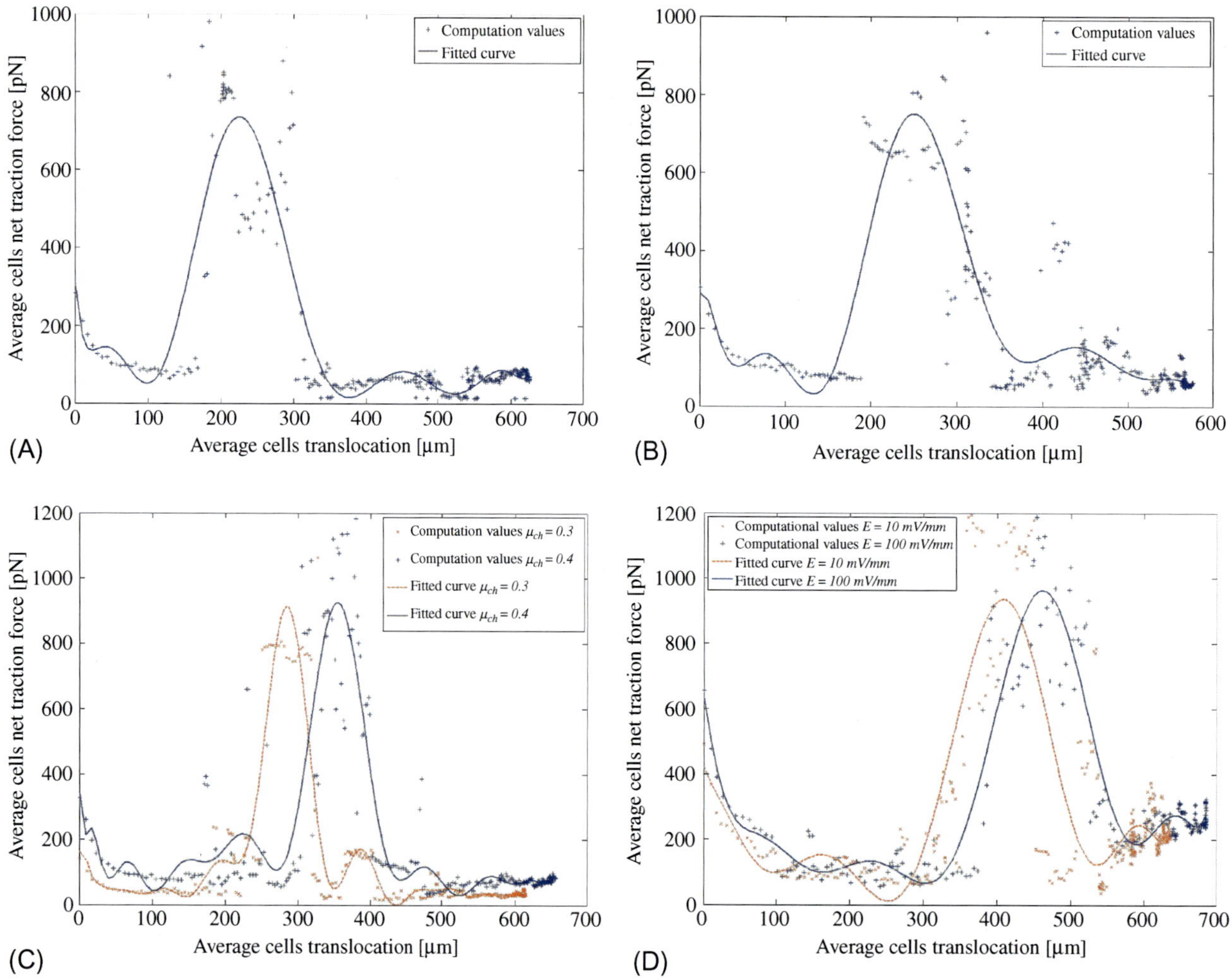


FIG. 15.10 Average cell net traction force versus average cell translocation during multicell migration within an ECM with (A) pure stiffness gradient and along with (B) thermal gradient, (C) chemical gradient, or (D) dcEF.

chemoattractant source or the cathode to locate at $x = 374 \pm 4$ μm (Fig. 15.11C) and $x = 383 \pm 2$ μm (Fig. 15.11D), respectively. The deviation of the obtained IEP location is due to the stochastic nature of cell migration between different runnings of simulations (random protrusion force).

Adding a new cue in the ECM displaces the IEP further toward the end of ECM and increases the cell elongation, $\epsilon_{elong}$, and CMI, as shown in Fig. 15.12. However, the most effective cue is electrotaxis.

### 15.3.3 Cell Differentiation and Proliferation Due to Mechanotaxis

Here, the aim is to study the lineage specification of MSCs in soft (with a stiffness of 0.1–1 kPa [28, 98, 153]), intermediate (with a stiffness of 20–25 kPa [98]), and hard (with a stiffness of 30–45 kPa [98]) ECM. Each classification resembles the stiffness of neuroblast, chondrocyte, and osteoblast tissues, respectively. In order to study the influence of ECM stiffness on each cell type lineage specification, for each classification, two different simulations with stiffnesses of the lower and upper bound are simulated. At the beginning, an MSC is located in the corner of the ECM with a potential to differentiate when the received signal is proper and maturation is achieved.

The differentiation of MSCs to neuroblast, chondrocyte, and osteogenic lineage specification in ECM with different stiffnesses is presented in Fig. 15.13. For each classification, during migration the cell gradually matures and once the MI = 1 (full maturation), it differentiates into neuroblasts, chondrocytes, and osteoblasts consistent with the ECM stiffness. The MSC is mature and differentiates into the osteoblast within ECMs with stiffnesses of 30 and 45 kPa after ~31 and ~24 h, respectively, as shown in Fig. 15.13A. Resident MSCs within ECMs with stiffnesses of 20 and 25 kPa are completely mature and start the differentiation process after ~44 and ~38 h, respectively (Fig. 15.13B). MSCs located

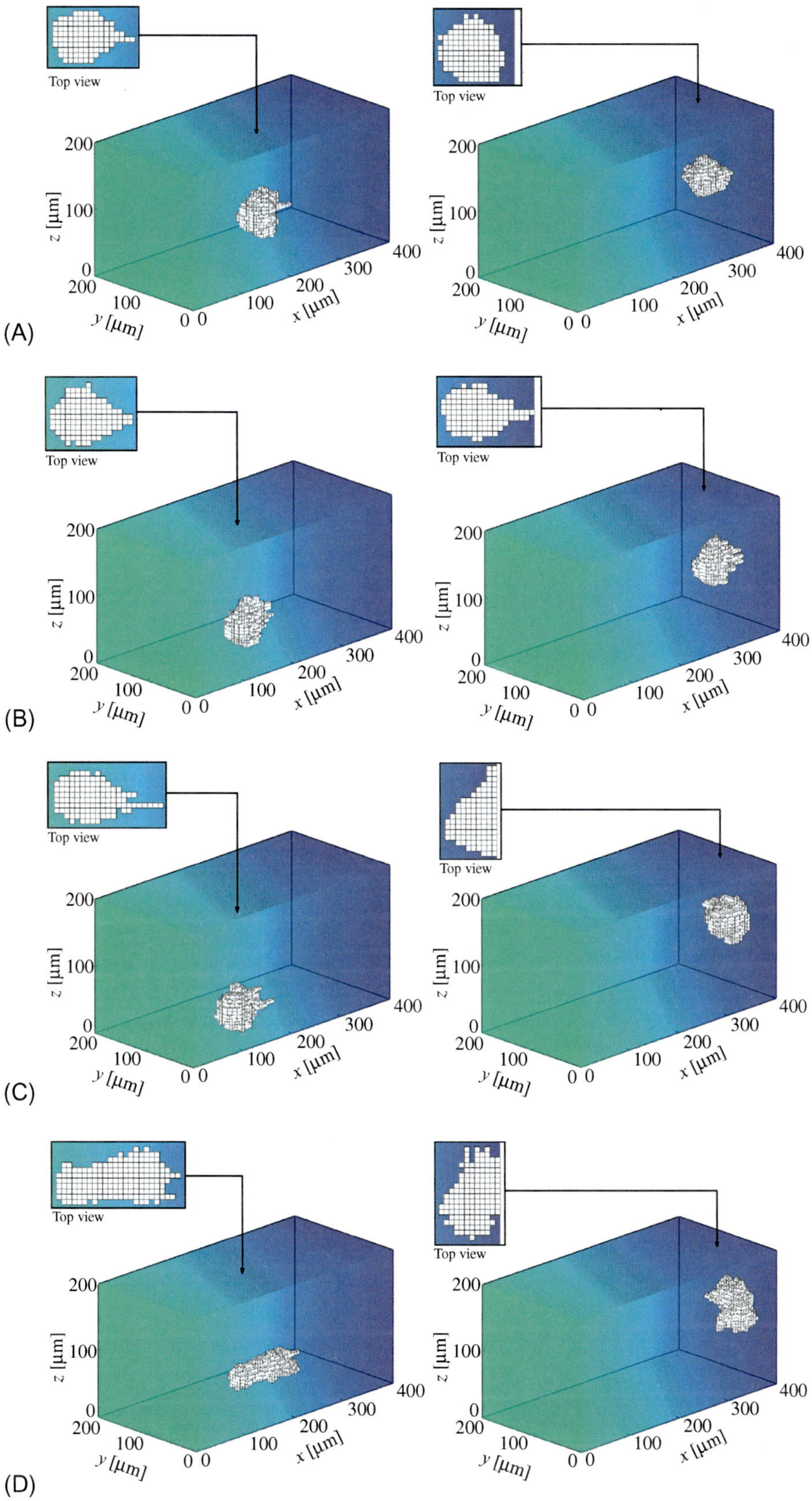


FIG. 15.11 Cell morphology in the presence of (A) pure stiffness gradient and along with (B) thermal gradient, (C) chemical gradient, or (D) dcEF.

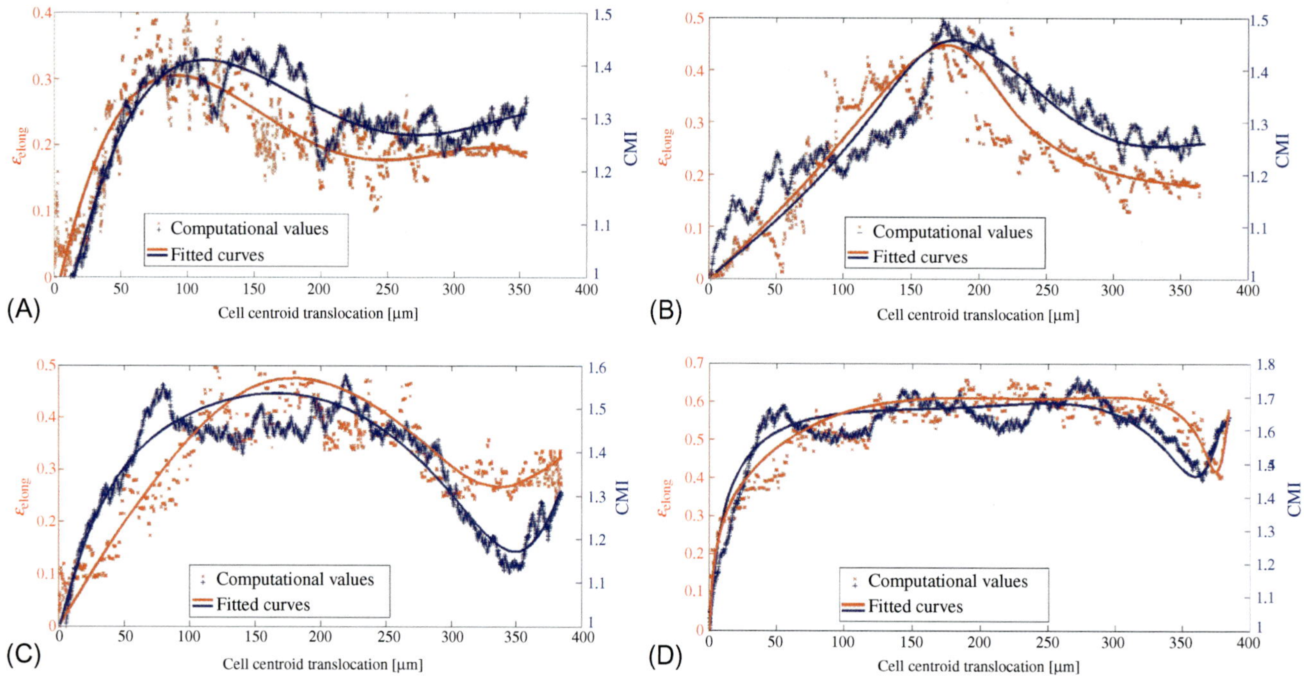


FIG. 15.12 Cell elongation, $\epsilon_{elong}$, and CMI versus the cell centroid translocation within an ECM with (A) pure stiffness gradient and along with (B) thermal gradient, (C) chemical gradient, or (D) dcEF.

within ECMs of 0.1 and 1 kPa stiffnesses are maturing and initiate to differentiate into neuroblasts after ~116 and ~75 h, respectively (Fig. 15.13C). Neuroblast, chondrocyte, and osteoblast lineage specifications of MSCs within ECMs with a stiffness equivalent to their natural tissue are supported by many experimental observations [28, 30, 154, 155].

For each classification studied here, an increase in the ECM stiffness advances MSC differentiation into the corresponding cell type, which is consistent with the findings of Fu et al. [154] and Evans et al. [156]. MSC differentiation into osteoblasts and chondrocytes instantly causes an increase in the magnitude of the net traction force (points A in Fig. 15.14a and b) while it leads to a sudden drop in the magnitude of the net traction force for neuroblast lineage specification (points A in Fig. 15.14c). Fu et al. [154] indicated that this is due to the strong correlation between the traction force and the ultimate lineage specification of MSCs. Zemel et al. [25] attributed this to the mechanical coupling between the ECM and internal CSK organization, indicating a perfect alignment of stress fibers in the direction of the cell polarization when the cell and ECM stiffness are similar due to the differentiation of MSCs into each cell type.

After MSC differentiation into neuroblasts, chondrocytes, or osteoblasts within ECMs with corresponding stiffnesses, new cell phenotypes can proliferate, depending on the strength of the mechanical signal received by the cell and its maturation state. Therefore, the first proliferation of the osteoblast within ECMs with stiffnesses of 30 and 45 kPa occurs after ~60 and ~45 h, respectively. These times are ~86 and ~74 h for chondrocytes within ECMs with stiffnesses of 20 and 25 kPa, respectively, while for neuroblasts the figures are ~230 and ~148 h within ECMs with stiffnesses of 0.1 and 1 kPa, respectively. Therefore, each new mature cell within the ECM can proliferate into many cells. Moreover, during cell proliferation, the average magnitude of the net traction force considerably increases for all cases (points B in Fig. 15.14) because of the cell-cell interaction, which causes an asymmetric nodal traction force distribution [100, 108]. The normalized density of each cell phenotype versus ECM stiffness is shown in Fig. 15.15 for identical times, indicating that the cell proliferation can be accelerated by an increase in the ECM stiffness of each classification, in agreement with the findings of Fu et al. [154].

## 15.3.4 Conclusions

In vitro, most cells are able to respond to mechanotaxis [24], thermotaxis [38], chemotaxis [44], and electrotaxis [42] by directional migration, shape change, proliferation, and/or differentiation. Here, a computational model is developed to predict the behavior of an individual cell as well as a cell population within a multisignaling environment. The numerical results presented here are qualitatively consistent with previous experimental observations from well-known literature [28, 30, 31, 39, 41–43, 47, 61, 124, 149, 150, 154, 155].

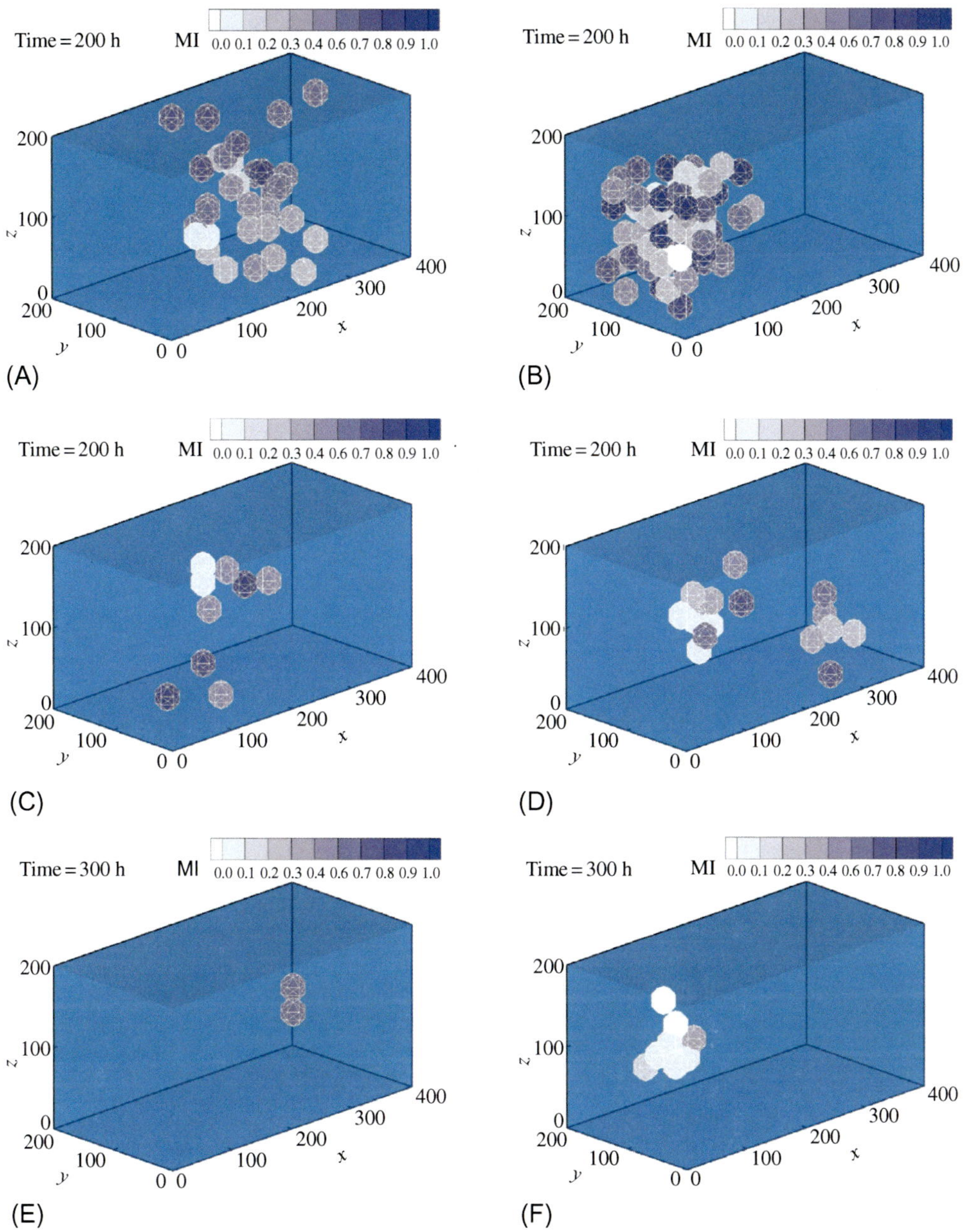


FIG. 15.13 MSC differentiation to osteoblast (A and B), chondrocyte (C and D), and neuroblast (E and F) as well as corresponding cell phenotype proliferation within hard, intermediate, and soft ECMs, respectively. ECM with stiffness of (A) 30 kPa, (B) 45 kPa, (C) 20 kPa, (D) 25 kPa, (E) 0.1 kPa, and (F) 1 kPa.

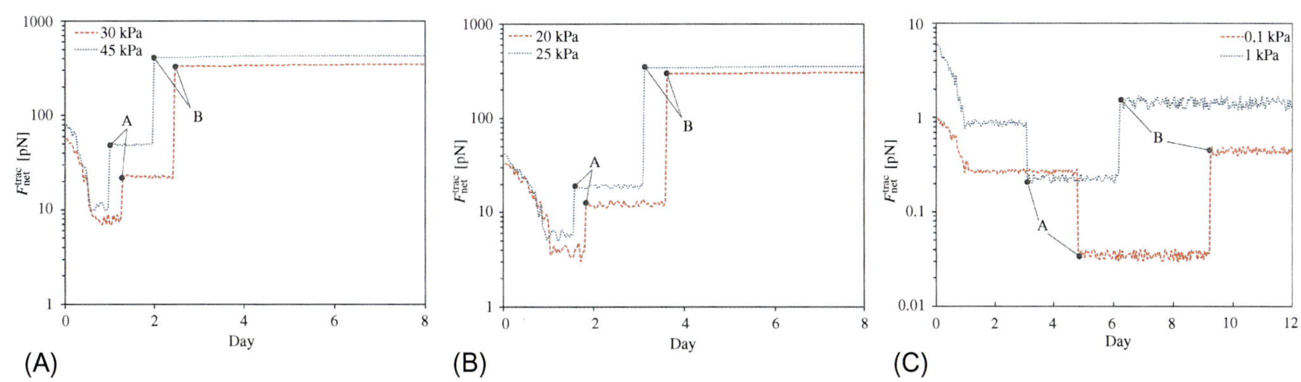


FIG. 15.14 Average cell traction force, $F_{net}^{trac}$, versus time within (a) hard (with a stiffness of 30–45), (b) intermediate (with a stiffness of 20–25 kPa), and (c) soft (with a stiffness of 0.1–1 kPa) ECMs during MSC differentiation and corresponding cell phenotype proliferation. Points A represents the instant of MSC differentiation to osteoblast (a), chondrocyte (b), and neuroblast (c) while points B are the initial instant of cell proliferation.

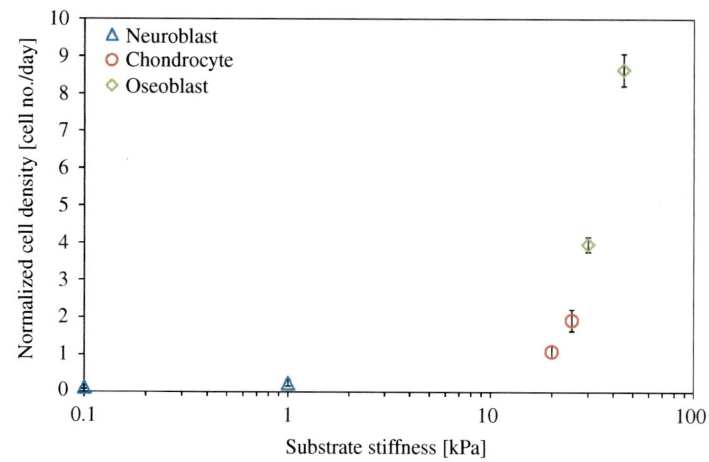


FIG. 15.15 Normalized density of each cell phenotype versus ECM stiffness during identical times as a consequence of MSC differentiation and proliferation of each cell phenotype. The *error bars* represent the mean standard deviation of different runs.

The model presented here can be considered a robust tool to study single or multicell behavior in a multisignaling ECM and it is sufficiently flexible to consider cell morphological changes. Besides, as the model allows the inclusion of different cell properties, it enables the investigation of the behavior of a wide range of cell types. The results obtained here demonstrate that alterations of the stimuli in the cell's environment can modify spontaneous cell behavior such as migration, morphology, and lineage specification, providing an insight into how a cell would potentially react to external stimuli. The model allows us to simultaneously consider cell-cell interactions in the multisignaling ECMs associated with different complex biological processes. For instance, this model can be a useful tool to develop a mechanism to enhance cell migration in the case of wound healing or to decrease cell invasion in the case of cancer growth. In addition, using this formulation and employing appropriate stimulus/stimuli within the ECM in the particular location/locations, physicians can disperse certain types of cells (for instance, tumor cells) or let these cells converge. These techniques can assist the physician to remove the tumor in such a way that the tumor might not seed out, or to prevent endothelial cells (such as blood capillaries) from migrating toward the tumor by chemo/mechanotaxis and/or changing the electrical environment locally. Although more sophisticated experiments are necessary to illustrate the accuracy of the results obtained here and to calibrate the model parameters, we believe that the computational model described here can be employed to design more efficient experiments for the prediction of the various aspects of single and multiple cell behavior within a multisignaling ECM.

## Acknowledgments

The authors gratefully acknowledge the financial support from the Spanish Ministry of Economy and Competitiveness (MINECO MAT2016-76039-C4-4-R, AEI/FEDER, UE), the Government of Aragon (DGA-T24-17R), and the Biomedical Research Networking Center in Bioengineering, Biomaterials, and Nanomedicine (CIBER-BBN). CIBER-BBN is financed by the Instituto de Salud Carlos III with assistance from the European Regional Development Fund.

## References

[1] A. Aman, T. Piotrowski, Cell migration during morphogenesis, Dev. Biol. 341 (1) (2010) 20–33.
[2] C. Cintron, H. Covington, C.L. Kublin, Morphogenesis of rabbit corneal stroma, Invest. Ophthalmol. Vis. Sci. 24 (5) (1983) 543–556.
[3] K.M. Hakkinen, J.S. Harunaga, A.D. Doyle, K.M. Yamada, Direct comparisons of the morphology, migration, cell adhesions, and actin cytoskeleton of fibroblasts in four different three-dimensional extracellular matrices, Tissue Eng. A 17 (5–6) (2011) 713–724.
[4] G.F. Oster, J.D. Murray, A.K. Harris, Mechanical aspects of mesenchymal morphogenesis, J. Embryol. Exp. Morphol. 78 (1983) 83–125.
[5] R.L. Trelstad, A.J. Coulombre, Morphogenesis of the collagenous stroma in the chick cornea, J. Cell. Sci. 50 (3) (1971) 840–858.
[6] N. Savill, P. Hogeweg, Modeling morphogenesis: from single cells to crawling slugs, J. Theor. Biol. 184 (1997) 229–235.
[7] T. Lecuit, P.F. Lenne, Cell surface mechanics and the control of cell shape, tissue patterns and morphogenesis, Nat. Rev. Mol. Cell Biol. 8 (2007) 633–644.
[8] M. Rauzi, P.F. Lenne, Cortical forces in cell shape changes and tissue morphogenesis, Curr. Top. Dev. Biol. 95 (2011) 93–144.
[9] Y. Gong, C. Mo, S.E. Fraser, Planar cell polarity signalling controls cell division orientation during zebrafish gastrulation, Nature 430 (7000) (2004) 689–693.
[10] R. McBeath, D.M. Pirone, C.M. Nelson, K. Bhadriraju, C.S. Chen, Cell shape, cytoskeletal tension, and RhoA regulate stem cell lineage commitment, Dev. Cell 6 (4) (2004) 483–495.
[11] M. Théry, M. Bornens, Cell shape and cell division, Curr. Opin. Cell Biol. 18 (6) (2006) 648–657.

[12] V.P. Hoppenreijs, E. Pels, G.F. Vrensen, J. Oosting, W.F. Treffers, Effects of human epidermal growth factor on endothelial wound healing of human corneas, Invest. Ophthalmol. Vis. Sci. 33 (6) (1992) 1946–1957.
[13] M. Matsuda, M. Sawa, H.F. Edelhauser, S.P. Bartels, A.H. Neufeld, K.R. Kenyon, Cellular migration and morphology in corneal endothelial wound repair, Invest. Ophthalmol. Vis. Sci. 26 (4) (1985) 443–449.
[14] E.K. Kirkeeide, I.F. Pryme, A. Vedeler, Morphological changes in Krebs II ascites tumour cells induced by insulin are associated with differences in protein composition and altered amounts of free, cytoskeletal-bound and membrane-bound polysomes, Mol. Cell Biochem. 118 (2) (1992) 131–140.
[15] D. Zink, A.H. Fischer, J.A. Nickerson, Nuclear structure in cancer cells, Nat. Rev. Cancer 4 (9) (2004) 677–687.
[16] J.J. Tomasek, G. Gabbiani, B. Hinz, Myofibroblasts and mechano-regulation of connective tissue remodelling, Nat. Rev. Mol. Cell Biol. 3 (5) (2002) 349–363.
[17] R.M. Castilho, C.H. Squarize, K. Leelahavanichkul, Rac1 is required for epithelial stem cell function during dermal and oral mucosal wound healing but not for tissue homeostasis in mice, PLoS ONE 5 (1) (2010) e10503.
[18] A.F. Chambers, A.C. Groom, I.C. MacDonald, Dissemination and growth of cancer cells in metastatic sites, Nat. Rev. Cancer 2 (8) (2002) 563–572.
[19] E. Fuchs, T. Tumbar, G. Guasch, Socializing with the neighbors: stem cells and their niche, Cell 116 (6) (2004) 769–778.
[20] M.M. Nava, M.T. Raimondi, R. Pietrabissa, Controlling self-renewal and differentiation of stem cells via mechanical cues, J. Biomed. Biotechnol. 2012 (2012) 797410, https://doi.org/10.1155/2012/797410.
[21] P. Friedl, K. Wolf, Plasticity of cell migration: a multiscale tuning model, J. Cell. Sci. 188 (1) (2009) 11–19.
[22] T. Lammermann, M. Sixt, Mechanical modes of amoeboid cell migration, Curr. Opin. Cell Biol. 21 (5) (2009) 636–644.
[23] D.A. Lauffenburger, A.F. Horwitz, Cell migration: a physically integrated molecular process, Cell 84 (3) (1996) 359–369.
[24] C.M. Lo, H.B. Wang, M. Dembo, Y.L. Wang, Cell movement is guided by the rigidity of the substrate, Biophys. J. 79 (1) (2000) 144–152.
[25] A. Zemel, F. Rehfeldt, A.B.D. Discher, S.A. Safran, Optimal matrix rigidity for stress fiber polarization in stem cells, Nat. Phys. 6 (2010) 468–473.
[26] P. Moreo, J.M. Garcia-Aznar, M. Doblaré, Modeling mechanosensing and its effect on the migration and proliferation of adherent cells, Acta Biomater. 4 (3) (2008) 613–621.
[27] M.H. Zaman, R.D. Kamm, P. Matsudaira, D.A. Lauffenburgery, Computational model for cell migration in three-dimensional matrices, Biophys. J. 89 (2005) 1389–1397.
[28] A.J. Engler, S. Sen, H.L. Sweeney, D.E. Discher, Matrix elasticity directs stem cell lineage specification, Cell 126 (4) (2006) 677–689.
[29] P.M. Gilbert, K.L. Havenstrite, Substrate elasticity regulates skeletal muscle stem cell self-renewal in culture, Science 329 (5995) (2010) 1078–1081.
[30] N. Huebsch, P.R. Arany, Harnessing traction-mediated manipulation of the cell/matrix interface to control stem-cell fate, Nat. Mater. 9 (2010) 518–526.
[31] E. Hadjipanayi, V. Mudera, R.A. Brown, Guiding cell migration in 3D: a collagen matrix with graded directional stiffness, Cell Motil. Cytoskelet. 66 (2009) 435–445.
[32] R.J. Pelham Jr, Y.L. Wang, Cell locomotion and focal adhesions are regulated by substrate flexibility, Proc. Natl Acad. Sci. USA 94 (24) (1997) 13661–13665.
[33] L. Bosgraaf, P.J.M. Van Haastert, Navigation of chemotactic cells by parallel signaling to pseudopod persistence and orientation, PLoS ONE 4 (8) (2009) e6842.
[34] V.S. Deshpande, R.M. McMeeking, A.G. Evans, A bio-chemo-mechanical model for cell contractility, PNAS 103 (38) (2006) 14015–14020.
[35] O. Hoeller, R.R. Kay, Chemotaxis in the absence of PIP3 gradients, Curr. Biol. 17 (9) (2007) 813–817.
[36] M.P. Neilson, D.M. Veltman, P.J.M. van Haastert, S.D. Webb, J.A. Mackenzie, R.H. Insall, Chemotaxis: a feedback-based computational model robustly predicts multiple aspects of real cell behaviour, PLoS Biol. 9 (5) (2011) e1000618.
[37] F.J. Vermolen, A. Gefen, A phenomenological model for chemico-mechanically induced cell shape changes during migration and cell-cell contacts, Biomech. Model Mechanobiol. 12 (2) (2013) 301–323.
[38] B.H. Choo, R.F. Donna, L.P. Kenneth, Thermotaxis of *Dictyostelium discoideum* amoebae and its possible role in pseudoplasmodial thermotaxis, Proc. Natl Acad. Sci. USA 80 (18) (1983) 5646–5649.
[39] A.A. Higazi, D. Kniss, J.M.E.S. Barnathan, D.B. Cines, Thermotaxis of human trophoblastic cells, Placenta 17 (8) (1996) 683–687.
[40] R.C. Gao, X.D. Zhang, Different roles of membrane potentials in electrotaxis and chemotaxis of *Dictyostelium* cells, Eukaryot. Cell 10 (9) (2011) 1251–1256.
[41] E.M. Maria, M.B.A. Djamgoz, Cellular mechanisms of direct-current electric field effects: galvanotaxis and metastatic disease, J. Cell Sci. 117 (9) (2004) 1631–1639.
[42] M. Zhao, Electrical fields in wound healing—an overriding signal that directs cell migration, Semin. Cell Dev. Biol. 20 (6) (2009) 674–682.
[43] P.J.M. Van Haastert, Chemotaxis: insights from the extending pseudopod, J. Cell Sci. 123 (18) (2010) 3031–3037.
[44] K.F. Swaney, C.H. Huang, P.N. Devreotes, Eukaryotic chemotaxis: a network of signaling pathways controls motility, directional sensing, and polarity, Annu. Rev. Biophys. 39 (2010) 265–289.
[45] M.E.A. Bahat, Sperm thermotaxis, Mol. Cell Endocrinol. 252 (1–2) (2006) 115–119.
[46] Y.X. Huang, X.J. Zheng, Quantum dots as a sensor for quantitative visualization of surface charges on single living cells with nano-scale resolution, Biosens. Bioelectron. 26 (8) (2011) 2114–2118.
[47] M.E. Mycielska, M.B. Djamgoz, Cellular mechanisms of direct current electric field effects: galvanotaxis and metastatic disease, J. Cell Sci. 117 (9) (2004) 1631–1639.
[48] K.Y. Nishimura, R.R. Isseroff, R. Nuccitelli, Human keratinocytes migrate to the negative pole in direct current electric fields comparable to those measured in mammalian wounds, J. Cell Sci. 109 (1) (1996) 199–207.
[49] E.K. Onuma, S.W. Hui, A calcium requirement for electric field-induced cell shape changes and preferential orientation, Cell Calcium 6 (3) (1985) 281–292.
[50] L.J. Shanley, P. Walczysko, M. Bain, D.J. MacEwan, M. Zhao, Influx of extracellular $Ca^{2+}$ is necessary for electrotaxis in *Dictyostelium*, J. Cell Sci. 119 (22) (2006) 4741–4748.
[51] M.J. Brown, L.M. Loew, Electric field-directed fibroblast locomotion involves cell surface molecular reorganization and is calcium independent, J. Cell Sci. 127 (1) (1994) 117–128.

[52] M.S. Cooper, R.E. Keller, Perpendicular orientation and directional migration of amphibian neural crest cells in DC electrical fields, Proc. Natl Acad. Sci. USA 81 (1) (1984) 160–164.
[53] C.A. Erickson, R. Nuccitelli, Embryonic fibroblast motility and orientation can be influenced by physiological electric fields, J. Cell Sci. 98 (1) (1984) 296–307.
[54] J. Lee, A. Ishihara, G. Oxford, B. Johnson, K. Jacobson, Regulation of cell movement is mediated by stretch-activated calcium channels, Nature 400 (6742) (1999) 382–386.
[55] E.K. Onuma, S.W. Hui, Electric field-directed cell shape changes, displacement, and cytoskeletal reorganization are calcium dependent, J. Cell Sci. 106 (1) (1988) 2067–2075.
[56] N. Orida, J.D. Feldman, Directional protrusive pseudopodial activity and motility in macrophages induced by extracellular electric fields, Cell Motil. 2 (3) (1982) 243–255.
[57] P.J.V. Haastert, P.N. Devreotes, Chemotaxis: signalling the way forward, Nat. Rev. Mol. Cell Biol. 5 (8) (2004) 626–634.
[58] O. Ilina, P. Friedl, Mechanisms of collective cell migration at a glance, J. Cell Sci. 122 (18) (2009) 3203–3208.
[59] R. Riahi, Y. Yang, D.D. Zhang, P.K. Wong, Advances in wound-healing assays for probing collective cell migration, J. Lab. Autom. 17 (1) (2012) 59–65.
[60] R. Nuccitelli, A role for endogenous electric fields in wound healing, Curr. Top. Dev. Biol. 58 (2003) 1–26.
[61] M. Ehrbar, A. Sala, P. Lienemann, A. Ranga, K. Mosiewicz, A. Bittermann, S.C. Rizzi, F.E. Weber, Elucidating the role of matrix stiffness in 3D cell migration and remodeling, Biophys. J. 100 (2) (2011) 284–293.
[62] E.F. Keller, L.A. Segel, Initiation of slide mold aggregation viewed as an instability, J. Theor. Biol. 26 (3) (1970) 399–415.
[63] E.F. Keller, L.A. Segel, Model for chemotaxis, J. Theor. Biol. 30 (2) (1971) 225–234.
[64] G. Oster, On the crawling of cells, J. Embryol. Exp. Morphol. 83 (1984) 329–364.
[65] G. Oster, A. Perelson, Cell spreading and motility: a model lamellipod, J. Math. Biol. 21 (1985) 383–388.
[66] D.C. Bottino, L.J. Fauci, A computational model of ameboid deformation and locomotion, Eur. Biophys. J. 27 (5) (1998) 532–539.
[67] D. Bottino, A. Mogilner, T. Roberts, M. Stewart, G. Oster, How nematode sperm crawl, J. Cell Sci. 115 (2) (2002) 367–384.
[68] H.G. Othmer, A. Stevens, Aggregation, blowup and collapse: the ABC's of generalized taxis, SIAM J. Appl. Math. 57 (4) (1997) 1044–1081.
[69] T. Hofer, J.A. Sherratt, P.K. Maini, *Dictyostelium discoideum*: cellular self-organisation in an excitable biological medium, Proc. Biol. Sci. 259 (1356) (1995) 249–257.
[70] M.A.J. Chaplain, A.M. Stuart, A model mechanism for the chemotactic response of endothelial cells to tumor angiogenesis factor, IMA J. Math. Appl. Med. Biol. 10 (3) (1993) 149–168.
[71] G.J. Pettet, H.M. Byrne, D.L.S. Mcelwain, J. Norbury, A model of wound-healing angiogenesis in soft tissue, Math. Biosci. 136 (1) (1996) 35–63.
[72] E. Palsson, A three-dimensional model of cell movement in multicellular systems, Futur. Gener. Comput. Syst. 17 (2001) 835–852.
[73] R. De, A. Zemel, S.A. Safran, Dynamics of cell orientation, Nat. Phys. 3 (2007) 655–659.
[74] D. Kabaso, R. Shlomovitz, K. Schloen, T. Stradal, N.S. Gov, Theoretical model for cellular shapes driven by protrusive and adhesive forces, PLoS Comput. Biol. 7 (5) (2011) e1001127.
[75] Y. Ni, M.Y.M. Chiang, Cell morphology and migration linked to substrate rigidity, Soft. Matter. 3 (2007) 1285–1292.
[76] K.T. Kang, J.H. Park, Study on differentiation of mesenchymal stem cells by mechanical stimuli and an algorithm for bone fracture healing, TERM 8 (4) (2011) 359–370.
[77] A.J. Stops, K.B. Heraty, M. Browne, F.J. O'Brien, P.E. McHugh, A prediction of cell differentiation and proliferation within a collagen-glycosaminoglycan scaffold subjected to mechanical strain and perfusive fluid flow, J. Biomech. 43 (4) (2010) 618–626.
[78] M.P. Neilson, J.A. Mackenzie, S.D. Webb, R.H. Insall, Modeling cell movement and chemotaxis using pseudopod-based feedback, SIAM J. Sci. Comput. 33 (3) (2011) 1035–1057.
[79] M. Yamao, H.N. Mail, S. Ishii, Multi-cellular logistics of collective cell migration, PLoS ONE 6 (12) (2011) e27950.
[80] R.F. van Oers, E.G. Rens, D.J. LaValley, C.A. Reinhart-King, R.M. Merks, Mechanical cell-matrix feedback explains pairwise and collective endothelial cell behavior in vitro, PLoS Comput. Biol. 10 (8) (2014) e1003774.
[81] A. Ducrot, F.L. Foll, H.M. Magal, P.J. Pasquier, G.F. Webb, An in vitro cell population dynamics model incorporating cell size, quiescence, and contact inhibition, Math. Models Methods Appl. Sci. 21 (1) (2011) 871–892.
[82] G. Gyllenberg, H. Heijmans, An abstract delay-differential equation modelling size dependent cell growth and division, SIAM J. Math. Anal. 18 (1) (1987) 74–88.
[83] J. Satulovsky, R. Lui, Y.L. Wang, Exploring the control circuit of cell migration by mathematical modeling, Biophys. J. 94 (9) (2008) 3671–3683.
[84] E. Hannezo, J. Prost, J.F. Joanny, Theory of epithelial sheet morphology in three dimensions, Proc. Natl Acad. Sci. 111 (1) (2014) 27–32.
[85] T.A. Ulrich, E.M.D.J. Pardo, S. Kumar, The mechanical rigidity of the extracellular matrix regulates the structure, motility, and proliferation of glioma cells, Cancer Res. 69 (10) (2009) 4167–4174.
[86] G. Albrecht-Buehler, Does blebbing reveal the convulsive flow of liquid and solutes through the cytoplasmic meshwork? Cold. Spring Harb. Symp. Quant. Biol. 46 (1) (1982) 45–49.
[87] C.C. Cunningham, Actin polymerization and intracellular solvent flow in cell surface blebbing, J. Cell Sci. 129 (6) (1995) 1589–1599.
[88] G.T. Charras, J.C. Yarrow, M.A. Horton, L. Mahadevan, T.J. Mitchison, Non-equilibration of hydrostatic pressure in blebbing cells, Nature 435 (2005) 365–369.
[89] S.J. Han, N.J. Sniadecki, Simulations of the contractile cycle in cell migration using a bio-chemical-mechanical model, Comput. Methods Biomech. Biomed. Eng. 14 (5) (2011) 459–468.
[90] D.R. Carter, P.R. Blenman, G.S. Beaupré, Correlations between mechanical stress history and tissue differentiation in initial fracture healing, J. Orthop. Res. 6 (5) (1988) 736–748.
[91] L.E. Claes, C.A. Heigele, Magnitudes of local stress and strain along bony surfaces predict the coarse and type of fracture healing, J. Biomech. 32 (3) (1999) 255–266.
[92] L. Geris, H.V. Oosterwyck, J.V. Sloten, J. Duyck, I. Naert, Assessment of mechanobiological models for the numerical simulation of tissue differentiation around immediately loaded implants, Comput. Methods Biomech. Biomed. Eng. 6 (5–6) (2003) 277–288.

[93] D.J. Kelly, P.J. Prendergast, Mechano-regulation of stem cell differentiation and tissue regeneration in osteochondral defects, J. Biomech. 38 (7) (2004) 1413–1422.

[94] D. Lacroix, P.J. Prendergast, A mechano-regulation model for tissue differentiation during fracture healing: analysis of gap size and loading, J. Biomech. 35 (9) (2002) 1163–1171.

[95] D. Lacroix, P.J. Prendergast, G. Li, D. Marsh, Biomechanical model of simulate tissue differentiation and bone regeneration: application to fracture healing, Med. Biol. Eng. Comput. 40 (1) (2002) 14–21.

[96] D.E. Ingber, I. Tensegrity, Cell structure and hierarchical systems biology, J. Cell Sci. 116 (7) (2003) 1157–1173.

[97] C.G. Penelope, P.A. Janmey, Cell type-specific response to growth on soft materials, J. Appl. Physiol. 98 (2005) 1547–1553.

[98] A. Buxboim, I.L. Ivanovska, D.E. Discher, Matrix elasticity, cytoskeletal forces and physics of the nucleus: how deeply do cells 'feel' outside and in? J. Cell Sci. 123 (3) (2010) 297–308.

[99] Y.T. Shiu, S. Li, W.A. Marganski, S. Uami, M.A. Schwartz, Y.L. Wang, M. Dembo, S. Chien, Rho mediates the shear-enhancement of endothelial cell migration and traction force generation, Biophys. J. 86 (4) (2004) 2558–2565.

[100] S.J. Mousavi, M.H. Doweidar, M. Doblaré, Computational modelling and analysis of mechanical conditions on cell locomotion and cell-cell interaction, Comput. Methods Biomech. Biomed. Eng. 17 (6) (2014) 678–693.

[101] S. Timoshenko, J.N. Goodier, Theory of Elasticity, McGraw-Hill Publishing Co, New York, NY, 1970.

[102] A.D. Bershadsky, N.Q. Balaban, B. Geiger, Adhesion-dependent cell mechanosensitivity, Annu. Rev. Cell Dev. Biol. 19 (2003) 677–695.

[103] G.W. Brofland, C.J. Wiebe, Mechanical effects of cell anisotropy on epithelia, Comput. Methods Biomech. Biomed. Eng. 7 (2) (2004) 91–99.

[104] D.L. Taylor, J. Heiple, Cellular and molecular aspects of amoeboid movement, Cold. Spring Harb. Symp. Quant. Biol. 46 (1) (1982) 101–111.

[105] S.J. Mousavi, M. Doblaré, M.H. Doweidar, Computational modelling of multi-cell migration in a multi-signalling substrate, Phys. Biol. 11 (2) (2014) 026002.

[106] A. Ponti, M. Machacek, S.L. Gupton, C.M. Waterman-Storer, G. Danuser, Two distinct actin networks drive the protrusion of migrating cells, Science 305 (5691) (2004) 1782–1786.

[107] D.W. James, J.F. Taylor, The stress developed by sheets of chick fibroblasts in vitro, Exp. Cell Res. 54 (1) (1969) 107–110.

[108] S.J. Mousavi, M.H. Doweidar, M. Doblaré, Cell migration and cell-cell interaction in the presence of mechano-chemo-thermotaxis, Mol. Cell Biomech. 10 (1) (2013) 1–25.

[109] S. Ramtani, Mechanical modelling of cell/ECM and cell/cell interactions during the contraction of a fibroblast-populated collagen microsphere: theory and model simulation, J. Biomech. 37 (11) (2004) 1709–1718.

[110] G.G. Stokes, On the effect of the inertial friction of fluids on the motion of pendulums, Trans. Cambr. Philos. Soc. 9 (1951) 8–106.

[111] I.V. Dokukina, M.E. Gracheva, A model of fibroblast motility on substrates with different rigidities, Biophys. J. 98 (2010) 2794–2803.

[112] E. Loth, Drag of non-spherical solid particles of regular and irregular shape, Powder Technol. 182 (2008) 342–353.

[113] S. Tran-Cong, M. Gay, E.E. Michaelides, Drag coefficients of irregularly shaped particles, Powder Technol. 139 (2004) 21–32.

[114] M. Dressel, Dynamic shape factors for particle shape characterization, Part. Charact. 2 (1985) 62–66.

[115] J.A. Jimenez, O.S. Madsen, A simple formula to estimate settling velocity of natural sediments, J. Waterw. Port Coast. Ocean Eng. 129 (2003), https://doi.org/10.1061/(ASCE)0733-950X(2003)129:2(70).

[116] M.S. Cooper, M. Shliwa, Motility of cultured fish epidermal cells in the presence and absence of direct current electric fields, J. Cell Sci. 102 (4) (1986) 1384–1399.

[117] D.M. Sheridan, R.R. Isseroff, R. Nuccitelli, Imposition of a physiologic DC electric field alters the migratory response of human keratinocytes on extracellular matrix molecules, J. Invest. Dermatol. 106 (1996) 642–646.

[118] G.L. Sulik, H.K. Soong, P.C. Chang, W.C. Parkinson, S.G. Elner, V.M. Elner, Effects of steady electric fields on human retinal pigment epithelial cell orientation and migration in culture, Acta Ophthalmol. (Copenhagen) 70 (1992) 115–122.

[119] B. Rapp, A. Boisfleury-Chevance, H. Gruler, Galvanotaxis of human granulocytes. Dose-response curve, Eur. Biophys. J. 16 (1988) 313–319.

[120] S.P. Fraser, J.K.J. Diss, M.E. Mycielska, R.C. Coombes, M.B.A. Djamgoz, Voltage-gated sodium channel expression in human breast cancer: possible functional role in metastasis, Breast Cancer Res. Treat. 76 (2002) S142.

[121] S.J. Mousavi, M.H. Doweidar, M. Doblaré, 3D computational modelling of cell migration: a mechano-chemo-thermo-electrotaxis approach, J. Theor. Biol. 329 (2013) 64–73.

[122] L. Armon, M. Eisenbach, Behavioral mechanism during human sperm chemotaxis: involvement of hyperactivation, PLoS ONE 6 (2011) e28359.

[123] C.B. Hong, D.R. Fontana, K.L. Poff, Thermotaxis of *Dictyostelium discoideum* amoebae and its possible role in pseudoplasmodial thermotaxis, Proc. Natl Acad. Sci. USA 80 (1983) 5646–5649.

[124] H. Long, G. Yang, Z. Wang, Galvanotactic migration of EA.Hy926 endothelial cells in a novel designed electric field bioreactor, Cell Biochem. Biophys. 61 (2011) 481–491.

[125] M.P. Sheetz, D. Felsenfeld, C.G. Galbraith, D. Choquet, Cell migration as a five-step cycle, Biochem. Soc. Symp. 65 (1997) 233–243.

[126] S.J. Mousavi, M.H. Doweidar, A novel mechanotactic 3D modeling of cell morphology, J. Phys. Biol. 11 (4) (2014) 046005.

[127] S.J. Mousavi, M.H. Doweidar, Three-dimensional numerical model of cell morphology during migration in multi-signaling substrates, PLoS ONE 10 (3) (2015) e0122094.

[128] D. Wessels, E. Voss, N.V. Bergen, R. Burns, J. Stites, D.R. Soll, A computer-assisted system for reconstructing and interpreting the dynamic three-dimensional relationships of the outer surface, nucleus, and pseudopods of crawling cells, Cell Motil. Cytoskelet. 41 (1998) 225–246.

[129] S.J. Gladman, R.E. Ward, A.T. Michael-Titus, M.M. Knight, J.V. Priestley, The effect of mechanical strain or hypoxia on cell death in subpopulations of rat dorsal root ganglion neurons in vitro, Neuroscience 171 (2) (2010) 577–587.

[130] N.C. Harrison, R.D. del Corral, B. Vasiev, Coordination of cell differentiation and migration in mathematical models of caudal embryonic axis extension, PLoS ONE 6 (2011) e22700.

[131] E.M. Kearney, P.J. Prendergast, V.A. Campbell, Mechanisms of strain-mediated mesenchymal stem cell apoptosis, J. Biomech. Eng. 130 (6) (2008) 061004.

[132] K. Kurpinski, J. Chu, C. Hashi, S. Li, Anisotropic mechanosensing by mesenchymal stem cells, Proc. Natl Acad. Sci. USA 103 (2006) 16095–16100.

[133] M.A. Wozniak, C.S. Chen, Mechanotransduction in development: a growing role for contractility, Nat. Rev. Mol. Cell Biol. 10 (2009) 34–43.

[134] S.J. Mousavi, M.H. Doweidar, Role of mechanical cues in cell differentiation and proliferation: a 3D numerical model, PLoS ONE 10 (5) (2015) e0124529.

[135] S.J. Mousavi, M.H. Doweidar, Numerical modeling of cell differentiation and proliferation in force-induced substrates via encapsulated magnetic nanoparticles, Comput. Methods Programs Biomed. 130 (2016) 106–117.

[136] P.A. Prokharaua, F.J. Vermolena, J.M. Garcia-Aznar, A mathematical model for cell differentiation, as an evolutionary and regulated process, Comput. Methods Biomech. Biomed. Eng. 17 (10) (2014) 1051–1070.

[137] R.M. Delaine-Smith, C.R. Gwendolen, Mesenchymal stem cell responses to mechanical stimuli, Muscles Ligaments Tendons J. 2 (3) (2012) 169–180.

[138] D.M. Cullinane, K.T. Salisbury, Y. Alkhiary, S. Eisenberg, Effects of the local mechanical environment on vertebrate tissue differentiation during repair: does repair recapitulate development, J. Exp. Biol. 206 (2003) 2459–2471.

[139] Q.Q. Wu, Q. Chen, Mechanoregulation of chondrocyte proliferation, maturation, and hypertrophy: ion-channel dependent transduction of matrix deformation signals, Exp. Cell Res. 256 (2) (2000) 383–391.

[140] S.J. Mousavi, M.H. Doweidar, Encapsulated piezoelectric nanoparticle-hydrogel smart material to remotely regulate cell differentiation and proliferation: a finite element model, Comput. Mech. 63 (3) (2018) 471–489, https://doi.org/10.1007/s00466-018-1604-7.

[141] S. Fouliarda, S. Benhamidaa, N. Lenuzzab, F. Xaviera, Modeling and simulation of cell populations interaction, Math. Comput. Model. 49 (11) (2009) 2104–2108.

[142] Hibbitt, Karlsson & Sorensen, Inc., ABAQUS-Theory Manual, sixth ed., 2011, pp. 11–13.

[143] S.K. Akiyama, K.M. Yamada, The interaction of plasma fibronectin with fibroblastic cells in suspension, J. Biol. Chem. 260 (7) (1985) 4492–4500.

[144] A. Schäfer, M. Radmacher, Influence of myosin II activity on stiffness of fibroblast cells, Acta Biomater. 1 (3) (2005) 273–280.

[145] K.R. Foster, H.P. Schwan, Dielectric properties of tissues and biological materials: a critical review, Crit. Rev. Biomed. Eng. 17 (1989) 25–104.

[146] K.L. Poff, M. Skokut, Thermotaxis by pseudoplasmodia of *Dictyostelium discoideum*, Proc. Natl Acad. Sci. USA 74 (1977) 2007–2010.

[147] K. Tawada, H. Miyamoto, Sensitivity of paramecium thermotaxis to temperature change, J. Protozool. 20 (1973) 289–292.

[148] R.R. Kay, P. Langridge, D. Traynor, O. Hoeller, Changing directions in the study of chemotaxis, Nat. Rev. Mol. Cell Biol. 9 (10) (2008) 455–463.

[149] N. Andrew, R.H. Insall, Chemotaxis in shallow gradients is mediated independently of Ptdlns 3-kinase by biased choices between random protrusions, Nat. Cell Biol. 9 (2007) 193–200.

[150] Y. Lee, J. Huang, Y. Wang, K. Lin, Three-dimensional fibroblast morphology on compliant substrates of controlled negative curvature, Integr. Biol. (Camb.) 5 (12) (2013) 1447–1455.

[151] S. Zhang, P.G. Charest, R.A. Firtel, Spatiotemporal regulation of Ras activity provides directional sensing, Curr. Biol. 18 (2008) 1587–1593.

[152] P.C.T. Chang, G.L. Sulik, H.G. Soong, W.C. Parkinson, Galvanotropic and galvanotactic responses of corneal endothelial cells, Formos. Med. Assoc. J. 95 (1996) 623–627.

[153] G.J. Hera, H.C. Wub, M.H. Chenc, M.Y. Chene, S.C. Change, T.W. Wanga, Control of three-dimensional substrate stiffness to manipulate mesenchymal stem cell fate toward neuronal or glial lineages, Acta Biomater. 9 (2) (2013) 5170–5180.

[154] J. Fu, Y.K. Wang, M.T. Yang, R.A. Desai, X. Yu, Z. Liu, C.S. Chen, Mechanical regulation of cell function with geometrically modulated elastomeric substrates, Nat. Methods 7 (9) (2010) 733–736.

[155] D.P. Burke, D.J. Kelly, Substrate stiffness and oxygen as regulators of stem cell differentiation during skeletal tissue regeneration: a mechanobiological model, PLoS ONE 7 (7) (2012) e40737.

[156] N.D. Evans, C. Minelli, Substrate stiffness affects early differentiation events in embryonic stem cells, Eur. Cell Mater. 18 (2009) 1–13.

C H A P T E R

# 16

# On the Simulation of Organ-on-Chip Cell Processes

## Application to an In Vitro Model of Glioblastoma Evolution

*Jacobo Ayensa-Jiménez*[*,†], *Mohamed H. Doweidar*[*,†,‡,§], *Teodora Randelovic*[*,†], *Luis. J. Fernández*[*,†,‡,§], *Sara Oliván*[*,†,‡,¶], *Ignacio Ochoa*[*,†,‡,¶], *Manuel Doblaré*[*,†,‡,§]

**[*]Aragón Institute of Engineering Research (I3A), University of Zaragoza, Zaragoza, Spain [†]Institute of Health Research of Aragon (IIS), Zaragoza, Spain [‡]Biomedical Research Networking Center in Bioengineering, Biomaterials and Nanomedicine (CIBER-BBN), Madrid, Spain [§]Mechanical Engineering Department, School of Engineering and Architecture (EINA), University of Zaragoza, Zaragoza, Spain [¶]Human Anatomy and Histology Department, Faculty of Medicine, University of Zaragoza, Zaragoza, Spain**

## 16.1 INTRODUCTION

Cancer is the leading cause of death in developed countries and second in the developing ones [1]. Out of the 14.1 million new cancer diagnoses in 2012 worldwide, there were 8.2 million deaths and, at that time, 32.6 million people living within 5 years of diagnosis. Furthermore, the Globoscan 2012 database [2] estimates a 27% overall increase in cancer incidence by 2035, whereas the death count will rise 78% by that year. This means that approximately 15–20 million people will die of cancer every year in the upcoming two decades. And all this, despite the immense research effort made in recent years. The main reason for this is the extreme complexity of the processes involved in cancer development. In fact, the cancer landscape has dramatically changed in recent years, demonstrating that it is much more complex than initially thought. Old paradigms focused only on tumor cells and genetics have now turned into a new scenario that integrates different cell populations, extracellular matrices, chemotactic gradients (oxygen or nutrients), and physical cues such as local deformation [3–6], all conforming a complex, dynamic, and multiple interactive tumor microenvironment (TME) [7].

Furthermore, tumor cells evolve differently within the same tumor, with different evolution paths and specific microenvironments for each type of cancer and tissue, thus explaining their high heterogeneity [8]. Whereas normal cells require specific physiological signals to proliferate, tumor cells can divide independently of these signals and ignore antigrowth ones. Besides, they acquire limitless replicative potential and are capable of evading apoptosis signals [9]. In this situation, exacerbated consumption of oxygen and nutrients leads to hypoxia and nutrient starvation, thus subjecting cells to extreme stress. This harsh environment favors survival of cells capable of resisting and adapting themselves to these highly demanding conditions. This lack of nutrients and oxygen also activates other mechanisms, such as autophagy or regulation of oxidative stress [10]. Moreover, tumor cells are also able to change their TME, for example, by promoting new blood vasculature that increases the supply of oxygen and nutrients to maintain their growth as well as colonizing new tissues through metastasis [11]. Finally, when a given treatment (surgery, chemo-, radio-, immune-, hormone, or combination therapy) is applied, the tumor and its TME undergo significant alterations.

Due to the complexity, heterogeneity, and dynamic changes that take place in the TME, it is difficult to investigate, in vivo, in a precise way, all interactions in the tumor and the surrounding stroma. A good reproduction in vitro of the

*Advances in Biomechanics and Tissue Regeneration*
https://doi.org/10.1016/B978-0-12-816390-0.00016-9

TME could help to better understand the evolution of the disease and would provide a tool for testing new drugs and reducing animal experiments. However, there is still an important lack of predictive power of currently available in vitro models, with this being one of the main reasons for the continuous drop of new drugs appearing per billion dollars invested [12, 13].

For example, many authors report strong changes in cell functions when cells are cultured in three dimensions (3D) compared with those in two dimensions (2D) (drug resistance, number, and specificity of focal adhesions) [14]. Despite this, cells are usually cultured on the traditional Petri dishes, where tumor complexity is mostly lost, which explains the efforts made to reproduce three-dimensionality in in vitro experiments [15]. Recently, microfluidics has arisen as a powerful tool to recreate the complex microenvironment that governs tumor dynamics [16, 17]. This technique allows reproducing important features of tumor evolution that were not seen before in 2D cultures as well as testing drugs in a much more reliable and efficient way. In particular, one of the most important fields of application of microfluidics in cell culture has been the study of chemotactic processes such as tumor-induced angiogenesis [18], tumor invasion [19], tumor intravasation [20], tumor extravasation [21], and metastasis [22], among others.

One particular case of cancer is the type that affects the central nervous system (CNS). According to the American Brain Tumor Association, the primary tumors of the CNS make up about 2% of all cancers in adults and 20% in children. Of these, different types of astrocytomas account for 76% of gliomas. Astrocytomas are classified in four different degrees to describe their level of abnormality and their degree of malignancy, with grade one having the best prognosis and grade 4, which is known as a glioblastoma (GBM), being the most aggressive. Survival of patients with this type of tumor who undergo the first-line standard treatments (local radiotherapy and chemotherapy with concomitant temozolomide) has a median of 14 months since diagnosis and has a 5-year survival rate of less than 10% [23, 24]. It is characterized by rapid growth and a high level of malignancy, being, unfortunately, the most frequent type of brain tumor with an incidence of 17% of all primary tumors.

GBM is a highly infiltrating and fast-progressing tumor, characterized by two main histopathological conditions: necrotic foci typically surrounded by areas of high cellularity known as pseudopalisades, and microvascular hyperplasia [25, 26]. In an early stage, GBM proliferation and secretion of procoagulant signals would cause thrombotic events, leading to hypoxia and nutrient depletion. As a consequence, GBM cells start to migrate toward enriched regions guided by this nutrient and oxygen gradient; this leads to the generation of the characteristic GBM pseudopalisades. These migrating cells would reach other healthy blood vessels, allowing GBM cell proliferation. Eventually, GBM cells will cause again the collapse of this blood vessel, restarting the process and creating an expanding wave of migrating tumor cells across the brain. Thereby, it has been proposed that one of the driving forces of glioma aggressiveness is the nutrient and oxygen starvation due to thrombotic events [27]. Recent histological studies have shown that proliferation in pseudopalisading areas is significantly lower while apoptosis is substantially larger than in neighboring regions. This evidence suggests that pseudopalisades are due to causes other than simply higher proliferation or survival rates [26]. In recent studies, it has been described that at least 50% of the pseudopalisades have a central occluded blood vessel and that GBM cells express several procoagulant factors [28] and hypoxia-induced factor-1 (HIF-1) [29], responsible for the vascular hyperplasia characteristic of GBM.

However, and despite these new experimental possibilities, the complexity and heterogeneity of the TME as well as of its dynamic interactions with tumor cells make it difficult to separate effects, check new hypotheses, and quantify the effect of every parameter for predicting the outcome in what if situations, even in these more realistic and controlled conditions. To get this, the only way is to combine the new possibilities of in vitro assays with the quantitative power and versatility of mathematical modeling and computational techniques [30, 31]. There have been many attempts to build mathematical models to describe how these tumors grow and respond to therapies [32–36]. In particular, a recent review [37], besides reviewing the mathematical models available to incorporate the main components of the TME, analyzes aspects such as the importance of the hypoxic environment for GBM progression and in the formation of cellular pseudopalisades [38]; tumor vasculature formation (including angiogenesis, and vessel cooption, remodeling and regression), the role of biophysical and biomechanical properties of the ECM in tumor cell invasion, the intravascular fluid microenvironment, tumor cell migration and dissemination through the lymphatic network, or the role of microenvironmental niches and sanctuaries in the emergence of acquired drug resistance in tumors. Also, in previous works in our group, we demonstrated the possibility of developing GBM pseudopalisades in vitro [39].

One of the main problems in these models is the lack of reliable values for the many parameters involved, which many times forces us to rely on approximations derived from very different situations, leading sometimes to unreliable conclusions.

In this chapter, we present a new mathematical framework to model the behavior of cell processes in vitro using microfluidic devices, especially for modeling the process of pseudopalisade formation in such devices. The first section describes the particular problem analyzed and the experiments performed in the microfluidic device as well as the main results obtained in such experiments. Next, we introduce the global mathematical framework, with the different equations, parameters, and interactions between the corresponding functions and variables. The following section presents the implementation of such equations in a finite element (FE) framework. The weak forms of the differential equations as well as the matrix components of the algebraic system resulting from the FE approximation are derived. Also, the corresponding results derived from the simulation results computed, after a process to identify the parameters that best fit some of those results, are presented and discussed. Most of these examples correspond to problems that can be assimilated to unidimensional, so the particularization of the global formulation to one dimension (1D) is used to get such results. As a proof of concept, and to observe the full potential of the proposed approach, another example is also solved and presented, now in 3D, although without experimental validation. Finally, the main conclusions of the work are stated and commented upon.

## 16.2 PROBLEM DESCRIPTION

Taking into account the poor prognosis and complex structure of the GBM described above, it is clear that the development of an accurate in vitro model for GBM research is very important. Three-dimensional cell cultures and microfluidic systems can give us a lot of new and useful information, as they can reproduce much better the physiological state and environment of a cell, which cannot be achieved in a standard 2D cell culture. As was said previously, the main characteristics of GBM are the appearance of necrotic foci surrounded by areas of high cellularity (pseudopalisades) and microvascular proliferation. In our in vitro models, we will focus on the process of necrotic core and pseudopalisade formation.

Uncontrolled proliferation of tumor cells and secretion of different factors induce occlusion of a surrounded blood vessel. This causes a decrease in nutrient and oxygen supply, hence the appearance of hypoxia in the perivascular region. Hypoxia provokes active cell migration away from this region and the formation of a hypercellular moving wave (pseudopalisade). Tumor cells that do not migrate activate a process of apoptosis or necrosis, creating in that way an enlarging necrotic zone. As the pseudopalisading cells are hypoxic, they have upregulated expression of the HIF, which induces overexpression of the vascular endothelial growth factor (VEGF) that is responsible for microvascular proliferation and hyperplasia. When the new blood vessels are formed, a pseudopalisade formation can be observed around them. Once the cells reach a functional blood vessel with appropriate environmental conditions, they start to proliferate at a high rate, and this can induce an occlusion of the vessel and restart the process [27].

Previous work in our laboratory showed that we are able to reproduce the necrotic core within the microfluidic device (see Fig. 16.1). If we seed a hydrogel with high cell density ($40 \times 10^6$ cells/mL) in the central microchamber, nutrients and oxygen cannot reach central parts of the device, mimicking the real case in which they cannot reach central parts of the tumor, far from functional blood vessels. This causes cell death and necrotic core formation [40].

On the other hand, we were also able to reproduce the pseudopalisade formation (see Fig. 16.2). In these experiments, cells are seeded within the central chamber at low density ($4 \times 10^6$ cells/mL) and growth medium is perfused just through one lateral channel while the other is sealed, simulating the functional and thrombotic blood vessels, respectively. Hypoxia occurs next to the sealed channel, which can be confirmed using a hypoxia-sensitive reagent that increases its fluorescence intensity as the concentration of oxygen decreases. This causes the cells to start to move toward the perfused channel with higher oxygen concentration, forming a migratory front and reaching the open channel (blood vessel) [39].

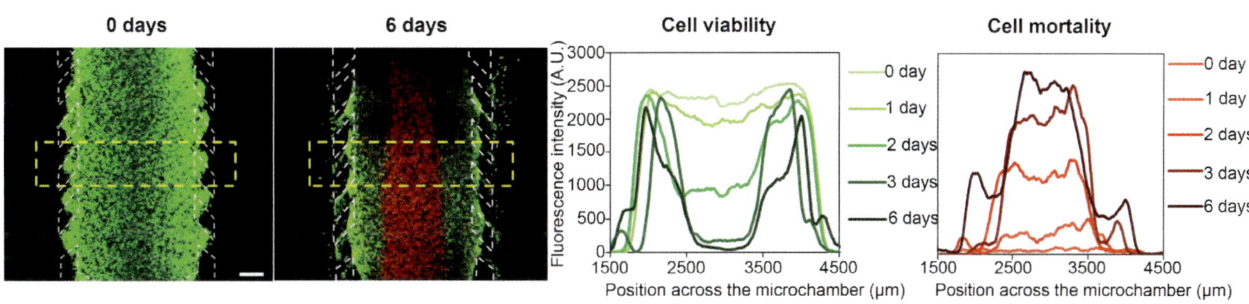


FIG. 16.1 Necrotic core formation. U251 cells were seeded at the concentration of $40 \times 10^6$ cells/mL within the central microchamber and growth medium was perfused every day through lateral channels. Cell viability is detected using calcein AM/propidium iodide (CAM/PI) staining, where CAM becomes fluorescent once it reaches the cytoplasm of the viable cell and PI stains dead cells with destroyed membranes [40]

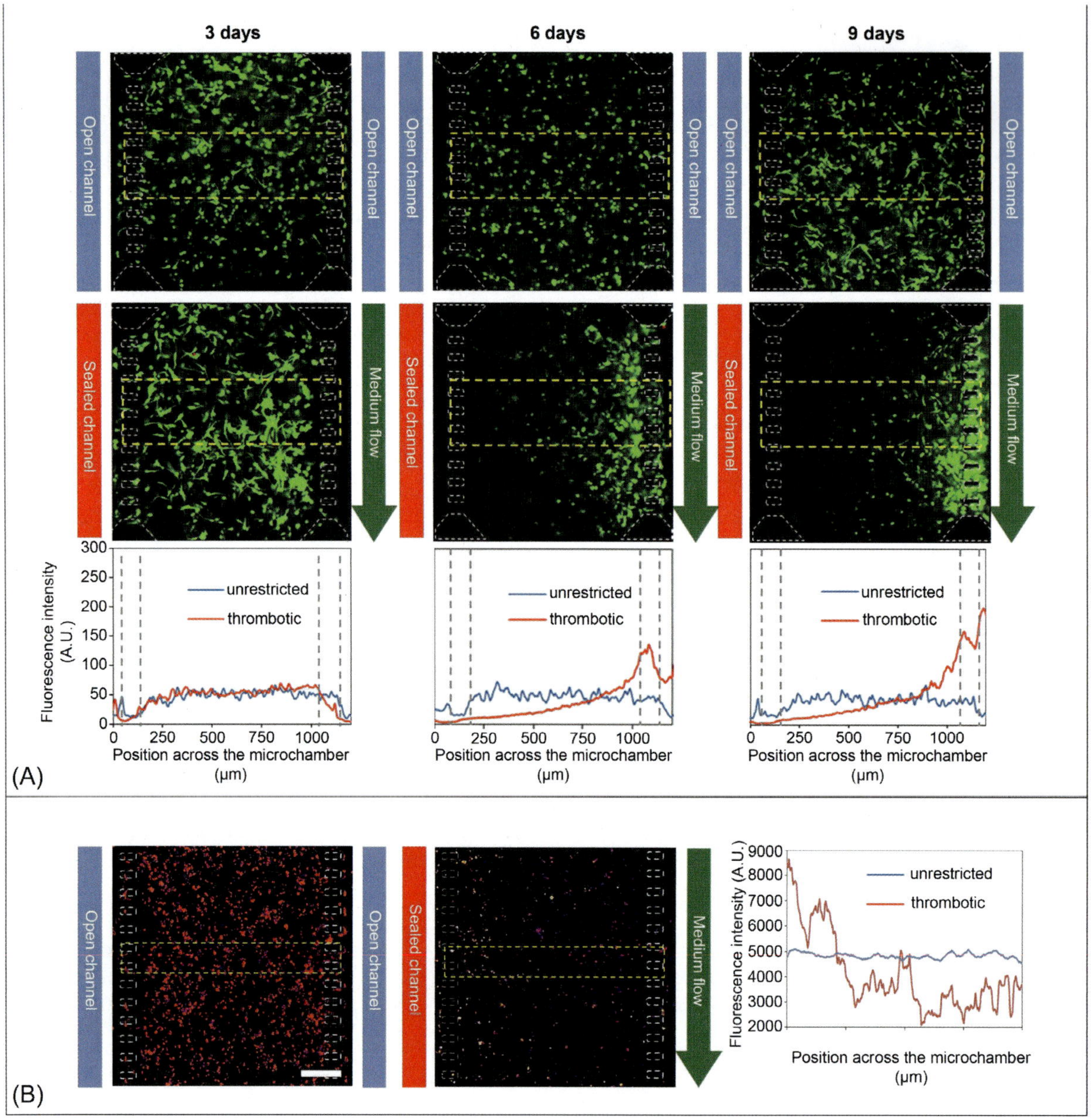


FIG. 16.2 Pseudopalisade formation. U-251 at 4 × $10^6$ cells/mL in collagen hydrogel were cultured within microdevices. Under unrestricted conditions, the medium was refreshed once a day. To mimic obstructed conditions, the medium flow was enabled only through the right microchannel. (A) Cell viability was evaluated after 3, 6, and 9 days using calcein (*green*) and propidium iodide (*red*). Graphs show the fluorescence intensity across the microchamber orthogonal view at 3, 6, and 9 days in obstructed and unrestricted conditions. (B) Oxygen profile was detected after 5 days in culture using Image-it Hypoxia reagent. Images are shown as heat-map hypoxia-induced fluorescence intensity. Hypoxia-induced fluorescence intensity across the microchamber revealed that the oxygen concentration was constant under unrestricted conditions, whereas an oxygen gradient was established under obstructed conditions. The graph shows the hypoxia-induced fluorescence intensity profile across the microchamber [39].

In the experiment that we shall simulate with our mathematical framework, we follow a slightly different approach. The same concentration of cells (4 × $10^6$ cells/mL) was seeded within the central chamber, but now the growth medium was perfused through both lateral microchannels, simulating two functional blood vessels. We wanted to see if we were able to maintain cell culture long enough to spontaneously create hypoxic conditions in the center of the chamber and induce cell migration and invasion of both lateral channels, that way mimicking double pseudopalisade formation.

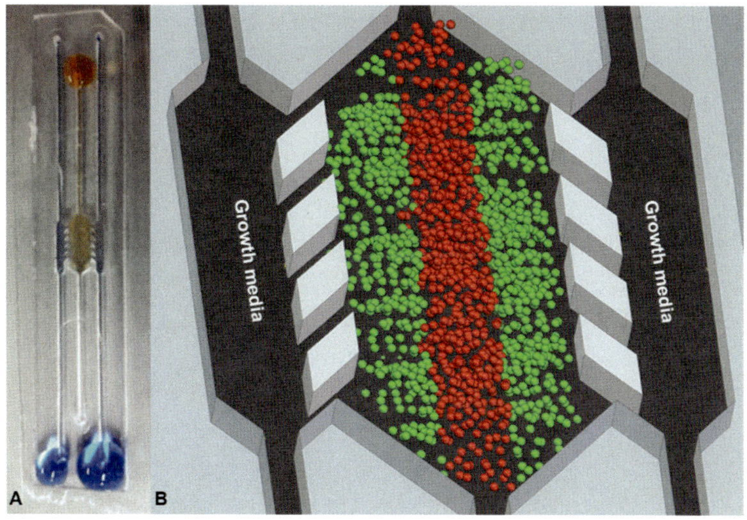


FIG. 16.3 (A) Microdevice confined with collagen hydrogel in the central chamber and *blue-colored* water perfused through the lateral channels. The droplets are left on the inlets to prevent evaporation. (B) Schematic view of the central region of the microdevice and necrotic core formation [40].

## 16.3 EXPERIMENT DESCRIPTION: MATERIALS AND METHODS

Microfluidic devices, made of cyclic olefin polymer, consist of a 2000-μm wide central chamber and two 700-μm wide lateral microchannels, interconnected with parallelogram-shaped pillars (see Fig. 16.3). The devices were fabricated by injection molding and attached to a Petri dish using a biocompatible adhesive.

In our experiments, the U251-MG human GBM cell line was used. The cells were transduced with a green fluorescent protein-expressing lentiviral vector [41], so while remaining alive, the cells produce fluorescent protein and can be visualized by microscopy. This permits following cellular behavior within the microfluidic device during long periods of time. A 3D cell culture was achieved by embedding cells into collagen type I hydrogel, which mimics the extracellular matrix. The concentration of cells used in experiments was $40 \times 10^6$ cells/mL for the necrotic core and $4 \times 10^6$ cells/mL for the pseudopalisade formation. The final concentration of collagen was 1.2 mg/mL. The mixture was injected into the central chamber of the microfluidic device using a micropipette and a 10-μL droplet was placed on the inlet to prevent evaporation. Earlier mentioned pillars prevented hydrogel leakage to lateral channels, so the cells are only located in the central microchamber. To obtain a homogeneous 3D cell distribution, the devices were turned up and down every 20 s for 3 min before being placed into an incubator for 15 min to promote collagen polymerization. Afterward, growth medium was added to the lateral microchannels, imitating blood vessels, and was refreshed every 24 h [40]. Cell culture was maintained for 6 and 21 days for the necrotic core and for the double pseudopalisade formation experiments, respectively.

Cell viability was detected using calcein AM/propidium iodide (CAM/PI) staining, where CAM becomes fluorescent once it reaches the cytoplasm of a viable cell and the PI stains dead cells, with a destroyed membrane.

Images were taken using confocal microscopy and the fluorescence intensity was measured using automated Fiji software. At the beginning of the experiment, cells were uniformly distributed through the central chamber, but as time passed and the oxygen level decreased in the center of the chamber, the cells started to migrate toward lateral microchannels, where the oxygen concentration is higher, and invade them, imitating the obstruction of blood vessels.

## 16.4 MATHEMATICAL FRAMEWORK

In this section, we describe the different equations that control the evolution of the number of cells for each phenotype and the concentration of each chemical substance present in the microfluidic device along time. In general, we shall consider the possible interdependence between the different cell types (dead cells are considered here as a particular phenotype without the capacity for proliferation, migration, or differentiation, but that can be dragged by the fluid flow) and the chemical species. In general, we shall consider also the influence of temperature, stiffness, strain, and electric potential, although in this first version and looking at the particular problems that will be simulated, we consider that this influence is uncoupled with the rest of the equations. We assume that the mechanical variables (fluid velocity, strain, and stiffness of the hydrogel) as well as temperature and electric potential are known or have been computed previously by the corresponding set of equations. This framework can be easily extended by coupling these latter variables with the cell number and species concentrations.

This framework is sufficiently general to model almost any possible biological problem. However, it is too complex in terms of the solution of the equations (highly nonlinear, strongly coupled, and very stiff equations), the number of parameters involved (many times unknown or difficult to measure), and the difficulty in validation. This is why this general framework is particularized and simplified for the problem of interest, keeping the most important influences and discarding those that can be considered comparatively negligible. Here, we shall see how to simplify it by focusing the research only on the most relevant aspects of our particular experiments. For example, as we will see, the dependence on external variables is not considered in our experiments, so these equations of the general framework will not be used.

## 16.4.1 Balance Equations for Cell Populations and Species

### 16.4.1.1 Cell Populations

The cell concentration (number of cells per unit volume) for each type of cell population is represented as continuous fields $C_i(\mathbf{x}, t)$, $i = 1, \ldots, n$ where $n$ is the number of cell populations dependent of space $\mathbf{x} \in \mathbb{R}^n$ and time $t \in \mathbb{I} \subset \mathbb{R}$. We note as $\mathrm{C}(\mathbf{x}, t) = (C_1, \ldots, C_n)^T$. The master equation that regulates cell population evolution is the transport equation with source terms considering the possible three standard reaction-convection-diffusion phenomena. For the $i$th cellular phenotype $C_i$, $i = 1, \ldots, n$, this equation is

$$\frac{\partial C_i}{\partial t} + (\mathbf{v} \cdot \nabla) C_i + \nabla \cdot \mathbf{q}_i = C_i F_i - \sum_{\substack{j=1 \\ j \neq i}}^{n} C_i F_{ij} + \sum_{\substack{j=1 \\ j \neq i}}^{n} C_j F_{ji}, \quad i = 1, \ldots, n \quad \text{in } \Omega \tag{16.1}$$

where $\Omega \subset \mathcal{R}^3$ represents the domain of study, v is the fluid velocity (convection term), $\mathbf{q}_i$ is the flux vector (number of cells at each point per unit surface and per unit time) associated with the migration of phenotype $i$ (diffusion and taxis terms), and $F_i$ is the source term corresponding to the population growth (number of new cells per unit cell and per unit time), $F_{ij}$ is the source (reaction) term corresponding to phenotype switching, that is, the number of cells that differentiates from phenotype $i$ to phenotype $j$ per unit cell and time. All biological phenomena that influence cell migration (different types of taxis) will be modeled by expressions affecting the flux vector $\mathbf{q}_i$.

Boundary conditions are defined in the boundary of the domain $\Omega$, $\partial\Omega = \Gamma_{D_i} \cup \Gamma_{R_i}$, where $\Gamma_{D_i}$ corresponds to the part of the boundary where the concentration $C_i$ is known (Dirichlet boundary part):

$$C_i = f_i, \quad i = 1, \ldots, n \quad \text{in } \Gamma_{D_i} \tag{16.2}$$

while $\Gamma_{R_i}$ corresponds to the Neumann boundary region, where the following general expression is fulfilled:

$$\kappa_i C_i + \frac{\partial C_i}{\partial \mathbf{n}} = g_i, \quad i = 1, \ldots, n \quad \text{in } \Gamma_{R_i} \tag{16.3}$$

### 16.4.1.2 Species Concentrations

Similarly, the transport equation for the $i$th chemical species $S_i$, $i = 1, \ldots, m$, including the reaction-convection-diffusion phenomena, is

$$\frac{\partial S_i}{\partial t} + (\mathbf{v} \cdot \nabla) S_i + \nabla \cdot \mathbf{q}'_i = \sum_{j=1}^{n} C_j F'_{ij}, \quad i = 1, \ldots, m \quad \text{in } \Omega \tag{16.4}$$

Again, $\mathbf{v}$ is the fluid velocity, $\mathbf{q}'_i$ is the flux vector associated with the chemical species $i$, and $F'_{ij}$ is the net source term corresponding to the production/consumption of species $i$ per unit cell of phenotype $j$. Chemical phenomena influencing species transport are again modeled using flux vectors.

The boundary conditions are defined again in $\partial\Omega = \Gamma_{D'_i} \cup \Gamma_{R'_i}$ with $S_i$ satisfying in the Dirichlet part, $\Gamma_{D'_i}$:

$$S_i = f'_i, \quad i = 1, \ldots, m \quad \text{in } \Gamma_{D'_i} \tag{16.5}$$

and in the Neumann part, $\Gamma_{R'_i}$, $S_i$:

$$\kappa'_i S_i + \frac{\partial S_i}{\partial \mathbf{n}} = g'_i, \quad i = 1, \ldots, m \quad \text{in } \Gamma_{R'_i} \tag{16.6}$$

### 16.4.2 Physical Models for Fluxes and Sources

In order to close the partial differential equation system (PDEs) given by Eqs. (16.1), (16.4), we need to model the biological transport phenomena, that is, to make explicit the relations between the flux vector fields $\mathbf{q}_i$ and $\mathbf{q}_i'$ and other physical variables that promote or inhibit cell migration (and/or substance diffusion) such as temperature (thermotaxis), electric potential (electrotaxis), substrate stiffness (durotaxis), strain (tensotaxis) (these latter two are usually included under the term mechanotaxis), or chemical species concentration (chemotaxis) as well as cell concentrations themselves (proper diffusion by random walk).

In addition to cell migration, growth, and differentiation of cellular phenotypes and consumption and production of chemical species, the reaction terms can also be dependent on similar variables, so they have to be expressed in terms of the same fields.

#### *16.4.2.1 Source Terms in Cell Population Equations*

First, the source terms in Eq. (16.1), which include proliferation $F_i$ and differentiation $F_{ij}$, will be analyzed.

##### 16.4.2.1.1 PROLIFERATION

The proliferation of the $i$th phenotype can be expressed using an equation of the type:

$$F_i = \mathcal{F}_i(C_1,\ldots,C_n,S_1,\ldots,S_m,\theta,p_1,\ldots,p_k), \quad i=1,\ldots,n \tag{16.7}$$

Here, $\theta$ is the temperature field and $p_1, \ldots, p_k$ are mechanical parameters characterizing the substrate (stiffness, anisotropy if required, strain, etc.). An example of proliferation models is the exponential growth model where $\mathcal{F}_i$ is written as

$$\mathcal{F}_i(C_1,\ldots,C_n,S_1,\ldots,S_m,\theta,p_1,\ldots,p_k) = r(\theta)$$

with $r$ the growth rate, depending only on the temperature, independent on the current concentration, and leading therefore to an exponential growth of the number of cells per unit volume. Another possible model is the so-called logistic growth model, where

$$\mathcal{F}_i(C_1,\ldots,C_n,S_1,\ldots,S_m,\theta,p_1,\ldots,p_k) = r_{max}(\theta)\left(1-\frac{\sum_{j=1}^{n} C_j}{C_{sat}}\right)$$

with $r_{max}$ the maximum growth rate, depending also on the temperature, and $C_{sat}$ the saturation parameter.

##### 16.4.2.1.2 DIFFERENTIATION

The differentiation of the $i$th phenotype to the $j$th phenotype can be modeled using a similar equation:

$$F_{ij} = \mathcal{F}_{ij}(C_1,\ldots,C_n,S_1,\ldots,S_m,\theta,p_1,\ldots,p_k), \quad i=1,\ldots,n \tag{16.8}$$

This very general expression is usually simplified to

$$\mathcal{F}_{ij}(C_1,\ldots,C_n,S_1,\ldots,S_m,\theta,p_1,\ldots,p_k) = \frac{1}{\tau}H(g(S_1,\ldots,S_m,\theta,p_1,\ldots,p_k)) \tag{16.9}$$

where $\tau$ is a characteristic time, $H$ is an activation function, for example, the Heaviside function ($H(x) = 0$ if $x \leq 0$ and $H(x) = 1$ if $x > 0$), or the sigmoid function ($H(x) = \frac{1}{2}(1 + \tanh(x))$), and $g$ is a function defining the domain of physiological behavior of a cell, that is, if $g(S_1, \ldots, S_m, \theta, p_1, \ldots, p_k) \leq 0$, the cell is in its physiological state without stressed or pathological behavior. A very simple example of these models is the one known as the threshold model, where

$$g(S_1,\ldots,S_m,\theta,p_1,\ldots,p_k) = \max\{S_l^{th} - S_l, 1 \leq l \leq m\} \tag{16.10}$$

In this model, the switch occurs when at least one of the chemical species influencing the metabolic behavior of the cell falls below a certain threshold. A more general approach, the $p$-norm threshold model, is stated as:

$$g(S_1,\ldots,S_m,\theta,p_1,\ldots,p_k) = \| \mathbf{S}^{th} - \mathbf{S} \|_p \tag{16.11}$$

where $\mathbf{S}^{th}$ is a vector of species thresholds. Analogous models can be formulated including other variables such as $\theta$ and $p_i$, $i = 1, \ldots, k$.

#### 16.4.2.2 Migration Terms in Cell Population Equations

Now, migration terms (biological related transport) are defined. The flux vector $q_i$ is postulated as a linear decomposition:

$$\mathbf{q}_i = \mathbf{q}_{D,i} + \mathbf{q}_{m,i} + \mathbf{q}_{s,i} + \mathbf{q}_{E,i} + \mathbf{q}_{T,i}, \quad i=1,\ldots,n \tag{16.12}$$

where the flux vectors $\mathbf{q}_{D,i}$, $\mathbf{q}_{m,i}$, $\mathbf{q}_{s,i}$, $\mathbf{q}_{E,i}$, and $\mathbf{q}_{T,i}$ are associated with diffusion, mechanotaxis, chemotaxis, electrotaxis, and thermotaxis phenomena, respectively.

##### 16.4.2.2.1 DIFFUSION

The common framework is used to model diffusion, that is

$$\mathbf{q}_{D,i} = -\mathbf{K}_{D,i}\nabla C_i, \quad i=1,\ldots,n \tag{16.13}$$

Here, matrix $\mathbf{K}_{D,\ i}$ is the diffusion matrix that can be expressed as

$$\mathbf{K}_{D,i} = \mathcal{K}_{D,i}(C_1,\ldots,C_n,\theta,p_1,\ldots,p_k), \quad i=1,\ldots,n \tag{16.14}$$

##### 16.4.2.2.2 MECHANOTAXIS

The mechanotaxis term can be expressed in a similar manner, using the expression

$$\mathbf{q}_{m,i} = \sum_{j=1}^{k}\mathbf{K}_{m,i,j}\nabla p_j, \quad i=1,\ldots,n \tag{16.15}$$

where $\mathbf{K}_{m,\ i,\ j}$ is the mechanotactic motility matrix of species $i$ with respect to parameter $p_j$ expressed as:

$$\mathbf{K}_{m,i,j} = \mathcal{K}_{m,i,j}(C_1,\ldots,C_n,p_j,\nabla p_j,\theta), \quad i=1,\ldots,n \tag{16.16}$$

##### 16.4.2.2.3 CHEMOTAXIS

The chemotaxis term is expressed as

$$\mathbf{q}_{s,i} = \sum_{j=1}^{m}\mathbf{K}_{s,i,j}\nabla S_j, \quad i=1,\ldots,n \tag{16.17}$$

Here, $\mathbf{K}_{s,\ i,\ j}$ is the chemotactic motility matrix with respect to the chemical species $j$. As before, it is possible to express the matrix as

$$\mathbf{K}_{s,i,j} = \mathcal{K}_{s,i,j}(C_1,\ldots,C_n,S_1,\ldots,S_m,\nabla S_1,\ldots,\nabla S_m,\theta,p_1,\ldots,p_k), \quad i=1,\ldots,n, \quad j=1,\ldots,m \tag{16.18}$$

It is common to use saturation models for chemotaxis [42], getting:

$$\mathcal{K}_{s,i,j}(C_1,\ldots,C_n,S_1,\ldots,S_m,\nabla S_1,\ldots,\nabla S_m,\theta,p_1,\ldots,p_k) = \frac{\chi_i C_i}{\sigma_j + \lambda_j \parallel \nabla S_j \parallel} \tag{16.19}$$

where $\sigma_j$ and $\lambda_j$ are parameters depending on the chemical species $j$ and $\chi_i$ is a sensitivity parameter depending on the cellular phenotype $i$.

##### 16.4.2.2.4 ELECTROTAXIS

For electrotaxis, a similar expression can be written:

$$\mathbf{q}_{E,i} = \mathbf{K}_{E,i}\nabla V, \quad i=1,\ldots,n \tag{16.20}$$

where $K_{E,\ i}$ is the electrostatic motility matrix and $V$ is the electric potential. As before,

$$\mathbf{K}_{E,i} = \mathcal{K}_{E,i}(C_1,\ldots,C_n,\theta,p_1,\ldots,p_k,V,\nabla V), \quad i=1,\ldots,n \tag{16.21}$$

##### 16.4.2.2.5 THERMOTAXIS

Thermotaxis can be modeled using analogous equations:

$$\mathbf{p}_{T,i} = \mathbf{K}_{T,i}\nabla\theta, \quad i=1,\ldots,n \tag{16.22}$$

where $\mathbf{K}_{T,\ i}$ is the thermotactic motility matrix and $\theta$ is the temperature field, with

$$\mathbf{K}_{T,i} = \mathcal{K}_{T,i}(C_1,\ldots,C_n,\theta,\nabla\theta,p_1,\ldots,p_k,), \quad i=1,\ldots,n \tag{16.23}$$

#### 16.4.2.3 Source Terms and Diffusion for Chemical Species

Source terms in Eq. (16.4) include production and consumption of species. The net production/consumption of the $i$th chemical species by the $j$th cell phenotype can be modeled using an equation of the kind:

$$F'_{ij} = \mathcal{F}'_{ij}(C_1,\ldots,C_n,S_1,\ldots,S_m,\theta), \quad i=1,\ldots,m, \quad j=1,\ldots,n \tag{16.24}$$

##### 16.4.2.3.1 DIFFUSION

As for cellular phenotypes, diffusion is stated as

$$\mathbf{q}'_i = -\mathbf{K}'_{D,i}\nabla S_i, \quad i=1,\ldots,m \tag{16.25}$$

Here, matrix $\mathbf{K}'_{D,i}$ is the diffusion matrix that will be expressed as

$$\mathbf{K}'_{D,i} = \mathcal{K}'_{D,i}(C_1,\ldots,C_n,\theta,p_1,\ldots,p_k), \quad i=1,\ldots,m \tag{16.26}$$

### 16.4.3 ECM Remodeling Coupling

One last step in this global framework corresponds to regeneration considerations. Mechanical parameters of the substrate $p_1, \ldots, p_k$ can be seen as constant parameters, such as elastic parameters (Young modulus $E$, Poisson coefficient $\nu$, etc.), or may be thought of as evolving parameters. In this latter approach, a dynamic approach to the problem is adopted, being necessary to define an evolution relationship:

$$\dot{p}_i = \mathcal{R}_i(C_1,\ldots,C_n,\theta), \quad i=1,\ldots,k \tag{16.27}$$

## 16.5 IMPLEMENTATION

### 16.5.1 3D Finite Element Implementation

#### 16.5.1.1 Weak Form

The weak form of Eqs. (16.1), (16.4) is derived. Let $\phi \in \mathcal{H}^1_0(\Omega), i=1, \ldots, n, j=1, \ldots, m$, being $\mathcal{H}^1_0(\Omega)$ the closure in $\mathcal{H}^1(\Omega)$ ($\mathcal{H}^1(\Omega) = W^{1,2}(\Omega)$), with this latter, the Sobolev space with respect to the $L^2$ norm of differentiable functions (in the weak sense) of order 1 of infinitely differentiable functions compactly supported in $\Omega$ or equivalently, the space of functions in $\mathcal{H}^1(\Omega)$ that vanish at the boundary $\partial\Omega$.

Eq. (16.1) can be multiplied by a test function $\phi \in \mathcal{H}^1_0(\Omega)$ and integrated in $\Omega$ to obtain

$$\int_\Omega \phi\frac{\partial C_i}{\partial t} + \phi\nabla\cdot\mathbf{A}_i d\Omega = \int_\Omega a_i\phi d\Omega, \quad i=1,\ldots,n \tag{16.28}$$

where we have defined

$$\begin{aligned} \mathbf{A}_i &= C_i\mathbf{v} + \mathbf{q}_i \\ a_i &= C_iF_i + \sum_{\substack{j=1 \\ j\neq i}}^{n} (C_jF_{ji} - C_iF_{ij}) \end{aligned}$$

Integrating by parts Eq. (16.28), splitting the boundary $\partial\Omega = \Gamma_{D_i} \cup \Gamma_{R_i}$ where, respectively, Dirichlet or Robin-Neumann boundary conditions are applied, and using that $\phi$ vanishes at the boundary, results in:

$$\int_\Omega \left(\phi\frac{\partial C_i}{\partial t} - \nabla\phi\cdot\mathbf{A}_i\right)d\Omega = \int_\Omega \phi a_i d\Omega - \int_{\Gamma_{R_i}} \phi\frac{\partial C_i}{\partial\mathbf{n}}d\Gamma, \quad i=1,\ldots,n \tag{16.29}$$

Multiplying Eq. (16.2) by a test function $\phi$ and integrating in $\Gamma_{D_i}$ we arrive to

$$\int_{\Gamma_{D_i}} \phi C_i d\Gamma = \int_{\Gamma_{D_i}} \phi f_i d\Gamma, \quad i=1,\ldots,n \tag{16.30}$$

and repeating it with Eq. (16.3) and integrating in $\Gamma_{R_i}$ we obtain

$$\int_{\Gamma_{R_i}} \kappa_i \phi C_i d\Gamma + \int_{\Gamma_{R_i}} \phi \frac{\partial C_i}{\partial \mathbf{n}} d\Gamma = \int_{\Gamma_{R_i}} \phi g_i d\Gamma, \quad i=1,\ldots,n \tag{16.31}$$

The same strategy is followed with Eq. (16.4) obtaining

$$\int_\Omega \phi \frac{\partial S_i}{\partial t} + \phi \nabla \cdot \mathbf{B}_i d\Omega = \int_\Omega b_i \phi d\Omega, \quad i=1,\ldots,m \tag{16.32}$$

where we have defined

$$\begin{aligned} \mathbf{B}_i &= S_i \mathbf{v} + \mathbf{q}_i' \\ b_i &= \sum_{j=1}^{m} C_j (F_{ji}' - F_{ij}') \end{aligned}$$

A new integration by parts of Eq. (16.32) leads to

$$\int_\Omega \left( \phi \frac{\partial S_i}{\partial t} - \nabla \phi \cdot \mathbf{B}_i \right) d\Omega = \int_\Omega \phi b_i d\Omega - \int_{\Gamma_{R_i'}} \phi \frac{\partial S_i}{\partial \mathbf{n}} d\Gamma, \quad i=1,\ldots,m \tag{16.33}$$

We do analogously with boundary conditions (16.5), (16.6) in order to obtain

$$\int_{\Gamma_{D_i'}} \phi S_i d\Gamma = \int_{\Gamma_{D_i'}} \phi f_i' d\Gamma, \quad i=1,\ldots,m \tag{16.34}$$

and repeating it with Eq. (16.3) and integrating in $\Gamma_{R_i'}$ we get

$$\int_{\Gamma_{R_i'}} \kappa_i' \phi S_i d\Gamma + \int_{\Gamma_{R_i'}} \phi \frac{\partial S_i}{\partial \mathbf{n}} d\Gamma = \int_{\Gamma_{R_i'}} \phi g_i' d\Gamma, \quad i=1,\ldots,m \tag{16.35}$$

#### *16.5.1.2 Spatial Discretization*

All scalar and vectorial fields involved in the problem are discretized using a finite basis of dimension $N$, $\mathcal{B} = \{\phi_r, r=1,\ldots,N\}$, that is,

$$\begin{aligned} C_i(\mathbf{x},t) &= \sum_{r=1}^{N} C_i^r(t)\phi_r(\mathbf{x}), \quad i=1,\ldots,n \\ S_i(\mathbf{x},t) &= \sum_{r=1}^{N} S_i^r(t)\phi_r(\mathbf{x}), \quad i=1,\ldots,m \\ \theta(\mathbf{x},t) &= \sum_{r=1}^{N} \theta^r(t)\phi_r(\mathbf{x}) \\ V(\mathbf{x},t) &= \sum_{r=1}^{N} V^r(t)\phi_r(\mathbf{x}) \\ p_k(\mathbf{x},t) &= \sum_{r=1}^{N} p_k^r(t)\phi_r(\mathbf{x}), \quad i=1,\ldots,k \\ \mathbf{v} &= \sum_{r=1}^{N} \mathbf{v}^r(t)\phi_r(\mathbf{x}) \end{aligned} \tag{16.36}$$

Note that all fields are approximated using the same basis and a separation representation in space time is postulated. Even if it is not the most general approach, the fact that Eqs. (16.29), (16.33) involve up to first-order derivatives is consistent with this FE approximation.

#### 16.5.1.2.1 CELL POPULATIONS

Plugging Eq. (16.36) into Eq. (16.29), we obtain

$$\int_{\Omega}\left(\phi\sum_{r=1}^{N}\frac{\partial C_i^r}{\partial t}\phi_r d\Omega-\int_{\Omega}\nabla\phi\cdot\mathbf{A}_i\right)d\Omega=\int_{\Omega}\phi a_i d\Omega-\int_{\Gamma_{R_i}}\phi\frac{\partial C_i}{\partial\mathbf{n}}d\Gamma,\ \ i=1,\dots,n \tag{16.37}$$

Using linear properties of $\nabla$ operator, Eqs. (16.13), (16.15), (16.17), (16.20), (16.22) may be written as

$$\begin{aligned}
\mathbf{q}_{D,i} &= -\sum_{r=1}^{N}(C_i^r\mathbf{K}_{D,i})\nabla\phi_r,\ \ i=1,\dots,n\\
\mathbf{q}_{m,i} &= \sum_{r=1}^{N}\left(\sum_{j=1}^{k}p_j^r\mathbf{K}_{m,i,j}\right)\nabla\phi_r,\ \ i=1,\dots,n\\
\mathbf{q}_{s,i} &= \sum_{r=1}^{N}\left(\sum_{j=1}^{n}S_j^r\mathbf{K}_{s,i,j}\right)\nabla\phi_r,\ \ i=1,\dots,n\\
\mathbf{q}_{E,i} &= \sum_{r=1}^{N}(\theta^r\mathbf{K}_{E,i})\nabla\phi_r,\ \ i=1,\dots,n\\
\mathbf{q}_{T,i} &= \sum_{r=1}^{N}(V^r\mathbf{K}_{T,i})\nabla\phi_r,\ \ i=1,\dots,n
\end{aligned} \tag{16.38}$$

So, as $A_i = C_i v + q_i$ it is obtained

$$\begin{aligned}
\mathbf{A}_i &= \left(\sum_{r=1}^{N}C_i^r\phi_r\right)\mathbf{v}\\
&+\sum_{r=1}^{N}\left(-C_i^r\mathbf{K}_{D,i}+\sum_{j=1}^{k}p_j^r\mathbf{K}_{m,i,j}+\sum_{j=1}^{m}S_j^r\mathbf{K}_{s,i,j}+\theta^r\mathbf{K}_{E,i}+V^r\mathbf{K}_{T,i}\right)\nabla\phi_r,\ \ i=1,\dots,n
\end{aligned} \tag{16.39}$$

And $\nabla\phi\cdot A_i$ writes

$$\begin{aligned}
\nabla\phi\cdot\mathbf{A}_i &= \sum_{r=1}^{N}C_i^r(\nabla\phi_i\cdot\phi_r\mathbf{v}-\nabla\phi_i\cdot(\mathbf{K}_{D,i}\nabla\phi_r))\\
&+\sum_{r=1}^{N}\nabla\phi\cdot\left(\sum_{j=1}^{k}p_j^r\mathbf{K}_{m,i,j}+\sum_{j=1}^{m}S_j^r\mathbf{K}_{s,i,j}+V^r\mathbf{K}_{i,E}+\theta^r\mathbf{K}_{i,T}\right)\nabla\phi_r,\ \ i=1,\dots,n
\end{aligned} \tag{16.40}$$

Besides, as $C_i^r = C_i^r(t)$, $\phi\frac{\partial C_i^r}{\partial t}$ writes

$$\phi\sum_{r=1}^{N}\frac{\partial C_i^r}{\partial t}\phi_r=\phi\sum_{r=1}^{N}\dot{C}_i^r\phi_r,\ \ i=1,\dots,n \tag{16.41}$$

Choosing $\phi = \phi_s$, $s = 1, \dots, N$ (Galerkin method) and reorganizing terms in Eq. (16.40), we may write

$$\nabla\phi_s\cdot\mathbf{A}_i=\sum_{r=1}^{N}\hat{\mathbb{K}}_{sr}^{(i)}C_i^r+\sum_{j=1}^{m}\sum_{r=1}^{N}\hat{\mathbb{K}}_{sr}^{(ij)}S_j^r+\hat{\mathbf{F}}_s^{(i)},\ \ i=1,\dots,n \tag{16.42}$$

with

$$\begin{aligned}
\hat{\mathbb{K}}_{sr}^{(i)} &= (\nabla\phi_s\cdot(\phi_r\mathbf{v})-\nabla\phi_s\cdot(\mathbf{K}_{D,i}\nabla\phi_r)),\ \ i=1,\dots,n\\
\hat{\mathbb{K}}^{(ij)} &= \nabla\phi_s\cdot\mathbf{K}_{s,i,j}\nabla\phi_r,\ \ i=1,\dots,n\\
\hat{\mathbf{F}}_s^{(i)} &= \nabla\phi_s\cdot\left[\sum_{r=1}^{N}\left(\left(\sum_{j=1}^{k}p_j^r\right)\mathbf{K}_{m,i,j}+V^r\mathbf{K}_{E,i}+\theta^r\mathbf{K}_{T,i}\right)\nabla\phi_r\right],\ \ i=1,\dots,n
\end{aligned} \tag{16.43}$$

And Eq. (16.41) becomes

$$\phi_s \sum_{r=1}^{N} \frac{\partial C_i^r}{\partial t} \phi_r = \sum_{r=1}^{N} \hat{\mathbb{M}}_{sr}^{(i)} C_i^r, \quad i=1,\ldots,n \tag{16.44}$$

where we have defined

$$\hat{\mathbb{M}}_{sr} = \phi_s \phi_r \tag{16.45}$$

Finally, Eq. (16.30) with $\phi = \phi_s$, $C_i = \sum_{r=1}^{N} C_i^r \phi_r$ and $f_i = \sum_{r=1}^{N} f_i^r \phi_r$ becomes

$$C_i^r = f_i^r, \quad i=1,\ldots,n \tag{16.46}$$

and Eq. (16.31) with $\phi = \phi_s$ and $C_i = \sum_{r=1}^{N} C_i^r \phi_r$:

$$\sum_{r=1}^{N} \left( \int_{\Gamma_{R_i}} \kappa_i \phi_s \phi_r d\Gamma \right) C_i^r + \int_{\Gamma_{R_i}} \phi_s \frac{\partial C_i}{\partial \mathbf{n}} d\Gamma = \int_{\Gamma_{R_i}} \phi_s g_i d\Gamma, \quad i=1,\ldots,n \tag{16.47}$$

#### 16.5.1.2.2 CHEMICAL SPECIES

In the same way as above, plugging now Eq. (16.36) into Eq. (16.33) we obtain

$$\int_{\Omega} \left( \phi \sum_{r=1}^{N} \frac{\partial S_i^r}{\partial t} \phi_r d\Omega - \int_{\Omega} \nabla\phi \cdot \mathbf{B}_i \right) d\Omega = \int_{\Omega} \phi b_i d\Omega - \int_{\Gamma_{R_i'}} \phi \frac{\partial S_i}{\partial \mathbf{n}} d\Gamma, \quad i=1,\ldots,m \tag{16.48}$$

Using again the linear properties of $\nabla$ operator, Eq. (16.25) writes

$$\mathbf{q}_i' = -\sum_{r=1}^{N} (S_i^r \mathbf{K}_{D,i}') \nabla\phi_r, \quad i=1,\ldots,m \tag{16.49}$$

So, as $\mathbf{B}_i = S_i \mathbf{v} + \mathbf{q}_i'$ it is obtained

$$\mathbf{B}_i = \left( \sum_{r=1}^{N} S_i^r \phi_r \right) \mathbf{v} - \sum_{r=1}^{N} (S_i^r \mathbf{K}_{D,i}') \nabla\phi_r, \quad i=1,\ldots,m \tag{16.50}$$

So, $\nabla\phi \cdot \mathrm{B}_i$ writes

$$\nabla\phi \cdot \mathbf{B}_i = \sum_{r=1}^{N} S_i^r \nabla\phi \cdot \phi_r \mathbf{v} - \nabla\phi \cdot (\mathbf{K}_{D,i}' \nabla\phi_r), \quad i=1,\ldots,m \tag{16.51}$$

Besides, as $S_i^r = S_i^r(t)$, $\phi \frac{\partial S_i^r}{\partial t}$ writes

$$\phi \sum_{r=1}^{N} \frac{\partial S_i^r}{\partial t} \phi_r = \phi \sum_{r=1}^{N} \dot{S}_i^r \phi_r, \quad i=1,\ldots,m \tag{16.52}$$

Choosing as before $\phi = \phi_s$, $s = 1, \ldots, N$ (Galerkin method) and reorganizing terms in Eq. (16.51) we may write

$$\nabla\phi_s \cdot \mathbf{B}_i = \sum_{r=1}^{N} \hat{\mathbb{K}}_{sr}^{\prime(i)} S_i^r, \quad i=1,\ldots,m \tag{16.53}$$

where we have defined

$$\hat{\mathbb{K}}_{sr}^{\prime(i)} = (\nabla\phi_s \cdot (\phi_r \mathbf{v}) - \nabla\phi_s \cdot \mathbf{K}_{D,i}' \nabla\phi_r), \quad i=1,\ldots,m \tag{16.54}$$

And Eq. (16.52) becomes

$$\phi_s \sum_{r=1}^{N} \frac{\partial S_i^r}{\partial t} \phi_r = \sum_{r=1}^{N} \hat{\mathbb{M}}_{sr}^{(i)} S_i^r, \quad i=1,\ldots,m \tag{16.55}$$

Finally, Eq. (16.34) with $= _s$, $S_i = \sum_{r=1}^{N} S_i^r \phi_r$, and $f_i' = \sum_{r=1}^{N} f_i'^r \phi_r$ becomes

$$S_i^r = f_i^r, \quad i = 1, \ldots, m \tag{16.56}$$

and Eq. (16.35) with $\phi = \phi_s$ and $S_i = \sum_{r=1}^{N} S_i^r \phi_r$:

$$\sum_{r=1}^{N} \left( \int_{\Gamma_{R_i'}} \kappa_i' \phi_s \phi_r d\Gamma \right) S_i^r + \int_{\Gamma_{R_i'}} \phi_s \frac{\partial S_i}{\partial \mathbf{n}} d\Gamma = \int_{\Gamma_{R_i'}} \phi_s g_i' d\Gamma, \quad i = 1, \ldots, m \tag{16.57}$$

#### 16.5.1.2.3 COMPACT FORM

For notation purposes, we define the vectors $\mathbf{C}_i = (C_i^1, \ldots, C_i^N)$, $\mathbf{S}_i = (S_i^1, \ldots, S_i^N)$, $\boldsymbol{\theta} = (\theta^1, \ldots, \theta^N)$, $V = (V^1, \ldots, V^N)$, $W = (v^1, \ldots, v^N)$, and $\mathbf{P}_k = (p_k^1, \ldots, p_k^N)$.

With these notations, Eq. (16.37) with $\phi = \phi_s$, and using Eq. (16.47), becomes

$$\mathbb{M}^{(i)} \dot{\mathbf{C}}_i + \mathbb{K}^{(i)} \mathbf{C}_i + \sum_{j=1}^{m} \mathbb{K}^{(ij)} \mathbf{S}_j = \mathbf{F}^{(i)}, \quad i = 1, \ldots, n \tag{16.58}$$

with

$$\begin{aligned}
\mathbb{K}_{sr}^{(i)} &= \int_{\Omega} \hat{\mathbb{K}}_{sr}^{(i)} d\Omega + \int_{\Gamma_{R_i}} \kappa_i \phi_s \phi_r d\Gamma, \quad i = 1, \ldots, n \\
\mathbb{K}_{sr}^{(ij)} &= \int_{\Omega} \hat{\mathbb{K}}_{sr}^{(ij)} d\Omega + \int_{\Gamma_{R_j}} \kappa_j \phi_s \phi_r d\Gamma, \quad i = 1, \ldots, n \\
\mathbb{M}_{sr}^{(i)} &= \int_{\Omega} \hat{\mathbb{M}}_{sr} d\Omega, \quad i = 1, \ldots, n \\
\mathbf{F}_s^{(i)} &= \int_{\Omega} f_i \phi_s d\Omega + \int_{\Gamma_{R_i}} \phi_s g_i d\Gamma - \int_{\Omega} \hat{\mathbf{F}}_s^{(i)} d\Omega, \quad i = 1, \ldots, n
\end{aligned} \tag{16.59}$$

Defining $C = (C_1, \ldots, C_n)$, $S = (S_1, \ldots, S_m)$, $P = (P_1, \ldots, P_k)$, $\mathbb{M}_C = \oplus_{i=1}^{n} \mathbb{M}^{(i)}$,

$$\mathbb{K}_C = \begin{bmatrix} \mathbb{K}^{(1)} & 0 & \cdots & 0 \\ 0 & \mathbb{K}^{(2)} & \cdots & 0 \\ \cdots & \cdots & \cdots & \cdots \\ 0 & 0 & \cdots & \mathbb{K}^{(nn)} \end{bmatrix} \tag{16.60}$$

$$\mathbb{K}_{CS} = \begin{bmatrix} \mathbb{K}^{(11)} & \mathbb{K}^{(12)} & \cdots & \mathbb{K}^{(1m)} \\ \mathbb{K}^{(21)} & \mathbb{K}^{(22)} & \cdots & \mathbb{K}^{(2m)} \\ \cdots & \cdots & \cdots & \cdots \\ \mathbb{K}^{(n1)} & \mathbb{K}^{(n2)} & \cdots & \mathbb{K}^{(nm)} \end{bmatrix} \tag{16.61}$$

and $F_C = (F^{(1)}, \ldots, F^{(n)})$, Eq. (16.58) for $i = 1, \ldots, n$ can be expressed in a compact form as:

$$\mathbb{M}_C \dot{\mathbf{C}} + \mathbb{K}_C \mathbf{C} + + \mathbb{K}_{CS} \mathbf{S} = \mathbf{F}_C \tag{16.62}$$

If we make explicit the functional dependencies we have

$$\mathbb{M}_C \dot{\mathbf{C}} + \mathbb{K}_C(\mathbf{C}, \mathbf{S}, \mathbf{P}, \mathbf{W}, \boldsymbol{\theta}) \mathbf{C} + \mathbb{K}_{CS} \mathbf{S}(\mathbf{C}, \mathbf{S}, \mathbf{P}, \boldsymbol{\theta}) = \mathbf{F}_C(\mathbf{C}, \mathbf{S}, \mathbf{P}, \mathbf{V}, \boldsymbol{\theta}) \tag{16.63}$$

Using the same vectorial notations, Eq. (16.48) becomes

$$\mathbb{M}'^{(i)} \dot{\mathbf{S}}_i + \mathbb{K}'^{(i)} \mathbf{S}_i = \mathbf{F}'^{(i)}, \quad i = 1, \ldots, m \tag{16.64}$$

where we have denoted

$$\mathbb{K}_{sr}'^{(i)} = \int_{\Omega} \hat{\mathbb{K}}_{sr}'^{(i)} d\Omega + \int_{\Gamma_{R_i'}} \kappa_i' \phi_s \phi_r d\Gamma, \quad i = 1, \ldots, m \tag{16.65}$$

$$\begin{aligned}
\mathbb{M}_{sr}'^{(i)} &= \int_{\Omega} \hat{\mathbb{M}}_{sr} d\Omega, \quad i = 1, \ldots, m \\
\mathbf{F}_s'^{(i)} &= \int_{\Omega} \hat{\mathbf{F}}_s'^{(i)} d\Omega + \int_{\Gamma_{R_i'}} \phi_s g_i' d\Gamma, \quad i = 1, \ldots, m
\end{aligned} \tag{16.66}$$

Defining $\mathbb{M}_S = \oplus_{i=1}^{m}\mathbb{M}'^{(i)}$, $\mathbb{K}_S = \oplus_{i=1}^{m}\mathbb{K}'^{(i)}$, and $\mathbf{F}_S = (\mathbf{F}'^{(1)},\ldots,\mathbf{F}'^{(m)})$ Eq. (16.58) for $i = 1, \ldots, m$ can be expressed in a compact form as:

$$\mathbb{M}_S\dot{\mathbf{S}} + \mathbb{K}_S\mathbf{S} = \mathbf{F}_S \tag{16.67}$$

If we explicit the functional dependences, we have

$$\mathbb{M}_S\dot{\mathbf{S}} + \mathbb{K}_S(\mathbf{C},\mathbf{S},\mathbf{P},\mathbf{W},\boldsymbol{\theta})\mathbf{S} = \mathbf{F}_S(\mathbf{C},\mathbf{S},\mathbf{P},\mathbf{V},\boldsymbol{\theta}) \tag{16.68}$$

Besides, using again vectorial notation, the regeneration term may be stated as

$$\dot{\mathbf{P}} = \mathbf{R}(\mathbf{C},\boldsymbol{\theta}) \tag{16.69}$$

Eqs. (16.63), (16.68) may be combined if we define U = (C, S), H = (W, V, $\boldsymbol{\theta}$), $\mathbb{M} = \mathbb{M}_C \oplus \mathbb{M}_S$, F = ($F_C$, $F_S$), and

$$\mathbb{K} = \begin{bmatrix} \mathbb{K}_C & \mathbb{K}_{CS} \\ 0 & \mathbb{K}_S \end{bmatrix} \tag{16.70}$$

arriving finally, to the equation:

$$\begin{aligned} \mathbb{M}\dot{\mathbf{U}} + \mathbb{K}(\mathbf{U},\mathbf{P},\mathbf{H})\mathbf{U} &= \mathbf{F}(\mathbf{U},\mathbf{P},\mathbf{H}) \\ \dot{\mathbf{P}} &= \mathbf{R}(\mathbf{U},\mathbf{H}) \end{aligned} \tag{16.71}$$

Eq. (16.71) shall be combined with Dirichlet boundary conditions, given in Eqs. (16.46), (16.56). For the former, $C_i = f_i(t)$ is known in nodes belonging to $\Gamma_{D_i}$ while for the later, $\mathbf{S}_i = \mathbf{f}_i'(t)$ is known in nodes belonging to $\Gamma_{D_i'}$. The first line of the compact Eq. (16.71), if $\overline{\mathbf{U}} = \overline{\mathbf{U}}(t)$ are the constrained variables, may be split symbolically in free and constrained variables:

$$\mathbb{M}_f\dot{\mathbf{U}}_f + \mathbb{M}_c\dot{\overline{\mathbf{U}}} + \mathbb{K}_f\mathbf{U}_f + \mathbb{K}_c\overline{\mathbf{U}} = \mathbf{F} \tag{16.72}$$

So, we finally obtain

$$\mathbb{M}^*\dot{\mathbf{U}}^* + \mathbb{K}^*\mathbf{U}^* = \mathbf{F}^* \tag{16.73}$$

where $\mathbb{M}^* = \mathbb{M}_f$, $\mathbb{K}^* = \mathbb{K}_f$, $\mathbf{F}^* = \mathbf{F} - \mathbb{M}_c\dot{\overline{\mathbf{U}}} - \mathbb{K}_c\overline{\mathbf{U}}$, and $U^* = U_f$ are the unknowns of the problem. Finally, Eq. (16.71) writes

$$\begin{aligned} \mathbb{M}^*\dot{\mathbf{U}}^* + \mathbb{K}^*(\mathbf{U}^*,\mathbf{P},\mathbf{H}(t))\mathbf{U}^* &= \mathbf{F}(\mathbf{U}^*,\mathbf{P},\mathbf{H}(t),\overline{\mathbf{U}}(t),\dot{\overline{\mathbf{U}}}(t)) \\ \dot{\mathbf{P}} &= \mathbf{R}(\mathbf{U}^*,\mathbf{H}(t),\overline{\mathbf{U}}(t)) \end{aligned} \tag{16.74}$$

#### *16.5.1.3 Time Integration*

In order to solve Eq. (16.71), that is, to compute $\mathbf{U}^* = \mathbf{U}^*(t)$ and $\mathbf{P} = \mathbf{P}(t)$, for $t \in [0;T]$, it is necessary to define the initial conditions $\mathbf{U}_o^* = \mathbf{U}^*(t = o)$ and $\mathbf{P}_0 = \mathbf{P}(t = 0)$ and to specify H = H($t$) (physical stimulus, including electrical stimulus, thermal stimulus, and flow stimulus) and $\overline{\mathbf{U}} = \overline{\mathbf{U}}(t)$ (boundary conditions) for $t \in [0;T]$.

This problem may be expressed in a standard form using the symbolic notation

$$\mathbf{x} = (\mathbf{U}^*,\mathbf{P}) \tag{16.75}$$

$$\mathbf{f}(\mathbf{x},t) = \begin{bmatrix} \mathbb{M}^{-1}(\mathbf{F}^*(\mathbf{U}^*,\mathbf{P},\mathbf{H}(t),\overline{\mathbf{U}}(t),\dot{\overline{\mathbf{U}}}(t)) - \mathbb{K}^*(\mathbf{U}^*,\mathbf{P},\mathbf{H}(t))) \\ \mathbf{R}(\mathbf{U}^*,\mathbf{H}(t),\overline{\mathbf{U}}(t)) \end{bmatrix} \tag{16.76}$$

$$\mathbf{x}_0 = (\mathbf{U}_0^*,\mathbf{P}_0) \tag{16.77}$$

thus to obtain

$$\begin{aligned} \frac{d\mathbf{x}}{dt} &= \mathbf{f}(\mathbf{x},t) \\ \mathbf{x}(t=0) &= \mathbf{x}_0 \end{aligned} \tag{16.78}$$

Eq. (16.78) is a nonlinear ordinary differential equation system that may be solved using different numerical schemes, accounting for the high nonlinearity, coupling between variables, and stiff behavior of the differential equation. Many numerical integrators have been implemented, including forward Euler, backward Euler, Midpoint, Adams-Bashforth, or adaptive Runge-Kutta solvers, such as Dormand-Prince scheme [43] or Bogacki-Shampine scheme [44].

#### 16.5.1.3.1 FORWARD EULER METHOD

Forward Euler is the simplest numerical integrator. Using forward Euler integration, Eq. (16.78) writes

$$\mathbf{x}_{t+1} = \mathbf{x}_t + \mathbf{f}_t \Delta t \tag{16.79}$$

where $x_t = x(t)$, $x_{t+1} = x(t + \Delta t)$, and $f_t = f(x_t, t)$. Eq. (16.79) is just an evaluation because it has an explicit nature. However, it is known that the forward Euler method can also be numerically unstable, especially for stiff equations, requiring very small time steps for obtaining accurate results.

#### 16.5.1.3.2 BACKWARD EULER COMBINED WITH THE BROYDEN METHOD

The backward Euler method is a numerical integrator that may work for greater time steps than forward Euler, due to its implicit nature. However, because of this, at each time-step, a multidimensional nonlinear equation must be solved. Eq. (16.78) discretized by means of the backward Euler method writes

$$\mathbf{x}_{t+1} = \mathbf{x}_t + \mathbf{f}_{t+1} \Delta t \tag{16.80}$$

where $x_t = x(t)$, $x_{t+1} = x(t + \Delta t)$, and $f_{t+1} = f(x_{t+1}, t + \Delta t)$. Eq. (16.80) is nonlinear in general and has to be solved iteratively. Computation of the tangent operator $\frac{\partial f}{\partial x}$ is computationally expensive and hard to derive from the expression of the function f. Instead of using the classical Newton method, based on the tangent operator, the multidimensional generalization of the secant method is going to be used, the Broyden method [45]. At each iteration, a secant operator, $J_n$, is going to be computed and an update of variables is performed as $\mathbf{x}_{t+1,n+1} = \mathbf{x}_{t+1,n} - \mathbf{J}_n^{-1} \mathbf{r}_{t+1,n}$, where we have defined the residual at iteration $n$ $r_{t+1,n} = x_{t+1,n} - x_{t,n} - f_{t+1,n} \Delta t$ and the secant operator is defined such as satisfying

$$\mathbf{J}_n \Delta \mathbf{x}_{t+1,n} = \Delta \mathbf{f}_{t+1,n} \tag{16.81}$$

where $\Delta x_{t+1,n} = x_{t+1,n+1} - x_{t+1,n}$ and $\Delta f_{t+1,n} = f_{t+1,n+1} - f_{t+1,n}$.

Of course, if the dimension of x and f are greater than 1, Eq. (16.81) is undetermined and further conditions shall be supplied. A possibility is to use the current estimate of the secant operator $J_{n-1}$ and improving upon it by taking the solution to the secant equation that is a minimal modification to $J_{n-1}$ in terms of the Frobenius norm, that is $\|J_n - J_{n-1}\|_F$ is minimal. Thus

$$\mathbf{J}_n = \mathbf{J}_{n-1} + \frac{\Delta \mathbf{f}_n - \mathbf{J}_{n-1} \Delta \mathbf{x}_n}{\| \Delta \mathbf{x}_n \|^2} \Delta \mathbf{x}_n^T \tag{16.82}$$

Using the Sherman-Morrison formula [46], the inverse of the secant operator may be updated, using Eq. (16.82) as follows

$$\mathbf{J}_n^{-1} = \mathbf{J}_{n-1}^{-1} + \frac{\Delta \mathbf{x}_n - \mathbf{J}_{n-1}^{-1} \Delta \mathbf{f}_n}{\Delta \mathbf{x}_n^T \mathbf{J}_{n-1}^{-1} \Delta \mathbf{f}_n} \Delta \mathbf{x}_n^T \mathbf{J}_{n-1}^{-1} \tag{16.83}$$

Finally, at each iteration, the residual $f_{t+1,n}$ is evaluated, where

$$\mathbf{r}_{t+1,n} = \mathbf{x}_{t+1,n} - \mathbf{x}_{t,n} - \mathbf{f}_{t+1,n} \Delta t \tag{16.84}$$

The iteration step stops when $\|r_{t+1,n}\| < TOL$. Note that with this approach, at each iteration, we have an evaluation of function $f_n$, that is, a construction of operators M, K, and f.

## 16.5.2 1D Finite Element Implementation

### *16.5.2.1 Unidimensional Equations*

Differential equations (16.1), (16.4) with boundary conditions (16.2), (16.3), (16.5), (16.6) may be expressed easily for unidimensional problems.

Let us define

$$\begin{aligned} u_i &= C_i, \quad i = 1, \ldots, n \\ u_{n+i} &= S_i, \quad i = 1, \ldots, m \end{aligned} \tag{16.85}$$

and $u = (u_0, \ldots, u_n, u_{n+1}, \ldots, u_{n+m})^T$.

Let us define

$$\begin{aligned} f_i &= -p_i - vC_i, \quad i=1,\dots,n \\ f_{n+i} &= -q_i - vS_i, \quad i=1,\dots,m \end{aligned} \tag{16.86}$$

and $f = (f_1, \dots, f_n, f_{n+1}, \dots, f_{n+m})^T$.

Finally let us define

$$\begin{aligned} s_i &= -C_i\frac{\partial v}{\partial x} + C_iF_i - \sum_{\substack{j=1 \\ j\neq i}}^{n} C_iF_{ij} + \sum_{\substack{j=1 \\ j\neq i}}^{n} C_jF_{ji}, \quad i=1,\dots,n \\ s_{n+i} &= -S_i\frac{\partial v}{\partial x} - \sum_{j=1}^{n} C_jG_{ij} + \sum_{j=1}^{n} C_jG_{ji}, \quad i=1,\dots,m \end{aligned} \tag{16.87}$$

and $s = (s_1, \dots, s_n, s_{n+1}, \dots, s_{n+m})^T$. With these notations, neglecting the ECM remodeling, it is possible to summarize the governing equations as:

$$\frac{\partial \mathbf{u}}{\partial t} = \frac{\partial \mathbf{f}}{\partial x} + \mathbf{s} \tag{16.88}$$

where $\mathbf{f} = \mathbf{f}\left(x, t, \mathbf{u}, \frac{\partial \mathbf{u}}{\partial x}\right)$ and $\mathbf{s} = \mathbf{s}(x, t, \mathbf{u})$. Eq. (16.88) has sense if and only if we define suitable boundary conditions and initial conditions. Boundary conditions are for each variable $u_i$, $i = 1, \dots, n$ given by Eq. (16.2) or (16.3) and for each variable $u_i$, $i = n + 1, \dots, m$ by Eq. (16.5) or Eq. (16.6) except for the fact that we replace directional derivatives $\frac{\partial}{\partial \mathbf{n}}$ by partial derivatives $\frac{\partial}{\partial x}$.

Finally, the initial conditions are

$$\mathbf{u} = \mathbf{u}^{\circ}(x) \tag{16.89}$$

where $u_i^{\circ} = C_i^{\circ}$ for $i = 1, \dots, n$ and $u_{n+i}^{\circ} = S_i^{\circ}$ for $i = 1, \dots, m$.

#### 16.5.2.2 Weak Form

The differential equation (16.88) with boundary conditions (16.2), (16.3), (16.6), (16.6) and initial condition (16.89) is a nonlinear parabolic differential equation in time and one space variable. We solve numerically this equation using a method based on a simple piecewise nonlinear Galerkin second-order accurate in space [47], which is compatible with this kind of nonlinear equation and boundary condition. Multiplying the PDE by a test function $\phi \in \mathcal{H}_o^1([\alpha;\beta])$ and integrating by parts in $[\alpha;\beta]$, we arrive to:

$$\phi(\beta)\mathbf{f}(\beta) - \phi(\alpha)\mathbf{f}(\alpha) - \int_\alpha^\beta \frac{\partial \phi}{\partial x}\mathbf{f}\, dx = \int_\alpha^\beta \phi \mathbf{Q}\, dx \tag{16.90}$$

where $\mathbf{Q} = \mathbf{Q}\left(x, t, \mathbf{u}, \frac{\partial \mathbf{u}}{\partial t}\right) = \frac{\partial \mathbf{u}}{\partial t} - \mathbf{s}(x, t, \mathbf{u})$. As a test function we select

$$\begin{aligned} \phi_\alpha(x) &- \frac{\beta - x}{\beta - \alpha} \\ \phi_\beta(x) &= \frac{x - \alpha}{\beta - \alpha} \end{aligned} \tag{16.91}$$

When using $\phi_\alpha$, Eq. (16.90) becomes

$$-\int_\alpha^\beta \mathbf{f}\frac{\partial \phi_\alpha}{\partial x} dx = \mathbf{f}(\alpha) + \int_\alpha^\beta \mathbf{Q}\phi_\alpha \; dx \tag{16.92}$$

and when using $\phi_\beta$, we get

$$\int_\alpha^\beta \mathbf{f}\frac{\partial \phi_\beta}{\partial x} dx = -\mathbf{f}(\beta) + \int_\alpha^\beta \mathbf{Q}\phi_\beta \; dx \tag{16.93}$$

#### 16.5.2.3 *Spatial Discretization*

After numerical quadrature we get, for $\xi \in [\alpha;\beta]$:

$$-\mathbf{f}\left(\xi,\mathbf{u}(\xi),\frac{\partial \mathbf{u}}{\partial x}(\xi)\right)\int_{\alpha}^{\beta}\frac{\partial \phi_{\alpha}}{\partial x}dx=\mathbf{f}(\alpha)+\mathbf{Q}\left(\xi,t,\mathbf{u}(\xi),\frac{\partial \mathbf{u}}{\partial t}(\xi)\right)\int_{\alpha}^{\beta}\phi_{\alpha}\,dx \tag{16.94}$$

Identifying $[\alpha;\beta]$ with $[x_{j-1};x_j]$ we get

$$\mathbf{f}_{j-1/2}=\mathbf{v}_{j-1}+(\xi_{j-1/2}-x_{j-1})(\dot{\mathbf{u}}_{j-1}-\mathbf{s}_{j-1/2}) \tag{16.95}$$

where $v_{j-1}=f_{j-1}$ is considered as a secondary variable of the problem. In a similar way, using $\phi_{\beta}$, we arrive to

$$-\mathbf{f}_{j-1/2}=-\mathbf{v}_{j}+(x_{j}-\xi_{j-1/2})(\dot{\mathbf{u}}_{j}-\mathbf{s}_{j-1/2}) \tag{16.96}$$

Adding Eq. (16.95) with $j+1$ and Eq. (16.96), we obtain

$$\mathbf{f}_{j+1/2}-\mathbf{f}_{j-1/2}=(\xi_{j+1/2}-x_j)(\dot{\mathbf{u}}_{j}-\mathbf{s}_{j+1/2})+(x_j-\xi_{j-1/2})(\dot{\mathbf{u}}_{j}-\mathbf{s}_{j-1/2}) \tag{16.97}$$

#### 16.5.2.4 *Time Integration*

Eqs. (16.95)–(16.97), including boundary conditions, form a system of differential-algebraic equations that is integrated using the MATLAB ODE suite [48]. For the time integrator, an absolute tolerance of $10^{-6}$ is fixed with a relative tolerance of $10^{-3}$.

## 16.6 SOME APPLICATIONS OF INTEREST

### 16.6.1 Reproducing In Silico Measurements of In Vitro Cell Cultures in Microfluidic Devices

We first particularize the general framework presented above to the particular examples we shall simulate. These correspond to the cell migration processes described above that can be approximated with reasonable accuracy by a 1D problem. Therefore, we particularize initially the mathematical models and fit all model parameters. In the notation presented in Section 16.4, this means establishing a proper functional relationship for $\mathcal{F}_i$, $\mathcal{F}_{ij}$, $\mathcal{K}_{D,i}$, $\mathcal{K}_{m,i,j}$, $\mathcal{K}_{s,i,j}$, $\mathcal{K}_{E,i}$, $\mathcal{K}_{T,i}$, $\mathcal{F}'_{ij}$, $\mathcal{K}'_{D,i}$, and $\mathcal{R}_i$. It is common to express these functions through empirical equations with some phenomenological meaning. This step, however, includes the definition of some phenomenological parameters that are, in fact, parameters of the model. The richer and more complex the model, the more parameters will be required. Here, we present an example of this strategy of modeling and parameter fitting.

#### 16.6.1.1 *Model and Parameters*

The considered model is a particular case of general equations presented in Section 16.4. Indeed, we consider the following equations, regulating the evolution of an alive cell population $C_n$, a dead cell population $C_d$, and oxygen pressure $O_2$:

$$\begin{aligned}\frac{\partial C_n}{\partial t}&=D_n\frac{\partial^2 C_n}{\partial^2 x}-\chi H_{go}(O_2,C_n)\frac{\partial}{\partial t}\left(C_n\frac{\partial O_2}{\partial x}\right)+\frac{1}{\tau_n}H_{gr}(O_2,C_n)C_n-\frac{1}{\tau_d}F_{12}(O_2)C_n\\ \frac{\partial C_d}{\partial t}&=\frac{1}{\tau_d}F_{12}(O_2)C_n\\ \frac{\partial O_2}{\partial t}&=D_{O_2}\frac{\partial^2 O_2}{\partial x^2}-\alpha F'_{11}(O_2)C_n\end{aligned} \tag{16.98}$$

Here, $D_n$ is the diffusion coefficient of the normoxic phenotype, $\chi$ is the chemotaxis coefficient, $\tau_n$ and $\tau_d$ are the growth and death characteristic times, respectively, and $\alpha$ is the oxygen consumption.

The functions $H_{go}(O_2)$, $H_{gr}$, $F_{12}$, and $F'_{11}$ are dimensionless activation functions that try to reproduce phenomenological behaviors observed in cell cultures and have the following meaning.

- *Go or grow dichotomy*: Functions $H_{go}$ and $H_{gr}$ are activation functions that try to reproduce the observed "go or grow" paradigm in a GBM microenvironment. GBM cells tend to proliferate in oxygenated areas, whereas they become much more migratory under hypoxic conditions. This behavior is almost exclusive, that is, when a cell invests energy resources in proliferative activities, it stops investing them in migratory activities. This is determined by a specific oxygenation threshold, $O_2^*$, the hypoxia threshold. Besides, cell growth and migration should be conditioned by the nutrient and spatial needs of the cell, so it is reasonable to use a logistic growth model including a maximum cell capacity $C_{sat}$ and to use the same capacity for avoiding cell migration. This suggests defining $H_{go}$ and $H_{gr}$ as:

$$H_{go}(O_2, C_n) = \phi_-(O_2; O_2^*)\phi_-(C_n; C_{sat}) \quad (16.99)$$

$$H_{gr}(O_2, C_n) = \phi_+(O_2; O_2^*)\rho(C_n; C_{sat}) \quad (16.100)$$

Here, we have used the following notation

$$\phi_-(x;\beta) = \begin{cases} 1 & \text{if} \quad x \le 0 \\ 1 - \dfrac{x}{\beta} & \text{if} \quad 0 \le x \le \beta \\ 0 & \text{if} \quad x > \beta \end{cases} \quad (16.101)$$

$$\phi_+(x;\beta) = \begin{cases} 0 & \text{if} \quad x \le 0 \\ \dfrac{x}{\beta} & \text{if} \quad 0 \le x \le \beta \\ 1 & \text{if} \quad x > \beta \end{cases} \quad (16.102)$$

$$\rho(x, \alpha) = 1 - \frac{x}{\alpha} \quad (16.103)$$

- *Cell death*: Cell death is a natural process depending on many factors and agents and has an inherent stochastic nature. Anoxia is the fundamental cause of cell death in the problem analyzed here, but due to the stochastic nature, the switch function should be a smooth function that can incorporate death cell variability depending on oxygenation conditions. Here, a two-parameter sigmoid model is used that is able to capture necrosis and apoptosis phenomena.

$$F_{12}(O_2) = \sigma_-(O_2; O_2^d, \delta O_2) \quad (16.104)$$

where $\sigma_-$ is the function:

$$\sigma_-(x;\beta,\delta) = \frac{1}{2}\left(1 - \tanh\left(\frac{x-\beta}{\delta}\right)\right) \quad (16.105)$$

Here, $\beta$ is a threshold parameter and $\delta$ is a sensitivity parameter. They can be seen as a pair of location-spread parameters summarizing the stochastic behavior of the considered phenomenon. The parameters $O_2^d$ and $\delta O_2$, associated with the oxygen level, define the limits for cell death, capturing the stochastic nature.

- *Oxygen consumption*: Oxygen consumption is a complex phenomenon related to the oxidative phosphorylation that occurs in the membrane of cellular mitochondria. Many authors have considered a zero-order consumption function, that is, a constant consumption rate independent of oxygen concentration $O_2$ [49–52]. A more realistic assumption is that the consumption function is described by the Michaelis-Menten model for enzyme kinetics [53, 54]. With this consideration we can define

$$F_{11}(O_2) = r(O_2; O_2^K) \quad (16.106)$$

with

$$r(x;K) = \frac{x}{x+K} \quad (16.107)$$

This type of equation was observed for the oxygen consumption rate in the late 1920s and early 1930s [55]. The Michaelis-Menten equation has a sigmoid shape that can be interpreted as an almost constant consumption rate for a high concentration of oxygen, followed by a rapid decrease when the oxygen concentration decreases. This equation describes more accurately the consumption at low oxygen concentrations and is compatible with previous constant consumption rate models, thus allowing the possibility of comparison with previous studies. The parameter $O_2^K$ is the oxygen concentration at which the reaction rate is half the rate in a fully oxygenated medium and therefore is related with oxidative phosphorylation kinetics, cell structure and morphology (size and number of mitochondria), and the diffusion process at the cytoplasm.

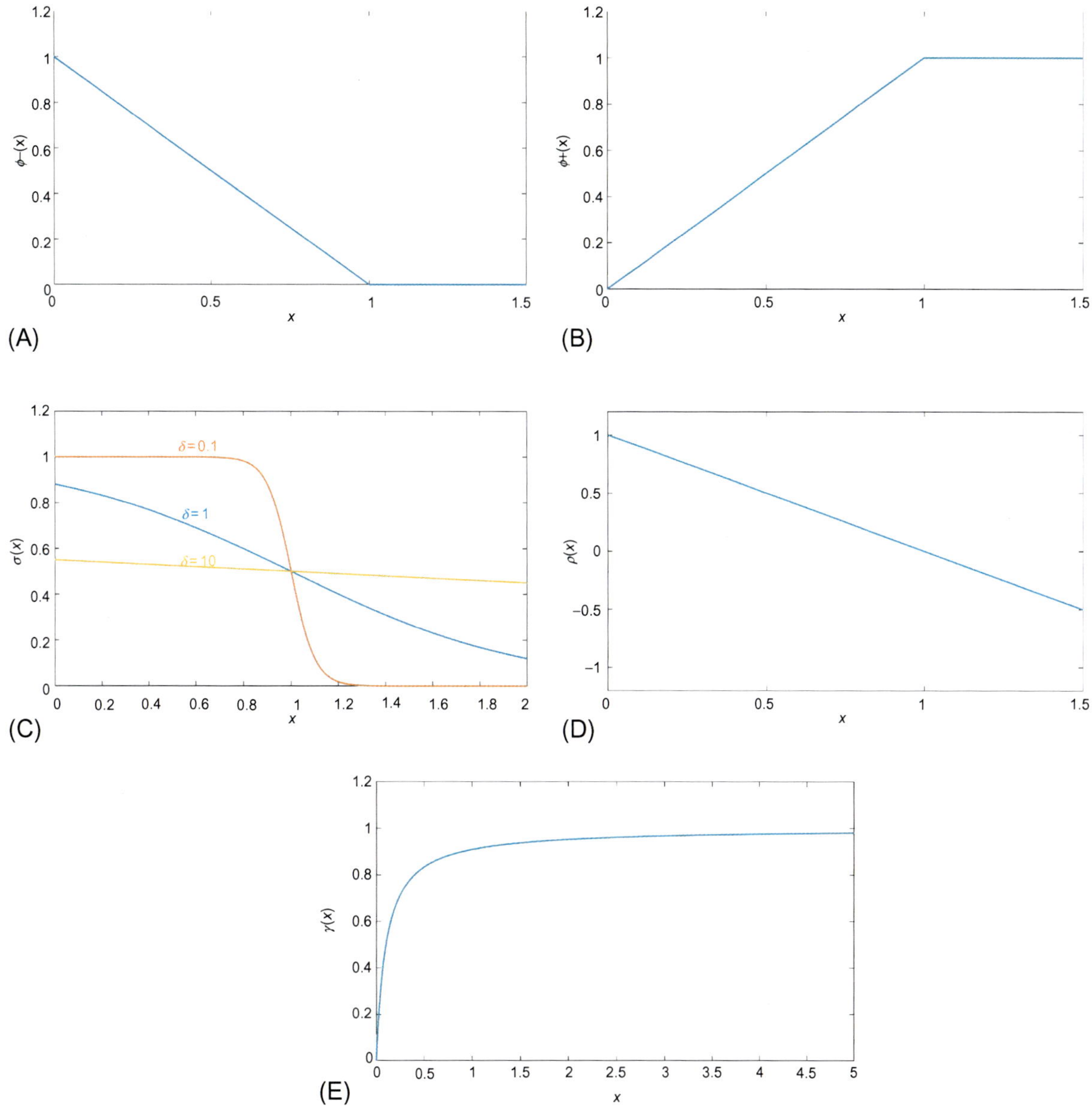


FIG. 16.4 Different activation functions for phenomenological models. (A) $\phi_-$ function. (B) $\phi_+$ function. (C) $\sigma_-$ function. (D) $\rho$ function. (E) $r$ function.

Fig. 16.4 shows functions $\phi_-$, $\phi_+$, $\rho$, and $r$ for $\beta = 1.0$ (for both $\phi_-$, $\phi_+$, and $\sigma_-$), $\delta = 0.1, 1, 10$, $\alpha = 1$, and $K = 0.1$. It is clear that these functions try to model cell adaptation to the environment. In principle, to define these functions, six parameters are needed ($\theta_1, \theta_2, \theta_3, \delta, C$, and $K$) but due to the "go or grow" assumption, it is assumed that $\theta_1 = \theta_2 = O_2^*$ so finally:

$$\begin{aligned} \beta_1 &= O_2^* \\ \beta_2 &= O_2^* \\ \beta_3 &= O_2^d \\ \delta &= \delta O_2 \\ \alpha &= C_{sat} \\ K &= O_2^K \end{aligned} \tag{16.108}$$

The entire set of parameters is listed in Table 16.1, where the bibliographic source of each parameter is also included.

TABLE 16.1 Parameters of the Model

| Symbol | Value | Reference |
|---|---|---|
| $D_{O_2}$ | $1.0 \times 10^{-5}$ cm$^2$/s | [56] |
| $D_n$ | $6.6 \times 10^{-12}$ cm$^2$/s | [54] |
| $C_{sat}$ | $3.5 \times 10^7$ cell/mL | [57] |
| $\chi$ | $1.5 \times 10^{-9}$ cm$^2$/mmHg ·s | [58] |
| $\tau_n$ | 300 h | [58] |
| $\tau_d$ | 72 h | [59] |
| $\alpha$ | $1.0 \times 10^{-9}$ cm$^3$ ·mmHg/cell ·s | [60] |
| $O_2^K$ | 2.5 mmHg | [56] |
| $O_2^*$ | 7.0 mmHg | [54] |
| $O_2^d$ | 0.7 mmHg | [54] |
| $\delta O_2$ | 0.1 mmHg | [54] |

#### 16.6.1.2 Boundary Conditions

Experimental practice shows that even if the microfluidic device is designed such that the cells are kept in the culture chamber, there is always cell leakage at the two sides of the chamber. Thus, it is natural to consider Robin boundary conditions at the two sides of the chip, as in Eq. (16.3). In order to do this, it is possible to define boundary conditions with the form:

$$K_n(x^*,t)(C_n - g_n(x^*,t)) + J_n(x^*,t)\left(\frac{\partial f_n}{\partial x} - h_n(x^*,t)\right) = 0 \quad (16.109)$$

where $f_n$ is cell flux and $k_n$, $J_n$, $g_n$, and $f_n$ are functions that in general are time-dependent and try to reproduce boundary cell permeability and boundary cell supplies ($x^* = 0$, $L$, where $L$ is the chip length). Note that if $K_n = 1$ and $J_n = 0$, Dirichlet boundary conditions are recovered, and if $K_n = 0$ and $J_n = 1$, Neumann boundary conditions are recovered. In our case, we consider, without loss of generality, $K_n = 1$, that there are neither cells nor cell flow at the boundaries at any moment, so $g_n(t) = f_n(t) = 0$. Therefore, $J_n = J$ is the only parameter that characterizes cell losses at the boundary of the chip.

With respect to oxygen, Dirichlet boundary conditions are considered where the oxygen concentration at both channels is assumed to be constant and equal to $O_2^S$, that is another parameter of the model because it is unknown. Actually, fresh cell culture media were perfused through lateral channels every 24 h, so the oxygen concentration in the channels is not constant in time but a small variation was assumed.

#### 16.6.1.3 Initial Conditions

As for the initial conditions, we assume that, at the beginning, there are no dead cells at the culture chamber and the concentration of alive cells is considered as known. Once cells are seeded and marked with fluorescence, the light emitted is captured by the microscope and, as the initial average concentration of cells is known $C_0 = 40 \times 10^6$ cell/mL, the fluorescence profile is normalized in order to obtain a concentration profile. Finally, it is assumed that oxygen pressure is homogeneous at $t = 0$ and equal to the oxygen concentration at the two supply channels, $O_2^S$.

#### 16.6.1.4 Results and Discussion

Fig. 16.5 shows the evolution of alive cells, which are marked with enhanced green fluorescent protein (EGFP), over time. Fig. 16.6 shows the evolution of the alive and dead cell profiles obtained, both experimentally and by numerical simulation. The estimated values for $O_2^S$ and $J$ were

$$\begin{aligned} O_2^S &= 7\,\text{mmHg} \\ J &= 5 \times 10^{16}\,\text{s/cm} \end{aligned} \quad (16.110)$$

Finally, in Fig. 16.7, the total number of cells in the chamber is computed for both the experimental and the computational cases.

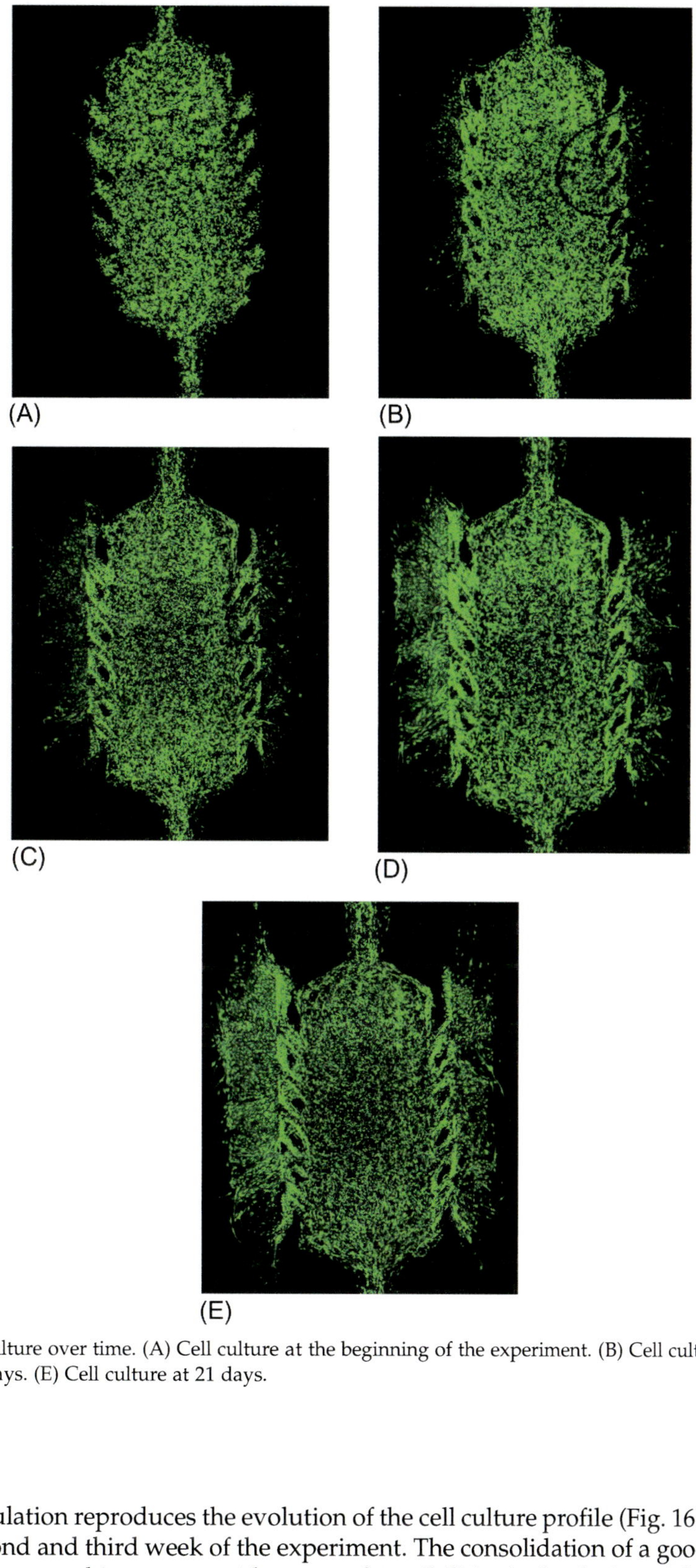


FIG. 16.5 Evolution of cell culture over time. (A) Cell culture at the beginning of the experiment. (B) Cell culture at 7 days. (C) Cell culture at 11 days. (D) Cell culture at 17 days. (E) Cell culture at 21 days.

As is observed, the simulation reproduces the evolution of the cell culture profile (Fig. 16.6) and the total amount of cells (Fig. 16.7) for the second and third week of the experiment. The consolidation of a good set of parameters able to reproduce culture evolution on a chip opens a wide range of possibilities in the design of drugs and therapies, as long as it allows the possibility of testing "what if" conditions that cannot be reproduced in actual experiments. For example, it permits the quantitative evaluation of GBM tissue response to vessel occlusion or collapse, the location of cell hyperdensity zones, and the necrotic cores depending on the vessel network. The inclusion of additional chemical

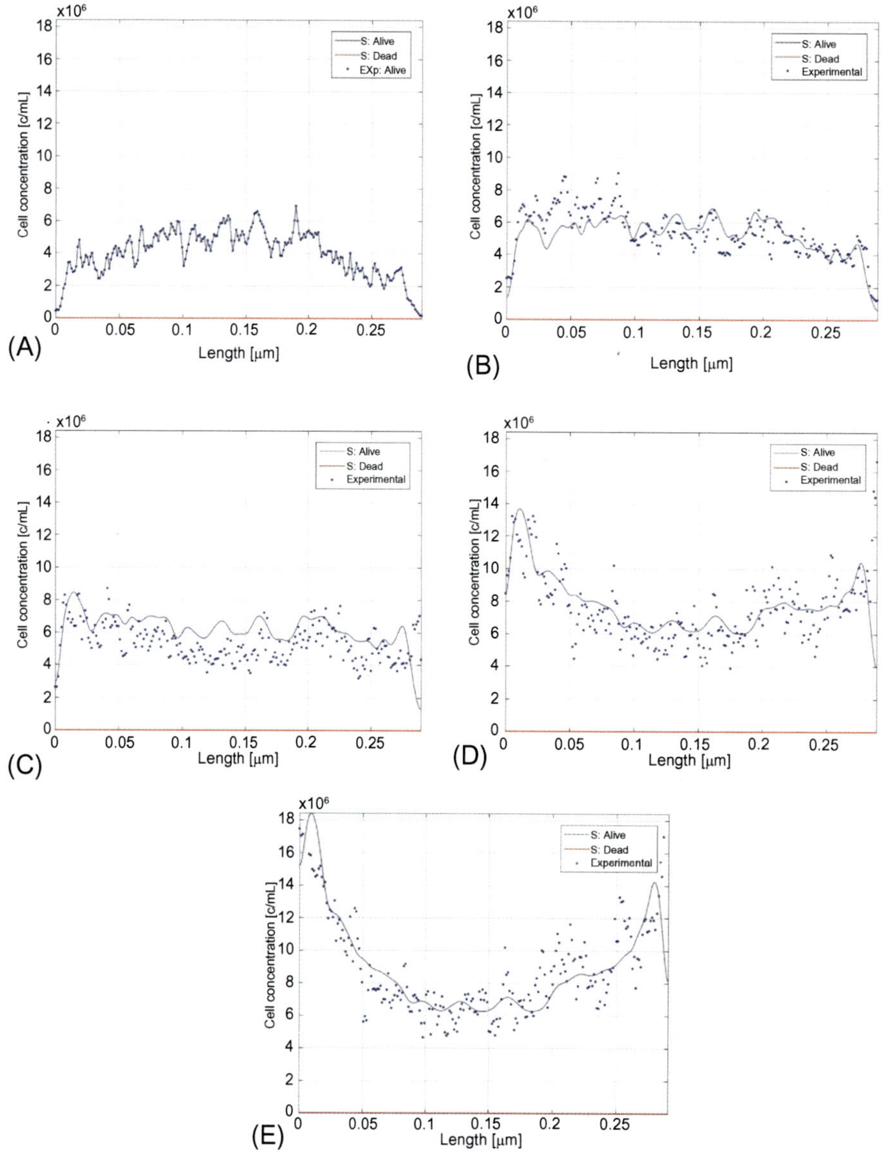


FIG. 16.6 Evolution of simulated and measured profiles over time. (A) Cell culture at the beginning of the experiment. (B) Cell culture at 7 days. (C) Cell culture at 11 days. (D) Cell culture at 17 days. (E) Cell culture at 21 days.

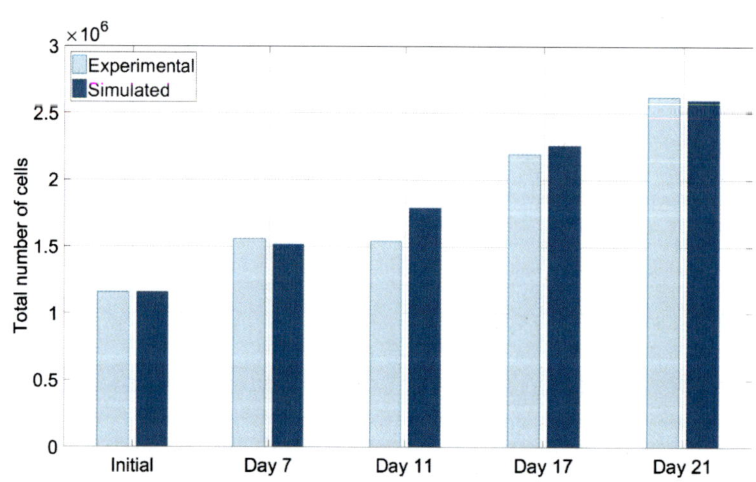


FIG. 16.7 Evolution of cell culture for the experimental and computational results.

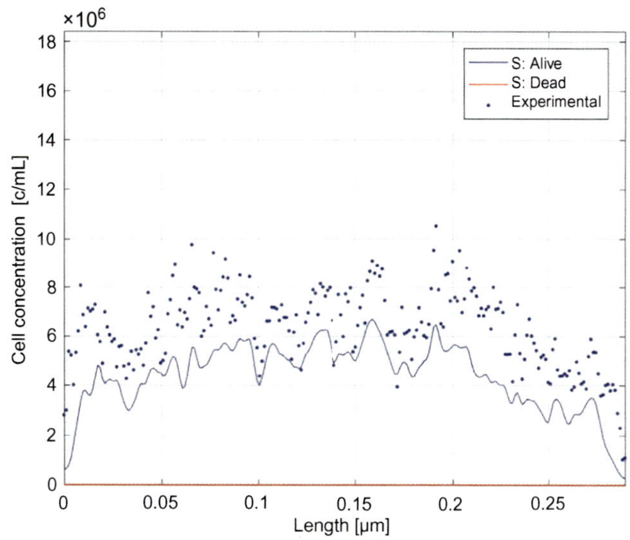


FIG. 16.8 Cell culture at 3 days.

species in the general framework, such as temozolomide (TMZ) or other drugs, may be used in order to evaluate the effect on GBM tissue evolution and to explore many other chemotherapy strategies.

Nevertheless, there are still many obstacles to overcome. First, there is an obvious need for a better characterization of boundary and initial conditions of the experiments. Cell leakage and oxygen supply as well as the initial oxygen profile have important impacts on cell evolution and therefore in parameter calibration. In the present work, boundary and initial conditions have been established in a reasonable manner, obtaining a good accordance between predicted and observed results but a perfect characterization is desirable.

Moreover, in order to make predictions, it is important to consider intercell culture parameter variability. As is usual in biological research, it is difficult to obtain a universal model able to reproduce GBM. With respect to applications, it is desirable to obtain patient-specific models, that is, parameters and geometries, for the simulations.

Finally, there is a lot of room in the short-term reaction of the cell for physiological stimuli. The presented model is able to capture long-term cell evolution features when genetic cell damage has already occurred. There is, however, the necessity of a better understanding of cell damage in the short term when the cell has not yet adapted its metabolism to new microenvironment conditions. How long does it take for a cell to become migratory? Is it instantaneous? Experiments demonstrate that it is not, but the mathematical model does not include this kind of feature, as is shown in Fig. 16.8, where we can conclude that the presented mathematical model underestimates cell proliferation in the first week.

### 16.6.2 In Silico Design and Quantification of Experiments in Microfluidic Devices

Three-dimensional simulations may be used in chip design and fabrication and in experiment set up. When a cell culture is going to be seeded on a chip, the objective is to reproduce, in the microfluidic device, the desired tumoral microenvironment, that is, the appropriate mechanical properties, nutrient and oxygen supplies, and gradients. The goal is to design properly the experiment in order to make suitable conclusions in a reasonable time. For that, it is necessary to define precise device geometry and boundary conditions so that cell cultures are subjected to the desired chemical and mechanical stimuli.

In order to illustrate all this, a simulation in a 3D chip-like geometry is going to be presented for a very fundamental model. Two cell phenotypes are going to be considered, alive and dead cells, $C_n$ and $C_d$, respectively, and the oxygen $O_2$ is going to be the driving chemoattractant.

#### 16.6.2.1 Model and Parameters

Equations of the model are the following:

$$\begin{aligned}
\frac{\partial C_n}{\partial t} &= D_n \nabla^2 C_n - \chi \nabla \cdot (C_n \nabla O_2) + \frac{1}{\tau_n}\left(1 - \frac{C_n}{C_{sat}}\right) - \frac{1}{\tau_d} S_{nd}(O_2) C_n \\
\frac{\partial C_n}{\partial t} &= \frac{1}{\tau_d} S_{nd}(O_2) C_n \\
\frac{\partial O_2}{\partial t} &= D_{O_2} \nabla^2 O_2 - \alpha \left(\frac{O_2}{O_2 + O_2^K}\right) C_n
\end{aligned} \tag{16.111}$$

Here, $D_n$ is the diffusion coefficient of the normoxic phenotype, $\chi$ is the chemotaxis coefficient, $\tau_n$ and $\tau_d$ are the growth and death characteristic times, $C_{sat}$ is the cell capacity, $\alpha$ is the oxygen consumption, and $O_2^K$ is the Michaelis-Menten constant of cellular respiration kinetics. The function $S_{nd}(O_2)$ is a step function that takes into account oxygen concentration, that is $S_{nd} = 1$ when $O_2 \leq O_2^d$ and $S_{nd} = 0$ when $O_2 > O_2^d$. In Table 16.2, the values of the parameters selected for the illustrative simulation are listed.

#### 16.6.2.2 Geometry

Fig. 16.9 shows the geometry and the mesh of the culture chamber of a microfluidic device. Geometry and dimensions are representative and of the order of hundreds of micrometers. This kind of simulation may permit the manufacturers to design a better chip in regard to the shape and dimensions. The geometry presented has a respective maximum width, length, and height of 600, 600, and 70 μm, respectively.

TABLE 16.2 Parameters of the Model

| Symbol | Meaning | Value |
|---|---|---|
| $D_{O_2}$ | Oxygen diffusion | $5.0 \times 10^{-5}$ cm²/s |
| $D_n$ | Cell diffusion | $3.3 \times 10^{-6}$ cm²/s |
| $C_{sat}$ | Cell capacity | $5.0 \times 10^{7}$ cell/mL |
| $\chi$ | Chemotaxis coefficient | $3.8 \times 10^{-5}$ cm²/mmHg ·s |
| $\tau_n$ | Growth characteristic time | 4 h |
| $\tau_d$ | Death characteristic time | 0.8 h |
| $\alpha$ | Oxygen consumption rate | $5.0 \times 10^{-8}$ cm³ ·mmHg/cell ·s |
| $O_2^K$ | Michaelis-Menten constant | 2.5 mmHg |
| $O_2^d$ | Anoxia threshold | 1.6 mmHg |

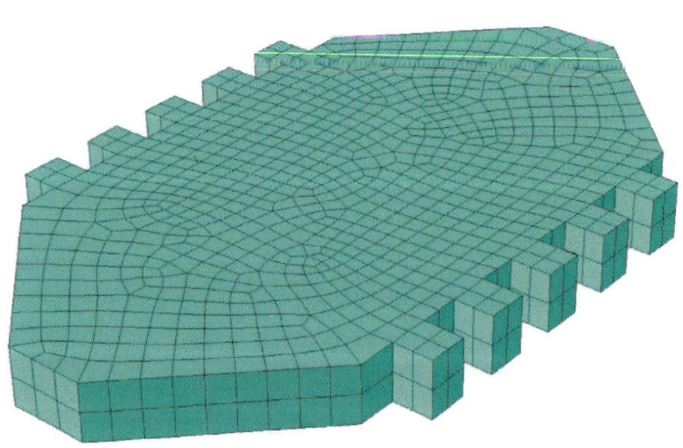

FIG. 16.9 Geometry and mesh of the culture chamber of a microfluidic device.

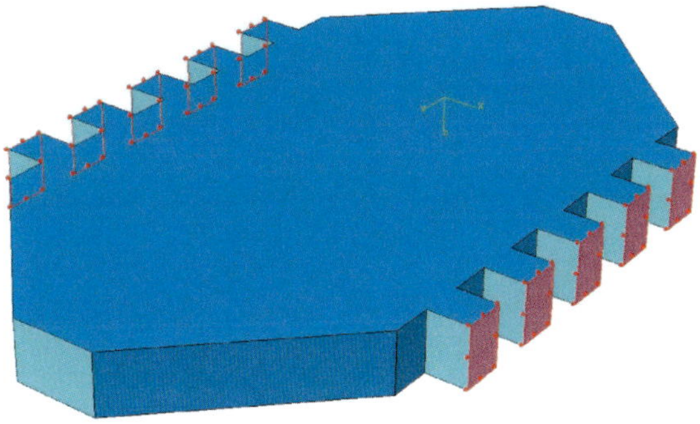

FIG. 16.10 Surfaces where Dirichlet boundary conditions are applied.

#### 16.6.2.3 *Boundary Conditions*

All faces of the chip are considered impervious to chemical species and cells, except for the supply surfaces that are marked in red in Fig. 16.10. Therefore, the boundary conditions are of the Neumann type in the rest of the contour of the chamber such that cell and oxygen flux are specified to be 0. In the red-marked surfaces, we assume Dirichlet boundary conditions. These boundary conditions depend on the desired microenvironment conditions (oxygen supply and oxygen gradient) and vary from one experiment to another. As an example, suppose that cell concentration is set to 0 for both alive and dead phenotypes and oxygen supply is fixed to $O_2^s = 2$ mmHg.

#### 16.6.2.4 *Initial Conditions*

As initial conditions, we assume that, at the beginning, there are no dead cells at the culture chamber, and the concentration of alive cells is homogeneous and equal to $C^0 = 1 \times 10^6$ cell/mL. Finally, it is assumed that oxygen pressure is homogeneous at $t = 0$ and equal $O_2^0 = 2$ mmHg.

#### 16.6.2.5 *Results and Discussion*

Fig. 16.11 shows the evolution over time of alive cells and Fig. 16.12 shows the evolution of dead cells on the culture chamber. As expected, the cell concentration remains high next to the supply channels and decreases in the central part of the chamber, where oxygen consumption induces anoxia. Therefore, once the oxygen threshold of $O_2^* = 1.6$ mmHg is achieved, cell death is promoted. This explains, analogously, why dead cell concentration increases at the same regions. However, cell chemotaxis explains why alive cells are even more concentrated at oxygen supply points: cells migrate in the direction of the oxygen gradient and when they arrive to a well-oxygenated point, proliferation occurs normally because the conditions are now favorable and the cell concentration is below the capacity limit.

The results of the simulation show how the cell culture is going to evolve during the virtual experiment. With the presented parameters and boundary conditions, the lack of oxygen diffusion along the culture chamber results in the fast appearance of a necrotic core occupying almost the entire chamber. The researcher should, therefore, consider whether the dimensions of the chip, the hydrogel diffusivity properties, or the initial cell concentration are appropriate for the experiment carried out and assess whether other conditions would be preferable. This kind of in silico prediction can be extrapolated to other cell populations and tissues, other geometries, and other mechanical frameworks relative to the experiment. Moreover, the simulation of the different processes allows access to all field value variables, which can, in turn, be interesting for the disclosure of correlation between phenomena or variables of clinical or physiological interest that would be inaccessible from an experimental point of view due to technical considerations (difficulty or impossibility of field variable monitorization).

It is important to note, however, that model and parameter characterization is always a very complicated task. Even if frequently simplified, the multiphysics nature of the TME is very complex: many different phenomena are coupled and many scales are involved, resulting in a hard nonlinear problem where even the semiquantitative analysis is often complicated. Too-simplistic models lead to the failure of predictive simulation models while complex and sophisticated ones result in a difficult parameter estimation (due to both numerical and experimental difficulties and validation). Moreover, highly nonlinear and coupled models in different physical scales may involve very expensive simulations from the computational point of view.

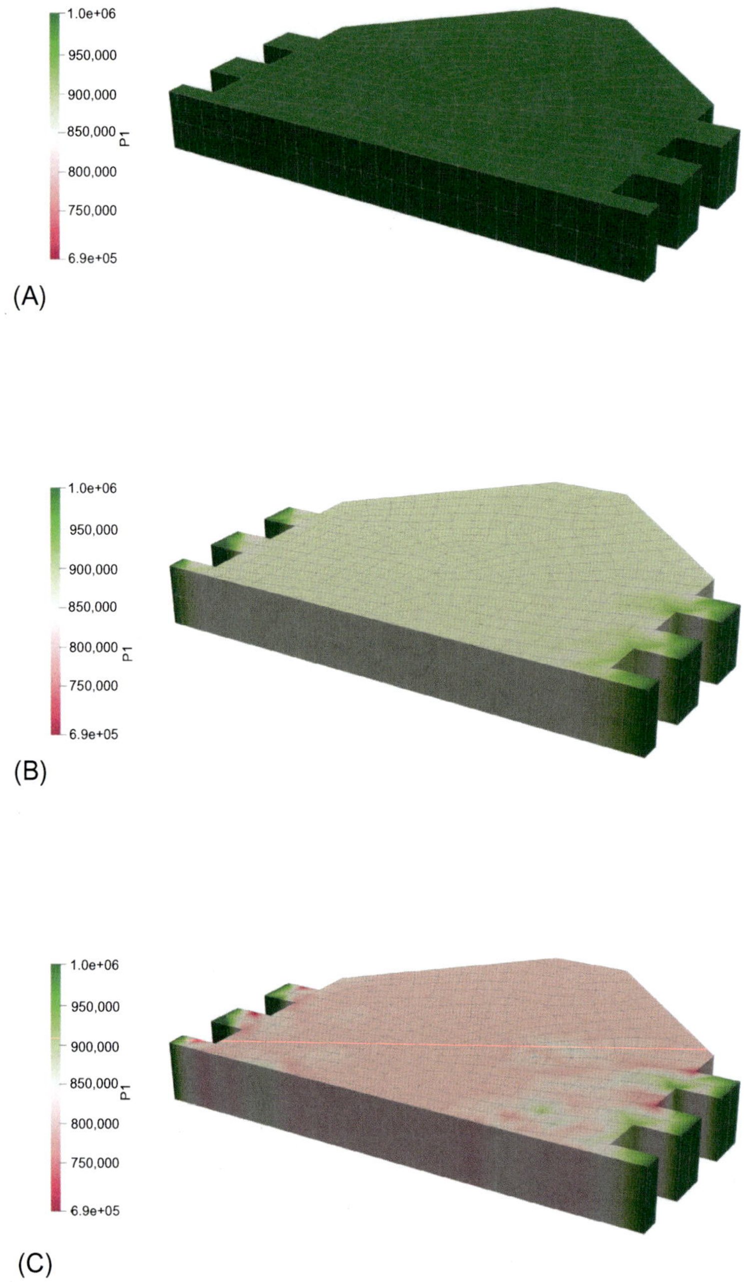


FIG. 16.11 Evolution of alive cells in the culture chamber (in cell/mL). (A) $t = 0$ s; (B) $t = 42$ h; (C) $t = 70$ h.

## 16.7 CONCLUSIONS

The combination of organ-on-chip devices and computational models is a perfect option to set up new complex biological models that include diffusion, advection, chemotaxis, mechanotaxis, electrotaxis, thermotaxis, proliferation, differentiation, and cell death as well as the interaction of the different cellular phenotypes with chemical species (such as nutrients or chemical cues) and ECM remodeling. Experimental campaigns are needed in order to define and calibrate proper mathematical models, but promising results have been obtained in the study of in vitro GBM models.

The main contribution of this work is the presentation of a framework integrating in vitro experiments in 3D biomimetic platforms, able to capture the enormous complexity of tumoral microenvironment biophysics, with in silico simulation models, which serve to extrapolate the conclusions to different pictures and to help the researcher in new hypotheses formulations and experimental campaign designs. Microfluidic devices offer flexible and realistic experimentation. The presented mathematical model is rich enough to capture all TME physics with a variable degree of complexity.

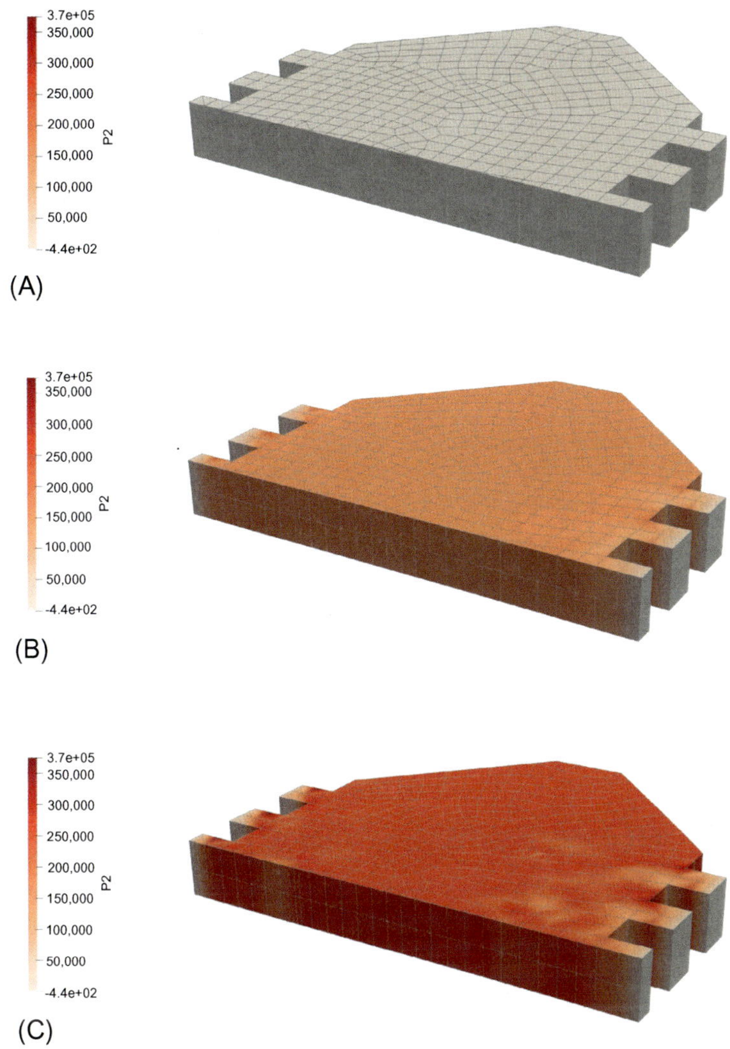


FIG. 16.12 Evolution of dead cells in the culture chamber (in cell/mL). (A) $t = 0$ s; (B) $t = 42$ h; (C) $t = 70$ h.

Coupling between organ-on-chip platforms and in silico simulations is a two-way knowledge generator. First, it offers the possibility of trying, once a tumor population has been sufficiently well characterized in terms of a parametric mathematical model, "what if" conditions, predicting tumor evolution and therefore patient prognosis, and exploring therapies (drugs, chemotherapy, radiotherapy) or surgical intervention. Second, it allows the biologist to speed up the experimental designs and set up, offering the possibility of an in silico design of the geometry and stabilizing the conditions of the experiment. In this chapter, both possibilities have been illustrated with two examples of application: an accurate characterization of cell culture evolution and a computational forecast of an experiment with a given set up.

However, the richness and complexity of the microenvironment physics results in the growing need for more specific devices and a major data assimilation from experiments, which in turn should feed in a proper way the computational models. Integration between data and simulations based on models that maintain the physics of the problem is a promising opportunity, which shows a trade-off between the power of data science techniques and the underlying knowledge of the universe that physics brings to us. In the rise of these techniques, which is a hot research area today, and the application to the clinical field, is the path of patient-specific medicine.

## Acknowledgments

The authors gratefully acknowledge the financial support from the Spanish Ministry of Economy and Competitiveness under the projects (MINECO MAT2016-76039-C4-4-R, AEI/FEDER, UE) and (MINECO BIO2016-79092-R, AEI/FEDER, UE), of the Government of Aragon (DGA-T24_17R) and of the Biomedical Research Networking Center in Bioengineering, Biomaterials and Nanomedicine (CIBER-BBN), financed by the Instituto de Salud Carlos III with assistance from the European Regional Development Fund.

## References

[1] K.D. Miller, R.L. Siegel, C.C. Lin, A.B. Mariotto, J.L. Kramer, J.H. Rowland, K.D. Stein, R. Alteri, A. Jemal, Cancer treatment and survivorship statistics, 2016, CA Cancer J. Clin. 66 (4) (2016) 271–289.

[2] J. Ferlay, I. Soerjomataram, R. Dikshit, S. Eser, C. Mathers, M. Rebelo, D.M. Parkin, D. Forman, F. Bray, Cancer incidence and mortality worldwide: sources, methods and major patterns in GLOBOCAN 2012, Int. J. Cancer 136 (5) (2015) E359–E386.

[3] A. Nagelkerke, J. Bussink, A.E. Rowan, P.N. Span, The mechanical microenvironment in cancer: how physics affects tumours, in: Seminars in Cancer Biology, vol. 35, Elsevier, 2015, pp. 62–70.

[4] S. Huang, D.E. Ingber, Cell tension, matrix mechanics, and cancer development, Cancer Cell 8 (3) (2005) 175–176.

[5] S.J. Mousavi, M.H. Doweidar, M. Doblaré, Cell migration and cell-cell interaction in the presence of mechano-chemo-thermotaxis, Mol. Cell Biomech. 10 (1) (2013) 1–25.

[6] S.J. Mousavi, M. Doblaré, M.H. Doweidar, Computational modelling of multi-cell migration in a multi-signalling substrate, Phys. Biol. 11 (2) (2014) 026002.

[7] D. Hanahan, R.A. Weinberg, Hallmarks of cancer: the next generation, Cell 144 (5) (2011) 646–674.

[8] D.F. Quail, J.A. Joyce, Microenvironmental regulation of tumor progression and metastasis, Nat. Med. 19 (11) (2013) 1423.

[9] K.J. Harrington, Biology of cancer, Medicine 39 (12) (2011) 689–692.

[10] V. Espina, L.A. Liotta, What is the malignant nature of human ductal carcinoma in situ? Nat. Rev. Cancer 11 (1) (2011) 68.

[11] P. Mehlen, A. Puisieux, Metastasis: a question of life or death, Nat. Rev. Cancer 6 (6) (2006) 449.

[12] J.W. Scannell, A. Blanckley, H. Boldon, B. Warrington, Diagnosing the decline in pharmaceutical R&D efficiency, Nat. Rev. Drug Discov. 11 (3) (2012) 191.

[13] G.P. Nolan, What's wrong with drug screening today, Nat. Chem. Biol. 3 (4) (2007) 187.

[14] R. Edmondson, J.J. Broglie, A.F. Adcock, L. Yang, Three-dimensional cell culture systems and their applications in drug discovery and cell-based biosensors, Assay Drug Dev. Technol. 12 (4) (2014) 207–218.

[15] C. Wang, Z. Tang, Y. Zhao, R. Yao, L. Li, W. Sun, Three-dimensional in vitro cancer models: a short review, Biofabrication 6 (2) (2014) 022001.

[16] E.K. Sackmann, A.L. Fulton, D.J. Beebe, The present and future role of microfluidics in biomedical research, Nature 507 (7491) (2014) 181.

[17] S.N. Bhatia, D.E. Ingber, Microfluidic organs-on-chips, Nat. Biotechnol. 32 (8) (2014) 760.

[18] J.A. Jiménez-Torres, S.L. Peery, K.E. Sung, D.J. Beebe, LumeNEXT: a practical method to pattern luminal structures in ECM gels, Adv. Healthcare Mater. 5 (2) (2016) 198–204.

[19] A. Boussommier-Calleja, R. Li, M.B. Chen, S.C. Wong, R.D. Kamm, Microfluidics: a new tool for modeling cancer-immune interactions, Trends Cancer 2 (1) (2016) 6–19.

[20] I.K. Zervantonakis, S.K. Hughes-Alford, J.L. Charest, J.S. Condeelis, F.B. Gertler, R.D. Kamm, Three-dimensional microfluidic model for tumor cell intravasation and endothelial barrier function, Proc. Natl Acad. Sci. USA 109 (34) (2012) 13515–13520.

[21] J.S. Jeon, S. Bersini, M. Gilardi, G. Dubini, J.L. Charest, M. Moretti, R.D. Kamm, Human 3D vascularized organotypic microfluidic assays to study breast cancer cell extravasation, Proc. Natl Acad. Sci. USA 112 (1) (2015) 214–219.

[22] S. Bersini, J.S. Jeon, G. Dubini, C. Arrigoni, S. Chung, J.L. Charest, M. Moretti, R.D. Kamm, A microfluidic 3D in vitro model for specificity of breast cancer metastasis to bone, Biomaterials 35 (8) (2014) 2454–2461.

[23] T.S. Deisboeck, G.S. Stamatakos, Multiscale Cancer Modeling, CRC Press, 2010, pp. 359–383.

[24] D.J. Brat, Glioblastoma: biology, genetics, and behavior, in: American Society of Clinical Oncology Educational Book, AmericanSociety of Clinical Oncology. Meeting, 2012, pp. 102–107. https://doi.org/10.14694/edbookam.2012.32.102.

[25] T. Oike, Y. Suzuki, K.-I. Sugawara, K. Shirai, S.-E. Noda, T. Tamaki, M. Nagaishi, H. Yokoo, Y. Nakazato, T. Nakano, Radiotherapy plus concomitant adjuvant temozolomide for glioblastoma: Japanese mono-institutional results, PLoS ONE 8 (11) (2013) e78943.

[26] D.J. Brat, A.A. Castellano-Sanchez, S.B. Hunter, M. Pecot, C. Cohen, E.H. Hammond, S.N. Devi, B. Kaur, E.G. Van Meir, Pseudopalisades in glioblastoma are hypoxic, express extracellular matrix proteases, and are formed by an actively migrating cell population, Cancer Res. 64 (3) (2004) 920–927.

[27] Y. Rong, D.L. Durden, E.G. Van Meir, D.J. Brat, "Pseudopalisading" necrosis in glioblastoma: a familiar morphologic feature that links vascular pathology, hypoxia, and angiogenesis, J. Neuropathol. Exp. Neurol. 65 (6) (2006) 529–539.

[28] D.J. Brat, E.G. Van Meir, Vaso-occlusive and prothrombotic mechanisms associated with tumor hypoxia, necrosis, and accelerated growth in glioblastoma, Lab. Invest. 84 (4) (2004) 397.

[29] C.-H. Hsieh, C.-H. Lee, J.-A. Liang, C.-Y. Yu, W.-C. Shyu, Cycling hypoxia increases U87 glioma cell radioresistance via ROS induced higher and long-term HIF-1 signal transduction activity, Oncol. Rep. 24 (6) (2010) 1629–1636.

[30] H.M. Byrne, T. Alarcon, M.R. Owen, S.D. Webb, P.K. Maini, Modelling aspects of cancer dynamics: a review, Philos. Trans. R. Soc. Lond. A Math. Phys. Eng. Sci. 364 (1843) (2006) 1563–1578.

[31] M.E. Katt, A.L. Placone, A.D. Wong, Z.S. Xu, P.C. Searson, In vitro tumor models: advantages, disadvantages, variables, and selecting the right platform, Front. Bioeng. Biotechnol. 4 (2016) 12.

[32] K.R. Swanson, E.C. Alvord, J.D. Murray, A quantitative model for differential motility of gliomas in grey and white matter, Cell Prolif. 33 (5) (2000) 317–329.

[33] E.L. Bearer, J.S. Lowengrub, H.B. Frieboes, Y.-L. Chuang, F. Jin, S.M. Wise, M. Ferrari, D.B. Agus, V. Cristini, Multiparameter computational modeling of tumor invasion, Cancer Res. 69 (10) (2009) 4493–4501.

[34] R.L. Jensen, Brain tumor hypoxia: tumorigenesis, angiogenesis, imaging, pseudoprogression, and as a therapeutic target, J. Neurooncol. 92 (3) (2009) 317–335, https://doi.org/10.1007/s11060-009-9827-2.

[35] C. Suarez, F. Maglietti, M. Colonna, K. Breitburd, G. Marshall, Mathematical modeling of human glioma growth based on brain topological structures: study of two clinical cases, PLoS ONE 7 (6) (2012) e39616.

[36] E. Scribner, O. Saut, P. Province, A. Bag, T. Colin, H.M. Fathallah-Shaykh, Effects of anti-angiogenesis on glioblastoma growth and migration: model to clinical predictions, PLoS ONE 9 (12) (2014) e115018.

[37] Y. Kim, H. Jeon, H. Othmer, The role of the tumor microenvironment in glioblastoma: a mathematical model, IEEE Trans. Biomed. Eng. 64 (3) (2017) 519–527.

[38] K.A. Rejniak, Systems Biology of Tumor Microenvironment, vol. 936, Springer, New York, NY, 2016.
[39] J.M. Ayuso, R. Monge, A. Martínez-González, M. Virumbrales-Muñoz, G.A. Llamazares, J. Berganzo, A. Hernández-Laín, J. Santolaria, M. Doblaré, C. Hubert, et al., Glioblastoma on a microfluidic chip: generating pseudopalisades and enhancing aggressiveness through blood vessel obstruction events, Neurooncology 19 (4) (2017) 503–513.
[40] J.M. Ayuso, M. Virumbrales-Muñoz, A. Lacueva, P.M. Lanuza, E. Checa-Chavarria, P. Botella, E. Fernández, M. Doblare, S.J. Allison, R.M. Phillips, et al., Development and characterization of a microfluidic model of the tumour microenvironment, Sci. Rep. 6 (2016) 36086.
[41] F. Morfoisse, A. Kuchnio, C. Frainay, A. Gomez-Brouchet, M.-B. Delisle, S. Marzi, A.-C. Helfer, F. Hantelys, F. Pujol, J. Guillermet-Guibert, et al., Hypoxia induces VEGF-C expression in metastatic tumor cells via a HIF-1$\alpha$-independent translation-mediated mechanism, Cell Rep. 6 (1) (2014) 155–167.
[42] Y. Kim, H. Jeon, H. Othmer, The role of the tumor microenvironment in glioblastoma: a mathematical model, IEEE Trans. Biomed. Eng. 64 (3) (2017) 519–527, https://doi.org/10.1109/TBME.2016.2637828.
[43] J.R. Dormand, P.J. Prince, A family of embedded Runge-Kutta formulae, J. Comput. Appl. Math. 6 (1) (1980) 19–26.
[44] P. Bogacki, L.F. Shampine, A 3 (2) pair of Runge-Kutta formulas, Appl. Math. Lett. 2 (4) (1989) 321–325.
[45] C.G. Broyden, A class of methods for solving nonlinear simultaneous equations, Math. Comput. 19 (92) (1965) 577–593.
[46] J. Sherman, W.J. Morrison, Adjustment of an inverse matrix corresponding to a change in one element of a given matrix, Ann. Math. Stat. 21 (1) (1950) 124–127.
[47] R.D. Skeel, M. Berzins, A method for the spatial discretization of parabolic equations in one space variable, SIAM J. Sci. Stat. Comput. 11 (1) (1990) 1–32.
[48] L.F. Shampine, M.W. Reichelt, J.A. Kierzenka, Solving index-1 DAEs in MATLAB and Simulink, SIAM Rev. 41 (3) (1999) 538–552.
[49] A. Krogh, The number and distribution of capillaries in muscles with calculations of the oxygen pressure head necessary for supplying the tissue, J. Physiol. 52 (6) (1919) 409–415.
[50] I.F. Tannock, Oxygen diffusion and the distribution of cellular radiosensitivity in tumours, Br. J. Radiol. 45 (535) (1972) 515–524.
[51] B.W. Pogue, J.A. O'Hara, C.M. Wilmot, K.D. Paulsen, H.M. Swartz, Estimation of oxygen distribution in RIF-1 tumors by diffusion model-based interpretation of pimonidazole hypoxia and Eppendorf measurements, Radiat. Res. 155 (1) (2001) 15–25.
[52] T.W. Secomb, R. Hsu, M.W. Dewhirst, B. Klitzman, J.F. Gross, Analysis of oxygen transport to tumor tissue by microvascular networks, Int. J. Radiat. Oncol. Biol. Phys. 25 (3) (1993) 481–489.
[53] A.A. Patel, E.T. Gawlinski, S.K. Lemieux, R.A. Gatenby, A cellular automaton model of early tumor growth and invasion: the effects of native tissue vascularity and increased anaerobic tumor metabolism, J. Theor. Biol. 213 (3) (2001) 315–331.
[54] A. Martínez-González, G.F. Calvo, L.A. Pérez Romasanta, V.M. Pérez-García, Hypoxic cell waves around necrotic cores in glioblastoma: a biomathematical model and its therapeutic implications, Bull. Math. Biol. 74 (12) (2012) 2875–2896, https://doi.org/10.1007/s11538-012-9786-1.
[55] P.-S. Tang, On the rate of oxygen consumption by tissues and lower organisms as a function of oxygen tension, Q. Rev. Biol. 8 (3) (1933) 260–274.
[56] A. Daşu, I. Toma-Daşu, M. Karlsson, Theoretical simulation of tumour oxygenation and results from acute and chronic hypoxia, Phys. Med. Biol. 48 (17) (2003) 2829.
[57] L. Hathout, B. Ellingson, W. Pope, Modeling the efficacy of the extent of surgical resection in the setting of radiation therapy for glioblastoma, Cancer Sci. 107 (8) (2016) 1110–1116.
[58] A. Agosti, C. Giverso, E. Faggiano, A. Stamm, P. Ciarletta, A personalized mathematical tool for neuro-oncology: a clinical case study, Int. J. Nonlinear Mech. 107 (2018) 170–181.
[59] H.B. Frieboes, J.S. Lowengrub, S. Wise, X. Zheng, P. Macklin, E.L. Bearer, V. Cristini, Computer simulation of glioma growth and morphology, Neuroimage 37 (2007) S59–S70.
[60] C.E. Griguer, C.R. Oliva, E. Gobin, P. Marcorelles, D.J. Benos, J.R. Lancaster Jr, G.Y. Gillespie, CD133 is a marker of bioenergetic stress in human glioma, PLoS ONE 3 (11) (2008) e3655.

CHAPTER

# 17

# Skin Mechanobiology and Biomechanics: From Homeostasis to Wound Healing


*Maria G. Fernandes*[*,†], *Lucília P. da Silva*[*,†], *and Alexandra P. Marques*[*,†,‡]

**[*]I3Bs—Research Institute on Biomaterials, Biodegradables and Biomimetics of University of Minho, Headquarters of the European Institute of Excellence on Tissue Engineering and Regenerative Medicine, University of Minho, Guimarães, Portugal [†]ICVS/3B's—PT Government Associate Laboratory, Guimarães, Portugal [‡]The Discoveries Centre for Regenerative and Precision Medicine, Headquarters at University of Minho, Guimarães, Portugal**


## 17.1 INTRODUCTION

The skin is the outmost and largest organ of the body constituting 6%–10% of the lean body mass. It is a layered tissue containing in total more than 20 different cell types that in a very coordinated way act to keep skin homeostasis and function. It is the first line of defense against the external environment, that is, external forces (tension, compression, and shear), external pathogens, temperature, and radiation. The outermost layer, epidermis, varies in thickness depending on its location and function. This layer consists of a stratified squamous epithelium of keratinocytes delimited by the basal membrane and contains melanocytes, Langerhans cells, and Merkel cells [1, 2]. The internal layer, dermis, is a connective tissue that represents most of the skin substance and structure. The dermis is composed of fibroblasts and extracellular matrix (ECM) enriched in collagen and elastin fibers and can be divided into two layers: the upper papillary and the thicker lower reticular dermis. The skin mechanical properties, strength, and elasticity are mostly owed to the composition and organization/orientation of the ECM in the dermis [3–5]. Lastly, beneath the dermis is the hypodermis, composed primarily of adipose cells used for fat storage, and is usually not regarded as part of the skin tissue.

Being a tissue constantly exposed to many external and endogenous factors, which disintegrate its structure and functions, skin requires intrinsic suitable mechanical properties that protect the body from suffering damage. While it is known that skin has high flexibility and is able to support large deformations, its mechanical properties are complex and difficult to describe or predict due to skin complexity. Moreover, the mechanical properties of skin not only differentiate between the different layers but also vary with skin anatomical region (heterogeneity), age, sex, pathology, body weight, and orientation (anisotropy) (Fig. 17.1). In addition to skin inherent biomechanics variability, the mechanical testing used to analyze skin biomechanics plays a pivotal role in the measurement [6–10]. Hence, it is not surprising that the evaluation of skin biomechanics has revealed inconsistent results.

The skin is a tensegrity tissue, and it is in passive tension at homeostasis. Once the mechanical properties of the skin are unable to support the external conditions or the tissue is removed, the skin tensegrity is compromised. After a breach, the skin responds with an orchestrated process to heal the wound and restore the integrity of the tissue. The wound healing process encompasses four interconnected and consecutive phases, namely, hemostasis, inflammation, proliferation, and remodeling. All of these phases are influenced by mechanical forces, and there is increasing evidence that mechanical influences regulate postinjury inflammation contributing to the closure of the wound [11] and the formation of fibrotic tissue [12, 13]. Human skin, as well as skin cells, reacts to mechanical forces and converts mechanical cues to biochemical signals that are crucial to the way the wound healing progresses [12, 14]. While the homeostatic and wound cellular and biochemical milieus have been largely studied to create better cues for regeneration, skin mechanical environment has not been as explored. Moreover, most of this knowledge has been provided by in vitro studies assessing the effect of tension over skin cells [15–24]. Mechanical stimuli have also been evaluated


https://doi.org/10.1016/B978-0-12-816390-0.00017-0

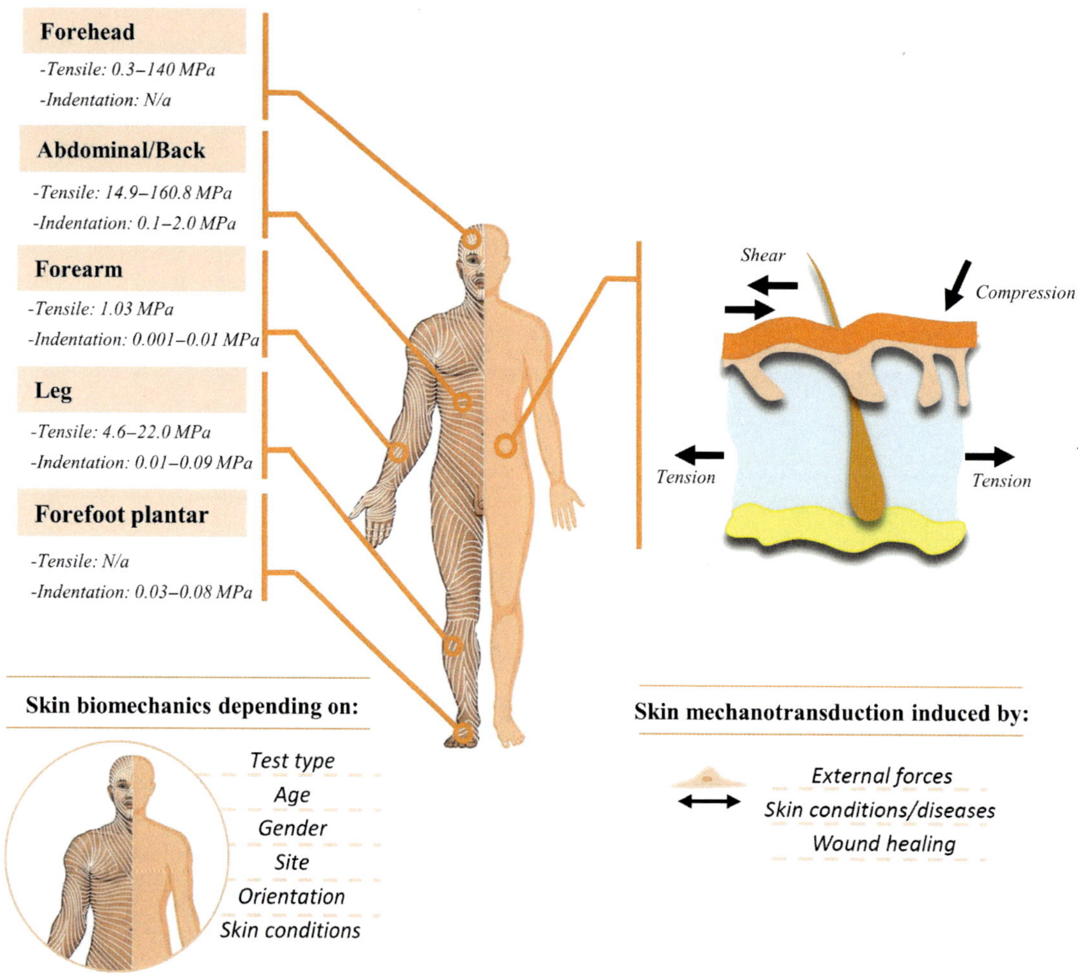


FIG. 17.1 Overview of human skin mechanical properties and schema of external mechanical forces that are transmitted across the skin tissue.

in vivo but mainly in rodent animal models that fail to represent skin biomechanics [25–27] and in pigs [28]. Thus, a better understanding of how healthy and wounded tissue deforms and how cells respond to this mechanical stress is critical for the creation of new therapies for scarless wound regeneration.

This chapter aims to elucidate the reader about skin biomechanics by focusing on the overall mechanical properties of the tissue and on its biomechanics. For this purpose, the main concepts and methodologies to measure skin mechanics as well as the mechanisms underlying skin mechanobiology and biomechanics are described. An overall analysis of this knowledge is presented, and a discussion on how it can be the basis to ameliorated wound healing approaches and relieve scarring through new therapies is provided.

## 17.2 BIOMECHANICS IN THE CONTEXT OF THE SKIN

As already mentioned the skin tissue is composed of different structural components, such as the collagen fibers (27%–39% by volume and 75%–80% of fat-free dry weight), elastin fibers (0.2%–0.6% by volume and 4% of the fat-free dry mass), and glycosaminoglycans (0.03%–0.35% by volume), in different combinations. Hence the properties of the skin are both dependent on the composition and their organizational direction [29]. Numerous studies were conducted over the last 40 years to characterize the mechanical behavior of the human skin. Most of them describe skin as a nonhomogeneous and anisotropic tissue with nonlinear and time-dependent mechanical behavior that can include viscoelastic response highly variable and sensitive to environmental conditions [4, 10, 30–32]. Accurate acknowledge of the skin structure and its constituents are important for determining its response to the different mechanical loads since each of these components has its own role in the mechanical properties of the skin tissue. Collagen (types I and III as the most prominent in human skin), as the main load-bearing and stiff component of the dermis, provides most of the mechanical strength of skin [33]; elastin fibers form a network in dermis and works like an energy storage device, bringing stretched collagen back to a relaxed position, which means they provide to the skin the ability to recoil after deformation [3]; and glycosaminoglycans, such as dermatan sulfate, hyaluronic acid, and chondroitin sulfate, provide to the skin their viscoelastic nature at low loads [34]. Glycosaminoglycans are covalently linked to peptide chains that form high-molecular-weight complexes called proteoglycans. The viscoelastic properties of connective tissue are strongly correlated with the type and the number of glycosaminoglycans [35].

Skin, as the outer shield of the human body, has often their mechanical integrity threatened and therefore needs to hold appropriate mechanical properties to respond correctly to the external mechanical forces. To understand the mechanical properties of skin, one must understand the main terms used to describe them. The following subsections highlight the basic definitions of the physics terms important to the understanding of the skin mechanics.

**Stress and Strain**

Most of the times materials are in a state of stress meaning that a force is being applied, and this will cause a change in its dimensions. Thus stress can be defined as the ratio of the applied force $F$ (in newtons, N) to cross-sectional area $A_0$ (in square meters, $m^2$) of a material. Depending on the force and the direction that is applied, three basic types of stress can be made on a material: tensile ($\sigma$), compressive ($\sigma$), and shear ($\tau$) stress (Fig. 17.2).

Strain, the change in material dimensions produced by a force that varies on its nature (tensile or compressive ($\varepsilon$), or shear ($\gamma$) force), is calculated by the ratio of the change in length ($\Delta L$) to the original length ($L_0$). Strain is dimensionless and sometimes represented as a percent change. The relation between the applied force and strain, or their rates, allows the assessment of material mechanical properties.

**Elasticity, Inelasticity, and Viscoelasticity**

Materials display different mechanical behaviors when an external load is applied and then removed (Fig. 17.3). Elasticity describes the material ability to deform instantaneously when a load is applied and to return immediately to its original state once the load is removed. On the other hand, inelasticity is the property of a material that keeps it permanently deformed by a force, which means that the material does not return to the original configuration when unloaded, remaining in the deformed state. Inelastic materials included viscous (fluids) and plastic materials, while viscoelastic ones combine the characteristics of both elastic solids and viscous fluids when undergoing deformation [36]. The viscoelastic materials exhibit time-dependent strain showing a "fading memory." Such behavior can be linear (stress and strain are proportional) or nonlinear.

**Stiffness, Young's Modulus, and Strength**

Stiffness describes the property of a material to resist the deformation under applied loads. The stiffness of a material is commonly defined by Young's or elastic modulus ($E$), determined by the slope of the linear region of a stress-strain curve when tested under tensile or compressive loads. Strength is a measure of the material resistance to failure by fracture or excessive deformation and can be defined as the maximum stress that the material supports before breaking. The strength of the materials varies with the nature of the applied force. For instance, there are materials with higher strength under compression but lower strength under tension.

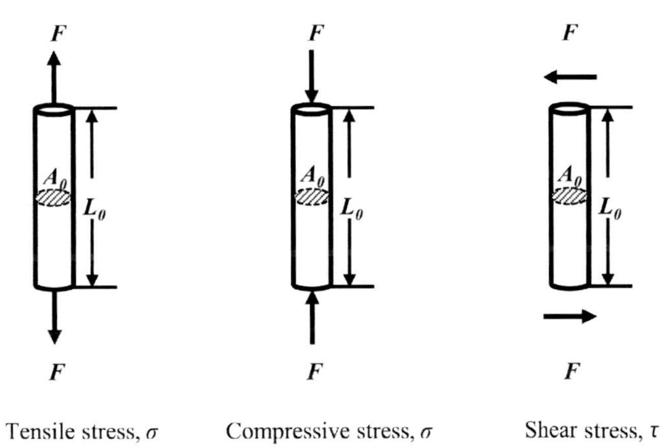


FIG. 17.2 Schematic description of the different types of applied stresses, expressed in newton per square meter ($N/m^2$) or Pascal (Pa), where $1\,Pa = 1\,N/m^2$ in the international system (SI). The units used for molecular, cell, or tissue levels are usually nanonewton per square micron ($nN/\mu m^2$) or kPa (where $1\ nN/\mu m^2 = 1\,kPa$).

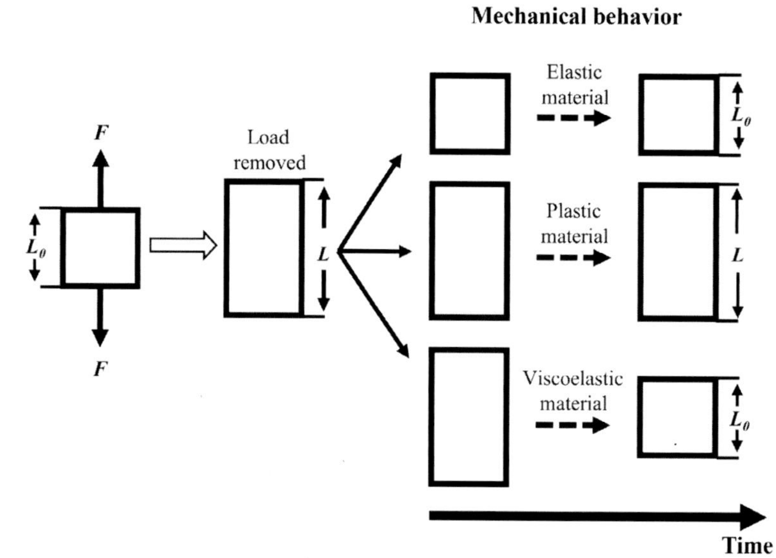


FIG. 17.3 Schematic representation of the different mechanical behavior along the time of materials subjected to a tensile load.

**Anisotropy and Nonhomogeneity**

Anisotropy is the property of being directionally dependent, which means that the physical properties of a material are different when measured along different axes (x, y, and z). As opposed to anisotropy, isotropy means homogeneity in all directions. Nonhomogeneity means that the material is made of the different material throughout. Because nonhomogeneous materials vary from point to point, it is important to note that the measured mechanical properties are also influenced by the structural characteristics of the material and their microstructural constituents.

## 17.2.1 Measuring Skin Mechanical Properties

Several methods have been followed to determine the mechanical properties of whole skin tissue in response to various loading conditions. These methods have been generally applied in in vivo and ex vivo samples. The majority of the approaches use either uniaxial/biaxial tensile testing, compression, indentation, or suction, subsequently coupled to imaging techniques such as motion analysis and digital image correlation to map the strain distribution in the skin tissue [9, 37–42]. Since skin mechanics is strongly dependent on active processes, the in vivo analyzes are the truly reliable methods to determine skin properties. Usually, a static or dynamic external force is applied on a specific area of the skin surface to provoke a stretch, compression, shear, or even torsion deformation and then assess the mechanical properties [38, 43–45]. However, these measurements can be restrictive since only small stresses can be applied and the boundary conditions cannot be fully controlled. In opposition, higher deformation along different skin directions can be applied in the ex vivo tests providing useful anisotropic data [46]. In addition, ex vivo experiments allow to conduct destructive tests and to obtain the skin mechanical behavior up to the failure point, as well as to analyze the skin layers' behavior separately.

### *17.2.1.1 Tensile Testing*

Considering all of the mechanical tests that are performed in skin tissue, tensile testing is probably the most common one. This is because, in normal physiological conditions, the skin is under a state of tension even in the absence of external loading [47, 48]. Tensile testing is performed by elongating a skin specimen under uniaxial or biaxial loading until failure. In uniaxial tests, the skin samples are tapered into a "dog bone" shape, while in the biaxial tests, the samples are cut into a square shape (Fig. 17.4A). From the representation of the results into a stress/strain curve, a variety of mechanical properties such as Young's modulus (stiffness), yield stress, ultimate stress, the ultimate tensile strength, and energy at failure can be extracted (Fig. 17.4B).

Some studies have shown that skin tissue under uniaxial tension experience three distinct stages (I, II, and III) (Fig. 17.5) [4, 49–51]. These stages are mainly related to the structural response of the dermal collagen and elastin fibers. In the first stage (up to 30%–40% of strain), the skin tissue is gradually stretched, and most of the mechanical response is carried out through the elastin fibers and the proteoglycan matrix. At this stage, the contribution of the crimped collagen fibers can be neglected [52]. When the skin is stretched to high strain levels (Stage II), the crimped collagen

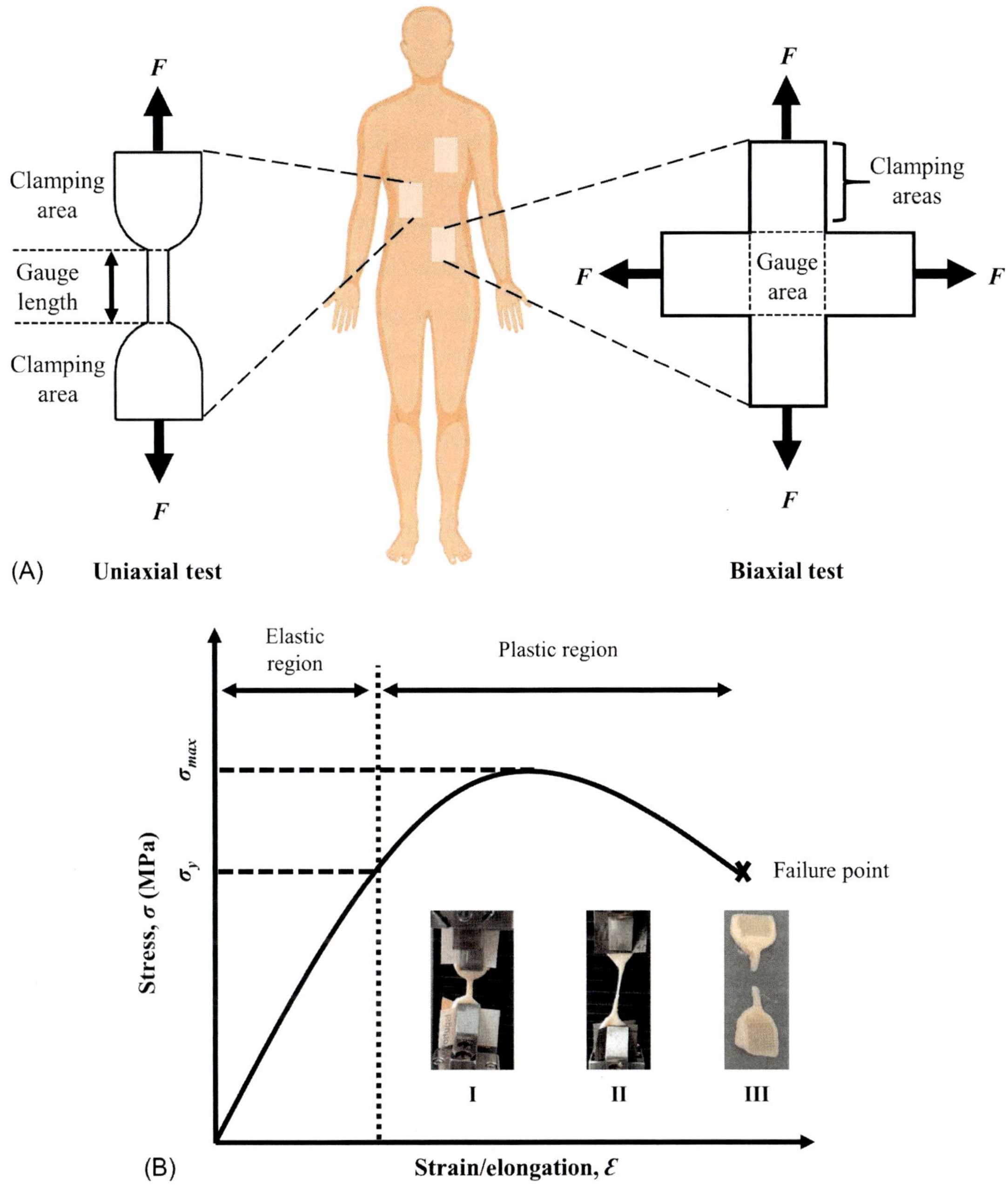


FIG. 17.4 Illustration of steps for mechanical tensile testing: (A) the samples are harvested and prepared into specific geometries; (B) typical stress-strain curve highlighting important measurement points. The typical stress-strain curve is characterized by an initial linear region before the yield point (called elastic region I) and a postyield nonlinear region (called plastic region II). In the elastic region, the loaded material will effectively return to its original length when the load is removed, that is, there is a negligible permanent extension. The slope of the stress-strain curve within the elastic region is defined by Hooke's law and Young's modulus can be determined. In the plastic region, the material begins to yield, which mean that the material does not recover to its original shape when the stress is removed. The yield point is called yield stress ($\sigma_y$) and corresponds to the transition from linear (elastic region) to nonlinear behavior (plastic region). The maximum stress ($\sigma_{max}$) is the highest tensile stress that the material can support before failure, which means the ultimate tensile strength (UTS) of the material. At the end of the curve the specimen fractures (III).

fibers gradually elongate and tend to align toward the load application direction [4]. The alignment of the collagen fibers leads to a high resistance the load, which makes the skin tissue to behave as a stiffer material. Thus, the stress-strain relation turns linear due to the stretching and slippage of the collagen fibers [53, 54]. The third stage begins after the yield point, for skin considered the strain above 60%–70%. Then the ultimate tensile strength is reached, and the rupture of the fibers starts to occur due to the loss of fibrillar structure [55]. At this stage, skin tissue loses the capacity to return to its original shape upon removal of the tensile stress.

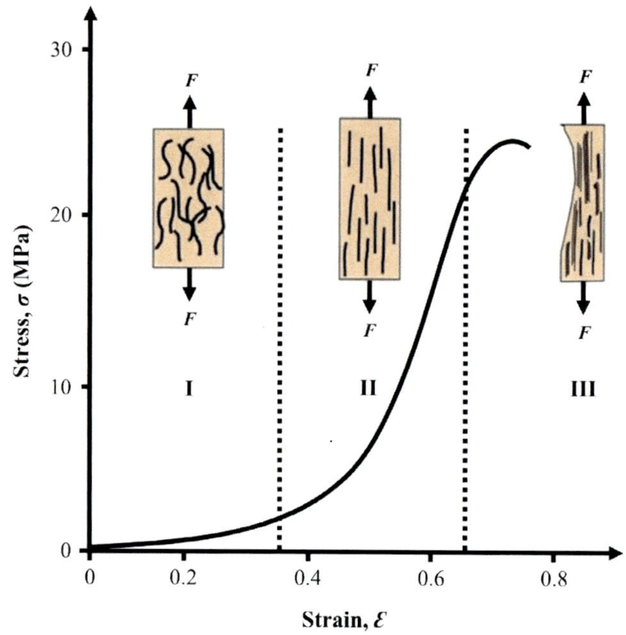


FIG. 17.5 General stress-strain curve for skin tissue under uniaxial tension and schematic drawings of the collagen fibers organization during the three stages of tensile loading: Stage I: the crimped collagen fibers begin to be oriented along the tensile axis, but its contribution can be neglected; Stage II: collagen fibers are straightening, and larger and larger amount of the fibrils reorient near to the tensile axis; Stage III: collagen fibers are fractured and curled back.

In the biaxial tests, loading is applied in two distinct directions. The first in vitro biaxial tension study of skin tissue was reported by Lanir and Fung [56]. They studied the biaxial mechanical properties of rabbit abdominal skin in two orthogonal directions. The results demonstrated the directional dependence of the stress-strain characteristics and the nonlinear response and the viscoelasticity of the skin. The comparison of the skin response under uniaxial and biaxial tension showed that the biaxial tensile loading results in the lateral compression of the stress-strain curve and the reduction in the strain before the entry into the linear region, which is due to the two-directional stretch of collagen fibers [50, 57]. While also suitable to understand the two-directional properties of the skin, the biaxial methods are not appropriated to determine the failure point since the hooks affixed to the edges of the sample tend to tear the skin.

Overall, testing of whole skin indicates that resultant Young's modulus from tensile tests varies from 0.1 to 160.8 MPa, depending on the tests conditions, sample orientation, location, and so on (Table 17.1) [58–62].

#### 17.2.1.2 Compression Testing

Although skin tissue is exposed to external compressive loads, the mechanical behavior of skin tissue under compression has been rarely studied. Shergold et al. [63] studied the uniaxial compressive response of pig back skin under a wide range of strain rates and showed that the stress-strain curves exhibit a similar profile to those obtained under the same conditions in tensile tests. Accordingly, the only difference is the strain levels at the transition to the different stages; the transitions in the compression tests occur at highest strain levels. Nonetheless, different studies confirmed that skin under compression is highly viscoelastic and shows a nonlinear and time-dependent mechanical behavior [63–66].

From our knowledge, the mechanical properties of human skin tissue under compression loadings are not clearly established yet. The few studies found in the literature are related to the animal skin.

#### 17.2.1.3 Indentation Testing

Tensile and compression tests provide an important comprehension of the mechanical behavior of the skin in response to extreme loading conditions but fail in providing any insight into the skin microenvironment. Micro- or nanoscale techniques such as indentation tests using atomic force microscopy (AFM) have proved useful to examine the skin mechanical properties at small length scales [60, 67–69]. The test consists in the use of a rigid cylindrical, pyramidal, or spherical indenter tip, to apply a perpendicular force on a target area of the skin tissue corresponding to a specific cell location (Fig. 17.6). The indenter tip is attached to a flexible microcantilever, which bends toward or away from the skin sample when attractive or repulsive forces are present, respectively [70, 71]. During the measurements, topographical images and the vertical displacement of the cantilever and its deflection are simultaneously recorded

TABLE 17.1 Young's Modulus of Human Skin Determined With Different Mechanical Tests

| Mechanical test | Skin location | Young's modulus, *E* (MPa) | Reference |
|---|---|---|---|
| Uniaxial tensile (ex vivo) | Abdomen | 14.96 | [78] |
| | Back (SC/E/D) | 160.8 (parallel to the Ll) | [62] |
| | | 121 (45° to the Ll) | |
| | | 70.6 (90° to the Ll) | |
| | Back (SC/E/D) | 112.47 (parallel to the Ll) | [10] |
| | | 63.75 (90° to the Ll) | |
| | Forehead (E/D) | 0.33 | [79] |
| | Forearm (E/D) | 1.03 | |
| | Submandibular neck (E/D) | 1.28 | |
| | Scalp | 22.74 | [80] |
| Uniaxial tensile (in vivo) | Leg/hand | 4.6–20 | [81, 82] |
| Indentation (ex vivo) | Abdominal (SC/E) | 1–2 | [60] |
| | Abdominal (Dp) | 0.1–0.25 | [83] |
| | Breast (Dp) | | |
| | Gluteus (Dp) | | |
| | – | 0.0258–1.18 | [74] |
| | Forearm | 0.014 | [84] |
| | Lower limb | 0.01–0.09 | [85] |
| | Forefoot plantar | 0.03–0.08 | [86] |
| Indentation (in vivo) | Forearm | 0.0045–0.008 | [45] |
| | Forearm | 0.0285 | [87] |
| | Forearm | 0.0011 | [88] |
| Suction (in vivo) | Forearm | 18–57 | [89] |
| | Forearm | 0.11–0.12 | [90] |
| | Forehead | 0.21–0.25 | [90] |

*SC*, stratum corneum; *E*, epidermis; *D*, dermis; *Dp*, papillary dermis; *Ll*, Langer lines.

and then converted to load–displacement curves [72]. For this equipment, both load and penetration depth can be controlled during the test.

From the AFM indentation measurements, the obtained values for human skin Young's modulus vary from 0.77 kPa to 322 kPa for dermal tissue, and some studies reported higher elastic modulus values in the mega Pascal range, potentially influenced by the collagen fibers in the probed area (Table 17.1) [61, 73, 74]. In fact, the contribution of the orientation of the collagen fibers to Young's modulus of the skin dermis and scar tissue was demonstrated by Grant et al. [74]. AFM imaging showed that the scarred skin has a higher degree of orientation of its collagen fibrils and displays stiffer behavior than the healthy intact skin and weaker viscoelastic creep and capability to dissipate energy at physiologically relevant frequencies.

#### *17.2.1.4 Suction Testing*

Suction test is one of the most widely used and accepted means of measurement of skin mechanical properties in vivo. Under this test, the tissue is elevated by applying a partial vacuum using a circular aperture, and the skin deformation is quantified by optical or ultrasound devices [75]. Generally, the obtained results only take into account the negative pressure applied, that is, suction, and the elevation of the dome of skin drawn up to deduce the properties of the skin. The suction device is used to measure the skin distensibility and the in vivo mechanical properties by measuring the skin elasticity as a percentage of skin retraction after the stretch [76]. This method has as a major disadvantage: the dependence on the experimental conditions used such as the size of the suction cup and the negative pressure

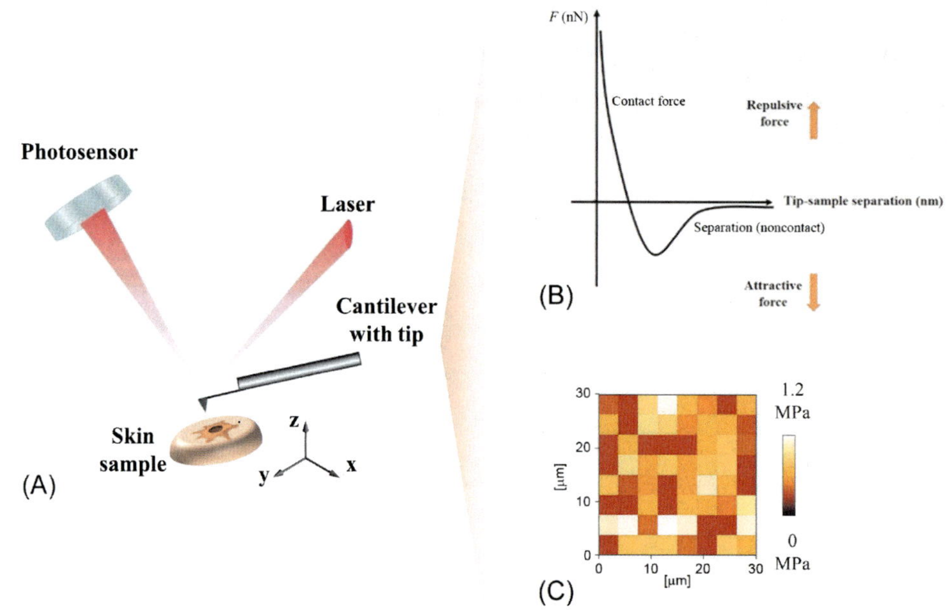


FIG. 17.6 Illustration of indentation test in and representative result plotting. (A) Bio-AFM setup in which skin sample is indented with the AFM tip with well-defined geometry; (B) sketch force-distance curve during force mapping; (C) AFM Young's modulus distribution in the analyzed skin area.

applied. Several authors have used this approach to calculate the mechanical properties of human skin [9, 58, 76, 77]. Young's Modulus measured by suction tests varies from 0.1 to 57 MPa (Table 17.1).

## 17.3 SKIN MECHANOBIOLOGY

### 17.3.1 Mechanosensing and Mechanotransduction

The field of mechanobiology studies the response of tissues and cells to mechanical signals that can be given by their surrounding environment [91]. Tissues and cells are continuously exposed to forces (*F*) of different types (compression, tension, and shear) and of varying magnitude, direction, and frequency that are able to change tissue/cell behaviors [92]. For instance, skin is continuously exposed to tension and compression forces resulting from ordinary daily actions like stretching. Mechanical forces also exist at the cellular level due to the tension forces of neighboring cells and/or surrounding ECM and contractile forces of cell cytoskeleton [92] (Fig. 17.7). Tissues and cells are able to sense those forces by the process of mechanosensing and convert this signal into a biological response through a process of mechanotransduction [91]. The process of mechanosensing is mediated by force-induced changes in the conformation of a protein, exposure of a peptide sequence of a protein, opening of ion channels, or receptor-ligand binding changes [93]. Integrins are the well-known transmembrane mechanoreceptors related with cell adhesion to the ECM, whereas cadherins, occludins, and connexins are the mechanoreceptors responsible for cell-cell mechanosensing [93, 94]. Other mechanoreceptors include G-protein-coupled receptors (GPCR, e.g., chemokine receptors (CXCR)), enzyme-like receptors (e.g., Discoidin domain receptors (DDR), ephrin receptors, and platelet endothelial cell adhesion molecule 1 (PECAM) receptors), ion channels (e.g., transient receptor potential (TRP) channels), lipid rafts, and glycocalyx [95].

FIG. 17.7 Cells are in a state defined as contractile tension due to the traction and tension forces applied to the cells. In equilibrium, tension and traction forces are in balance, and cell mechanotransduction is not activated. Mechanotransduction is activated once the forces are unbalanced. Cell cytoskeletons elongate once ECM-cell tension forces are higher than cell cytoskeleton contraction forces, whereas the cytoskeleton relaxes once cell cytoskeleton contraction forces are higher than ECM-cell tension forces.

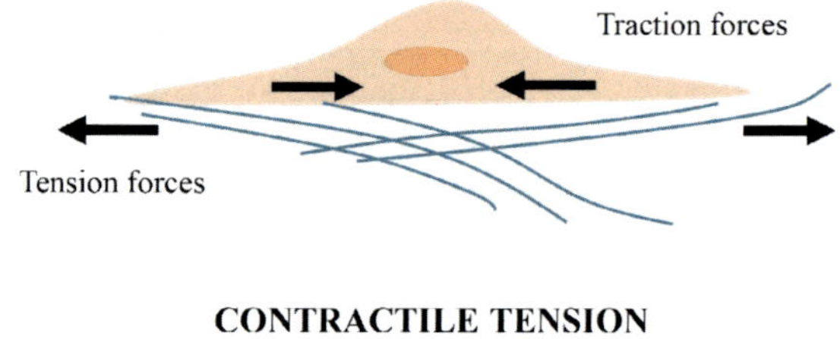

Mechanosensing is then followed by the process of mechanotransduction that occurs through alterations at the subcellular and molecular levels through intracellular proteins (e.g., vinculin and talin), kinases and phosphatases (e.g., focal adhesion kinase (FAK)), cytoskeletal components (i.e., actin filament and microtubule), or intermediate filaments (e.g., keratin, dermin, plectin, lamin, and vimentin) [93]. Thus it is clear why the mechanosensing force is mechanotransduced into a biological response related with the activated signaling pathway that can be, for instance, adhesion, survival, apoptosis, proliferation, migration, or differentiation [93, 96, 97].

The effect of the stiffness and topography over skin cells has been largely reviewed, but the result of mechanical forces on skin cells has only been considered recently and will be the focus of this section. In the skin, different cells are able to mechanosense mechanical forces. The most distinctive mechanoreceptors are neurons that are able to sense touch, itch, pain, and temperature stimuli. Merkel cells, the Ruffini corpuscle, and the Meissner's corpuscle both at the epidermis and dermis of the skin are also able to mechanosense external forces and transmit the signal to sensory neurons called low-threshold mechanoreceptors (LTMRs) that in turn send the signal to the central nervous system [98, 99]. Other cells existing in the skin, including endothelial cells [100], adipocytes [101], stem cells [102], Langerhans cells [103], and melanocytes [104, 105], also sense mechanical stimuli, but the effect of mechanical loading over skin cells has only been explored for the main skin resident cells—fibroblasts and keratinocytes. This is probably because these cells, in addition to the external stimuli, are continuously exposed to internal passive tension forces resulting from cell-cell (keratinocytes and fibroblasts) and cell (fibroblast)-ECM interactions. External forces exerted at air-epidermis interface enhance the active tension in the skin and on both cell types due to the stretching of cell junctions and collagen fibrils. The mechanical stimuli are transmitted along the air-epidermis interface throughout the epidermis due to hemidesmosomes existing in the epidermal-dermal junction that maintain the mechanical continuity along the two skin layers [106]. Animal models have been used to understand the effect of forces over the different skin cells [25–27]. However, these studies are mainly performed using rodent skin that is both anatomically and mechanically different from human skin [107]. Moreover, human skin is a complex organ, and the continuous exposition to different stimuli, including other than mechanical ones such as chemical, radiation, electric, etc., renders difficult the understanding of isolated effects. Hence, the effect of external forces over keratinocytes and fibroblasts has been mainly explored in vitro by the application of external forces onto cells.

### 17.3.2 Effect of Forces Over Fibroblasts and Keratinocytes

The first study of the effect of forces over skin cells was in 1990. Görmar et al. showed that cyclic compression forces of pestle-shaped weights over keratinocytes induced their differentiation [108]. Similar studies using compression forces conducted in dermal skin fibroblasts showed increased production of metalloproteinases (MMPs) in response to the stimulus [109]. To our knowledge, these were the only studies that explored the effect of compression forces over skin cells. This may be due to the limited significance of compression forces, as keratinocytes and fibroblasts in the skin are only subjected to tension forces. Thus skin mechanobiology studies have focused on the effect of tension forces of different orientations, magnitudes, frequencies, and periods over keratinocytes (Table 17.2) and fibroblasts (Tables 17.3 and 17.4).

In the late 90's, researchers started developing in-house mechanical loading units to study the effect of tension on cells cultured onto silicone substrates [110, 111]. Keratinocyte mechanobiology was empowered with the arising of mechanical stretching units for sell [16, 112–115]. Tension forces showed to affect keratinocytes by activating the epithelial-mesenchymal transition [115], promoting proliferation [112, 113, 115–118] through the activation of ERK signaling pathway [111, 116, 119], and by activating antiapoptotic mechanisms [120, 121] mediated by the activation of the Akt signaling pathway [117, 120, 121]. Moreover, tension forces induced keratinocytes migration [122, 123], basement membrane ECM production [16, 118], expression of basal layer-associated keratins [16, 116], and decreased expression of upper layer-associated ones [116]. These responses seem to be mediated by integrin β1 [111, 119, 123]. These studies revealed that tension mechanical loading has a prominent effect on keratinocytes and that the observed responses are independent of force orientation, magnitude, frequency, and period of the loading (Fig. 17.8).

The first observations regarding the mechanobiology of fibroblasts refer to 1983 when Allen and Schor detected the contraction of free-floating collagen gels loaded with dermal fibroblasts [127, 128]. In the absence of gel anchorage (no mechanical loading), the contractile tension existing between dermal fibroblasts and collagen gels becomes unbalanced prevailing fibroblasts traction forces that lead to gel contraction. This did not occur if gels were anchored because the mechanical loading due to anchoring maintained a balanced between the cell traction and the cell-ECM forces [127] (Fig. 17.9). The dermal fibroblasts in the anchored gels showed an elongated morphology and produced higher amounts of ECM in contrast with the round morphology and the reduced synthesis of the mediators of ECM synthesis observed with fibroblasts in free-floating gels [127, 129–131] (Fig. 17.9, Table 17.3). These responses

TABLE 17.2 Mechanotransduced Response of Keratinocytes Subjected to Tension Loading

| Force | Substrate | Cell | Mechanotransduced response | Reference |
|---|---|---|---|---|
| Uniaxial<br>Cyclic<br>Strain: 14%–33% | N/A | KCs | ↑IL-1α<br>↑IL-1R antagonist<br>↑*E*-selectin in human vascular endothelial cells mediated by media conditioned by mechanically stimulated KCs<br>Activate vascular endothelium | [110] |
| Uniaxial<br>Static/cyclic<br>Strain: 10%<br>Frequency: 0.17 Hz<br>Period: 15 min–7 days | Type I collagen | hNFKCs | ↑Proliferation ↑DNA synthesis ↑Elongation ↑Protein synthesis<br>↓cAMP ↓PKA ↓PGE2<br>Regulated by Il-1 release | [112, 113] |
| Uniaxial<br>Static<br>Strain: 10%<br>Period: 5 min–24 h | Arginine | HaCaT | Activation of ERK signaling pathway<br>Regulated by β1-integrins | [111] |
| Uniaxial<br>Static<br>Strain: 20%<br>Period: 24 h | Type I collagen | NHK | ↑Calcium influx<br>↑Proliferation<br>Activation of ERK1/2 and Akt signaling pathways<br>↑K6 ↓K10 | [116] |
| Uniaxial<br>Static<br>Strain: 10%<br>Period: 5 min | Arginine | HaCaT | Regulated by angiotensin II type 1 receptor.<br>Activation of Akt signaling pathway<br>Antiapoptotic | [121] |
| Uniaxial<br>Static<br>Strain: 20%<br>Period: 15–30 min | Type I collagen | NHK | Activation of Akt signaling pathway<br>Antiapoptotic | [120] |
| Uniaxial<br>Static<br>Strain: 10%<br>Period: 5 min–48 h | Type IV collagen, fibronectin, laminin, arginine, fetal calf serum | HaCaT | ↑ Substrate adhesion to fibronectin and collagen type IV but not to laminin<br>Activation of ERK signaling pathway<br>Regulated by β1-integrins | [119] |
| Uniaxial<br>Static/cyclic<br>Strain: 10%–30%<br>Frequency: 1 Hz<br>Period: 3 min–24 h | Type I/IV collagen | NHKs | ↓ uPA protease (static stretch)<br>↑ uPA protease (cyclic stretch) | [114] |
| Uniaxial<br>Static/cyclic<br>Strain: 10%<br>Frequency: 0.17 Hz<br>Period: 72 h | Type I collagen | HaCaT | Orientation perpendicular to stress<br>Reduced area<br>↑ MMP-9 protease | [124] |

| | | | | |
|---|---|---|---|---|
| Uniaxial<br>Cyclic<br>Intermittent (tension +relaxation)<br>Strain: 10%<br>Frequency: 0.17 Hz | Type I collagen | NFKCs | ↑AKT<br>↑BAD (I)<br>↓Proliferation (after 2 days of I) | [117] |
| Uniaxial<br>Cyclic<br>Strain: 10%–20%<br>Frequency: 1 Hz<br>Period: 24–48 h | Type I collagen | NHKs<br>HF | ↑Endothelin 1 | [125] |
| Uniaxial<br>Static<br>Strain: 15%–30%<br>Period: 6 days | PDMS | HaCaT<br>HF | Asymmetric migration of KCs and HF in coculture<br>Regulated by EGF derived from HF, not KCs | [122] |
| Uniaxial<br>Cyclic<br>Strain: 20%<br>Frequency: 0.2 Hz<br>Period: 12–72 h | Type I collagen | HaCaT hESCs | ↑ E-cadherin, catenin β1, connexin 43, desmoglein 1, endothelin 1, integrin α6, IL-α1, keratin 1/6/10/14, KGFR, laminin α5, fibronectin 11, MMP-9 | [16] |
| Uniaxial<br>Cyclic<br>Strain: 10%<br>Frequency: 0.5 Hz<br>Period: 1–72 h | Type I collagen | Neonatal or adult foreskin KCs | ↑Proliferation ↓Differentiation | [115] |
| Uniaxial<br>Cyclic<br>Strain: 10%<br>Frequency: 0.03 Hz<br>Period: 5 days | Type I collagen | HSE<br>(NHK+HF) | Thicker epidermal layer<br>↑Laminin 5<br>↑Collagen IV/VII<br>Development of basement membrane | [118] |
| Uniaxial<br>Static<br>Strain: 20%<br>Period: 6 days | Type I collagen | HaCaT<br>HF | KC migration mediated by β1-integrin | [123] |
| Uniaxial<br>Static<br>Strain: 10%<br>Period: 5 min | Arginine | HaCaT | Desmosomes and keratins involved in mechanosensing | [126] |
| Uniaxial<br>Static/cyclic<br>Strain: 20%<br>Frequency: 1 Hz<br>Period: 6 days | Type I collagen or fibronectin | HaCaT | ↓Proliferation (cyclic)<br>Regulated by β1-CASK signal pathway | [20] |

*BAD*, BCL2-antagonist of cell death; *cAMP*, cyclic adenosine monophosphate; *EGF*, epidermal growth factor; *ERKs*, extracellular signal-regulated kinases; *HaCaT*, keratinocyte cell line from human adult skin; *hESCs*, keratinocytes derived from human embryonic stem cells; *HF*, (primary) human fibroblasts; *hNFKCs*, human neonatal foreskin keratinocytes; *HSE*, human skin equivalents; *IL*, interleukin; *KCs*, keratinocytes; *KGFR*, keratinocyte growth factor receptor; *MMP*, metalloproteinase; *N/A*, not available; *NF-κB*, nuclear factor kappa-light-chain enhancer of activated B cells; *NHKs*, (primary) normal human keratinocytes; *PDMS*, poly-dimethylsiloxane; *PGE2*, prostaglandin E2; *PKA*, protein kinase A; *PKB*, proto-oncogene protein kinase B; *TGF*, transforming growth factor; *uPA*, urokinase-type plasminogen activator.

**TABLE 17.3** Mechanotransduced Response of Human Dermal Fibroblasts Loaded into Free-Floating or Anchorage Type I Collagen Gels

| Cell | Mechanotransduced response | Reference |
|---|---|---|
| HNFFbs | Cell migration<br>Gel contraction | [127] |
| HDFbs | ↓Procollagen type I and III<br>↓Carboxy-procollagen peptidases<br>↓Lysyl oxidase<br>(relaxation) | [129] |
| HDFbs (fetal) | ↓Contractile forces ↓Stress fibers ↓Collagen contraction, modulated by PGE2 through cAMP-dependent mechanism through the EP2 receptor (tension) | [130] |
| HDFbs | ↓hDFb size<br>↓Mechanical force<br>↓TGF-β2 receptor<br>↓SMAD3 pathway<br>↓ECM production<br>(relaxation) | [131] |
| HDFbs | ↑p130Cas<br>Phosphorylation of Src family kinase | [132] |

*hNFFbs*, human neonatal foreskin fibroblasts; *hDFbs*, human dermal fibroblasts; *PGE*, protaglandin E synthase; *cAMP*, cyclic adenosine monophosphate; *EP*, prostaglandin E receptor; *TGF*, transforming growth factor; *ECM*, extracellular matrix.

were then associated with the activation of the p130Cas phosphorylation of Src family kinase [132], of the PGE2 through cAMP-dependent mechanism in the EP2 receptor [130] and of the TGF-β/SMAD3 pathway [131].

Further studies on human dermal fibroblasts were conducted by exposing the cells to tension forces (Table 17.4). The most evident result after the application of tensile forces on fibroblasts loaded in flexible silicon membranes was the increased production of ECM synthesis mediators [133] and of ECM proteins, especially collagen and elastin [17–24, 133–138]. Moreover, collagen fibrils [134, 136, 137, 139, 140] were aligned in agreement with the alignment of the cells but perpendicular to the stretch [137, 139–141]. Interestingly, tension was also found to regulate levels of metalloproteinases (MMPs) and tissue inhibitors of metalloproteinases (TIMPs) secreted by fibroblasts, which impacts ECM degradation [19, 23, 138, 140, 142, 143]. The secretion of some growth factors, including vascular endothelial growth factor (VEGF), TGF-β1, TGF-β2, TGF-β3, connective tissue growth factor (CTGF), Cyr61, nerve growth factor (NGF), and stromal cell-derived factor (SDF)-1α was also found to be upregulated after subjecting fibroblasts to tension forces [137, 138, 144]. Moreover, tension leads to increased cell migration [140] and proliferation [24, 137, 138] and reduce cell apoptosis [18, 140]. All these results suggest that fibroblasts present a more active "synthetic" phenotype that coincides with myofibroblast phenotype as shown by the expression of α-SMA [19, 135, 145, 146]. These mechanotransduced responses are mediated by different signaling pathways and mediators, including integrin β1 [24, 137, 141], focal adhesion kinase [141], Rho GTPases [141], TGF-β pathway [24, 133, 135, 137, 147], p38 pathway [21, 135, 141, 148], ERK pathway [18, 21, 135, 148], Jnk pathway [18], Akt pathway [18, 148], Wnt pathway [140], SMAD [147], and P130cas [24, 137]. Nevertheless, in some studies, tension was found to decrease fibroblast proliferation [20, 148], collagen production [20, 149], and the release of CTFG [149, 150].

## 17.4 BIOMECHANICS AND MECHANOBIOLOGY IN THE CONTEXT OF SKIN WOUND HEALING

Healthy skin tissue is normally under tensile stress, but upon a small incisional wound, the skin relaxes. Cell-cell and cell-matrix forces are disrupted at the wound margins breaking the cell stress-shielding cap. A wound of a greater diameter than the incision wound that tends to elongate in the direction of the greatest stress is formed. Wound elongation after a surgical incision was originally demonstrated by the 19th century by the German anatomist Karl Langer [5, 151]. Langer thrust conical spikes through the skin of cadavers producing multiple splits over the entire human body and observed that the wound transformed into an elliptic form. By joining the major axes of these ellipses, he drew and catalogued the pattern of tension lines on the body, producing what we now term Langer's lines [5]. These

TABLE 17.4 Mechanotransduced Response of Dermal Fibroblasts After Tension Loading

| Force | Substrate | Cell | Mechanotransduced response | Reference |
|---|---|---|---|---|
| Cyclic<br>Biaxial<br>Strain: 20%<br>Frequency: 1 Hz<br>Period: 48 h | Type I collagen | hDFbs | ↑Procollagen<br>↑Proteinase C<br>Activated by TGF-β- | [133] |
| Uniaxial<br>Force: 120 dynes<br>Cyclic<br>Strain: 10%<br>Frequency: 11 dynes/min<br>Period: 10 min–11 h | Collagen Lattices<br>Collagen Sponges | hDFbs | ↑MMP-2, MMP-9, and PLAT<br>↓MMP-3 and uPA | [142, 143] |
| Equibiaxial cyclic<br>Strain: 20%<br>Frequency: 0.1 Hz<br>Period: 24 h | Type I collagen | hDFbs | ↑Cell proliferation transcripts<br>↑Connective tissue synthesis<br>↓ECM degradation ↓Inflammatory mediators<br>↑VEGF, TGF-β1, TGF-β3, CTFG, Cyr61 transcripts | [138] |
| Uniaxial<br>N/a | Rat tail<br>Type I collagen | hDFbs<br>(healthy and striae) | ↑ α-SMA<br>↑ Contractile force | [145] |
| Equibiaxial<br>Cyclic<br>Strain: 16%<br>Frequency:<br>0.2 Hz<br>Period:8 days | Fibrin | hDFbs | ↑Stronger gels ↑Denser ↑Thinner ↑Collagen | [17] |
| Cyclic<br>Strain: 10–24%<br>Frequency: 0.1–0.17 Hz<br>Period: 24 h | Type I collagen | hDFbs | ↓Proliferation<br>Activation of p38 and ERK1/2 (repetitive)<br>Activation of AKT and BAD pathways (intermittent) | [148] |
| Uniaxial cyclic<br>Strain: 20%<br>Frequency: 0.16 Hz | Type I collagen | hDFbs | ↓CTFG<br>↓COL1A2<br>↑Heparan sulfate proteoglycan 2 | [149] |
| N/a | PGA fibers | hDFbs | Spindle shape cells<br>Aligned collagen fibers<br>↑Collagen fibril diameter<br>↑Tensile strength | [134] |
| Axial<br>Cyclic<br>Strain: 4–12%<br>Frequency: 0.1 Hz<br>Period: 30 min–12 h | Type I collagen | hNFFbs | Cell orientation<br>Activation of FAK, p38, and Rho<br>Regulated in part by integrin *β*1 | [141] |
| Uniaxial<br>Cyclic<br>Strain: 10%<br>Frequency: 2 mm/min<br>Period: 7 w | Fibrin-based tubular constructs | hNDFbs | ↑Collagen ↑Elastin ↑ECM deposition and disorganization<br>↑α-SMA and SMAD2/3<br>Activation of TGF-β1 and p38<br>Inhibition ERK | [135] |
| Biaxial<br>Cyclic<br>Strain: 5%<br>Frequency: 0.3 mm/s<br>Period: 60–90 min | Fibrin gels | hDFbs | Fiber alignment<br>Fibrin degradation<br>↑Elastin and collagen, in a geometry-dependent manner | [136] |
| Equibiaxial<br>Cyclic<br>Strain: 15%<br>Frequency: 0.1 Hz<br>Period: 24 h | Type I Collagen | hDFbs | ↓α-SMA, CTGF, and ET-1<br>α-SMA and CTGF modulated by ET-1 levels | [150] |

*Continued*

TABLE 17.4 Mechanotransduced Response of Dermal Fibroblasts After Tension Loading—cont'd

| Force | Substrate | Cell | Mechanotransduced response | Reference |
|---|---|---|---|---|
| Uniaxial<br>Cyclic<br>Strain: 10%<br>Frequency: 0.5 Hz<br>Period:1 h | N/A | hDFbs | ↑NGF and TGF-β2 | [144] |
| Uniaxial<br>Cyclic<br>Strain: 120%<br>Frequency: 0.17 Hz<br>Period: 24 h | N/A | hDFbs | ↑Migration<br>Cell orientation<br>↓Apoptosis<br>↑Collagen degradation<br>Regulated by Wnt and integrin | [140] |
| Uniaxial<br>Static/cyclic<br>Strain: 5%<br>Frequency: 0.5 Hz<br>Period: 3–4 w | Tubular cell sheets | hNDFbs | Cell and collagen alignment<br>↑ Stiffness | [139] |
| Cyclic/static<br>Uniaxial<br>Strain: 10%<br>Frequency: 1 Hz<br>Period: 96 h | Type I collagen | hDFbs | ↑Type I procollagen<br>↑TIMP-1<br>↓MMP-1<br>Differentiation | [19] |
| Cyclic<br>Frequency: 1 Hz | Type I collagen | hDFbs | ↓Apoptosis<br>↑Cell survival<br>↑Adhesion<br>↑ERK ↑JNKs ↑AKT<br>↑Synthesis ECM<br>↑Denser focal adhesion<br>↑VEGF ↑SDF-1α | [18] |
| Uniaxial<br>Cyclic<br>Strain: 10%<br>Frequency: 0.1 Hz<br>Period: 48 h | N/A | hDFbs (from two different skin sites) | ↑Proliferation<br>↑Integrin β1, p130Cas, and TGF-β1 | [24] |
| Uniaxial<br>Static<br>Strain: 10%<br>Period: 24 h | Rat tail<br>Type I collagen | hDFbs<br>hNFFbs | ↑ MMP-2, TIMP-2, and collagen type III in HDFs (hDFbs but not hNFFbs) | [23] |
| Static/cyclic<br>Strain: 20 Hz<br>Frequency: 1 Hz<br>Period: 6 d | N/A | hDFbs | ↓Proliferation (cyclic)<br>↓Collagen I (cyclic)<br>↑Fibronectin (cyclic)<br>↑Collagen I (static)<br>↓Fibronectin (static) | [20] |
| Uniaxial<br>Cyclic<br>Strain: 10%–20%<br>Frequency: 0.1 Hz<br>Period:24 h | N/A | hDFbs (hypertrophic and normal skin) | ↑ Cell proliferation<br>Cell Orientation<br>↑TGF-β1<br>↑Collagen<br>Regulated by integrin β1 and P130Cas | [137] |
| Uniaxial<br>Cyclic<br>Strain: 15%<br>Frequency: 0.5 Hz<br>Period: 15 min–2 weeks | Fibrin gels | hDFbs | ↑Collagen deposition<br>Activation of ERK1/2 and p38 | [21] |

*CTFG*, connective tissue growth factor; *ECM*, extracellular matrix; *ERK*, extracellular signal-regulated kinases; *ET-1*, endothelin 1; *FAK*, focal adhesion kinase; *hDFbs*, human dermal fibroblasts; *hNDFbs*, human neonatal dermal fibroblasts; *hNFFbs*, human neonatal foreskin fibroblasts; *JNK*, c-Jun N-terminal kinase; *MMP*, metalloproteinase; *N/A*, not available; *NGF*, nerve growth factor; *PLAT*, plasminogen activator; *SDF-1*, stromal cell-derived factor 1; *TGF*, transforming growth factor; *TIMP*, metalloproteinase inhibitor; *uPA*, urokinase-type plasminogen activator; *VEGF*, vascular endothelial growth factor; *α-SMA*, α smooth muscle actin.

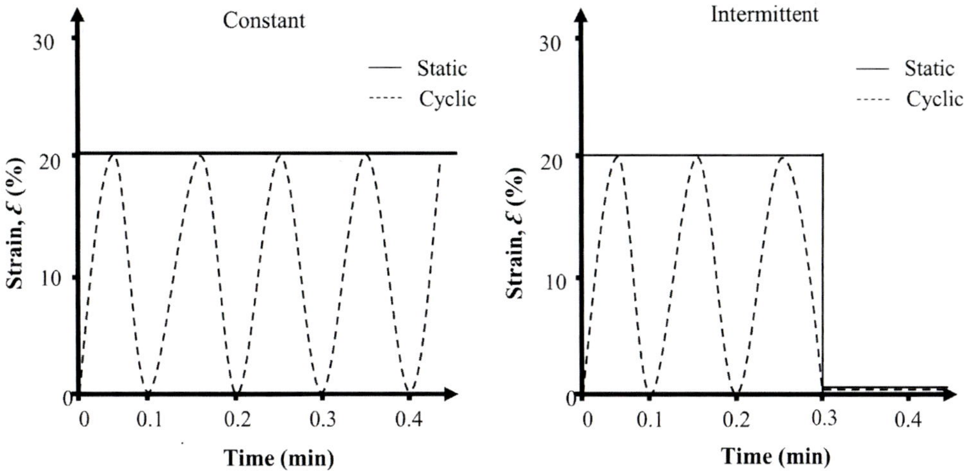


FIG. 17.8 Tension forces applied into cells.

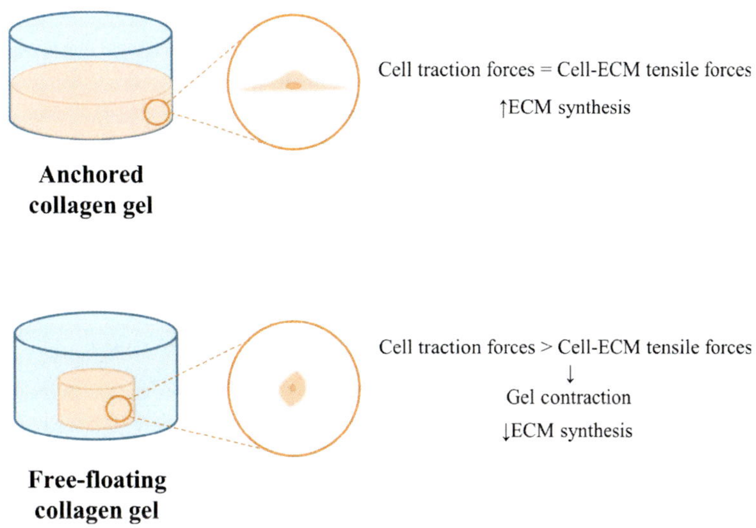


FIG. 17.9 Mechanotransduced response of human dermal fibroblasts loaded into free-floating or anchorage type I collagen gels.

lines run parallel to the main collagen fibers in the dermis but do not always follow the line of wrinkle [10]. Through these studies, it was possible to demonstrate the anisotropic nature of skin and to map the natural lines of great tension that occur within the skin tissue. Hence, surgical incisions that are performed perpendicular to the lines tend to pull open in the direction of the greatest stress and close with the formation of a scar tissue [152]. This evidence was further confirmed in a pig model that holds more significance in terms of skin anatomy and biomechanics [28, 153]. This has motivated prevention actions during surgical procedures to reduce scarring, that is, incision procedures are now being performed along the Langer's lines to reduce the scar area. Moreover, to prevent scar formation during the healing of incisional wounds, care products that relieve wound tension have emerged. Silicone bandages that contract the wound holding the incision margins together and bandages with zip that are able to distribute the forces during the skin closure through the action of a zip system are some examples. These bandages compress the wound along axial incision extension, and the magnitude of the force can be controlled, resulting in a reduced scar area. These therapeutic approaches are however inefficient in the treatment of excisional wounds as wound margins cannot be connected due to the lack of skin tissue. The healing of excisional wounds is associated with scarring due to the contraction of the dermis mediated by myofibroblasts. Myofibroblasts are specialized contractile fibroblasts differentiated from fibroblasts in response to transforming growth factor β1 (TGF-β1) [11]. Myofibroblasts express α-SMA and synthesize type I collagen that contribute to the reorganization of the ECM. These specialized contracting cells exert increased traction forces that are able to pull the ECM shortening the surrounding collagen network. Due to their "synthetic" phenotype, myofibroblasts synthesize new collagen that stabilizes the surrounding collagen network. This phenomenon leads to not only the remodeling of the damaged ECM but also contraction of the dermis. After wound repair, TGF-β1 is reduced, and the activity of myofibroblasts is reduced due to myofibroblast dedifferentiation to fibroblasts [11]. These findings demonstrate that the focus of new therapeutics for the treatment of excisional wounds should go beyond skin mechanics and interfere with myofibroblasts mechanobiology. In fact, Kun et al. have recently shown accelerated healing and lower scar formation of mice excisional wounds by inhibiting FAK, one of the main mediators of mechanobiology [154]. This response was associated with the reduced action of myofibroblasts and reduced collagen deposition. Despite the increasing knowledge observed in the last years in the fields of skin mechanobiology and biomechanics, the mechanobiological processes occurring in wound healing are still far from being totally understood. Better knowledge of these processes will be a breakthrough in the development of new therapies for a scarless wound healing.

## 17.5 FINAL REMARKS

The growing knowledge of skin biomechanics and skin mechanobiology has greatly contributed to understand the importance of mechanical stress in skin homeostasis and in skin wound healing. Knowledge of the importance of mechanical stress in wound healing has potentiated the creation of new therapies to counteract the wound healing forces and improve the normal skin tensegrity. In addition to the mechanical off-loading therapies, other therapies that focus on skin mechanobiology events are a pathway to follow in the future. It is clear that strategies that aim for scarless wound regeneration should consider not only the cellular and biochemical milieu of the wound healing process but also the mechanical environment. Nonetheless, the future successes will greatly depend on a better understanding of the molecular mechanisms involved in the skin response to the wound intrinsic mechanical forces as the stimuli and underlying mechanisms that drive wound healing to be elucidated.

## References


[1] E.A. Gantwerker, D.B. Hom, Clin. Plast. Surg. 39 (2012) 85–97.
[2] R.A. Kamel, J.F. Ong, E. Eriksson, J.P. Junker, E.J. Caterson, J. Am. Coll. Surg. 217 (2013) 533–555.
[3] H. Oxlund, J. Manschot, A. Viidik, J. Biomech. 21 (1988) 213–218.
[4] F.H. Silver, J.W. Freeman, D. DeVore, Skin Res. Technol. 7 (2001) 18–23.
[5] K. Langer, Br. J. Plast. Surg. 31 (1978) 93–106.
[6] M.H. Alireza Karimi Ahmad Shojaei, M. Navidbakhsh, A.M. Haghi, S.J.A. Sadati, Proc. Inst. Mech. Eng. Pt L J. Mater. Des. Appl. 0 (2015) 1–8.
[7] G. Boyer, C. Pailler Mattei, J. Molimard, M. Pericoi, S. Laquieze, H. Zahouani, Med. Eng. Phys. 34 (2012) 172–178.
[8] S. Girardeau, S. Mine, H. Pageon, D. Asselineau, Exp. Dermatol. 18 (2009) 704–711.
[9] M. Hendriks, D. Brokken, W. Oomens, L. Bader, P. Baaijens, Med. Eng. Phys. 28 (2006) 259–266.
[10] A. Ni Annaidh, K. Bruyere, M. Destrade, M.D. Gilchrist, M. Ottenio, J. Mech. Behav. Biomed. Mater. 5 (2012) 139–148.
[11] J.J. Tomasek, G. Gabbiani, B. Hinz, C. Chaponnier, R.A. Brown, Nat. Rev. Mol. Cell Biol. 3 (2002) 349–363.
[12] R. Agha, R. Ogawa, G. Pietramaggiori, D.P. Orgill, J. Surg. Res. 171 (2011) 700–708.
[13] V.W. Wong, S. Akaishi, M.T. Longaker, G.C. Gurtner, J. Invest. Dermatol. 131 (2011) 2186–2196.
[14] L.A. Barnes, C.D. Marshall, T. Leavitt, M.S. Hu, A.L. Moore, J.G. Gonzalez, M.T. Longaker, G.C. Gurtner, Adv. Wound Care (New Rochelle) 7 (2018) 47–56.
[15] W. Wang, L. Guo, Y. Yu, Z. Chen, R. Zhou, Z. Yuan, J. Biomed. Mater. Res. Part A 103 (2015) 1703–1712.
[16] T. Cherbuin, M.M. Movahednia, W.S. Toh, T. Cao, Stem Cell Rev. Rep. 11 (2015) 460–473.
[17] J.L. Balestrini, K.L. Billiar, J. Biomech. 39 (2006) 2983–2990.
[18] E.A. Lee, D.Y. Kim, E. Chung, E.A. Lee, K.S. Park, Y. Son, Cell Transplant. 23 (2014) 285–301.
[19] G. Rolin, D. Binda, M. Tissot, C. Viennet, P. Saas, P. Muret, P. Humbert, J. Biomech. 47 (2014) 3555–3561.
[20] M. Jiang, J. Qiu, L. Zhang, D. Lü, M. Long, L. Chen, X. Luo, Exp. Ther. Med. 12 (2016) 1542–1550.
[21] J.B. Schmidt, K. Chen, R.T. Tranquillo, Cell. Mol. Bioeng. 9 (2016) 55–64.
[22] Q. Shu, J. Tan, V.D. Ulrike, X. Zhang, J. Yang, S. Yang, X. Hu, W. He, G. Luo, J. Wu, Sci. Rep. 6 (2016) 1–12.
[23] J.S. Suwandi, R.E.M. Toes, T. Nikolic, B.O. Roep, Clin. Exp. Rheumatol. 33 (2015) 97–103.
[24] R. Kuang, Z. Wang, Q. Xu, S. Liu, W. Zhang, Int. J. Clin. Exp. Med. 8 (2015) 7641–7647.
[25] M.S. Chin, L. Lancerotto, D.L. Helm, P. Dastouri, M.J. Prsa, M. Ottensmeyer, S. Akaishi, D.P. Orgill, R. Ogawa, Plast. Reconstr. Surg. 124 (2009) 102–113.
[26] S.B. Zhou, J. Wang, C.A. Chiang, L.L. Sheng, Q.F. Li, Stem Cells 31 (2013) 2703–2713.
[27] S. Aarabi, K.A. Bhatt, Y. Shi, J. Paterno, E.I. Chang, S.A. Loh, J.W. Holmes, M.T. Longaker, H. Yee, G.C. Gurtner, FASEB J. 21 (2007) 3250–3261.
[28] G.C. Gurtner, R.H. Dauskardt, V.W. Wong, K.A. Bhatt, K. Wu, I.N. Vial, K. Padois, J.M. Korman, M.T. Longaker, Ann. Surg. 254 (2011) 217–225.
[29] J.E. Sanders, B.S. Goldstein, D.F. Leotta, J. Rehabil. Res. Dev. 32 (1995) 214–226.
[30] J.E. Bischoff, E.M. Arruda, K. Grosh, J. Biomech. 33 (2000) 645–652.
[31] M.G. Dunn, F.H. Silver, D.A. Swann, J. Invest. Dermatol. 84 (1985) 9–13.

[32] M.J. Munoz, J.A. Bea, J.F. Rodriguez, I. Ochoa, J. Grasa, A.P. del Palomar, P. Zaragoza, R. Osta, M. Doblare, J. Biomech. 41 (2008) 93–99.
[33] C.H. Daly, J. Invest. Dermatol. 79 (1982) S17–S20.
[34] Y. Lanir, Skin mechanics, in: Handbook of Bioengineering, McGraw-Hill Book Company, New York, 1987.
[35] M.H. Flint, A.S. Craig, H.C. Reilly, G.C. Gillard, D.A. Parry, Connect. Tissue Res. 13 (1984) 69–81.
[36] D. Gutierrez-Lemini, Engineering Viscoelasticity, 2014.
[37] A. Delalleau, G. Josse, J.M. Lagarde, H. Zahouani, J.M. Bergheau, Skin Res. Technol. 14 (2008) 152–164.
[38] S.L. Evans, C.A. Holt, J. Strain Anal. Eng. Des. 44 (2009) 337–345.
[39] R.B. Groves, S.A. Coulman, J.C. Birchall, S.L. Evans, Comput. Methods Biomech. Biomed. Eng. 15 (2012) 73–82.
[40] E. Jacquet, G. Josse, F. Khatyr, C. Garcin, Skin Res. Technol. 14 (2008) 1–7.
[41] K.H. Lim, C.M. Chew, P.C. Chen, S. Jeyapalina, H.N. Ho, J.K. Rappel, B.H. Lim, J. Biomech. 41 (2008) 931–936.
[42] J. Mahmud, C.A. Holt, S.L. Evans, J. Biomech. 43 (2010) 1002–1006.
[43] C. Flynn, A. Taberner, P. Nielsen, Biomech. Model. Mechanobiol. 10 (2011) 27–38.
[44] X. Liang, S.A. Boppart, IEEE Trans. Biomed. Eng. 57 (2010) 953–959.
[45] C. Pailler-Mattei, S. Bec, H. Zahouani, Med. Eng. Phys. 30 (2008) 599–606.
[46] R.B. Groves, S.A. Coulman, J.C. Birchall, S.L. Evans, J. Mech. Behav. Biomed. Mater. 18 (2013) 167–180.
[47] H. Frederick, Mechanosensing and Mechanochemical Transduction in Extracellular Matrix, first ed., Springer-Verlag US, 2006.
[48] W. Li, Biomed. Eng. Lett. (2015).
[49] G.L. Wilkes, I.A. Brown, R.H. Wildnauer, CRC Crit. Rev. Bioeng. 1 (1973) 453–495.
[50] X. Feng, L. Tianjian, Springer (Ed.), Introduction to Skin Biothermomechanics and Thermal Pain, 2011, pp. 85–97.
[51] P. Fratzl, K. Misof, I. Zizak, G. Rapp, H. Amenitsch, S. Bernstorff, J. Struct. Biol. 122 (1998) 119–122.
[52] H. Oxlund, J. Manschot, A. Viidik, J. Biomech. 21 (1988) 213–218.
[53] T. Gibson, R.M. Kenedi, J.E. Craik, Br. J. Surg. 52 (1965) 764–770.
[54] W. Yang, V.R. Sherman, B. Gludovatz, E. Schaible, P. Stewart, R.O. Ritchie, M.A. Meyers, Nat. Commun. 6 (2015) 6649.
[55] A. Serafinifracassini, Nature 332 (1988) 794.
[56] Y. Lanir, Y.C. Fung, J. Biomech. 7 (1974) 171–182.
[57] W.A. Wan Abas, Biomed. Mater. Eng. 4 (1994) 473–486.
[58] S. Diridollou, F. Patat, F. Gens, L. Vaillant, D. Black, J.M. Lagarde, Y. Gall, M. Berson, Skin Res. Technol. 6 (2000) 214–221.
[59] M. Gasior-Glogowska, M. Komorowska, J. Hanuza, M. Maczka, A. Zajac, M. Ptak, R. Bedzinski, M. Kobielarz, K. Maksymowicz, P. Kuropka, S. Szotek, J. Mech. Behav. Biomed. Mater. 18 (2013) 240–252.
[60] M. Geerligs, L. van Breemen, G. Peters, P. Ackermans, F. Baaijens, C. Oomens, J. Biomech. 44 (2011) 1176–1181.
[61] A.P. Kao, J.T. Connelly, A.H. Barber, J. Mech. Behav. Biomed. Mater. 57 (2016) 14–23.
[62] M. Ottenio, D. Tran, A. Ni Annaidh, M.D. Gilchrist, K. Bruyere, J. Mech. Behav. Biomed. Mater. 41 (2015) 241–250.
[63] O.A. Shergold, N.A. Fleck, D. Radford, Int. J. Impact Eng. 32 (2006) 1384–1402.
[64] K. Sakata, G. Parfitt, K.L. Pinder, Biorheology 9 (1972) 173–184.
[65] E.S. Wilkins, K.L. Pinder, Physiol. Chem. Phys. 11 (1979) 23–35.
[66] J.Z. Wu, R.G. Dong, W.P. Smutz, A.W. Schopper, Biomed. Mater. Eng. 13 (2003) 373–385.
[67] M.L. Crichton, X. Chen, H. Huang, M.A. Kendall, Biomaterials 34 (2013) 2087–2097.
[68] J.A. Last, S.J. Liliensiek, P.F. Nealey, C.J. Murphy, J. Struct. Biol. 167 (2009) 19–24.
[69] P.-H. Wu, D.R.-B. Aroush, A. Asnacios, W.-C. Chen, M.E. Dokukin, B.L. Doss, P. Durand-Smet, A. Ekpenyong, J. Guck, N.V. Guz, P.A. Janmey, J.S.H. Lee, N.M. Moore, A. Ott, Y.-C. Poh, R. Ros, M. Sander, I. Sokolov, J.R. Staunton, N. Wang, G. Whyte, D. Wirtz, Nat. Methods (2018).
[70] S. Cardenas-Perez, J.J. Chanona-Perez, J.V. Mendez-Mendez, G. Calderon-Dominguez, R. Lopez-Santiago, I. Arzate-Vazquez, Innov. Food Sci. Emerg. Technol. 34 (2016) 234–242.
[71] S. Vahabi, B. Nazemi Salman, A. Javanmard, Iran J. Med. Sci. 38 (2013) 76–83.
[72] N. Gavara, Microsc. Res. Tech. 80 (2017) 75–84.
[73] M.L. Crichton, B.C. Donose, X. Chen, A.P. Raphael, H. Huang, M.A. Kendall, Biomaterials 32 (2011) 4670–4681.
[74] C.A. Grant, P.C. Twigg, D.J. Tobin, Acta Biomater. 8 (2012) 4123–4129.
[75] H. Joodaki, M.B. Panzer, Proc. Inst. Mech. Eng. H 232 (2018) 323–343.
[76] G.B. Jemec, M. Gniadecka, B. Jemec, Skin Res. Technol. 2 (1996) 164–166.
[77] G. Boyer, N. Lachmann, G. Bellemere, C. Belilovsky, C. Baudouin, Skin Res. Technol. (2018).
[78] J. Ankersen, A.E. Birkbeck, R.D. Thomson, P. Vanezis, Proc. Inst. Mech. Eng. Part H J. Eng. Med. 213 (1999) 493–501.
[79] M.F. Griffin, B.C. Leung, Y. Premakumar, M. Szarko, P.E. Butler, J. Otolaryngol. Head Neck Surg. 46 (2017) 33.
[80] L. Falland-Cheung, M. Scholze, P.F. Lozano, B. Ondruschka, D.C. Tong, P.A. Brunton, J.N. Waddell, N. Hammer, J. Mech. Behav. Biomed. Mater. 84 (2018) 188–197.
[81] F. Hendriks, Biomed. Eng. 4 (1969) 322–327.
[82] J.F. Manschot, A.J. Brakkee, J. Biomech. 19 (1986) 511–515.
[83] L. Penuela, C. Negro, M. Massa, E. Repaci, E. Cozzani, A. Parodi, S. Scaglione, R. Quarto, R. Raiteri, Exp. Dermatol. 27 (2018) 150–155.
[84] K. Elleuch, R. Elleuch, H. Zahouani, Polym. Eng. Sci. 46 (2006) 1715–1720.
[85] Y. Zheng, A.F. Mak, IEEE Trans. Rehabil. Eng. 7 (1999) 257–267.
[86] C.Y.L. Chao, Y.-P. Zheng, Y.-P. Huang, G.L.Y. Cheing, Clin. Biomech. (Bristol, Avon) 25 (2010) 594–600.
[87] J. Isaza, J. Ramirez, 4th International Conference Tissue Engineering ICTE2015, An ECCOMAS Thematic Conference, vol. 110, (2015) pp. 45–50.
[88] D.L. Bader, P. Bowker, Biomaterials 4 (1983) 305–308.
[89] R. Grahame, P.J. Holt, Gerontologia 15 (1969) 121–139.
[90] S. Diridollou, V. Vabre, M. Berson, L. Vaillant, D. Black, J.M. Lagarde, J.M. Gregoire, Y. Gall, F. Patat, Int. J. Cosmet. Sci. 23 (2001) 353–362.
[91] K.A. Jansen, D.M. Donato, H.E. Balcioglu, T. Schmidt, E.H.J. Danen, G.H. Koenderink, Biochim. Biophys. Acta Mol. Cell Res. 1853 (2015) 3043–3052.

[92] C.S. Chen, J. Cell Sci. 121 (2008) 3285–3292.
[93] Y. Chen, L. Ju, M. Rushdi, C. Ge, C. Zhu, Mol. Biol. Cell 28 (2017) 3134–3155.
[94] S.J. Mousavi, M.H. Doweidar, M. Doblaré, Comput. Methods Biomech. Biomed. Eng. (2014).
[95] A.N. Gasparski, K.A. Beningo, Arch. Biochem. Biophys. 586 (2015) 20–26.
[96] S.J. Mousavi, M.H. Doweidar, PLoS One (2015).
[97] S.J. Mousavi, M.H. Doweidar, Comput. Methods Prog. Biomed. (2016).
[98] E.A. Lumpkin, M.J. Caterina, Nature 445 (2007) 858–865.
[99] A. Zimmerman, L. Bai, D.D. Ginty, Science (80-) 346 (2014) 950–954.
[100] V.Z. Luu, B. Chowdhury, M. Al-omran, D.A. Hess, S. Verma, Atherosclerosis 275 (2018) 196–204.
[101] N. Shoham, A. Gefen, J. Biomech. 45 (2012) 1–8.
[102] F. Guilak, D.M. Cohen, B.T. Estes, Cell Stem Cell 5 (2009) 17–26.
[103] M.P. Ruutu, X. Chen, O. Joshi, M.A. Kendall, I.H. Frazer, Exp. Dermatol. 20 (2011) 534–536.
[104] S. Kippenberger, A. Bernd, S. Loitsch, J. Muller, M. Guschel, R. Kaufmann, Pigment Cell Res. 12 (1999) 246–251.
[105] S. Kippenberger, S. Loitsch, J. Müller, M. Guschel, A. Ramirez-Bosca, R. Kaufmann, A. Bernd, Pigment Cell Res. 13 (2000) 278–280.
[106] F.H. Silver, L.M. Siperko, G.P. Seehra, Skin Res. Technol. 9 (2003) 3–23.
[107] L. Dunn, H.C.G. Prosser, J.T.M. Tan, L.Z. Vanags, M.K.C. Ng, C.A. Bursill, J. Vis. Exp. (2013) 1–6.
[108] F.E. Görmar, A. Bernd, J. Bereiter-Hahn, H. Holzmann, Arch. Dermatol. Res. 282 (1990) 22–32.
[109] M. Hara, T. Fujii, R. Hashizume, Y. Nomura, Cytotechnology 66 (2014) 723–728.
[110] R.T. Lee, W.H. Briggs, G.C. Cheng, H.B. Rossiter, P. Libby, T. Kupper, J. Immunol. 159 (1997) 5084–5088.
[111] S. Kippenberger, A. Bernd, S. Loitsch, M. Guschel, J. Müller, J. Bereiter-Hahn, R. Kaufmann, J. Invest. Dermatol. 114 (2000) 408–412.
[112] T. Takei, C. Rivas-Gotz, C.A. Delling, J.T. Koo, I. Mills, T.L. Mccarthy, M. Centrella, B.E. Sumpio, J. Cell. Physiol. 173 (1997) 64–72.
[113] T. Takei, H. Kito, W. Du, I. Mills, B.E. Sumpio, J. Cell. Biochem. 69 (1998) 95–103.
[114] N. Bhadal, I.B. Wall, S.R. Porter, S. Broad, G.E. Lindahl, S. Whawell, M.P. Lewis, Br. J. Dermatol. 158 (2008) 396–398.
[115] J. Zhou, J. Wang, N. Zhang, Y. Zhang, Q. Li, Am. J. Transl. Res. 7 (2015) 2187–2198.
[116] S. Yano, M. Komine, M. Fujimoto, H. Okochi, K. Tamaki, J. Invest. Dermatol. 122 (2004) 783–790.
[117] K. Nishimura, P. Blume, S. Ohgi, B.E. Sumpio, J. Surg. Res. 155 (2009) 125–131.
[118] E. Tokuyama, Y. Nagai, K. Takahashi, Y. Kimata, K. Naruse, PLoS One 10 (2015) 1–12.
[119] Y. Knies, A. Bernd, R. Kaufmann, J. Bereiter-Hahn, S. Kippenberger, Exp. Dermatol. 15 (2006) 347–355.
[120] S. Yano, M. Komine, M. Fujimoto, H. Okochi, K. Tamaki, Exp. Dermatol. 15 (2006) 356–361.
[121] S. Kippenberger, S. Loitsch, M. Guschel, J. Müller, Y. Knies, R. Kaufmann, A. Bernd, J. Biol. Chem. 280 (2005) 3060–3067.
[122] D. Lü, X. Liu, Y. Gao, B. Huo, Y. Kang, J. Chen, S. Sun, L. Chen, X. Luo, M. Long, PLoS One 8 (2013).
[123] D. Lï, Z. Li, Y. Gao, C. Luo, F. Zhang, L. Zheng, J. Wang, S. Sun, M. Long, Biomed. Eng. Online 15 (2016) 605–620.
[124] F. Reno, V. Traina, M. Cannas, Cell Adhes. Migr. 3 (2009) 239–242.
[125] M. Kurita, M. Okazaki, T. Fujino, A. Takushima, K. Harii, Biochem. Biophys. Res. Commun. 409 (2011) 103–107.
[126] S. Kippenberger, J. Kleemann, M. Meissner, K. Steinhorst, J. Müller, C.C. Zouboulis, R. Kaufmann, N. Zöller, J. Dermatol. Sci. 89 (2018) 241–247.
[127] T.D. Allen, S.L. Schor, J. Ultrasructure Res. 83 (1983) 205–219.
[128] S. Manzano, R. Moreno-Loshuertos, M. Doblaré, I. Ochoa, M.H. Doweidar, Med. Biol. Eng. Comput. (2015).
[129] C.A. Lambert, A.C. Colige, C.M. Lapière, B.V. Nusgens, Eur. J. Cell Biol. 80 (2001) 479–485.
[130] A. Parekh, V.C. Sandulache, T. Singh, S. Cetin, M.S. Sacks, J.E. Dohar, P.A. Hebda, Wound Repair Regen. 17 (2009) 88–98.
[131] G.J. Fisher, Y. Shao, T. He, Z. Qin, D. Perry, J.J. Voorhees, T. Quan, Aging Cell 15 (2016) 67–76.
[132] M. Takeya, Y. Okumura, T. Nikawa, J. Physiol. Sci. 67 (2017) 613–622.
[133] M. Parsons, E. Kessler, G.J. Laurent, R.A. Brown, J.E. Bishop, Exp. Cell Res. 252 (1999) 319–331.
[134] D. Deng, W. Liu, F. Xu, Y. Yang, G. Zhou, W.J. Zhang, L. Cui, Y. Cao, Biomaterials 30 (2009) 6724–6730.
[135] Z.H. Syedain, R.T. Tranquillo, J. Biomech. 44 (2011) 848–855.
[136] E.A. Sander, V.H. Barocas, R.T. Tranquillo, Ann. Biomed. Eng. 39 (2011) 714–729.
[137] R. Kuang, Z. Wang, Q. Xu, X. Cai, T. Liu, Ann. Plast. Surg. 76 (2016) 388–393.
[138] D. Kessler, S. Dethlefsen, I. Haase, M. Plomann, F. Hirche, T. Krieg, B. Eckes, J. Biol. Chem. 276 (2001) 36575–36585.
[139] N.K. Weidenhamer, R.T. Tranquillo, Tissue Eng. Part C Methods 19 (2013) 386–395.
[140] C. Huang, K. Miyazaki, S. Akaishi, A. Watanabe, H. Hyakusoku, R. Ogawa, J. Plast. Reconstr. Aesthetic Surg. 66 (2013) e351–e361.
[141] H. Wen, P.A. Blume, B.E. Sumpio, Int. Wound J. 6 (2009) 149–158.
[142] R.T. Prajapati, M. Eastwood, R.A. Brown, Wound Repair Regen. 8 (2000) 238–246.
[143] R.T. Prajapati, B. Chavally-Mis, D. Herbage, M. Eastwood, R.A. Brown, Wound Repair Regen. 8 (2000) 226–237.
[144] M. Kim, D.W. Shin, H. Shin, M. Noh, J.H. Shin, Exp. Dermatol. 22 (2013) 72–74.
[145] C. Viennet, J. Bride, V. Armbruster, F. Aubin, A.C. Gabiot, T. Gharbi, P. Humbert, Arch. Dermatol. Res. 297 (2005) 10–17.
[146] A. Sarasa-Renedo, V. Tunç-Civelek, M. Chiquet, Exp. Cell Res. 312 (2006) 1361–1370.
[147] A.S. Peters, G. Brunner, T. Krieg, B. Eckes, Arch. Dermatol. Res. 307 (2015) 191–197.
[148] K. Nishimura, P. Blume, S. Ohgi, B.E. Sumpio, Wound Repair Regen. 15 (2007) 646–656.
[149] Y. Kanazawa, J. Nomura, S. Yoshimoto, T. Suzuki, K. Kita, N. Suzuki, M. Ichinose, Connect. Tissue Res. 50 (2009) 323–329.
[150] A.S. Peters, G. Brunner, K. Blumbach, D.J. Abraham, T. Krieg, B. Eckes, Exp. Dermatol. 21 (2012) 765–770.
[151] K. Langer, Br. J. Plast. Surg. 31 (1978) 3–8.
[152] M.D. Ridge, V. Wright, J. Invest. Dermatol. 46 (1966) 341–346.
[153] A. Summerfield, F. Meurens, M.E. Ricklin, Mol. Immunol. 66 (2015) 14–21.
[154] A.A. Kun Ma, S.H. Kwon, J. Padmanabhan, D. Duscher, G.C. Trotsyuk, Y. Dong, M. Inayathullah, J. Rajadas, Gurtner, J. Clin. Invest. (2018) 1–9.

CHAPTER

# 18

# Cartilage Regeneration and Tissue Engineering

María Sancho-Tello*, Lara Milián*, Manuel Mata Roig*, José Javier Martín de Llano*, Carmen Carda*,†

*Department of Pathology, University of Valencia and INCLIVA Health Research Institute, Valencia, Spain
†Biomedical Research Networking Center—Bioengineering, Biomaterials and Nanomedicine (CIBER-BBN), Valencia, Spain

## Nomenclature

| | |
|---|---|
| **3D** | three-dimensional |
| **ACI** | autologous chondrocyte implantation |
| **ADSC** | adipose stromal/stem cell |
| **BM** | bone marrow |
| **BMP** | bone morphogenetic protein |
| **CB** | cord blood |
| **CHT** | chitosan |
| **CHI3L1** | chitinase 3-like-1 |
| **CSPC** | cartilage stem/progenitor cell |
| **DPSC** | dental pulp stem cell |
| **ECM** | extracellular matrix |
| **FGF** | fibroblast growth factor |
| **GAG** | glycosaminoglycan |
| **HA** | hyaluronic acid |
| **IGF** | insulin-like growth factor |
| **MACI** | matrix-induced autologous chondrocyte implantation |
| **MMP** | matrix metalloproteinase |
| **MSC** | mesenchymal stem cell |
| **PB** | peripheral blood |
| **PCL** | poly(L-ε-caprolactone) |
| **PCM** | pericellular matrix |
| **PGA** | poly(glycolic acid) |
| **PLCL** | poly(L-lactide-*co*-ε-caprolactone) |
| **PLGA** | poly(lactic-*co*-glycolic acid) |
| **PLLA** | poly(L-lactic) acid |
| **PRP** | platelet-rich plasma |
| **rER** | rough endoplasmic reticulum |
| **SF** | synovial fluid |
| **SM** | synovial membrane |
| **TGF** | transforming growth factor |
| **VEGF** | vascular endothelial growth factor |
| **WJ** | Wharton's jelly |

## 18.1 CARTILAGE TISSUE [1, 2]

Cartilage is a special type of connective tissue, containing a single type of specialized cells (chondrocytes) embedded in an extensive extracellular matrix (ECM) exhibiting specific characteristics that soften impacts and loads. Usually, cartilage is surrounded by a perichondrium (a fibrous connective tissue layer with undifferentiated cells involved

*Advances in Biomechanics and Tissue Regeneration*
https://doi.org/10.1016/B978-0-12-816390-0.00018-2

in appositional cartilage growth). However, articular cartilage lacks a perichondrium, exhibiting, rather, smooth, lubricated, synovial joint surfaces. Unlike most connective tissues, cartilage lacks vascularization; nutrients and metabolites must reach cells via diffusion through the ECM. The avascular nature of cartilage is attributable to the fact that its biochemical composition prevents vascular invasion; breakdown of the antiangiogenic cartilage barrier triggers unwanted vascular invasion and irreversible cartilage degeneration [3].

There are three types of cartilage that differ in both appearance and mechanical properties and are characterized by their matrices: hyaline cartilage, elastic cartilage, and fibrocartilage. Hyaline cartilage is the most abundant in humans; this cartilage covers the articular surfaces of most synovial joints.

### 18.1.1 Cartilage Cells [1]

Cartilage cells are derived from mesenchymal stem cells, which are small undifferentiated cells with thin processes exhibiting a high rate of proliferation; the cells can differentiate into chondroblasts and other types of cells. Chondroprogenitor mesenchymal cells aggregate and differentiate into chondroblasts, which secrete the diverse components of the cartilage matrix; when cells become completely surrounded by the matrix material that they have secreted, they are termed chondrocytes and occupy small spaces in the ECM, termed lacunae (Fig. 18.1). Therefore, chondroblasts and chondrocytes are the same cellular type, but they differ in terms of their state of activity; these are the only cell types present in articular cartilage. Chondrocytes divide by mitosis and originate two cells occupying a lacuna. Such cellular clusters are termed isogenous groups; the presence of such groups indicates that cells have recently divided; this is the basis of interstitial cartilage growth. Thus daughter cells initially occupy the same lacuna, but as they secrete new matrix, a partition is formed between them, and each cell comes to occupy its own lacuna; the cells move farther apart as matrix secretion continues, as the chondral matrix is distensible.

Chondrocyte mitosis is prominent during bodily development and growth; immature cartilage exhibits some regeneration capacity, but this appears to be lost with increasing age [4]. Although mature chondrocytes can proliferate and exhibit chondrogenic potential, they rarely divide by mitosis, and therefore, their ability to engage in tissue repair is limited [5].

The morphology of the chondrocyte cytoplasm varies by the extent of activity. Cells actively producing matrix exhibit typical features of protein-secreting cells, such as a patent nucleolus, a well-developed Golgi apparatus (observed as a clear cytoplasmic area with optical microscope), and a large amount of rough endoplasmic reticulum (rER) that renders the cytoplasm basophilia, thus staining with basic dyes such as hematoxylin; the cells also contain secretory granules, vesicles, and a cytoskeleton. However, less active cells have reduced amount of the Golgi and rER organelles and increased amounts of stored energy reserves such as lipid droplets and glycogen inclusions.

When cartilage is observed with transmission electron microscopy after standard histological procedures, the chondrocytes exhibit retraction and shrinkage and appear to be surrounded by partially empty pericellular lacunae; the plasmalemma is partially retracted but maintains focal contacts with the surrounding matrix via small cellular

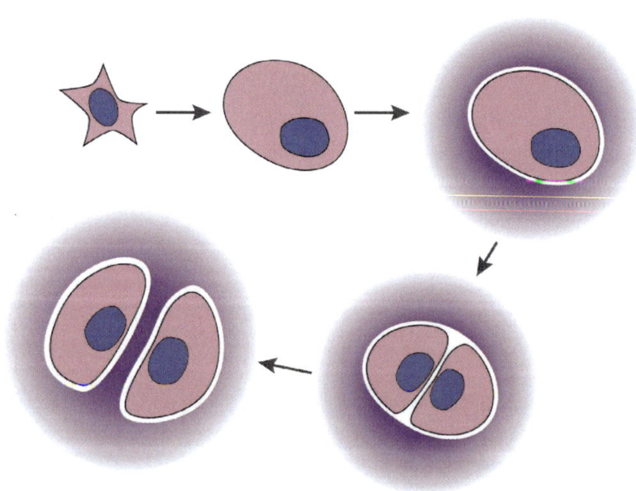

FIG. 18.1 Chondrocytes differentiation from mesenchymal stem cells. Mesenchymal stem cell differentiates to chondroblast, then chondrocyte in a lacuna surrounded by extracellular matrix, isogenic group when the chondrocyte divides by mitosis, and finally isolated chondrocytes in their respective lacunae.

processes. However, Hunziker et al. [6] showed that when cartilage is processed using specific methods, such as high-pressure freezing, freeze-substitution, and low-temperature embedding, the chondrocytes remain in an expanded state and empty spaces around cells are lacking.

The chondrocyte density is $1.4 \times 10^7$ cells/cm$^3$ in mature cartilage [7], indicating that cells occupy less than 5% of the total cartilage volume but are essential for ECM production and maintenance throughout life. The cells are metabolically active; have high nutritional requirements; and, although suited to low-oxygen environments because of the absence of blood vessels, in fact, consume oxygen (although at lower rates than other cell types) and are susceptible to damage caused by oxidative stress [8, 9].

### 18.1.2 Hyaline Cartilage Extracellular Matrix [1, 2, 10]

Hyaline cartilage exhibits a homogeneous amorphous matrix when stained and observed with standard light microscopy. ECM components are secreted locally and then assembled into an organized meshwork closely associated with the surfaces of chondroblasts and chondrocytes that produce them. The ECM contains many components that can modify matrix behavior; the three major macromolecules are the fibrous proteins collagen, proteoglycans, and glycoproteins (Fig. 18.2), embedded in a highly hydrated gel-like ground substance that confers physical and functional properties to cartilage and, therefore, the ability to resist compressive forces because of the strength of collagen while permitting rapid diffusion of nutrients, metabolites, and waste products.

#### *18.1.2.1 ECM Components*

Collagens are a family of fibrous proteins including fibril-forming, fibril-associated, and network-forming proteins, present in different proportions in the diverse connective tissues. The collagen molecule has a long, stiff, triple-stranded helical structure; three collagen polypeptide chains (α-chains) extremely rich in proline and glycine residues are wrapped around one another to form a ropelike superhelix that is highly resistant to tension. These collagen molecules assemble into higher-order polymers termed collagen fibrils (10–300 nm in diameter and many hundreds of micrometers long) that exhibit characteristic cross striations every 68 nm observed with electron microscopy (Fig. 18.2C), attributable to the regularly staggered packing of individual collagen molecules. In hyaline cartilage, type II collagen, found only in cartilage, accounts for 90%–95% of all collagens in the ECM [3] and forms a three-dimensional (3-D) meshwork of fibrils that are thinner than those of the typical type I collagen present in other connective tissues that contains different types of α-chains.

Proteoglycan monomers are formed from abundant glycosaminoglycan (GAG) chains; these are linear unbranched polymers of repeating sulfated disaccharides covalently attached to a core protein (Fig. 18.2B). Sulfated GAGs are basophilic when stained with hematoxylin or other basic dyes and also exhibit metachromasia. The major proteoglycan of the cartilage ECM is aggrecan, a cartilage-specific proteoglycan. Each aggrecan monomer contains about 100 chains of chondroitin sulfate and 60 of keratan sulfate, featuring three globular domains that bind noncovalently to hyaluronic acid (HA), matrix proteins, and other unknown substances [11]. The chondroitin and keratan sulfate

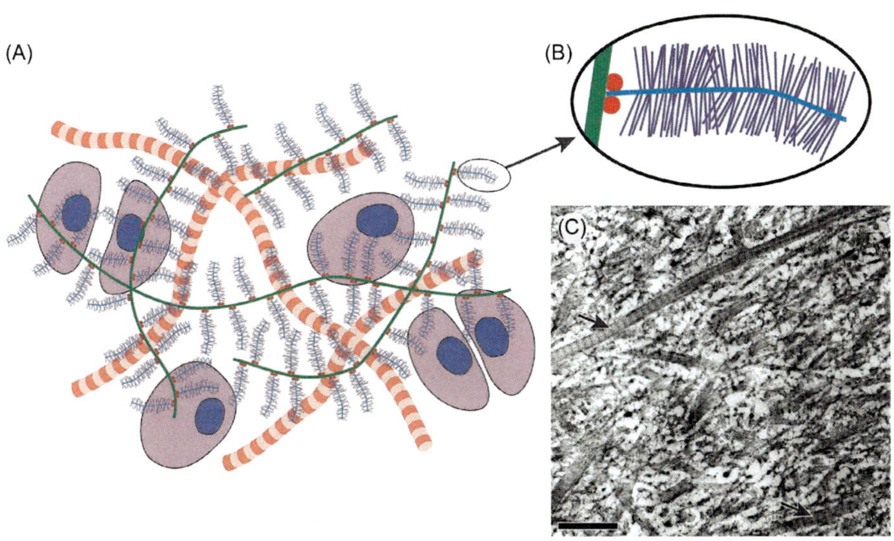


FIG. 18.2 Chondral extracellular matrix. (A) Chondrocytes surrounded by collagen fibrils and aggregates of proteoglycans. (B) Proteoglycan monomer with abundant glycosaminoglycan chains attached to a core protein, which is bound to hyaluronic acid by means of link proteins. (C) Enlarged detail showing transmission electron micrograph of the extracellular matrix, where type II collagen fibrils can be identified *(arrows)*. Bar represents 500 nm.

contents correlate strongly with the cartilage creep modulus, indicating that compressive stiffness is determined by the GAGs rather than collagen [12]. Other proteoglycans of the chondral matrix are biglycan, decorin, fibromodulin, and perlecan [13]. Proteoglycans of various types associate to form larger ECM polymeric complexes termed aggregates; about 300 proteoglycan molecules attach to a linear molecule of HA (a huge GAG containing repeats of up to 25,000 nonsulfated disaccharide units) via linkage proteins, forming the large proteoglycan aggregates mentioned earlier.

Sulfated aggrecan GAGs are highly negatively charged and therefore strongly hydrophilic, forming highly hydrated gels; in fact, water represents 70%–80% of cartilage net weight. Water does not flow out of articular cartilage when walking [6], but transient changes in water content do occur during joint movement and when joints are subjected to pressure; this allows, on the one hand, rapid diffusion of water-soluble molecules from blood vessels in surrounding tissues toward chondrocytes and, on the other hand, creation of a swelling pressure or turgor allowing the ECM to withstand compressive forces. The low-friction surface explains ECM resilience, and the 3-D meshwork of tension-resisting collagen fibrils allows the swelling pressure imparted by proteoglycans to be withstood; cartilage is well adapted to high intermittent pressures and efficiently resists shear stress imposed on synovial joints [14].

Glycoproteins are adhesive proteins featuring a short, branched oligosaccharide chain with multiple domains, each of which specifically binds other macromolecules of the ECM or receptors on the cell surface. These glycoproteins are important in terms of cellular interactions, organizing and assembling other ECM components, and helping cells attach to the ECM.

Within the chondral ECM, various glycoproteins such as laminin or fibronectin are detected at different concentrations during the life span and in periods of good health or illness [13]. The glycoprotein tenascin is abundant during cartilage development, but the level decreases during maturation; this glycoprotein disappears almost completely in adult articular cartilage, although it has been detected in both the cartilage and synovium of osteoarthritis patients [15].

The glycoproteins lubricin and chitinase 3-like-1 (CHI3L1) are present in normal articular cartilage, produced by both articular chondrocytes and other cell types. Lubricin is abundant in normal cartilage but decreases in patients with osteoarthritis; this is one of the major joint lubricants and is considered to be chondroprotective, preventing cartilage wear and reducing the amount of friction on the surface of articular cartilage [3]. However, glycoprotein CHI3L1 has the opposite effect, being associated with mediators of inflammation and cartilage damage during the pathogenesis of osteoarthritis [16].

#### *18.1.2.2 ECM Territories*

ECM components are not uniformly distributed in chondral ECM; different regions are observed, defined principally by the relative concentrations of sulfated proteoglycans and, therefore, their staining properties. The ECM features pericellular, territorial, and interterritorial matrices (Fig. 18.3).

The pericellular matrix (PCM) is a narrow space surrounding individual cells. The PCM and the enclosed cell are termed a chondron; this is the primary structural, functional, and metabolic unit of hyaline cartilage [17]. This matrix is distinct from other regions of the ECM in all of biochemical composition and ultrastructural and biomechanical properties and plays important roles in protecting cells from mechanical stress and exposure to certain

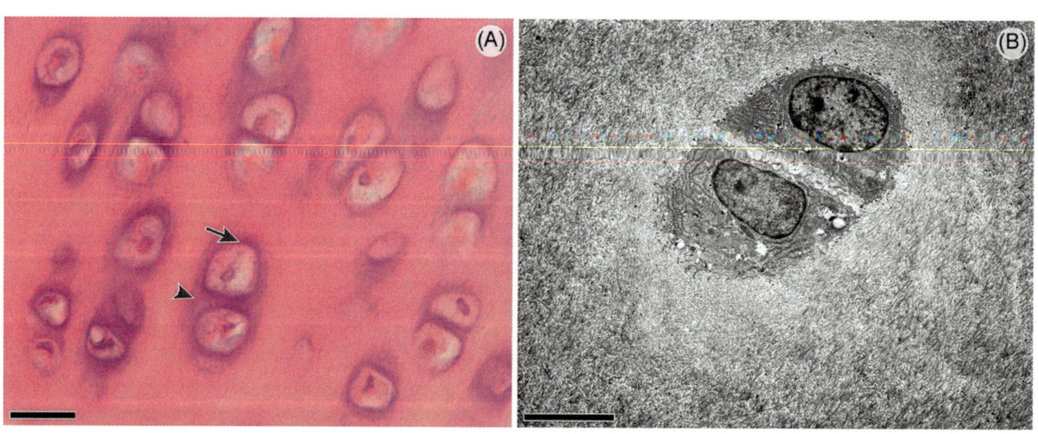


FIG. 18.3 Extracellular matrix territories of hyaline cartilage. (A) Pericellular matrix, stained in dark purple, is a thin band surrounding each chondrocyte (arrow); territorial matrix is a moderately stained area around pericellular matrix and isogenous groups (arrowhead); interterritorial matrix is a more eosinophil area that occupies the space between territorial matrices. Sample stained with hematoxylin-eosin. Bar represents 20 μm. (B) Transmission electron micrograph of an isogenous group surrounded by extracellular matrix with different electrodensities. Bar represents 5 μm.

molecules, ensuring that articular cartilage is not degraded [13, 18]. The PCM stains densely with basic dyes because it contains high concentrations of sulfated GAGs and glycoproteins, but only some loose collagen fibrils, creating a meshlike capsular ultrastructure that modifies the mechanical load imparted to the chondrocyte; in fact, the elastic modulus is significantly lower than that of the bulk ECM [17, 18]. Aggrecan is anchored to both cell surfaces and HA, and this is of particular importance within the PCM; Knudson et al. [19] concluded that both aggrecan and HA contribute to the establishment and maintenance of cell-cell spaces between chondrocytes. PCM contains type VI and IX collagens but lacks type II collagen; thus, under normal conditions, type II collagen fibrils are not exposed to chondrocytes [13]. PCM also contains perlecan, a large proteoglycan unique to the PCM, which colocalizes with type VI collagen [13].

The second ECM territory is the territorial matrix that surrounds the PCM and isogenous groups; this matrix stains moderately. The matrix contains abundant type II collagen fibrils that are irregularly distributed and GAGs and proteoglycans (but at lower concentrations than the PCM).

Finally, the interterritorial matrix occupies the space between territorial matrices. This matrix stains slightly with basic dyes and contains large amounts of type II collagen fibrils, glycoproteins, and aggrecan, associated with HA (as in other territories), thus supporting biomechanical function [19].

### 18.1.3 Synovial Joints

Diarthroses are bone articulations that permit wide joint movements. The joints have capsules that join both bone ends. The capsules have two layers; the outermost is formed from dense connective tissue containing blood vessels and nerves in continuity with the periosteum of the bones joined at the edge of the articular cartilage. The inner layer is a synovial membrane that covers the synovial or articular cavity. The inner surface of the synovial membrane is lined with one or two layers of synovial cells covering a loose connective tissue with abundant fenestrated capillaries. There are two types of synovial cells: type A synoviocytes are macrophage-like, and type B synoviocytes are fibroblast-like.

The articular cavity contains the synovial fluid, a liquid that reduces friction between the hyaline cartilages covering both articular surfaces. Synovial liquid is produced by the synovial cells and is an ultrafiltrated blood plasma containing abundant HA (which serves as a lubricant) [10]. The liquid contains leukocytes and glycoproteins such as lubricin (secreted into the synovial cavity by both type B synoviocytes and superficial chondrocytes); this glycoprotein reduces friction within the joint [13].

Articular cartilage is a hyaline cartilage lacking a perichondrium; the surface is bathed by synovial fluid and is in continuity with subchondral bone in the deeper regions. This cartilage features a characteristic organization of collagen fibers into overlapping arches that support the mechanical tension at articular surfaces and can be divided into four zones [3, 9, 20] (Fig. 18.4):

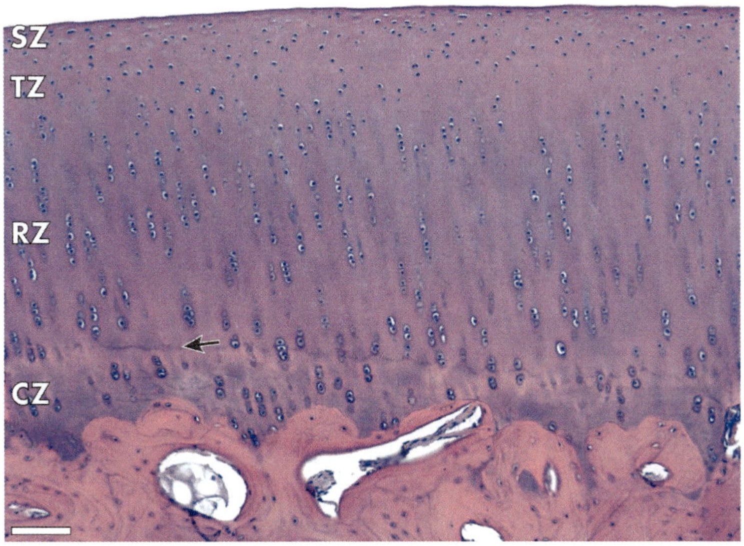


FIG. 18.4 Articular cartilage zones. Superficial zone (SZ), transitional zone (TZ), radial zone (RZ), and calcified zone (CZ). *Arrow* indicates the tidemark. Sample stained with hematoxylin-eosin. Bar represents 100 μm.

- *Upper superficial or tangential zone* (SZ): This zone, in contact with the synovial cavity, is thin and contains small flattened chondrocytes surrounded by an ECM featuring abundant type II collagen fibrils arranged in parallel to the surface, with lower levels of proteoglycans and keratan sulfate GAGs than the deeper zones.
- *Intermediate or transitional zone* (TZ): This zone contains round chondrocytes that are apparently randomly distributed and collagen fibrils arranged obliquely to the surface and the highest levels of proteoglycans.
- *Radial or deep zone* (RZ): This zone occupies the greatest area; short columns of round large chondrocytes are arrayed perpendicular to the surface, together with collagen fibrils that parallel the chondrocyte columns.
- *Calcified zone* (CZ): This zone contains a partly mineralized matrix and smaller chondrocytes and serves as the transition zone between cartilage and underlying subchondral bone.
- *Tidemark*: This is a heavily calcified undulating line marking the transition between the noncalcified radial zone and the calcified zone.

Articular cartilage exhibits crucial, diverse biomechanical functions in synovial joints, absorbing and distributing loads to the opposing bony shafts, thus allowing frictionless movement [6]. Articular chondrocytes require specific balancing of mechanical loads, potentially differing within each of the four layers; Chen et al. [3] showed that biomechanical stimulation in vitro was required to sustain functional tissue-specific cell populations. Although mature chondrocytes exhibit few mitoses, Muiños-Lopez et al. [5] reported that biopsies of the superficial and deep zones of articular cartilage exhibited proliferative and chondrogenic potential, particularly of deep zone chondrocytes. A subpopulation of slow-cycling cells has been identified on the surface of articular cartilage [21].

Apart from hyaline articular cartilage, the capsule, and synovial fluid, some synovial joints contain additional structures such as menisci, tendons, or ligaments that improve mobility and function.

### 18.1.4 Cartilage ECM Turnover [2, 10]

Cartilage ECM undergoes continuous internal remodeling throughout life since cells replace degraded matrix components. Normal matrix turnover requires that chondrocytes detect changes in matrix composition and respond by synthesizing appropriate new molecules; chondrocytes are thus engaged in continual matrix turnover [19]. Stimuli (such as pressure) applied to articular cartilage create mechanical, electrical, and chemical signals directing chondrocyte synthetic activities. However, as the body ages, chondrocytes lose the ability to respond to such stimuli, and the matrix composition changes gradually.

Constant turnover of ECM molecules throughout life is required for tissue repair. However, renewal of mature articular cartilage is very slow because certain ECM components such as the collagen network are very stable, exhibiting negligible turnover after adolescence [22]. Matrix molecules are degraded by matrix metalloproteinases (MMPs) ($Zn^{2+}$- or $Ca^{2+}$-dependent enzymes secreted by chondrocytes but present at only low levels in healthy articular cartilage). The MMPs include a collagenase that specifically degrades ECM collagen fibrils, allowing cells to expand, to become repositioned within growing isogenous groups. The aggrecan and HA turnover rates are similar [19]; both are continuously synthesized by cartilage and have half-lives of the order of weeks reflecting, in part, endocytosis by chondrocytes. Knudson et al. [19] showed that cleavage of aggrecan and further degradation to a threshold size were required for HA internalization. Both proteoglycan synthesis and breakdown decrease with increasing age [23].

The primary role of articular cartilage is load transmission/distribution; mechanical loading influences cartilage composition. The reduced mechanical load in paraplegic patients diminished articular cartilage thickness over 1 year; a lack of mechanical stimulation caused atrophy, with formation of thinner and softer cartilage that was more susceptible to trauma [23, 24].

## 18.2 ARTICULAR CARTILAGE AGING AND SENESCENCE

During aging, articular cartilage undergoes many changes in molecular structure in terms of ECM component levels and qualities; the changes resemble those noted during degeneration associated with osteoarthritis. Anabolic activity and water content fall, as does fibrillation, associated with fragmentation of core protein of aggrecan and chondroitin sulfate GAGs, altered sulfation of GAG chains, and changes in cross-linking between collagen molecules of reduced tensile strength and stiffness [9, 25]. Some changes of the articular surface and focal defects such as superficial fibrillation are common and are spontaneously repaired but tend to accumulate with age [14].

Mammals accumulate senescent cells as they age, contributing to aging per se and age-related disease [26]. In articular cartilage, senescence may predispose joint tissues to osteoarthritis development and/or progression; senescent cells have been observed near osteoarthritic lesions but not in intact cartilage from the same patient [26, 27].

Cellular senescence is mediated via signal transduction; cells enter stable growth arrest but remain metabolically active. However, senescent cell persistence and accumulation impair cartilage function [26]. Aging chondrocytes exhibit signs of senescence, losing the ability to divide, which is the major factor contributing to deterioration of cartilage homeostasis and function [9, 28]. Senescence is associated with diminished mitotic activity and telomere shortening [25, 29], but other factors that do not affect telomere length are also in play; Martin et al. [29] showed that in vitro exposure to mechanical loading increased chondrocyte oxidative stress that ultimately caused senescence without any reduction in telomere length.

Therefore, increasing age triggers decline in chondrocyte function caused by senescence and, hence, diminished remodeling and maintenance capacities. The lack of tissue turnover and renewal triggers end-product accumulation, increasing stiffness caused by fibrillar cross-linking, followed by declines in articular cartilage quality and deformation capacity and, ultimately an increased susceptibility to destruction. Thus age is a major predisposing factor for the development of osteoarthritis [23].

The cell density of articular cartilage decreases with age, because apoptosis increases [30, 31]. Moreover, articular chondrocytes exhibit reduced proteoglycan synthesis and altered proteoglycan composition, modifying the biomechanical properties of the ECM [14]. Thus the articular cartilage of elderly individuals exhibits a diminished capacity for deformation in vivo compared with that of younger individuals [32]. Also, senescent cells act synergistically with inflammation to drive further cellular senescence [26].

Structural changes in collagen fibers also develop with age, contributing to weakening of fibrillar elasticity and, therefore, of biomechanical properties: type II collagen fibers become larger in diameter, increasing stiffness and decreasing the resistance of articular cartilage to tension, mainly because of fibrillar cross-linking with advanced glycation end products [14, 33]. With increasing age, type II collagen synthesis decreases, the expression and activity of MMPs and several collagenases increase, and increased proteolytic fragmentation of collagen molecules is apparent, promoting matrix remodeling and reducing tensile strength, in turn causing articular fibrillation and erosion [34, 35]. These changes commence via disruption of the superficial zone of articular cartilage, progressing later to deeper zones in the degenerative disease termed osteoarthritis.

Holmes et al. [36] showed that the total amount of aggrecan did not change with age; however, these results are controversial since the sizes of large proteoglycan aggregates decreased with age [14], accompanied by shortening of keratan sulfate and chondroitin sulfate chains, as well as the size of HA. In contrast, the proportion of HA in the ECM increased fourfold from birth to 90 years of age [34]. Aggrecanase proteolytic activity increased with age; aggrecan fragments were released into synovial fluid, leaving HA binding domains free for occupation by other molecules, thus limiting formation of fully functional aggregates [37]. Therefore, proteoglycan synthesis and level fall with age [13], reducing the ECM intercellular water content, finally altering the biomechanical properties of articular cartilage in a manner compromising function.

In addition, age-related changes in subchondral bone, which often undergoes significant remodeling [38], probably affect joint oxygen tensions, as does increased cartilage calcification, associated with advancing age [9].

## 18.3 CARTILAGE REPAIR AND OSTEOARTHRITIS

Cartilage changes and loss of cartilage thickness in synovial joints with aging contribute to the development of osteoarthritis, the most common disease of synovial joints, causing progressive reduction of mobility and increased pain on joint movement [16]. Several risk factors are associated with the development of osteoarthritis: gender (females are at higher risk), genetic predisposition, obesity, and advancing age [14, 39, 40]. Small injuries can trigger osteoarthritis in younger subjects because, although articular cartilage can tolerate intense repetitive stress, cartilage manifests a striking inability to heal even after very minor injury, given its avascularity, chondrocyte immobility, and the limited ability of mature chondrocytes to proliferate [1].

Osteoarthritis features various changes in synovial joint components, including progressive degeneration of articular cartilage, formation of bony peripheral outgrowths (osteophytes), changes in subchondral bone, thickening of both the synovium and ligaments, and (in many cases) synovial inflammation (synovitis) [26].

Moderate loading exerts a beneficial effect on osteoarthritis, associated with cartilage hypertrophy and maintenance of articular cartilage quality via increases in lubricin levels, regardless of age [23]. However, joint overloading (as in overweight subjects) affects cartilage composition; depending on loading intensity, the earliest damage is softening without visible collagen loss, followed by collagen loss and, finally, irreversible destruction [41]. In vitro, mechanical stress accelerated chondrocyte senescence via increased oxidative stress, damaging mitochondria and triggering either apoptosis or senescence, thus reducing the number of functional chondrocytes [29, 35].

Although mature chondrocytes exhibit little mitotic activity, a subpopulation of slow-cycling cells in the superficial zone exhibited increased proliferation and rearrangement at the onset of osteoarthritis, contributing to cartilage repair under pathological conditions [21, 42]. Seol et al. [43] reported that chondrogenic progenitor cells of the superficial zone could migrate to damaged areas and proliferate therein to cover damaged cartilage with a continuous coat of lubricin; the cells were thus involved in the early stages of cartilage repair.

During osteoarthritis development, chondrocytes undergo many phenotypic changes, often influenced by aging, including changes in responses to external stimuli promoting abnormal ECM remodeling [14]. Chondrocytes secrete increased levels of many inflammatory cytokines, growth factors, and other factors that are together termed the senescence-messaging secretome [44], which suggests that cell senescence may play a pathological role in osteoarthritis [26]. One such factor is vascular endothelial growth factor (VEGF), a signaling protein promoting blood vessel formation, which may contribute to dysregulated osteogenesis and osteophyte formation. Matrix-degrading proteases induced by proinflammatory cytokines also increase in level; these degrade type II collagen lattices. Other changes include telomere attrition, activation of the DNA damage response, and secretion of reactive oxygen species [26]. Oxidative stress is a major contributor to cartilage breakdown in osteoarthritis and age-associated disease [45]; advanced oxidation protein products have been detected in the synovial fluid and plasma of osteoarthritis patients and used as biomarkers of disease progression [46].

Articular chondrocytes are responsible for the secretion and turnover of ECM components and are crucial in terms of ECM homeostasis; in osteoarthritis, the balance between synthesis and degradation of matrix components shifts in favor of catabolic events, thus promoting pathological tissue remodeling and, eventually, breakdown [14, 47]. As the articular cartilage is destroyed in osteoarthritis, its ECM components (such as type II collagen and proteoglycan) are degraded [13]; the pronounced loss of aggrecan observed in osteoarthritis causes a dramatic increase in tissue hydraulic permeability [6].

The PCM surrounding chondrocytes plays an important role in joint protection from osteoarthritis. The lack of one or more PCM components disrupts matrix structure; the chondrocytes are then less protected from mechanical stress and more exposed to molecules from the territorial and interterritorial matrices (which would normally not be encountered). In particular, when type II collagen binds to chondrocyte membranes, it activates the tyrosine kinase receptor and its intracellular signaling pathway, in turn inducing expression of MMPs that degrade type II collagen and proteoglycans, thus advancing osteoarthritis [13].

Currently, no pharmacological therapy effectively treats osteoarthritis, preventing or reversing progressive joint damage in most patients. The only therapeutic approaches are pain management and joint replacement in the most severe cases; new treatments are required.

## 18.4 ARTICULAR CARTILAGE AND TISSUE ENGINEERING

Various surgical methods including subchondral drilling, microfracture, or nanofracture have often failed to trigger functional hyaline cartilage regeneration [48]; in this context, tissue engineering seems to be promising in terms of repairing, regenerating, and/or enhancing movement of arthritic joints [14]. The first efforts in this area were made in the 1970s when Green [49] transplanted rabbit chondrocytes were grown ex vivo into cartilage defects (allografts). In 1994 cartilage tissue engineering was tested in patients with deep cartilage defects of the knee; healthy chondrocytes from the same patients were isolated, cultured, and injected into the affected areas, and the results were promising [50]. In the time since these early attempts, tissue engineering has sought to create articular cartilage as similar as possible to native tissue, using different approaches involving, principally, three components: cells that can differentiate into chondrocytes and maintain the chondrocyte phenotype, scaffolds providing adequate 3-D environments, and growth factors inducing cell growth and differentiation (Fig. 18.5).

### 18.4.1 Cells

Various sources of cells generating neocartilage in both scaffold-free and scaffold-based systems are available. Such cells must be readily accessible and exhibit high proliferative rates and stable phenotypes. Regenerated cartilage must be rich in type II collagen and aggrecan, nonimmunogenic, and nontumorigenic [51]. Thus the most obvious cell source

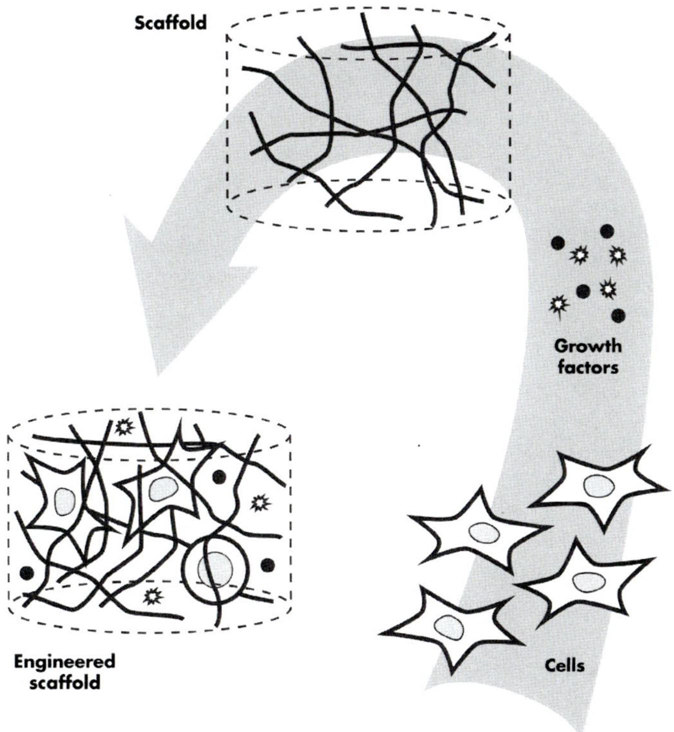


FIG. 18.5 Main components used in tissue engineering. Cells of different origins, biomolecules to induce cell growth and differentiation, and scaffolds to provide an adequate 3-D environment.

is cartilage per se, but its low growth rate and dedifferentiation during in vitro expansion encouraged exploration of the chondrogenic potentials of various stem cells [52–54].

#### 18.4.1.1 Cartilage-Derived Cells

Chondrocytes obtained from uninjured articular cartilage regions of the same patients have been extensively used in tissue engineering because the cells synthesize cartilage-specific ECM after appropriate stimulation in specific culture medium and a 3-D environment [55]. Chondrocytes can survive under the hypoxic conditions found in implant areas. However, the cells exhibit several limitations, including a poor proliferation rate and dedifferentiation during expansion [56]. Although it is possible to redifferentiate the cells using appropriate growth factors in a 3-D environment, the native phenotype is not fully restored, resulting in fibrocartilage formation in most cases [57, 58]. Finally, it is necessary to use fibrin glue, sutures, or other materials to fix engineered tissue into lesions; the tissue adheres poorly to host cartilage, compromising integration [59].

Recently, cartilage stem/progenitor cells (CSPCs) have attracted attention [60]. Jiang et al. [61] suggested that fully differentiated chondrocytes possessed a "reserved stemness" that could be activated under specific culture conditions (such as 2-D, low-density, and low-glucose culture), promoting expression of CD166 (an early-stage mesenchymal stem cell (MSC) marker) in vitro. Cartilage-derived chondroprogenitor cells can create tissue with the characteristics of hyaline cartilage when grown under specific culture conditions. These cells exhibit a phenotype similar to that of bone marrow mesenchymal stromal/stem cells (BM-MSCs) but have a higher chondrogenic potential and do not induce hypertrophy (unlike BM-MSCs) [60].

Also, nasal septal cartilage is a promising source of chondrocytes because of its inherent chondrogenic potential. Chondrogenic cells in the surface zone of the nasal septa are positive for MSC markers, maintain chondrogenic ability in vitro to passage 35, and proliferate faster than articular chondrocytes. Interestingly, such cells also synthesize sulfated GAGs, large amounts of type II collagen, and (to a lesser extent) collagen type I during pellet culture, without the addition of transforming growth factor-β (TGF-β) or bone morphogenetic proteins (BMPs) to the culture medium [62].

Although significant advances have been made in terms of the use of chondrocytes for cartilage repair, several issues remain. First, two surgeries are necessary; the first to obtain undamaged cartilage from a joint and the second for defect implantation of chondrocytes expanded in vitro; morbidity is thus high. Also, in elderly patients, both the quantity and quality of chondrocytes are less than in younger patients; fewer functional cells are obtained on culture in agarose [63].

Another important question is the chondrocyte density required to regenerate articular cartilage. Human chondrocytes exhibit a poor proliferation rate, which often limits their use in cartilage regeneration. Finally, Roberts et al. [9] observed no reduction in terms of clinical benefit from autologous chondrocyte implantation (ACI), which related to the age of the patient. For these reasons, other cells exhibiting chondrogenic potential but higher proliferation rates are required.

#### *18.4.1.2 Mesenchymal Stem Cells*

Mesenchymal stem cells (MSCs) are a valuable cell source and a useful alternative for cartilage engineering because of their chondrogenic potential and the fact that they can be isolated from various tissues (Fig. 18.6), as described later.

##### 18.4.1.2.1 BONE MARROW

Bone marrow (BM) is one of the best-known sources of MSCs that have been widely used for cartilage regeneration either alone or seeded onto scaffolds facilitating implantation into defects. Both approaches have been successful in the laboratory and the clinic; hyaline cartilage-like tissue part-filled defects and were associated with improvements in the quality of life and joint function [60]. BM-MSCs are used when the microfracture technique is applied; the cells flow toward the defect site and then differentiate into chondrocytes if the microenvironment and mechanical stimulation are appropriate, as we showed in an in vivo animal model [64, 65].

However, some studies found no remarkable benefits when using BM-MSCs to repair cartilage defects; the new cartilage was fibrocartilaginous in nature or even calcified. Also, the proportion of MSCs in BM is low (0.01%–0.001%), and their differentiation potential falls with increased expansion in vitro. Finally, morbidity associated with cell harvesting points to the need for other MSC sources [60].

##### 18.4.1.2.2 ADIPOSE TISSUE

Adipose stromal/stem cells (ADSCs) can differentiate into chondrocytes, adipocytes, or osteoblasts, among other cell types [66]. ADSCs are readily accessible as they can be isolated from adipose tissue routinely available after liposuction surgery and expand quickly in vitro [67, 68]. Another advantage of such MSCs is yield; the number of MSCs obtained from 100g of adipose tissue is 300-fold that from 100mL of BM. Several in vitro studies have found that ADSCs exhibit chondrogenic differentiation when first cultured under appropriate 2-D conditions and then in

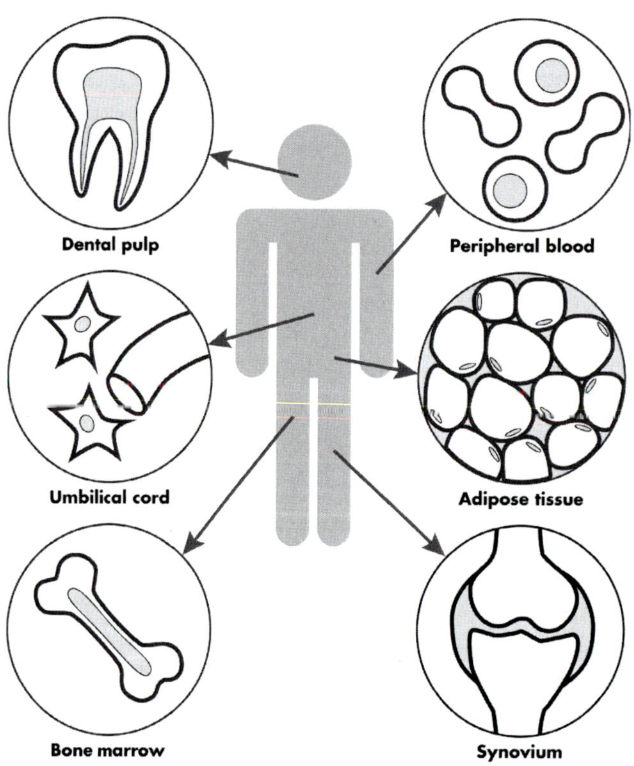


FIG. 18.6 Sources of mesenchymal stem cells. Diverse sources are used to differentiate cells into chondrocytes.

a 3-D environment with specific chondrogenic factors including TGF-β1, insulin-like growth factor-1 (IGF-1), and BMP-6 [60]. However, in terms of chondrogenic potential, the data are controversial compared with those for BM-MSCs. Thus Afizah et al. [69] reported that better results were obtained in vitro using BM-MSCs; histological, immunohistochemical, and GAG analyses showed that type II collagen and proteoglycans were synthesized by BM-MSCs but not ADSCs. However, other authors found no significant differences between the two cell types [70]. Recent clinical studies have highlighted the use of ADSCs, and phase I and II trials have shown that intraarticular injection of ADSCs yielded glossy white cartilage similar to native cartilage, although the optimal number of injected cells remains unclear [71].

#### 18.4.1.2.3 UMBILICAL CORD

The umbilical cord is an important source of stem cells affording promising results in terms of cartilage tissue engineering. In recent years, several researchers have used this new source of stem cells for cartilage regeneration. Umbilical cord blood–derived mesenchymal stem cells (CB-MSCs) are principally used to this end, both in vitro and in vivo [62]. In terms of culture conditions, the effects of hypoxia or normoxia on hypertrophy of chondrocytes derived from CB-MSCs are controversial. Gómez-Leduc et al. [72] determined that initial normoxia followed by a period of hypoxia aided expression of chondrocyte-specific markers, as did the addition of TGF-β1 and BMP-2. Initial normoxia favored chondral differentiation and hypoxia helped to stabilize the chondrocyte phenotype. Other studies also emphasized the utility of hypoxia, which improved the expression of chondrocytic markers such as Sox-9 and type II collagen [73]. However, Desancé et al. [74] found no differences in hypertrophic chondrocyte marker (collagen X and MMP-13) levels on normoxic or hypoxic culture.

In vivo studies using CB-MSCs combined with HA hydrogels have been performed both in animal models and clinically in osteoarthritic patients. GAG-rich cartilage developed in the defects; the amount was greater than when BM-MSCs were used, and collagens of type X and II were also present [75]. Also, arthroscopy revealed firm hyaline cartilage [76].

Another interesting study using a rabbit osteochondral defect model employed human umbilical cord Wharton's jelly–derived mesenchymal stem cells (hWJ-MSCs), which are less immunogenic than other adults MSCs. Hyaline-like, completely integrated neocartilage was obtained using undifferentiated hWJ-MSCs; the cartilage was of better quality than that obtained using TGF-β-induced, differentiated hWJ-MSCs [77].

#### 18.4.1.2.4 DENTAL PULP

Human dental pulp stem cells (hDPSCs) are self-renewing MSCs located within the perivascular niche of the pulp [78]. DPSCs are easily obtained and, under specific conditions, can differentiate in vitro into a variety of cell types, including chondrocytes [79, 80]. Some studies showed that DPSCs cultured in chondrogenic medium expressed type I and II collagen [81] along with other chondrogenic markers such as aggrecan, Sox9, and alkaline phosphatase [82]. In a recent study, significant upregulation of chondrogenic genes was evident upon culture in 3-D hydrogel scaffolds composed of a composite of methacrylated gelatin–HA; thus, a 3-D environment was required for chondrogenic differentiation of not only DPSCs but also other chondrogenic cells.

We recently found that after 6 weeks of culture under chondrogenic medium, hDPSCs expressed both aggrecan and low levels of type I and II collagen. In fact, we sought to repair osteochondral defects in rabbits using either hDPSCs or primary rabbit chondrocytes embedded in 3% ($w/v$) alginate. After 12 weeks, we observed improved tissue regeneration and a smoother articular surface when hDPSCs rather than primary chondrocytes were used, which may reflect the antiinflammatory effects of hDPSCs [54].

#### 18.4.1.2.5 PERIPHERAL BLOOD

Peripheral blood (PB) is an easily accessible source of MSCs useful in cartilage regeneration. Granulocyte-colony stimulating factor efficiently enriches hematopoietic stem/progenitor cells in peripheral blood [83].

Several studies have compared the chondrogenic potential of PB-MSCs with that of BM-MSCs. In terms of morphology and GAG production in vitro, both types of cells exhibited similar behavior, but PB-derived cells exhibited a greater chondrogenic differentiation potential than BM-MSCs [84]. Using an in vivo defect model in rabbits, similar findings were noted.

Clinical studies in patients have confirmed the effectiveness of PB-MSCs injected in combination with HA [85, 86]. Histological evaluation of regenerated cartilage and magnetic resonance imaging data showed that the results were better than those afforded by HA alone.

As PB-MSCs can be obtained using a minimally invasive method and as their chondrogenic differentiation potential seems to be comparable with or even higher than that of BM-MSCs, PB may be a promising source of cells for articular cartilage regeneration.

#### 18.4.1.2.6 SYNOVIUM

MSCs derived from synovial membranes (SM-MSCs) afford several advantages in terms of cartilage regeneration. Such cells can be expanded in vitro for prolonged periods; their multilineage differentiation capability is not affected by donor age or cell passage number. Such cells exhibit higher chondrogenic potential than BM-MSCs in vitro, and thus, although the surface epitopes and proliferation potential are similar, cartilage pellets derived from synovium were significantly larger than those derived from BM in patient-matched comparisons under optimal culture conditions (pellet culture with TGF-β, dexamethasone, and BMP-2) [60, 87].

MSCs are present in the synovial fluid (SF-MSCs) of normal knee joints and increase in number after injury and as osteoarthritis progresses [60, 88]; this is very useful when treating osteoarthritis patients. SF-MSCs share more MSC antigens and genes with SM-MSCs than BM-MSCs. SF-MSCs recently exhibited chondrogenic potential in a rabbit cartilage defect repair model; hyaline-like cartilage was detected macroscopically and histologically [60, 89].

## 18.4.2 Scaffolds

As previously mentioned, 2-D culture expansion of chondrocytes triggers dedifferentiation, but a 3-D environment facilitates redifferentiation. To optimize chondrogenic characteristics, different 3-D approaches have been trialed (pellet culture, or the use of scaffolds to create structures that are physically and mechanically stable).

An optimal scaffold for cartilage tissue engineering should exhibit the following properties:

- *Biocompatibility*: The material must allow cells to proliferate and differentiate into chondrocytes with the characteristic round morphology and create a microenvironment favoring the synthesis of a chondral matrix rich in type II collagen and aggrecan. Also, immune reactions to the scaffold material must be minimized following in vivo scaffold implantation [90].
- *Porosity*: Pores of appropriate size are essential for cell growth and flow transport of nutrients and metabolic waste [91].
- *Bioresorbability*: Degradation and resorption must be controllable to match cell/tissue growth rates in vitro and/or in vivo [57].
- *Mechanical issues*: The mechanical properties must be similar to those of native cartilage; this is crucial if neocartilage is to be functional [92, 93].

A wide variety of materials are used to fabricate scaffolds; some of them are natural, and some others are synthetic.

### *18.4.2.1 Natural Materials*

Natural materials are attractive because they are biochemically similar to cartilage and can be enzymatically degraded. Both protein- and polysaccharide-based biomaterials are used. The first group features collagens, gelatin, and fibrin, and the second group agarose, alginate, HA, and chitosan (CHT).

#### 18.4.2.1.1 PROTEIN-BASED SCAFFOLDS

Collagen-based scaffolds may recreate environments that are rather physiological; collagen is the principal ECM protein, allowing seeded cells to proliferate, differentiate, and generate cartilage-specific ECM [94].

The several collagen-based scaffolds include type I and II collagen hydrogels. Although the type I collagen used in cartilage tissue engineering is of some utility in terms of cartilage defect repair, type II collagen more efficiently promotes chondrogenic differentiation of embedded MSCs and supports a more homogeneous cell distribution throughout the matrix [94, 95].

Studies combining both types of collagens have reported induction of chondrocyte-like morphologies and cartilage-like matrices of seeded MSCs [96]. However, an important limitation is the poor mechanical properties of collagen-based scaffolds. Therefore, collagens are often mixed with other biomaterials such as CHT, poly(L-lactic) acid (PLLA), sodium alginate, and HA, perhaps mimicking the mechanical properties of native cartilage more accurately [97, 98].

Type I/III or I/II collagen membranes are clinically used for matrix-induced autologous chondrocyte implantation (MACI) [99, 100]. Although the membranes yielded promising results in terms of cartilage quality, their superiority compared with ACI in the absence of membranes, or microfracture, requires corroboration in long-term follow-up studies.

Of protein-based materials, fibrin and gelatin also create environments favorable for maintenance of the chondrocytic phenotype and chondrogenic differentiation of MSCs [101]. However, their mechanical properties are even poorer than those of collagen; the materials are thus often used in combination with other materials to fabricate scaffolds with appropriate mechanical properties [95, 102]. Both fibrin and gelatin can contribute to scaffold structure or encapsulation of growth factors to be later released to improve adhesion, proliferation, and chondrogenic differentiation of cells contained within the scaffolds [103].

##### 18.4.2.1.2 POLYSACCHARIDE-BASED SCAFFOLDS

Alginate and agarose were the first materials used for cartilage tissue engineering. Both polysaccharides are derived from marine algae and exhibit useful gelation and cell encapsulation properties. Several studies found that these materials allowed chondrocytes and MSCs of various origins to express the characteristic ECM components of hyaline cartilage at both the RNA and protein levels [104]. However, although alginate gels easily in the presence of divalent ions, elution of such ions by the surrounding environment causes gel degradation and loss of the soft mechanical properties [102]. Recent studies have combined alginate with other polymers (natural or synthetic) to improve the mechanical properties of scaffolds [105, 106]. Also, agarose, despite being biocompatible and nonimmunogenic and exhibiting mechanical properties closer to those of hyaline cartilage, is not biodegradable by humans [107].

HA is a component of both articular cartilage and synovial fluid and supports chondrocyte phenotype retention and matrix deposition and MSC chondrogenesis [108, 109]. Compared with other hydrogels formed of polyethylene glycol or fibrin, cartilage formation was enhanced by HA hydrogels, emphasizing the important roles of biochemical cues during cartilage formation. However, although scaffold biomaterials must be biodegradable, hyaluronidase action must be controlled as this can compromise the strength of the newly formed matrix. Chemical modifications enhancing or diminishing hyaluronidase activity may be required [95]. Also, HA is of low mechanical strength, as is also true of other natural polymers mentioned earlier. HA is often combined with other polymers to improve mechanical properties [110, 111]; the properties of HA can also be enhanced by increasing the extent of cross-linking [102].

Chitosan (CHT) is a copolymer derived via alkaline deacetylation of chitin. CHT shares certain structural characteristics with the GAGs of hyaline cartilage; CHT is thus an ideal scaffold for articular cartilage engineering. Another advantage is that CHT is degraded by lysozyme to nontoxic products [112]. The life span of CHT is longer than that of HA, favoring ECM deposition, and resorption can be modulated via chemical modification. CHT induces chondrogenesis of both chondrocytes and MSCs [113].

Despite these advantages, limitations include poor cell adhesion [112], which is greatly improved on addition of bioactive materials such as gelatin, collagen, or HA. We and others have observed that various growth factors favor chondrogenesis within CHT-based scaffolds [113, 114].

#### *18.4.2.2 Synthetic Materials*

The use of materials of biological origin for cartilage tissue engineering is associated with risks of immunological reactions, disease transmission, and limited availability. Synthetic materials lack these disadvantages, also allowing control of mechanical properties (via chemical modifications), hydrophilia (a common problem, rendering cell nesting difficult), and biodegradability [115, 116].

Poly(glycolic acid) (PGA), poly(L-lactic acid) (PLLA), and poly(L-ε-caprolactone) (PCL), as well as the copolymers poly(lactic-*co*-glycolic acid) (PLGA) and poly(L-lactide-*co*-ε-caprolactone) (PLCL), have been widely used as synthetic scaffolds for cartilage regeneration.

PCL exhibits elastic properties close to those of native cartilage [117]. PLGA is a degradable synthetic polymer widely used in tissue engineering because of excellent biocompatibility, biodegradability, and mechanical strength [118], in addition to low cytotoxicity and immunogenicity, compared with protein-based polymers [119]. PLCL is degraded much more slowly than PLLA, avoiding the abrupt falls in pH reported in PLLA scaffolds [120]. To overcome the shortcomings of synthetic materials, combinations with natural materials are commonly used to enhance wettability and bioactivity. For example, PCL better supports chondrocyte growth when combined with CHT to enhance the hydrophilia of PCL [121].

Therefore, scaffolds combining both natural and synthetic materials are of great interest, because they feature the desirable properties of both materials.

### 18.4.3 Growth Factors

Although an appropriate scaffold and cell source are essential for tissue engineering, the use of growth and differentiation factors inducing differentiation and maintenance of the chondrocytic phenotype are also essential. We will focus on four of the principal factors: those of the transforming growth factor-β superfamily (TGF-β), bone morphogenetic proteins subfamily (BMPs), fibroblast growth factors family (FGFs), and insulin-like growth factor family (IGFs).

Of the five members of the TGF-β superfamily, TGF-β1, TGF-β2, and TGF-β3 stimulate proteoglycan and type II collagen synthesis by chondrocytes and induce chondrogenic differentiation of MSCs in vitro [122, 123]; types 1 and 3 are the most effective in terms of synthesis of ECM markers [124].

Although in vivo studies have shown that TGF-β1 induces MSCs to differentiate chondrally, other studies found that direct injection into joints triggered calcification [122, 125], suggesting that the use thereof must be regulated.

BMPs are members of the TGF-β superfamily that induce bone and cartilage formation. About 15 members have been identified to date; BMP-2, BMP-4, BMP-6, BMP-7, and BMP-9 have been best studied in terms of chondrogenic potential and effectively stimulated expression of chondrocyte-specific markers when added to isolated chondrocytes or various MSCs [126, 127]. Also, scaffolds chemically loaded with BMP-2 or genetically modified MSCs of various origins expressing BMP-2, BMP-4, BMP-6, and BMP-7 have proved useful for repair of osteochondral defects [126].

Microfracture triggers articular cartilage repair via the induction of fibrous cartilage. Some authors combined microfracture with administration of BMP-2 or BMP-7, affording a more hyaline-like cartilage repair [126].

BMP-2, BMP-4, and BMP-6 enhance stem cell chondrogenesis because they induce expression of the TGF-β receptor [128]. BMP-2 seems to be the most effective in this context [129]. However, a disadvantage is that higher levels of BMP may trigger ectopic bone formation in vivo; the spectrum of the tissue regenerative dose is thus narrow [130].

Although TGF-β and the BMPs are the most important factors in terms of chondral differentiation, other factors (such as FGFs and IGFs) can also be used. Of the FGFs, types 2 and 18 are interesting because of their MSC chondrogenic potentials. FGF-18 exerts anabolic effects on cartilage, and FGF-2 participates in cartilage homeostasis [123]. Both FGFs exhibited chondrogenic potential in vitro and in vivo, promoting matrix deposition and the expression of chondrogenic markers of stimulated MSCs when used either alone or in combination with other growth factors such as TGF-β1 or BMP-2 [131–133].

IGF-1 is an essential mediator of cartilage homeostasis, stimulating proteoglycan synthesis and promoting chondrocyte survival and proliferation [134]. IGF-1 enhanced cartilage formation in vitro in tissue-engineered cartilage constructs and IGF-1-transfected MSC pellets [135] and also promoted cell-based repair of articular cartilage defects in vivo [115, 136].

Recently, platelet-rich plasma (PRP) has been investigated as a chondrogenic inducer of various types of stem cells [137]. PRP is an autologous platelet concentrate obtained from fresh blood via centrifugation and contains various growth factors and bioactive proteins [138]. We recently used PRP in combination with biomaterials to enhance chondrogenic differentiation to an extent greater than that in the absence of PRP [114].

In summary, cartilage tissue engineering has been extensively researched, using various cell sources, scaffold materials, and growth factors. However, determining the optimal combination of these components, yielding optimal functional, regenerated articular cartilage, requires further work.

## Acknowledgments

Supported by projects from the Spanish Ministry of Economy and Competitiveness (project no. MAT2016-76039-C4-2-R) and from CIBER-BBN Valorization program (JOINTCART project). CIBER-BBN is an initiative funded by the VI National R&D&I Plan 2008-2011, Iniciativa Ingenio 2010, and Consolider Program. CIBER Actions are financed by the Instituto de Salud Carlos III with assistance from the European Regional Development Fund. The authors would like to acknowledge William V. Barber for the drawings and graphics support.

## References

[1] M.H. Ross, W. Pawlina, Histology: A Text and Atlas, sixth ed., Lippincott Williams & Wilkins, Baltimore, 2011 (Chapter 7).

[2] A.L. Kierszenbaum, L.L. Tres, Histology and Cell Biology: An Introduction to Patology, third ed., Elsevier Saunders, Philadelphia, 2012 (Chapter 4).

[3] S. Chen, P. Fu, H. Wu, M. Pei, Meniscus, articular cartilage and nucleus pulposus: a comparative review of cartilage-like tissues in anatomy, development and function, Cell Tissue Res. 370 (1) (2017) 53–70.

[4] R.A. Calandruccio, W. Gilmer, W. Scott, Proliferation, regeneration, and repair of articular cartilage of immature animals, J. Bone Joint Surg. Am. 44 (3) (1962) 431–455.

[5] E. Muiños-López, M.E. Rendal-Vázquez, T. Hermida-Gómez, I. Fuentes-Boquete, S. Díaz-Prado, F.J. Blanco, Cryopreservation effect on proliferative and chondrogenic potential of human chondrocytes isolated from superficial and deep cartilage, Open Orthop. J. 6 (2012) 150–159.

[6] E.B. Hunziker, K. Lippuner, N. Shintani, How best to preserve and reveal the structural intricacies of cartilaginous tissue, Matrix Biol. 39 (2014) 33–43.

[7] M.D. Buschmann, Y.A. Gluzband, A.J. Grodzinsky, J.H. Kimura, E.B. Hunziker, Chondrocytes in agarose culture synthesize a mechanically functional extracellular matrix, J. Orthop. Res. 10 (6) (1992) 745–758.
[8] S. Zhou, Z. Cui, J.P. Urban, Factors influencing the oxygen concentration gradient from the synovial surface of articular cartilage to the cartilage-bone interface: a modeling study, Arthritis Rheum. 50 (12) (2004) 3915–3924.
[9] S. Roberts, P. Colombier, A. Sowman, C. Mennan, J.H. Rölfing, J. Guicheux, et al., Ageing in the musculoskeletal system, Acta Orthop. 87 (eSuppl. 363) (2016) 15–25.
[10] B. Alberts, A. Johnson, J. Lewis, D. Morgan, M. Raff, K. Roberts, et al., Molecular Biology of the Cell, sixth ed., Garland Science, New York, 2015 (Chapter 19).
[11] D. Heinegård, Proteoglycans and more—from molecules to biology, Int. J. Exp. Pathol. 90 (6) (2009) 575–586.
[12] G.E. Kempson, H. Muir, S.A. Swanson, M.A. Freeman, Correlations between stiffness and the chemical constituents of cartilage on the human femoral head, Biochim. Biophys. Acta 215 (1) (1970) 70–77.
[13] C. Liu, B. Wang, L. Xiao, Y. Li, L. Xu, Z. Zhao, et al., Protective effects of the pericellular matrix of chondrocyte on articular cartilage against the development of osteoarthritis, Histol. Histopathol. 33 (2018) 757–764.
[14] M. Rahmati, G. Nalesso, A. Mobasheri, M. Mozafari, Aging and osteoarthritis: central role of the extracellular matrix, Ageing Res. Rev. 40 (2017) 20–30.
[15] M. Hasegawa, T. Yoshida, A. Sudo, Role of tenascin-C in articular cartilage, Mod. Rheumatol. 28 (2) (2018) 215–220.
[16] M.A. Szychlinska, F.M. Trovato, M. Di Rosa, L. Malaguarnera, L. Puzzo, R. Leonardi, et al., Co-expression and co-localization of cartilage glycoproteins CHI3L1 and lubricin in osteoarthritic cartilage: morphological, immunohistochemical and gene expression profiles, Int. J. Mol. Sci. 17 (3) (2016) 359, https://doi.org/10.3390/ijms17030359.
[17] C.A. Poole, Articular cartilage chondrons: form, function and failure, J. Anat. 191 (1997) 1–13.
[18] E.M. Darling, R.E. Wilusz, M.P. Bolognesi, S. Zauscher, F. Guilak, Spatial mapping of the biomechanical properties of the pericellular matrix of articular cartilage measured in situ via atomic force microscopy, Biophys. J. 98 (12) (2010) 2848–2856.
[19] W. Knudson, S. Ishizuka, K. Terabe, E.B. Askew, C.B. Knudson, The pericellular hyaluronan of articular chondrocytes, Matrix Biol. (2018), https://doi.org/10.1016/j.matbio.2018.02.005.
[20] J. Becerra, J.A. Andrades, E. Guerado, P. Zamora-Navas, J.M. López-Puertas, A.H. Reddi, Articular cartilage: structure and regeneration, Tissue Eng. Part B Rev. 16 (6) (2010) 617–627.
[21] B. Rolauffs, J.M. Williams, M. Aurich, A.J. Grodzinsky, K.E. Kuettner, A.A. Cole, Proliferative remodeling of the spatial organization of human superficial chondrocytes distant from focal early osteoarthritis, Arthritis Rheum. 62 (2) (2010) 489–498.
[22] K.M. Heinemeier, P. Schjerling, J. Heinemeier, M.B. Møller, M.R. Krogsgaard, T. Grum-Schwensen, et al., Radiocarbon dating reveals minimal collagen turnover in both healthy and osteoarthritic human cartilage, Sci. Transl. Med. 8 (346) (2016) 346ra90.
[23] A.E.M. Jørgensen, M. Kjær, K.M. Heinemeier, The effect of aging and mechanical loading on the metabolism of articular cartilage, J. Rheumatol. 44 (4) (2017) 410–417.
[24] B. Vanwanseele, F. Eckstein, H. Knecht, A. Spaepen, E. Stüssi, Longitudinal analysis of cartilage atrophy in the knees of patients with spinal cord injury, Arthritis Rheum. 48 (12) (2003) 3377–3381.
[25] J.A. Martin, J.A. Buckwalter, The role of chondrocyte senescence in the pathogenesis of osteoarthritis and in limiting cartilage repair, J. Bone Joint Surg. Am. 85-A (Suppl. 2) (2003) 106–110.
[26] K. McCulloch, G.J. Litherland, T.S. Rai, Cellular senescence in osteoarthritis pathology, Aging Cell 16 (2) (2017) 210–218.
[27] J.S. Price, J.G. Waters, C. Darrah, C. Pennington, D.R. Edwards, S.T. Donell, et al., The role of chondrocyte senescence in osteoarthritis, Aging Cell 1 (1) (2002) 57–65.
[28] W.S. Toh, M. Brittberg, J. Farr, C.B. Foldager, A.H. Gomoll, J.H. Hui, et al., Cellular senescence in aging and osteoarthritis, Acta Orthop. 87 (2016) 6–14. eSuppl. 363.
[29] J.A. Martin, T.D. Brown, A.D. Heiner, J.A. Buckwalter, Chondrocyte senescence, joint loading and osteoarthritis, Clin. Orthop. Relat. Res. 427 (Suppl) (2004) S96–103.
[30] M.M. Temple, W.C. Bae, M.Q. Chen, M. Lotz, D. Amiel, R.D. Coutts, et al., Age- and site-associated biomechanical weakening of human articular cartilage of the femoral condyle, Osteoarthr. Cartil. 15 (9) (2007) 1042–1052.
[31] C.S. Adams, W.E. Horton, Chondrocyte apoptosis increases with age in the articular cartilage of adult animals, Anat. Rec. 250 (4) (1998) 418–425.
[32] M. Hudelmaier, C. Glaser, J. Hohe, K.H. Englmeier, M. Reiser, R. Putz, et al., Age-related changes in the morphology and deformational behavior of knee joint cartilage, Arthritis Rheum. 44 (11) (2001) 2556–2561.
[33] J.A. Buckwalter, S.L. Woo, V.M. Goldberg, E.C. Hadley, F. Booth, T.R. Oegema, et al., Soft-tissue aging and musculoskeletal function, J. Bone Joint Surg. Am. 75 (10) (1993) 1533–1548.
[34] V.M. Dejica, J.S. Mort, S. Laverty, J. Antoniou, D.J. Zukor, M. Tanzer, et al., Increased type II collagen cleavage by cathepsin K and collagenase activities with aging and osteoarthritis in human articular cartilage, Arthritis Res. Ther. 14 (3) (2012) R113, https://doi.org/10.1186/ar3839.
[35] W. Wu, R.C. Billinghurst, I. Pidoux, J. Antoniou, D. Zukor, M. Tanzer, et al., Sites of collagenase cleavage and denaturation of type II collagen in aging and osteoarthritic articular cartilage and their relationship to the distribution of matrix metalloproteinase 1 and matrix metalloproteinase 13, Arthritis Rheum. 46 (8) (2002) 2087–2094.
[36] M.W. Holmes, M.T. Bayliss, H. Muir, Hyaluronic acid in human articular cartilage. Age-related changes in content and size, Biochem. J. 250 (2) (1988) 435–441.
[37] L.S. Lohmander, P.J. Neame, J.D. Sandy, The structure of aggrecan fragments in human synovial fluid. Evidence that aggrecanase mediates cartilage degradation in inflammatory joint disease, joint injury, and osteoarthritis, Arthritis Rheum. 36 (9) (1993) 1214–1222.
[38] R.F. Loeser, Age-related changes in the musculoskeletal system and the development of osteoarthritis, Clin. Geriatr. Med. 26 (3) (2010) 371–386.
[39] S. Manzano, M. Doblaré, M.H. Doweidar, Altered mechano-electrochemical behavior of articular cartilage in populations with obesity, Appl. Sci. 6 (7) (2016) 186.
[40] S. Manzano, R. Manzano, M. Doblaré, M.H. Doweidar, Altered swelling and ion fluxes in articular cartilage as a biomarker in osteoarthritis and joint immobilization: a computational analysis, J. R. Soc. Interface 12 (102) (2015) 20141090.

[41] S.M. Hosseini, M.B. Veldink, K. Ito, C.C. van Donkelaar, Is collagen fiber damage the cause of early softening in articular cartilage? Osteoarthr. Cartil. 21 (1) (2013) 136–143.

[42] G.P. Dowthwaite, J.C. Bishop, S.N. Redman, I.M. Khan, P. Rooney, D.J. Evans, et al., The surface of articular cartilage contains a progenitor cell population, J. Cell Sci. 117 (6) (2004) 889–897.

[43] D. Seol, D.J. McCabe, H. Choe, H. Zheng, Y. Yu, K. Jang, et al., Chondrogenic progenitor cells respond to cartilage injury, Arthritis Rheum. 64 (11) (2012) 3626–3637.

[44] T. Kuilman, D.S. Peeper, Senescence-messaging secretome: SMS-ing cellular stress, Nat. Rev. Cancer 9 (2) (2009) 81–94.

[45] P. Lepetsos, A.G. Papavassiliou, ROS/oxidative stress signaling in osteoarthritis, Biochim. Biophys. Acta 1862 (4) (2016) 576–591.

[46] U. Ahmed, A. Anwar, R.S. Savage, P.J. Thornalley, N. Rabbani, Protein oxidation, nitration and glycation biomarkers for early-stage diagnosis of osteoarthritis of the knee and typing and progression of arthritic disease, Arthritis Res. Ther. 18 (1) (2016) 250.

[47] L.J. Sandell, T. Aigner, Articular cartilage and changes in arthritis. An introduction: cell biology of osteoarthritis, Arthritis Res. 3 (2) (2001) 107–113.

[48] P.C. Kreuz, M.R. Steinwachs, C. Erggelet, S.J. Krause, G. Konrad, M. Uhl, et al., Results after microfracture of full-thickness chondral defects in different compartments in the knee, Osteoarthr. Cartil. 14 (11) (2006) 1119–1125.

[49] W.T. Green, Articular cartilage repair: behavior of rabbit chondrocytes during tissue culture and subsequent allografting, Clin. Orthop. Relat. Res. (124) (1977) 237–250.

[50] M. Brittberg, A. Lindahl, A. Nilsson, C. Ohlsson, O. Isaksson, L. Peterson, Treatment of deep cartilage defects in the knee with autologous chondrocyte transplantation, N. Engl. J. Med. 331 (14) (1994) 889–895.

[51] T.A.E. Ahmed, M.T. Hincke, Mesenchymal stem cell-based tissue engineering strategies for repair of articular cartilage, Histol. Histopathol. 29 (6) (2014) 669–689.

[52] M. Schnabel, S. Marlovits, G. Eckhoff, I. Fichtel, L. Gotzen, V. Vécsei, et al., Dedifferentiation-associated changes in morphology and gene expression in primary human articular chondrocytes in cell culture, Osteoarthr. Cartil. 10 (1) (2002) 62–70.

[53] E.M. Darling, K.A. Athanasiou, Rapid phenotypic changes in passaged articular chondrocyte subpopulations, J. Orthop. Res. 23 (2) (2005) 425–432.

[54] M. Mata, L. Milian, M. Oliver, J. Zurriaga, M. Sancho-Tello, J.J. Martin de Llano, et al., In vivo articular cartilage regeneration using human dental pulp stem cells cultured in an alginate scaffold: a preliminary study, Stem Cells Int. 2017 (2017) 8309256.

[55] B.J. Huang, J.C. Hu, K.A. Athanasiou, Cell-based tissue engineering strategies used in the clinical repair of articular cartilage, Biomaterials 98 (2016) 1–22.

[56] Y. Zhang, W. Guo, M. Wang, C. Hao, L. Lu, S. Gao, et al., Co-culture systems-based strategies for articular cartilage tissue engineering, J. Cell. Physiol. 233 (3) (2018) 1940–1951.

[57] D.J. Huey, J.C. Huand, K.A. Athanasiou, Unlike bone, cartilage regeneration remains elusive, Science 338 (6109) (2012) 917–921.

[58] S. Roberts, J. Menage, L.J. Sandell, Immunohistochemical study of collagen types I and II and procollagen IIA in human cartilage repair tissue following autologous chondrocyte implantation, Knee 16 (5) (2009) 398–404.

[59] K. Shimomura, W. Ando, H. Fujie, D.A. Hart, H. Yoshikawa, N. Nakamura, Scaffold-free tissue engineering for injured joint surface restoration, J. Exp. Orthop. 5 (1) (2018) 2.

[60] C.R. Fellows, C. Matta, R. Zakany, I.M. Khan, A. Mobasheri, Adipose, bone marrow and synovial joint-derived mesenchymal stem cells for cartilage repair, Front. Genet. 7 (2016) 213.

[61] Y. Jiang, Y. Cai, W. Zhang, Z. Yin, C. Hu, T. Tong, et al., Human cartilage-derived progenitor cells from committed chondrocytes for efficient cartilage repair and regeneration, Stem Cells Transl. Med. 5 (6) (2016) 733–744.

[62] J.N. Fisher, I. Tessaro, T. Bertocco, G.M. Peretti, L. Mangiavini, The application of stem cells from different tissues to cartilage repair, Stem Cells Int. 2017 (2017) 2761678.

[63] N. Tran-Khanh, C.D. Hoemann, M.D. McKee, J.E. Henderson, M.D. Buschmann, Aged bovine chondrocytes display a diminished capacity to produce a collagen-rich, mechanically functional cartilage extracellular matrix, J. Orthop. Res. 23 (6) (2005) 1354–1362.

[64] M. Sancho-Tello, F. Forriol, P. Gastaldi, A. Ruiz-Saurí, J.J. Martín de Llano, E. Novella-Maestre, et al., Time evolution of in vivo articular cartilage repair induced by bone marrow stimulation and scaffold implantation in rabbits, Int. J. Artif. Organs 38 (4) (2015) 210–223.

[65] M. Sancho-Tello, F. Forriol, J.J. Martín de Llano, C.M. Antolinos-Turpín, J.A. Gómez-Tejedor, J.L. Gómez Ribelles, et al., Biostable scaffolds of polyacrylate polymers implanted in the articular cartilage induce hyaline-like cartilage regeneration in rabbits, Int. J. Artif. Organs 40 (7) (2017) 350–357.

[66] P.A. Zuk, M. Zhu, H. Mizuno, J.W. Futrell, A.J. Katz, P. Benhaim, et al., Multilineage cells from human adipose tissue: implications for cell-based therapies, Tissue Eng. 7 (2) (2001) 211–228.

[67] H.A. Awad, M.Q. Wickham, H.A. Leddy, Chondrogenic differentiation of adipose-derived adult stem cells in agarose, alginate, and gelatin scaffolds, Biomaterials 25 (16) (2004) 3211–3222.

[68] S.C. Wu, J.K. Chang, C.K. Wang, Enhancement of chondrogenesis of human adipose derived stem cells in a hyaluronan-enriched microenvironment, Biomaterials 31 (4) (2010) 631–640.

[69] H. Afizah, Z. Yang, J.H. Hui, H.W. Ouyang, E.H. Lee, A comparison between the chondrogenic potential of human bone marrow stem cells (BMSCs) and adipose-derived stem cells (ADSCs) taken from the same donors, Tissue Eng. 13 (4) (2007) 659–666.

[70] M.C. Ronziere, E. Perrier, F. Mallein-Gerin, A.M. Freyria, Chondrogenic potential of bone marrow- and adipose tissue-derived adult human mesenchymal stem cells, Biomed. Mater. Eng. 20 (3) (2010) 145–158.

[71] C.H. Jo, Y.G. Lee, W.H. Shin, Intra-articular injection of mesenchymal stem cells for the treatment of osteoarthritis of the knee: a proof-of-concept clinical trial, Stem Cells 32 (5) (2014) 1254–1266.

[72] T. Gómez-Leduc, M. Desancé, M. Hervieu, F. Legendre, D. Ollitrault, C. de Vienne, et al., Hypoxia is a critical parameter for chondrogenic differentiation of human umbilical cord blood mesenchymal stem cells in type I/III collagen sponges, Int. J. Mol. Sci. 18 (9) (2017) E1933.

[73] A. Marmotti, S. Mattia, F. Castoldi, A. Barbero, L. Mangiavini, D.E. Bonasia, et al., Allogeneic umbilical cord-derived mesenchymal stem cells as a potential source for cartilage and bone regeneration: an in vitro study, Stem Cells Int. 2017 (2017) 1732094.

[74] M. Desancé, R. Contentin, L. Bertoni, T. Gomez-Leduc, T. Branly, S. Jacquet, et al., Chondrogenic differentiation of defined equine mesenchymal stem cells derived from umbilical cord blood for use in cartilage repair therapy, Int. J. Mol. Sci. 19 (2) (2018) E537.

[75] J.L. White, N.J. Walker, J.C. Hu, D.L. Borjesson, K.A. Athanasiou, A comparison of bone marrow and cord blood mesenchymal stem cells for cartilage self-assembly, Tissue Eng. Part A 24 (15–16) (2018) 1262–1272.

[76] Y.B. Park, C.W. Ha, C.H. Lee, Y.G. Park, Restoration of a large osteochondral defect of the knee using a composite of umbilical cord blood-derived mesenchymal stem cells and hyaluronic acid hydrogel: a case report with a 5-year follow-up, BMC Musculoskelet. Disord. 18 (1) (2017) 59.

[77] S. Liu, Y. Jia, M. Yuan, W. Guo, J. Huang, B. Zhao, et al., Repair of osteochondral defects using human umbilical cord Wharton's jelly-derived mesenchymal stem cells in a rabbit model, Biomed. Res. Int. 2017 (2017) 8760383.

[78] S. Gronthos, J. Brahim, W. Li, L.W. Fisher, N. Cherman, A. Boyde, et al., Stem cell properties of human dental pulp stem cells, J. Dent. Res. 81 (8) (2002) 531–535.

[79] K. Chen, H. Xiong, N. Xu, Y. Shen, Y. Huang, C. Liu, Chondrogenic potential of stem cells from human exfoliated deciduous teeth in vitro and in vivo, Acta Odontol. Scand. 72 (8) (2014) 664–672.

[80] J. Dai, J. Wang, J. Lu, The effect of co-culturing costal chondrocytes and dental pulp stem cells combined with exogenous FGF9 protein on chondrogenesis and ossification in engineered cartilage, Biomaterials 33 (31) (2012) 7699–7711.

[81] S.H. Zainal Ariffin, S. Kermani, R. Megat Abdul Wahab, S. Senafi, Z. Zainal Ariffin, M. Abdul Razak, In vitro chondrogenesis transformation study of mouse dental pulp stem cells, Sci. World J. 2012 (2012) 827149.

[82] C.L. Nemeth, K. Janebodin, A.E. Yuan, J.E. Dennis, M. Reyes, D.H. Kim, Enhanced chondrogenic differentiation of dental pulp stem cells using nanopatterned PEG-GelMA-HA hydrogels, Tissue Eng. Part A 20 (21 – 22) (2014) 2817–2829.

[83] W.L. Fu, C.Y. Zhou, J.K. Yu, A new source of mesenchymal stem cells for articular cartilage repair: MSCs derived from mobilized peripheral blood share similar biological characteristics in vitro and chondrogenesis in vivo as MSCs from bone marrow in a rabbit model, Am. J. Sports Med. 42 (3) (2014) 592–601.

[84] P.P. Chong, L. Selvaratnam, A.A. Abbas, T. Kamarul, Human peripheral blood derived mesenchymal stem cells demonstrate similar characteristics and chondrogenic differentiation potential to bone marrow derived mesenchymal stem cells, J. Orthop. Res. 30 (4) (2012) 634–642.

[85] K.Y. Saw, A. Anz, C. Siew-Yoke Jee, S. Merican, R. Ching-Soong Ng, S.A. Roohi, et al., Articular cartilage regeneration with autologous peripheral blood stem cells versus hyaluronic acid: a randomized controlled trial, Arthroscopy 29 (4) (2013) 684–694.

[86] K.Y. Saw, A. Anz, C.S. Jee, R.C. Ng, N. Mohtarrudin, K. Ragavanaidu, High tibial osteotomy in combination with chondrogenesis after stem cell therapy: a histologic report of 8 cases, Arthroscopy 31 (10) (2015) 1909–1920.

[87] R. Chijimatsu, M. Kobayashi, K. Ebina, T. Iwahashi, Y. Okuno, M. Hirao, et al., Impact of dexamethasone concentration on cartilage tissue formation from human synovial derived stem cells in vitro, Cytotechnology 70 (2) (2018) 819–829.

[88] Y. Matsukura, T. Muneta, K. Tsuji, H. Koga, I. Sekiya, Mesenchymal stem cells in synovial fluid increase after meniscus injury, Clin. Orthop. Relat. Res. 472 (5) (2014) 1357–1364.

[89] Z. Jia, Q. Liu, Y. Liang, X. Li, X. Xu, K. Ouyang, et al., Repair of articular cartilage defects with intra-articular injection of autologous rabbit synovial fluid-derived mesenchymal stem cells, J. Transl. Med. 16 (1) (2018) 123.

[90] Y. Cai, J. Li, C.K. Poh, H.C. Tan, E. San Thian, J.Y. Hsi Fuh, et al., Collagen grafted 3D polycaprolactone scaffolds for enhanced cartilage regeneration, J. Mater. Chem. B 1 (43) (2013) 5971–5976.

[91] F. Berthiaume, T.J. Maguire, M.L. Yarmush, Tissue engineering and regenerative medicine: history, progress, and challenges, Annu. Rev. Chem. Biomol. Eng. 2 (2) (2011) 403–430.

[92] J.A. Hendriks, L. Moroni, J. Riesle, The effect of scaffold-cell entrapment capacity and physico-chemical properties on cartilage regeneration, Biomaterials 34 (17) (2013) 4259–4265.

[93] S. Manzano, M. Doblaré, M.H. Doweidar, Parameter-dependent behavior of articular cartilage: 3D mechano-electrochemical computational model, Comput. Methods Prog. Biomed. 122 (3) (2015) 491–502.

[94] C.H. Wu, C.S. Ko, J.W. Huang, Effects of exogenous glycosaminoglycans on human chondrocytes cultivated on type II collagen scaffolds, J. Mater. Sci. Mater. Med. 21 (2) (2010) 725–729.

[95] K.L. Spiller, S.A. Maher, A.M. Lowman, Hydrogels for the repair of articular cartilage defects, Tissue Eng. Part B Rev. 17 (4) (2011) 281–299.

[96] C. Sanjurjo-Rodríguez, A.H. Martínez-Sánchez, T. Hermida-Gómez, I. Fuentes-Boquete, S. Díaz-Prado, F.J. Blanco, Differentiation of human mesenchymal stromal cells cultured on collagen sponges for cartilage repair, Histol. Histopathol. 31 (11) (2016) 1221–1239.

[97] X. Yang, Z. Lu, H. Wu, W. Li, L. Zheng, J. Zhao, Collagen-alginate as bioink for three-dimensional (3D) cell printing based cartilage tissue engineering, Mater. Sci. Eng. C Mater. Biol. Appl. 83 (2018) 195–201.

[98] F. Mohammadi, S. Mohammadi Samani, N. Tanideh, F. Ahmadi, Hybrid scaffolds of hyaluronic acid and collagen loaded with prednisolone: an interesting system for osteoarthritis, Adv. Pharm. Bull. 8 (1) (2018) 11–19.

[99] P. Cherubino, F.A. Grassi, P. Bulgheroni, M. Ronga, Autologous chondrocyte implantation using a bilayer collagen membrane: a preliminary report, J. Orthop. Surg. 11 (1) (2003) 10–15.

[100] L. Han, Z.W. Zhang, B.H. Wang, Z.K. Wen, Construction and biocompatibility of a thin type I/II collagen composite scaffold, Cell Tissue Bank. 19 (1) (2018) 47–59.

[101] T.A.E. Ahmed, M.T. Hincke, Strategies for articular cartilage lesion repair and functional restoration, Tissue Eng. Part B Rev. 16 (3) (2010) 305–329.

[102] W. Zhao, X. Jin, Y. Cong, Degradable natural polymer hydrogels for articular cartilage tissue engineering, J. Chem. Technol. Biotechnol. 88 (3) (2013) 327–339.

[103] H. Park, J.S. Temenoff, Y. Tabata, Effect of dual growth factor delivery on chondrogenic differentiation of rabbit marrow mesenchymal stem cells encapsulated in injectable hydrogel composites, J. Biomed. Mater. Res. Part A 88A (4) (2009) 889–897.

[104] L. Kim, R.L. Mauck, J.A. Burdick, Hydrogel design for cartilage tissue engineering: a case study with hyaluronic acid, Biomaterials 32 (34) (2011) 8771–8782.

[105] P. Stagnaro, I. Schizzi, R. Utzeri, E. Marsano, M. Castellano, Alginate-polymethacrylate hybrid hydrogels for potential osteochondral tissue regeneration, Carbohydr. Polym. 185 (2018) 56–62.

[106] H. Yu, G. Cauchois, N. Louve, Y. Chen, R. Rahouadj, C. Huselstein, Comparison of MSC properties in two different hydrogels, impact of mechanical properties, Biomed. Mater. Eng. 28 (s1) (2017) S193–S200.

[107] N.C. Hunt, L.M. Grover, Cell encapsulation using biopolymer gels for regenerative medicine, Biotechnol. Lett. 32 (6) (2010) 733–742.
[108] C. Chung, I.E. Erickson, R.L. Mauck, Differential behavior of auricular and articular chondrocytes in hyaluronic acid hydrogels, Tissue Eng. Part A 14 (7) (2008) 1121–1131.
[109] C. Chung, J.A. Burdick, Influence of three-dimensional hyaluronic acid microenvironments on mesenchymal stem cell chondrogenesis, Tissue Eng. Part A 15 (2) (2009) 243–254.
[110] E. Amann, P. Wolff, E. Breel, M. van Griensven, E.R. Balmayor, Hyaluronic acid facilitates chondrogenesis and matrix deposition of human adipose derived mesenchymal stem cells and human chondrocytes co-cultures, Acta Biomater. 52 (2017) 130–144.
[111] X. Lin, W. Wang, W. Zhang, Z. Zhang, G. Zhou, Y. Cao, et al., Hyaluronic acid coating enhances biocompatibility of nonwoven PGA scaffold and cartilage formation, Tissue Eng. Part C Methods 23 (2) (2017) 86–97.
[112] W.Y. Xia, W. Liu, L. Cui, Tissue engineering of cartilage with the use of chitosan-gelatin complex scaffolds, J Biomed Mater Res B Appl Biomater 71B (2) (2004) 373–380.
[113] R.A. Muzzarelli, F. Greco, A. Busilacchi, V. Sollazzo, A. Gigante, Chitosan, hyaluronan and chondroitin sulfate in tissue engineering for cartilage regeneration: a review, Carbohydr. Polym. 89 (3) (2012) 723–739.
[114] M. Sancho-Tello, S. Martorell, M. Mata Roig, L. Milián, M.A. Gámiz-González, J.L. Gómez Ribelles, et al., Human platelet-rich plasma improves the nesting and differentiation of human chondrocytes cultured in stabilized porous chitosan scaffolds, J. Tissue Eng. 8 (2017) 1–6.
[115] M.B. Gugjoo, G.T. Amarpal, H.P. Sharma, P.K. Aithal, Cartilage tissue engineering: role of mesenchymal stem cells along with growth factors & scaffolds, Indian J. Med. Res. 144 (3) (2016) 339–347.
[116] S. Poveda-Reyes, T. Gamboa-Martínez, S. Manzano, M.H. Doweidar, J.L. Gómez Ribelles, I. Ochoa, et al., Engineering interpenetrating polymer networks of poly(2-hydroxyethyl acrylate) as *ex vivo* platforms for articular cartilage regeneration, Int. J. Polym. Mater. 64 (14) (2015) 745–754.
[117] S. Martinez-Diaz, N. Garcia-Giralt, M. Lebourg, In vivo evaluation of 3-dimensional polycaprolactone scaffolds for cartilage repair in rabbits, Am. J. Sports Med. 38 (3) (2010) 509–519.
[118] P. Duan, Z. Pan, L. Cao, The effects of pore size in bilayered poly(lactide-co-glycolide) scaffolds on restoring osteochondral defects in rabbits, J. Biomed. Mater. Res. A 102 (1) (2014) 180–192.
[119] J. Shi, X. Zhang, X. Zeng, J. Zhu, Y. Pi, C. Zhou, et al., One-step articular cartilage repair: combination of in situ bone marrow stem cells with cell-free poly(L-lactic-co-glycolic acid) scaffold in a rabbit model, Orthopedics 35 (5) (2012) e665–e671.
[120] V. Kesireddy, J. Ringe, M. Endres, S. Stich, D. Klose, R. Bodmeier, et al., Development of growth factor/insulin release for articular cartilage regeneration in PLGA scaffolds, Tissue Eng. 12 (4) (2006) 1073.
[121] S.C. Neves, L.S. Moreira Teixeira, L. Moroni, Chitosan/poly(epsilon-caprolactone) blend scaffolds for cartilage repair, Biomaterials 32 (4) (2011) 1068–1079.
[122] C. Vinatier, D. Mrugala, C. Jorgensen, J. Guicheux, D. Noël, Cartilage engineering: a crucial combination of cells, biomaterials and biofactors, Trends Biotechnol. 27 (5) (2009) 307–314.
[123] E. Augustyniak, T. Trzeciak, M. Richter, J. Kaczmarczyk, W. Suchorska, The role of growth factors in stem cell-directed chondrogenesis: a real hope for damaged cartilage regeneration, Int. Orthop. 39 (5) (2015) 995–1003.
[124] W. Wang, D. Rigueur, K.M. Lyons, TGFβ signaling in cartilage development and maintenance, Birth Defects Res. C Embryo Today 102 (1) (2014) 37–51.
[125] C. Scotti, B. Tonnarelli, A. Papadimitropoulos, Recapitulation of endochondral bone formation using human adult mesenchymal stem cells as a paradigm for developmental engineering, Proc. Natl. Acad. Sci. U. S. A. 107 (16) (2010) 7251–7256.
[126] Z.H. Deng, Y.S. Li, X. Gao, G.H. Lei, J. Huard, Bone morphogenetic proteins for articular cartilage regeneration, Osteoarthr. Cartil. 26 (9) (2018) 1153–1161.
[127] T. Grunder, C. Gaissmaier, J. Fritz, Bone morphogenetic protein (BMP)-2 enhances the expression of type II collagen and aggrecan in chondrocytes embedded in alginate beads, Osteoarthr. Cartil. 12 (7) (2004) 559–567.
[128] C.H. Lu, T.S. Yeh, C.L. Yeh, Regenerating cartilages by engineered ASCs: prolonged TGF-beta3/BMP-6 expression improved articular cartilage formation and restored zonal structure, Mol. Ther. 22 (1) (2014) 186–195.
[129] I. Sekiya, B.L. Larson, J.T. Vuoristo, R.L. Reger, D.J. Prockop, Comparison of effect of BMP-2, -4, and -6 on in vitro cartilage formation of human adult stem cells from bone marrow stroma, Cell Tissue Res. 320 (2) (2005) 269–276.
[130] M. Fujioka-Kobayashi, M. Abd El Raouf, N. Saulacic, E. Kobayashi, Y. Zhang, B. Schaller, et al., Superior bone-inducing potential of rhBMP9 compared to rhBMP2, J. Biomed. Mater. Res. A 106 (6) (2018) 1561–1574.
[131] L. Huang, L. Yi, C. Zhang, Y. He, L. Zhou, Y. Liu, et al., Synergistic effects of FGF-18 and TGF-β3 on the chondrogenesis of human adipose-derived mesenchymal stem cells in the pellet culture, Stem Cells Int. (2018) (2018) 7139485.
[132] C. Shu, S.M. Smith, C.B. Little, J. Melrose, Use of FGF-2 and FGF-18 to direct bone marrow stromal stem cells to chondrogenic and osteogenic lineages, Future Sci. OA 2 (4) (2016) FSO142.
[133] T. Pizzute, J. Li, Y. Zhang, M.E. Davis, M. Pei, Fibroblast growth factor ligand dependent proliferation and chondrogenic differentiation of synovium-derived stem cells and concomitant adaptation of Wnt/mitogen-activated protein kinase signals, Tissue Eng. Part A 22 (15–16) (2016) 1036–1046.
[134] L.C. Davies, E.J. Blain, S.J. Gilbert, B. Caterson, V.C. Duance, The potential of IGF-1 and TGFb1 for promoting 'adult' articular cartilage repair: an in vitro study, Tissue Eng. Part A 14 (2008) 1251–1261.
[135] Y. Ikeda, M. Sakaue, R. Chijimatsu, D.A. Hart, H. Otsubo, K. Shimomura, et al., IGF-1 gene transfer to human synovial MSCs promotes their chondrogenic differentiation potential without induction of the hypertrophic phenotype, Stem Cells Int. 2017 (2017) 5804147.
[136] Z. Zhang, L. Li, W. Yang, Y. Cao, Y. Shi, X. Li, et al., The effects of different doses of IGF-1 on cartilage and subchondral bone during the repair of full-thickness articular cartilage defects in rabbits, Osteoarthr. Cartil. 25 (2) (2017) 309–320.
[137] Y. Qian, Q. Han, W. Chen, J. Song, X. Zhao, Y. Ouyang, et al., Platelet-rich plasma derived growth factors contribute to stem cell differentiation in musculoskeletal regeneration, Front. Chem. 5 (2017) 89.
[138] J. Van Den Dolder, R. Mooren, A.P.G. Vloon, Platelet-rich plasma: quantification of growth factor levels and the effect on growth and differentiation of rat bone marrow cells, Tissue Eng. 12 (11) (2006) 3067–3073.

CHAPTER

# 19

# Impact of Mechanobiological Perturbation in Cartilage Tissue Engineering

*Zheng Yang*,†, Yingnan Wu*,†, Lu Yin*,†, Hin Lee Eng*,†*

*NUS Tissue Engineering Program, Life Sciences Institute, National University of Singapore, Singapore

†Department of Orthopaedic Surgery, National University of Singapore, Singapore

## 19.1 INTRODUCTION

Articular cartilage at the diarthrodial joints serves a critical mechanical role with a smooth, lubricated surface, allowing joint articulation that withstands cyclic loading deformation while minimizing wear. Cartilage is a highly hydrated tissue, with high content of water drawn in by the highly negatively charged sulfated proteoglycans (PG), embedded within the fibrillar type II collagens (Col II), endowing the cartilage with its high compressive strength. Despite its simple appearance, articular cartilage exhibits significant heterogeneity, comprising of superficial, middle (transitional), and deep zones with distinct chondrogenic phenotypes (Fig. 19.1). The density, morphology, and metabolic activity of the cells, as well as the composition and structural arrangement of the extracellular matrix (ECM) components, vary greatly across these zones [1, 2]. The superficial zone (constituting the top 10%–15% of total cartilage thickness) contains flattened chondrocytes with collagen fibrils aligned parallel to the articulating surface. Chondrocytes in the superficial zone produce superficial zone protein (SZP, also known as proteoglycan 4 (PRG4) and lubricin), which acts as a lubricant for efficient gliding motion during joint movement [3]. The middle zone (40%–50% of total cartilage thickness) contains more rounded chondrocytes, thicker collagen fibrils with random orientation. The deep zone (30%–40% of total cartilage thickness) is made up of large, spherical chondrocytes in a columnar arrangement, embedded in a dense extracellular matrix rich in proteoglycans (PGs), with thick collagen fibrils aligned perpendicularly to the articulating surface. While PG concentration increases along the cartilage depth, collagen content (per wet weight) does not change significantly, but displays depth-dependent increases in collagen hydroxylysine and hydroxylysyl pyridinoline cross-links [4]. The presence of other minor collagens isoforms such as type IX and XI and cartilage oligomeric matrix protein (COMP) play critical roles in the regulation of fibril size, inter fibril cross-linking, and interactions with PGs [5, 6], conferring the characteristic compressive strength and dimensional stability of the articular cartilage tissue.

The mechanical property and function of the articular cartilage are defined by the ECM composition and organization. On the other hand, mechanical signals experienced by the tissue chondrocytes play a key role in cartilage homeostasis and in shaping stem cell chondrogenic differentiation [7]. Understanding how chondrocytes and mesenchymal stem cells (MSCs) respond to mechanical signals is a major focus of research that has important implications for cartilage tissue engineering and regeneration. Mechanical signals can be derived passively from the base properties of the materials to which cells adhere or actively from extrinsically applied mechanical deformations. Interaction with ECM molecules through membrane proteins such as integrins and adhesion molecules, perturbation of ion channels, and cytoskeletal components are believed to play key roles in the complex mechanotransduction mechanisms that govern how cells sense and transmit mechanical signals. This chapter will summarize our current understanding of how mechanical cues control chondrocytes and direct MSCs towards the chondrogenic lineage, including (i) the influence of extracellular substrate mechanics (stiffness and topography) in regulating cell shape/cytoskeleton tension and (ii) the effects of various extrinsic mechanical signals (e.g., compression, hydrostatic pressure, tension, and fluid

*Advances in Biomechanics and Tissue Regeneration*
https://doi.org/10.1016/B978-0-12-816390-0.00019-4

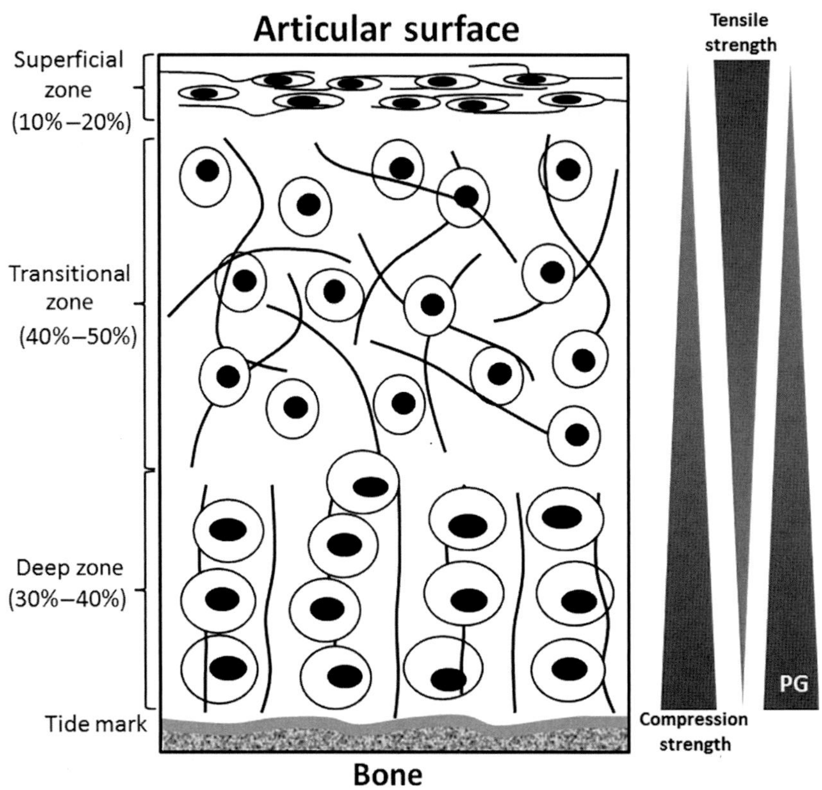


FIG. 19.1 Schematic diagram showing zonal differences in mature articular cartilage. Variations in chondrocyte morphology, collagen fiber arrangement, proteoglycan content, tensile, and compressive strength are illustrated. PG = proteoglycans.

flow), with an emphasis on the phenotype of the cartilage generated. The role of various cellular components, in particular the involvement of the nucleus as a mechanosensor, that takes part in the mechanotransduction of these different mechanical cues will be described.

## 19.2 MECHANOTRANSDUCTION OF MECHANICAL SIGNALS

Cells are inextricably linked to their extracellular environment primarily through integrin molecules at the plasma membrane, which act as a bridge to the complex interpenetrating cytoskeletal networks, and through this anchorage, mechanical signals can be transmitted into the cell [8, 9]. Stretch-activated ion channels, adhesion complexes, cell-cell junctions, and cytoskeletal and nuclear components have all been identified as mechanosensitive elements (Fig. 19.2) that can activate cellular signaling pathways, such as the mitogen-activated protein kinase–extracellular signal-regulated kinase (MAPK-ERK), YAP/TAZ, and MKL1, that ultimately result in the expression of mechanoresponsive genes. The tension generated by the cytoskeleton depends on substrate stiffness and topography, type, and density of the ligand [10], which determines cell morphology and affects the activity of focal adhesions and cell-cell junctions [11, 12]; all play prominent roles in MSC differentiation [13]. When mechanical signals are transmitted to integrins from the ECM, large cytoskeletal protein complexes form at cells' periphery, known as focal adhesions (FA), which triggers signaling cascades within the cell that are transmitted to the nucleus. Intracellularly, integrins help to form large FA protein complexes [14] that serve as anchorage sites for the actin cytoskeleton [15, 16]. The assembly of focal adhesions is a key regulator of MSC differentiation as this acts to stabilize integrin binding [17], which in turn can regulate cell shape, and activate numerous signaling cascades including tyrosine kinases such as FAK and paxillin [18], Rho-family GTPases [19], serine-threonine kinases such as MAPK [20], and intracellular calcium concentration [21], as well as the nuclear translocation of the mechanosensitive transcriptional regulators YAP/TAZ [22, 23], all of which or the inhibition of some plays a critical role in mechanotransduction and regulation of chondrogenesis.

Integrins communication with the ECM also triggers cadherins at the plasma membrane to participate in cell-cell interaction and mechanotransduction [24, 25]. Cadherins are calcium-dependent molecules that bind in a homophilic manner with the extracellular domain of cadherins from adjacent cells. Activation of N-cadherin to initiate cell-cell junctions is also known to be stabilized by actin cytoskeleton, through complex formation with catenins [26, 27]. Cellular

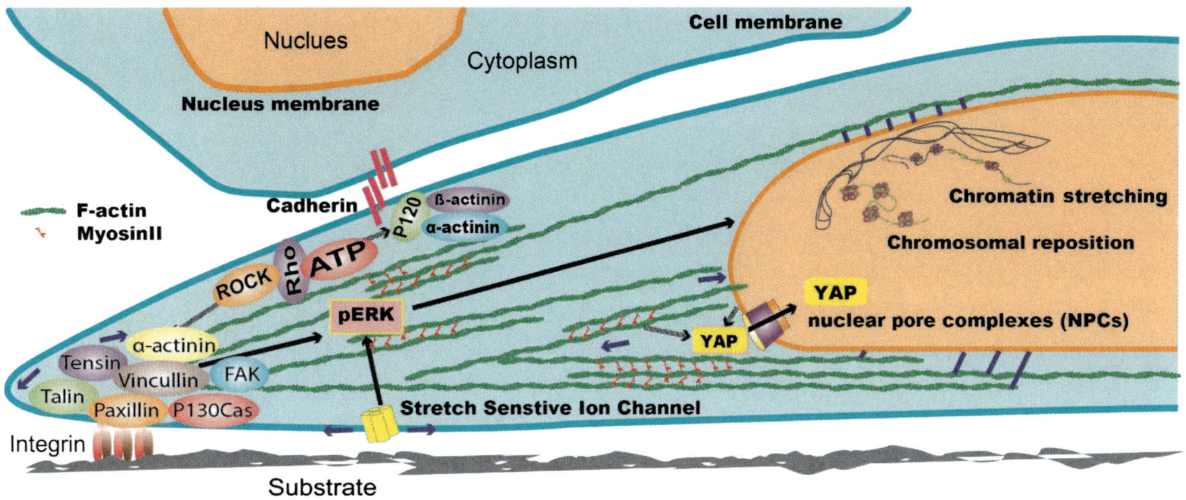


FIG. 19.2 Schematic illustration of the routes of mechanotransduction. Mechanoactivation occurs via substrate/matrix-cell and cell-cell interaction that induce cytoskeleton tension generating forces that get transmitted to the nucleus. The process involves the interaction of cytoskeletal elements and stretching of the nuclear membrane, which results in the modulation of mechanosensitive transcriptional and translational activities.

condensation of MSCs during development through cadherin binding is essential for the initiation of chondrogenic differentiation [28]. In transforming growth factor (TGF)-β-induced chondrogenesis of MSCs, activation of N-cadherin expression and the accumulation of β-catenin in the nucleus and subsequent promotion of β-catenin-activated transcriptional activity are significantly implicated in the commitment of MSCs to chondrogenic lineage [29–32].

Forces acting at the cell surface get carried through the cytoskeleton and transmitted to the interior of the nucleus. Increasingly, the nucleus is being recognized as a mechanosensor in itself [33]. Nuclei in MSCs have been found to deform in response to both intrinsic and extrinsic mechanical signals [34], and a significant increase in nuclear deformation also occurs during mesenchymal condensation [35]. Nucleus deformation can alter chromatin architecture including chromatin stretching [36] and reposition chromatin domains, which can directly affect transcription [37]. Force is transmitted to the nucleus across the nuclear envelope, which is known to interact with cytoskeleton actin and microtubules through binding proteins known collectively as the linker of nucleoskeleton and cytoskeleton (LINC) complex [37, 38]. These proteins then interact with lamins inside the nucleus, which in turn provide support for several nuclear proteins involved in DNA replication, transcription, and posttranslational modification [33, 37, 39]. Mechanical forces can also increase nuclear membrane tension and modulate nuclear pore complex (NPC) permeability, promoting nuclear entry of mechanosensitive transcription factors such as YAP/TAZ [40]. Nevertheless, many questions remain, including to what extent the nucleus itself responds to mechanical forces, instead of indirectly reacting through downstream cytoplasmic signaling pathways.

## 19.3 INFLUENCE OF EXTRACELLULAR CUES

Microenvironments are critical in influencing cell response and function. Passive mechanical inputs provided from the base properties of the materials to which cells adhered dictate cellular and cytoskeletal orientation, and facilitate mechanical perturbation of the attached cells [10]. Substratum of specific mechanical stiffness and topographical features (Fig. 19.3), through engineering specific receptor-ligand interactions, can modulate the extent of chondrocyte phenotypic conversion and MSC chondrogenesis.

### 19.3.1 Stiffness

Chondrocytes, on two-dimensional (2-D) substrates, has been shown to deposit more cartilaginous ECM with a substrate stiffness that is similar to that of articular cartilage, through a ROCK-dependent mechanism [41]. Chondrocytes are less sensitive to 3-D substrate stiffness as compared with MSCs [42]. However, matrix elasticity does influence the chondrogenic phenotype of chondrocytes [43], with softer gels of 4 kPa supporting higher type II collagen and aggrecan and lower type I collagen expression, while stiffer gels induced in an organization of the actin cytoskeleton correlating with the loss of chondrocyte phenotype [43, 44].

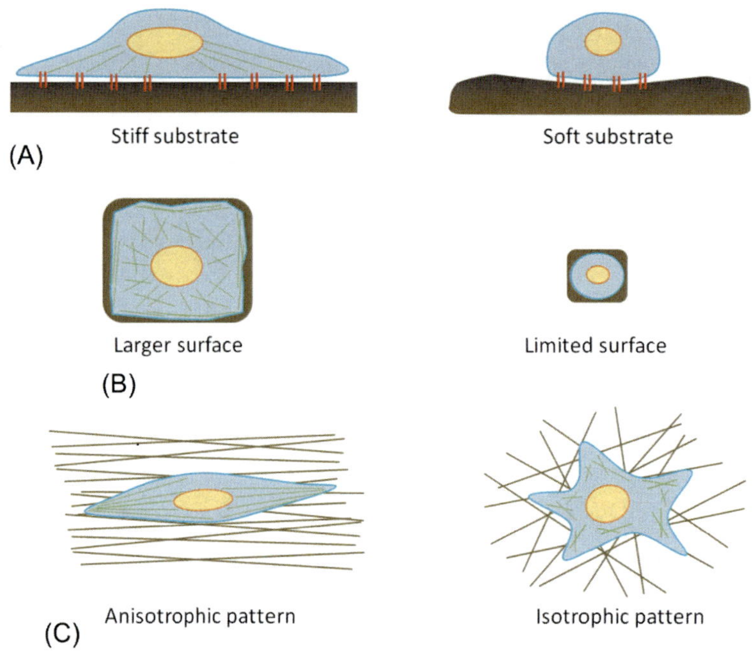


FIG. 19.3 Schematic diagrams of the response of cells to substrates with different stiffness (A), surface area (B), and topography (C).

Stem cells are exquisitely sensitive to the biochemical and mechanical properties of the extracellular matrix, which has been shown to regulate the differentiation of MSCs towards specific lineages [45]. When cultured on 2-D substrates that mimicked the stiffness of physiologic tissue environments, MSCs adopted a phenotype corresponding to the tissue stiffness, as demonstrated by cellular morphology, transcript markers, and protein production. In 2-D culture systems, substrate stiffness generally affects cellular morphology. Compared with myogenic and osteogenic differentiation, MSC chondrogenesis preferred softer substratum that permits adoption of spherical cell morphology, with lower cell adhesion strength and decreased F-actin stress fiber formation [46–48]. Investigation of both substrate stiffness and adhesivity (through provision of Arg-Gly-Asp peptide, RGD) on electrospun hyaluronic acid (HA) fibers indicated that MSC spreading and focal adhesion formation were dependent on RGD density, with traction force increased with more adhesive fibers [49]. The expression of chondrogenic markers, unlike trends in cell spreading and cytoskeletal organization, was influenced by both fiber mechanics and adhesivity in which softer fibers and lower RGD densities enhanced chondrogenesis.

In 3-D hydrogels, MSCs have been shown to retain a spherical morphology irrespective of the hydrogel stiffness [50, 51]. A study in 3-D hydrogel found that modulus-driven differentiation of MSCs was independent of actin polymerization, ROCK signaling, or NMM II [51]. In spite of this, the fate of encapsulated MSCs is still dependent on the stiffness of the hydrogel, with a decrease of chondrogenesis with increasing gel stiffness [42, 47, 52, 53]. Development of HA-based scaffolds with tunable mechanical and rheological properties has allowed the identification of an optimal, lower cross-linked matrix elasticity (Young's modulus at ~3–6 kPa) to favor hyaline cartilage formation [53]. A shift in MSC differentiation towards the fibrocartilage and fibrous tissue formation [52] or induction of calcification [53] was reported with increasing cross-linking and matrix stiffness of HA hydrogels. Substrates with compliant mechanical cues can also facilitate matrix-induced cell-cell interactions that are essential for the recapitulation of precartilage mesenchymal condensation [32, 54]. Engagement with integrin was shown to be essential for MSC chondrogenesis as both collagen and RGD-modified hydrogel enhanced differentiation [55–57]. Prior to condensation, cell-matrix interactions in hyaluronan and type I collagen (Col I)–rich ECM mediate aggregation of mesenchymal progenitors [58]. Provision of a 3-D environment with RGD, hyaluronic acid, and/or type I collagen, synergized with a softer hydrogel to support MSC condensation, enhanced cartilaginous development [57]. Mechanosensing computational models have been designed to relate the role of substratum mechanical cues in directing MSC proliferation and differentiation in 3-D hydrogel and could be employed to predict essential aspects of cell maturation, differentiation, proliferation and apoptosis during regenerative events [59–61]. Such computational models could be favorably exploited to identify substratum mechano-specificity in tissue engineering applications.

### 19.3.2 Cell Shape and Dynamic Morphological Changes

While primary chondrocytes are typically round shaped, they undergo a dedifferentiation process upon cultivation on 2-D substrates when cells adopt an elongated, fibroblastic morphology (with the formation of stress actin fibers), which is accompanied by a drastic loss in chondrogenic phenotype with the downregulation of chondrogenic transcription factor, SOX9, and Col II and concomitant upregulation of type I collagen I (Col I) [62–64]. Dedifferentiation is reversible with fibroblastic chondrocytes reexpressing the chondrogenic phenotype upon regaining a rounded cell shape by cultivation in 3-D hydrogel [64]. Treatment of fibroblastic chondrocytes with pharmacological agents (e.g., dihydrocytochalasin B, cytochalasin D, and staurosporine) that interfere with the integrity of the actin cytoskeleton, resulting in rounding of dedifferentiated chondrocytes, can induce the reexpression of the chondrogenic phenotype [65–68]. Inhibition of myosin/actin contractility by blebbistatin further enhanced staurosporine-induced redifferentiation that involved PI3K, PKC, and mitogen-activated protein kinase (MAPK)–signaling pathways [67]. These studies indicate that traction force generated from actin polymerization and myosin/actin contractility associated with the development of fibroblastic morphology account for loss of chondrocyte phenotype during 2-D culture.

The role of cell shape on MSC chondrogenic differentiation was explored by subjecting transforming growth factor (TGF)-β3 stimulated MSCs to fibronectin-coated micropatterned island that allowed cells to either flatten and spread on large islands or maintain a rounded cell morphology on small islands. MSCs kept rounded were committed to a chondrogenic lineage, while MSCs that were allowed to spread proceeded down a myogenic lineage [69]. Inhibition of Rac1, a member of the Rho GTPase family associated with cytoskeleton tension, resulted in the inhibition of myogenesis while upregulating Sox9, indicating that structural changes to the cytoskeleton play key roles in determining MSC lineage commitment. Although hydrogels maintain the encapsulated cells in spherical morphology and enhanced MSC chondrogenesis has been observed with the provision of biochemical cues to improve cell-matrix interactions, the inherent limitation with cells in hydrogel is that they are subjected to a "locked" morphology within the gel and have limited direct cell-cell interactions, which are essential for early mesenchymal condensation process and initiation of MSC chondrogenesis [28]. ECM deposited by differentiated MSCs in poly(ethylene glycol) or HA hydrogel was found to be restricted to the pericellular domain, even after prolonged culture period [70–72]. Robust chondrogenesis of MSCs requires dynamic morphological changes, initiated through integrin engagement that induces the association of their cytoplasmic domains with the actin cytoskeleton, adopting a fibroblastic morphology. Such integrin-driven changes in cell morphology facilitate enhanced cell-cell interactions, through N-cadherin expression during condensation events [25, 73]. Condensation is followed by a shift in cell morphology, involving cytoskeletal rearrangement to form cortical actin, with transient expression of NCAD and N-CAM molecules [29, 58], a process regulated by Rho kinase (ROCK)–driven actomyosin contractions and myosin II-generated differential cell cortex tension [35]. The importance of providing an orderly sequence of cell/matrix-induced cell-cell interaction in promoting robust MSC chondrogenic differentiation has been recognized in several studies [32, 56]. In the study employing oriented Col I ligand in interfacial polyelectrolyte complexation (IPC)–based hydrogels, Raghothaman et al. demonstrated that proximal integrin-mediated cell-matrix contacts initiated early morphological dynamics and the onset of N-cadherin/β-catenin-mediated chondrogenic induction resulted in superior chondrogenesis and the generation of mature hyaline neocartilage in comparison with scaffold-free pellet culture (in which initial matrix-cell interaction is not provided), and MSC in the conventional collagen hydrogel (in which cell-cell interaction is lacking) [32].

### 19.3.3 Substrate Topography

Cells can be forced to adopt specific morphology and cytoskeletal orientation by interaction with substrate topography, especially with engineered surface patterns of nanoscale features that mimic native ECM. MSCs on nonaligned nanofibers are well spread, with actin-rich processes extending isotropically. In contrast, cells on aligned nanofibers are fibroblastic, extending along the fiber direction. Hyaline cartilage matrix production was preferentially upregulated on randomly oriented nanofibers, whereas structurally anisotropic, aligned nanofibrous scaffold promotes fibrochondrogenesis [74–76]. The effect of specific nanotopography in directing MSC chondrogenic differentiation was further demonstrated using thermal imprinted nanotopographical patterns [77]. MSCs on nanograting topography formed actin stress fiber organization and were induced into a fibrocartilaginous or superficial zone–like neocartilage formation, while MSCs on nanopillars formed round morphology with their F-actin organized at the cell cortex, readily underwent cell aggregation reminiscent of condensation, and were induced to form hyaline-like neocartilage tissue [77]. By varying the stiffness of substratum nanotopography, Wu et al. further explored the combined effect of substrate topography and stiffness on MSC chondrogenesis, demonstrating that stiffer patterns inhibit the generation of the hyaline and fibrocartilage phenotype on the pillar and grating topography, respectively [78]. Formation of polygonal morphology on the

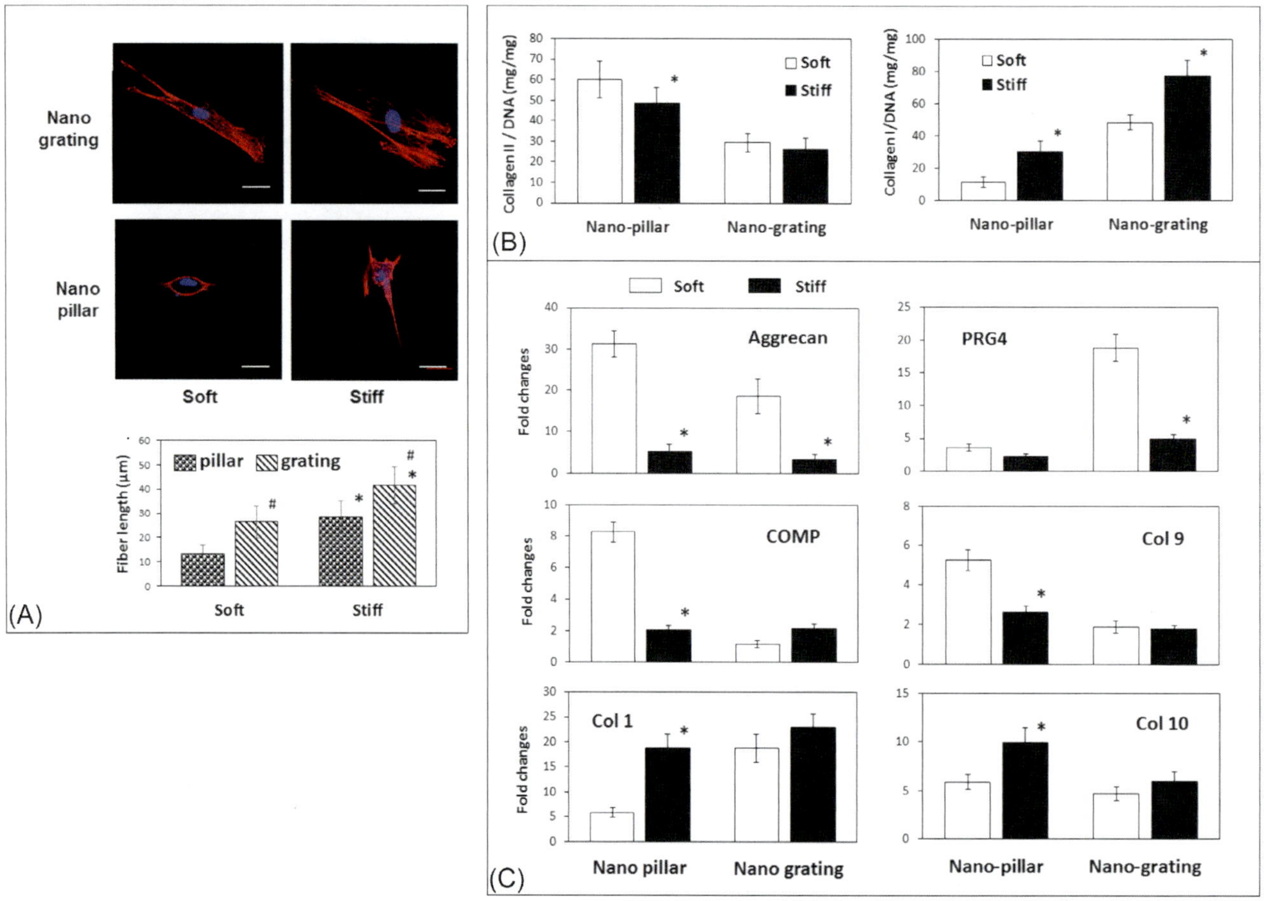


FIG. 19.4 (A) Cell morphology analysis by F-actin phalloidin staining on different nanopatterned surfaces after 3 days of culture under chondrogenic condition. Fluorescence analysis of F-actin fiber length by ImageJ. #Significant different fiber length in cells on nanopillar to nanograting of the same polymeric material. *Significant different fiber length in cells on stiff nanopillar or nanograting, compared with those on soft materials of the same pattern. (B) Quantification of type I and II collagen expression. (C) Real-time PCR analysis of mRNA expression levels of cartilaginous genes. *Significant different in stiff samples compared with soft samples of the same pattern. Modified from Y. Wu, et al., The combined effect of substrate stiffness and surface topography on chondrogenic differentiation of mesenchymal stem cells, Tissue Eng. Part A 23 (1–2) (2017) 43–54.

stiffer nanopillar instead of the rounded morphology on softer pillar induced increased stress fiber length and resulted in reduced chondrogenesis and directed MSCs towards a mixed hyaline/fibro/hypertrophic cartilage. On stiffer nanograting surfaces, MSCs did not undergo chondrogenesis (Fig. 19.4). These studies demonstrate the sensitivity of MSC differentiation and the possibility of refining the phenotype of regenerated cartilage by manipulating the material stiffness and surface nanotopographies of the scaffolding substrate. Employing sequential electrospinning technology, microstructural, trilaminar scaffolds were fabricated with depth-dependent variations in orientation and fiber sizes in a continuous construct, which resulted in the formation of neotissue mimicking some organizational characteristics of native cartilage [79]. Also, a bilayered polymeric scaffold with overlaid aligned microfiber layers enhanced the mechanical and surface properties of the underlying macroporous scaffold [80]. Zonal analysis of these scaffolds yielded region-specific variations in chondrocyte number, GAG-rich ECM, and chondrocytic gene expression, demonstrating the potential of multiphasic structural organization for the regeneration of a hierarchically organized articular cartilage tissue.

## 19.3.4 Intracellular Mechanotransduction

Substratum stiffness and topography influence on chondrocyte and stem cell chondrogenesis are heavily influenced by the integrity of cytoskeleton contractility. Accumulated studies indicate that compliant substrates that yield reduction of actin stress fiber, that is low cytoskeleton tension, are conducive for chondrogenic phenotype and the induction of MSC chondrogenesis. Expression of Rho GTPase, which is intrinsically associated with cytoskeleton contractile generation and regulates focal adhesions and organization of actin, was found to correlate with conversion of round primary chondrocytes into a fibroblast morphology and the loss of chondrogenic phenotype [81]. Rho/ROCK and Rac1 have been shown to generally inhibit MSC chondrogenesis by regulating Sox9 expression [69, 82, 83], indicating that structural changes to the cytoskeleton and reduction of cytoskeleton tension are critical for chondrogenesis. Nuclear morphology of MSCs, compared with terminally differentiated fibrochondrocytes, was shown to be much sensitive to mechanical cues. Traction force generated from interaction with ECM cues that get transmitted through the cytoskeleton actin would have translated to differences in nuclear geometry and organization that could directly alter transcriptional events [34, 84] and regulate collagen expression [37, 77, 85]. Reduced cytoskeletal tension, when cells are cultured on soft substrates, results in nuclear envelope proteins posttranslational modification and leads to the modulation of transcriptional activity [33]. The different traction forces generated by actin cytoskeleton of MSCs dictated by specific surface topography and material stiffness can also regulate the nuclear translocation of the mechanosensitive transcription factor, Yes-associated protein (YAP), through altering the permeability of the nuclear pore complexes (NPC) [40]. YAP has been identified as sensor and mediator of mechanical cues instructed by ECM rigidity and cell shape, which relays actomyosin cytoskeleton tension through a Rho GTPase–dependent translocation to the nucleus [22, 86]. The phenotypic variation of chondrocytes in response to substrate stiffness was shown to be concomitant with the changes in YAP localization, with YAP silencing of chondrocytes on stiffer substrate reversing the loss of chondrocyte phenotype [87]. YAP is a negative regulator of chondrogenesis in mesenchymal stem cells, which is downregulated during MSC chondrogenesis, while overexpression of YAP inhibits chondrogenic differentiation [88]. The nuclear translocation and total expression level of YAP/TAZ in adipose stem cells on aligned fibers was shown to be notably elevated compared with those on random fibers, suggesting regulation of YAP/TAZ by surface topography at transcriptional and translational levels [89]. While the exact mechanism of enhanced fibro/superficial zone-like chondrogenic development by unidirectional topography and substrate stiffness [77, 78] remains unclear, it is likely that the internal tension generated by the augmented stress fibers could lead to YAP nuclear translocation and activation, which in turn regulate chondrogenesis, and modulate the phenotype of the derivative chondrocytes.

## 19.4 EFFECT OF EXTERNAL MECHANICAL SIGNALS

During physiologic cartilage loading, water in this highly hydrated tissue is gradually squeezed out, causing direct strains at the tissue, cellular, and nuclear levels [7]. Concomitantly, mechanical compressive loading of cartilage generates secondary biophysical signals, such as hydrostatic pressures and shear force caused by the fluid flow. The oscillating sliding motion of the articulating joint also imparts shearing tension to the surface of the articular cartilage. Physiologic mechanical loading is a pivotal factor for the development of zonally defined cartilage in mature animals. Cartilage tissues formed during embryonic chondrogenesis consist of neither multiple zones nor the unique Benninghoff collagen architecture associated with articular cartilage. Such features are absent during birth and develop during skeletal maturation in postnatal development [90, 91], shaped by articular motions exerting a combination of compressive, tensile, and shear loading. Thus externally applied mechanical force can also significantly influence the chondrocyte ECM biosynthesis and chondrogenic differentiation of MSCs during articular cartilage repair. The type, frequency, magnitude and duration of such cues have all been shown to affect chondrocyte response and MSC differentiation [92]. The following subsections will summarize the response of chondrocytes and MSCs to the different forms of mechanical signals.

### 19.4.1 Compression

Compression by direct loading or the nondeforming hydrostatic pressure (HP) of chondrocytes and MSCs encapsulated in hydrogels has been found to be a strong prochondrogenic stimulus. Chondrocytes respond to physiologic magnitudes of dynamic compression (DC; ~10%–20%) with enhanced synthesis of ECM molecules, including PGs, collagens, and COMP [93–95]. The responses of chondrocytes to mechanical loading are highly dependent on loading frequency, amplitude, strain rate, and loading history, with superphysiologic magnitudes of loading (>20%) failing to enhance matrix production [93], while static or very low frequency loading [96] inhibits matrix synthesis. Chondrocytes respond to dynamic hydrostatic pressure similarly to DC [97, 98].

For MSCs, both DC [99–102] and HP [103–105] were shown to induce chondrogenic gene expression in MSCs in the absence of exogenous growth factor stimulation, by increasing autocrine TGFβ1 production [104, 106], suggesting that compression and exogenous TGFβ stimulation activate similar pathways. However, unlike the effect on chondrocytes, there were some uncertainties as to the impact of DC/HP stimulation on MSCs, with some studies demonstrating that the simultaneous application of TGFβ and mechanical loading inhibits chondrogenesis [107, 108]. Delayed application of dynamic loading provided after a preculture period of chondrogenic induction was instead shown to be beneficial for MSC chondrogenesis [109, 110]. The formation of sufficient pericellular matrix (PCM) by chondro-differentiating MSCs was speculated to be an important factor in determining MSCs' response to compression. PCM is both biochemically and biomechanically distinct from the surrounding ECM that can influence the stress-strain microenvironment of the cells and serves as a direct transducer of biochemical and biomechanical signals to the cells [111]. MSCs seeded in the stiffer hydrogels, with increased PCM formation, displayed a more prochondrogenic response to the application of DC [112] and hydrostatic pressure [113], compared with those seeded in softer hydrogels with lower level of PCM formation. Cytoskeletal organization and focal adhesion formation were observed to be altered in the stiff gels relative to the soft gels in which cell-matrix interaction via integrin could have played a critical role in distributing and transmitting the exogenous mechanical signal [112, 114].

Stabilization of chondrogenic phenotype by DC [110, 115] and HP [116] has also been reported. Inhibition of hypertrophic development by mechanical compression correlated with the outcomes of an in vivo study involving orthotopic transplantation of MSC, wherein collagen type X–positive chondrocytes were found only at the osteochondral interface [117]. Study with elastomeric poly-L-lactide-*co*-ε-caprolactone (PLCL) scaffolds suggested that compression led to hypertrophic development (Zhang et al., 2015). However, long-term dynamic loading has also been shown to induce hypertrophy development [108]. These discrepancies in compression associated-hypertrophy development could depend on the loading strain, frequency, and also the existence of pericellular matrix deposition to provide proximal matrix-cell interaction that is essential for mechanotransduction [113].

### 19.4.2 Shear Stress

Apart from compression loading, articular cartilage experiences mechanical shear deformation at the tissue surface [118, 119]. The combined interplay between load and articular oscillation shearing and their relationship to chondrocytes and MSC responses were explored with the employment of multiaxial bioreactor, which superimposed surface shearing on cyclic axial compression, mimicking in vivo articulation [120]. Adding a component of shear to compressive loading was shown to be superior to when only compressional loading was provided at inducing matrix biosynthesis in chondrocytes. Specifically, expression of the superficial zone–specific PRG4 was induced [72, 119]. The application of shear superimposed upon dynamic compression also led to significant increases in MSC chondrogenic gene expression, including PRG4, resulting in a tissue of improved tensile modulus [102, 121, 122], with the phenotype of the derived neocartilage tissue dependent on the frequency and amplitude of the compression and shear stress.

### 19.4.3 Fluid Flow

With prolonged and/or higher loading magnitude, interstitial fluid within the extracellular matrix is eventually exuded, generating secondary biophysical signals in the form of fluid shear stress and changes in osmolarity across the cellular membrane. Fluid flow in response to joint loading is complex and challenging to recapitulate for isolated chondrocytes in in vitro three-dimensional cultures. Controlled medium flow has been used to culture chondrocyte-seeded constructs, as culture with dynamic fluid flow provides several advantages over static culture including enhanced mass transport, improved nutrient delivery, and suitable hydrodynamic stimuli. Perfusion and rotating wall bioreactors have been shown to enhance extracellular matrix accumulation by chondrocytes seeded in porous polymeric constructs [123–125]. Perfusion of chondrocytes induced superficial zone–specific phenotype [126, 127] that was dependent on intracellular calcium elevation [126, 128]. Perfusion and oscillating bioreactors have similarly been found to enhance the biochemical and functional properties of MSC-seeded scaffolds in TGFβ-supplemented chondrogenic medium [129–131]. However, the effect of fluid flow is dependent on flow rate and the porosity of the tissue construct. Mechanical signaling induced by fluid flow likely acts through calcium response that is dependent on the fluid flow rate [128]. At high fluid flow rate, reduced chondrogenesis with decreased GAG and diminished Col II was reported [132]. Shear stress as a result of increased fluid flow has also been associated with increased hypertrophy [133].

### 19.4.4 Mechanotransduction of External Mechanical Stimulation

A number of key transduction mechanisms have been identified that facilitate the mechanically driven enhancement of cartilage ECM biosynthesis and functional properties, including mechanosensitive ion channels [7] and signaling through integrins [134]. Chondrogenesis is known to be modulated by calcium signaling cascades of specific temporal sensitivity [135, 136]. Calcium influx via membrane-associated cation channels is a key event in initiating chondrogenesis, potentially mediated by either the transient receptor potential (TRP) channels or voltage-gated calcium channels (VGCC) [136]. As a ubiquitous second messenger, mechanically induced calcium signaling has been shown to act as a regulator of mechanotransduction in multiple signaling pathways, including nuclear factor of activated T lymphocytes (NFAT), protein kinase C, nuclear factor kappa-light-chain enhancer of activated B cells (NF-κB), c-Jun N-terminal kinase 1 (JNK1), and cyclic adenosine monophosphate (cAMP) response element-binding protein (CREB) [137]. In addition, Sox9, the master transcription factor of chondrogenesis, is subject to $Ca^{2+}$-calmodulin regulation [138]. Calcium signaling, mediated by transient receptor potential vanilloid 4 (TRPV4), an osmo-mechanosensitive ion channel highly expressed in articular chondrocytes [139], has been shown to play a primary role in promoting chondrogenesis [140] and transducing the mechanical signals that support cartilage extracellular matrix maintenance and joint health [141–143]. Intracellular calcium signaling is also involved in the shear stress–driven enhancement of PRG4 expression in superficial zone chondrocytes in which purinergic ATP/P2X7 and PKA/CREB signaling pathways were also implicated [126]. Other studies have further indicated TGF-β/Smad, Erk1/2, p38, and ciliary signaling [144, 145], as well as integrin/FAK signaling [110, 144], in mediating the responses of chondrocytes to loading. Dynamic compression study with elastomeric polymeric scaffolds suggested that compression-driven hypertrophic development involves cross talk between TGFβ/SMAD2/3 signaling and integrin-ECM interactions, regulating the suppression of the BMP/GDP and integrin/FAK/ERK signaling [110].

Mechanical stress can directly induce nuclear conformational changes, exerting transcriptional and posttranslational modification [33, 34]. Apart from the response of YAP expression and activity to passive mechanical force [22], the role of YAP in transducing extrinsic mechanical force stimuli, in relation to chondrocytes and MSC chondrogenesis, is being established. Using an integrated microfluidic perfusion device that subjects cells to controlled fluid shear stresses (FSS), Zhong et al. showed that the levels of YAP nuclear distribution in both chondrocyte and MSC increase, accompanying formation of fibroblastic morphology, with the increased strength of FSS [146]. Chondrocytes under FSS undergo phenotypic conversion with the loss of Col II, while Col I expression increases. Treatment with cytochalasin D reverted the FSS-induced morphological change, YAP distribution, and phenotypic conversion, revealing the connection between YAP and MSC/chondrocyte fates in a fluid flow–induced mechanical microenvironment [146]. The combined effect of external mechanical treatment such as FSS [147] or tensile stress [34, 86] on chondrocytes and MSCs already subjected to passive mechanical signal, directed by attachment to aligned fibers, was investigated. Chondrocytes and MSCs on aligned fibrous meshes elicited further alterations in cell and nuclear morphology, dependent on the direction of force application in relation to the fiber direction [86, 147], which resulted in generation of fibrochondrocyte when force is parallel to the fiber direction, while yielding a more hyaline phenotype when force is perpendicular to the fiber direction. The effect of applied tension is mediated by filamentous actin cytoskeleton and can be curtailed with ROCK inhibitor and knockdown of YAP/TAZ [34, 147]. These studies suggest that the direction between applied force and the orientation of cells on the substrate have profound effects on intracellular mechanotransduction and play a significant role in regulating chondrogenesis. Of similar note, when comparing the response of MSCs seeded in either agarose (spherical MSC morphology) or fibrin (spread MSC morphology with clear stress fiber formation) hydrogels to the application of HP, it was demonstrated that while agarose provided a stronger pro-chondrogenic environment, a more robust response to the application of HP was observed in fibrin hydrogels [114], again demonstrating that cells with differing cytoskeleton contractility will alter their response to external mechanical stimulation.

## 19.5 FUTURE DIRECTIONS

The impact of both mechanical environment and extrinsic mechanical force in directing chondrocyte phenotype and MSC chondrogenic differentiation has been well recognized. Recent studies have been performed to decipher the intracellular mechanotransduction mechanism in transmitting the various mechanical signals. Nonetheless, many questions remain, including the specific microenvironment cues best suited for imparting desirable mechanical signaling for chondrocytes or MSCs, given that the differences in the developmental status of MSCs and the terminally differentiated mature chondrocytes have been shown to prefer different microenvironment cues [42, 148–150]. The majority of studies exploring the mechanobiology of chondrocytes and MSCs have been performed using in vitro systems, which have helped to improve our understanding on how cells respond to specific stimuli. Increasing studies have started exploring the effect of multiple cues, including biophysical and biochemical factors, in combination with externally applied force, to mimic the interplay between intrinsic and extrinsic mechanical cues in the in vivo environment. Computational modeling has also been utilized to provide predictive evaluation to investigate the complex relationships between external mechanical forces, the cell, and the surrounding substrate [59, 151–154]. Further research is needed to fully elucidate how chondrocytes and MSCs sense and respond to the complex sets of mechanical stimuli applied in tandem. It should also be recognized that the creation of a translational platform to encompass all the relevant biological, biochemical, and mechanical cues could present a significant technical challenge. Thus the onus will be on the identification of the vital elements from the vast scientific landscape for the optimal application to cartilage tissue engineering and regeneration.

## References


[1] A.J. Hayes, et al., Macromolecular organization and in vitro growth characteristics of scaffold-free neocartilage grafts, J. Histochem. Cytochem. 55 (8) (2007) 853–866.

[2] T.J. Klein, et al., Tissue engineering of articular cartilage with biomimetic zones, Tissue Eng. Part B Rev. 15 (2) (2009) 143–157.

[3] G.D. Jay, K.A. Waller, The biology of lubricin: near frictionless joint motion, Matrix Biol. 39 (2014) 17–24.

[4] R.A. Bank, et al., Ageing and zonal variation in post-translational modification of collagen in normal human articular cartilage. The age-related increase in non-enzymatic glycation affects biomechanical properties of cartilage, Biochem. J. 330 (Pt 1) (1998) 345–351.

[5] K. Blumbach, et al., Combined role of type IX collagen and cartilage oligomeric matrix protein in cartilage matrix assembly: cartilage oligomeric matrix protein counteracts type IX collagen-induced limitation of cartilage collagen fibril growth in mouse chondrocyte cultures, Arthritis Rheum. 60 (12) (2009) 3676–3685.

[6] H. Haleem-Smith, et al., Cartilage oligomeric matrix protein enhances matrix assembly during chondrogenesis of human mesenchymal stem cells, J. Cell. Biochem. 113 (4) (2012) 1245–1252.

[7] J. Sanchez-Adams, et al., The mechanobiology of articular cartilage: bearing the burden of osteoarthritis, Curr. Rheumatol. Rep. 16 (10) (2014) 451.

[8] A.J. Maniotis, C.S. Chen, D.E. Ingber, Demonstration of mechanical connections between integrins, cytoskeletal filaments, and nucleoplasm that stabilize nuclear structure, Proc. Natl. Acad. Sci. U. S. A. 94 (3) (1997) 849–854.

[9] N. Wang, J.D. Tytell, D.E. Ingber, Mechanotransduction at a distance: mechanically coupling the extracellular matrix with the nucleus, Nat. Rev. Mol. Cell Biol. 10 (1) (2009) 75–82.

[10] F. Guilak, et al., Control of stem cell fate by physical interactions with the extracellular matrix, Cell Stem Cell 5 (1) (2009) 17–26.

[11] Z. Liu, et al., Mechanical tugging force regulates the size of cell-cell junctions, Proc. Natl. Acad. Sci. U. S. A. 107 (22) (2010) 9944–9949.

[12] H. Wolfenson, et al., Actomyosin-generated tension controls the molecular kinetics of focal adhesions, J. Cell Sci. 124 (Pt 9) (2011) 1425–1432.

[13] A. Woods, G. Wang, F. Beier, Regulation of chondrocyte differentiation by the actin cytoskeleton and adhesive interactions, J. Cell. Physiol. 213 (1) (2007) 1–8.

[14] K. Burridge, et al., Focal adhesions: transmembrane junctions between the extracellular matrix and the cytoskeleton, Annu. Rev. Cell Biol. 4 (1988) 487–525.

[15] B. Geiger, et al., Transmembrane crosstalk between the extracellular matrix-cytoskeleton crosstalk, Nat. Rev. Mol. Cell Biol. 2 (11) (2001) 793–805.

[16] R. Zaidel-Bar, et al., Functional atlas of the integrin adhesome, Nat. Cell Biol. 9 (8) (2007) 858–867.

[17] R. McBeath, et al., Cell shape, cytoskeletal tension, and RhoA regulate stem cell lineage commitment, Dev. Cell 6 (4) (2004) 483–495.

[18] K. Burridge, C.E. Turner, L.H. Romer, Tyrosine phosphorylation of paxillin and pp125FAK accompanies cell adhesion to extracellular matrix: a role in cytoskeletal assembly, J. Cell Biol. 119 (4) (1992) 893–903.

[19] A. Hall, Small GTP-binding proteins and the regulation of the actin cytoskeleton, Annu. Rev. Cell Biol. 10 (1994) 31–54.

[20] Q. Chen, et al., Integrin-mediated cell adhesion activates mitogen-activated protein kinases, J. Biol. Chem. 269 (43) (1994) 26602–26605.

[21] T.J. Jones, S.M. Nauli, Mechanosensory calcium signaling, Adv. Exp. Med. Biol. 740 (2012) 1001–1015.

[22] S. Dupont, et al., Role of YAP/TAZ in mechanotransduction, Nature 474 (7350) (2011) 179–183.

[23] T. Panciera, et al., Mechanobiology of YAP and TAZ in physiology and disease, Nat. Rev. Mol. Cell Biol. 18 (12) (2017) 758–770.

[24] A. Ganz, et al., Traction forces exerted through N-cadherin contacts, Biol. Cell. 98 (12) (2006) 721–730.

[25] G.F. Weber, M.A. Bjerke, D.W. DeSimone, Integrins and cadherins join forces to form adhesive networks, J. Cell Sci. 124 (Pt 8) (2011) 1183–1193.

[26] H. Aberle, H. Schwartz, R. Kemler, Cadherin-catenin complex: protein interactions and their implications for cadherin function, J. Cell. Biochem. 61 (4) (1996) 514–523.

[27] M. Smutny, A.S. Yap, Neighborly relations: cadherins and mechanotransduction, J. Cell Biol. 189 (7) (2010) 1075–1077.

[28] A.M. DeLise, L. Fischer, R.S. Tuan, Cellular interactions and signaling in cartilage development, Osteoarthr. Cartil. 8 (5) (2000) 309–334.

[29] A.M. Delise, R.S. Tuan, Analysis of N-cadherin function in limb mesenchymal chondrogenesis in vitro, Dev. Dyn. 225 (2) (2002) 195–204.

[30] R. Tuli, et al., Transforming growth factor-beta-mediated chondrogenesis of human mesenchymal progenitor cells involves N-cadherin and mitogen-activated protein kinase and Wnt signaling cross-talk, J. Biol. Chem. 278 (42) (2003) 41227–41236.

[31] R. Modarresi, et al., N-cadherin mediated distribution of beta-catenin alters MAP kinase and BMP-2 signaling on chondrogenesis-related gene expression, J. Cell. Biochem. 95 (1) (2005) 53–63.

[32] D. Raghothaman, et al., Engineering cell matrix interactions in assembled polyelectrolyte fiber hydrogels for mesenchymal stem cell chondrogenesis, Biomaterials 35 (9) (2014) 2607–2616.

[33] T.J. Kirby, J. Lammerding, Emerging views of the nucleus as a cellular mechanosensor, Nat. Cell Biol. 20 (4) (2018) 373–381.

[34] A.S. Nathan, et al., Mechano-topographic modulation of stem cell nuclear shape on nanofibrous scaffolds, Acta Biomater. 7 (1) (2011) 57–66.

[35] P. Ray, S.C. Chapman, Cytoskeletal reorganization drives mesenchymal condensation and regulates downstream molecular signaling, PLoS One 10 (8) (2015) e0134702.

[36] A. Tajik, et al., Transcription upregulation via force-induced direct stretching of chromatin, Nat. Mater. 15 (12) (2016) 1287–1296.

[37] K.N. Dahl, A.J. Ribeiro, J. Lammerding, Nuclear shape, mechanics, and mechanotransduction, Circ. Res. 102 (11) (2008) 1307–1318.
[38] M. Crisp, et al., Coupling of the nucleus and cytoplasm: role of the LINC complex, J. Cell Biol. 172 (1) (2006) 41–53.
[39] Y. Shibata, et al., Mechanisms determining the morphology of the peripheral ER, Cell 143 (5) (2010) 774–788.
[40] A. Elosegui-Artola, et al., Force triggers YAP nuclear entry by regulating transport across nuclear pores, Cell 171 (6) (2017) 1397–1410 e14.
[41] J.L. Allen, M.E. Cooke, T. Alliston, ECM stiffness primes the TGFbeta pathway to promote chondrocyte differentiation, Mol. Biol. Cell 23 (18) (2012) 3731–3742.
[42] Z. Yang, et al., Influence of fibrinogen concentration on mesenchymal stem cells and chondrocytes chondrogenesis in fibrin hydrogels, J. Biomater. Tissue Eng. 7 (11) (2017) 1136–1145.
[43] E. Schuh, et al., Effect of matrix elasticity on the maintenance of the chondrogenic phenotype, Tissue Eng. Part A 16 (4) (2010) 1281–1290.
[44] P. Sanz-Ramos, et al., Response of sheep chondrocytes to changes in substrate stiffness from 2 to 20 Pa: effect of cell passaging, Connect. Tissue Res. 54 (3) (2013) 159–166.
[45] A.J. Engler, et al., Matrix elasticity directs stem cell lineage specification, Cell 126 (4) (2006) 677–689.
[46] S. Ghosh, et al., In vitro model of mesenchymal condensation during chondrogenic development, Biomaterials 30 (33) (2009) 6530–6540.
[47] J.S. Park, et al., The effect of matrix stiffness on the differentiation of mesenchymal stem cells in response to TGF-beta, Biomaterials 32 (16) (2011) 3921–3930.
[48] J. Nam, et al., Modulation of embryonic mesenchymal progenitor cell differentiation via control over pure mechanical modulus in electrospun nanofibers, Acta Biomater. 7 (4) (2011) 1516–1524.
[49] I.L. Kim, et al., Fibrous hyaluronic acid hydrogels that direct MSC chondrogenesis through mechanical and adhesive cues, Biomaterials 34 (22) (2013) 5571–5580.
[50] N. Huebsch, et al., Harnessing traction-mediated manipulation of the cell/matrix interface to control stem-cell fate, Nat. Mater. 9 (6) (2010) 518–526.
[51] S.H. Parekh, et al., Modulus-driven differentiation of marrow stromal cells in 3D scaffolds that is independent of myosin-based cytoskeletal tension, Biomaterials 32 (9) (2011) 2256–2264.
[52] W.S. Toh, et al., Modulation of mesenchymal stem cell chondrogenesis in a tunable hyaluronic acid hydrogel microenvironment, Biomaterials 33 (15) (2012) 3835–3845.
[53] L. Bian, et al., The influence of hyaluronic acid hydrogel crosslinking density and macromolecular diffusivity on human MSC chondrogenesis and hypertrophy, Biomaterials 34 (2) (2013) 413–421.
[54] C.F. Chang, et al., Three-dimensional collagen fiber remodeling by mesenchymal stem cells requires the integrin-matrix interaction, J. Biomed. Mater. Res. A 80 (2) (2007) 466–474.
[55] T. Re'em, O. Tsur-Gang, S. Cohen, The effect of immobilized RGD peptide in macroporous alginate scaffolds on TGFbeta1-induced chondrogenesis of human mesenchymal stem cells, Biomaterials 31 (26) (2010) 6746–6755.
[56] Y.Y. Li, et al., Scaffold composition affects cytoskeleton organization, cell-matrix interaction and the cellular fate of human mesenchymal stem cells upon chondrogenic differentiation, Biomaterials 52 (2015) 208–220.
[57] B. Carrion, et al., The synergistic effects of matrix stiffness and composition on the response of chondroprogenitor cells in a 3D precondensation microenvironment, Adv. Healthc. Mater. 5 (10) (2016) 1192–1202.
[58] M.B. Goldring, K. Tsuchimochi, K. Ijiri, The control of chondrogenesis, J. Cell. Biochem. 97 (1) (2006) 33–44.
[59] D.P. Burke, D.J. Kelly, Substrate stiffness and oxygen as regulators of stem cell differentiation during skeletal tissue regeneration: a mechanobiological model, PLoS One 7 (7) (2012) e40737.
[60] S.J. Mousavi, M.H. Doweidar, Numerical modeling of cell differentiation and proliferation in force-induced substrates via encapsulated magnetic nanoparticles, Comput. Methods Prog. Biomed. 130 (2016) 106–117.
[61] S.J. Mousavi, M.H. Doweidar, Role of mechanical cues in cell differentiation and proliferation: a 3D numerical model, PLoS One 10 (5) (2015) e0124529.
[62] K. von der Mark, et al., Relationship between cell shape and type of collagen synthesised as chondrocytes lose their cartilage phenotype in culture, Nature 267 (5611) (1977) 531–532.
[63] P.D. Benya, S.R. Padilla, M.E. Nimni, Independent regulation of collagen types by chondrocytes during the loss of differentiated function in culture, Cell 15 (4) (1978) 1313–1321.
[64] D.G. Stokes, et al., Regulation of type-II collagen gene expression during human chondrocyte de-differentiation and recovery of chondrocyte-specific phenotype in culture involves Sry-type high-mobility-group box (SOX) transcription factors, Biochem. J. 360 (Pt 2) (2001) 461–470.
[65] P.D. Benya, P.D. Brown, S.R. Padilla, Microfilament modification by dihydrocytochalasin B causes retinoic acid-modulated chondrocytes to reexpress the differentiated collagen phenotype without a change in shape, J. Cell Biol. 106 (1) (1988) 161–170.
[66] P.D. Brown, P.D. Benya, Alterations in chondrocyte cytoskeletal architecture during phenotypic modulation by retinoic acid and dihydrocytochalasin B-induced reexpression, J. Cell Biol. 106 (1) (1988) 171–179.
[67] M. Rottmar, et al., Interference with the contractile machinery of the fibroblastic chondrocyte cytoskeleton induces re-expression of the cartilage phenotype through involvement of PI3K, PKC and MAPKs, Exp. Cell Res. 320 (2) (2014) 175–187.
[68] J. Parreno, et al., Interplay between cytoskeletal polymerization and the chondrogenic phenotype in chondrocytes passaged in monolayer culture, J. Anat. 230 (2) (2017) 234–248.
[69] L. Gao, R. McBeath, C.S. Chen, Stem cell shape regulates a chondrogenic versus myogenic fate through Rac1 and N-cadherin, Stem Cells 28 (3) (2010) 564–572.
[70] B. Sharma, et al., Designing zonal organization into tissue-engineered cartilage, Tissue Eng. 13 (2) (2007) 405–414.
[71] N.S. Hwang, et al., Regulation of osteogenic and chondrogenic differentiation of mesenchymal stem cells in PEG-ECM hydrogels, Cell Tissue Res. 344 (3) (2011) 499–509.
[72] L.H. Nguyen, et al., Engineering articular cartilage with spatially-varying matrix composition and mechanical properties from a single stem cell population using a multi-layered hydrogel, Biomaterials 32 (29) (2011) 6946–6952.
[73] S. Gronthos, et al., Integrin-mediated interactions between human bone marrow stromal precursor cells and the extracellular matrix, Bone 28 (2) (2001) 174–181.

[74] B.M. Baker, et al., The influence of an aligned nanofibrous topography on human mesenchymal stem cell fibrochondrogenesis, Biomaterials 31 (24) (2010) 6190–6200.
[75] J.K. Wise, et al., Chondrogenic differentiation of human mesenchymal stem cells on oriented nanofibrous scaffolds: engineering the superficial zone of articular cartilage, Tissue Eng. Part A 15 (4) (2009) 913–921.
[76] W.J. Li, et al., Engineering controllable anisotropy in electrospun biodegradable nanofibrous scaffolds for musculoskeletal tissue engineering, J. Biomech. 40 (8) (2007) 1686–1693.
[77] Y.N. Wu, et al., Substrate topography determines the fate of chondrogenesis from human mesenchymal stem cells resulting in specific cartilage phenotype formation, Nanomedicine 10 (7) (2014) 1507–1516.
[78] Y. Wu, et al., The combined effect of substrate stiffness and surface topography on chondrogenic differentiation of mesenchymal stem cells, Tissue Eng. Part A 23 (1–2) (2017) 43–54.
[79] S.D. McCullen, et al., Anisotropic fibrous scaffolds for articular cartilage regeneration, Tissue Eng. Part A 18 (19–20) (2012) 2073–2083.
[80] J.A. Steele, et al., Combinatorial scaffold morphologies for zonal articular cartilage engineering, Acta Biomater. 10 (5) (2014) 2065–2075.
[81] D. Kumar, A.B. Lassar, The transcriptional activity of Sox9 in chondrocytes is regulated by RhoA signaling and actin polymerization, Mol. Cell. Biol. 29 (15) (2009) 4262–4273.
[82] A. Woods, F. Beier, RhoA/ROCK signaling regulates chondrogenesis in a context-dependent manner, J. Biol. Chem. 281 (19) (2006) 13134–13140.
[83] A. Woods, G. Wang, F. Beier, RhoA/ROCK signaling regulates Sox9 expression and actin organization during chondrogenesis, J. Biol. Chem. 280 (12) (2005) 11626–11634.
[84] Y. Wang, et al., Orientation and repositioning of chromosomes correlate with cell geometry-dependent gene expression, Mol. Biol. Cell 28 (14) (2017) 1997–2009.
[85] C.H. Thomas, et al., Engineering gene expression and protein synthesis by modulation of nuclear shape, Proc. Natl. Acad. Sci. U. S. A. 99 (4) (2002) 1972–1977.
[86] T.P. Driscoll, et al., Cytoskeletal to nuclear strain transfer regulates YAP signaling in mesenchymal stem cells, Biophys. J. 108 (12) (2015) 2783–2793.
[87] W. Zhong, et al., YAP-mediated regulation of the chondrogenic phenotype in response to matrix elasticity, J. Mol. Histol. 44 (5) (2013) 587–595.
[88] A. Karystinou, et al., Yes-associated protein (YAP) is a negative regulator of chondrogenesis in mesenchymal stem cells, Arthritis Res. Ther. 17 (2015) 147.
[89] S. Wan, et al., FAK- and YAP/TAZ dependent mechanotransduction pathways are required for enhanced immunomodulatory properties of adipose-derived mesenchymal stem cells induced by aligned fibrous scaffolds, Biomaterials 171 (2018) 107–117.
[90] J. Rieppo, et al., Changes in spatial collagen content and collagen network architecture in porcine articular cartilage during growth and maturation, Osteoarthr. Cartil. 17 (4) (2009) 448–455.
[91] P. Julkunen, et al., Maturation of collagen fibril network structure in tibial and femoral cartilage of rabbits, Osteoarthr. Cartil. 18 (3) (2010) 406–415.
[92] A.J. Steward, D.J. Kelly, Mechanical regulation of mesenchymal stem cell differentiation, J. Anat. 227 (6) (2015) 717–731.
[93] M. Wong, M. Siegrist, X. Cao, Cyclic compression of articular cartilage explants is associated with progressive consolidation and altered expression pattern of extracellular matrix proteins, Matrix Biol. 18 (4) (1999) 391–399.
[94] R.L. Mauck, et al., Functional tissue engineering of articular cartilage through dynamic loading of chondrocyte-seeded agarose gels, J. Biomech. Eng. 122 (3) (2000) 252–260.
[95] K.W. Ng, et al., Duty cycle of deformational loading influences the growth of engineered articular cartilage, Cell. Mol. Bioeng. 2 (3) (2009) 386–394.
[96] F. Guilak, et al., The effects of matrix compression on proteoglycan metabolism in articular cartilage explants, Osteoarthr. Cartil. 2 (2) (1994) 91–101.
[97] R.L. Smith, et al., Time-dependent effects of intermittent hydrostatic pressure on articular chondrocyte type II collagen and aggrecan mRNA expression, J. Rehabil. Res. Dev. 37 (2) (2000) 153–161.
[98] R.L. Smith, et al., In vitro stimulation of articular chondrocyte mRNA and extracellular matrix synthesis by hydrostatic pressure, J. Orthop. Res. 14 (1) (1996) 53–60.
[99] C.Y. Huang, P.M. Reuben, H.S. Cheung, Temporal expression patterns and corresponding protein inductions of early responsive genes in rabbit bone marrow-derived mesenchymal stem cells under cyclic compressive loading, Stem Cells 23 (8) (2005) 1113–1121.
[100] J.K. Mouw, et al., Dynamic compression regulates the expression and synthesis of chondrocyte-specific matrix molecules in bone marrow stromal cells, Stem Cells 25 (3) (2007) 655–663.
[101] D. Pelaez, C.Y. Huang, H.S. Cheung, Cyclic compression maintains viability and induces chondrogenesis of human mesenchymal stem cells in fibrin gel scaffolds, Stem Cells Dev. 18 (1) (2009) 93–102.
[102] Z. Li, et al., Chondrogenesis of human bone marrow mesenchymal stem cells in fibrin-polyurethane composites is modulated by frequency and amplitude of dynamic compression and shear stress, Tissue Eng. Part A 16 (2) (2010) 575–584.
[103] P. Angele, et al., Cyclic hydrostatic pressure enhances the chondrogenic phenotype of human mesenchymal progenitor cells differentiated in vitro, J. Orthop. Res. 21 (3) (2003) 451–457.
[104] K. Miyanishi, et al., Effects of hydrostatic pressure and transforming growth factor-beta 3 on adult human mesenchymal stem cell chondrogenesis in vitro, Tissue Eng. 12 (6) (2006) 1419–1428.
[105] K. Sakao, et al., Induction of chondrogenic phenotype in synovium-derived progenitor cells by intermittent hydrostatic pressure, Osteoarthr. Cartil. 16 (7) (2008) 805–814.
[106] Z. Li, et al., Mechanical load modulates chondrogenesis of human mesenchymal stem cells through the TGF-beta pathway, J. Cell. Mol. Med. 14 (6A) (2010) 1338–1346.
[107] A.H. Huang, et al., Long-term dynamic loading improves the mechanical properties of chondrogenic mesenchymal stem cell-laden hydrogel, Eur. Cell. Mater. 19 (2010) 72–85.
[108] M.G. Haugh, et al., Temporal and spatial changes in cartilage-matrix-specific gene expression in mesenchymal stem cells in response to dynamic compression, Tissue Eng. Part A 17 (23–24) (2011) 3085–3093.

[109] J. Li, et al., The influence of delayed compressive stress on TGF-beta1-induced chondrogenic differentiation of rat BMSCs through Smad-dependent and Smad-independent pathways, Biomaterials 33 (33) (2012) 8395–8405.
[110] T. Zhang, et al., Cross-talk between TGF-beta/SMAD and integrin signaling pathways in regulating hypertrophy of mesenchymal stem cell chondrogenesis under deferral dynamic compression, Biomaterials 38 (2015) 72–85.
[111] R.E. Wilusz, J. Sanchez-Adams, F. Guilak, The structure and function of the pericellular matrix of articular cartilage, Matrix Biol. 39 (2014) 25–32.
[112] A.J. Steward, D.R. Wagner, D.J. Kelly, Exploring the roles of integrin binding and cytoskeletal reorganization during mesenchymal stem cell mechanotransduction in soft and stiff hydrogels subjected to dynamic compression, J. Mech. Behav. Biomed. Mater. 38 (2014) 174–182.
[113] A.J. Steward, D.R. Wagner, D.J. Kelly, The pericellular environment regulates cytoskeletal development and the differentiation of mesenchymal stem cells and determines their response to hydrostatic pressure, Eur. Cell. Mater. 25 (2013) 167–178.
[114] A.J. Steward, et al., Cell-matrix interactions regulate mesenchymal stem cell response to hydrostatic pressure, Acta Biomater. 8 (6) (2012) 2153–2159.
[115] L. Bian, et al., Dynamic compressive loading enhances cartilage matrix synthesis and distribution and suppresses hypertrophy in hMSC-laden hyaluronic acid hydrogels, Tissue Eng. Part A 18 (7–8) (2012) 715–724.
[116] T. Vinardell, et al., Hydrostatic pressure acts to stabilise a chondrogenic phenotype in porcine joint tissue derived stem cells, Eur. Cell. Mater. 23 (2012) 121–132 (discussion 133–134).
[117] E. Steck, et al., Mesenchymal stem cell differentiation in an experimental cartilage defect: restriction of hypertrophy to bone-close neocartilage, Stem Cells Dev. 18 (7) (2009) 969–978.
[118] G.E. Nugent, et al., Dynamic shear stimulation of bovine cartilage biosynthesis of proteoglycan 4, Arthritis Rheum. 54 (6) (2006) 1888–1896.
[119] S. Grad, et al., Effects of simple and complex motion patterns on gene expression of chondrocytes seeded in 3D scaffolds, Tissue Eng. 12 (11) (2006) 3171–3179.
[120] S.D. Waldman, et al., Multi-axial mechanical stimulation of tissue engineered cartilage: review, Eur. Cell. Mater. 13 (2007) 66–73 (discussion 73–74).
[121] O. Schatti, et al., A combination of shear and dynamic compression leads to mechanically induced chondrogenesis of human mesenchymal stem cells, Eur. Cell. Mater. 22 (2011) 214–225.
[122] A.H. Huang, et al., Sliding contact loading enhances the tensile properties of mesenchymal stem cell-seeded hydrogels, Eur. Cell. Mater. 24 (2012) 29–45.
[123] G. Vunjak-Novakovic, et al., Bioreactor cultivation conditions modulate the composition and mechanical properties of tissue-engineered cartilage, J. Orthop. Res. 17 (1) (1999) 130–138.
[124] G. Vunjak-Novakovic, et al., Bioreactor studies of native and tissue engineered cartilage, Biorheology 39 (1–2) (2002) 259–268.
[125] D. Pazzano, et al., Comparison of chondrogensis in static and perfused bioreactor culture, Biotechnol. Prog. 16 (5) (2000) 893–896.
[126] H. Ogawa, et al., Mechanical motion promotes expression of Prg4 in articular cartilage via multiple CREB-dependent, fluid flow shear stress-induced signaling pathways, Genes Dev. 28 (2) (2014) 127–139.
[127] L. Yu, et al., Tubular perfusion system for chondrocyte culture and superficial zone protein expression, J. Biomed. Mater. Res. A 103 (5) (2015) 1864–1874.
[128] S. Degala, et al., Calcium signaling in response to fluid flow by chondrocytes in 3D alginate culture, J. Orthop. Res. 30 (5) (2012) 793–799.
[129] A. Goncalves, et al., Effect of flow perfusion conditions in the chondrogenic differentiation of bone marrow stromal cells cultured onto starch based biodegradable scaffolds, Acta Biomater. 7 (4) (2011) 1644–1652.
[130] M.L. Alves da Silva, et al., Chondrogenic differentiation of human bone marrow mesenchymal stem cells in chitosan-based scaffolds using a flow-perfusion bioreactor, J. Tissue Eng. Regen. Med. 5 (9) (2011) 722–732.
[131] P.K. Valonen, et al., In vitro generation of mechanically functional cartilage grafts based on adult human stem cells and 3D-woven poly(epsilon-caprolactone) scaffolds, Biomaterials 31 (8) (2010) 2193–2200.
[132] L.M. Kock, et al., Flow-perfusion interferes with chondrogenic and hypertrophic matrix production by mesenchymal stem cells, J. Biomech. 47 (9) (2014) 2122–2129.
[133] J.C. Bernhard, et al., Perfusion enhances hypertrophic chondrocyte matrix deposition, but not the bone formation, Tissue Eng. Part A 24 (11 – 12) (2018) 1022–1033.
[134] S.J. Millward-Sadler, D.M. Salter, Integrin-dependent signal cascades in chondrocyte mechanotransduction, Ann. Biomed. Eng. 32 (3) (2004) 435–446.
[135] C. Matta, et al., Cytosolic free Ca2+ concentration exhibits a characteristic temporal pattern during in vitro cartilage differentiation: a possible regulatory role of calcineurin in Ca-signalling of chondrogenic cells, Cell Calcium 44 (3) (2008) 310–323.
[136] C. Matta, R. Zakany, Calcium signalling in chondrogenesis: implications for cartilage repair, Front. Biosci. (Schol. Ed.) 5 (2013) 305–324.
[137] R.E. Dolmetsch, et al., Differential activation of transcription factors induced by Ca2+ response amplitude and duration, Nature 386 (6627) (1997) 855–858.
[138] A. Argentaro, et al., A SOX9 defect of calmodulin-dependent nuclear import in campomelic dysplasia/autosomal sex reversal, J. Biol. Chem. 278 (36) (2003) 33839–33847.
[139] M.N. Phan, et al., Functional characterization of TRPV4 as an osmotically sensitive ion channel in porcine articular chondrocytes, Arthritis Rheum. 60 (10) (2009) 3028–3037.
[140] S. Muramatsu, et al., Functional gene screening system identified TRPV4 as a regulator of chondrogenic differentiation, J. Biol. Chem. 282 (44) (2007) 32158–32167.
[141] C.J. O'Conor, et al., TRPV4-mediated mechanotransduction regulates the metabolic response of chondrocytes to dynamic loading, Proc. Natl. Acad. Sci. U. S. A. 111 (4) (2014) 1316–1321.
[142] F. Guilak, H.A. Leddy, W. Liedtke, Transient receptor potential vanilloid 4: the sixth sense of the musculoskeletal system? Ann. N. Y. Acad. Sci. 1192 (2010) 404–409.
[143] S.V. Eleswarapu, K.A. Athanasiou, TRPV4 channel activation improves the tensile properties of self-assembled articular cartilage constructs, Acta Biomater. 9 (3) (2013) 5554–5561.

[144] W. Liang, et al., Periodic mechanical stress stimulates the FAK mitogenic signal in rat chondrocytes through ERK1/2 activity, Cell. Physiol. Biochem. 32 (4) (2013) 915–930.
[145] A.K. Wann, et al., Primary cilia mediate mechanotransduction through control of ATP-induced Ca2+ signaling in compressed chondrocytes, FASEB J. 26 (4) (2012) 1663–1671.
[146] W. Zhong, et al., Mesenchymal stem cell and chondrocyte fates in a multishear microdevice are regulated by yes-associated protein, Stem Cells Dev. 22 (14) (2013) 2083–2093.
[147] W. Zhong, et al., Regulation of fibrochondrogenesis of mesenchymal stem cells in an integrated microfluidic platform embedded with biomimetic nanofibrous scaffolds, PLoS One 8 (4) (2013) e61283.
[148] J. Zhang, et al., The influence of scaffold microstructure on chondrogenic differentiation of mesenchymal stem cells, Biomed. Mater. 9 (3) (2014) 035011.
[149] J. Zhang, et al., Cells behave distinctly within sponges and hydrogels due to differences of internal structure, Tissue Eng. Part A 19 (19–20) (2013) 2166–2175.
[150] D. Raghothaman, et al., Cell type dependent morphological adaptation in polyelectrolyte hydrogels governs chondrogenic fate, Biomed. Mater. 11 (2) (2016) 025013.
[151] L.A. McMahon, et al., Regulatory effects of mechanical strain on the chondrogenic differentiation of MSCs in a collagen-GAG scaffold: experimental and computational analysis, Ann. Biomed. Eng. 36 (2) (2008) 185–194.
[152] M.T. Raimondi, et al., Comparative chondrogenesis of human cells in a 3D integrated experimental-computational mechanobiology model, Biomech. Model. Mechanobiol. 10 (2) (2011) 259–268.
[153] S.J. Mousavi, M.H. Doweidar, M. Doblare, Computational modelling and analysis of mechanical conditions on cell locomotion and cell-cell interaction, Comput. Methods Biomech. Biomed. Engin. 17 (6) (2014) 678–693.
[154] S.J. Mousavi, M.H. Doweidar, Three-dimensional numerical model of cell morphology during migration in multi-signaling substrates, PLoS One 10 (3) (2015) e0122094.

CHAPTER

# 20

# Biomechanical Analysis of Bone Tissue After Insertion of Dental Implants Using Meshless Methods: Stress Analysis and Osseointegration

M.M.A. Peyroteo*,†, H.I.G. Gomes†, Jorge Belinha*,‡, Renato M. Natal Jorge†

*Institute of Science and Innovation in Mechanical and Industrial Engineering (INEGI), Porto, Portugal †Faculty of Engineering of University of Porto (FEUP), Porto, Portugal ‡School of Engineering of Polytechnic of (ISEP), Porto, Portugal

## 20.1 INTRODUCTION

Edentulism is the designation used for partial or total absence of teeth. Loss of teeth can be influenced by several factors, mainly related to oral health. Some examples include unhealthy eating habits, poor oral hygiene, smoking, and alcoholism. Consequently, several problems arise, such as functional limitations (i.e., mastication difficulties) or psychological and social stigmas [1]. Moreover a set of pathologies can be developed and ultimately cause edentulism. The major pathologies are dental caries and periodontitis, which affect about 3053 million and 743 million people worldwide in 2010, respectively [2]. Overall, edentulism affected about 158 million people worldwide in 2010, which corresponded to 2.3% of the world's population. In the future the socioeconomic impact of edentulism will tend to increase due to the aging of the population.

Nowadays, there are several solutions available, such as removable dental prosthesis, fixed partial dentures (i.e., dental bridges), and dental implants. Functional rehabilitation can be achieved with the placement of a removable dental prosthesis, since it is capable of returning the masticatory, phonetic, and aesthetic functions to a partially edentulous patient. However, this solution presents several drawbacks, such as difficulties and discomfort during mastication, damages to the remaining surrounding natural tissues and teeth, displacement of the teeth abutment, and bone loss in the edentulous regions [3, 4]. Fixed partial dentures are also not an ideal solution, since dental bridges are usually fixed to the teeth with dental cement, which over time deteriorates. Consequently a bacterial infection can affect the remaining teeth and increase the risk of tooth loss. The placement of dental implants is currently a valid treatment that has a high success rate [5]. Its use has increased exponentially in recent years, with >5 million implants inserted per year in the United States of America, which is equivalent to €800 billion [6]. A dental implant is a biocompatible device placed on the maxilla or mandible bone to provide support for a prosthetic reconstruction. It consists of four components—crown, screw, abutment, and implant. All these components require rigorous control of design and construction to ensure that mastication forces are properly transferred to the bone [7]. Different types of implants of distinct materials can be classified according to their size, shape, design, and surface of the threads. Testing implants with different characteristics is an effective approach to find the ideal combination of features that allows primary stability and osseointegration [8]. Immediately after implant's placement the implant should remain fixed to withstand the functional loads on teeth. This is accomplished through osseointegration in which the bone and the surface of the implant interact to form a direct bond. Branemark et al. [9–11] were the first to observe this phenomenon with the bone and a titanium implant. Osseointegration usually occurs in the periimplant region within the first 3–6 months after surgery. During this period the implant becomes increasingly more stable through bone remodeling. After a


https://doi.org/10.1016/B978-0-12-816390-0.00020-0

certain healing period, a state of equilibrium is achieved in which bone loss is minimal and implant failure rate is low. Ideally a dental implant should have a biocompatible chemical composition to avoid adverse tissue reactions, excellent corrosion resistance within the physiological limits, high wear resistance, and an elasticity modulus close to the bone. In this way, bone resorption around the implant would be minimized, and stress shielding would be avoided. In this work a standard dental implant of a titanium alloy with a conical thread is used [12].

Its impact on the trabecular morphology of the mandible is studied with the remodeling algorithm proposed in previous works [13, 14]. With this mechanical model the adaptation and remodeling of the trabecular structure around the implant are numerically simulated using two distinct numerical approaches—finite element method (FEM) and Natural Neighbor Radial Point Interpolation Method (NNRPIM). To study this adaptation process, two computational models are created. The following sections explain in detail the computational simulation conducted and the trabecular morphologies obtained.

## 20.2 COMPUTATIONAL MODEL

This section describes the construction of a computational model of a single dental implant inserted in a patch of mandible bone. Two distinct two-dimensional models are designed, and a set of loading conditions are applied to reproduce the physiological loading scenario of a tooth.

### 20.2.1 Single Dental Implant

The mandible was sectioned according to two distinct analysis planes—*Oxy* and *Oyz*—as schematically represented in Fig. 20.1. Model 1 is the geometric model obtained from *Oxy* plane, and model 2 is the one obtained from *Oyz* plane. Model 1 was discretized with an irregular triangular mesh of 3893 nodes and 7500 elements, while model 2 has 4157 nodes and 8013 elements. Both models were analyzed separately and are presented in Fig. 20.2. The materials considered in both models are a titanium alloy implant, a trabecular bone, and a thin layer of cortical bone (thickness of 1 mm). The dental implant used was a 4.1 × 12 mm ITI solid-screw implant [16]. All materials were considered linear elastic, isotropic, and homogeneous. Their mechanical properties are summarized in Table 20.1.

### 20.2.2 Boundary Conditions

Afterward, natural and essential boundary conditions are defined for each model. The boundary conditions of model 1 were based on the work of Meijer et al. [17]. Therefore three loads are applied to the implant system: a horizontal load, $F_h$, with a magnitude of 10 N; a vertical load, $F_v$, with a magnitude of 35 N; and an oblique load, $F_o$, with a magnitude of 70 N with 120 degrees of inclination with respect to the horizontal axis $x$. In this work a fourth loading condition, $F_t$, is included to simulate the effect of mandibular twisting [18]. In regard to the essential boundary conditions, model 1 is restricted on the base along $x$- and $y$-directions. In Fig. 20.3A the boundary conditions of model 1 are schematically represented.

Using these four load cases, two analyses are performed. Firstly, it is assumed that all loads have the same impact (equal number of load cycles). But in a second analysis the torsion load, $F_t$, is assumed as independent, while the three bite forces are weighted according to the number of cycles presented in Table 20.2. Knowing that, per day, three masticatory events, of 15 min each, take place with a chewing rate of 60 cycles/min (1 Hz), 2700 chewing cycles occur per day [17].

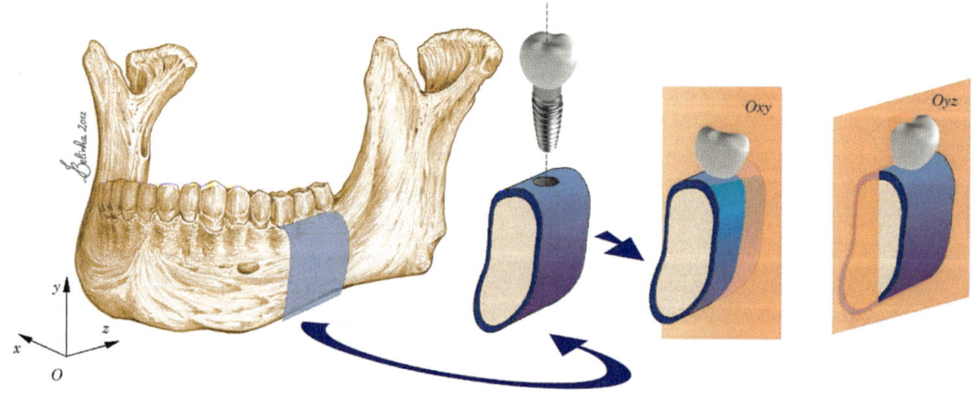


FIG. 20.1 Patch of mandible bone with a dental implant and its two analysis planes: *Oxy* and *Oyz* [15].

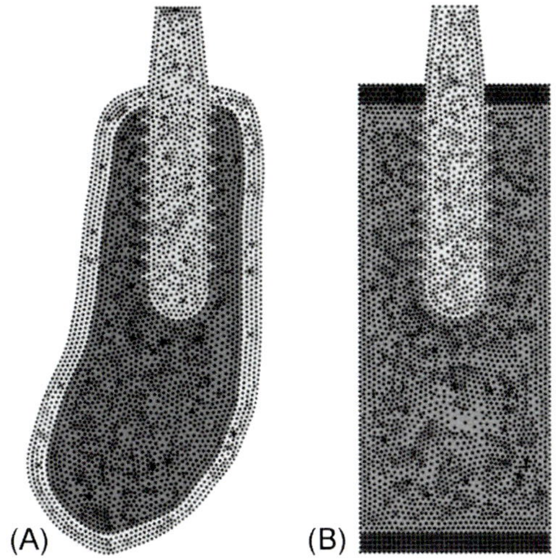


FIG. 20.2 Geometric model (A) 1 and (B) 2.

TABLE 20.1 Mechanical Properties of the Considered Materials

| Material | Elasticity modulus (*E*)(MPa) | Poisson's coefficient (*ν*) |
|---|---|---|
| Titanium alloy | 110,000 | 0.32 |
| Cortical bone | 13,700 | 0.3 |
| Trabecular bone | 1000 | 0.3 |

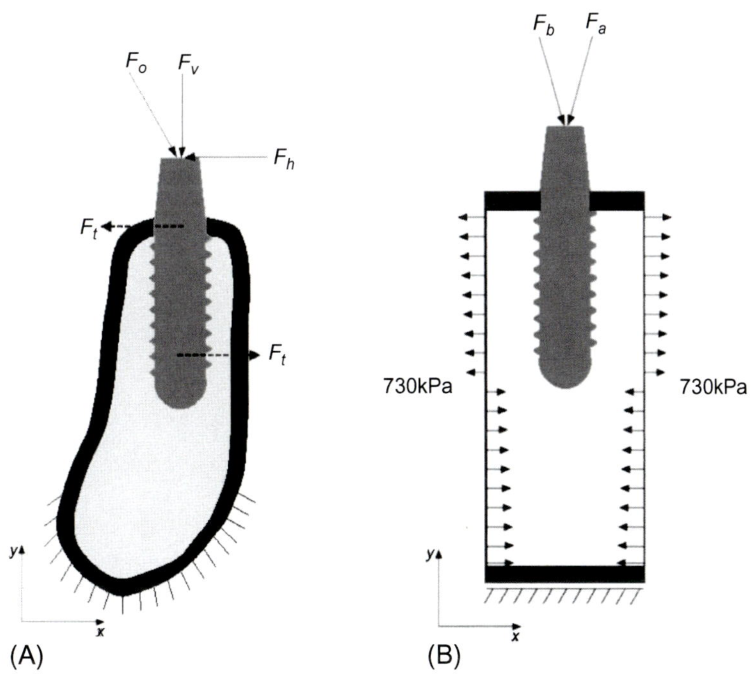


FIG. 20.3 Natural and essential boundary conditions applied on (A) model 1 and (B) model 2.

In model 2, two loads were simultaneously applied to the implant system. Both with a magnitude of 100 N, $F_a$ had an orientation of 79 degrees, and $F_b$ had a 101 degrees of orientation with respect to the horizontal axis. A distributed pressure along the vertical limits of the model was also applied to simulate the stress induced by mandibular flexion and the internal pressure of surrounding fluids. Model 2 was again restricted on the nodes at the base along $x$- and $y$-directions. The boundary conditions of model 2 are schematically presented in Fig. 20.3B.

TABLE 20.2 Specification of the Number of Load Cycles for Each Load Case Tested

| | Number of load cycles | | | |
|---|---|---|---|---|
| **Load cases** | $F_h$ | $F_v$ | $F_o$ | **Total** |
| LC1 | 540 | 540 | 1620 | 2700 |
| LC2 | 540 | 1620 | 540 | 2700 |
| LC3 | 1620 | 540 | 540 | 2700 |
| LC4 | 900 | 900 | 900 | 2700 |

# 20.3 ALGORITHM DESCRIPTION

## 20.3.1 Numerical Discretization

The algorithm starts with a discretization of the geometric domain under analysis. Although nodal distribution is directly obtained from medical images, nodal connectivity has to be imposed according to the used numerical technique. In FEM, nodes are connected using elements. Therefore, for the same nodal distribution, different nodal connectivities can be obtained, depending on the type of the used element. For this work an irregular triangular mesh was used. However, NNRPIM uses a distinct approach since it is a meshless method. From the nodal distribution a Voronoï diagram is constructed, dividing the spatial domain into several Voronoï cells. As presented in Fig. 20.4A, each node possesses its respective Voronoï cell. Then, nodal connectivity is imposed using either first-order or second-order influence cells. For this work, second-order influence cells are used, which means that a certain node, $n_i$, is connected with its first and second neighbor cells, as depicted in Fig. 20.4B. This meshless approach is more organic since nodal connectivity can change during the simulation, while FEM's approach preestablishes a constant connectivity.

Afterward, since the problem will not be analyzed at the nodes but at the integration points the next step is to define their spatial localization. Following the quadrature scheme of Gauss-Legendre [19], FEM sets one integration point inside each element. NNRPIM first divides each Voronoï cell into several quadrilateral subcells and then imposes one integration point inside each cell using the Gauss-Legendre quadrature scheme. This process is schematically represented in Fig. 20.4C.

Lastly a set of shape functions has also to be defined. Since triangular elements are used, FEM's shape functions are isoparametric interpolation functions. The shape function of NNRPIM is constructed using the radial point interpolation technique, which combines a polynomial basis with a radial basis function. Additional information regarding shape functions can be found in the literature [19, 20].

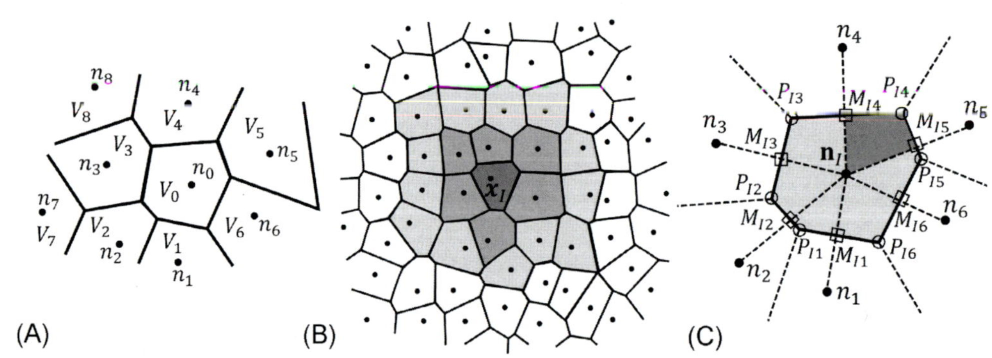


FIG. 20.4 NNRPIM's formulation: (A) Voronoï diagram; (B) second-order influence cells; (C) Voronoï cell division into quadrilateral integration subcells.

### 20.3.2 Mechanical Analysis

For each time instant, $t_j$, a mechanical analysis is performed, considering each load case, $k$, separately and sequentially. So, for each load case, $k$, and integration point, $\boldsymbol{x}_I$, a local stiffness matrix, $\boldsymbol{K}_I$, is constructed using the following expression:

$$\boldsymbol{K}_I = \int_{\Omega_I} \boldsymbol{B}_I^T \, \boldsymbol{c}_I \, \boldsymbol{B}_I d\Omega_I = \hat{w}_I \boldsymbol{B}_I^T \, \boldsymbol{c}_I \, \boldsymbol{B}_I \tag{20.1}$$

in which $\boldsymbol{B}_I$ is the deformation matrix and $\boldsymbol{c}_I$ is the material constitutive matrix, both for integration point, $\boldsymbol{x}_I$. The physical volume occupied by integration point, $\boldsymbol{x}_I$, is denoted by $\Omega_I$ and its numerical representation by $\hat{w}_I$. By assembling all local stiffness matrices, $\boldsymbol{K}_I$, into a single one, the global stiffness matrix, $\boldsymbol{K}$, is constructed. Then the essential boundary conditions for each load case, $k$, are imposed in $\boldsymbol{K}$. Consequently, knowing $\boldsymbol{K}$ and the force vector, $\boldsymbol{f}^k$, for the respective load case, $k$, the equation system $\boldsymbol{K}^k \, \boldsymbol{u}^k = \boldsymbol{f}^k$ can be solved, determining the displacement field, $\boldsymbol{u}^k$. Then, using Hooke's law [13], the strain, $\boldsymbol{\varepsilon}^k$, and stress fields, $\boldsymbol{\sigma}^k$, are obtained. These two fields allow the calculus of the principal stresses field, $\boldsymbol{\sigma}(\boldsymbol{x}_I)_i^k$, with the following:

$$\det\left(\begin{bmatrix} \sigma_{xx}(\boldsymbol{x}_I)^k & \sigma_{xy}(\boldsymbol{x}_I)^k \\ \sigma_{xy}(\boldsymbol{x}_I)^k & \sigma_{yy}(\boldsymbol{x}_I)^k \end{bmatrix} - \sigma(\boldsymbol{x}_I)_i^k \begin{bmatrix} 1 & 0 \\ 0 & 1 \end{bmatrix}\right) = 0 \tag{20.2}$$

and principal directions field, $\boldsymbol{n}(\boldsymbol{x}_I)_i^k$, with

$$\left(\begin{bmatrix} \sigma_{xx}(\boldsymbol{x}_I)^k & \sigma_{xy}(\boldsymbol{x}_I)^k \\ \sigma_{xy}(\boldsymbol{x}_I)^k & \sigma_{yy}(\boldsymbol{x}_I)^k \end{bmatrix} - \sigma(\boldsymbol{x}_I)_i^k \begin{bmatrix} 1 & 0 \\ 0 & 1 \end{bmatrix}\right) \begin{Bmatrix} n_x(\boldsymbol{x}_I)_i^k \\ n_y(\boldsymbol{x}_I)_i^k \end{Bmatrix} = 0 \tag{20.3}$$

Moreover the von Mises effective stress field, $\overline{\sigma}(\boldsymbol{x}_I)^k$, is also determined by the following expression:

$$\overline{\sigma}(\boldsymbol{x}_I)^k = \sqrt{\frac{1}{2}\left(\left(\sigma(\boldsymbol{x}_I)_1^k - \sigma(\boldsymbol{x}_I)_2^k\right)^2 + \left(\sigma(\boldsymbol{x}_I)_2^k - \sigma(\boldsymbol{x}_I)_3^k\right)^2 + \left(\sigma(\boldsymbol{x}_I)_3^k - \sigma(\boldsymbol{x}_I)_1^k\right)^2\right)} \tag{20.4}$$

At the end of this analysis, the obtained variable fields, $\lambda = \{\boldsymbol{u}_j^k, \boldsymbol{\varepsilon}_j^k, \boldsymbol{\sigma}_j^k, \boldsymbol{\sigma}(\boldsymbol{n})_j^k, \boldsymbol{n}_j^k\}$, are weighted at each time instant, $t_j$, and load case, $k$, using the following expression:

$$\lambda = \sum_{k=1}^{n_k} \frac{d_k \lambda^k}{\sum_{k=1}^{n_k} d_k} \tag{20.5}$$

So the variable field, $\lambda$, is weighted according to the number of load cases, $n_k$, and the corresponding number of load cycles, $d_k$, that corresponds to the average number of times a certain load occurs daily.

### 20.3.3 Bone Remodeling

The premise of this algorithm is that bone gradually changes its apparent density during bone remodeling. However, to ensure a continuous and progressive process, only a portion of the integration points will undergo remodeling and consequently update its apparent density. These remodeling points are chosen according to the used remodeling criterion. In this work, principal stress, $\sigma_{11}$, is the remodeling criterion used to divide the integration points into three distinct groups—resorption, formation, and lazy groups. The resorption group, $R(\boldsymbol{x}_I)$, is given by.

$$R(\boldsymbol{x}_I) \in [\sigma_{11\,min}, \sigma_{11\,min} + \alpha \cdot \triangle\sigma_{11}[, \quad \forall R(\boldsymbol{x}_I) \in \mathbb{R} \tag{20.6}$$

and the formation group, $F(\boldsymbol{x}_I)$, by

$$F(\boldsymbol{x}_I) \in ]\sigma_{11\,max} - \beta \cdot \triangle\sigma_{11}, \sigma_{11\,max}], \quad \forall F(\boldsymbol{x}_I) \in \mathbb{R} \tag{20.7}$$

being $\sigma_{11min}=\min(\sigma_{11})$, $\sigma_{11max}=\max(\sigma_{11})$, and $\triangle\sigma_{11}=\sigma_{11max}-\sigma_{11min}$. In Eqs. (20.6), (20.7), it is shown that the resorption group includes the integration points with the lowest values of $\sigma_{11}$ and the formation group has the integration points with the highest values of $\sigma_{11}$. Only the integration points belonging to $R(\boldsymbol{x}_I)$ or $F(\boldsymbol{x}_I)$ will have a new apparent density, being parameters $\alpha$ and $\beta$ the decay and growth rate of apparent density, respectively. The remaining integration points are included in the lazy group in which their apparent density will remain the same during that iteration step.

The next step is to update the apparent density using the phenomenological law proposed by Belinha et al. [21]. This law classifies bone architecture according to the apparent density, $\rho$, considering trabecular bone when $\rho \leq 1.3\ \text{g/cm}^3$ and cortical bone when $\rho > 1.3\ \text{g/cm}^3$. Based on experimental works performed by Zioupos et al. [22], the following mathematical expressions were proposed by Belinha et al. [21]:

$$\overline{\sigma}=\sum_{j=0}^{3} a_j \cdot \rho^j \tag{20.8}$$

$$E=\begin{cases} \sum_{j=0}^{3} b_j \cdot \rho^j & \text{if} \quad \rho \leq 1.3\,\text{g/cm}^3 \\ \sum_{j=0}^{3} c_j \cdot \rho^j & \text{if} \quad \rho > 1.3\,\text{g/cm}^3 \end{cases} \tag{20.9}$$

The von Mises effective stress, $\overline{\sigma}$, and the elasticity modulus, $E$, are expressed in MPa, whereas apparent density, $\rho$, is expressed in $\text{g/cm}^3$. The values of coefficients present in Eqs. (20.8), (20.9) are presented in Table 20.3.

The remodeling points will update their apparent density, $\rho$, using Eq. (20.8), since the von Mises effective stress field, $\overline{\sigma}$, has been already determined during a mechanical analysis. Then the elasticity modulus, $E$, of these points will also be updated, since they possess a new apparent density, $\rho$, using Eq. (20.9). The remodeling process is then reproduced with this phenomenological law in which the apparent density and the mechanical properties of the remodeling points are updated at each iteration.

At the end of each iteration, the mean apparent density of the model, $\rho_{med}$, is determined with

$$\rho_{med}=\frac{1}{Q}\sum\nolimits_{I=1}^{Q}\rho(\boldsymbol{x}_I) \tag{20.10}$$

being $Q$ the total number of integration points and $\rho(\boldsymbol{x}_I)$ the apparent density at the integration point $\boldsymbol{x}_I$.

Thereafter the process progresses to the next iteration step, $t_{j+1}$, performing again a mechanical analysis. Remodeling ends when $\rho_{med}$ reaches a value determined by the user or when two consecutive iteration steps have the same $\rho_{med}$ (i.e., $\Delta\rho/\Delta t=0$).

## 20.4 COMPUTATIONAL ANALYSIS OF BONE REMODELING

### 20.4.1 Model 1

The four loading case scenario of model 1 are analyzed with two distinct approaches. The results, obtained when assuming the same impact for all four load cases, are presented in Figs. 20.5 and 20.6, containing the isomaps of the final trabecular architecture, von Mises effective stress, and principal stresses obtained with FEM and NNRPIM. Analyzing the final trabecular architecture, it is possible to depict that, below the implant, the algorithm predicts high apparent density regions oriented horizontally, connecting the opposite cortical layers. However, in the region, just

TABLE 20.3 Coefficient's Values of Belinha's Law

| Coefficient | $j=0$ | $j=1$ | $j=2$ | $j=3$ |
|---|---|---|---|---|
| $a_j$ | 0.000E+00 | 0.000E+00 | 2.680E+01 | 2.035E+01 |
| $b_j$ | 0.000E+00 | 7.216E+02 | 8.059E+02 | 0.000E+00 |
| $c_j$ | −1.770E+05 | 3.861E+05 | −2.798E+05 | 6.836E+04 |

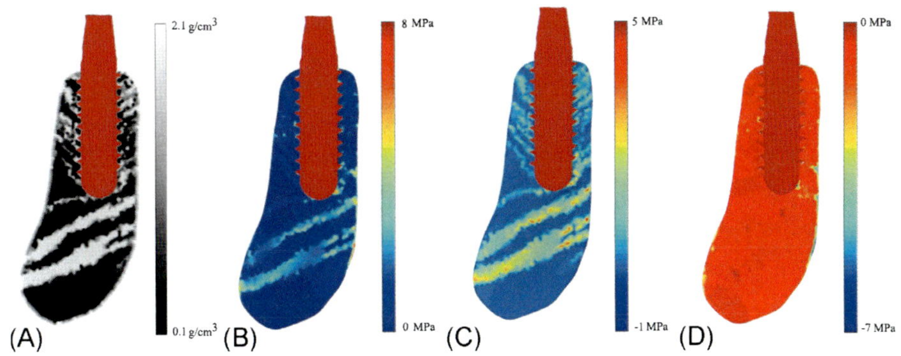


FIG. 20.5 FEM solution using model 1: final isomaps of (A) apparent density, $\rho$; (B) von Mises effective stress, $\overline{\sigma}$; and principal stresses (C) $\sigma_{11}$ and (D) $\sigma_{22}$.

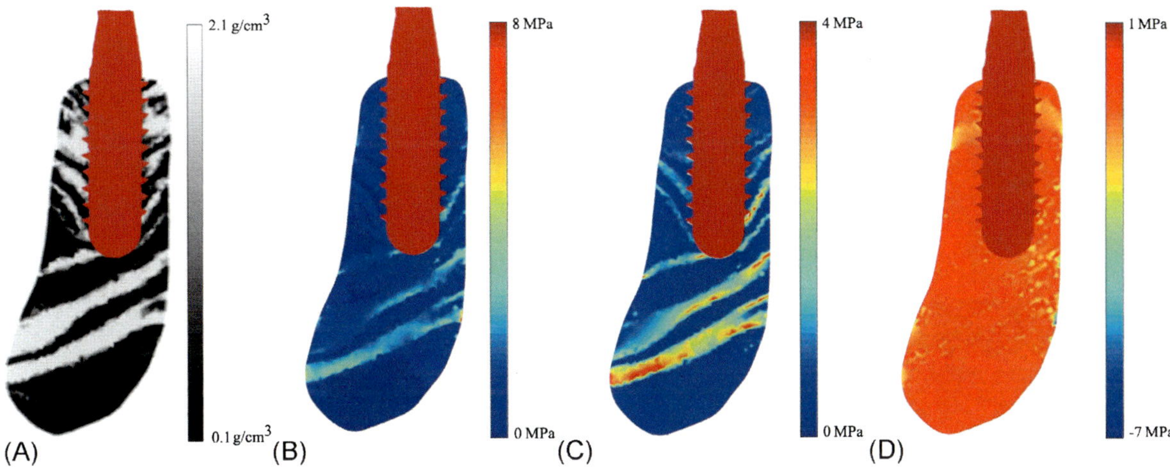


FIG. 20.6 NNRPIM solution using model 1: final isomaps of (A) apparent density, $\rho$; (B) von Mises effective stress, $\overline{\sigma}$; and principal stresses (C) $\sigma_{11}$ and (D) $\sigma_{22}$.

below the implant (i.e., apical region), a more intensive bone resorption has occurred. The isomaps of principal stresses $\sigma_{11}$ and $\sigma_{22}$ show that the model is subjected to higher compression than tensile stresses, since the applied loads are only compressive loads.

Fig. 20.7A presents the intersection between FEM and NNRPIM solutions. White nodes mean that both methods predicted the existence of bone, whereas black nodes depict incoherent predictions between both methods. After this spatial intersection the main trabecular structures of the mandibular bone remained present, confirming a good agreement between both numerical techniques.

Moreover, Fig. 20.7B presents the real trabecular distribution of a baboon's mandible with a dental implant [23]. These clinical results and the predictions obtained in this work have two common regions: trabecular morphology surrounding the implant in region A and the low apparent density area in the apical region of the implant marked by region B in Fig. 20.7B.

The second analysis with model 1, when each load has distinct impacts, is presented in Figs. 20.8 and 20.9. Here, it is possible to depict that LC1 is the one that leads to trabecular distributions more similar to clinical data (Fig. 20.7B). Therefore the results suggest that the oblique force, $F_o$, has a greater contribution during bone remodeling, since LC1 considers a greater impact (greater number of load cycles) for $F_o$.

## 20.4.2 Model 2

The bone remodeling simulation predicted with model 2 was very similar with the results obtained with model 1. The isomaps of the final trabecular architecture, von Mises effective stress, and principal stresses of model 2 are presented in Figs. 20.10 and 20.11. Trabecular distributions predict again bone resorption immediately below the implant

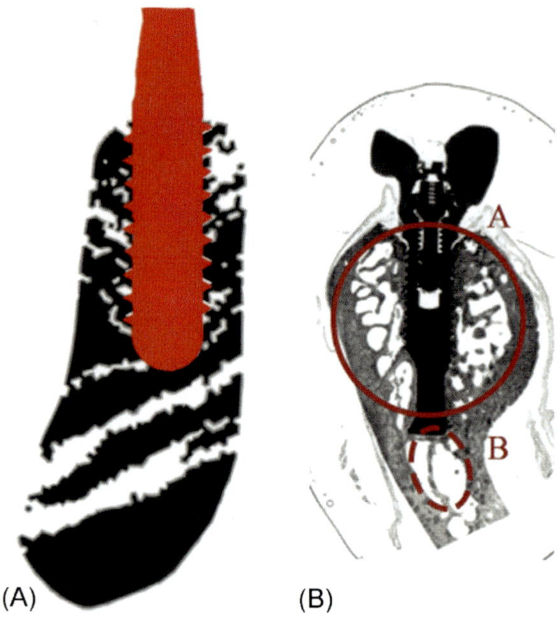


FIG. 20.7 (A) Intersection of FEM and NNRPIM solutions using model 1; (B) trabecular morphology of a baboon's mandible after insertion of a dental implant.

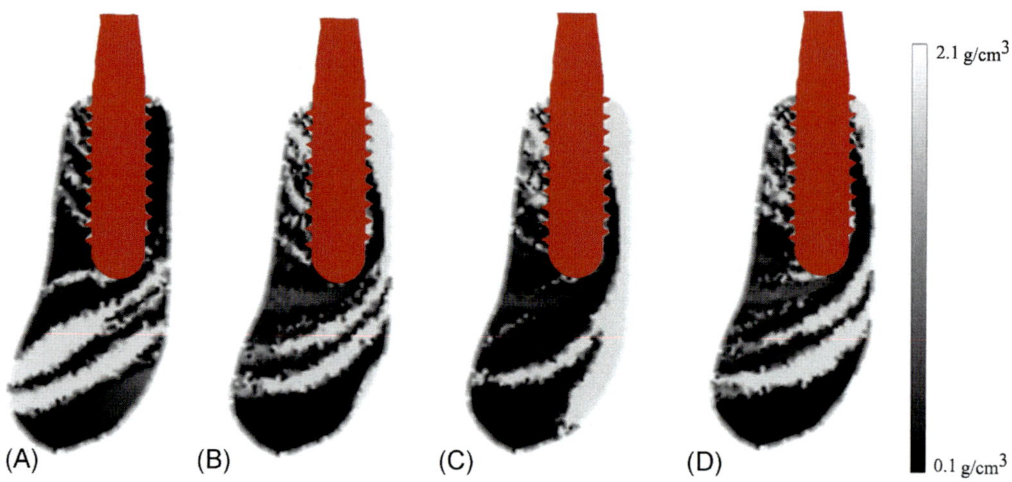


FIG. 20.8 FEM solution: trabecular morphology obtained with distinct loading cases—(A) LC1, (B) LC2, (C) LC3, and (D) LC4.

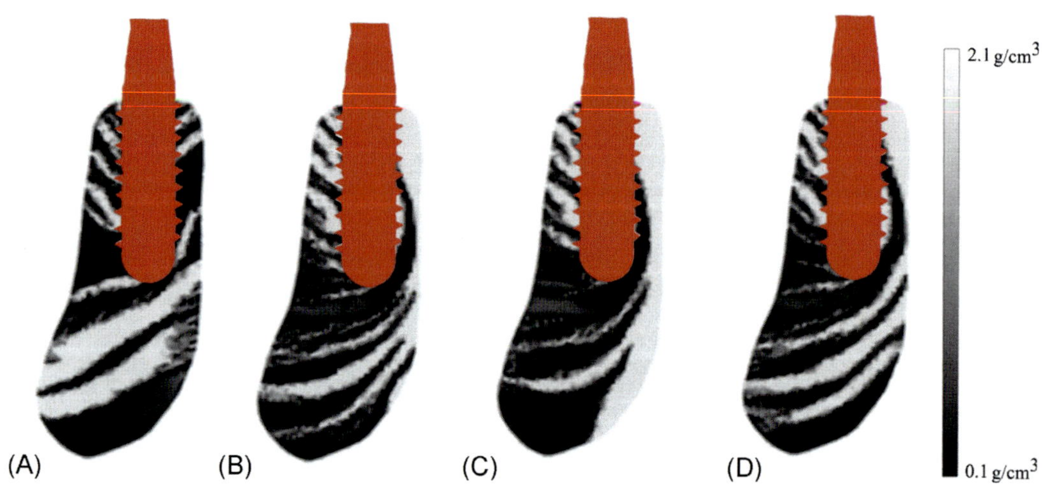


FIG. 20.9 NNRPIM solution: trabecular morphology obtained with distinct loading cases—(A) LC1, (B) LC2, (C) LC3, and (D) LC4.

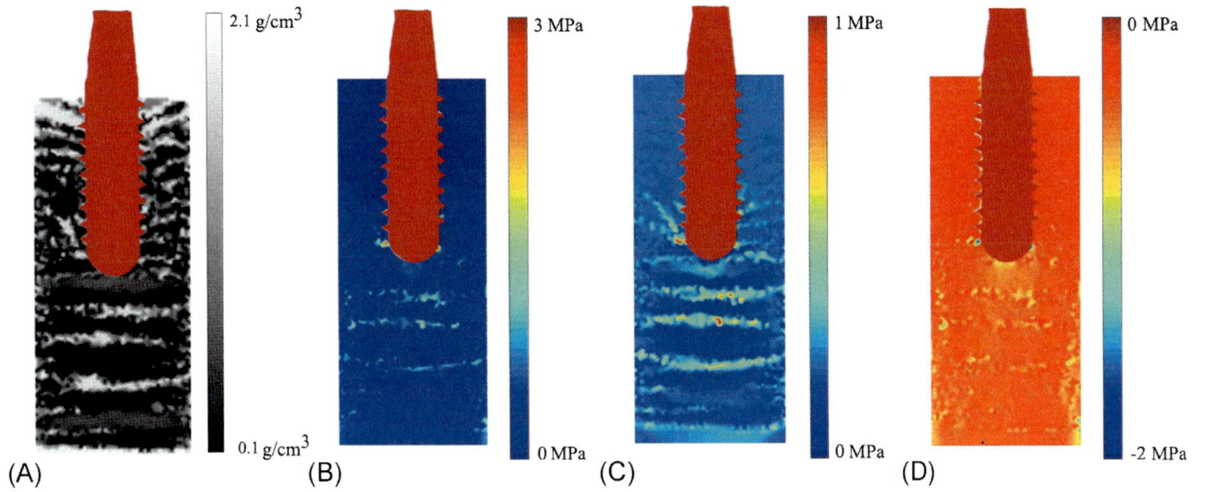


FIG. 20.10 FEM solution using model 2: final isomaps of (A) apparent density, $\rho$; (B) von Mises effective stress, $\overline{\sigma}$; and principal stresses (C) $\sigma_{11}$ and (D) $\sigma_{22}$.

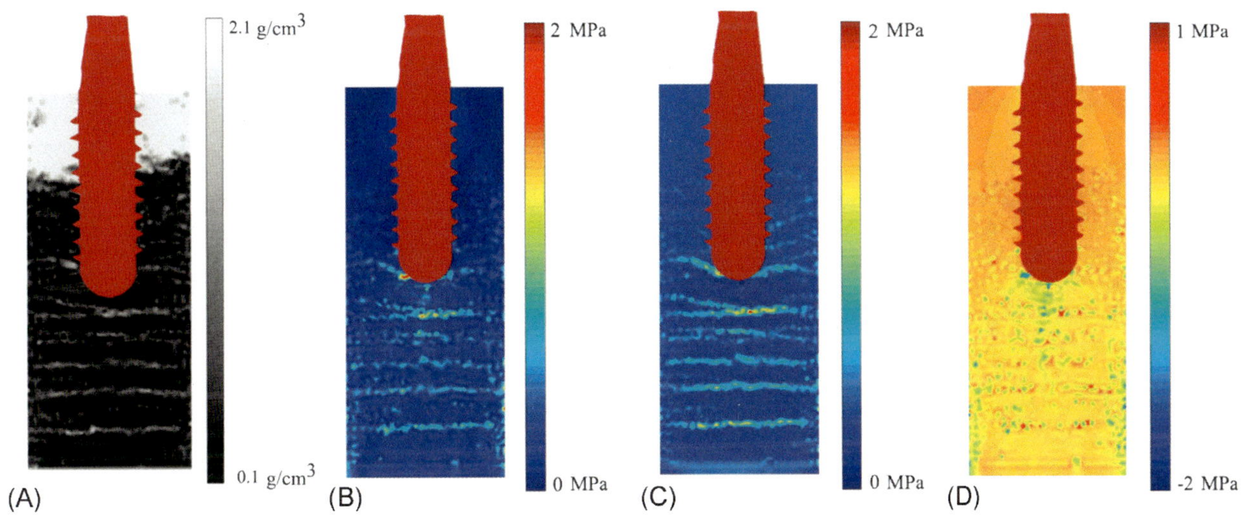


FIG. 20.11 NNRPIM solution using model 2: final isomaps of (A) apparent density, $\rho$; (B) von Mises effective stress, $\overline{\sigma}$; and principal stresses (C) $\sigma_{11}$ and (D) $\sigma_{22}$.

and well-defined horizontal trabeculae. However, NNRPIM's solution, due to its meshless formulation, produces smoother results when compared with FEM.

Once again an intersection of the bone apparent density distribution map of both numerical solutions was constructed. As presented in Fig. 20.12, solutions are coherent with each other preserving the main trabecular structures of the mandibular bone.

## 20.5 CONCLUSIONS

Studying and predicting bone remodeling of the mandible, after insertion of a dental implant, are an important approach to extend our knowledge about implant's characteristics and develop strategies to increase its integration rate. The survival and effectiveness of the implant were studied in this work through an adaptive bone remodeling algorithm, which includes a phenomenological law capable of correlating the apparent density of bone tissue with its mechanical properties. This algorithm was combined with two different numerical techniques (i.e., FEM and NNRPIM). For both methods the algorithm was able to accurately predict the main trabecular structures of the mandible. However, due to its meshless formulation, NNRPIM's stress maps were more accurate, leading to smoother trabecular distributions when compared with FEM.

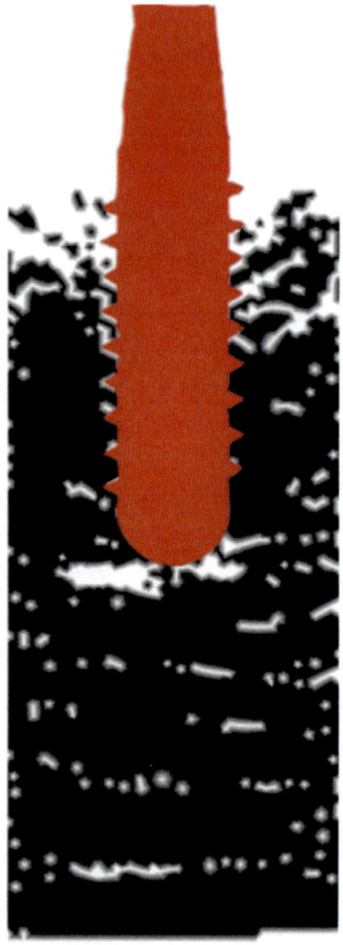

FIG. 20.12 Intersection of FEM and NNRPIM solutions using model 2.

The model considers only the influence of a mechanical stimulus. Thus, in the future, it would be interesting to use an algorithm that would simultaneously combine bone cell dynamics with distinct loading scenarios.

## Acknowledgments

The authors truly acknowledge the funding provided by Ministério da Ciência, Tecnologia e Ensino Superior, Fundação para a Ciência e a Tecnologia (Portugal), under grants SFRH/BD/133105/2017 and by project funding MIT-EXPL/ISF/0084/2017. Additionally the authors gratefully acknowledge the funding of Project NORTE-01-0145-FEDER-000022, SciTech, Science and Technology for Competitive and Sustainable Industries, cofinanced by Programa Operacional Regional do Norte (NORTE2020), through Fundo Europeu de Desenvolvimento Regional (FEDER).

## References

[1] E. Emami, R.F. de Souza, M. Kabawat, J.S. Feine, The impact of edentulism on oral and general health, Int. J. Dent. 2013 (2013) 1–7, https://doi.org/10.1155/2013/498305.

[2] T. Vos, A.D. Flaxman, M. Naghavi, R. Lozano, et al., Years lived with disability (YLDs) for 1160 sequelae of 289 diseases and injuries 1990–2010: a systematic analysis for the global burden of disease study 2010, Lancet 380 (9859) (2012) 2163–2196, https://doi.org/10.1016/S0140-6736(12)61729-2.

[3] L. Rissin, J.E. House, C. Conway, E.R. Loftus, H.H. Chauncey, Effect of age and removable partial dentures on gingivitis and periodontal disease, J. Prosthet. Dent. 42 (2) (1979) 217–223, https://doi.org/10.1016/0022-3913(79)90178-1.

[4] M. Saito, K. Notani, Y. Miura, T. Kawasaki, Complications and failures in removable partial dentures: a clinical evaluation, J. Oral Rehabil. 29 (7) (2002) 627–633, https://doi.org/10.1046/j.1365-2842.2002.00898.x.

[5] J. Zupnik, S. Kim, D. Ravens, N. Karimbux, K. Guze, Factors associated with dental implant survival: a 4-year retrospective analysis, J. Periodontol. 82 (10) (2011) 1390–1395, https://doi.org/10.1902/jop.2011.100685.

[6] S. Jivraj, W. Chee, Rationale for dental implants, Br. Dent. J. 200 (12) (2006) 661–665, https://doi.org/10.1038/sj.bdj.4813718.

[7] S. Raikar, P. Talukdar, S. Kumari, S.K. Panda, V.M. Oommen, A. Prasad, Factors affecting the survival rate of dental implants: a retrospective study, J. Int. Soc. Prev. Commun. Dent. 7 (6) (2017) 351–355, https://doi.org/10.4103/jispcd.JISPCD_380_17.

[8] F.W. Neukam, T.F. Flemmig, Working Group 3, Local and systemic conditions potentially compromising osseointegration, Clin. Oral Implants Res. 17 (S2) (2006) 160–162, https://doi.org/10.1111/j.1600-0501.2006.01359.x.
[9] R. Adell, U. Lekholm, B. Rockler, P.I. Brånemark, A 15-year study of osseointegrated implants in the treatment of the edentulous jaw, Int. J. Oral Surg. 10 (6) (1981) 387–416, https://doi.org/10.1016/S0300-9785(81)80077-4.
[10] P.I. Branemark, Osseointegration and its experimental background, J. Prosthet. Dent. 50 (3) (1983) 399–410, https://doi.org/10.1016/S0022-3913(83)80101-2.
[11] T. Albrektsson, T. Jansson, U. Lekholm, Osseointegrated dental implants, Dent. Clin. N. Am. 30 (1) (1986) 151–174.
[12] S.S. Al-Johany, M.D. Al Amri, S. Alsaeed, B. Alalola, Dental implant length and diameter: a proposed classification scheme, J. Prosthodont. 26 (3) (2017) 252–260, https://doi.org/10.1111/jopr.12517.
[13] J. Belinha, Meshless Methods in Biomechanics—Bone Tissue Remodelling Analysis, Springer International Publishing, Switzerland, 2014.
[14] M.M.A. Peyroteo, J. Belinha, S. Vinga, L.M.J.S. Dinis, R.M. Natal Jorge, Mechanical bone remodelling: comparative study of distinct numerical approaches, Eng. Anal. Bound. Elem. (2018), https://doi.org/10.1016/J.ENGANABOUND.2018.01.011.
[15] J. Belinha, L.M.J.S. Dinis, R.M. Natal Jorge, The mandible remodeling induced by dental implants: a meshless approach, J. Mech. Med. Biol. 15 (04) (2015) 1550059, https://doi.org/10.1142/S0219519415500591.
[16] O. Kayabaşı, E. Yüzbasıoğlu, F. Erzincanlı, Static, dynamic and fatigue behaviors of dental implant using finite element method, Adv. Eng. Softw. 37 (10) (2006) 649–658, https://doi.org/10.1016/J.ADVENGSOFT.2006.02.004.
[17] H.J. Meijer, F.J. Starmans, W.H. Steen, F. Bosman, A three-dimensional, finite-element analysis of bone around dental implants in an edentulous human mandible, Arch. Oral Biol. 38 (6) (1993) 491–496, https://doi.org/10.1016/0003-9969(93)90185-O.
[18] H.-Y. Chou, J.J. Jagodnik, S. Müftü, Predictions of bone remodeling around dental implant systems, J. Biomech. 41 (6) (2008) 1365–1373, https://doi.org/10.1016/j.jbiomech.2008.01.032.
[19] K.-J. Bathe, Finite Element Procedures, Prentice Hall, Pearson Education, Inc., New Jersey, 1996.
[20] J.G. Wang, G.R. Liu, A point interpolation meshless method based on radial basis functions, Int. J. Numer. Methods Eng. 54 (11) (2002) 1623–1648, https://doi.org/10.1002/nme.489.
[21] J. Belinha, R.M.N. Jorge, L.M.J.S. Dinis, A meshless microscale bone tissue trabecular remodelling analysis considering a new anisotropic bone tissue material law, Comput. Methods Biomech. Biomed. Engin. 16 (11) (2013) 1170–1184, https://doi.org/10.1080/10255842.2012.654783.
[22] P. Zioupos, R.B. Cook, J.R. Hutchinson, Some basic relationships between density values in cancellous and cortical bone, J. Biomech. 41 (9) (2008) 1961–1968, https://doi.org/10.1016/j.jbiomech.2008.03.025.
[23] G. Watzak, W. Zechner, C. Ulm, S. Tangl, G. Tepper, G. Watzek, Histologic and histomorphometric analysis of three types of dental implants following 18 months of occlusal loading: a preliminary study in baboons, Clin. Oral Implants Res. 16 (4) (2005) 408–416, https://doi.org/10.1111/j.1600-0501.2005.01155.x.

CHAPTER

# 21

# Numerical Assessment of Bone Tissue Remodeling of a Proximal Femur After Insertion of a Femoral Implant Using an Interpolating Meshless Method

*M.M.A. Peyroteo**,†, A.T.A. Castro†, Jorge Belinha*,‡, Renato M. Natal Jorge†*

*Institute of Science and Innovation in Mechanical and Industrial Engineering (INEGI), Porto, Portugal †Faculty of Engineering of University of Porto (FEUP), Porto, Portugal ‡School of Engineering Polytechnic of Porto (ISEP), Porto, Portugal

## 21.1 INTRODUCTION

Total hip arthroplasty (THA) is considered the "surgery of the century," being one of the most successful surgical replacement procedures in orthopedics. In THA the natural hip joint is replaced with an artificial prosthesis, aiming to restore the joint function and consequently increase the quality of life of the patient. A variety of pathologies can lead to the necessity for hip joint replacement, being the three main diseases primary osteoarthritis (76%); rheumatoid arthritis (6%); and conditions that can evolve to fracture (11%), such as osteoporosis [1].

To take on the missing physiological functions, a hip prosthesis must assure three important requirements [2]. The set of materials constituting the prosthesis must have adequate mechanical properties to withstand millions of charge cycles without fracturing. Besides this structural requirement, artificial joint must not compromise the movement of the musculoskeletal system and withstand the corrosive physiological bone environment.

There are two types of arthroplasty: primary arthroplasty and revision arthroplasty. Primary arthroplasty is the replacement of the natural hip joint with an artificial one. However, when the artificial joint fails, a second surgical procedure is required. The revision arthroplasty is then the removal of the original implant and the replacement with a new implant. Revision arthroplasties are often more difficult to perform compared with primary arthroplasties, since the patient has less bone tissue volume.

In the United States of America, between 2003 and 2013, primary and revision THAs increased up to 174%–130%, respectively [3]. Moreover a recent study predicted that, by 2030, the number of primary and revision THAs will increase, causing an increase in costs from $8.43 billion in 2003 to $22.7 billion in 2030 [4]. The aging of the population and the reduced useful lifetime of the prosthesis components are some of the factors that explain these statistics [2].

Although the use of implants has helped many patients to restore their joint function, complications can occur, since integration of a foreign body into a highly corrosive physiological environment can be exceptionally challenging. Therefore an early revision arthroplasty is required when some factors such as infection, displacement, stress shielding, or aseptic loosening occur [5, 6]. Implant failure due to stress shielding is explained by a severe bone resorption and consequently weakening of bone as a result of decreased physiological loading of the bone. The release of debris from wear of the prosthesis materials is the main cause of aseptic loosening. This process induces again bone resorption and, subsequently, detachment of the implant. The main outcomes of implant failure are fracture or total displacement of the implant.


https://doi.org/10.1016/B978-0-12-816390-0.00021-2

Excessive bone resorption is a common factor in implant failure, which reflects an unbalanced process of bone remodeling after placement of the prosthesis. Thus this work proposes a bone remodeling study after placement of a femoral implant. Bone remodeling is a synchronized process of the removal of damaged bone and formation of new bone tissue. Through this process, bone is able to adapt to different internal and external stimuli, such as biochemical disturbances or new loading conditions. After THA the implant becomes the receiving structure of functional loads, and so the remaining bone has to functionally adapt to this new condition and guarantee osseointegration through bone remodeling. However, different implants promote different osseointegration processes, depending on their material, geometry, or surface texture [1]. A large set of different implants are currently available on the market. The most widely used materials are cobalt-chromium alloys and titanium alloys; however, different materials have emerged, combining polymer properties with metal alloys or even combining these metal alloys with carbon fibers.

The ideal combination remains a subject of debate. Many simulation studies analyzed the influence of different geometries and materials, concluding that the ideal choice should be studied for each specific patient.

This work uses the M.E. Müller Straight Stem prosthesis of titanium alloy and applies the in silico remodeling model initially proposed by Belinha et al. [7] and extended by Peyroteo et al. [8]. The application of this mathematical model has been validated in the literature, studying bone remodeling of the femur [9], the calcaneus [9], and natural teeth and dental implants [10]. Moreover, this remodeling model has been successfully combined with different numerical methods, namely, the finite element method (FEM), the radial point interpolation method (RPIM), and the natural neighbor radial point interpolation method (NNRPIM). Thus, in the following sections, a complete description of the remodeling model is included. Also, using a real case of THA as a numerical example, simulation results using FEM and RPIM are analyzed and compared with the X-ray images captured after the insertion of the femoral implant.

## 21.2 BONE REMODELING MODEL

The remodeling algorithm applied in this work is a meshless adaptation of Carter's model [11–13]. For an easy understanding of its functioning, the algorithm was divided in four steps—preprocessing, mechanical analysis, remodeling points, and phenomenological law. Each of these steps is explained in the following sections.

### 21.2.1 Preprocessing

Using X-ray images the problem domain is segmented and discretized with a nodal mesh. Then, nodal connectivity is imposed according to the numerical technique used and the integration mesh. For both numerical techniques a triangular background integration mesh is used, defining the spatial localization of the integration points with the Gauss-Legendre quadrature Scheme [9, 14]. However, the nodal connectivity differs from FEM and RPIM. FEM uses elements, while RPIM considers influence domains. Afterward the shape functions for each integration point, $x_I$, are constructed, $\varphi(x_I)$. Lastly the boundary conditions and the mechanical properties of the different materials of the computational model are imposed.

### 21.2.2 Mechanical Analysis

Then an iterative process takes place, beginning with a mechanical analysis performed at each instant $t_j$. For each iterative step, $t_j$, each load case, $k$, $(k=1, 2, \ldots, l)$ is analyzed sequentially and separately. The local stiffness matrix, $\boldsymbol{K}_I$, for each integration point, $\boldsymbol{x}_I$, is defined, using the deformation matrix, $\boldsymbol{B}_I$, and the constitutive material matrix, $\boldsymbol{c}_I$, as the following:

$$\boldsymbol{K}_I = \int_{\Omega_I} \boldsymbol{B}_I^T \boldsymbol{c}_I \boldsymbol{B}_I d\Omega_I = \hat{w}_I \boldsymbol{B}_I^T \boldsymbol{c}_I \boldsymbol{B}_I \tag{21.1}$$

being $\Omega_I$ the physical volume occupied by integration point, $\boldsymbol{x}_I$, and $\hat{w}_I$ its numerical representation. Then, all local stiffness matrices, $\boldsymbol{K}_I$, are assembled into a global stiffness matrix, $\boldsymbol{K}$. For each load case, $k$, the essential boundaries are imposed in the stiffness matrix, $\boldsymbol{K}_k$, as well as the respective force vector, $\boldsymbol{f}_k$. Consequently the system of equations $\boldsymbol{K}_k \boldsymbol{u}_k = \boldsymbol{f}_k$ is constructed, where $\boldsymbol{u}_k$ is the unknown displacement field for load case $k$. After determining the displacement field, $\boldsymbol{u}_k$, it is possible to obtain the strain field, $\boldsymbol{\varepsilon}_k$, and the stress field, $\boldsymbol{\sigma}_k$, with the Hooke law [9]. These two fields are required to calculate the strain energy density (SED) field, $\boldsymbol{U}_k$, and the von Mises effective stress field, $\bar{\boldsymbol{\sigma}}_k$, using Eqs. (21.2), (21.3), respectively.

$$U(x_I)_k = \frac{1}{2}\int_{\Omega_I} \{\sigma_{xx}(x_I)_k\ \sigma_{yy}(x_I)_k\ \sigma_{xy}(x_I)_k\} \begin{Bmatrix} \varepsilon_{xx}(x_I)_k \\ \varepsilon_{yy}(x_I)_k \\ \gamma_{xy}(x_I)_k \end{Bmatrix} d\Omega_I \tag{21.2}$$

$$\overline{\sigma}(x_I)_k = \sqrt{\frac{1}{2}\left(\left(\sigma_1(x_I)_k - \sigma_2(x_I)_k\right)^2 + \left(\sigma_2(x_I)_k - \sigma_3(x_I)_2\right)^2 + \left(\sigma_3(x_I)_k - \sigma_1(x_I)_k\right)^2\right)} \tag{21.3}$$

In Eq. (21.3), $\sigma_i(x_I)_k$ is the principal stress of the integration point, $x_I$, of load case $k$, being $i=\{1,2,3\}$. Lastly, at each time instant, $j$, the variable fields obtained for each load case, $k$, are weighted using the following expression:

$$\{u^j, \varepsilon^j, \sigma^j, U^j, \overline{\sigma}^j\} = \sum_{k=1}^{l} \frac{m_k\{u^j, \varepsilon^j, \sigma^j, U^j, \overline{\sigma}^j\}}{\sum_{s=1}^{l} m_s} \tag{21.4}$$

being $l$ the number of load cases and $s$ the corresponding number of load cycles.

## 21.2.3 Remodeling Points

A particular feature of this model is the fact that only a portion of integration points is selected to suffer bone remodeling. Therefore, with an optimization algorithm, only the integration points with SED values are belonging to the following chosen interval:

$$M(x_I) \in [U_{min}, U_{min} + \alpha \cdot \Delta U[\vee]U_{max} - \beta \cdot \Delta U, U_{max}],\ \ \forall M(x_I) \in \mathbb{R} \tag{21.5}$$

in which $U_{min} = \min(\boldsymbol{U})$, $U_{max} = \max(\boldsymbol{U})$, and $\Delta U = U_{max} - U_{min}$. This approach then considers that only integration points that are under extreme levels of mechanical stimulation will be actively remodeled. The ratio of integration points included in the low stimulus group is given by $\alpha$, while $\beta$ defines the ratio of integration points composing the high stimulus group. The low stimulus group is named the "resorption group" since the bone apparent density of the points in that group will potentially decrease. In turn the high stimulus group is denominated "formation group" since its points will potentially increase their apparent density.

## 21.2.4 Phenomenological Law

In this step, each of the selected remodeling points will update their bone apparent density using the phenomenological law proposed by Belinha et al. [7]. Belinha's law is a mathematical proposal capable to describe the experimental results obtained by Lotz et al. [15] and Zioupos et al. [16]. Correlating bone apparent density, $\rho$, with the ultimate compression stress in the axial, $\sigma^c_{axial}$, and transverse, $\sigma^c_{trans}$, directions, a new value of $\rho$ is determined by back substitution in Eqs. (21.6), (21.7).

$$\sigma^c_{axial} = \sum_{j=0}^{3} a_j.\rho^j \tag{21.6}$$

$$\sigma^c_{trans} = \sum_{j=0}^{3} b_j.\rho^j \tag{21.7}$$

A new $\rho$ field leads to an update of the mechanical properties of the chosen remodeling points. Thus, using again the anisotropic Belinha's law, the elasticity modulus in the axial, $E_{axial}$, and transverse, $E_{trans}$, directions is determined with the following expressions:

$$E_{axial} \begin{cases} \sum_{j=0}^{3} c_j.\rho^j \ \text{ if } \ \rho \le 1.3\,\text{g/cm}^3 \\ \sum_{j=0}^{3} d_j.\rho^j \ \text{ if } \ \rho > 1.3\,\text{g/cm}^3 \end{cases} \tag{21.8}$$

$$E_{trans} = \sum_{j=0}^{3} e_j.\rho^j \tag{21.9}$$

In Belinha's law the apparent density is expressed in g/cm$^3$, while $E_{axial}$, $E_{trans}$, $\sigma^c_{axial}$, and $\sigma^c_{trans}$ are expressed in MPa. The values of the coefficients in Eqs. (21.6)–(21.9) are presented in Table 21.1.

The final step is to determine the mean apparent density of the computational model, $\rho^{med}$, as described in the following expression:

TABLE 21.1 Coefficients of Belinha's Law

| Coefficients | $j=0$ | $j=1$ | $j=2$ | $j=3$ |
|---|---|---|---|---|
| $a_j$ | 0.000E+00 | 0.000E+00 | 2.680E+01 | 2.035E+01 |
| $b_j$ | 0.000E+00 | 0.000E+00 | 2.501E+01 | 1.247E+00 |
| $c_j$ | 0.000E+00 | 7.216E+02 | 8.059E+02 | 0.000E+00 |
| $d_j$ | −1.770E+05 | 3.861E+05 | −2.798E+05 | 6.836E+04 |
| $e_j$ | 0.000E+00 | 0.000E+00 | 2.004E+03 | −1.442E+02 |

$$\rho^{med} = \frac{1}{Q}\sum_{I=1}^{Q}\rho(\boldsymbol{x}_I) \tag{21.10}$$

in which $Q$ is the number of integration points and $\rho(\boldsymbol{x}_I)$ the bone apparent density of integration point $\boldsymbol{x}_I$.

The algorithm moves on to the next iterative step, $j$, performing a new mechanical analysis. It should be noted that a new material constitutive matrix, $\boldsymbol{c}(\boldsymbol{x}_I)_{j+1}$, is constructed for the next iteration given by $j+1$.

The remodeling process ends when $\rho^{med}$ reaches a value determined by the user or if two consecutive iteration steps have the same $\rho^{med}$ $(\Delta\rho^{med}/\Delta t=0)$.

# 21.3 BONE REMODELING AFTER THA

## 21.3.1 Computational Model

The goal of this work is to analyze the remodeling of bone tissue after THA. Using X-ray images as the one presented in Fig. 21.1A, a computational model is created and discretized in an irregular triangular mesh with 4473 nodes and 8591 elements, as shown in Fig. 21.1B and C.

To simulate the daily loading history of the femur, the three loading cases proposed by Beaupré et al. [17–19] are imposed. Each load case consists of a parabolic distributed load applied in the femoral head and another in the great trochanter. An essential boundary condition is also imposed, considering that all degrees of freedom are constrained at

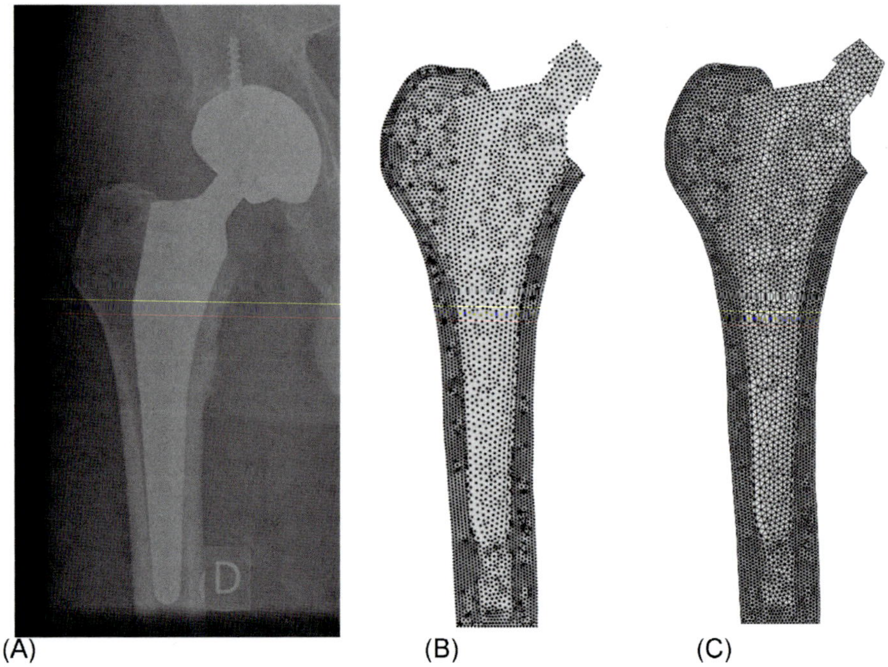


FIG. 21.1 (A) Anteroposterior X-ray of the right femur; geometric model discretized with (B) 4473 nodes and (C) 8591 elements.

the base of the model, preventing its movement in $x$- and $y$-directions. The aforementioned natural and essential boundary conditions are represented schematically in Fig. 21.2A, while the magnitude and inclination of each load are presented in Table 21.2.

Regarding material properties, the bone is considered isotropic with an initial uniform apparent density distribution equal to 2.1 g/cm$^3$ and a Poisson's coefficient of 0.3. The titanium implant has an elasticity modulus of 110 GPa and a Poisson's coefficient of 0.32. This study does not consider interface elements between the implant and the bone tissue. The remodeling parameters $\alpha$ and $\beta$ are assumed as 0.0 and 0.03, respectively.

## 21.3.2 Prediction of Bone Remodeling

Bone remodeling after implant placement is reproduced using the algorithm previously described. In Figs. 21.3 and 21.4, the isomaps of the final trabecular architecture, von Mises effective stress, and principal stresses obtained with FEM and RPIM are respectively presented.

Analyzing the trabecular architecture obtained, all numerical techniques present a trabecular morphology in agreement with Fig. 21.2B. However, only RPIM's solution is able to predict the compressive trabeculae, even though less dense than expected. The tensile zones are more difficult to depict since the load is applied in the implant and not directly in the femoral bone. In Figs. 21.3C and 21.4C, the principal stress isomap $\sigma_{11}$ reveals the impact of the tensile load applied in the great trochanter.

When comparing the normal trabecular morphology of a healthy femur, as obtained previously by Peyroteo et al. [8], with this model after THA, a significant difference is found. Because of stress shielding the obtained trabecular architecture is poorer when compared with a healthy case. Since the implant is stiffer than the bone, the applied loads and the consequent stress levels are backed up by the implant. As a result, the bone is not mechanically stimulated enough, and thus bone resorption occurs.

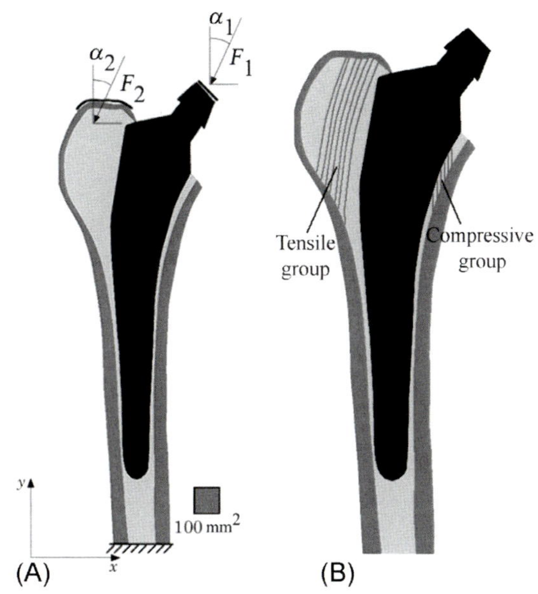


FIG. 21.2 (A) Natural and essential boundary conditions and (B) schematic representation of the major trabecular groups after THA.

TABLE 21.2 Load Cases Specifications

| Load case | $F_1$ (N) | $\alpha_1$ (°) | $F_2$ (N) | $\alpha_2$ (°) | Load cycles |
|---|---|---|---|---|---|
| LC1 | 2317 | 24° | 703 | 28° | 6000 |
| LC2 | 1158 | −15° | 351 | −8° | 2000 |
| LC3 | 1548 | 56° | 468 | 35° | 2000 |

To analyze, with more detail, the spatial apparent density variation throughout the numerical model, the isomap was divided into the seven Gruen zones. The Gruen zones provide a reference system based on the anatomy of the femur. This division system is usually used to visualize the zones of interest of the X-ray and thence to evaluate the outcome of the surgical intervention. Thus the numerical model was divided as shown in Fig. 21.5A. In zone 1, known as a zone of high resorption, only a small group of trabeculae is formed as result of the tensile load applied in the great trochanter. On the opposite side of the model, in zone 7, FEM predicts a severe bone loss, while RPIM's solution presents higher apparent density levels. Then, it is possible to conclude that, unlike FEM's solution, RPIM can predict the compressive group specified in Fig. 21.2B. The compressive group is also depicted in zone 6. For the other zones, both numerical solutions agree. Since a greater resorption occurs in the proximal-medial zone of the femur, a distal redistribution of the load is expected [20]. Therefore, as observed in Fig. 21.5A, zones 2 and 5 are zones of medium resorption, whereas in zones 3 and 4, distal bone hypertrophy is observed. This load's redistribution is depicted in the von Mises effective stress isomaps (Figs. 21.3B and 21.4B). Higher stress concentration occurs in the lower zone of the diaphysis, with higher compressive stresses at the left side and higher tensile stresses on the right side of the prosthesis.

Comparing the X-ray image in Fig. 21.1A with the predicted trabecular morphology, a good agreement is obtained at the proximal-medial regions. Unfortunately, it is not possible to compare at the hypertrophic region around the

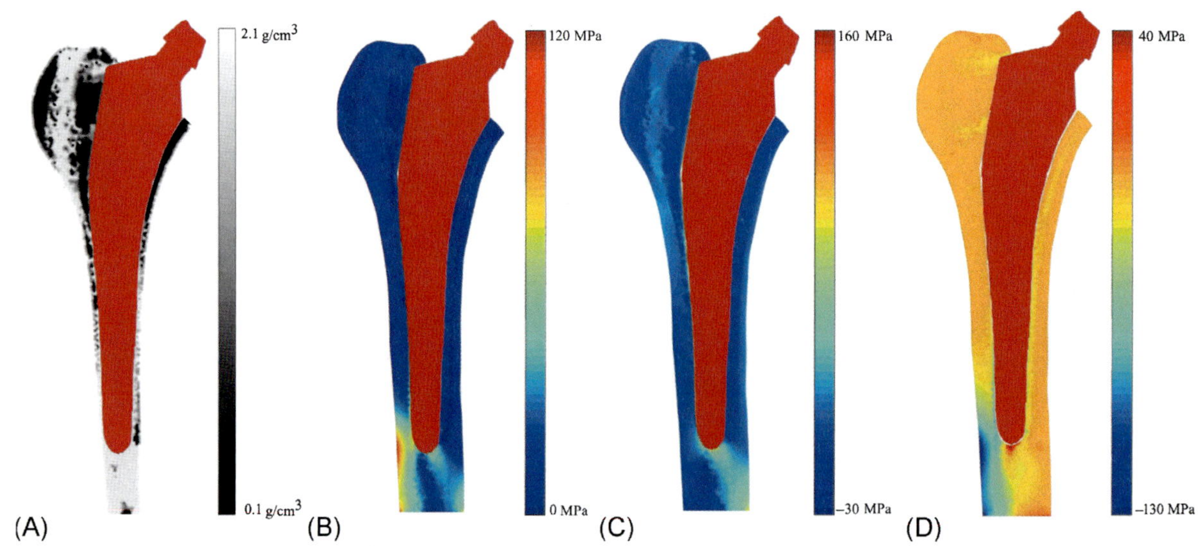


FIG. 21.3 FEM solution: final isomaps of (A) apparent density, $\rho$; (B) von Mises effective stress, $\bar{\sigma}$; and principal stresses (C) $\sigma_{11}$ and (D) $\sigma_{22}$.

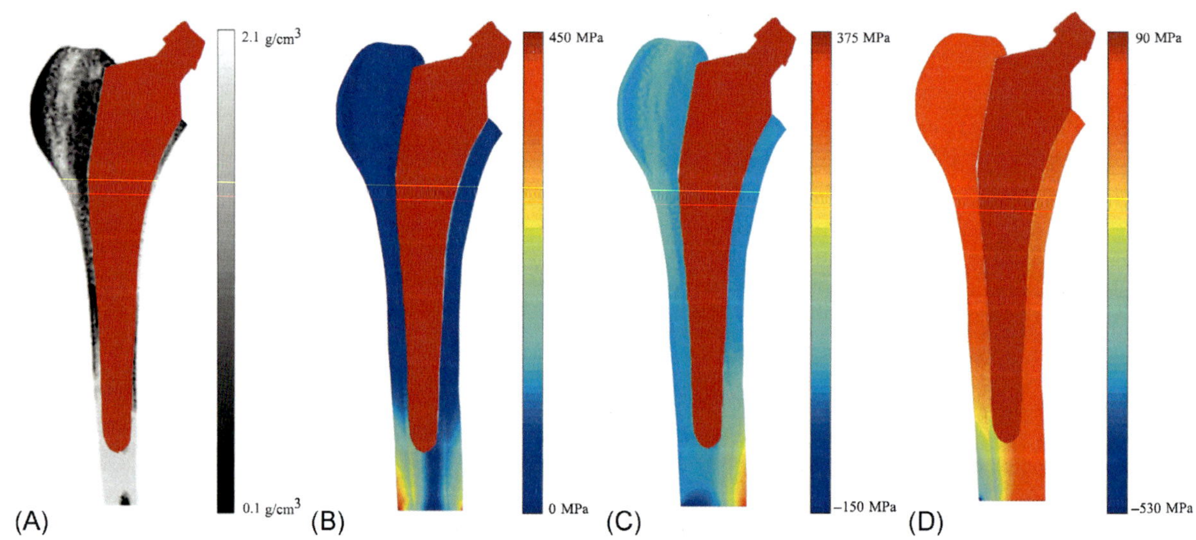


FIG. 21.4 RPIM solution: final isomaps of (A) apparent density, $\rho$; (B) von Mises effective stress, $\bar{\sigma}$; and principal stresses (C) $\sigma_{11}$ and (D) $\sigma_{22}$.

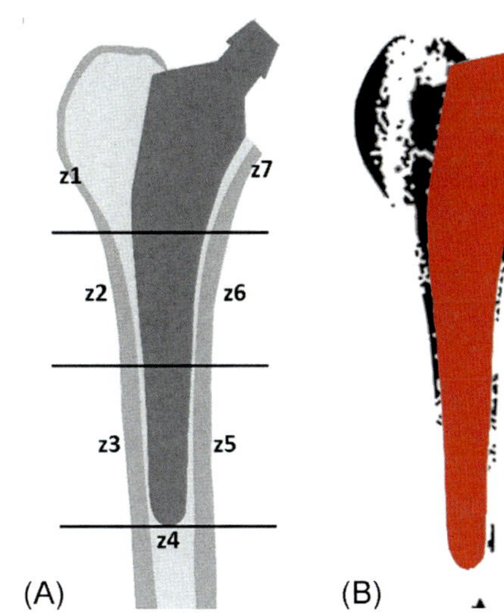


FIG. 21.5 (A) Gruen zones and (B) intersection of FEM and RPIM solutions.

distal stem, since the X-ray image does not include this region. FEM and RPIM's solutions were also compared. In Fig. 21.5B the black-and-white map indicates the spatial intersection of the two solutions. White nodes mean that both solutions predicted bone, whereas black nodes reflect contradictory results or nonbone zones predicted by both methods. With this qualitative analysis, it is possible to depict that FEM and RPIM produce similar results.

## 21.4 CONCLUSIONS

The mechanical model adopted in this work produced a trabecular distribution similar to the X-ray image, proving the efficiency of the remodeling algorithm. Moreover the success of the approach was independent of the numerical method used. However, unlike FEM, RPIM's solution was able to predict the compressive trabecular group and produce more accurate and smoother apparent density distribution maps. As future work a three-dimensional (3-D) model should be used to simulate/predict the 3-D structural interaction between femur and implant. Nonetheless, using this remodeling algorithm to predict bone remodeling after implant insertion is a promising approach with an important impact on implants' evaluation and success of THA.

## Acknowledgments

The authors truly acknowledge the funding provided by Ministério da Ciência, Tecnologia e Ensino Superior, Fundação para a Ciência e a Tecnologia (Portugal), under Grants SFRH/BD/133105/2017 and by project funding MIT-EXPL/ISF/0084/2017. Additionally the authors gratefully acknowledge the funding of Project NORTE-01-0145-FEDER-000022—SciTech—Science and Technology for Competitive and Sustainable Industries, cofinanced by Programa Operacional Regional do Norte (NORTE2020), through Fundo Europeu de Desenvolvimento Regional (FEDER).

## References

[1] S. Affatato, Perspectives in Total Hip Arthroplasty: Advances in Biomaterials and Their Tribological Interactions, Woodhead Publishing, Oxford, UK, 2014.

[2] A.D. Woolf, B. Pfleger, Burden of major musculoskeletal conditions, Bull. World Health Organ. 81 (9) (2003) 646–656, https://doi.org/10.1371/journal.pone.0090633.

[3] S. Kurtz, K. Ong, E. Lau, F. Mowat, M. Halpern, Projections of primary and revision hip and knee arthroplasty in the United States from 2005 to 2030, J. Bone Jt. Surg. 89 (4) (2007) 780, https://doi.org/10.2106/JBJS.F.00222.

[4] R.L. Barrack, Economics of revision total hip arthroplasty, Clin. Orthop. Relat. Res. (319) (1995) 209–214, https://doi.org/10.1016/J.CUOR.2006.02.007.

[5] C.K. Ledford, T.S. Watters, S.S. Wellman, D.E. Attarian, M.P. Bolognesi, Risk versus reward: total joint arthroplasty outcomes after various solid organ transplantations, J. Arthroplast. 29 (8) (2014) 1548–1552, https://doi.org/10.1016/j.arth.2014.03.027.

[6] M.R. Abdul Kadir, Computational Biomechanics of the Hip Joint, Springer Berlin Heidelberg, Berlin, Heidelberg, 2014.

[7] J. Belinha, R.M.N. Jorge, L.M.J.S. Dinis, A meshless microscale bone tissue trabecular remodelling analysis considering a new anisotropic bone tissue material law, Comput. Methods Biomech. Biomed. Engin. 16 (11) (2013) 1170–1184, https://doi.org/10.1080/10255842.2012.654783.

[8] M.M.A. Peyroteo, J. Belinha, S. Vinga, L.M.J.S. Dinis, R.M. Natal Jorge, Mechanical bone remodelling: comparative study of distinct numerical approaches, Eng. Anal. Bound. Elem. (2018), https://doi.org/10.1016/J.ENGANABOUND.2018.01.011.

[9] J. Belinha, R.M.N. Jorge, L.M.J.S. Dinis, Bone tissue remodelling analysis considering a radial point interpolator meshless method, Eng. Anal. Bound. Elem. 36 (11) (2012) 1660–1670, https://doi.org/10.1016/j.enganabound.2012.05.009.
[10] J. Belinha, L.M.J.S. Dinis, R.M.N. Jorge, The bone tissue remodelling analysis in dentistry using a meshless method, in: Proceedings of the III International Conference on Biodental Engineering, 2014, pp.213 – 220.
[11] D.R. Carter, D.P. Fyhrie, R.T. Whalen, Trabecular bone density and loading history: regulation of connective tissue biology by mechanical energy, J. Biomech. 20 (8) (1987) 785–794, https://doi.org/10.1016/0021-9290(87)90058-3.
[12] R.T. Whalen, D.R. Carter, C.R. Steele, Influence of physical activity on the regulation of bone density, J. Biomech. 21 (10) (1988) 825–837, https://doi.org/10.1016/0021-9290(88)90015-2.
[13] D.R. Carter, T.E. Orr, D.P. Fyhrie, Relationships between loading history and femoral cancellous bone architecture, J. Biomech. 22 (3) (1989) 231–244, https://doi.org/10.1016/0021-9290(89)90091-2.
[14] K.-J. Bathe, Finite Element Procedures, Prentice Hall, Pearson Education, Inc., New Jersey, 1996.
[15] J.C. Lotz, T.N. Gerhart, W.C. Hayes, Mechanical properties of metaphyseal bone in the proximal femur, J. Biomech. 24 (5) (1991) 317–329, https://doi.org/10.1016/0021-9290(91)90350-V.
[16] P. Zioupos, R.B. Cook, J.R. Hutchinson, Some basic relationships between density values in cancellous and cortical bone, J. Biomech. 41 (9) (2008) 1961–1968, https://doi.org/10.1016/j.jbiomech.2008.03.025.
[17] G.S. Beaupré, T.E. Orr, D.R. Carter, An approach for time-dependent bone modeling and remodeling-application: a preliminary remodeling simulation, J. Orthop. Res. 8 (5) (1990) 662–670, https://doi.org/10.1002/jor.1100080507.
[18] G.S. Beaupré, T.E. Orr, D.R. Carter, An approach for time-dependent bone modeling and remodeling. Theoretical development, J. Orthop. Res. 8 (5) (1990) 651–661, https://doi.org/10.1002/jor.1100080506.
[19] G. Bergmann, et al., Hip contact forces and gait patterns from routine activities, J. Biomech. 34 (7) (2001) 859–871, https://doi.org/10.1016/S0021-9290(01)00040-9.
[20] J.H. Keyak, T.S. Kaneko, J. Tehranzadeh, H.B. Skinner, Predicting proximal femoral strength using structural engineering models, Clin. Orthop. Relat. Res. (437) (2005) 219–228, https://doi.org/10.1097/01.blo.0000164400.37905.22.

# Index

Note: Page numbers followed by *f* indicate figures and *t* indicate tables.

Made in the USA
Las Vegas, NV
13 October 2025

32269626R00259